MECHANOBIOLOGY: CARTILAGE AND CHONDROCYTE

Biomedical and Health Research

Volume 61

Recently published in this series:

ISSN 0929-6743

Mechanobiology: Cartilage and Chondrocyte

Volume 3

Edited by

J.-F. Stoltz

Hématologie-Hémorhéologie, UMR CNRS 7563, Faculté de Médecine, Université Henri Poincaré, Vandœvre-lès-Nancy, France

IOS Press

Amsterdam • Berlin • Oxford • Tokyo • Washington, DC

ISBN 1 58603 436 7
Library of Congress Control Number: 2004109133

This is the book edition in the BHR book series of the journal *Biorheology*, Volume 41, Nos 3,4 (2004), ISSN 0006-355X

Publisher
IOS Press
Nieuwe Hemweg 6B
1013 BG Amsterdam
The Netherlands
fax: +31 20 620 3419
e-mail: order@iospress.nl

Distributor in the UK and Ireland
IOS Press/Lavis Marketing
73 Lime Walk
Headington
Oxford OX3 7AD
England
fax: +44 1865 750079

Distributor in the USA and Canada
IOS Press, Inc.
4502 Rachael Manor Drive
Fairfax, VA 22032
USA
fax: +1 703 323 3668
e-mail: iosbooks@iospress.com

PRINTED IN THE NETHERLANDS

Contents

Part II: Mechanobiology and chondrocytes

Part III : Subchondral bone and osteoarthritis

Part IV: Cartilage remodeling and engineering

Biorheology 41 (2004) 153
IOS Press

Foreword

I am pleased to introduce the proceedings of the third International Symposium on Mechanobiology of Cartilage and Chondrocyte (Brussels, 16–17 May 2003).

Mechanobiology is now a vigorous branch of biomechanics and biorheology mainly concerned with the study of the influence of mechanical forces on cells and tissues and their clinical or therapeutical applications. As we are now in the age of proteomics, genomics and of cell micromechanical approaches using methods like laser tweezers or confocal microscopy, mechanobiology brings new challenges. With such new researches, mechanobiology has the promise of leading to new diagnostic and therapeutic approaches.

The main aim of this third symposium is to focus on more recent research on cartilage and chondrocyte. I am very pleased that the speakers and session chairmen kindly accepted my invitation. I also thank Doctor Martine Burger for her investment in the preparation of this symposium and the firm NEGMA-LERADS for its support.

In order to ensure the continuity of the first three symposia (Nice 1999, Paris 2001, Brussels 2003) a fourth one will be organized in May 2005 in Budapest, where it is again proposed to present awards to young scientist investigators.

J.F. Stoltz
Nancy, May 2004

Biorheology 41 (2004) 155–156
IOS Press

Introduction

Adaptation concept, tissue remodeling, mechanobiology and tissue engineering: A survey

J.F. Stoltz
Cell and Tissue Mechanics and Engineering group – LEMTA UMR CNRS 7563 and IFR 111, Bioengineering- (CNRS-UHP-INPL-CHU) Faculté de Médecine and Cell and Tissue Therapy, 54500 Vandoeuvre les Nancy, France

Man's interest in biomechanics dates back to ancient Greece (Aristotle 384–322 BC) and perhaps even earlier than that, but it is with Leonardo de Vinci (1452–1519), Galileo (1564–1632) and Harvey (1578–1652) that work reveals the importance of mechanics in physiology. For example, in his work relating to blood circulation and which presented the existence of microcirculation, Harvey wrote: "*How does the blood manage to traverse the porosities of the flesh on its way from arteries to the vein?*"

The term Biomechanics, however, did not come into use before 1887 with Benedikt (über Mathematische Morphologie und Biomechanick).

The 1980's witnessed the birth of new biomechanics with parallel progress in biology, physics and chemistry. Indeed, we stepped into the era of genomics and proteomics about ten years ago, thanks to new technology allowing us to study microstructures or to measure microforces (e.g. confocal microscopy, magnetic bead cytometry, laseroptical tweezers). Comparing the data arising from these new experimental approaches allowed a better understanding of the relation between applied forces, gene or receptor expression, and synthesis of autocrine factors. This new biomechanics (mechanobiology) led to a better grasp of the tissue remodeling principles observed in some pathologies. Remodeling corresponds to the biological dynamics of a tissue. It can be physiological and ensure the stability of tissue homeostasis or it can evolve and lead to pathological modifications. Remodeling induces mobilization in two different tissue components, i.e. cells that constitute tissue, and the extra-cellular matrix. In more complex mechanisms, the influence of local mechanical forces often plays a predominant role. Tissue remodelling combines modifications in cell expression and the possible mobilization of other cell types like inflammatory cells or progenitors. The deregulation of tissue homeostasis that leads to pathological remodeling are chronic phenomena that often result from diseases related to age.

The first observations regarding the mechanical adaptation and the structure/strain relation of the tissues were made a long time ago (Meyer, 1866; Wolff, 1892). Indeed, Meyer in a paper on the structure of cancerous bone wrote (1866) "*the spongiasa showed a well-motivated architecture which is closely connected with the statics and mechanics of bone.*" Culman, a mathematician who was attending the conference, then pointed out that the shape of a cantilver beam in civil engineering was the same as the head

0006-355X/04/$17.00

of a femur and undertook a first attempt at modeling. Meyer then organised his ideas into three questions: "*is it possible that structures like the observed ones are formed by static (force–equilibrium) conditions? What is the internal metamorphosis that makes that structures so "fit to service?" Can these structures be understood if one adds to the external loads the mechanical influence of the traction of muscles and ligaments?*" Some years later (1864), Wolff, a German surgeon, wrote "*every change in the form and function of bone or of their functions alone is followed by certain definitive changes in their internal architecture and equally definitive secondary alterations in their external conformation in accordance with mathematical laws*", thus defining the phenomenon of tissue mechanical remodeling (law of bone transformation) without putting it in writing. At the same time, Wilhelm Roux (1881) introduced the concepts of functional adaptation which he described as a quantitative regulating auto-mechanism controlled by functional stimuli. He then wrote (1881): "*many cells are influenced by functional stimuli, the nerve, muscle, gland cells by the related electric pulse, bone cells and connective tissue cells by pressure and tension. It is plausible, that, if this influence is disavantageous for some cells of a tissue or some parts of it, then these cells will disappear during the process of physiological regeneration. If however, variations appear for which the functional stimulus is favorable, then in the course of time these cells will replace all the other cells, which are indifferent to the functional stimulus.*" This sentence described the mechanotransduction phenomena that are the subject of many studies today.

Current research in mechanobiology finds its application in many domains and particularly in the vascular, cardiac and osteoarticular fields. Cartilage especially, is a tissue model that enables us to consider all the approaches of mechanobiology, in spite of its complexity and thanks to its autosynthetised matrix, from the most fundamental mechanical approaches (poroelasticity) to pharmacological studies or to the concept of biotissues.

The most recent work shows that the incidence of mechanical forces is specific to the system under scrutiny and that stresses are implicated in tissue physiology, for example by the production of the extracellular matrix (i.e. cartilage), secretions (i.e. production of NO and prostaglandins by endothelial cells), or for the induction of specific functions via intercellular communication; hence the interest from pharmacology in studies on new molecules. Moreover, these new findings have led to the development of tissue engineering, that is the concept of substitute tissue developed *in vitro*, from bioresorbable or non-bioresorbable scaffolds and from cells harvested in a physiologic mechanical environment such as from cartilage, bone and vessels. At the same time, the problem of cell grafting in tissue repair and especially the use of stem cells have led to new therapeutic fields.

Economic forecasts predict that these new approaches will soar with an estimated turnover of more than 50 billion euros within 5 years.

References

[1] Y.C. Fung, *Biomechanics: Mechanical Properties of Living Tissues*, Springer Verlag Publ., N.Y., Berlin, 1981, 433 pp.
[2] Y.C. Fung, *Biomechanics: Motion, Flow, Stress and Growth*, Springer Verlag Publ., N.Y., Berlin, 1990, 569 pp.
[3] K.J. Gooch and C.J. Tennant, *Mechanical Forces: Their Effects on Cells and Tissues*, Springer Verlag Publ., N.Y., Berlin, 1997, 182 pp.
[4] M.C.H. van der Meulen and R. Huiskes, Why mechanobiology? A survey article, *Journal of Biomechanics* **35** (2002), 401–414.
[5] J.F. Stoltz and X. Wang, De la biomécanique à la mécanobiologie. Conférence plénière, 15ème congrès français de Mécanique, 2001, pp. 1–7.
[6] J.F. Stoltz, *Mechanobiology: Cartilage and Chondrocyte* (multi authors books), Vol. 2, 2002, 288 pp.
[7] X. Wang and G.A. Dumas, Simulation of bone adaptative remodeling using a stochastic process as loading history, *Journal of Biomechanics* **35** (2002), 375–380.

Part I

Cartilage mechanics: Theoretical aspects

Biorheology 41 (2004) 159–166
IOS Press

Biomechanical properties of human articular cartilage under compressive loads

Federica Boschetti [a,*], Giancarlo Pennati [a], Francesca Gervaso [a], Giuseppe M. Peretti [b] and Gabriele Dubini [a]

[a] *Laboratory of Biological Structure Mechanics, Department of Structural Engineering and Department of Bioengineering, Politecnico di Milano, Italy*

[b] *Orthopaedic Department, San Raffaele Hospital, Milan, Italy*

Abstract. The function of articular cartilage is to support and distribute loads and to provide lubrication in the diarthrodial joints. Cartilage function is described by proper mechanical and rheological properties, strain and depth-dependent, which are not completely assessed. Unconfined and confined compression are commonly used to evaluate the Young's modulus (E) and the aggregate modulus (H_A), respectively. The Poisson's ratio (ν) can be calculated indirectly from the equilibrium compression data, or using the biphasic indentation technique; it has recently been optically evaluated by using video microscopy during unconfined compression. The transient response of articular cartilage during confined compression depends on its permeability k; a constant value of k can be easily identified by a simple analytical model of confined compression tests, whereas more complex models or direct measurements (permeation tests) are needed to study the permeability dependence on deformation. A poroelastic finite element model of articular cartilage was developed for this purpose. The elastic parameters (E, ν) of the model were evaluated performing unconfined compression creep tests on human articular cartilage disks, whereas k was identified from the confined test response. Our combined experimental and computational method can be used to identify the parameters that define the permeability dependence on deformation, as a function of depth from articular surface.

Keywords: Femoral head, creep experiments, permeability, poroelastic model

1. Introduction

Articular cartilage is composed of a charged solid matrix phase consisting of charged proteoglycan macromolecules and collagen fibres, an interstitial fluid phase, and an ion phase [19]. The function of articular cartilage is to support and distribute loads and to provide lubrication in the diarthrodial joints [18]. Articular cartilage is a heterogeneous tissue, having a depth dependent composition and structure, and showing anisotropic and nonlinear properties in compression and tension. This explains the anisotropic and nonlinear properties in compression and tension. Biomechanical models of increasing complexity have been developed to interpret and quantify the mechanical properties of this tissue, from the simpler monophasic linear elastic model [20] to the more complex and realistic FEM models [13]. Experimental investigation of cartilage biomechanics during compression is typically performed by using confined or unconfined compression, or indentation. The equilibrium response of articular cartilage is satisfactorily described, for small deformations, by the homogeneous isotropic elastic model [6]. In particular, unconfined and confined compression tests are commonly used to evaluate the Young's modulus (E)

*Address for correspondence: Federica Boschetti, PhD, LaBS, Dipartimento di Ingegneria Strutturale, Politecnico di Milano, Piazza L. da Vinci 32, 20133 Milano, Italy. E-mail: federica.boschetti@polimi.it.

and the aggregate modulus (H_A), respectively. Recently, video microscopy has been used during unconfined compression for the optical evaluation of the Poisson's ratio [9,22]. This parameter has also been calculated indirectly from the equilibrium compression data [10], or using the biphasic indentation technique [16]. The transient response of cartilage during compression has been described by the biphasic model [17], which introduces the hydraulic permeability, k, as a parameter regulating the velocity of fluid exudation from the tissue. Due to the compaction of the solid matrix during compression, the permeability of articular cartilage is not a constant property of the tissue; in fact, it decreases for increasing deformation [14], according to an exponential curve [12]. Nevertheless a constant value of k can be identified by fitting the experimental curve to the analytical solution from confined compression experiments [11], or indentation tests [16]. On the other hand, the strain dependent permeability can be directly evaluated performing permeation tests. These measurements are conceptually easy, but difficult to be accurately performed, both at low strains (i.e. low applied pressures), due to the very low flow through the cartilage sample, and at high strains (i.e. high applied pressures), due to the difficulties in guaranteeing the seal. Furthermore, a complete permeation study, taking into account several strain levels, can take up to two days: because of the very low permeability of articular cartilage, it can take more than 1 hour for 1 microliter to flow at 0.02 MPa [1].

It is well known that the composition [15] and structure [8] of articular cartilage vary through the depth of the tissue. As a consequence, the biomechanical parameters are expected to vary with depth. Indeed, a number of experimental studies have shown variations of H_A, E, ν, and k with depth [2,3, 9,21,22]. Nonetheless, only a few studies are related to human articular cartilage. To the best of our knowledge, none of these studies measured all the above parameters for the same samples. In order to evaluate the values and depth dependence of H_A, E, ν, and k for the human articular cartilage, we performed confined and unconfined creep experiments on disks of human articular cartilage representing consecutive layers of the full cartilage thickness. Furthermore, we implemented a numerical poroelastic model of articular cartilage based on the measured elastic parameters. The possibility to identify the strain-dependent permeability of human cartilage by experimental curve fitting was examined. Confined compression experiments were simulated both using idealised and actual testing geometry, in order to assess the effects of simplifications on the estimate of the biomechanical parameters.

2. Materials and methods

2.1. Biomechanical tests

The cartilage samples were harvested from the femoral head, derived from 3 patients (age and sex 73-M, 75-M, 81-F), who suffered from hip fracture and who underwent hip endoprosthesis as treatment. Only those samples free from osteoarthritis were utilised for this study. Using a drill we obtained cylindrical plugs of bone and cartilage of approximately 9 mm in diameter. While drilling, the articular surface was moistened with physiological solution. The cartilage plugs were stored at $-26°$C prior testing. On the day of testing, the cartilage plug was cut into three slices, having roughly the same thickness of approximately 650 μm, starting from the superficial zone, using a custom made device. The cartilage slices were thawed at room temperature and equilibrated in phosphate buffered saline (PBS: 0.15 mmol/l NaCl; 1.54 mmol/l NaH_2PO_4; 2.71 mmol/l Na_2HPO_4, purchased from Sigma-Aldrich Srl, Milan, Italy) containing protease inhibitors (PI: 1 mmol/l phenylmethyl-sulfonyl-fluoride; 2 mmol/l EDTA; 5 mmol/l benzamidine; 10 mmol/l N-ethylmaleimide, purchased from Sigma-Aldrich Srl, Milan, Italy) for about

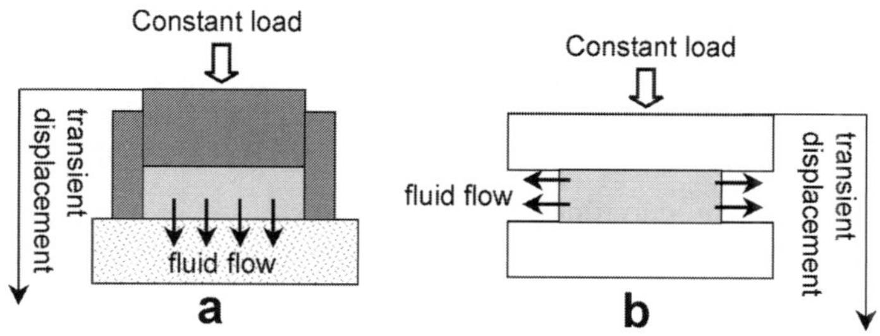

Fig. 1. (a) Confined compression; (b) unconfined compression.

30 minutes. Each slice was punched to obtain a 5 mm disk, used in the tests. The thickness of the uncompressed cartilage disks was measured optically using a stereomicroscope (SMZ800, Nikon). The samples first underwent the confined compression creep test, and then, after re-equilibration, the unconfined compression creep test. The total testing time for each sample was typically 1–2 h.

The confined test was performed by placing the specimen in a confining chamber (5 mm in diameter) over a stainless steel porous filter (Aegis Advanced Materials Ltd., Bewdley, UK, average pore size $\sim$40 μm) allowing free fluid flow (Fig. 1a). Compressive loading (15 kPa) was applied to the cartilage specimen by a non-porous stainless steel indenter. The clearance between the confining chamber and the indenter was 0.1 mm. The indenter was connected to a LVDT (W1E, Hottinger Baldwin Messtechnik Gmbh, Darmstadt, Germany, with 1 μm resolution), which measured the indenter vertical displacement for a total time of approximately 1800 seconds, or until the equilibrium was reached. The sampling frequency was 200 Hz for the first minute after application of the load and 2 Hz for the remaining creep response. Using the equilibrium stress-strain data, we calculated the aggregate modulus H_A; we identified a constant value of permeability ($\bar{k}$) fitting the experimental creep curve to the analytical solution of the confined compression creep problem [11]. Since it is well known that the permeability is strain-dependent [12], the identified value represents the averaged value of the true k over the time and depth dependent deformation.

The unconfined test was performed by placing the specimen between two plexiglass platens (Fig. 1b) and applying a load resulting in 15 kPa compression for a total time of approximately 1800 seconds. The superior platen was connected to a LVDT (W1E, Hottinger Baldwin Messtechnik Gmbh, Darmstadt, Germany, with 1 μm resolution). Using the equilibrium stress-strain data, we calculated the Young's modulus E. Furthermore, the top surface of the cartilage disk was observed using a stereo-microscope with a magnification of 2.6× connected to a PC by a digital camera (M50, JAI Corporation, USA), and digital images were acquired during the testing to allow optical evaluation of the radial deformation and, hence, the Poisson's ratio. Throughout the measurements, cartilage disks were immersed in PBS containing PI.

The statistical significance of the parameters variation among layers was assessed by paired t-test.

2.2. *Poroelastic model*

A coupled pore fluid diffusion and stress analysis was performed using the commercial code ABAQUS (Hibbit Karlsson and Sorenses, Inc., Pawtucket, RI, USA), reproducing the creep-compression experimental tests. An axial-symmetric poroelastic model of the cartilage disks was developed, in which the cartilage was described as a multiphase material constituted of an incompressible, isotropic, linear, elastic solid phase, and an incompressible fluid phase [17]. The model parameters were Young's modulus (E), Poisson's ratio (ν), hydraulic permeability (k) and porosity (ϕ). E and ν were obtained by the compression equilibrium data measured on cartilage disks of human articular cartilage as described above. The porosity, ϕ, was evaluated from the water content, determined by the difference between the weights of the hydrated and dried cartilage sample. Cartilage samples were dried in an oven at 80°C for 24 h.

The model was used to simulate a confined compression experiment of 5 mm-cartilage disks. Model simulations were used to identify the permeability of the tested cartilage layers. The permeability was considered strain (ε) dependent according to the formulation $k = k_0 e^{-M\varepsilon}$ suggested by Lai and Mow [12]. Since ABAQUS requires k as a function of porosity, we assumed for the porosity a linear variation with deformation [5]: $\phi = \phi_0 + (1 - \phi_0) \cdot \varepsilon$, being ϕ_0 the initial porosity.

The parameters k_0 and M were evaluated by fitting the experimental confined compression displacement to the numerical solution, using three methods. The first method identifies k_0 and M independently, so they are both changed by guessing; conversely, the second and third methods assume a relationship between k_0 and M, based on the averaged value of k during creep (*Averaged Method*), or on the initial slope of the creep response (*Initial Slope Method*), respectively.

The averaged value of k is the one ($\bar{k}$) identified using the analytical solution for the confined compression creep problem. Therefore, the Averaged Method assumes that k_0 and M are related as follows:

$$\bar{k} = \frac{1}{\varepsilon_f} \int_0^{\varepsilon_f} k_0 \cdot e^{-M\varepsilon}\, d\varepsilon, \tag{1}$$

where ε_f is the final deformation.

For the first few moments after the load is applied, a linear relation $\varepsilon = \beta\sqrt{t}$ exists between the deformation ε and the square root of time (t) [7], where β, for small strains, is:

$$\beta = \frac{2\varepsilon_f}{h}\left(1 - \frac{\varepsilon_f M}{\pi}\right)\sqrt{\frac{k_0 H_A}{\pi}}, \tag{2}$$

where h is the cartilage disk thickness. So the Initial Slope Method assumes that k_0 and M are related through Eq. (2), where β is determined by fitting with a line the first $\varepsilon - \sqrt{t}$ points of the creep response. We considered the points relative approximately to the first 0.03 seconds.

3. Results and discussion

From the femoral heads we obtained 13 cylindrical plugs of bone and cartilage; a total of 33 disks were tested for confined compression, unconfined compression, or both. The layers obtained, though consecutive, are not representative of the real anatomical superficial, middle and deep layer. They were identified as "layer 1" ("superficial", 0.796 ± 0.262 mm), layer 2 ("middle", 0.599 ± 0.175 mm), and layer 3 ("deep", 0.552 ± 0.147 mm). Results of the experiments are shown in Fig. 2. The parameters H_A, E, and v increase, while $\bar{k}$ decreases with depth, reflecting the well-known variability in tissue composition and structure with depth. The increase in H_A was statistically significant between layer 1 and 2 and between layer 2 and 3 ($p < 0.05$). The increase in E was significant ($p < 0.05$) between layer 1 and 3 and layer 1 and 2; no significant differences were observed between layer 2 and 3, probably because of the little number of samples tested in unconfined compression and high variability between samples. The increase in v and the decrease in $\bar{k}$ were never statistically significant.

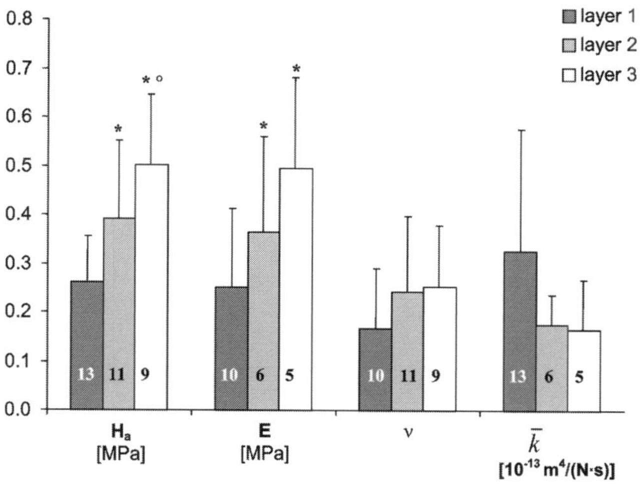

Fig. 2. Experimental results of confined (H_a and $\bar{k}$) and unconfined (E, ν) compression for different layers of cartilage. Mean values with standard deviations are reported for the tested disks (sample numbers are indicated on the bars). *Significant differences vs. layer 1 ($p < 0.05$), °significant differences vs. layer 2 ($p < 0.05$).

Figure 3 compares a typical experimental confined compression plot to the simulation results from the poroelastic model. The simulations related to Fig. 3a were obtained using the constant permeability $\bar{k}$, and the strain dependent permeability with parameters k_0 and M evaluated independently (by best-fitting the experimental curve). The comparison clearly shows that a constant value of permeability, as the one identified by confined compression experiments, is not able to precisely describe the transient response of articular cartilage during compression. Since the permeability decreases with strain [14], a constant value underestimates the true permeability for small strains and, conversely, the permeability is overestimated for strains approaching the equilibrium. On the other hand, the use of the exponential dependence on strain for permeability results in quite a good fitting, provided that proper values of k_0 and M are adopted; nevertheless, the identification by guessing of two independent parameters is very time-consuming. Figures 3b and 3c report three simulated curves using k_0 and M evaluated according to the Averaged and Initial Slope Methods: the experimental curve could not be fitted to the simulated ones for any couple of k_0 and M. Reasons for failure of those latter methods remain to be investigated. The Averaged Method may be incorrect since $\bar{k}$ is calculated on the basis of the displacement of the superior surface of the cartilage disk, whereas the strain varies not only during the transient phase, but also throughout the cartilage disk thickness [23]. On the other hand, the calculation of β may require a more precise acquisition of the displacement during the first instants of the creep experiments. Furthermore, the experimental confined compression curve is never obtainable in "ideal" conditions. In fact, a gap between the indenter and the confining chamber is necessary to guarantee a correct indenter movement. In our experimental set-up the gap corresponds to 4% of the chamber radius, a value smaller than that used and discussed by other researchers [4]. The poroelastic model that we adopted was used to investigate whether the presence of a gap produced significant differences in the creep response curve and, hence, in the evaluated permeability and H_A parameters. An additional model geometry representing the actual gap was developed and compared with the ideal model. For these latter simulations, typical values were adopted for the cartilage model parameters ($E = 0.5$ MPa, $\nu = 0.25$, $k = 1.0\mathrm{e}^{-14}$ m^4/(N·s), $\phi = 75\%$); a pressure of 0.025 MPa was applied to a 600 μm-disk of cartilage. Results (Fig. 4) indicate that the creep response is different between ideal and actual confined geometries: the presence of the gap

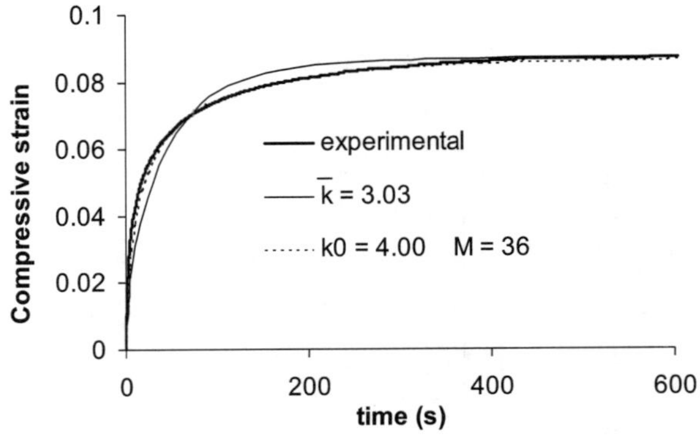

a)

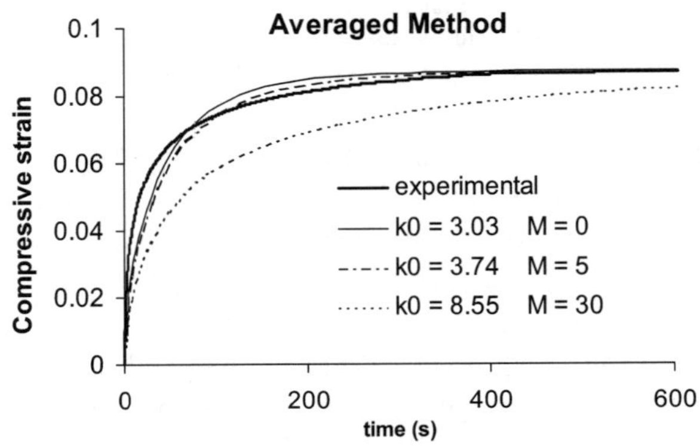

b)

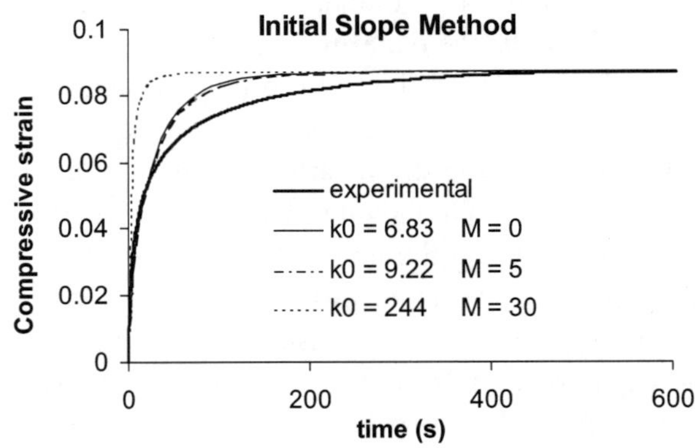

c)

Fig. 3. Comparison between experimental and simulated curves of confined compression. In the simulations, different permeability values (constant $\bar{k}$ or strain-dependent $k_0 e^{-M\varepsilon}$) were adopted according to three methods (see text). Permeability values (k and k_0) are expressed in 10^{-14} $m^4/(N{\cdot}s)$.

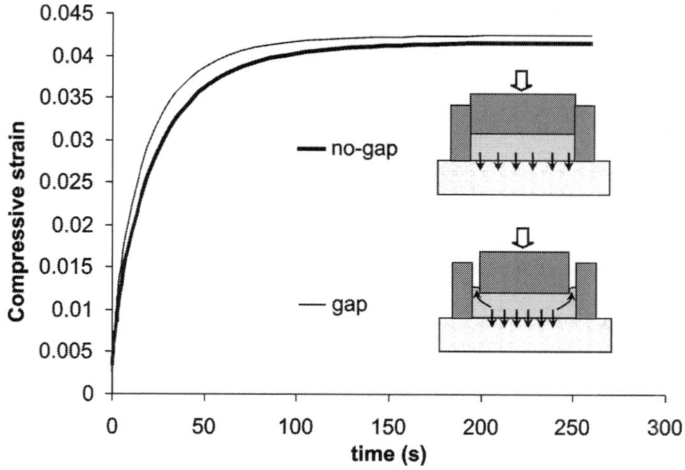

Fig. 4. Simulated creep response of a cartilage disk (radius = 2.5 mm, thickness = 600 μm) during confined compression using ideal (no-gap) and actual (gap) model configurations ($E = 0.5$ MPa, $\nu = 0.25$, $k = 1.0\text{e}^{-14}$ $\text{m}^4/(\text{N}\cdot\text{s})$, $\phi = 75\%$, pressure = 0.025 MPa).

allows flow exudation and tissue extrusion around the indenter, thus leading to underestimate H_A and to overestimate $\bar{k}$. For the adopted model parameters in those last simulations, errors are negligible for H_A (2%), but relevant for $\bar{k}$ (~30%).

In conclusion: (i) the analytical, uni-dimensional, approach to the confined creep test seems inadequate to describe the actual transient response; therefore, the permeability cannot be identified using the Averaged or Initial Slope Methods; (ii) the identification of the permeability parameters requires FEM models based on actual geometry and iterative and independent change of k_0 and M, requiring considerable computational time; (iii) the adopted approach based on experimental and computational methods can be used to identify the parameters that define the permeability dependence on deformation as a function of depth from articular surface.

Acknowledgements

Authors are grateful to Dr. Maurizio Colombo and Mr. Floriano Scienza for technical support during the biomechanical tests.

References

[1] F. Boschetti, M. Colombo, G.M. Peretti, G. Fraschini and R. Pietrabissa, Direct and indirect measurement of human articular cartilage permeability, *Trans. Orthop. Res. Soc.* **28** (2003), 293.

[2] A.C. Chen, W.C. Bae, R.M. Schinagl and R.L. Sah, Depth- and strain-dependent mechanical and electromechanical properties of full-thickness bovine articular cartilage in confined compression, *J. Biomech.* **34** (2001), 1–12.

[3] S.S. Chen, Y.H. Falcovitz, R. Schneiderman, A. Maroudas and R.L. Sah, Depth-dependent compressive properties of normal aged human femoral head articular cartilage: relationship to fixed charge density, *Osteoarthr. Cartil.* **9** (2001), 561–569.

[4] M.R. DiSilvestro and J.-K.F. Suh, A cross-validation of the biphasic poroviscoelastic model of articular cartilage in unconfined compression, indentation, and confined compression, *J. Biomech.* **34** (2001), 519–525.

[5] W.Y. Gu, W.M. Lai and V.C. Mow, Transport of fluids and ions through a porous-permeable charged-hydrated tissue, and streaming potential data on normal bovine articular cartilage, *J. Biomech.* **26** (1993), 709–723.

[6] W.C. Hayes, L.M. Keer, G. Herrmann and L.F. Mockros, A mathematical analysis for indentation tests of articular cartilage, *J. Biomech.* **5** (1972), 541–551.
[7] M.H. Holmes, Comparison theorems and similarity solution approximations for a nonlinear diffusion equation arising in the study of soft tissue, *SIAM J. Appl. Math.* **44** (1984), 545–556.
[8] E.B. Hunziker, Articular structure in humans and experimental animals, in: *Articular Cartilage and Osteoarthritis*, K.E. Kuettner, R. Schleyerbach, J.G. Peyron and V.C. Hascall, eds, Raven Press, New York, 1992, pp. 183–199.
[9] J.S. Jurvelin, M.D. Buschmann and E.B. Hunziker, Optical and mechanical determination of Poisson's ratio of adult bovine humeral articular cartilage, *J. Biomech.* **30** (1997), 235–241.
[10] R.K. Korhonen, M.S. Laasanen, J. Toyras, J. Rieppo, J. Hirvonen, H.J. Helminen and J.S. Jurvelin, Comparison of the equilibrium response of articular cartilage in unconfined compression, confined compression and indentation, *J. Biomech.* **35** (2002), 903–909.
[11] M.K. Kwan, W.M. Lai and V.C. Mow, Fundamentals of fluid transport through cartilage in compression, *Ann. Biomed. Eng.* **12** (1984), 537–58.
[12] W.M. Lai and V.C. Mow, Drag-induced compression of articular cartilage during a permeation experiment, *Biorheology* **17** (1980) 111–123.
[13] L.P. Li, M.D. Buschmann and A. Shirazi-Adl, A fibril reinforced nonhomogeneous poroelastic model for articular cartilage: inhomogeneous response in unconfined compression, *J. Biomech.* **33** (2000), 1533–1541.
[14] J.M. Mansour and V.C. Mow, The permeability of articular cartilage under compressive strain and at high pressures, *J. Bone Joint Surg.* **58A** (1976), 509–516.
[15] A. Maroudas, Physico-chemical properties of articular cartilage, in: *Adult Articular Cartilage*, M.A.R. Freeman, ed., Pitman Medical, Tunbridge Wells, England, 1979, pp. 215–290.
[16] V.C. Mow, M.C. Gibbs, W.M. Lai, W.B. Zhu and K.A. Athanasiou, Biphasic indentation of articular cartilage–II. A numerical algorithm and an experimental study, *J. Biomech.* **22** (1989), 853–861.
[17] V.C. Mow, S.C. Kwei, W.M. Lai and C.G. Armstrong, Biphasic creep and stress relaxation of articular cartilage in compression: theory and experiments, *J. Biomech. Eng.* **102** (1980), 73–84.
[18] V.C. Mow and L.A. Setton, Mechanical properties of normal and osteoarthritic articular cartilage, in: *Osteoarthritis*, K.D. Brandt, M. Doherty and L.S. Lohmander, eds, Oxford University Press, Oxford, UK, 1998, pp. 108–122.
[19] V.C. Mow, C.C. Wang and C.T. Hung, The extracellular matrix, interstitial fluid and ions as a mechanical signal transducer in articular cartilage, *Osteoarth. Cartil.* **7** (1999), 41–58.
[20] L. Sokoloff, Elasticity on aging cartilage, *Fed. Proc. Am. So. Exp. Bio.* **25** (1966), 1089–1095.
[21] S. Treppo, H. Koepp, E.C. Quan, A.A. Cole, K.E. Kuettner and A.J. Grodzinsky, Comparison of biomechanical and biochemical properties of cartilage from human knee and ankle pairs, *J. Orthop. Res.* **18** (2000), 739–748.
[22] C.C.-B.Wang, J.-M. Deng, G.A. Athesian and C.T. Hung, An automated approach for direct measurement of two-dimensional strain distributions within articular cartilage under unconfined compression, *J. Biomech. Eng.* **124** (2002), 557–567.
[23] C.C.-B. Wang, C.T. Hung and V.C. Mow, An analysis of the effects of depth-dependent aggregate modulus on articular cartilage stress-relaxation behavior in compression, *J. Biomech.* **34** (2001), 75–84.

Biorheology 41 (2004) 167–179
IOS Press

Mechano-acoustic determination of Young's modulus of articular cartilage

Simo Saarakkala [a,b], Rami K. Korhonen [c], Mikko S. Laasanen [b,d], Juha Töyräs [c], Jarno Rieppo [d] and Jukka S. Jurvelin [b,c,*]

[a] *Department of Nuclear Medicine, Etelä-Savo Hospital District, Mikkeli Central Hospital, Porrassalmenkatu 35-37, 50100 Mikkeli, Finland*

[b] *Department of Clinical Physiology and Nuclear Medicine, Kuopio University Hospital and University of Kuopio, POB 1777, 70211 Kuopio, Finland*

[c] *Department of Applied Physics, University of Kuopio, POB 1627, 70211 Kuopio, Finland*

[d] *Department of Anatomy, University of Kuopio, POB 1627, 70211 Kuopio, Finland*

Abstract. The compressive stiffness of an elastic material is traditionally characterized by its Young's modulus. Young's modulus of articular cartilage can be directly measured using unconfined compression geometry by assuming the cartilage to be homogeneous and isotropic. In isotropic materials, Young's modulus can also be determined acoustically by the measurement of sound speed and density of the material. In the present study, acoustic and mechanical techniques, feasible for *in vivo* measurements, were investigated to quantify the static and dynamic compressive stiffness of bovine articular cartilage *in situ*. Ultrasound reflection from the cartilage surface, as well as the dynamic modulus were determined with the recently developed ultrasound indentation instrument and compared with the reference mechanical and ultrasound speed measurements in unconfined compression ($n = 72$). In addition, the applicability of manual creep measurements with the ultrasound indentation instrument was evaluated both experimentally and numerically. Our experimental results indicated that the sound speed could predict 47% and 53% of the variation in the Young's modulus and dynamic modulus of cartilage, respectively. The dynamic modulus, as determined manually with the ultrasound indentation instrument, showed significant linear correlations with the reference Young's modulus ($r^2 = 0.445$, $p < 0.01$, $n = 70$) and dynamic modulus ($r^2 = 0.779$, $p < 0.01$, $n = 70$) of the cartilage. Numerical analyses indicated that the creep measurements, conducted manually with the ultrasound indentation instrument, were sensitive to changes in Young's modulus and permeability of the tissue, and were significantly influenced by the tissue thickness. We conclude that acoustic parameters, i.e. ultrasound speed and reflection, are indicative to the intrinsic mechanical properties of the articular cartilage. Ultrasound indentation instrument, when further developed, provides an applicable tool for the *in vivo* detection of cartilage mechano-acoustic properties. These techniques could promote the diagnostics of osteoarthrosis.

Keywords: Articular cartilage, Young's modulus, ultrasound, osteoarthrosis, finite element analysis

1. Introduction

Articular cartilage is both structurally and mechanically highly complex tissue. Basic constituents of cartilage, i.e. proteoglycans, collagen fibrils and water, are inhomogenously distributed in tissue (Figs 1 and 2). As a result of this, in the view of biomechanics, cartilage is inhomogeneous, layered poroelastic

*Address for correspondence: Professor Jukka Jurvelin, Ph.D., Department of Applied Physics, University of Kuopio, P.O.B. 1627, FIN-70211, Kuopio, Finland. Tel.: +358 17 162530; Fax: +358 17 162585; E-mail: jukka.jurvelin@uku.fi; URL: www.luotain.uku.fi.

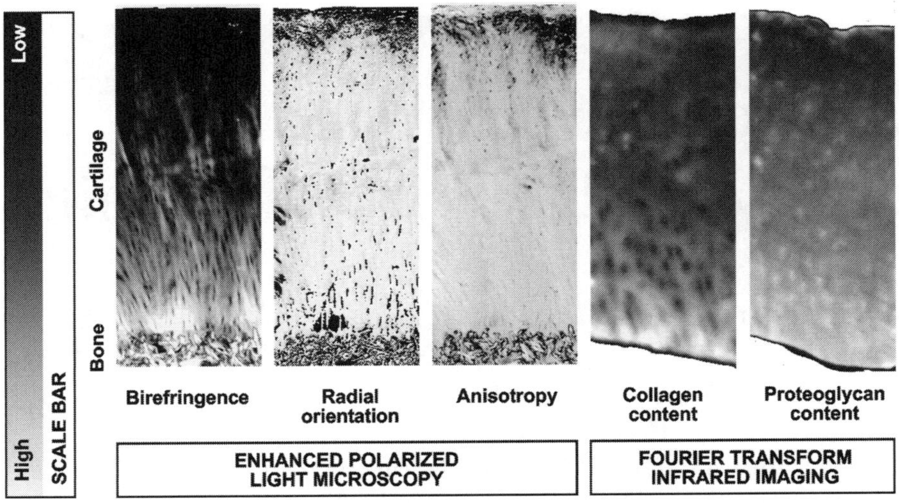

Fig. 1. Representative enhanced polarized light microscopic images showing birefringence, orientation and anisotropy of human cartilage collagen fibrils. Images confirm the classical collagen network arrangement described by Benninghoff [5]. Collagen fibrils are highly organized in the superficial and deep layers whereas the middle layer shows more randomly arranged arcading fibrils. Fourier transform infrared spectroscopic images visualize the spatial collagen and proteoglycan content of extracellular matrix. Both collagen and proteoglycan concentrations show increasing gradient toward cartilage–bone interface.

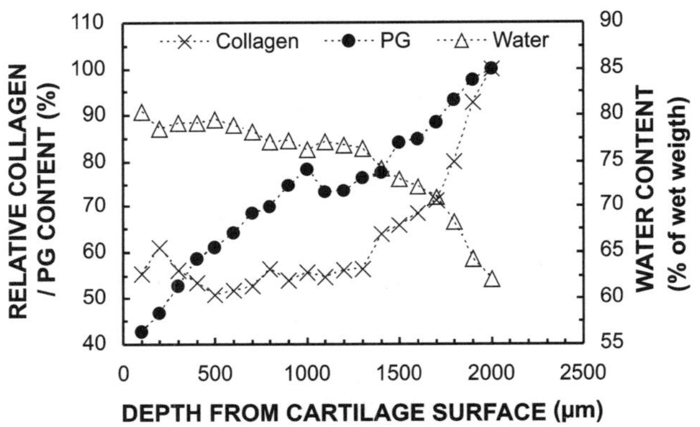

Fig. 2. Spatial distribution of human cartilage collagen, proteoglycan (PG) and water contents. Bulk water content was determined by weighting the wet and dry weight of the cartilage tissue whereas Fourier transform infrared imaging (FTIRI) was used for measuring spatial collagen (absorption at 1710–1610 cm^{-1}) and proteoglycan (absorption at 1145–980 cm^{-1}) concentrations. Spatial distribution of solid matrix could be calculated by assuming that it contains $1/3$ of PG and $2/3$ of collagen. Both collagen and proteoglycan concentrations increase while water content decreases toward the cartilage–bone junction.

tissue with nonlinear and anisotropic mechanical properties [7,19,37]. For an anisotropic single-phasic elastic material, the stress–strain relation, known as Hooke's law, has the form:

$$\sigma_{ij} = C_{ijkl}\varepsilon_{kl}, \tag{1}$$

where σ_{ij} is the stress tensor, C_{ijkl} is the elastic stiffness matrix and ε_{kl} is the strain tensor. To fully characterize the mechanical behaviour of anisotropic material altogether 21 stiffness components (elastic constants, C_{ijkl}) are needed. For the material with mutually perpendicular planes of elastic symmetry (orthotropic material), nine elastic constants are needed to characterize the mechanical behaviour. By further simplification, i.e. by assuming the same properties in one plane (e.g., the x–y plane) and different properties in the direction normal to this plane (e.g., the z-axis), the transversely isotropic material is described by five independent elastic constants. For a perfectly isotropic material, there are two independent elastic constants: the Young's modulus (E) and Poisson's ratio (ν). Then, the Young's modulus, aggregate modulus (H_A) and shear modulus (μ) are related as follows:

$$H_A = \frac{1-\nu}{(1+\nu)(1-2\nu)}E, \tag{2}$$

$$\mu = \frac{E}{2(1+\nu)}. \tag{3}$$

Due to the complex structure and properties of articular cartilage, even the use of linear orthotropic elasticity seems not to capture all essential features, including different responses in compression and tension [37]. Several simplifying assumptions have been made in order to describe the mechanical behaviour of tissue with a fewer material parameters, extractable from experimental measurements. In the classical linear biphasic model cartilage is assumed to consist of a perfectly elastic and isotropic solid matrix with homogeneously distributed fluid phase [23]. Thus, the material parameters of the model are Young's modulus (E), Poisson's ratio (ν) and permeability (κ). The isotropic biphasic model has been used extensively for characterizing the mechanical behaviour of cartilage. As a further modification of the isotropic biphasic model, several modeling approaches including matrix transverse isotropy [10], fibril reinforcement [33] and viscoelasticity [11,20] have been developed to improve the model predictions especially under dynamically varying loads.

Under laboratory conditions, isotropic E for cartilage is directly measured using unconfined stress relaxation or creep test. After the full relaxation or creep, E is calculated as a ratio of the axial stress and strain. E depends only on the elastic stiffness of solid matrix, as suggested by the biphasic model [23]. Alternatively, for the calculation of E from indentation measurements, value of ν has to be assumed, independently determined or derived using a model analysis of the experimental measurement [13,21]. Since the indentation measurements do not require isolated cartilage plugs, they are also applicable to *in vivo* applications. Higher values of E for the cartilage have been reported in indentation geometry than in unconfined compression geometry, but the modulus values are highly related between measurement geometries [14]. Larger values of E have systematically been reported for the deep layers of cartilage than for the surface layers [7,17,32]. In addition, values of E vary between different animal species [4,29] and measurement sites [3,4,16]. Importantly, decrease of E has been found to relate sensitively to cartilage degeneration [18,30].

High frequency ultrasound has been used for the imaging and quantitative analysis of structural and functional properties of articular cartilage. In the isotropic elastic solid, the speed of longitudinal sound wave, c, is given by

$$c = \sqrt{\frac{E(1-\nu)}{\rho_0(1+\nu)(1-2\nu)}}, \tag{4}$$

where ν and ρ are Poisson's ratio and density of the material, respectively. When sound penetrates in a narrow bar the modulus and speed are related as

$$E = \rho c^2. \tag{5}$$

Thereby, provided that ρ is known, E can be directly determined through the measurement of c. Ultrasound reflection from an acoustic interface is related to the acoustic impedances of the materials before (Z_1) and after (Z_2) the reflecting interface as follows:

$$R = \left(\frac{Z_2 - Z_1}{Z_2 + Z_1}\right)^2, \tag{6}$$

where R is the reflection coefficient describing the ratio of the reflected and incident sound amplitude. Furthermore, in the isotropic elastic solid, Z is linearly related to the material density and speed of sound:

$$Z = \rho c. \tag{7}$$

By combining the Eqs (4) and (7) we obtain:

$$Z_i = \rho_i \sqrt{\frac{E_i(1 - \nu_i)}{\rho_i(1 + \nu_i)(1 - 2\nu_i)}}, \quad i = 1, 2, \tag{8}$$

where 1 and 2 refers to the material before (1) and after (2) the reflective interface. By substituting Eq. (8) in Eq. (6), we obtain the relation between the ultrasound reflection coefficient and E, ν and ρ. Thereby, provided that ρ and ν are known, E can be determined also through the simple measurement of R. However, direct acoustic determination of the cartilage Young's modulus is more difficult due to inhomogeneous multiphasic structure of cartilage, and, e.g., in our previous study ultrasound reflection was only weakly related to the tissue Young's modulus [35]. As far as we know there are no theoretical models that have successfully described sound penetration in cartilage. In addition, designing simple measurement geometries for acoustic testing of cartilage may be difficult. However, several experimental studies have shown that acoustic parameters derived from relatively simple measurements significantly correlate with the cartilage structure and moduli. These measurements include ultrasound reflection and backscattering from the cartilage surface, sound speed and attenuation in cartilage. Similarly as for mechanical parameters, values of acoustic parameters vary in different animal species and measurement sites, and are significantly affected by the tissue integrity.

Combination of the mechanical and acoustic measurements has been suggested to serve as a technique capable for *in vivo* measurement of cartilage properties. Recently, we have presented an ultrasound indentation instrument which consists of an ultrasound transducer mounted on a tip of a commercial arthroscopic indentation instrument [15]. Dynamic indentation modulus (E_{dyn}) of cartilage and ultrasound reflection (R) from the cartilage surface can be quantified with the instrument. We have demonstrated that the instrument is suitable for detecting topographical differences in bovine knee joint *in situ* [16] as well as changes in the mechano-acoustic properties of spontaneously degenerated [30] or enzymatically degraded [15] bovine cartilage.

In the present study, we used both mechanical and acoustic techniques to quantify the material parameters, i.e. Young's modulus and dynamic modulus, for bovine articular cartilage. This study is an extension of our three earlier studies [15,16,30]. The materials used in those studies were now combined and the results analyzed again. First, we clarified experimentally the capability of acoustic measurements to predict Young's modulus and dynamic modulus of bovine articular cartilage *in vitro*. For the same samples, we then used combined mechanical and acoustic measurements to test whether the prediction of tissue moduli could be improved. We further investigated, both experimentally and numerically, the feasibility of manual creep measurements with the ultrasound indentation instrument. Finally, the interoperator reproducibility of the ultrasound indentation measurements was determined in the present study.

2. Materials and methods

2.1. Sample processing

Bovine knee joints, obtained from the local abattoir, were opened and altogether 72 samples were prepared [15,16,30]. Osteochondral samples were prepared from the medial tibial plateau (MTM, $n = 6$), lateral patello-femoral groove (LPG, $n = 6$), medial femoral condyle (MFC, $n = 6$) and lateral upper quadrant of patella (PAT, $n = 54$). Mechano-acoustic measurements were conducted for all samples with the ultrasound indentation instrument (Fig. 3). Furthermore, manual creep measurements were conducted for the four patellar samples showing visually different degenerative stages. Subsequently, reference measurements were conducted with the high-resolution mechano-acoustic material tester for all samples [17]. Finally, samples used in the creep measurements were processed for the histologic evaluation of cartilage quality [22,36].

2.2. Measurements with the ultrasound indentation instrument

Sample thickness (h), dynamic modulus (E_{dyn}) and ultrasound reflection coefficient from the surface (R) were quantified with the ultrasound indentation instrument (Fig. 3). The h was determined with the instrument by assuming a constant ultrasound speed for the cartilage, as described in the study by Töyräs et al. [36]. Measurement and calculation of E_{dyn} and R have been presented in detail in our recent study [30]. For the determination of R, the ultrasound transducer was kept at a constant distance from the cartilage surface and, subsequently, reflected (and backscattered) signal from the cartilage surface was collected.

Protocol for manual creep measurements consisted of a 1 N pre-stress for 5 seconds, followed by the test load to induce instantaneously a 4% strain. The subsequent creep was registered for 20 seconds. The creep rate (k), defined as the slope of indentation compliance vs. logarithmic time (compliance = indentation strain-to-stress ratio), was calculated in order to evaluate the time-dependent behaviour of samples [27].

2.3. Reference measurements

Biomechanical reference measurements of the samples were conducted with the high-resolution mechano-acoustic material testing device [36]. The measurements were conducted by using stress relaxation protocol in unconfined compression geometry [15,16,30]. Reference dynamic modulus ($E_{\mathrm{ref}}^{\mathrm{dyn}}$) was calculated as the maximum stress–strain ratio whereas Young's modulus (E_{ref}) indicated the equilibrium stress-to-strain ratio. Measurement protocol and calculation of sound speed (c) is presented with full details in our recent study [36].

2.4. Histologic evaluation

The samples were histologically evaluated using the Mankin score method [22]. Degenerative grade of each sample was evaluated by three of the authors and compared with the experimental and numerical results from mechano-acoustic tests [30].

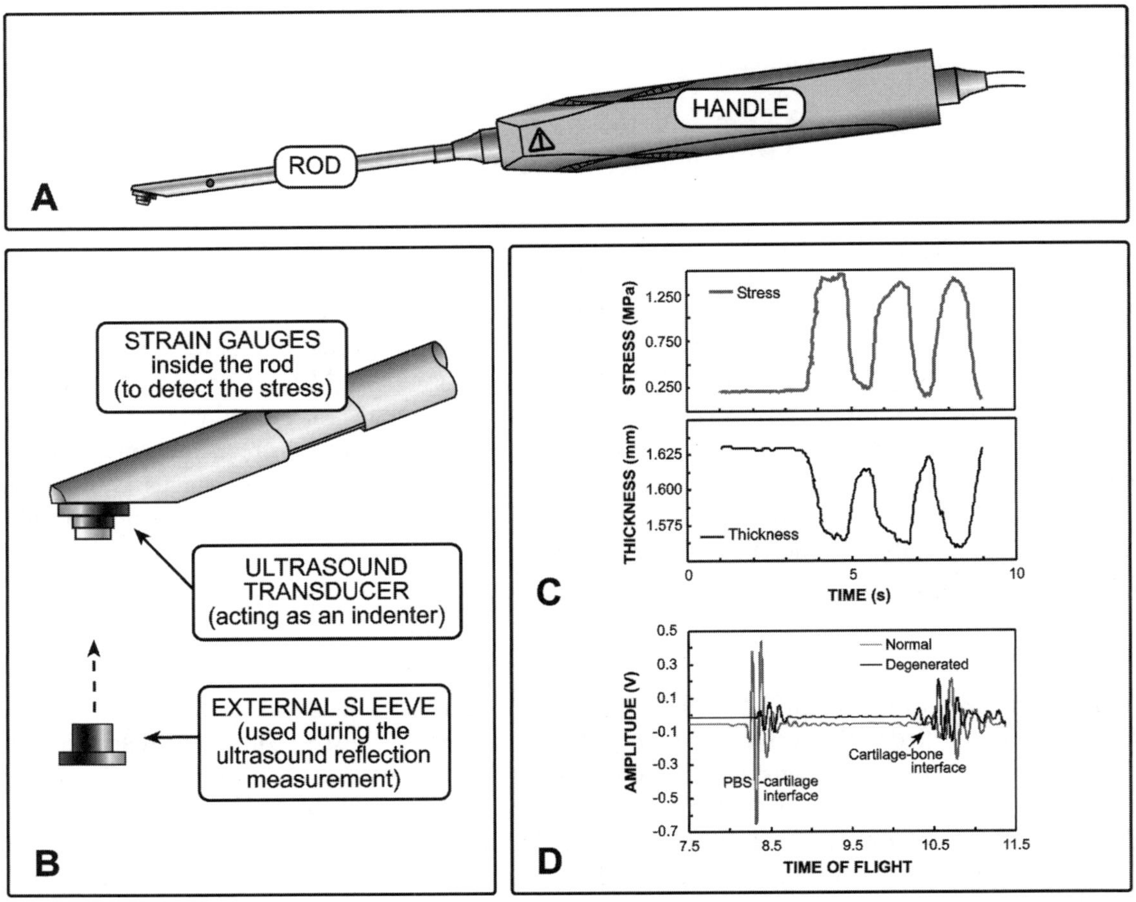

Fig. 3. (A) Schematic presentation of the design of the ultrasound indentation instrument. Instrument consists of a handle, measurement rod and the ultrasound transducer mounted on the instrument tip (Artscan 200, Artscan Oy, Helsinki, Finland). (B) An unfocused miniature 10 MHz contact ultrasound transducer was used as an indenter, and the stress was simultaneously detected using a strain gauge inside the rod. In ultrasound reflection measurements, an external plastic sleeve (length = 3.5 mm) was attached over the ultrasound transducer in order to keep the transducer at a constant distance from the articular surface. (C) Typical ultrasound indentation measurement. Instantaneous compressions were conducted to the predefined strain and the stress was recorded and analysed. (D) Typical ultrasound reflection measurement. Backscattered signal consisted of two interfaces: PBS–cartilage and cartilage–bone. Amplitude of the reflected signal from the PBS–cartilage interface was determined. Significant decrease in the surface reflection signal was observed in degenerated cartilage samples as compared to the normal samples.

2.5. Numerical analyses of the manual creep measurements with the ultrasound indentation instrument

The manual creep measurements were modeled numerically by using a finite element (FE) code of ABAQUS 6.1 (Hibbitt, Karlsson & Sorensen Inc., Pawtucket, RI, USA). The axially symmetric transversely isotropic poroelastic model of cartilage consisted of 8-node biquadratic elements (Fig. 4). The material parameters of the model were Young's moduli (E_1 and E_3), Poisson's ratios (ν_{31} and ν_{12}), shear modulus (μ_{13}) and permeability (κ). The plane of isotropy and loading direction (perpendicular to articular surface) are indicated by 1–2 and 3, respectively. The initial void ratio, i.e. ratio of fluid to solid content, was 3.5 [24]. The diameter of the indenter (transducer) and the radius of the cartilage (diameter = 3.0 mm, radius = 9.5 mm) were similar to those used in the actual instrument. Fluid was

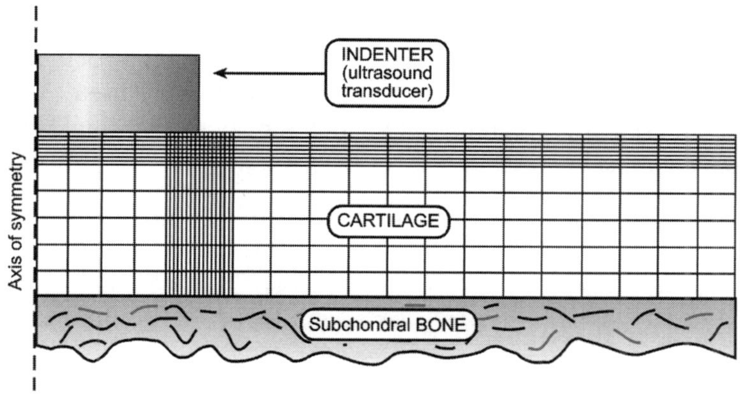

Fig. 4. A part of the axisymmetric finite element mesh for the ultrasound indentation geometry. In the numerical simulations, the cartilage was modeled as a transversely isotropic poroelastic material. The diameter of the cartilage was 19 mm.

Table 1

Material parameters for the transversely isotropic poroelastic finite element model and experimentally determined parameters. E_1 and ν_{12} are in-plane Young's modulus and Poisson's ratio, E_3 and ν_{31} are out-of-plane Young's modulus and Poisson's ratio, μ_{13} is shear modulus and κ is permeability

	Numerical							Experimental		
	Thickness (mm)	E_1 (MPa)	E_3 (MPa)	ν_{31}	ν_{12}	μ_{13} (MPa)	$\kappa \times 10^{-15}$ (m^4/Ns)	E_{dyn} (MPa)	k (1/kPa)	Mankin score
Simulation	0.50–5.00	5.8	0.46	0	0.5	0.37	5.1	–	–	–
Sample 1	1.54	16.0	0.44	0	0.5	3.20	1.2	15.23	11	0
Sample 2	1.64	5.5	0.28	0	0.5	1.10	1.8	2.36	18	2
Sample 3	2.19	3.1	0.27	0	0.5	0.62	5.1	4.32	32	5
Sample 4	1.81	0.5	0.03	0	0.5	0.10	15.0	1.41	95	10

allowed to flow through the cartilage surface and through the lateral edge, but not through the indenter or fixed cartilage–bone interface. At the axis of symmetry, neither fluid flow nor displacement was allowed in the horizontal direction. The indenter was modeled as an analytical rigid body, and the contact between the indenter and the cartilage surface was set to be adhesive [2,34]. The FE mesh was adjusted to be fine in the superficial layer and below the indenter edge, and coarser elsewhere (Fig. 4). Altogether 650 elements were used in the model. First, effect of the cartilage thickness on the long-term behaviour of cartilage was investigated using 84.4 kPa stress [34]. Material parameters for the model (Table 1) were adopted from the literature [9], and the measurements were modeled with five different values of cartilage thickness (1, 2, 3, 4 and 5 mm). Subsequently in the model, the ultrasound transducer (Fig. 3B) was pressed against the cartilage surface with a predefined creep protocol. Results of the experimental creep measurements were compared with those of the numerical analyses.

2.6. *Statistical analyses*

Pearson's correlation analysis was used for interrelating the experimental parameters. The interoperator reproducibility test for the ultrasound indentation instrument was conducted by three of the authors. Intraclass correlation coefficient (Rho and 95% confidence interval) [1] was calculated for the ultrasound indentation parameters (h, E_{dyn} and R). Visually healthy bovine patellar cartilage samples ($n = 8$) were used for the interoperator reproducibliity measurements. Reproducibility of the instrument was also described by calculating the coefficient of variation (CV) and standardized coefficient of variation (sCV) [6,12,26]. In calculation of CV, mean and SD of three operators were used as a result for individual sample. To establish a representative reference population for calculating single user sCV, the data from the same operator were combined ($n = 56$).

3. Results

Significant linear correlations ($p < 0.01$) were established between the ultrasound parameters (c, R) and the cartilage reference mechanical properties (E_{ref}, $E_{\mathrm{ref}}^{\mathrm{dyn}}$) (Fig. 5). Specifically, sound speed, c, predicted cartilage moduli better than the ultrasound reflection, R (c: $r^2 = 0.467$–0.533 and R: $r^2 = 0.162$–0.316, Fig. 5). A significant linear correlation was also obtained between c and R ($r^2 = 0.388$).

E_{dyn}, as measured with the ultrasound indentation instrument, predicted cartilage reference dynamic modulus ($E_{\mathrm{ref}}^{\mathrm{dyn}}$) better than it predicted the reference Young's modulus (E_{ref}) ($r^2 = 0.779$ vs. $r^2 = 0.445$,

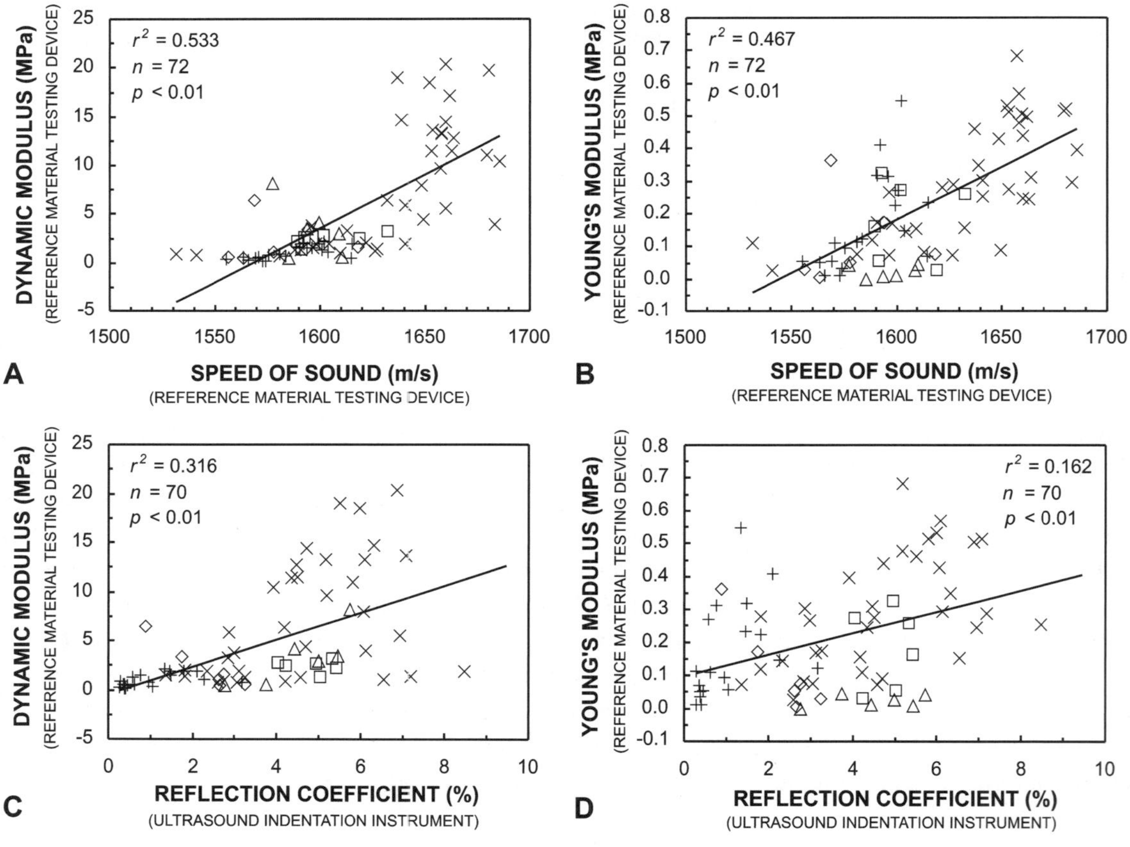

Fig. 5. Relations between (A) reference dynamic modulus ($E_{\mathrm{ref}}^{\mathrm{dyn}}$) and speed of sound ($c$), (B) Young's modulus ($E_{\mathrm{ref}}$) and speed of sound ($c$), (C) reference dynamic modulus ($E_{\mathrm{ref}}^{\mathrm{dyn}}$) and ultrasound reflection coefficient from the surface (R) and (D) Young's modulus (E_{ref}) and ultrasound reflection coefficient from the surface (R). Significant ($p < 0.01$) linear correlations were established between the ultrasound parameters and the cartilage reference mechanical parameters. Symbols used in the figure: visually normal samples (×), visually degenerated samples (+), and samples digested with chondroitinase ABC (□), collagenase (◇) or trypsin (△) [15,16,30].

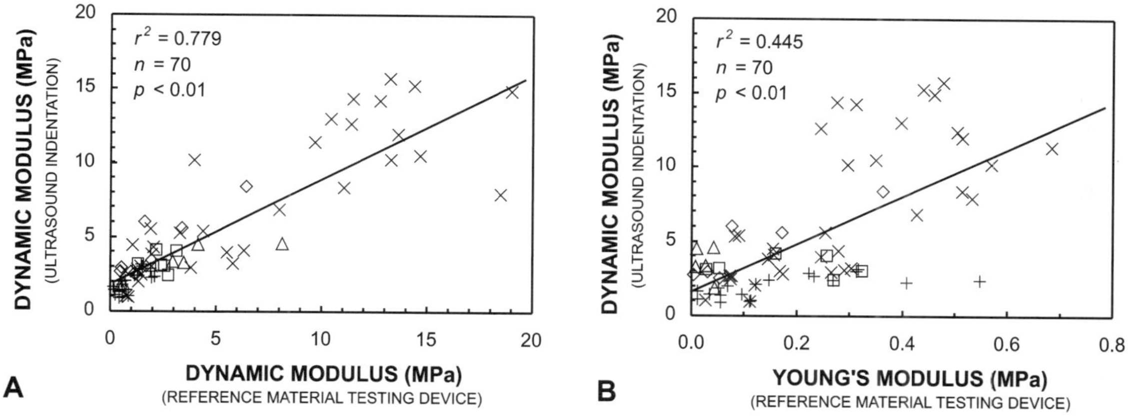

Fig. 6. Relations between (A) dynamic modulus (E_{dyn}) and reference dynamic modulus (E_{ref}^{dyn}) and (B) dynamic modulus (E_{dyn}) and Young's modulus (E_{ref}). E_{dyn}, measured with the ultrasound indentation instrument, predicted well the cartilage reference dynamic modulus (E_{ref}^{dyn}). Symbols used in the figure: visually normal samples ($\times$), visually degenerated samples ($+$), and samples digested with chondroitinase ABC ($\Box$), collagenase ($\Diamond$) or trypsin ($\triangle$) [15,16,30].

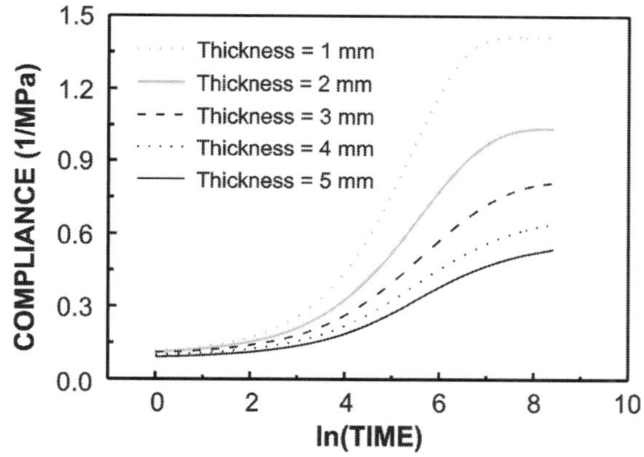

Fig. 7. Long-term creep measurements, analyzed numerically using the poroelastic transversely isotropic model (cartilage thickness 1–5 mm). Material parameters of the model (Table 1) were adopted from the literature [9]. Numerical analyses demonstrated the significant dependence of the tissue thickness on the creep measurements, as conducted with the ultrasound indentation instrument.

$p < 0.01$, Fig. 6). Linear correlation between the ultrasound reflection coefficient (R) and the dynamic modulus (E_{dyn}), both measured with the ultrasound indentation instrument, was significant ($r^2 = 0.285$, $p < 0.01$). Significant correlation was also established between E_{dyn} and c ($r^2 = 0.584$, $p < 0.01$).

Numerical analyses revealed that the tissue thickness had a significant effect on the time-dependent indenter force, measured with the ultrasound indentation instrument (Fig. 7). Material parameters of the experimental creep measurements are presented in Table 1. Numerical results agreed well with the experimental findings (Fig. 8). Importantly, significant differences in the material parameters were observed between the samples with different degenerative grades (Table 1). Young's modulus (E_1 or E_3) and shear modulus (μ_{13}) decreased and permeability (κ) increased as the degenerative grade (Mankin score) increased. Experimental findings demonstrated the same trend, i.e. E_{dyn} decreased and k increased as degeneration progressed (Table 1).

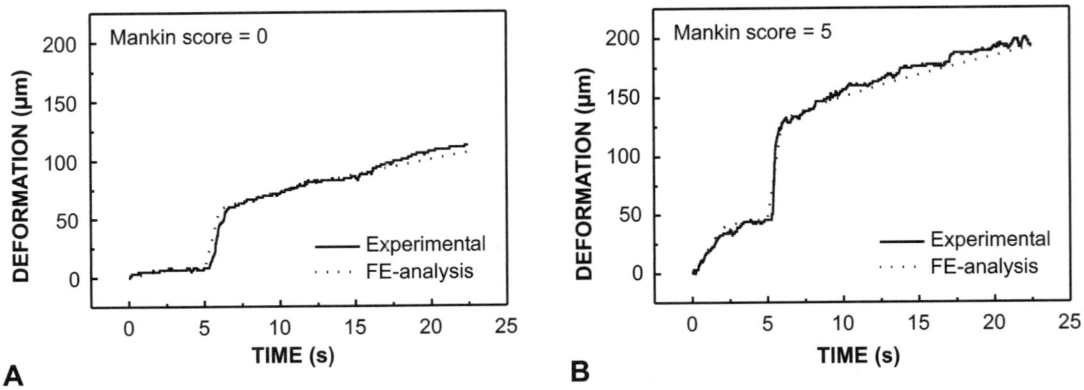

Fig. 8. Experimental ultrasound indentation creep measurements and numerical analyses of the (A) healthy cartilage sample (Mankin score = 0) and (B) degenerated cartilage sample (Mankin score = 5). Poroelastic transversely isotropic FE-model revealed the strong effect of changes in the values of Young's moduli, shear modulus and permeability on the creep measurements. Numerical analyses agreed well with the experimental measurements. Material parameters of the model are presented in Table 1.

As a measure for reproducibility of the ultrasound indentation measurements, intraclass correlation coefficients (Rho) were 0.905, 0.669 and 0.786 for thickness, dynamic modulus and reflection coefficient, respectively. 95% confidence intervals were 0.72–0.98, 0.27–0.91 and 0.46–0.95 for thickness, dynamic modulus and reflection coefficient, respectively. *CV* values, calculated for all three operators, were 5.0%, 17.2% and 5.5% for the h, E_{dyn} and R, respectively. Similarly, *sCV* values were 0.5%, 4.3% and 0.4% for the h, E_{dyn} and R, respectively.

4. Discussion and conclusion

In the present study, mechano-acoustic techniques were used to characterize the mechanical properties of normal and degenerated cartilage in compression. Especially, we were interested in techniques that are potentially applicable to *in vivo* diagnostics. For calculation of the Young's modulus of articular cartilage, the solid matrix was assumed to be isotropic and linearly elastic. The ability of the acoustic parameters, i.e. sound speed and surface reflection, to predict the tissue Young's modulus and dynamic modulus was investigated experimentally in bovine articular cartilage *in vitro*. Furthermore, the feasibility of the ultrasound indentation instrument [15] to determine cartilage intrinsic mechanical properties was investigated both experimentally and numerically. We found that the sound speed was significantly related to tissue compressive stiffness and integrity. Using the ultrasound indentation technique, we could derive a true dynamic modulus that was consistent with the one obtained from the reference measurements in unconfined compression. Although information on the dynamic modulus is essential in order to judge the functional competence of the cartilage under physiological joint loads, it had a limited capacity to predict the true material stiffness, i.e. the Young's modulus at equilibrium. Articular cartilage, which is mechanically unique tissue, shows a composite structure with different load carrying mechanisms under static and dynamic compressive and tensile loads. Consequently, compressive stiffness, when determined in the form of Young's modulus, is only one indicator for the functional integrity of the tissue.

In the present study, we combined results from our previous studies on normal [16], enzymatically degraded [15] and spontaneously degenerated [30] cartilage. Earlier, we demonstrated in bovine cartilage that sound speed is significantly related to tissue composition and Young's modulus: in proteoglycan rich tissue with high Young's modulus, sound speed is higher than in proteoglycan depleted, soft tissue [36]. The current results, derived for the combined material ($n = 72$), confirm these findings. In clinical situation, however, designing a simple and easy-to-use technique for the thickness measurement, necessary for the determination of sound speed, may be difficult and threaten its successful diagnostic use.

The ultrasound reflection from the cartilage surface, as well as the backscatter from the superficial tissue layer, is closely related to collagen network and provides a sensitive measure of the integrity of superficial tissue [8,25,35]. Although the reflection measurements are only related to properties of the superficial tissue they served as a moderate indicator of cartilage stiffness, as revealed by statistically significant linear correlations between the ultrasound surface reflection and the reference moduli in our recent study ($n = 30$, $r^2 = 0.745$ and 0.332 for $E_{\mathrm{ref}}^{\mathrm{dyn}}$ and E_{ref}, respectively) [30]. For the combined data of the current study, the linear correlations were lower ($n = 70$, $r^2 = 0.316$ and 0.162 for $E_{\mathrm{ref}}^{\mathrm{dyn}}$ and E_{ref}, respectively) although statistically significant. The current results included also the data from the cartilage samples digested with chondroitinase ABC. These samples show proteoglycan depletion [38] and are mechanically soft but demonstrate highly normal acoustic backscatter [15,28,31,35]. Based on the present and previous data we conclude that the reflection measurements provide a sensitive measure of the integrity of collagenous network in the superficial tissue but are not an unambiguous indicator of tissue compressive stiffness.

Ultrasound indentation measurements revealed the dynamic stiffness consistent with the dynamic modulus measured in the reference unconfined compression geometry. These results are encouraging and let us believe that the ultrasound indentation technique may be capable of determining mechanical properties of cartilage *in vivo*. The interoperator reproducibility of the measurements was determined and the intraclass correlation coefficients were found to be reasonable for all parameters. However, large confidence intervals diminish the reproducibility of some parameters. *CV* and *sCV* values indicate moderate or good reproducibilities for parameters. However, the final significance of the reproducibility values depend on the variation of parameter values encountered in true clinical populations and can be judged after that is revealed.

In our earlier study, we conducted manual creep measurements with the ultrasound indentation instrument before and after enzymatic degradation of bovine cartilage samples [15]. In that study, we suggested that our instrument is capable of quantifying the proteoglycan depletion of cartilage, known to be of high importance for the time-dependent mechanical properties. However, time-dependent relaxation process of cartilage under an impervious indenter is a complicated process, and has not been solved analytically. In the present study, the suitability of creep measurements was further investigated both experimentally and numerically. Poroelastic transversely isotropic FE-model revealed the strong effect of changes in the values of Young's moduli, shear modulus and permeability on the creep behavior. This was consistent with the experimental measurements showing a decrease of Young's modulus and an increase of creep as the degenerative grade (Mankin score) increased. Numerical results demonstrated the significant dependence of the tissue thickness on the creep measurements. Further theoretical analysis with an anisotropic, layered cartilage model should be conducted in order to successfully judge, and even eliminate the thickness dependence. In the operation theater, all measurements should be easy, quick and reproducible. Especially, successful manual creep measurements with the instrument prototype require a significant amount of training from the operator. It may be necessary to develop an additional facilitation if the creep measurements are applied clinically.

To conclude, the present results suggest that the mechano-acoustic techniques presented in this study are feasible for the estimation of the structural and functional properties of cartilage. Acoustic and mechanical data are complementary and may improve diagnostics of cartilage-related diseases. To exploit all possibilities of acoustic measurements, theoretical analyses focusing on the physical sound–tissue interactions in cartilage are needed. To address the changes in acoustic properties during tissue degeneration, these analyses should include realistic multiphasic cartilage structure. Ultrasound indentation lets us determine cartilage thickness as well as time dependent stress and strain and, thereby, enable objective measurement of cartilage material parameters, such as Young's modulus. However, further *in situ* and *in vivo* studies with the human cartilage are needed before these techniques can be applied clinically.

Acknowledgements

Financial support from the National Technology Agency (TEKES, projects 40714/01 and 70061/02), Finland; Kuopio University Hospital (EVO, projects 5173 and 5203), Kuopio, Finland; Etelä-Savo Hospital District (EVO), Mikkeli, Finland; The Finnish Cultural Foundation of Northern Savo, Finland; The Academy of Finland; The Finnish Biomaterial Graduate School and The Finnish Graduate School for Musculoskeletal Diseases is acknowledged.

References

[1] R.C. Appleyard, M.V. Swain, S. Khanna and G.A. Murrell, The accuracy and reliability of a novel handheld dynamic indentation probe for analysing articular cartilage, *Phys. Med. Biol.* **46** (2001), 541–50.

[2] C.G. Armstrong, W.M. Lai and V.C. Mow, An analysis of the unconfined compression of articular cartilage, *J. Biomech. Eng.* **106** (1984), 165–73.

[3] J.P. Arokoski, M.M. Hyttinen, H.J. Helminen and J.S. Jurvelin, Biomechanical and structural characteristics of canine femoral and tibial cartilage, *J. Biomed. Mater. Res.* **48** (1999), 99–107.

[4] K.A. Athanasiou, M.P. Rosenwasser, J.A. Buckwalter, T.I. Malinin and V.C. Mow, Interspecies comparisons of in situ intrinsic mechanical properties of distal femoral cartilage, *J. Orthop. Res.* **9** (1991), 330–40.

[5] A. Benninghoff, Form und bau der gelenkknorpel in ihren beziehungen zur function. Erste mitteilung: die modellierenden und formerhaltenden faktoren des knorpelreliefs, *Z. Anat.* **76** (1925), 46–63.

[6] G.M. Blake, H.W. Wahner and I. Fogelman, *The Evaluation of Osteoporosis: Dual Energy X-Ray Absorptiometry and Ultrasound in Clinical Practice*, 2nd edn, Martin Dunitz Ltd., London, 1999.

[7] A.C. Chen, W.C. Bae, R.M. Schinagl and R.L. Sah, Depth- and strain-dependent mechanical and electromechanical properties of full-thickness bovine articular cartilage in confined compression, *J. Biomech.* **34** (2001), 1–12.

[8] E. Chérin, A. Saïed, P. Laugier, P. Netter and G. Berger, Evaluation of acoustical parameter sensitivity to age-related and osteoarthritic changes in articular cartilage using 50-MHz ultrasound, *Ultrasound Med. Biol.* **24** (1998), 341–54.

[9] B. Cohen, T.R. Gardner and G. Ateshian, The influence of transverse isotropy on cartilage indentation behaviour. A study of the human humeral head, *Trans. Orthop. Res. Soc.* **18** (1993), 185.

[10] B. Cohen, W.M. Lai and V.C. Mow, A transversely isotropic biphasic model for unconfined compression of growth plate and chondroepiphysis, *J. Biomech. Eng.* **120** (1998), 491–6.

[11] M.R. DiSilvestro and J.K. Suh, A cross-validation of the biphasic poroviscoelastic model of articular cartilage in unconfined compression, indentation, and confined compression, *J. Biomech.* **34** (2001), 519–25.

[12] C.C. Glüer, G. Blake, Y. Lu, B.A. Blunt, M. Jergas and H.K. Genant, Accurate assessment of precision errors: how to measure the reproducibility of bone densitometry techniques, *Osteoporos. Int.* **5** (1995), 262–70.

[13] W.C. Hayes, L.M. Keer, G. Herrmann and L.F. Mockros, A mathematical analysis for indentation tests of articular cartilage, *J. Biomech.* **5** (1972), 541–51.

[14] R.K. Korhonen, M.S. Laasanen, J. Töyräs, H.J. Helminen and J.S. Jurvelin, Comparison of the equilibrium response of articular cartilage in unconfined compression, confined compression and indentation, *J. Biomech.* **35** (2002), 903–9.

[15] M.S. Laasanen, J. Töyräs, J. Hirvonen, S. Saarakkala, R.K. Korhonen, M.T. Nieminen et al., Novel mechano-acoustic technique and instrument for diagnosis of cartilage degeneration, *Physiol. Meas.* **23** (2002), 491–503.

[16] M.S. Laasanen, S. Saarakkala, J. Töyräs, J. Hirvonen, J. Rieppo, R.K. Korhonen et al., Ultrasound indentation of bovine knee articular cartilage in situ, *J. Biomech.* **36** (2003), 1259–67.

[17] M.S. Laasanen, J. Toyras, R.K. Korhonen, J. Rieppo, S. Saarakkala, M.T. Nieminen et al., Biomechanical properties of knee articular cartilage, *Biorheology* **40** (2003), 133–40.

[18] J.M. Lane, E. Chisena and J. Black, Experimental knee instability: early mechanical property changes in articular cartilage in a rabbit model, *Clin. Orthop.* (1979), 262–5.

[19] L.P. Li, A. Shirazi-Adl and M.D. Buschmann, Alterations in mechanical behaviour of articular cartilage due to changes in depth varying material properties – a nonhomogeneous poroelastic model study, *Comput. Methods Biomech. Biomed. Eng.* **5** (2002), 45–52.

[20] A.F. Mak, The apparent viscoelastic behavior of articular cartilage – the contributions from the intrinsic matrix viscoelasticity and interstitial fluid flows, *J. Biomech. Eng.* **108** (1986), 123–30.

[21] A.F. Mak, W.M. Lai and V.C. Mow, Biphasic indentation of articular cartilage. I. Theoretical analysis, *J. Biomech.* **20** (1987), 703–14.

[22] H.J. Mankin, H. Dorfman, L. Lippiello and A. Zarins, Biochemical and metabolic abnormalities in articular cartilage from osteo-arthritic human hips. II. Correlation of morphology with biochemical and metabolic data, *J. Bone Joint Surg. Am.* **53** (1971), 523–37.

[23] V.C. Mow, S.C. Kuei, W.M. Lai and C.G. Armstrong, Biphasic creep and stress relaxation of articular cartilage in compression: Theory and experiments, *J. Biomech. Eng.* **102** (1980), 73–84.

[24] V.C. Mow, D.C. Fithian and M.A. Kelly, Fundamentals of articular cartilage and meniscus biomechanics, in: *Articular Cartilage and Knee Joint Function: Basic Science and Arthroscopy*, J.W. Ewing, ed., Raven Press Ltd., New York, 1990, pp. 1–18.

[25] H.J. Nieminen, J. Töyräs, J. Rieppo, M.T. Nieminen, J. Hirvonen, R. Korhonen et al., Real-time ultrasound analysis of articular cartilage degradation in vitro, *Ultrasound Med. Biol.* **28** (2002), 519–525.

[26] C.F. Njeh, D. Hans, J. Li, B. Fan, T. Fuerst, Y.Q. He et al., Comparison of six calcaneal quantitative ultrasound devices: precision and hip fracture discrimination, *Osteoporos. Int.* **11** (2000), 1051–62.

[27] J.R. Parsons and J. Black, The viscoelastic shear behavior of normal rabbit articular cartilage, *J. Biomech.* **10** (1977), 21–9.

[28] B. Pellaumail, A. Watrin, D. Loeuille, P. Netter, G. Berger, P. Laugier et al., Effect of articular cartilage proteoglycan depletion on high frequency ultrasound backscatter, *Osteoarthritis Cartilage* **10** (2002), 535–41.

[29] J. Rieppo, E.P. Halmesmäki, U. Siitonen, M.S. Laasanen, J. Töyräs, I. Kiviranta et al., Histological differences of human, bovine and porcine cartilage, *Trans. Orthop. Res. Soc.* **28** (2003).

[30] S. Saarakkala, M.S. Laasanen, J.S. Jurvelin, K. Törrönen, M.J. Lammi, R. Lappalainen et al., Ultrasound indentation of normal and spontaneously degenerated bovine articular cartilage, *Osteoarthritis Cartilage* **11** (2003), 697–705.

[31] S. Saarakkala, J. Töyräs, J. Hirvonen, M.S. Laasanen, R. Lappalainen and J.S. Jurvelin, Ultrasonic quantitation of superficial degradation of articular cartilage, *Ultrasound Med. Biol.* (in press) (2004).

[32] R.M. Schinagl, D. Gurskis, A.C. Chen and R.L. Sah, Depth-dependent confined compression modulus of full-thickness bovine articular cartilage, *J. Orthop. Res.* **15** (1997), 499–506.

[33] J. Soulhat, M.D. Buschmann and A. Shirazi-Adl, A fibril-network-reinforced biphasic model of cartilage in unconfined compression, *J. Biomech. Eng.* **121** (1999), 340–7.

[34] J.K. Suh and R.L. Spilker, Indentation analysis of biphasic articular cartilage: nonlinear phenomena under finite deformation, *J. Biomech. Eng.* **116** (1994), 1–9.

[35] J. Töyräs, J. Rieppo, M.T. Nieminen, H.J. Helminen and J.S. Jurvelin, Characterization of enzymatically induced degradation of articular cartilage using high frequency ultrasound, *Phys. Med. Biol.* **44** (1999), 2723–33.

[36] J. Töyräs, M.S. Laasanen, S. Saarakkala, M.J. Lammi, J. Rieppo, J. Kurkijärvi et al., Speed of sound in normal and degenerated bovine articular cartilage, *Ultrasound Med. Biol.* **29** (2003), 447–54.

[37] C.C. Wang, N.O. Chahine, C.T. Hung and G.A. Ateshian, Optical determination of anisotropic material properties of bovine articular cartilage in compression, *J. Biomech.* **36** (2003), 339–53.

[38] T. Yamagata, H. Saito, O. Habuchi and S. Suzuki, Purification and properties of bacterial chondroitinases and chondrosulfatases, *J. Biol. Chem.* **243** (1968), 1523–1535.

The role of viscoelasticity of collagen fibers in articular cartilage: Theory and numerical formulation

L.P. Li * and W. Herzog
Human Performance Lab, Faculty of Kinesiology, University of Calgary, Alberta, Canada

Abstract. The relative importance of fluid-dependent and fluid-independent transient mechanical behavior in articular cartilage was examined for tensile and unconfined compression testing using a fibril reinforced model. The collagen matrix of articular cartilage was modeled as viscoelastic using a quasi-linear viscoelastic formulation with strain-dependent elastic modulus, while the proteoglycan matrix was considered as linearly elastic. The collagen viscoelastic properties were obtained by fitting experimental data from a tensile test. These properties were used to investigate unconfined compression testing, and the sensitivity of the properties was also explored. It was predicted that the stress relaxation observed in tensile tests was not caused by fluid pressurization at the macroscopic level. A multi-step tensile stress relaxation test could be approximated using a hereditary integral in which the elastic fibrillar modulus was taken to be a linear function of the fibrillar strain. Applying the same formulation to the radial fibers in unconfined compression, stress relaxation could not be simulated if fluid pressurization were absent. Collagen viscoelasticity was found to slightly weaken fluid pressurization in unconfined compression, and this effect was relatively more significant at moderate strain rates. Therefore, collagen viscoelasticity appears to play an import role in articular cartilage in tensile testing, while fluid pressurization dominates the transient mechanical behavior in compression. Collagen viscoelasticity plays a minor role in the mechanical response of cartilage in unconfined compression if significant fluid flow is present.

Keywords: Cartilage mechanics, cartilage tensile properties, collagen fibril reinforcement, nonlinear biomechanics, strain rates, unconfined compression

1. Introduction

Articular cartilage is known to exhibit strong transient mechanical behavior [40,43]. However, the mechanism of this strain-rate dependent response is not completely understood. While some studies have shown evidence of a fluid-flow dominant mechanism of the transient behavior [29,30,34,37,49,53], others indicate a significant influence from intrinsic viscoelasticity (referred to as viscoelasticity hereafter) on the response [7–9,19,20,26,31,32,38]. Theories have experienced difficulties in determining the roles of fluid-flow and viscoelastic properties in the mechanical behavior of articular cartilage [35].

Collagen fibers in rat tail tendon, or type I collagen in other soft tissues, have been extensively investigated, and were found to be viscoelastic and nonlinear [4,14–16,44,45,47,48,52,59,61]. The importance of collagen fibers (mainly type II collagen, e.g., [13]) in the transient response of cartilage has been suggested in experimental studies [3,33,36], and it is known that the cross-links in the collagen fibers make a significant contribution to the tensile properties of cartilage [5,58]. The viscoelasticity of cartilage has

*Address for correspondence: LePing Li, Faculty of Kinesiology, University of Calgary, 2500 University Drive, N.W., Calgary, Alberta, Canada T2N 1N4. Tel.: +1 403 220 3449; Fax: +1 403 284 3553; E-mail: leping@kin.ucalgary.ca.

also been observed in tensile testing [46,50,64], where fluid pressure is negligible, and the contribution from the proteoglycan matrix is likely low because of its small tensile stiffness compared to that of the fibrillar matrix. Therefore, there must be a dominant role of collagen for the observed viscoelastic properties in tensile testing. The intrinsic viscoelasticity of proteoglycans is not well documented. The viscoelastic shear modulus of cartilage has been reported to increase with collagen cross-links [17], indicating that the collagen fibers account for the viscoelasticity of cartilage. Furthermore, fluid flow and pressurization were found to be the dominant mechanisms of load support in confined compression [49], for which the collagen fibers are in compression, and thus are unlikely to play a significant role in the mechanical response. Therefore, it was assumed in the current work that collagen fibers in articular cartilage are viscoelastic, but the proteoglycan matrix is elastic [11].

Collagen viscoelasticity was incorporated into a fibril reinforced model [28] using a quasi-linear viscoelastic formulation [12]. The distinct tensile properties of cartilage [1,2,6,21–25,33,36,41,46,51,63] were recognized by assuming a zero compressive modulus and a strain-dependent tensile modulus for the fibrillar matrix. Using a mechanism of fluid-driven fibril reinforcement, the fibril reinforced models were able to explain the strain-rate dependent mechanical response of cartilage in compression [29]. The objective of the present study was to find the possible role of fiber viscoelasticity in the transient mechanical response of articular cartilage.

2. Methods

The fibrillar matrix is defined by three strain-dependent tensile moduli in the coordinate directions, $E_r^f(\varepsilon_r)$, $E_\theta^f(\varepsilon_\theta)$ and $E_z^f(\varepsilon_z)$. All other fibrillar properties are taken as zero. If in any direction, for example in the radial (r) direction, the fibers are in tension, the fibrillar stress is then determined by the hereditary integral [12] as

$$\sigma_r^f(t) = \sigma_r^f(0) + \int_0^t G_r(t-\tau)E_r^f(\varepsilon_r)\dot{\varepsilon}_r\,\mathrm{d}\tau, \tag{1}$$

where the relaxation function is represented by a discrete spectrum approximation [42,56]

$$G_r(t) = 1 + \sum_m g_m \exp(-t/\lambda_m). \tag{2}$$

Here λ_m are characteristic times for the viscoelastic dissipation. In many cases, taking three terms of the exponentials is sufficient for numerical analysis [56]. For the purpose of numerical computation, Eq. (1) is rewritten as

$$\begin{aligned}\sigma_r^f(t) = \sigma_r^f(t-\Delta t) &+ \int_{t-\Delta t}^t G_r(t-\tau)E_r^f(\varepsilon_r)\dot{\varepsilon}_r\,\mathrm{d}\tau \\ &+ \int_0^{t-\Delta t}\big[G_r(t-\tau) - G_r(t-\Delta t-\tau)\big]E_r^f(\varepsilon_r)\dot{\varepsilon}_r\,\mathrm{d}\tau.\end{aligned} \tag{3}$$

Supposing that the stress is known at time t_{k-1} ($k \geqslant 1$ and $t_0 = 0$), we find the stress at t_k as

$$\sigma_r^f(t_k) \approx \sigma_r^f(t_{k-1}) + G_r(\Delta t_k/2)E_r^f\left(\frac{\varepsilon_r(t_{k-1}) + \varepsilon_r(t_k)}{2}\right)\Delta\varepsilon_r^k$$

$$+\sum_m g_m\big(1-\exp(\Delta t_k/\lambda_m)\big)\exp(-t_k/\lambda_m)I^r_m(t_{k-1}), \tag{4}$$

where $\Delta t_k = t_k - t_{k-1}$, $\Delta\varepsilon^k_r = \varepsilon^k_r - \varepsilon^{k-1}_r$, and

$$I^r_m(t) = \int_0^t \exp(\tau/\lambda_m)E^f_r(\varepsilon_r)\dot{\varepsilon}_r\,\mathrm{d}\tau. \tag{5}$$

Equation (5) can be evaluated in a step-by-step fashion, and only the value from the previous step needs to be stored in a computer procedure. However, for very slow relaxation, t can become so large that the calculation of the exponential in (5) may cause a floating overflow in computer memory. For such cases, we calculate the following instead

$$\Psi^r_m(t_k) \doteq \exp(-t_k/\lambda_m)I^r_m(t_{k-1}). \tag{6}$$

Obviously, $\Psi^r_m(t_1) = 0$. For $k \geqslant 2$,

$$\begin{aligned}\Psi^r_m(t_k) &= \exp(-t_k/\lambda_m)\left[I^r_m(t_{k-2}) + \int_{t_{k-2}}^{t_{k-1}} \exp(\tau/\lambda_m)E^f_r(\varepsilon_r)\dot{\varepsilon}_r\,\mathrm{d}\tau\right]\\ &\approx \exp(-\Delta t_k/\lambda_m)\big[\Psi^r_m(t_{k-1}) + E^f_r\big(\varepsilon_r(t_{k-1})\big)\Delta\varepsilon^{k-1}_r\big].\end{aligned} \tag{7}$$

In order to find the partial derivative of the stress to the strain, $\partial\sigma^f_r/\partial\varepsilon_r$, we use an alternative form of the stress calculation obtained from (1) and (2)

$$\sigma^f_r(t) = \sigma^f_r(0) + \int_0^t E^f_r(\varepsilon_r)\dot{\varepsilon}_r\,\mathrm{d}\tau + \sum_m g_m\exp(-t/\lambda_m)I^r_m(t). \tag{8}$$

The integral in (8) represents the elastic stress, whose derivative to the strain is the modulus, $E^f_r(\varepsilon_r)$. Using its definition, we also obtain

$$\frac{\partial I^r_m(t)}{\partial\varepsilon_r} = \exp(t/\lambda_m)E^f_r(\varepsilon_r). \tag{9}$$

Therefore, the derivative takes the form

$$\frac{\partial\sigma^f_r}{\partial\varepsilon_r} = G_r(0)E^f_r(\varepsilon_r). \tag{10}$$

A user-defined fibrillar element (Appendix A) was incorporated into the commercial finite element code ABAQUS, which was used to work with the existing porous element for the nonfibrillar matrix. Finite element solutions were obtained for tensile and unconfined compression testing. Analytical solutions were found (Appendix B) and used to test the finite element results when fluid pressurization was not considered.

3. Results

For convenience of derivation of the closed form solutions used to test the finite element results, the effect of finite deformation was not included in Figs 1–3. Finite deformation theory was used for all other cases (Figs 4–9).

Consider a 5-step tensile relaxation test. In each step, 2% strain was applied in the direction parallel to the articular surface at a rate of 0.15%/s, and then sufficient time was given to allow for full relaxation (Charlebois, McKee and Buschmann, *J. Biomech. Eng.* **126** (2004), 129–137). We have verified that the fluid has negligible impact on the mechanical response in tension using a fibril reinforced model (results not shown). Therefore, in our simulations, the tensile stress was determined analytically (in Appendix B) using the hereditary integral (1), once the elastic stress of the nonfibrillar matrix ($E_m\varepsilon$) had been added. The elastic properties (given in the caption of Fig. 1) were obtained using the data measured at equilibrium. The parameters for the relaxation function (2) (three exponentials were taken) were extracted by fitting the model to the experimental data (legend in Fig. 1).

The same elastic and viscoelastic properties determined for the tensile test (for the horizontal direction) were then applied to the horizontal fibers in the analysis of unconfined compression (closed form solution given by (B.4)), and surprisingly, no viscoelastic response was visible when fluid pressurization was absent (Fig. 2). The transient response shown in Fig. 2 was almost all produced by fluid flow/pressure. Increasing values of Poisson's ratio were examined to determine if Poisson's ratio could account for the loss of viscoelasticity in compression (Fig. 3a). The ratio of peak and equilibrium stresses did not increase significantly with increased compression rates (case iii, Fig. 3b), or with greater collagen viscoelastic relaxation (case ii, Fig. 3b), or with faster collagen viscoelastic dissipation (case iv, Fig. 3b).

The impact of collagen viscoelasticity on the fluid pressurization was then examined for strain rates of 0.15 and 15%/s, respectively. In both cases, the viscoelastic dissipation in the radial fibers resulted

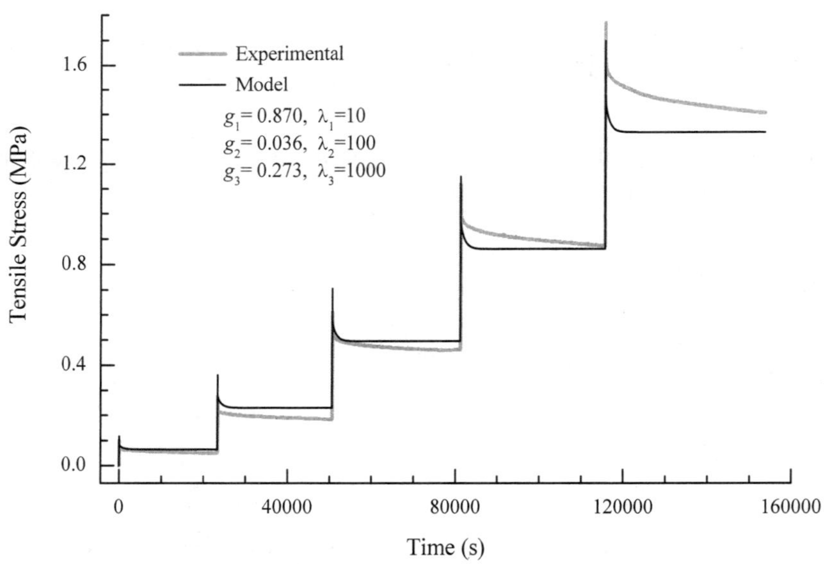

Fig. 1. Simulation of a 5-step tensile relaxation test using the viscoelastic model. 2% tensile strain was applied at each step at a rate of 0.15%/s. The Young's modulus of the nonfibrillar matrix was $E_m = 0.25$ MPa. The fibrillar modulus, obtained by fitting the tensile stress at equilibrium, was $E_r^f = (0.5 + 250\varepsilon_r)$ MPa (the instantaneous modulus $G_r(0)E_r^f = 2.179E_r^f$). The effect of fluid flow was not considered in this case, and thus the analytical solution could be obtained using Eqs (B.2) and (B.3).

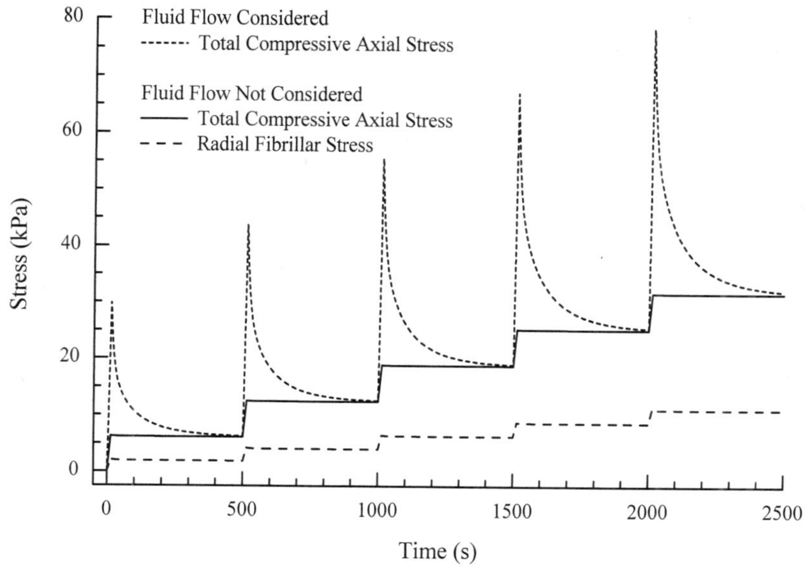

Fig. 2. Compressive axial stress and radial fibrillar stress in unconfined compression predicted by the viscoelastic model. The transient response could be neglected if fluid pressurization in the tissue were absent. 2% compressive axial strain was applied at each step at a rate of 0.15%/s. The moduli were the same as those for Fig. 1. In addition, Poisson's ratio of the nonfibrillar matrix was $\nu_m = 0.3$. The permeability was $k = 0.003\exp(10 \times \text{dilatation})$ mm^4/Ns [27], and the void ratio was 3.5 when applicable. The radius and thickness of the disk were 1.5 and 1.0 mm, respectively.

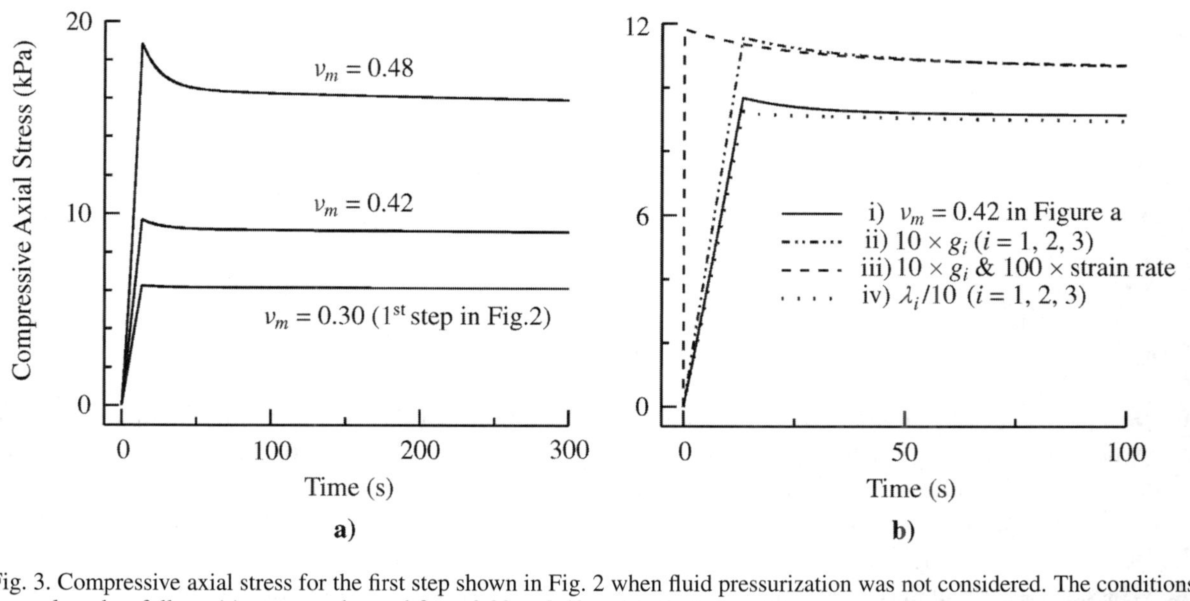

Fig. 3. Compressive axial stress for the first step shown in Fig. 2 when fluid pressurization was not considered. The conditions were altered as follow: (a) ν_m was changed from 0.30 to 0.42 and 0.48; (b) conditions were altered in reference to case (i) as follow: (ii) g_i was increased by 10 times ($G_r(0) = 12.79$ instead of 2.179, c.f. Eq. (2)); (iii) g_i was increased by 10 times and the strain rate was increased by 100 times; (iv) the characteristic times λ_i were reduced to one tenth of that for the other cases.

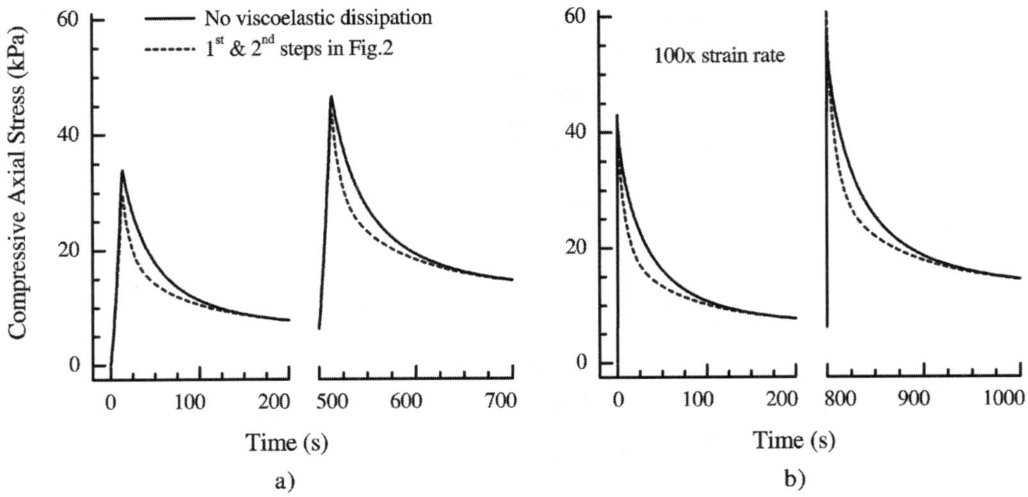

Fig. 4. Impact of collagen viscoelasticity on the axial stress in the presence of fluid pressurization. Comparisons were made with cases when collagen viscoelastic dissipation was ignored, i.e. $G_r(t) \equiv G_r(0)$. For clarity, only the data for early relaxation are shown. (a) The strain rate was 0.15%/s, as for the case shown in Fig. 2; (b) The strain rate was increased to 15%/s.

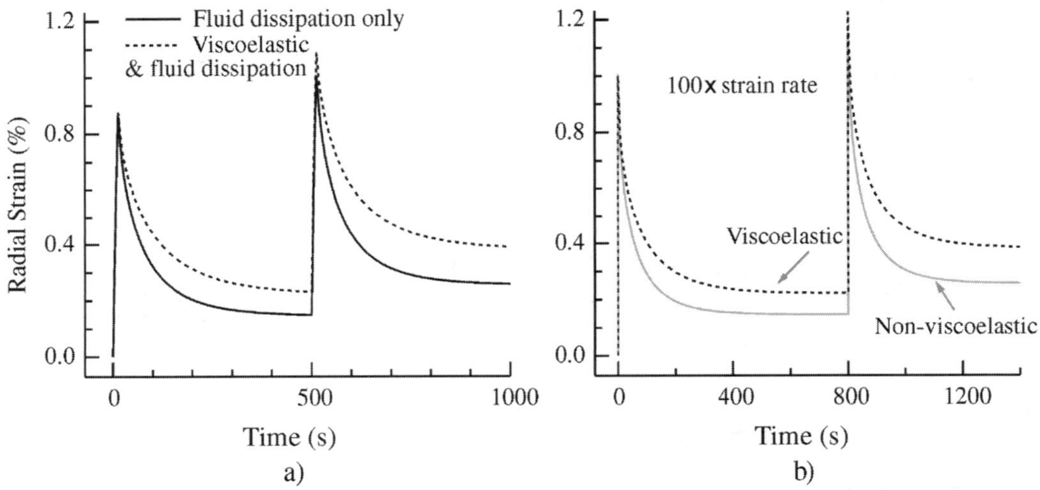

Fig. 5. Radial strain at $r = 0.62R$ for the cases shown in Fig. 4 (R refers to the disk radius). Cartilage strain was larger when collagen viscoelasticity was included. In the non-viscoelastic case, it was assumed that $G_r(t) \equiv G_r(0)$.

in a faster decrease in the axial stress during early relaxation (Fig. 4). During ramp loading, collagen viscoelasticity had an influence on the stress at low strain rates (Fig. 4a). Note that the actual fibrillar modulus at equilibrium was smaller when viscoelasticity was present [$G_r(\infty)E_r^f(\varepsilon_r)$ for the viscoelastic case, as compared to $G_r(0)E_r^f(\varepsilon_r)$ for the non-viscoelastic case], resulting in a greater radial strain in the former case (Fig. 5). This explains why the peak stress of the 2nd step shown in Fig. 4b is higher for the viscoelastic case than for the non-viscoelastic case (because $E_r^f(\varepsilon_r)$ is greater in the former case).

The stiffness of cartilage in unconfined compression is shown in Fig. 6 for various strain rates. Here the stiffness is defined as the ratio of the nominal axial stress averaged over the cross-section of the disk and the nominal strain. In all cases, the stiffness for a given strain rate was slightly lowered by the viscoelastic dissipation (Fig. 6a), which was more evident at a moderate strain rate (around 0.5%/s).

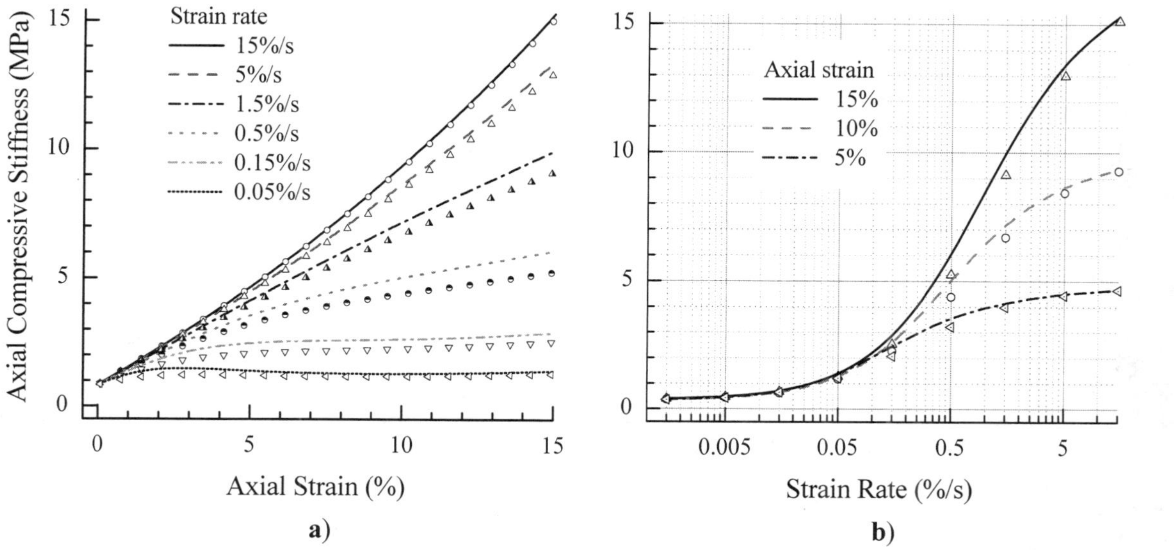

Fig. 6. Axial compressive stiffness of the cartilage disk as a function of axial strain and strain rate. It was always assumed that the tissue was compressed from the unloaded state to a nominal axial strain of 15% at a given strain rate. $E_m = 0.25$ MPa, $\nu_m = 0.3$, and $E_r^f = (0.5 + 250\varepsilon_r)$ MPa. The permeability was $k = 0.003\exp(10 \times \text{dilatation})$ mm^4/Ns, and the void ratio was 3.5. The parameters for the relaxation function are shown in Fig. 1. The lines correspond to the cases when collagen viscoelasticity was neglected. The symbols refer to the cases when collagen viscoelasticity and fluid flow were considered. (a) Stiffness as a function of strain for various strain rates; (b) Stiffness as a function of strain rate (on $\log_{10}$ axis) for axial strains of 5, 10 and 15%, respectively.

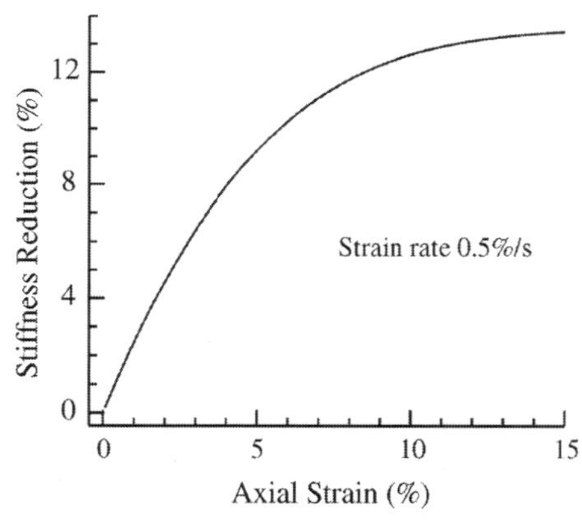

Fig. 7. Stiffness reduction (in percentage) of cartilage produced by viscoelastic dissipation, for a strain rate of 0.5%/s, plotted using the data shown in Fig. 6a. The maximum stiffness reduction shown is 13.46%. The impact of viscoelasticity in other cases shown in Fig. 6a is less significant.

Similarly, the viscoelastic dissipation did not significantly change the stiffness as a function of strain rate (Fig. 6b). The relative importance of viscoelastic dissipation to the stiffness of cartilage for a strain rate of 0.5%/s is shown in Fig. 7 (maximum 13.5%, which is close to the asymptotic value).

The fibrillar modulus used for the non-viscoelastic cases (Figs 4–7) was $G_r(0)E_r^f = (1.0895 + 544.75\varepsilon_r)$ MPa, which is the instantaneous modulus obtained from the tensile test shown in Fig. 1. In a previous study, the modulus was assumed to range from $(3 + 800\varepsilon_r)$ to $(3 + 1600\varepsilon_r)$ MPa [29].

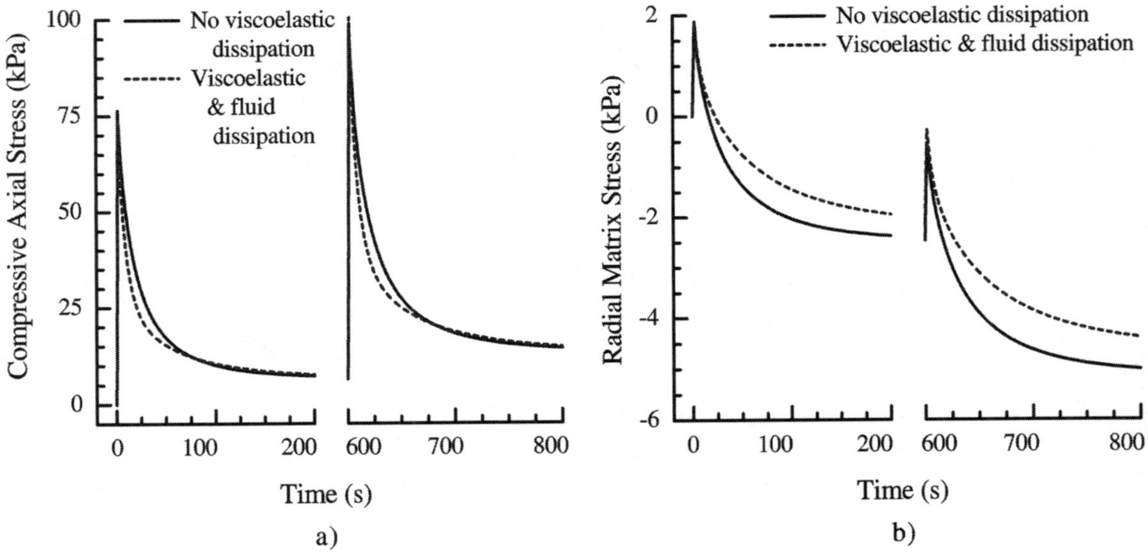

Fig. 8. Effect of increasing fibrillar modulus: the conditions were the same as those for Fig. 4, except that the fibrillar modulus was doubled (i.e. $E_r^f = (1 + 500\varepsilon_r)$ MPa) and the strain rate was 1%/s. (a) Compressive axial stress (nominal average stress for the cross section of the disk); (b) Radial stress in the nonfibrillar matrix at $r = 0.62R$ (which is the same for the viscoelastic and non-viscoelastic cases during the first ramp compression).

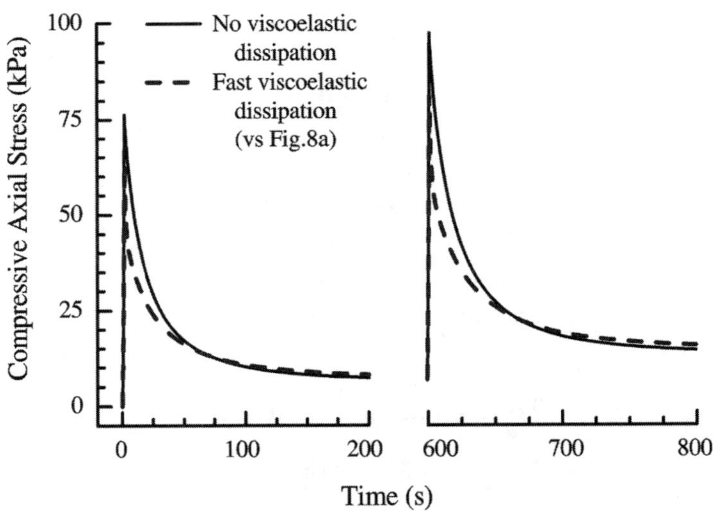

Fig. 9. Effect of reduced characteristic times: the conditions were the same as for Fig. 8a, except that all three characteristic times (λ_i) were reduced to one tenth of the values previously used. The peak stress in the first and second steps was reduced by 17.0% and 16.6% respectively, compared with that for the non-viscoelastic case.

Therefore, it would be helpful to see the viscoelastic effects when E_r^f is greater. For the cases shown in Figs 8 and 9, $G_r(0)E_r^f = (2.179 + 1089.5\varepsilon_r)$ MPa (doubled), and the strain rate was 1%/s, a strain rate at which the viscoelastic effect might be more evident. No qualitative change was observed as can be seen when comparing Fig. 8a with Fig. 4. The viscoelastic dissipation lowered the compressive radial stress in the nonfibrillar matrix during relaxation (Fig. 8b), in response to the strain change (the strain for this case was similar to that shown in Fig. 5). If the viscoelastic dissipation time was reduced by 90%

(which is an unlikely scenario according to the data shown in Fig. 1), there would be a significant (17% at peak stress) change in the stiffness compared to the non-viscoelastic case (Fig. 9).

4. Discussion

The impact of fluid pressurization is negligible in tensile testing (actually negative pressure). The transient response observed in tensile tests is primarily (intrinsically) viscoelastic, although it might have partial contributions from fluid flow in collagen fibers (i.e., fluid flow in smaller scale of "pores", rather than interstitial "pores" that are assumed to exist between fibers), as suggested by some investigators [52]. In contrast, collagen viscoelasticity plays a small role in unconfined compression of articular cartilage, when compared to the effect of fluid pressurization. Furthermore, the influence of collagen viscoelasticity in unconfined compression would be negligible, if fluid flow/pressure were absent (Fig. 2): collagen viscoelasticity plays a minor role through fluid flow. This finding might explain why some investigators considered cartilage as poroelastic [34], while others believed that the transient behavior was associated with fluid-flow and intrinsic viscoelastic properties of the tissue matrix [20]. The results presented here indicate that a fibril reinforced model without collagen viscoelasticity approximates the load response of cartilage well, provided that the instantaneous fibrillar modulus ($G_r(0)E_r^f$) is used.

If fluid pressurization were absent, using the same material properties and loading protocol (for tensile and unconfined compression), we would obtain a rather strong transient response in tensile testing (Fig. 1), but almost no transient response in unconfined compression (Fig. 2). The Poisson's effect might play a role in this event. However, only a Poisson's ratio close to 0.5 produces a significant transient response (Fig. 3a). (In the extreme case, when $\nu_m = 0.5$, which is unlikely, the ratio of the peak axial stress and the equilibrium stress is only 1.477, compared to 1.810 in tensile testing.) The amplitude of the relaxation function [$G_r(0)$ vs $G_r(\infty)$], the characteristic times of viscoelastic dissipation, and the strain rate do not account for the loss of the transient response in unconfined compression (Fig. 3b). In order to understand this observation, it is necessary to investigate the fibrillar stress as a function of its strain and relaxation time. The actual fibrillar stiffness decreases with time, resulting in an increased fibrillar strain, which increases the elastic stress of the fibers. Therefore, the actual viscoelastic stress of the fibers depends on the relative decrease in fibrillar stiffness and increase in elastic stress (Eq. (1)). This observation may explain the loss of the transient response in unconfined compression in the absence of fluid pressurization.

Now consider the physiologic case, when fluid pressurization is present. Collagen viscoelasticity has no impact on the instantaneous response of cartilage since there is no time for viscoelastic dissipation. Therefore, collagen viscoelasticity has little influence on the mechanical response at high strain-rates. However, at equilibrium (zero fibrillar strain rate), the fibrillar modulus for the viscoelastic case is $G_r(\infty)E_r^f = E_r^f$, but for the elastic case is $G_r(0)E_r^f = 2.179E_r^f$. The question now arises, why is the axial stress virtually the same in both cases (Figs 4, 8a and 9)? The answer is, because fibril reinforcement in the radial direction only has a small effect on the axial stress *at equilibrium*, as has been found previously. (The fibril reinforcement is mainly produced by fluid pressure which is zero at equilibrium. The Poisson effect in fibril reinforcement is secondary.) Therefore, collagen viscoelasticity does not influence the axial stress in cartilage significantly in unconfined compression at high and low strain rates. Collagen viscoelasticity has relatively more impact on the mechanical behavior of cartilage in compression at moderate strain-rates (0.5 and 1.5%/s, Fig. 6a). In other words, collagen viscoelasticity plays a minor role if (and only if) significant fluid flow is present.

The results appear to be more sensitive to the characteristic times λ_i (Fig. 9) than to other viscoelastic parameters. The characteristic times determined for Fig. 1 were actually too small for the forth and fifth steps in the tensile test. Even if the characteristic times were further reduced, the role of collagen viscoelasticity in the mechanical behavior of articular cartilage would still be secondary (Fig. 9), especially for the cases of high strain rates.

In many studies, the proteoglycan and collagen matrices are treated as one constituent. Therefore, it was not clear if possible viscoelastic properties of the matrix were contributed by the fibrillar matrix and/or nonfibrillar matrix. In the present study, collagen viscoelasticity was explicitly considered, while proteoglycan viscoelasticity was neglected. It is not clear if the transient response, or a significant part of the transient response observed in shear and torsion tests [17,18,54], is produced by the proteoglycan matrix. First, the effect of collagen fibers is not excluded in such tests. Second, fluid flow/pressure might contribute to the transient response, since the tissue is not isotropic and homogeneous, and thus volume changes may occur in shear deformation. Here, we assumed that the viscoelasticity of proteoglycans is negligible. However, this assumption needs thorough testing and awaits verification.

Further work is required to generalize the present model to a more realistic representation of cartilage properties and deformations. The directional differences observed in articular cartilage [21,60,62,63] can be considered when three-dimensional elements with orthotropic material properties are used. A hyperelastic formulation may be required to consider larger deformations than currently involved. More measured data are required to find typical values for the viscoelastic properties of cartilage in tension, in order to understand the mechanical response of cartilage in unconfined compression. Although unknown at present, the quasi-linear viscoelastic representation of the fibrillar stress might be insufficient to capture its essential mechanical properties of articular cartilage.

Acknowledgements

We would like to acknowledge the financial support from the Alberta Heritage Foundation for Medical Research and CIHR (Canada). The experimental data shown in Fig. 1 were obtained from M. Charlebois, M.D. McKee and M.D. Buschmann, and we would like to thank them for providing the data prior to publication.

Appendix A. Finite element formulation for the fibrillar matrix

The fibrillar element was incorporated into ABAQUS using the USER ELEMENT option. For simplicity, only small deformations are considered here. The strain energy for a fibrillar element produced by the virtual displacements δu_r and δu_z can be written as

$$\int_{V_0} [\sigma_r^f \delta\varepsilon_r + \sigma_\theta^f \delta\varepsilon_\theta + \sigma_z^f \delta\varepsilon_z]\, \mathrm{d}V_0 = \int_{V_0} \left[\left(\sigma_r^f N_{n,r} + \frac{\sigma_\theta^f}{r} N_n \right) \delta u_r^n + \sigma_z^f N_{n,z} \delta u_z^n \right] \mathrm{d}V_0, \tag{A.1}$$

where N_n denotes the interpolation (shape) function for node n. This equation is compatible with the weak forms used in viscoelastic analyses [10,57], while it is different from convolution type variational functionals [39,55].

For convenience, it was assumed that external loads are applied only on the porous elements. Hence, equilibrium of the element requires (A.1) to be zero, resulting in

$$F_r^i(\mathbf{u}_r) = \int_{V_0} \left(\sigma_r^f N_{i,r} + \frac{\sigma_\theta^f}{r} N_i \right) \mathrm{d}V_0 = 0,$$
$$F_z^i(\mathbf{u}_z) = \int_{V_0} \sigma_z^f N_{i,z}\, \mathrm{d}V_0 = 0, \tag{A.2}$$

where $\mathbf{u}_r$ and $\mathbf{u}_z$ are the vectors containing the nodal displacements. Supposing $\tilde{\mathbf{u}}_z$ is an approximation for $\mathbf{u}_z$, we need to find $\Delta\tilde{\mathbf{u}}_z$ so that

$$F_z^i(\tilde{\mathbf{u}}_z + \Delta\tilde{\mathbf{u}}_z) = 0. \tag{A.3}$$

Therefore, letting $\mathrm{d}F_z^i$ be equal to ΔF_z^i results in

$$\frac{\partial F_z^i}{\partial u_z^j}(\tilde{\mathbf{u}}_z)\Delta\tilde{u}_z^j \approx -F_z^i(\tilde{\mathbf{u}}_z). \tag{A.4}$$

Here $-F_z^i$ are elements of the residual vector, and $k_{ij}^z = \partial F_z^i / \partial u_z^j$ are elements of the Jacobian matrix, both of which must be evaluated in the user subroutine. If the fibrils are in tension in the axial direction,

$$\begin{aligned} k_{ij}^z &= \int_{V_0} \frac{\partial \sigma_z^f}{\partial \varepsilon_z} \frac{\partial \varepsilon_z}{\partial u_z^j} N_{i,z}\, \mathrm{d}V_0, \\ &= \int_{V_0} \frac{\partial \sigma_z^f}{\partial \varepsilon_z} N_{i,z} N_{j,z}\, \mathrm{d}V_0. \end{aligned} \tag{A.5}$$

Similarly, if the fibrils are in tension in the radial direction (and thus in the circumferential direction),

$$k_{ij}^r = \int_{V_0} \left[\frac{\partial \sigma_r^f}{\partial \varepsilon_r} N_{i,r} N_{j,r} + \frac{\partial \sigma_\theta^f}{\partial \varepsilon_\theta} \frac{N_i N_j}{r^2} \right] \mathrm{d}V_0 \tag{A.6}$$

and $-F_r^i$ defined in (A.2) are elements of the residual vector.

Appendix B. Viscoelastic stresses in uniaxial tensile and compression testing in the absence of fluid pressurization

A multi-step relaxation test can be defined by the strain and strain rate as

$$\varepsilon = \begin{cases} \varepsilon_{k-1} & \text{at } t = 0, \\ \varepsilon_k & \text{at } t = t_k, \end{cases} \qquad \dot{\varepsilon} = \begin{cases} c & \text{if } t < t_k, \\ 0 & \text{if } t \geqslant t_k, \end{cases} \tag{B.1}$$

where c is a constant, t_k is the loading time of step k. Let the long term Young's modulus be a linear function of the fibrillar strain, i.e. $E = E^0 + E^\varepsilon \varepsilon$. The fibrillar stress at relaxation ($t \geqslant t_k$) for tensile testing can be obtained by the hereditary integral as

$$\begin{aligned}\sigma^f(t) &= \sigma^f(0) + \int_0^{t_k} G(t-\tau)E(\varepsilon)c\,\mathrm{d}\tau \\ &= \sigma^f_{\text{elastic}}(t_k) + c\sum_m \lambda_m g_m \mathrm{e}^{-(t-t_k)/\lambda_m}\big[E^\varepsilon\big(\varepsilon(t_k) - \varepsilon_{k-1}\mathrm{e}^{-t_k/\lambda_m}\big) \\ &\quad + \big(E^0 - c\lambda_m E^\varepsilon\big)\big(1-\mathrm{e}^{-t_k/\lambda_m}\big)\big].\end{aligned} \tag{B.2}$$

The stress for ramp tension ($t < t_k$) can be obtained by using the above equation after t_k is replaced by t, resulting in

$$\sigma^f(t) = \sigma^f_{\text{elastic}}(t) + c\sum_m \lambda_m g_m\big[E^\varepsilon\big(\varepsilon - \varepsilon_{k-1}\mathrm{e}^{-t/\lambda_m}\big) + \big(E^0 - c\lambda_m E^\varepsilon\big)\big(1-\mathrm{e}^{-t/\lambda_m}\big)\big]. \tag{B.3}$$

Now consider the case of unconfined compression, for which ε_z is known. Equilibrium in the radial direction requires $\sigma_r^f = -\sigma_r^m$ (the fibrillar stress is identical to the negative of the radial stress in the nonfibrillar matrix) since fluid pressure is absent. Applying Hooke's law to the nonfibrillar matrix gives

$$\varepsilon_r + \nu_m\varepsilon_z + (1+\nu_m)(1-2\nu_m)\sigma_r^f/E_m = 0. \tag{B.4}$$

σ_r^f is related to ε_r by the hereditary integral. The Newton method was used to solve this nonlinear equation.

References

[1] S. Akizuki, V.C. Mow, F. Muller, J.C. Pita, D.S. Howell and D.H. Manicourt, Tensile properties of human knee joint cartilage: I. Influence of ionic conditions, weight bearing, and fibrillation on the tensile modulus, *J. Orthopaedic Research* **4** (1986), 379–392.

[2] D.L. Bader, G.E. Kempson, A.J. Barrett and W. Webb, The effects of leucocyte elastase on the mechanical properties of adult human articular cartilage in tension, *Biochimica et Biophysica Acta* **677** (1981), 103–108.

[3] R.A. Bank, M. Soudry, A. Maroudas, J. Mizrahi and J.M. TeKoppele, The increased swelling and instantaneous deformation of osteoarthritic cartilage is highly correlated with collagen degradation, *Arthritis and Rheumatism* **43** (2000), 2202–2210.

[4] D.F. Betsch and E. Baer, Structure and mechanical properties of rat tail tendon, *Biorheology* **17** (1980), 83–94.

[5] A.C. Chen, M.M. Temple, D.M. Ng, C.D. Richardson, J. DeGroot, N. Verzijl, J.M. teKoppele and R.L. Sah, AGE-related crosslinking alters tensile propertiets of articular cartilage, *Transactions of the Orthopaedic Research Society* **26** (2001), 128.

[6] S. Cusack and A. Miller, Determination of the elastic constants of collagen by Brillouin light scattering, *Journal of Molecular Biology* **135** (1979), 39–51.

[7] M.R. DiSilvestro and J.K.F. Suh, A cross-validation of the biphasic poroviscoelastic model of articular cartilage in unconfined compression, indentation, and confined compression, *Journal of Biomechanics* **34** (2001), 519–525.

[8] M.R. DiSilvestro, Q. Zhu and J.K.F. Suh, Biphasic poroviscoelastic simulation of the unconfined compression of articular cartilage: II – Effect of variable strain rates, *ASME J. Biomechanical Eng.* **123** (2001), 198–200.

[9] M.R. DiSilvestro, Q. Zhu, M. Wong, J.S. Jurvelin and J.K.F. Suh, Biphasic poroviscoelastic simulation of the unconfined compression of articular cartilage: I – Simultaneous prediction of reaction force and lateral displacement, *ASME J. Biomechanical Eng.* **123** (2001), 191–197.

[10] W. Ehlers and B. Markert, A linear viscoelastic biphasic model for soft tissues based on the theory of porous media, *ASME J. Biomechanical Eng.* **123** (2001), 418–424.

[11] S.M. Elmore, L. Sokoloff, G. Norris and P. Carmeci, Nature of "imperfect" elasticity of articular cartilage, *Journal of Applied Physiology* **18** (1963), 393–396.

[12] Y.C. Fung, *Biomechanics: Mechanical Properties of Living Tissues*, 2nd edn, Springer-Verlag, New York, 1993.
[13] E.M. Hasler, W. Herzog, J.Z. Wu, W. Muller and U. Wyss, Articular cartilage biomechanics: theoretical models, material properties, and biosynthetic response, *Critical Reviews in Biomedical Engineering* **27** (1999), 415–488.
[14] R.C. Haut, Age-dependent influence of strain rate on the tensile failure of rat-tail tendon, *ASME J. Biomechanical Eng.* **105** (1983), 296–299.
[15] T.L. Haut and R.C. Haut, The state of tissue hydration determines the strain-rate-sensitive stiffness of human patellar tendon, *Journal of Biomechanics* **30** (1997), 79–81.
[16] R.C. Haut and R.W. Little, A constitutive equation for collagen fibers, *Journal of Biomechanics* **5** (1972), 423–430.
[17] W.C. Hayes and A.J. Bodine, Flow-independent viscoelastic properties of articular cartilage matrix, *Journal of Biomechanics* **11** (1978), 407–419.
[18] W.C. Hayes and L.F. Mockros, Viscoelastic properties of human articular cartilage, *Journal of Applied Physiology* **31** (1971), 562–568.
[19] C.Y. Huang, V.C. Mow and G.A. Ateshian, The role of flow-independent viscoelasticity in the biphasic tensile and compressive responses of articular cartilage, *ASME Journal of Biomechanical Engineering* **123** (2001), 410–417.
[20] C.Y. Huang, M.A. Soltz, M. Kopacz, V.C. Mow and G.A. Ateshian, Experimental verification of the roles of intrinsic matrix viscoelasticity and tension–compression nonlinearity in the biphasic response of cartilage, *ASME J. Biomech. Eng.* **125** (2003), 84–93.
[21] S. Kamalanathan and N.D. Broom, The biomechanical ambiguity of the articular cartilage, *Journal of Anatomy* **183** (1993), 567–578.
[22] G.E. Kempson, Relationship between the tensile properties of articular cartilage from the human knee and age, *Annals of the Rheumatic Diseases* **41** (1982), 508–511.
[23] G.E. Kempson, Age-related changes in the tensile properties of human articular cartilage: a comparative study between the femoral head of the hip joint and the talus of the ankle joint, *Biochimica et Biophysica Acta* **1075** (1991), 223–230.
[24] G.E. Kempson, M.A.R. Freeman and S.A.V. Swanson, Tensile properties of articular cartilage, *Nature* **220** (1968), 1127–1128.
[25] G.E. Kempson, H. Muir, C. Pollard and M. Tuke, The tensile properties of the cartilage of human femoral condyles related to the content of collagen and glycosaminoglycans, *Biochimica et Biophysica Acta* **297** (1973), 456–472.
[26] P.S. Khalsa and S.R. Eisenberg, Compressive behaviour of articular cartilage is not completely explained by proteoglycan osmotic pressure, *J. Biomech.* **30** (1997), 589–594.
[27] W.M. Lai and V.C. Mow, Drag induced compression of articular cartilage during a permeation experiment, *Biorheology* **17** (1980), 111–123.
[28] L.P. Li, M.D. Buschmann and A. Shirazi-Adl, The role of fibril reinforcement in the mechanical behavior of cartilage, *Biorheology* **39** (2002), 89–96.
[29] L.P. Li, M.D. Buschmann and A. Shirazi-Adl, Strain-rate dependent stiffness of articular cartilage in unconfined compression, *ASME J. Biomechanical Eng.* **125** (2003), 161–168. Erratum: **125** (2003), 566.
[30] T. Macirowski, S. Tepic and R.W. Mann, Cartilage stresses in the human hip joint, *ASME Journal of Biomechanical Engineering* **116** (1994), 10–18.
[31] A.F. Mak, The apparent viscoelastic behavior of articular cartilage – The contributions from the intrinsic matrix viscoelasticity and interstitial fluid flows, *ASME Journal of Biomechanical Engineering* **108** (1986), 123–130.
[32] A.F. Mak, Unconfined compression of hydrated viscoelastic tissues: a biphasic poroviscoelastic analysis, *Biorheology* **23** (1986), 371–383.
[33] A. Maroudas, Balance between swelling pressure and collagen tension in normal and degenerate cartilage, *Nature* **260** (1976), 808–809.
[34] C.W. McCutchen, Cartilage is poroelastic, not viscoelastic (Including an exact theorem about strain energy and viscous loss, and an order of magnitude relation for equilibrium time), *Journal of Biomechanics* **15** (1982), 325–327.
[35] K. Miller, Modelling soft tissue using biphasic theory – A word of caution, *Computer Methods in Biomechanics and Biomedical Engineering* **1** (1998), 261–263.
[36] J. Mizrahi, A. Maroudas, Y. Lanir, I. Ziv and T.J. Webber, The "instantaneous" deformation of cartilage: Effects of collagen fiber orientation and osmotic stress, *Biorheology* **23** (1986), 311–330.
[37] N. Mukherjee and J.S. Wayne, Load sharing between solid and fluid phases in articular cartilage: II – Comparison of experimental results and u-p finite element predictions, *ASME Journal of Biomechanical Engineering* **120** (1998), 620–624.
[38] T. Nguyen and A. Oloyede, Predictive rheological models for the consolidation behaviour of articular cartilage under static loading, *Proceedings of the Institution of Mechanical Engineers, Part H, Journal of Engineering in Medicine* **215** (2001), 565–577.
[39] J.T. Oden and J.N. Reddy, *Variational Methods in Theoretical Mechanics*, Springer-Verlag, New York, 1976.
[40] A. Oloyede, R. Flachsmann and N.D. Broom, The dramatic influence of loading velocity on the compressive response of articular cartilage, *Connective Tissue Research* **27** (1992), 211–244.

[41] G.D. Pins, E.K. Huang, D.L. Christiansen and F.H. Silver, Effects of static axial strain on the tensile properties and failure mechanisms of self-assembled collagen fibers, *Journal of Applied Polymer Science* **63** (1997), 1429–1440.

[42] M.A. Puso and J.A. Weiss, Finite element implementation of anisotropic quasi-linear viscoelasticity using a discrete spectrum approximation, *ASME J. Biomechanical Eng.* **120** (1998), 62–70.

[43] E.L. Radin, I.L. Paul and M. Lowy, A comparison of the dynamic force transmitting properties of subchondral bone and articular cartilage, *Journal of Bone and Joint Surgery* **52-A** (1970), 444–456.

[44] K.H. Rajini, R. Usha, V. Arumugam and R. Sanjeevi, Fracture behaviour of cross-linked collagen fibres, *Journal of Materials Science* **36** (2001), 5589–5592.

[45] B.J. Rigby, N. Hirai, J.D. Spikes and H. Eyring, The mechanical properties of rat tail tendon, *Journal of General Physiology* **43** (1959), 265–283.

[46] V. Roth and V.C. Mow, The intrinsic tensile behaviour of the matrix of bovine articular cartilage and its variation with age, *J. Bone and Joint Surgery* **62** (1980), 1102–1117.

[47] R. Sanjeevi, A viscoelastic model for the mechanical properties of biological materials, *Journal of Biomechanics* **15** (1982), 107–109.

[48] R. Sanjeevi, N. Somanathan and D. Ramaswamy, A viscoelastic model for collagen fibres, *Journal of Biomechanics* **15** (1982), 181–183.

[49] L.A. Setton, W. Zhu and V.C. Mow, The biphasic poroviscoelastic behavior of articular cartilage: Role of the surface zone in governing the compressive behavior, *Journal of Biomechanics* **26** (1993), 581–592.

[50] F.H. Silver, G. Bradica and A. Tria, Viscoelastic behavior of osteoarthritic cartilage, *Connective Tissue Research* **42** (2001), 223–233.

[51] F.H. Silver, G. Bradica and A. Tria, Elastic energy storage in human articular cartilage: estimation of the elastic modulus for type II collagen and changes associated with osteoarthritis, *Matrix Biology* **21** (2002), 129–137.

[52] F.H. Silver, A. Ebrahimi and P.B. Snowhill, Viscoelastic properties of self-assembled type I collagen fibers: Molecular basis of elastic and viscous behaviors, *Connective Tissue Research* **43** (2002), 569–580.

[53] M.A. Soltz and G.A. Ateshian, Experimental verification and theoretical prediction of cartilage interstitial fluid pressurization at an impermeable contact interface in confined compression, *Journal of Biomechanics* **31** (1998), 927–934.

[54] A.A. Spirt, A.F. Mak and R.P. Wassell, Nonlinear viscoelastic properties of articular cartilage in shear, *Journal of Orthopaedic Research* **7** (1989), 43–49.

[55] P. Stucky and W. Lord, Finite element modeling of transient ultrasonic waves in linear viscoelastic media, *IEEE Trans. on Ultrasonics, Ferroelectrics, and Frequency Control* **48** (2001), 6–16.

[56] J.K. Suh and M.R. DiSilvestro, Biphasic poroviscoelastic behavior of hydrated biological soft tissue, *ASME Journal of Biomechanical Engineering* **66** (1999), 528–535.

[57] P. Le Tallec, *Numerical Analysis of Viscoelastic Problems*, Springer-Verlag, Berlin, 1990.

[58] N. Verzijl, J. DeGroot, C.B. Zaken, O. Braun-Benjamin, A. Maroudas, R.A. Bank, J. Mizrahi, C.G. Schalkwijk, S.R. Thorpe, J.W. Baynes, J.W.J. Bijlsma, F.P.J.G. Lafeber and J.M. TeKoppele, Crosslinking by advanced glycation end products increases the stiffness of the collagen network in human articular cartilage – A possible mechanism through which age is a risk factor for osteoarthritis, *Arthritis and Rheumatism* **46** (2002), 114–123.

[59] A. Viidik, A rheological model for uncalcified parallel-fibred collagenous tissue, *Journal of Biomechanics* **1** (1968), 3–11.

[60] C.C.-B. Wang, N.O. Chahine, C.T. Hung and G.A. Ateshian, Optical determination of anisotropic material properties of bovine articular cartilage in compression, *Journal of Biomechanics* **36** (2003), 339–353.

[61] J.L. Wang, M. Parnianpour, A. Shirazi-Adl and A.E. Engin, Failure criterion of collagen fiber: viscoelastic behavior simulated by using load control data, *Theoretical and Applied Fracture Mechanics* **27** (1997), 1–12.

[62] S.L.Y. Woo, W.H. Akeson and G.F. Jemmott, Measurements of nonhomogeneous, directional mechanical properties of articular cartilage in tension, *Journal of Biomechanics* **9** (1976), 785–791.

[63] S.L.Y. Woo, P. Lubock, M.A. Gomez, G.F. Jemmott, S.C. Kuei and W.H. Akeson, Large deformation nonhomogeneous and directional properties of articular cartilage in uniaxial tension, *Journal of Biomechanics* **12** (1979), 437–446.

[64] S.L.Y. Woo, B.R. Simon, S.C. Kuei and W.H. Akeson, Quasi-linear viscoelastic properties of normal articular cartilage, *ASME J. Biomechanical Eng.* **102** (1980), 85–90.

Biorheology 41 (2004) 195–202
IOS Press

Quantitative electrical impedance analysis of cartilage degradation

E. Frank, R. Evans, C. Lee, S. Treppo, M. Spector and A. Grodzinsky
Continuum Electromechanics Group, Center for Biomedical Engineering, Massachusetts Institute of Technology, Cambridge, MA, USA

1. Introduction

Degradation of articular cartilage involves several physicochemical changes in the extracellular matrix, such as loss of highly charged glycosaminoglycan (GAG) molecules and increased water content. A variety of physical and biochemical techniques are currently being investigated for detection of cartilage degeneration or the staging cartilage repair tissue following surgery. Radiographic and MRI imaging give an overall assessment of cartilage loss, but few methods are available to probe the loss of physical properties of the cartilage. Lyyra et al. [3] developed a probe that measures the mechanical indentation properties of cartilage *in situ*, as a means of assessing cartilage stiffness.

We have shown electrical impedance to be a sensitive indicator of biochemical composition and tissue integrity [5] using a minimally invasive, hand-held surface impedance device. Tissue electrical impedance is directly dependent on electrolyte content of a given tissue and is particularly appropriate to cartilage due to its high water content. Since impedance measurements are made using high frequency, they can be used to gather data on the state of a joint cartilage surface very quickly. The impedance probe described below uses multiple electrodes which may be reconfigured in such a way as to probe different depths into the tissue in order to assess surface fibrillation. In the studies presented here, we measured the electrical impedance of the articulating surface of intact joints to gauge (1) enzymatic degradation *in vitro* bovine articular cartilage, (2) OA degradation in human patellae, and (3) canine model graft repair tissue, by comparing impedance measurements with biomechanical measurements, biochemical analyses, and histology.

2. Methods

2.1. *Surface impedance device*

Silver sheet (25 μm thick) was laminated onto an insulating substrate and patterned as 4 electrodes using standard photofabrication techniques [4]. The structure was backed and mounted on the end of a 6.4 mm diameter stainless tube (Fig. 1). The electrodes had nominally equal areas, 4 mm $\times$ 0.75 mm, with a center to center spacing of 1.05 mm. The individual electrodes allowed the excitation spatial wavelength, defined as 2 times the spacing between (+) and (−) electrode area centroids, to be varied

0006-355X/04/$17.00

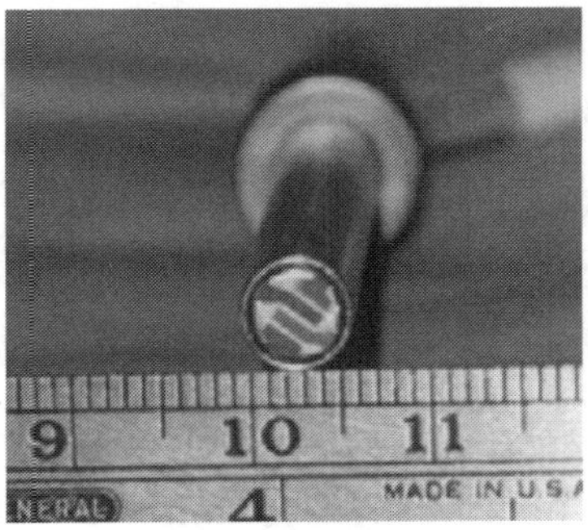

Fig. 1. Surface impedance probe.

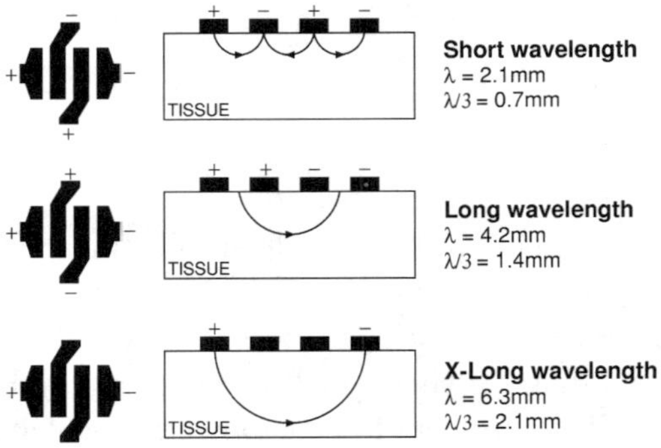

Fig. 2. Electrical wavelength configuration showing relative penetration depth.

among short, long, and extra-long wavelength configurations (Fig. 2). Tissue impedance was calculated as the measured electrode voltage divided by the driving current, and was normalized to the impedance of the bathing solution. For these studies, the impedance was measured in the 250–1,000 Hz frequency range; it is known that the electrode–electrolyte interfacial impedance is minimized at high frequency. To ensure electrode contact, the probe was held flush to the tissue surface using a constant-force spring to apply a small 50 kPa tare stress. A sinusoidal current density of 1 mA/cm^2 was then applied to the tissue. The Student's t-test and the Fisher Exact test were used, with the criterion of $p < 0.05$ for significance.

2.2. *Articular cartilage*

Studies described here which used human donor tissue were approved by the MIT committee on the use of humans as experimental subjects. Animal experiments were approved by the Brockton/West Roxbury VA Animal Care Committee.

2.3. *Bovine calf patellae*

Bovine calf patellae were harvested and the impedance measured in 0.15 M phosphate buffered saline at 4 sites on each specimen using three spatial wavelengths. The specimens were cut in two and one section of each was trypsin digested (1 mg/ml) at the articulating surface for four hours. Impedance tests on the trypsin digested sites were then repeated. Normal and digested sites were analyzed for proteoglycan (PG) content (DMMB dye binding). Adult bovine cartilage surfaces were treated with

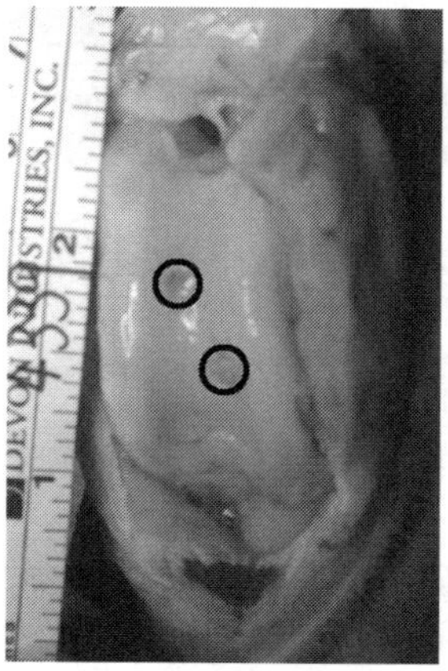

Fig. 3. Autologous cartilage implant sites.

10 μg/ml of activated rhMMP-13 (collagenase-3) for 24 hours. Impedance tests were then performed on normal and rhMMP-13 treated samples.

2.4. *Human patellae*

Human patellae were obtained from the Regional Organ Bank of Illinois and kept frozen. Donor ages ranged between 13 to 70 years and were visually graded [1] either Collins 0/1 (normal) or 2 (mild degradation) prior to testing. Joint surfaces were tested in a 0.1 M phosphate buffered solution with 0.01 M EDTA, pH 6.5. Impedance was measured at four sites on each patella and normalized to solution as above. Tested sites were cored and the top 1 mm layer of the articular surface and the subsequent cartilage layer to the bone surface removed for analysis of PG content.

2.5. *Autologous articular cartilage*

Autologous articular cartilage was harvested from the trochlear ridge of the left knees of six mature hound dogs [2]. A strip of approximately 20–25 mm by 3–5 mm tissue was removed down to the tide-mark. The chondrocytes were extracted and expanded in monolayer up to third passage and seeded at 10^6 cells/disk into EDAC cross-linked, porcine type-II collagen matrices. After four weeks, matrices were implanted in partial-thickness 4-mm diameter surgically constructed defects in the proximal and distal patellar groove of the right knees of each dog. Animals were sacrificed sixteen weeks after implantation. *Short-wavelength* impedance measurements at 100–1000 Hz were made *in situ* at the two defect sites (Fig. 3, circled) in the patellar groove and then on adjacent visually normal tissue, as well as at three sites on the medial and lateral halves of the excised right and left patellae.

3. Results

Trypsin digestion of bovine cartilage resulted in a 66% loss of PG content from the top 1 mm of tissue as measured by DMMB assay (data not shown). This is consistent with the known effect of trypsin-induced GAG loss. Impedance from calf patellae sites significantly increased after trypsin digestion (Fig. 4). In contrast to trypsin treatment, a *decrease* in impedance ($p < 0.05$) was observed in MMP-13 treated samples compared with normal bovine cartilage (Fig. 5).

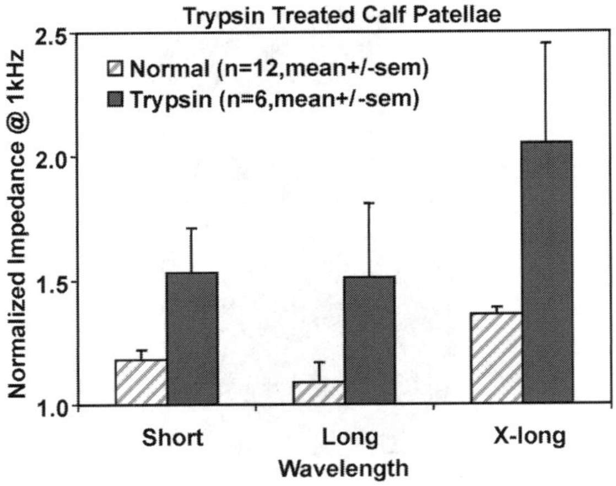

Fig. 4. Normalized impedance (NI) from the average of calf patellae sites significantly increased after trypsin digestion at all frequencies and spatial wavelengths (both $p < 0.05$).

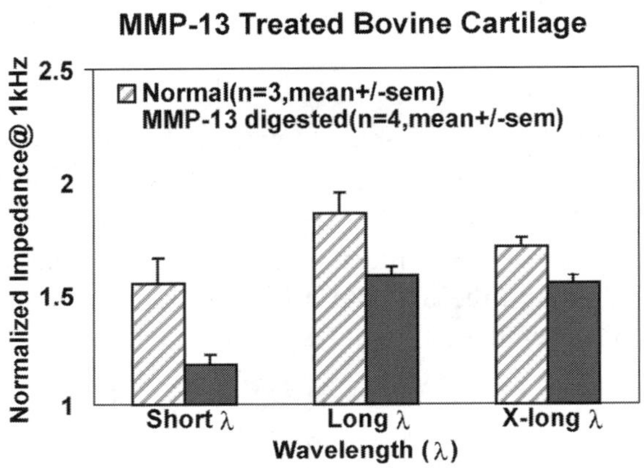

Fig. 5. Adult bovine cartilage surfaces were treated with 10 g/ml of activated rhMMP-13 (collagenase-3) for 24 hours which cleaves superficial zone collagen fibrils and releases ~10% of total tissue hydroxyproline and ~20% PG [5].

Collins Grade 2 human patellae, characterized by collagen fibrillation as well as loss of superficial PG, had a significantly lower GAG content (Fig. 6) and higher water content (hydration, Fig. 7) within the top 1 mm of the tissue. There were no significant differences found in deeper regions of the tissue.

Collins Grade 2 human patellae had a significantly lower impedance compared to normal (Collins 0&1) patellae. Significant differences were seen using the long wavelength configuration at 250, 500, and 1000 Hz ($p < 0.05$, Fig. 8), using an independent t-test with equal variance as determined by an F-test (95% confidence). A power analysis showed that the sample size gave a power of 80% with $\alpha = 0.01$. To highlight differences between grades, we took the ratio of short to long wavelength normalized impedance (NI). Since the short wavelength has a shallow penetration depth compared to the long wavelength measurement, the ratio is an indicator of surface degradation.

Autologous implant model: Histomorphometric evaluation of tissue in surgically-induced defects revealed predominately fibrous tissue and fibrocartilage. *In vitro* mechanical tests revealed that the graft was much less stiff ($p < 0.01$) than the control. Normalized impedance of graft sites in a canine repair model [2] was significantly less than that in the adjacent the visually-normal tissue ($p < 0.01$).

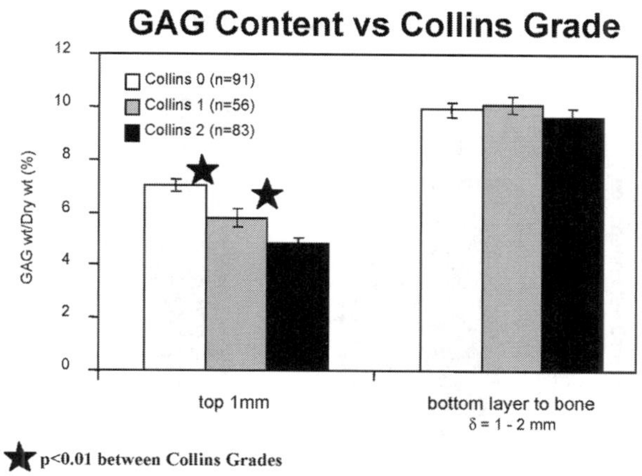

Fig. 6. Gag content in the top 1 mm layer of human patellae decreased significantly with higher Collins grade ($p < 0.01$). However deeper zones had relatively high GAG content with no difference between grades.

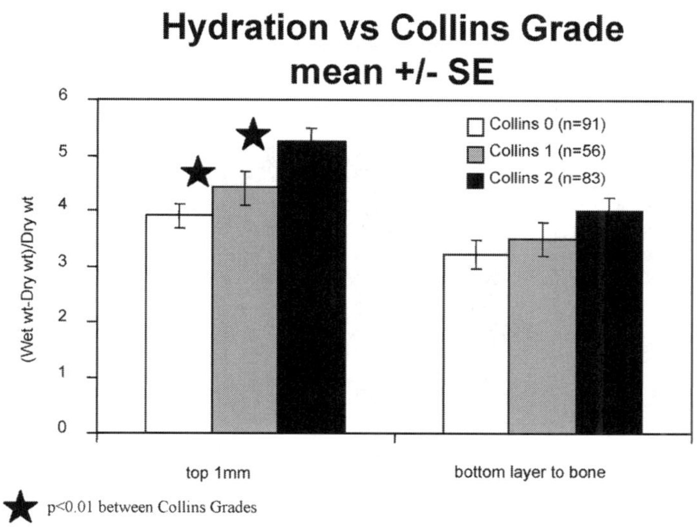

Fig. 7. Hydration, similarly, was increased significantly with higher Collins grade ($p < 0.01$) in the top 1 mm, but was lower with small differences between grades in the deeper zones.

Repair tissue in the grafts was compared to adjacent visually normal tissue on the femoral patellar groove. Tissue on the right patellae, articulating against the graft sites, was compared with that on the left patellae, articulating against the harvest sites. Impedance in the graft sites on the femoral patellar groove was significantly less than that in the adjacent the visually-normal tissue ($p < 0.01$). The impedance in the proximal repair tissue was less than in the distal, corresponding with visual inspection at necroscopy, where it was noted that the surface of the proximal repair tissue was slightly uneven while the distal repair tissue was smooth. *In vitro* mechanical tests revealed that the distal graft was much less stiff ($p < 0.01$) than the control.

Untreated patellar tissue: There were no gross abnormalities distant from the defect and harvest sites at necroscopy. Electrical and biochemical analysis could detect no significant changes between the patellae articulating against the graft and harvest sites. No difference was found in the hydration or GAG per wet

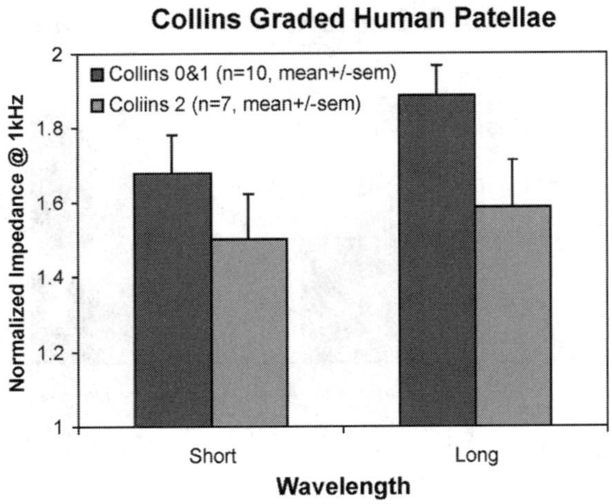

Fig. 8. Degraded tissue, characterized by increased swelling, shows a decrease in impedance, similar to treatment by MMP-13 which involved damage to the collagen matrix.

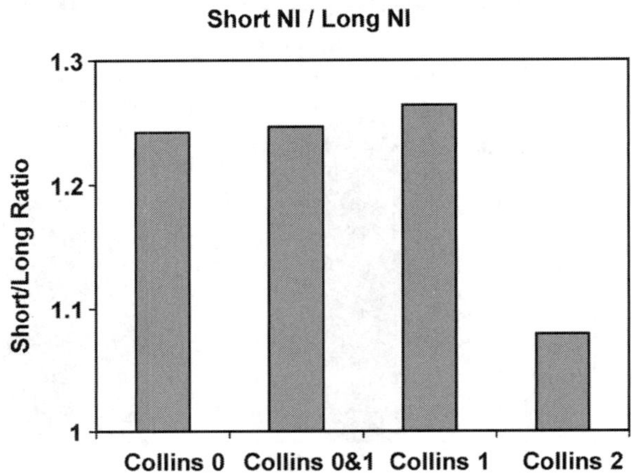

Fig. 9. The ratio of short to long wavelength impedance highlights the differences between grades. Since the short wavelength has a shallow penetration depth compared to the long wavelength measurement, the ratio is an indicator of surface degradation.

weight content of the graft articulating vs. the harvest-articulating patella tissue. These results suggest that surgically induced tissue repair did not affect the uninvolved tissue in this model.

4. Discussion

Loss of GAG alone by trypsin digestion was found to result in an increase in tissue impedance (Fig. 4). (Trypsin is known to leave the collagen matrix largely unaffected.) As the highly charged GAG molecules are primarily responsible for tissue swelling, loss of GAG generally results in a loss of hydration. At the same time, the decrease in fixed charge sites within the tissue would, in accordance with Donnan theory, result in a decrease of the total intratissue mobile ion content. Both these effects would contribute to an increase in tissue impedance.

In contrast, MMP-13 (collagenase-3) cleaves the helical type II collagen molecule at specific sites, leading to degradation of the collagen matrix, increased permeability, and swelling. Treatment of MMP-

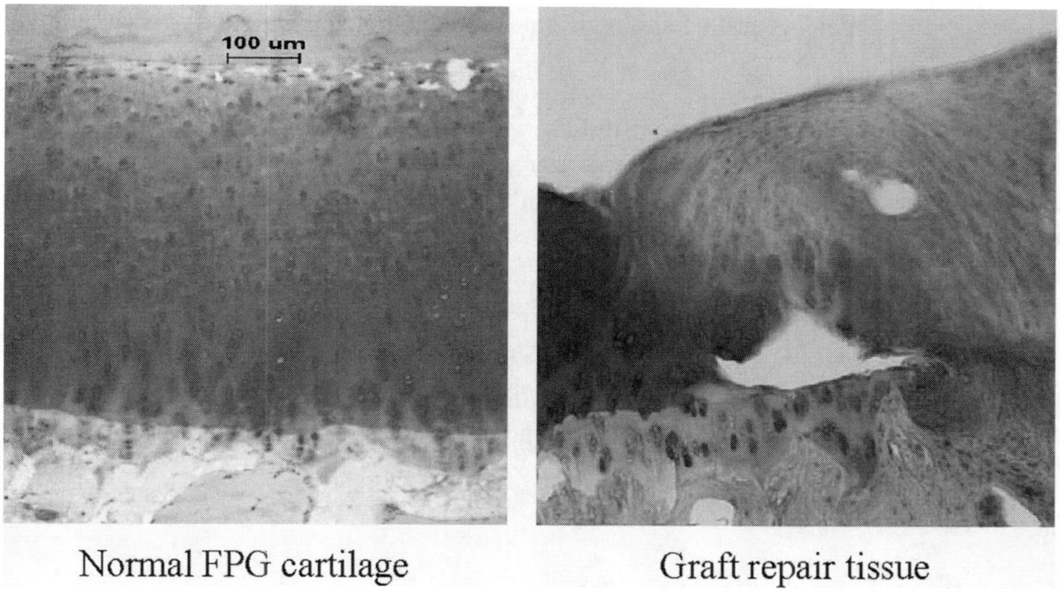

Fig. 10. *Histology:* Defects were 90–100% filled with repair tissue that was grossly distinct from adjacent uninvolved tissue. The repair tissue was white in color, but there were visible borders around the defect.

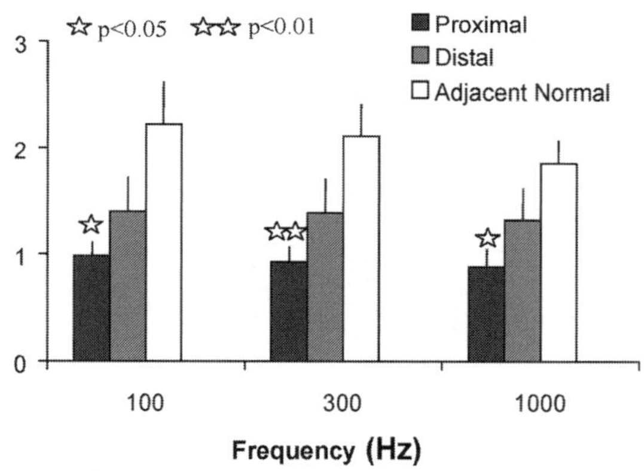

Fig. 11. *In situ* short-wavelength normalized impedance of proximal and distal defect sites compared to that of adjacent visually normal tissue.

13 for 24 h was found to release ∼10% of total tissue hydroxyproline and ∼20% of tissue proteoglycan [5]. The loss of GAG from the cartilage was not as extensive nor as rapid as in the case of trypsin, which can result in greater than 90% loss of GAG from the tissue. The loss of collagen matrix integrity coupled with the relatively intact GAG content would result in an increase in hydration and thereby a decrease in tissue impedance, as observed.

In characterizing normal (Collins grade 0), slightly degraded (grade 1), and moderately degraded (grade 2) human patellar tissue, we observe a progressive loss of GAG (Fig. 6) coupled with a general increase in hydration (Fig. 7). These observations are consistent with the view of progressive fibrillation of the collagen matrix leading to pathological degeneration in human tissue. Although we saw no difference between Collins grade 0 and 1, we found a significant decrease in tissue impedance in moderately degraded tissue (Collins grade 2, Fig. 8). We also saw a significantly lower impedance in fibrous graft tissue (Fig. 10), which generally has a high water content, compared to that of surrounding normal tissue.

Since the short wavelength has a shallow penetration depth compared to the long wavelength measurement (Fig. 2), we interpret the ratio of short-to-long wavelength measurements as an indicator of surface degradation (Fig. 9). Since these measurements can be made in rapid succession, the ratio may be used in a practical application as a single “figure of merit”. The ratio is also a “self-normalizing” parameter in the sense that we are normalizing the surface measurement to that of the tissue as a whole.

Acknowledgement

This research was partially supported by NIH SBIR Phase II Grant No. 2 R44 AR42285-02A1 under subcontract to Daniel H. Wagner, Associates Inc., and VA Grant No. A2012PA.

References

[1] Collins et al., *The Pathology of Articular and Spinal Diseases*, 1949, p. 74.
[2] C.R. Lee, A.J. Grodzinsky, H.-P. Hsu, S.D. Martin and M. Spector, Harvest and selected cartilage repair procedures affect physical and biochemical properties of uninvolved articular cartilage in the canine knee, *JOR* **18** (2000), 790–798.
[3] T. Lyyra, J. Jurvelin, P. Pitkänen, U. Väätäinen and I. Kiviranta, Indentation instrument for the measurement of cartilage stiffness under arthroscopic control, *Med. Eng. Phys.* **17** (1995), 395–399.
[4] E.C. Quan, S. Treppo, E. Butts, E.H. Frank, H. Koepp, A.A. Cole and A.J. Grodzinsky, Detection of cartilage degradation in human patellae via tissue impedance measurements using a hand-held surface spectroscopy probe, *Trans. ORS*, 1994.
[5] S. Treppo, S.I. Berkenblit, D.L. Bombard, E.H. Frank and A.F. Grodzinsky, Physical diagnosis of cartilage degradation, in: *Advances of Osteoarthritis*, Springer-Verlag, Tokyo, 1999, pp. 59–73.

Biorheology 41 (2004) 203–213
IOS Press

Zonal and directional variations in tensile properties of bovine articular cartilage *with special reference to strain rate variation*

A. Verteramo and B.B. Seedhom
Division of Bioengineering, Academic Unit of Musculoskeletal and Rehabilitation Medicine, Leeds School of Medicine, University of Leeds, Leeds, UK

Abstract. *The aims* of this study were: (i) to investigate the variation in the tensile properties of articular cartilage with depth through cartilage thickness and fibre orientation; (ii) to determine the effect of strain rate on tensile properties of articular cartilage.

Materials and method: All experimental work was performed on cartilage specimens taken from two bovine knee joints. Osteochondral plugs 12 mm in diameter were harvested with a special reamer from the femur and the tibial plateaux of each knee. Slices (0.2 mm thick), of articular cartilage were cut from the plug with a microtome. The predominant orientation of the collagen fibres on the cartilage surface was determined using the pinpricking technique. Each specimen used for the tensile test was cut, so as to produce a dumbbell shape, with a gauge length of 6 mm. Uniaxial tensile tests were performed on each specimen in order to determine the tensile Young's modulus, and ultimate tensile strength (UTS). In this investigation, these tensile tests were carried out at different strain rate: 1, 20, 50 and 70%/sec.

Results: As regards the zonal properties, it was found that tensile stiffness was greater in the superficial layer than in deep layer. However, a few specimens from the deep layer displayed similar or greater stiffness compared to the superficial layer. With respect to the directional properties, the specimens oriented parallel to the predominant alignment of collagen, were stiffer than those, which were perpendicular to it in each layer. However, only the results regarding the deep layer can be considered statistically significant. In regard to the variation of modulus with the strain-rate, the results showed that there is no significant increase of the modulus with increasing strain rate from 20 to 50% per second. However, at 70% per second, articular cartilage stiffness considerably increased by up to one order of magnitude greater than that determined at lower strain rates in both the superficial and deep layer. Moreover, the UTS of cartilage specimens tested at 70% per second showed a significant rise, reaching values of four to five times that of those measured at 1, 20 or 50% per second.

Conclusion: The steep increases in both the stiffness and ultimate tensile strength of cartilage at high strain rates point to the existence in cartilage of a mechanism for its protection from damage by stresses arising in trauma, which are usually applied at high rates. This mechanism needs to be elucidated. The reduced anisotropy found in the present study pointed out that collagen is likely to be less organized in bovine cartilage than in the human and therefore, a study of its ultra-structure would be appropriate.

1. Introduction

Articular cartilage (AC) is a thin layer lining the surfaces of the stiffer bone ends of synovial joints. Throughout a whole life time joints are typically subjected to between 70 and 300 million load cycles. Most of these cycles occur during walking when the load transmitted across the knee joint is up to 2–4 times body weight [4–10]. As a consequence, cartilage surfaces are continually subjected to deformation resulting into both compressive and tensile stresses. Cartilage must therefore act as attenuating excessive stresses, which otherwise would lead to bone destruction, by deforming under load. Moreover, cartilage deformation compensates the incongruities and asperities of bones, increasing the contact area and levelling the load evenly over the bone surfaces, resulting in lower stress [3–7]. When load is applied in synovial joint, instantaneous deformations of cartilage occur as a reflection of the proteoglycan gel to swell under the pressure of the gradients within cartilage. It seems clear that only the collagen fibres tend to restrict this swelling, experiencing increased tensile stresses. Flachmann and Broom [2] have experimentally shown that compressive stresses similar to that occurring during locomotion and running resulted in surface tensile strains and fracture of cartilage.

While there are many studies into the behaviour and properties of cartilage under compressive load, few studies exist on its tensile properties. The relationship between the tensile properties and the orientation of the specimen relative to the predominant direction of collagen was examined by Kempson [6] and later on by Woo [11]. The results showed that the specimens oriented parallel to the predominant collagen direction were stiffer than those perpendicular to it. Besides, it was found that the tensile stiffness of AC decreases significantly with depth from the articular surface, showing a ratio of stiffness of 4–6. Those early investigations into cartilage tensile properties were made testing specimens with a very low strain rate ($\sim$0.01 sec^{-1}), to avoid a significant flow-generated stiffening effect [6–8]. Applying such a low rate, the water, forming 80% of cartilage, was not able to resist the applied load and to harden the specimen. Therefore, the measurement of stiffness regarded mostly the intrinsic tensile modulus of cartilage excluding the contribution of the viscous phase due to the fluid effect. Furthermore, testing cartilage at the very low strain rates in the previous studies is not representative of any physiological conditions. Loads arising in sport activities and trauma are applied on cartilage at high strain rate and they are responsible of injuries to articular cartilage. Besides, as mechanical factors have been implicated in the development of osteoarthritis [1–9] it is essential to study cartilage properties and modes of failure to understand the role of mechanical factors in the development of disease and also to benefit research into cartilage repair.

Because collagen failure is primarily due to tensile stress the aims of this study were: (i) to investigate the variation in the tensile properties of articular cartilage with depth through cartilage thickness and fibre orientation; (ii) to discover if the rate of tensile strain affects the tensile modulus and ultimate tensile strength (UTS) of articular cartilage.

2. Materials and method

2.1. Osteochondral plug preparation

All the experimental work was performed on cartilage specimens obtained from osteochondral plugs, which were harvested from two bovine knee joints, A and B. Intact knee joins were procured from the slaughterhouse and, once acquired, they were immediately stored in a deep freezer at −18°C. Before use, the joint was thawed up overnight at room temperature. The joint was then opened and the two bones, femur and tibia were separated. Osteochondral plugs were so harvested from the femur and the tibial plateaux with a reamer mounted on a drill that allowed cutting and extraction of cylindrical specimens, 12 mm in diameter. A total of 26 plugs were harvested from different regions of knee A: 16 from medial and lateral femoral condylar surfaces, 8 from tibial surfaces covered by menisci and 2 tibial surfaces not covered by menisci. The sites were labelled as shown in Fig. 1. A total of 12 plugs were taken from the femur condyles of knee B. Once all the osteochondral plugs have been removed, they were re-frozen at −18°C until the specimen preparation.

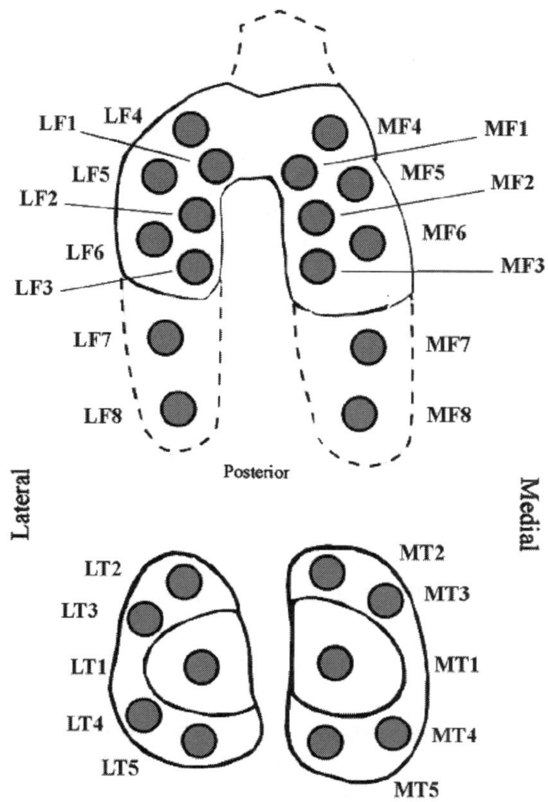

Fig. 1. Each osteochondral plug was identified with a label showing the original site, where F and T stand for femur and tibia, while L and M stand for lateral and medial, respectively.

2.2. Specimen preparation

The specimens used for the tensile tests were prepared from osteochondral plugs harvested from bovine knees. One plug at a time was thawed at room temperature.

Before slicing the cartilage from each plug, the predominant orientation of the collagen fibres on the cartilage surface was investigated by using the pinpricking technique [5]. Slices of articular cartilage 0.2 mm thick were cut from the plug with a microtome. Once the cartilage surface was shaved away, the information regarding the predominant collagen fibres orientation determined by the pinpricking pattern could be lost. To avoid such eventuality, the other slices of the cartilage were marked with lines of Haematoxylin acid to preserve such information. Since cartilage thickness was approximately 1.5–2 mm, the cutting procedure enabled to obtain three different layers of cartilage: superficial, middle and deep layers. Then, the specimen used for the tensile test was cut from each slice with a specially designed cutter, so as to produce a dumbbell shape with a gauge length of 6 mm. All the slices obtained were labeled and refrozen until required for testing. Before being placed in the test chamber and stretched, the specimen was then glued to plastic blocks, using a special device that enabled the specimen to be positioned in the right alignment. The blocks were beforehand scratched with a scalpel to roughen their surfaces, in order to prevent the specimen from slipping during tensile tests. Then the specimen was ready to be clamped in the special designed apparatus. In this investigation 65 specimens (41 from knee A and 24 from knee B) were tested with the collagen fibres parallel to the gauge length and 20 specimens (from

Table 1

The number of cartilage specimens categorized by site, collagen direction and strain rate

	No. of plugs	Site	Strain rate (%/s)	No. of specimens	
				Parallel	Perpendicular
Knee A	26	Femur + tibia	1	41	20
Knee B	12	Femur	20	8	–
			50	8	–
			70	8	–

knee A) with the fibres perpendicular to it. Table 1 shows how specimens were harvested and sorted according to site, orientation of collagen fibres and strain rate applied in the different groups of test specimens.

2.3. *Methodology*

Uniaxial tension has been carried out on cartilage to evaluate the mechanical properties. Each specimen was subjected to a continuously increasing tensile force in the direction of the gauge length. All the specimens were tested till fracture, maintaining a defined constant rate of strain. Performing tensile tests, a clear relation between load (N), applied to the specimen, and displacement (mm) was obtained. These parameters were measured simultaneously at a constant rate and a frequency of sampling of 100 Hz, and then displayed on a screen. Using mechanical notations, load and displacement were normalized to the cross-sectional area and initial length, thus obtaining stress and strain. According to the stress–strain graphs of articular cartilage, it was possible to differentiate three different parts, which characterized cartilage response: initial response, linear response and creep response. The Young's modulus (E) was evaluated by the gradient to the stress/strain curve of the linear response. This evaluation was based on international standards ISO 527-1, 527-2 for tensile testing of plastics with gauge length being less than 10 mm.

2.4. *Variation of strain rate*

All the tensile tests were carried out at a defined constant rate of strain, in order to compare results maintaining a fixed mechanical parameter. In the first part of this investigation, to avoid stiffening effects and compare the results with previous investigations, the tensile tests were carried out at a strain rate of 1% per second on knee A. Once the mechanical properties were determined at this rate, new tests were performed only on cartilage specimens in the direction parallel to the collagen orientation at higher strain rates: 20, 50 and 70%/sec. These specimens were taken from knee B.

2.5. *Thickness measurement*

In this study the measurement of cartilage specimen thickness was carried out using a special apparatus, consisting of a digital micrometer (0.5 μm accuracy) with a needle mounted on it and an electrical circuit. By using the electrical conductivity of the cartilage, it was possible to measure its thickness.

2.6. *Application of tensile load*

The apparatus used for articular cartilage tensile tests was the ELF 3200, a machine manufactured by EnduraTEC Inc. and specially designed for biomaterials and tissue engineered medical products. This

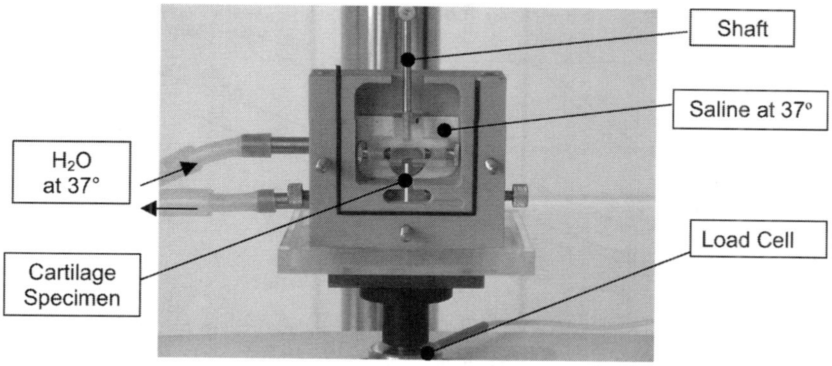

Fig. 2. Environmental chamber used to maintain the specimen hydrated during the tensile test. The specimen was surrounded by saline solution throughout the tensile test at the temperature of 36÷37°C.

machine enables small specimens to be tested at a small range of loads with much improved control on both load and displacement, and also with greater facilities for data collection. The uniaxial testing machine comprises four assemblies: the frame, which has a rigid steel structure; the electro-dynamic linear actuator, which enables stretching the specimen along the vertical axis; the instrumentation consisting of load and displacement transducers, which respectively measure the load applied to cartilage and the corresponding movement of the actuator. The measurement of the displacement was monitored using a transducer of the linear variable differential transformer (LVDT) type with a range of ±6.35 mm. The load applied to the cartilage was measured using a bonded strain gauge load transducer, with a capacity of ±225 N.

A special test chamber (Fig. 2) was designed to keep cartilage hydrated with saline solution at 36÷37°C. This temperature in the chamber was maintained by circulating warm water in a jacket mounted in the chamber.

3. Results

A total of 62 specimens from knee A and 24 from knee B were subjected to uniaxial tensile load. Almost every specimen was able to reach fracture within 12 mm of elongation, which was the maximum travel limit imposed by the apparatus.

3.1. Zonal properties

Because of the variability of middle layer data, comparison of tensile properties was made between superficial and deep layer only. Figures 3 and 4 show the variation in tensile properties of articular cartilage with depth during the linear response phase. Figure 3 shows the results for specimens tested with the long axis of the specimen being parallel to the predominant alignment of the collagen while Fig. 4 illustrates the corresponding results for specimens tested perpendicularly to the predominant alignment. The results suggested that tensile stiffness, measured by the gradient to the stress/strain curve, was greater in superficial layer than in deep layer. The pattern of variation was different between the two orientations, since the stiffness in parallel-orientated specimens allowed sudden changes of slope, while specimens orientated perpendicularly showed gradual patterns. Furthermore, in Fig. 3 a few specimens from the deep layer displayed similar or greater stiffness compared to the superficial layer.

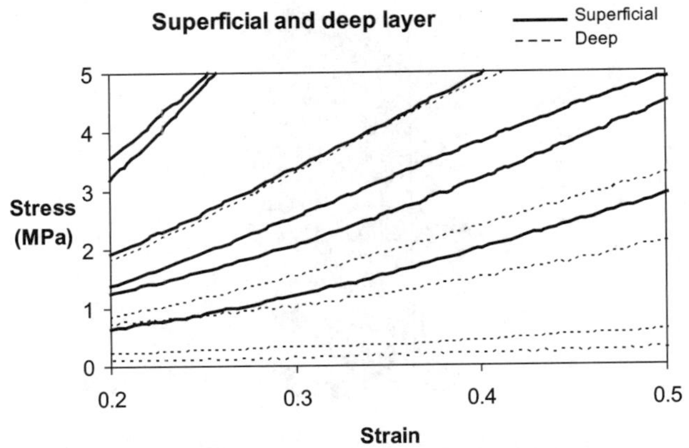

Fig. 3. Tensile stress/strain curves of superficial and deep layer; the specimens were tested in the direction parallel to the predominant alignment.

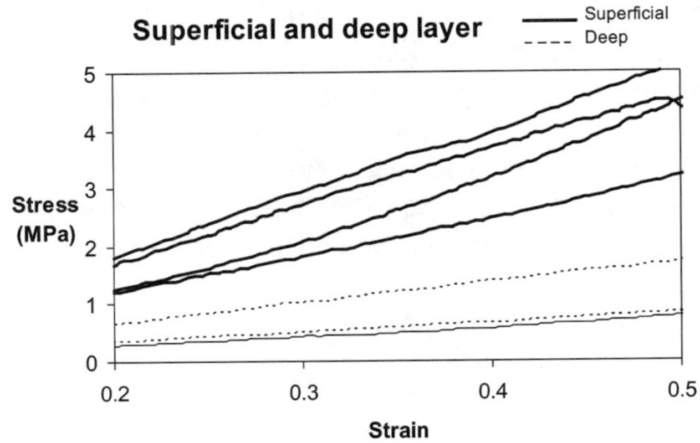

Fig. 4. Tensile stress/strain curves of superficial and deep layer; the specimens were tested in the direction perpendicular to the predominant alignment.

3.2. Directional properties

As already mentioned, the pin-pricking technique enabled to determine the collagen orientation and accordingly to classify the specimens as parallel or perpendicular to the predominant. Figures 5, 6 and 7 display tensile stress/strain curves for the superficial, middle and deep layers. It can be readily seen that the specimens, of which the gauge length was oriented parallel to the predominant alignment of collagen, were stiffer than those which were perpendicular to it. However, only the deep layer contained reliable values that were consistent with the above-mentioned statement, since the scatter of the data revealed that this difference was not significant.

3.3. Variation of stiffness with the rate of strain

All the tensile tests in this part of the investigation were carried out on cartilage specimens with the collagen fibres oriented parallel to the gauge length at the strain rates of 20, 50 and 70% per second. To compare stiffness at different strain rates, we used 4 groups, each comprising 3 plugs. These three plugs were harvested as closely as possible, in order to minimise regional variations of cartilage properties

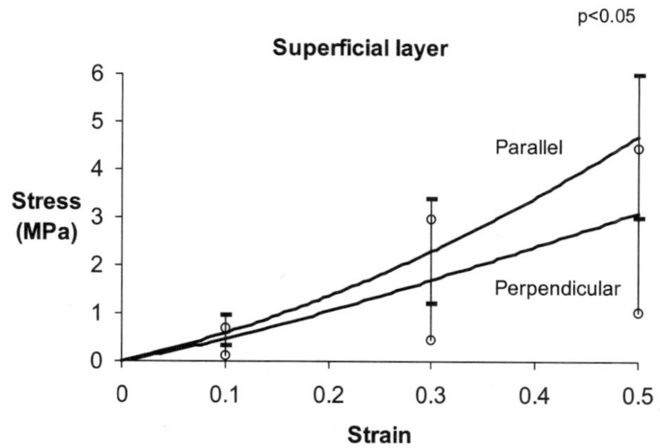

Fig. 5. Tensile stress/strain curves of superficial layer. Specimens were compared when tested parallel and perpendicular to the predominant alignment of collagen in the superficial layer.

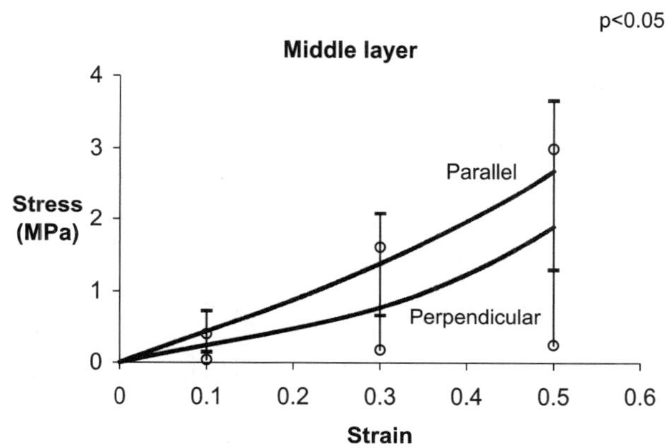

Fig. 6. Tensile stress/strain curves of middle layer. Specimens were compared when tested parallel and perpendicular to the predominant alignment of collagen.

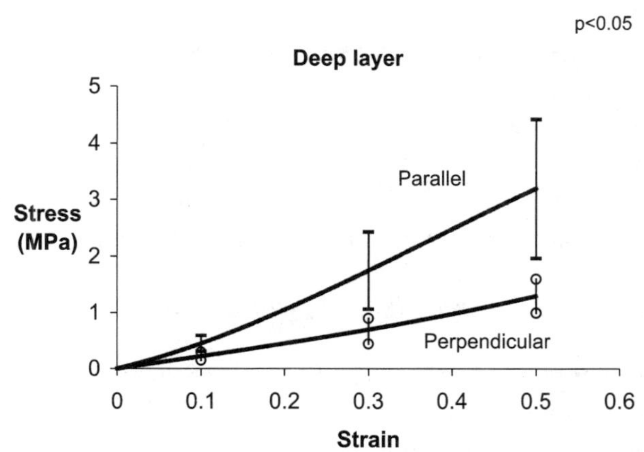

Fig. 7. Tensile stress/strain curves of deep layer. Specimens were compared when tested parallel and perpendicular to the predominant alignment of collagen.

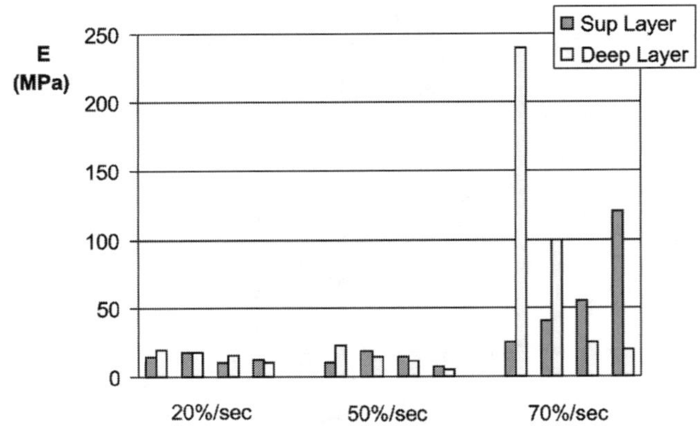

Fig. 8. Variation of Young's modulus of articular cartilage when tested in tension at strain rates of 20, 50 and 70% per second.

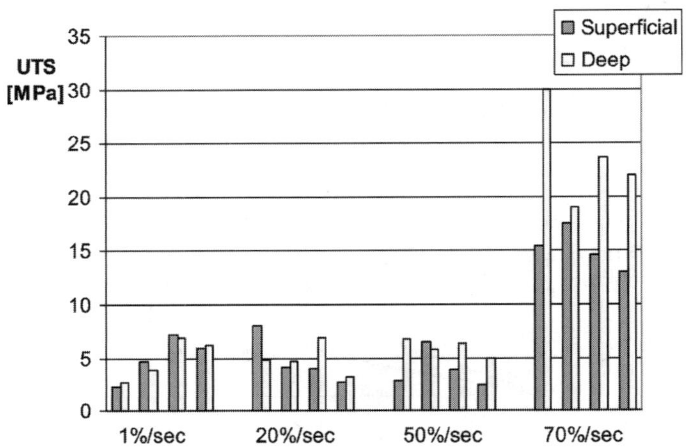

Fig. 9. Variation of the UTS of articular cartilage when tested in tension at strain rates of 1, 20, 50 and 70% per second.

within the individual group. The results obtained from these are shown in Fig. 8. It can be observed that there was no significant difference between stiffness and depth; in fact using such strain rates the superficial layer was not stiffer than the deep one, though it was expected to be because of the previous data obtained at a strain rate of 1% per second. As the strain rate increased up to 50% this lack of significant difference between the two layers difference remained. However at a strain rate of 70% per second the difference between the stiffness of cartilage from the two layers became much pronounced but inconsistent, i.e. in half of the specimens the deep layer had the stiffer cartilage while in the other half the reverse was observed. Further, specimens tested at 70% per second showed a dramatic increase of the elastic modulus of up to one order of magnitude in comparison with that measured at the lower strain rates.

3.4. Variation of the Ultimate Tensile Strength (UTS) with strain rate

The capability to endure an applied force is given by the UTS, which is the ultimate value of the stress before the material collapses under tensile loading. Values of UTS have been taken from stress–strain graphs of cartilage specimens tested at strain rates of 1, 20, 50 and 70% per second, and are shown in Fig. 9. The specimens tested at 1% per second were taken from the femoral condyles of the knee A and with the gauge length parallel to the predominant direction. It is clearly shown in Fig. 9 that there is no significant variation in the UTS of specimens tested up to 50% per second, displaying values between 2.3 and 8 MPa. However, carrying out tensile tests at the strain rate of 70% per second, the specimens exhibit a considerable rise in the UTS, which varies between 13 and 30 MPa. Furthermore, the graph indicates that at this strain rate the UTS of the deep layers is greater than that of the superficial layers, showing values which are almost doubled (from 15.5 to 30 MPa). Nevertheless, this feature is repeated at strain rates of 20 and 50% per second, though less pronounced, whereas it is not present in specimens tested at 1% per second.

4. Discussion

Articular cartilage layers are distinguishable by their different structure that also confers a specific mechanical feature on each layer. The response of cartilage to load clearly depends on the tensile properties of the tissue from the articular surface to the subchondral bone interface. Cartilage can be analyzed as a composite material made of a fibre, which is embedded in a matrix, where the fibre is the collagen while the matrix is the surrounding part. The mechanical properties of composite materials depend on the distribution and alignment of fibres in the matrix. The combination of these two components interacts to obtain structures with differentiated stiffness, which characterizes the heterogeneity and anisotropy of a material. Although the fibres act as reinforcement, they need the support of the matrix, which holds the fibres together, absorbs energy by extending under stress and increasing toughness of the material, and finally gives compressive strength to the structure. The fibres in the superficial layer are distributed parallel to the surface, while the fibres in the deep layer are aligned perpendicular to the subchondral plate. Analysing the deep layer, it can be noticed that the fibres are cut and work as particles in order to reinforce the material. This structure results in increasing the stiffness, which sometimes was greater than that from the superficial layers.

A typical stress–strain diagram of cartilage, when tested at 1% per second, is demonstrated in Fig. 10 where the areas represent the total energy absorbed by the material. Assuming that the initial response shown in Fig. 10 was due to internal energy change, it was possible to identify and trace out the area concerning the energy due to the increase in entropy during the process of stretching of the specimen. It was found the average amount of internal energy being about 60% of the total energy, whereas the entropic energy was about 40%. This latter energy was noticed to change while applying different strain rates; in fact, the area due to the entropy tended to decrease with the increase of the strain rate.

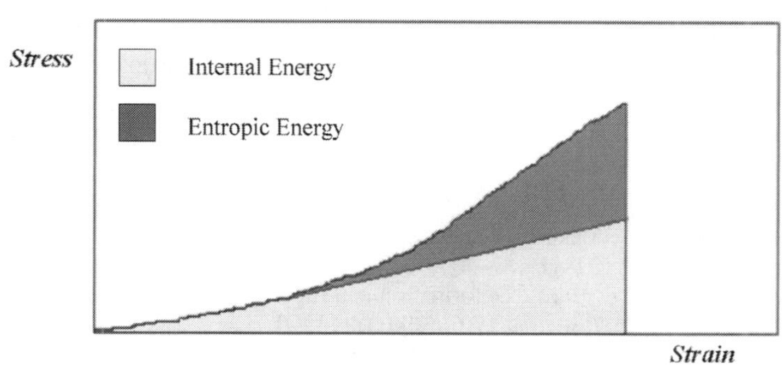

Fig. 10. Typical stress–strain curve of cartilage, which clearly shows the different absorptions of energy. The average amount of internal energy is about 60%, whereas the entropic energy is about 40%.

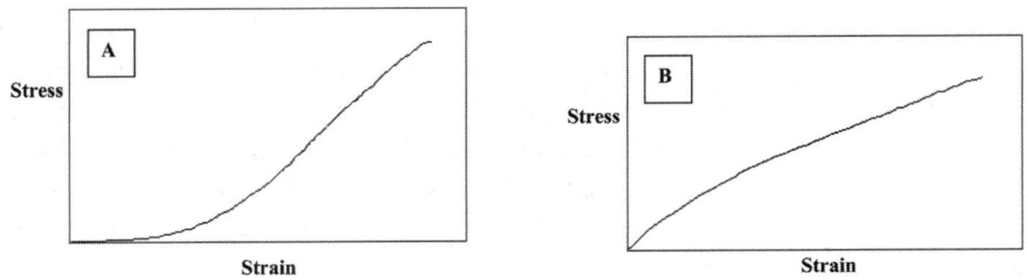

Fig. 11. Stress–strain curves of cartilage specimens, which show different mechanical responses. Curve A: typical viscoelastic behaviour. Curve B: typical elasto-plastic behaviour.

The second part of the study was carried out on cartilage at different strain rates: 20, 50 and 70% per second. The stress–strain graphs at these rates showed basically two different behaviours of cartilage. The first one was the typical behaviour found by testing specimens at a strain rate of 1% per second (Fig. 11A); the second one was characterized by an initial high slope decreasing with the increase of strain until fracture (Fig. 11B). The cartilage response shown in curve B was mostly observed when testing the deep layer at the strain rates of 20, 50 and 70% per second. Moreover, graph B displays a stress–strain relation, which is typical of particles reinforced materials. That is confirmed by assuming that fibres in the deep layer can be considered as particles rather than continuous lines as in the superficial layer. It was noticed that, evaluating the stiffness measured in stress–strain graphs, specimens performing type B curve were stiffer than type A curve. This examination was consistent with the results showing more stiffness in the deep layer when compared to the superficial one.

The drastic increase in both of the Young's modulus and ultimate tensile strength, exhibited by cartilage at the higher strain rate of 70%/sec, suggested that this biological composite material posses an extraordinary and very diverse mode of responding to loads, which was dependent on the rate of strain applied. This behaviour points out to an ability of cartilage to protect itself from traumatic situations, which comes into operation at high rates of strain, by both stiffening and strengthening the structure. This mechanism enables cartilage to preserve its thin layer from damage and to transfer the load to the subchondral bone.

Acknowledgements

The present study was supported by a scholarship from EPSRC. Thanks to the Enduratec Inc. for the testing machine and for the useful support provided throughout this work. Thanks are also due to Mike Pullan and Brian Whitham, technicians at the Bioengineering Division, University of Leeds for manufacturing the apparatus used in this study and for their considerable help.

References

[1] J.A. Buckwalter and H.J. Mankin, Articular cartilage: degeneration and osteoarthritis, repair, regeneration, and transplantation. AAOS, *Instr. Course Lect.* **47** (1984), 487–504.

[2] R. Flachsmann, N.D. Broom and A.E. Hardy, Deformation and rupture of the articular surface under dynamic and static compression, *Journal of Orthopaedic Research* **19**(6) (2001), 1131–9.

[3] Y.C. Fung, *Biomechanics*, Y.C. Fung, ed., 2nd edn, Springer-Verlag, New York, 1993.

[4] I.J. Harrington, A bioengineering analysis of force actions at the knee in normal and pathological gait, *Journal of Biomedical Engineering* **11** (1976), 167–172.

[5] J.W. Hultkrantz, Über die Spaltrichtungen der Gelenkknorpel, Verh. Anat. Ges., 1898.

[6] G.E. Kempson, The tensile properties of articular cartilage and their relevance to the development of osteoarthrosis, *Orthopaedic Surgery and Traumatology* (1972), 44–55.

[7] G.E. Kempson, Mechanical properties of articular cartilage, in: *Adult Articular Cartilage*, M.A.R. Freeman, ed., Pitman Medical Ltd, London, 1979, pp. 333–414.

[8] V.C. Mow, S. Akizuki, F. Muller, J.C. Pita, D.S. Howell and D.H. Manicourt, Tensile properties of human knee cartilage: I. Influence of ionic conditions, weight bearing, and fibrillation on the tensile modulus, *Journal of Orthopaedic Research* **4**(4) (1986), 397–392.

[9] C.W. Slemenda, The epidemiology of osteoarthritis of the knee, *Curr. Opin. Rheumatol.* **4** (1992), 546–51.

[10] S.J. Taylor, P.S. Walker, J.S. Perry, S.R. Cannon and R. Woledge, The forces in the distal femur and the knee during walking and other activities measured by telemetry, *Journal of Arthroplasty* **13**(4) (1998), 428–37.

[11] S.L.Y. Woo, W.H. Akeson and G.F. Jemmott, Measurements of nohomogenous directional mechanical properties of articular cartilage in tension, *Journal of Biomechanics* **9** (1976), 785–791.

Biorheology 41 (2004) 215–222
IOS Press

Macroscale modeling of cartilage: Mixture theory versus homogenization

Christian Moyne [a] and Márcio A. Murad [b]
[a] *LEMTA (UMR 7563) CNRS-INPL-UHP, 2, Avenue de la Forêt de Haye, 54504 Vandoeuvre lès Nancy Cedex, France*
[b] *Laboratório Nacional de Computação Científica LNCC/MCT, Av Getúlio Vargas 333, 25651-070 Petrópolis, RJ, Brazil*

Abstract. The equations governing the electro-chemo-hydro-mechanical processes in cartilage are derived using the periodic homogenization technique. First the equations at the microscale are recalled. Then the homogenization technique is applied to propagate the physics to the macroscale. The results of this work and of the thermodynamical approach used by various authors are finally compared.

1. Introduction

During the past few decades a significant amount of research has been developed toward the derivation of models capable of capturing coupled electro-chemo-hydro-mechanical effects in charged hydrated and swelling tissue and especially in cartilage. In addition to classical phenomenological approaches developed at the macroscale, coupling phenomena can naturally be described within the framework of the Mixture Theory and Thermodynamics of Irreversible Processes [3,5–10,13,14,18–20]. In this approach, the interaction between matter flux, electric charge and chemical osmosis can be properly embedded in the framework of Onsager's reciprocity relations. Another point of view is to derive the equations from an up-scaling procedure: assuming that electrostatic forces are dominant at the microscopic level, use of classical continuum mechanics allows a complete description of the chemo-electro-hydrodynamics problem. The electric potential is given by a Poisson equation, the movement of the fluid phase by Stokes equations supplemented by an electric force term, the transport of ions by Nernst–Planck equations and finally the solid phase is supposed to be isotropic elastic. The homogenization technique is then applied to propagate information available at the pore scale to the macroscale where the following equations are derived: a Darcy modified equation supplemented by both a chemical osmosis term induced by a salt concentration gradient and an electro-osmotic term induced by a streaming potential gradient; two convection-diffusion equations for the cations and anions transport; Biot poroelasticity equations with a Terzaghi principle modified by an additive swelling pressure term to incorporate the physicochemical effects. Given a spatial periodic structure this method allows to calculate the coefficients of the model. They depend in particular of the double layer potential, solution of the Poisson–Boltzmann equation, which has to be solved at the microscopic scale in a representative unit cell of the medium. The aim of this presentation is to compare both approaches in order to derive an unified theory for water flow, ions transport and mechanical equilibrium in cartilage.

2. Microscopic description of the problem

Consider at the microscopic scale the porous medium (cartilage) composed of a solid phase "s" and a liquid one "f". For simplicity the liquid phase is considered a mixture of water and ions of an entirely dissociated monovalent salt: anions "−" and cations "+" with volumetric concentrations c^- and c^+. If F designates the Faraday constant, $\tilde{\varepsilon}\tilde{\varepsilon}_0$ the permittivity of the mixture, the electric potential Φ is given by the Poisson equation (1) (with $\epsilon = 1$). The movement of the fluid phase (assumed Newtonian of dynamic viscosity μ_f) is given by the Stokes equations ((2), (3)) supplemented by an electric force term represented in terms of the divergence of the Maxwell tensor [11]. Here $\mathbf{v}$ is the velocity of the fluid, p the pressure and $\mathbf{E} = -\boldsymbol{\nabla}\Phi$ denotes the electric field. The movement of the ions is due to a convective movement and a diffusion one. The diffusion term is the sum of a Fickian term ($\mathcal{D}_{w\pm}$ are the binary water–ions diffusion coefficients, R the perfect gas constant and T the temperature) and a forced diffusion one accounting for the effect of the electric forces acting on the ions (4). The solid phase is supposed to be isotropic elastic with Lamé parameters λ_s and μ_s and its deformation is denoted by $\mathcal{E}$. The equation for the displacement $\mathbf{u}$ is given by (5). Boundary conditions are prescribed on the solid–fluid interfaces (A_{fs}). Since the solid surface is charged with a surface charge density σ, the electric field at the interface is given by (6) where $\mathbf{n}_{fs}$ is the normal to the interface oriented from the fluid to the solid phase. The boundary condition for the Stokes problem (7) expresses the non-slip condition at the interface. For the transport of the ions, the solid is assumed impervious (8). Finally, the stress continuity is given by (9).

3. Order of magnitude analysis

In order to derive the equations governing the phenomena at the macroscopic scale, the homogenization method is used [22]. Two length scales are introduced, a microscopic one ℓ of the order of the pore size and a macroscopic one L. Their ratio $\epsilon \equiv \ell/L$ is a small parameter. In what follows the microscopic governing equations are written in dimensionless form thoroughly using references values in the spirit of [1]. The order of magnitude of the electric field is given by the imposed value for the normal component on the solid–fluid interface, $E_{\text{ref}} = \sigma/(\tilde{\varepsilon}\tilde{\varepsilon}_0)$. As the Debye's length (see below) is assumed to be of the order of magnitude of ℓ, $\Phi_{\text{ref}} = \ell E_{\text{ref}}$. The electric charge is supposed to counterbalance the surfaces charge and $c_{\text{ref}} = \sigma/(F\ell)$. The classical theory based on Darcy's law allows to assume $v_{\text{ref}} = \ell^2 p_{\text{ref}}/(\mu_f L)$. Three dimensionless quantities appear: the Péclet number $\text{Pe} = v_{\text{ref}}L/\mathcal{D}$; $\text{N} = F\sigma\ell/(\tilde{\varepsilon}\tilde{\varepsilon}_0 RT)$ which is the ratio of electric energy to the thermal energy of an ion; $\text{M} = (\tilde{\varepsilon}\tilde{\varepsilon}_0/2)E_{\text{ref}}^2 L/u_{\text{ref}}(3\lambda_s + 2\mu_s)$ which is a measure of the importance of the electric effects on the stress acting on the solid phase. When swelling effects are important, it seems reasonable to consider that $\text{M} = \mathcal{O}(1)$ and $\text{N} = \mathcal{O}(1)$. The Péclet number is assumed $\mathcal{O}(\epsilon^m)$ with $m \geqslant 0$. If $m = 1$, the convection terms for the ions transport disappear at the macroscale [15,16]. To retain these terms, we postulate here that $m = 0$ [12]. Using the macroscopic scale L to normalize the operator $\boldsymbol{\nabla}$ and the time scale $t_{\text{ref}} = L^2/\mathcal{D}$ for the time derivatives, the following equations are listed below in dimensional form with a formal ϵ^n factor to indicate the order of magnitude of each term.

– Poisson equation

$$\epsilon^2 \nabla^2 \Phi = -\frac{(c^+ - c^-)F}{\tilde{\varepsilon}\tilde{\varepsilon}_0}. \tag{1}$$

– Stokes equation

$$0 = \nabla . \mathbf{v}, \tag{2}$$

$$0 = \nabla . \left[-p\boldsymbol{I} + \frac{\tilde{\varepsilon}\tilde{\varepsilon}_0}{2}(2\mathbf{EE} - E^2\boldsymbol{I}) \right] + \epsilon^2 \mu_f \nabla^2 \mathbf{v}. \tag{3}$$

– Transport equation for the ions

$$\frac{\partial c^{\pm}}{\partial t} + \epsilon^{m-1} \nabla . (c^{\pm}\mathbf{v}) = \nabla . \left[\mathcal{D}_{w\pm}\left(\nabla c^{\pm} \pm \frac{Fc^{\pm}}{RT} \nabla \Phi \right) \right]. \tag{4}$$

– Elasticity equation for the solid phase

$$\mu_s \nabla^2 \mathbf{u} + (\lambda_s + \mu_s)\nabla(\nabla . \mathbf{u}) = 0. \tag{5}$$

The boundary conditions on the solid–fluid interface (A_{fs}) are:

$$\epsilon \nabla \Phi . \mathbf{n}_{fs} = \frac{\sigma}{\tilde{\varepsilon}\tilde{\varepsilon}_0}, \tag{6}$$

$$\mathbf{v} - \frac{\partial \mathbf{u}}{\partial t} = 0, \tag{7}$$

$$-\mathcal{D}_{w\pm}\left(\nabla c^{\pm} \pm \frac{Fc^{\pm}}{RT} \nabla \Phi \right) . \mathbf{n}_{fs} = 0, \tag{8}$$

$$\left[-p\boldsymbol{I} + \frac{\tilde{\varepsilon}\tilde{\varepsilon}_0}{2}(2\mathbf{EE} - E^2\boldsymbol{I}) + \epsilon^2(\nabla \mathbf{v} + \nabla \mathbf{v}^T) \right] . \mathbf{n}_{fs} = \boldsymbol{\sigma}_s . \mathbf{n}_{fs}, \tag{9}$$

where $\boldsymbol{\sigma}_s$ is the stress tensor in the solid phase. According to the classical homogenization method for periodic media, two independent variables are introduced $\boldsymbol{x}$ for the macroscopic scale and $\boldsymbol{y}$ for the microscopic scale, the medium being spatially periodic in $\boldsymbol{y}$. With the choice of the macroscopic length as reference value L, $\nabla = 1/\epsilon \nabla_y + \nabla_x$. The homogenization method consists in searching a development in series of the small parameter ϵ for each variable $\psi = \sum \epsilon^k \psi^k$ of the problem.

4. Homogenization

4.1. Poisson–Boltzmann equation

We introduce (at the zeroth order) a bulk solution with the concentration $c_b^0(\boldsymbol{x}, t)$ and with the electric potential $\psi_b^0(\boldsymbol{x}, t)$ in local thermodynamical equilibrium with the solution inside the porous material. It is possible to prove that the zeroth order electric potential can be split into $\boldsymbol{\Phi}^0(\boldsymbol{x}, \boldsymbol{y}, t) = \varphi^0(\boldsymbol{x}, \boldsymbol{y}, t) + \psi_b^0(\boldsymbol{x}, t)$ where ψ_b^0 is the so-called *streaming potential* and φ^0 is a *double layer potential* satisfying the Poisson–Boltzmann equation

$$\nabla^2 \varphi^0 = \frac{1}{L_D^2} \sinh(F\varphi^0/RT) \quad \text{in } Y_f, \tag{10}$$

$$\nabla_y \varphi^0 . \mathbf{n}_{fs} = \sigma/\tilde{\varepsilon}\tilde{\varepsilon}_0 \qquad \text{on } \partial Y_{fs}, \tag{11}$$

where $L_D = \sqrt{\tilde{\varepsilon}\tilde{\varepsilon}_0 RT/2F^2 c_b^0}$ is the Debye's length [15]. The ions obey a Boltzmann distribution $c^{\pm 0} = c_b \exp(\mp\bar{\varphi})$ with $\bar{\varphi} = F\varphi/RT$.

4.2. Transport of ions

In the case $m = 0$, the homogenization procedure leads to the following macroscopic equation

$$\frac{\partial}{\partial t}(n_f G_\pm^* c_b^0) + \boldsymbol{\nabla}_x \cdot \boldsymbol{J}_\pm^0 = 0 \tag{12}$$

with

$$\boldsymbol{J}_\pm^0 \equiv n_f G_\pm^* c_b^0 \mathbf{v}_\pm^{*0} - n_f(\boldsymbol{D}_\pm^C \boldsymbol{\nabla}_x c_b^0 + \boldsymbol{D}_\pm^E c_b^0 \boldsymbol{\nabla}_x \psi_b^0 + \boldsymbol{D}_\pm^P \boldsymbol{\nabla}_x p_b^0) \tag{13}$$

and

$$G_\pm^* = \langle \exp(\mp\overline{\varphi}^0)\rangle^f; \qquad G_\pm^* \mathbf{v}_\pm^{*0} = \langle \exp(\mp\overline{\varphi}^0)\mathbf{v}^0\rangle^f, \tag{14}$$

where $\langle\ \rangle^f$ designates the volume average over the fluid phase of a unit cell, n_f is the volume fraction of the fluid phase and $\boldsymbol{D}_\pm^C$, $\boldsymbol{D}_\pm^E$ and $\boldsymbol{D}_\pm^P$ are diffusion tensors which can be determined solving *closure* problems on the unit cell of the periodic medium [17].

4.3. Fluid movement

Introducing

$$\begin{aligned} p_b^0 &= p^0 + \int_0^{\varphi^0} F(c^{+0} - c^{-0})\,\mathrm{d}\varphi = p^0 - 2Fc_b^0 \int_0^{\varphi^0} \sinh\bar{\varphi}^0\,\mathrm{d}\varphi^0 \\ &= p^0 - 2RTc_b^0(\cosh\bar{\varphi}^0 - 1) = p^0 - RT(c^{+0} + c^{-0} - 2c_b^0) = p^0 - \pi^0, \end{aligned} \tag{15}$$

where π^0 is the Donnan osmotic pressure [4], the Stokes equations show that p_b^0 is only a function of $\boldsymbol{x}$ and t. Following standard arguments used in the homogenization of the Stokes problem (with $m = 0$) leads to the following modified form of Darcy's law

$$\mathbf{v}_D^0 = \langle \mathbf{v}^0 \rangle - n_f \frac{\partial \mathbf{u}^0}{\partial t} = -\boldsymbol{K}_P \cdot \boldsymbol{\nabla}_x p_b^0 - \boldsymbol{K}_C \cdot \boldsymbol{\nabla}_x c_b^0 - \boldsymbol{K}_E \cdot \boldsymbol{\nabla}_x \psi_b^0, \tag{16}$$

where $\langle\ \rangle = n_f \langle\ \rangle^f$ is the average value over the unit cell. $\boldsymbol{K}_{P,C,E}$ are respectively the permeability, chemico-osmotic and electro-osmotic tensors. They can be determined solving a *closure* on the unit cell [17].

4.4. *Elasticity equation for the solid phase* [21]

At the order $\mathcal{O}(\epsilon^0)$, the problem solution for $\mathbf{u}^0$ shows that $\mathbf{u}^0$ depends only on x and t. The problem at order $\mathcal{O}(\epsilon)$ is the following:

$$\mu_s \nabla_y^2 \mathbf{u}^1 + (\lambda_s + \mu_s)\boldsymbol{\nabla}_y(\boldsymbol{\nabla}_y \cdot \mathbf{u}^1) = 0 \tag{17}$$

with the boundary condition on ∂Y_{fs}:

$$\begin{aligned} &-\left[p_b^0(\boldsymbol{x},t)\boldsymbol{I} + \boldsymbol{\Pi}_d^0(\boldsymbol{x},\boldsymbol{y},t)\right] \cdot \mathbf{n}_{fs} \\ &\quad = \left[\lambda_s(\boldsymbol{\nabla}_x \cdot \mathbf{u}^0 + \boldsymbol{\nabla}_y \cdot \mathbf{u}^1)\boldsymbol{I} + 2\mu_s(\boldsymbol{\mathcal{E}}_x(\mathbf{u}^0) + \boldsymbol{\mathcal{E}}_y(\mathbf{u}^1))\right] \cdot \mathbf{n}_{fs}, \end{aligned} \tag{18}$$

where the disjoining pressure $\boldsymbol{\Pi}_d^0$ is the sum of the chemico-osmotic pressure π^0 and the electrical Maxwell stresses $\boldsymbol{\tau}_{\mathbf{M}}^0$:

$$\boldsymbol{\Pi}_d^0 = \pi^0 \boldsymbol{I} - \boldsymbol{\tau}_{\mathbf{M}}^0 \quad \text{with} \quad \pi^0 = 2RTc_b^0(\cosh\bar{\varphi}^0 - 1) \quad \text{and} \quad \boldsymbol{\tau}_{\mathbf{M}}^0 = \frac{\tilde{\varepsilon}\tilde{\varepsilon}_0}{2}(2\mathbf{E}\mathbf{E} - E^2\boldsymbol{I}). \tag{19}$$

Three non-homogeneous terms appear in the boundary condition (18) depending on p_b^0, $\boldsymbol{\nabla}_x\mathbf{u}^0$ and on the "disjoining pressure" term $\boldsymbol{\Pi}_d^0$. Due to the linearity of the problem, the solution is searched in the form:

$$\mathbf{u}^1 = \boldsymbol{\zeta}(\boldsymbol{y})p_b^0(\boldsymbol{x},t) + \boldsymbol{\xi}(\boldsymbol{y}) : \mathcal{E}_x(\mathbf{u}^0) + \mathbf{u}_\pi^1(\boldsymbol{x},\boldsymbol{y},t) + \hat{\mathbf{u}}^1(\boldsymbol{x},t). \tag{20}$$

The problems for the vector $\boldsymbol{\zeta}$ and the third order tensor $\boldsymbol{\xi}$ are classical [2]. $\mathbf{u}_\pi^1$ satisfies

$$\mu_s \nabla_y^2 \mathbf{u}_\pi^1 + (\lambda_s + \mu_s)\boldsymbol{\nabla}_y(\boldsymbol{\nabla}_y \cdot \mathbf{u}_\pi^1) = 0 \quad \text{in } Y_s \tag{21}$$

and

$$-\boldsymbol{\Pi}_d^0 \cdot \mathbf{n}_{fs} = \left[\lambda_s \boldsymbol{\nabla}_y \cdot \mathbf{u}_\pi^1 \boldsymbol{I} + 2\mu_s \boldsymbol{\mathcal{E}}_y(\mathbf{u}_\pi^1)\right] \cdot \mathbf{n}_{fs} \quad \text{on } \partial Y_{fs}. \tag{22}$$

Introducing the total stress tensor $\boldsymbol{\sigma}_T^0 = \langle \boldsymbol{\sigma}_f^0 \rangle$ in Y_f and $\langle \boldsymbol{\sigma}_s^0 \rangle$ in Y_s, the following macroscopic results are obtained

$$\boldsymbol{\nabla}_x \cdot \boldsymbol{\sigma}_T^0 = 0, \tag{23}$$

$$\boldsymbol{\sigma}_T^0 = -\boldsymbol{\alpha} p_b^0 + \boldsymbol{C}_s \cdot \boldsymbol{\mathcal{E}}_x(\mathbf{u}^0) - \boldsymbol{\Pi}^0 \tag{24}$$

with

$$\boldsymbol{\Pi}^0 = -(1 - n_f)\langle \boldsymbol{c}_s \cdot \boldsymbol{\mathcal{E}}_y(\mathbf{u}_\pi^1)\rangle^s + n_f\langle \boldsymbol{\Pi}_d^0 \rangle, \tag{25}$$

where $\boldsymbol{\alpha} = -n_f\boldsymbol{I} + \langle \boldsymbol{c}_s \cdot \boldsymbol{\mathcal{E}}_y(\boldsymbol{\zeta})\rangle$ is the Biot coefficient for the solid phase; $\boldsymbol{c}_s$ and $\boldsymbol{C}_s = \langle \boldsymbol{c}_s \cdot (\boldsymbol{I} + \boldsymbol{\mathcal{E}}_y(\boldsymbol{\xi}))\rangle$ are respectively the microscopic elastic modulus of the solid phase and the macroscopic elastic modulus for the equivalent medium; $\boldsymbol{\Pi}^0$ is the *swelling stress tensor*.

4.5. *Overall mass balance*

From the mass conservation for the fluid at order $\mathcal{O}(\epsilon)$, an integration over the unit cell leads to

$$\nabla_x \cdot \mathbf{v}_D^0 = (-n_f \boldsymbol{I} + \langle \nabla_y \cdot \boldsymbol{\xi} \rangle) : \frac{\partial \boldsymbol{\mathcal{E}}_x(\mathbf{u}^0)}{\partial t} + \langle \nabla_y \cdot \boldsymbol{\zeta} \rangle \frac{\partial p_b^0}{\partial t} + \frac{\partial \gamma_\pi}{\partial t}, \tag{26}$$

where $\gamma_\pi = \langle \nabla_y \cdot \mathbf{u}_\pi^1 \rangle$.

4.6. *Fluid mass conservation*

To close the system it remains to derive a mass balance equation for the fluid phase in order to compute the porosity n_f. Averaging the mass balance for the incompressible fluid flow leads to (neglecting the convective effect introduced by $\partial \mathbf{u}^0/\partial t$):

$$\frac{\partial n_f}{\partial t} + \nabla_x \cdot \mathbf{v}_D^0 + n_f \nabla_x \cdot \frac{\partial \mathbf{u}^0}{\partial t} = 0. \tag{27}$$

4.7. *Summary of the Two-Scale Model*

The macroscopic two-scale problem contains several unknowns: the total tensor $\boldsymbol{\sigma}_T^0$, the bulk pressure p_b^0, the global solid displacement $\mathbf{u}^0$, the Darcy's velocity $\mathbf{v}_D^0$, the bulk concentration c_b^0, the streaming potential ψ^0 and the porosity n_f. To compute all the unknowns, we have to solve the mechanical equilibrium using the Terzaghi's principle, the total volume conservation using the Darcy's law, the two conservation equation for the cations and the anions and finally the fluid volume conservation equation:

$$\begin{cases} \nabla_x \cdot \boldsymbol{\sigma}_T^0 = 0, \\ \boldsymbol{\sigma}_T^0 = -\boldsymbol{\alpha} p_b^0 + \boldsymbol{C}_s \boldsymbol{\mathcal{E}}_x(\mathbf{u}^0) - \boldsymbol{\Pi}^0, \\ \nabla_x \cdot \mathbf{v}_D^0 + \boldsymbol{\alpha} : \dfrac{\partial}{\partial t} \boldsymbol{\mathcal{E}}_x(\mathbf{u}^0) = \beta \dfrac{\partial p_b^0}{\partial t} + \dfrac{\partial \gamma_\pi}{\partial t}, \\ \mathbf{v}_D^0 = -\boldsymbol{K}_P \nabla_x p_b^0 - \boldsymbol{K}_C \nabla_x c_b^0 - \boldsymbol{K}_E \nabla_x \psi_b^0, \\ \dfrac{\partial}{\partial t}(n_f G_\pm^* c_b^0) + \nabla_x \cdot \boldsymbol{J}_\pm^0 = 0, \\ \boldsymbol{J}_\pm^0 = n_f G_\pm^* c_b^0 \mathbf{v}_*^{\pm 0} - n_f (\boldsymbol{D}_\pm^C \nabla_x c_b^0 \pm \boldsymbol{D}_\pm^E c_b^0 \nabla_x \psi_b^0 + \boldsymbol{D}_\pm^P \nabla_x p_b^0), \\ \dfrac{\partial n_f}{\partial t} + \nabla_x \cdot \mathbf{v}_D^0 + n_f \nabla_x \cdot \dfrac{\partial \mathbf{u}^0}{\partial t} = 0. \end{cases} \tag{28}$$

5. Conclusion: "mixture theory" versus "homogenization"

The aim of this section is to compare the models derived using the thermodynamic method and our model using the homogenization technique.

Deriving the Fick's law, the thermodynamic models [7–10,14] gives two equations for the *average* concentration of the cations and the anions $\langle c^\pm \rangle^f$. Thus three unknowns (the two concentrations and the streaming potential ψ_b^0) appear for only two transport equations. An electroneutrality condition at the macroscale (anions charge + cations charge + solid fixed charge $= 0$) is added to remove this indetermination. The homogenization method demonstrates that the electroneutrality condition is nothing else but the integration on the fluid part of the unit cell of the Poisson–Boltzmann equation (10). Therefore only a bulk concentration c_b^0 is introduced. More all the *bulk* variables introduced in the homogenization method (bulk pressure p_b^0, concentration c_b^0 and (streaming) potential ψ_b^0) are continuous variables at the frontier with the bulk phase.

The homogenization procedure derives a Darcy's law (16) in which three driving forces appear: the bulk pressure gradient, the bulk concentration gradient and the streaming potential gradient. In the case of a small Péclet number ($m = 1$), it is possible to prove that the permeability tensor $\boldsymbol{K}_P$ is the classical one which arises when homogenizing the Stokes equations [15]. In the case $m = 0$ used here due to a coupling between the double layer potential and the fluid flow, this coefficient incorporates an electro-viscous effect coming from the additional electric drag induced by the distorsion of the electric double layer [12,17]. The two other driving forces are the chemico-osmotic and electro-osmotic components which quantify the flow driven by concentration and streaming potential gradients.

The last point is the generalization of the Terzaghi's principle. Equation (25) clearly shows that the *swelling pressure* (in fact *swelling tensor*) comes from the chemico-osmotic pressure π^0 and the electrical forces given by the Maxwell's tensor.

References

[1] J.L. Auriault, Heterogeneous media: is an equivalent homogeneous description always possible?, *Int. J. Engrg. Sci.* **29** (1991), 785–795.

[2] J.L. Auriault and E. Sanchez-Palencia, Etude du comportement macroscopique d'un milieu poreux saturé déformable, *Journal de Mécanique* **16** (1977), 575–603.

[3] L.S. Bennethum, M.A. Murad and J.H. Cushman, Macroscale thermodynamics and the chemical potential for swelling porous media, *Transport in Porous Media* **39** (2000), 187–225.

[4] F.G. Donnan, The theory of membrane equilibria, *Chemical Review* **1** (1924), 73–90.

[5] L. Dormieux, P. Barboux, O. Coussy and P. Dangla, A macroscopic model of the swelling phenomenon of a saturated clay, *Eur. Mech. A/Solids* **14** (1995), 981–1004.

[6] L. Dormieux, E. Lemarchand and O. Coussy, Macroscopic and micromechanical approaches to the modelling of the osmotic swelling in clays, *Transport in Porous Media* **50** (2003), 75–91.

[7] W.Y. Gu, W.M. Lai and V.C. Mow, A triphasic analysis of negative osmotic flows through charged hydrated tissues, *J. Biomechanics* **30** (1998), 71–78.

[8] W.Y. Gu, W.M. Lai and V.C. Mow, A mixture theory for charged–hydrated soft tissues containing multi-electrolytes: Passive transport and swelling behaviors, *J. Biomech. Engrg.* **120** (1998), 169–180.

[9] J.M. Huyghe and J.D. Janssen, Quadriphasic mechanics of swelling incompressible porous media, *Int. J. Engrg. Sci.* **25** (1997), 793–802.

[10] W.M. Lai, J.S. Hou and V.C. Mow, A triphasic theory for the swelling and deformation behaviors of articular cartilage, *J. Biomech. Engrg.* **113** (1991), 245–258.

[11] L.D. Landau and E.M. Lifchitz, *Electrodynamics of Continuous Media*, Pergamon Press, Oxford, 1960.

[12] T. Lemaire, C. Moyne, D. Stemmelen and M.A. Murad, Electro-chemo-mechanical couplings in swelling clays derived by homogenization: electroviscous effects and Onsager's relations, in: *2nd Biot Conference on Poromechanics*, Auriault et al., eds, Balkema Publishers, 2002, pp. 489–494.

[13] M.E. Levenston, S.R. Eisenberg and A.J. Grodzinsky, A variational formulation for coupled physicochemical flows during finite deformation of charged porous media, *International J. Solids and Structures* **35** (1998), 4999–5019.

[14] V.C. Mow, A. Ateshian, W. Lai and W. Gu, Effects of fixed charges on the stress–relaxation behavior of hydrated soft tissues in a confined compression problem, *International J. Solids and Structures* **35** (1998), 4945–4962.

[15] C. Moyne and M.A. Murad, Electro-chemo-mechanical couplings in swelling clays derived from a micro/macro homogenization procedure, *International J. Solids and Structures* **39** (2002), 6159–6190.

[16] C. Moyne and M.A. Murad, Macroscopic behavior of swelling porous media derived from micromechanical analysis, *Transport in Porous Media* **50** (2003), 127–151.

[17] C. Moyne and M.A. Murad, A two-scale model for coupled electro-chemo-mechanical phenomena and Onsager's reciprocity relations in expansive clays: I. Homogenization analysis. II. Computational validation, *Submitted* (2003).
[18] M.A. Murad, A thermomechanical model of hydration swelling in smectite clays, I. Two-scale mixture-theory approach, *International J. for Numerical and Analytical Methods in Geomechanics* **23** (1999), 73–696.
[19] M.A. Murad and J.C. Cushman, A multiscale theory of swelling porous media: II. Dual porosity models for consolidation of clays incorporating physicochemical effects, *Transport in Porous Media* **28** (1997), 69–108.
[20] M.A. Murad and J.C. Cushman, Thermomechanical theories for swelling porous media with microstructure, *Int. J. Engrg. Sci.* **38** (2000), 517–564.
[21] M.A. Murad and C. Moyne, Micromechanical computational modeling of expansive porous media, *C. R. Mecanique* **330** (2002), 865–870.
[22] E. Sanchez-Palencia, *Non-Homogeneous Media and Vibration Theory, Lectures Notes in Physics*, Springer-Verlag, 1980.

Role of cell-associated matrix in the development of free-swelling and dynamically loaded chondrocyte-seeded agarose gels

Terri-Ann N. Kelly [a], Christopher C.-B. Wang [a], Robert L. Mauck [a], Gerard A. Ateshian [b] and Clark T. Hung [a,c]

[a] *Cellular Engineering Laboratory, Department of Biomedical Engineering, Columbia University, New York, NY 10027, USA*

[b] *Musculoskeletal Biomechanics Laboratory, Department of Mechanical Engineering, Columbia University, New York, NY 10027, USA*

[c] *351 Engineering Terrace, MC 8904, 1210 Amsterdam Avenue, Department of Biomedical Engineering, Columbia University, New York, NY 10027, USA*
Tel.: +1 212 854 6542; Fax: +1 212 854 8725; E-mail: cth6@columbia.edu

Abstract. Chondrocytes embedded in agarose and subjected to dynamic deformational loading produce a functional matrix with time in culture, but there is usually a delay in the development of significant differences compared to free swelling. In this study, we hypothesized that the initial presence of a cell-associated matrix would expedite construct development in response to dynamic deformational loading. Seeded samples with enzymatically isolated chondrocytes and chondrons (the chondrocyte and its pericellular matrix) and examined the effects of seeding density and dynamic loading on the development of tissue properties. At 60 million/ml, dynamic loading significantly augmented the development of material properties in chondrocyte- and chondron-seeded constructs. Biochemical content and histological analysis indicated that the deposition of GAG, link protein and collagens are affected by the pre-existing cell-associated matrix of the chondron-seeded samples. The pericellular matrix associated with the chondrons did not expedite the development of material properties in response to deformational loading, disproving our hypothesis. The relative concentration and distribution of matrix proteins may play a major role in the disparate responses observed for the chondrocyte- and chondron-seeded cultures. In further support of these findings, culturing chondrocytes in agarose for two weeks prior to the application of deformational loading also did not exhibit expedited construct development.

Keywords: Chondron, functional tissue engineering, material properties, collagens, link protein, glycosaminoglycans

1. Introduction

The extracellular matrix (ECM) of articular cartilage is composed of a dense, strong network of collagen fibers (60% dry weight), primarily type II collagen (⩾80% of the total collagen) interwoven with proteoglycan (PG) aggregates (25–35% dry weight) and other non-collagenous proteins (15–20% dry weight) [2,14,30]. These components form a functional ECM that provides articular cartilage with its unique load-bearing properties, and disruption of this ECM leads to the development of osteoarthritis [2,14,30]. Accordingly, there have been significant research efforts utilizing a variety of *in vitro* models (alginate [43], agarose [1], high density monolayer cultures [11]), for studying cartilage matrix development by isolated chondrocytes. As chondrocytes are subjected to a plethora of physical stimuli arising from joint loading-induced tissue deformation, there has also been interest in understanding chondrocyte and cartilage mechanotransduction and response to these physical forces [2,14,29,30].

In vitro studies of chondrocytes have suggested that the presence of a cell-associated matrix modulates the temporal development of further matrix elaboration [6,12,18], growth factor sensitivity [39, 40], and biosynthetic activity in response to physical loading [13,15]. In support of this notion, Lee and co-workers have reported that chondrocytes in pellet cultures possessing a native pericellular matrix (i.e., chondrons) elaborate ECM more rapidly than pellet cultures of isolated matrix-free chondrocytes maintained in free-swelling culture [12,13,18]. The chondron, referred to in the literature as the basic functional biomechanical unit of articular cartilage [31–35], is composed of one or more chondrocytes and their associated pericellular matrix. This study investigated if the presence of such a native pericellular matrix has a similar influence on ECM development of seeded agarose hydrogel cultures. Agarose hydrogels have been used extensively in chondrocyte basic science studies [1,3,4,6,19,20,42], and for some cartilage tissue engineering applications (e.g., [7,36]). Through these studies, we aim to elucidate the cause of an initial 2–3 week delay that is typically observed before the effects of applied deformational loading on chondrocyte-seeded agarose construct material and biochemical properties become apparent [24,25,27].

In the present study, we hypothesized that the retention of a cell-associated matrix after isolation would dramatically expedite construct development in chondrocyte-seeded agarose constructs maintained in free-swelling culture or subjected to daily dynamic deformational loading. This hypothesis was tested by examining the effects of free-swelling culture (Study 1) and dynamic deformational loading culture (Study 2) on the development of material properties and biochemical composition of agarose constructs seeded with enzymatically isolated chondrocytes and chondrons over six weeks in culture. Early work from our laboratory has demonstrated that constructs seeded at 60 million/ml develop material properties more rapidly and achieve higher stiffness than 10 million/ml constructs cultured over the same period; constructs seeded at these densities were adopted for these studies [24,25,27]. In a related study (Study 3), we repeated Study 2 (60 million/ml) using chondrocyte-seeded agarose constructs that were maintained under free-swelling culture for two weeks, permitting elaboration of a cell-associated matrix, before the initiation of applied deformational loading.

2. Materials and methods

2.1. *Culturing chondrocytes and chondrons in agarose gels*

Full thickness articular cartilage was harvested from the carpometacarpal or patellofemoral joints (Fresh Farms Beef, Rutland, VT) of 2 to 3 month old calves. For each study, tissue was harvested from four animals and pooled before digestion. Chondrons were enzymatically isolated from minced cartilage via digestion in 0.2% (w/v) dispase (Life Technologies, Rockville, MD) and 0.05% (w/v) collagenase (Sigma Chemicals, St. Louis, MO) for 5 hours with mild agitation [21]. Chondrocytes were isolated via digestion in 0.25% (w/v) pronase (Calbiochem, San Diego, CA) for 1 hour followed by a 4-hour digestion in 0.05% (w/v) collagenase, and the samples were stirred for the duration of the digestions to maximize cell yield.

2.2. *Study 1: Chondrocytes and chondrons in agarose gels at different seeding densities*

Chondrocytes and chondrons were encapsulated at a final concentration of 10 million/ml or 60 million/ml in 2% (w/v) low-melt agarose (type VII, Sigma Chemicals) in phosphate buffered saline (PBS). Agarose slabs were cast at 2.3 mm thickness between glass plates, and cylindrical constructs cored with a 4.76 mm diametedermal punch. The constructs were allowed to equilibrate for 3 days in 30 ml of serum-free media, where day 0 was taken to be the final day of this period. The constructs were then cultured in 30 ml of DMEM (Dulbecco's Modification of Eagle's Medium, Life Technologies) containing 10% (for 10 million/ml constructs) or 20% (for 60 million/ml constructs) fetal bovine serum (FBS, Sigma Chemicals), buffers, antibiotics, antimycotics, and fresh 50 μg/ml ascorbic acid (Sigma Chemicals) [25,27]. The constructs were maintained aseptically in free-swelling culture with daily media changes. Samples were removed for mechanical testing on days 0, 14 and 28.

2.3. *Study 2: Dynamic deformational loading of chondrocytes and chondrons in agarose gels*

Chondrocytes and chondrons were encapsulated at final concentration of 10 million/ml or 60 million/ml in 2% (w/v) low-melt agarose in PBS. Agarose slabs were cast, cored, equilibrated, and then cultured in 30 ml of DMEM (containing 10% or 20% FBS, buffers, antibiotics, antimycotics, and 50 μg/ml ascorbic acid), as described above. Constructs were either maintained under free-swelling conditions or dynamically loaded (Fig. 1A), as described previously [25], through a 6-week culture period with daily

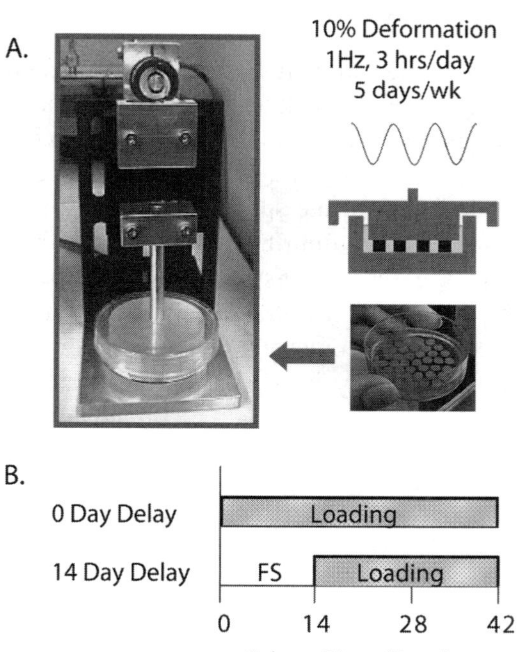

Fig. 1. (A): Schematic representation of dynamic deformational loading protocol. 10% peak-to-peak compression was applied to samples in an unconfined compression configuration at 1 Hz (sinusoidal) for 3 hours/day, 5 days/week over the course of a six-week culture period. (B): Schematic representation of load delay experimental protocol. The shaded segments represent the periods over which the samples were maintained in dynamic loading culture. 0 Day Delay is for Study 2 and Study 3; 14 Day Delay is for Study 3; FS represents free-swelling culture.

media changes. Constructs were loaded in unconfined compression, between two impermeable platens, using a custom deformational loading bioreactor. A sinusoidal deformation (10% surface-to-surface displacement at a frequency of 1 Hz) was superimposed on a 2% tare strain (based on day 0 dimensions). This loading regimen was carried out five days a week, with dynamic deformational loading applied for three hours (continuous) each day. Custom agarose templates with 5.76 mm diameter wells, cast into Falcon 60 mm petri dishes (Fisher Scientific, Pittsburgh, PA), were utilized to ensure that the constructs were properly seated. The samples were maintained in 5 ml of culture medium during dynamic loading. Free-swelling controls were also transferred to agarose templates and maintained in parallel with loaded samples. After loading, samples were returned to petri dishes containing 30 ml of fresh culture medium. Samples were removed for mechanical testing on days 0, 14, 28, and 42.

2.4. *Study 3: Dynamic deformational loading of chondrocytes with pre-elaborated matrix*

Chondrocytes were encapsulated at 60 million/ml and maintained as described in Study 1. Samples were divided into 2 free-swelling and 2 dynamically loaded groups. For dynamic loading groups, loading was initiated on day 0 or after 14 days of free-swelling culture (Fig. 1B). Free-swelling controls were maintained for both dynamically loaded groups to control for the effects of transferring and reduced media conditions during loading. Samples were collected and analyzed on days 0, 14, 28 and 42. The effect of loading was compared by normalizing the construct stiffness by the magnitude of the construct stiffness achieved in free-swelling at the time point when loading was initiated.

2.5. *Mechanical testing of agarose constructs*

For Studies 1–3, the equilibrium Young's modulus (E_Y) of the constructs was measured using a custom unconfined compression device [37]. Before each test, the disk thickness and diameter were measured using a digital caliper, and the specimens were equilibrated in creep, under a tare load of 1 gram. A stress-relaxation test was then performed, consisting of an applied ramp-and-hold displacement, with a ramp speed of 1 μm/s to 10% of the thickness of the specimen after tare load. Constructs were allowed to relax to equilibrium (2000 seconds). This equilibrium reaction force and sample geometry were used to calculate the E_Y. For study 2 and 3, a dynamic displacement of 40 microns was applied at 1, 0.5, 0.1, 0.05, 0.01, and 0.005 Hz and the respective unconfined dynamic modulus (G^*), was calculated from the load/displacement profiles and construct geometry. Following mechanical testing the constructs were either frozen at -30°C for biochemical analysis or processed for histological analysis.

2.6. *Histological analysis*

For histology, samples were fixed at 4°C for 24 hours in a solution containing 5% acetic acid, 3.7% formaldehyde and 70% ethanol, serially dehydrated in ethanol, embedded in paraffin (Fisher Scientific), sectioned to 8 μm, and mounted onto microscope slides. The samples were then de-waxed, rehydrated, and stained as follows. Sections were digested in 0.5 mg/ml of testicular hyaluronidase, swollen in 0.5 M of acetic acid, blocked in 10% normal goat serum (NGS) and labeled with 10% NGS containing monoclonal primary antibody for link protein-2 and types II, VI and IX collagens (9/30/8-A-4, II-II6B3, 5C6, and D1-9, respectively; Developmental Studies Hybridoma Bank, Iowa City, IA). Non-immune controls were incubated in 10% NGS alone. Alexa 488-conjugated goat anti-mouse secondary antibody labeling (Molecular Probes, Eugene, OR) and propidium iodide nuclear counterstaining (Molecular Probes) were performed to visualize the ECM and cells, respectively. After staining, the slides were coverslipped and sections were analyzed using an inverted microscope with an Olympus Fluoview confocal system (New York/New Jersey Scientific, Middlebush, NJ) with dual wavelengths excitation at 488 and 568 nm (20× to 100×-oil objective lens).

2.7. Biochemical analysis

The samples were thawed, weighed wet, lyophilized, reweighed dry, and papain digested for 16 hours at 60°C in 0.1 M sodium acetate buffer (pH 5.6) containing 0.05 M EDTA and 0.01 M cysteine-HCl (Sigma Chemical). The DNA content was determined using the PicoGreen double strand DNA quantification assay (Molecular Probes), which had been modified for use with a microplate reader. Calf thymus DNA (0 to 2000 ng/ml) was used as a standard [28]. The GAG content was measured with the 1,9-dimethylmethylene blue (Sigma Chemicals) dye-binding assay modified for use with a microplate reader. Shark chondroitin sulfate (0 to 50 μg/ml) was used as a standard [10]. The overall collagen content was measured using the orthohydroxyproline (OHP) colorimetric assay adapted for use with a microplate reader. OHP content was determined by reaction of hydrolyzed papain digest with dimethyaminobenzaldehyde and chloramine T. Bovine OHP (0 to 10 μg/ml) was used as a standard [38]. Collagen content was determined by assuming a mass ratio of 1 part OHP per 10 part collagen [41].

2.8. Statistical analysis

Statistica software package (Statsoft, Inc, Tulsa, OK) was used to perform two-way analysis of variance (ANOVA), with culture time and loading as independent variables. Fisher's LSD post-hoc test with $\alpha = 0.05$ were used for statistical comparisons. For Study 1, data was pooled from two independent experiments conducted with constructs derived from independent isolations yielding $n = 6$–9 per group. Study 2 and 3 were performed using cells derived from one isolation each, yielding $n = 3$ constructs per group. Data is presented as the mean and standard deviation.

3. Results

Immediately upon encapsulation, chondron-seeded constructs (Fig. 2B, D) exhibited a local pericellular matrix rich in type VI collagen similar to that found in articular cartilage (Fig. 2E), which was absent in chondrocyte-seeded constructs (Fig. 2A, C). The initial DNA content of chondrocyte- and chondron-seeded constructs was similar, and increased 300–400% for the lower seeding density constructs. For 60 million/ml groups, the DNA content increased by 51% for the chondrocyte-seeded samples and 81% for the chondron-seeded samples. The DNA content was similar for free-swelling and dynamically loaded samples.

3.1. Study 1: Chondrocytes and chondrons in free-swelling culture

E_Y increased over time in culture for all groups (Fig. 3A). These increases were significant at day 28 for 10 million/ml constructs ($p < 0.005$), and by day 14 for both 60 million/ml constructs ($p < 0.005$). At 10 million/ml, the chondrocyte and chondron groups were not significantly different from one another. However, at 60 million/ml, the chondron-seeded samples were significantly stiffer than the chondrocyte-seeded controls by day 14 ($p < 0.01$). Furthermore, by day 28, both 60 million/ml groups were significantly stiffer than the 10 million/ml groups ($p < 0.005$).

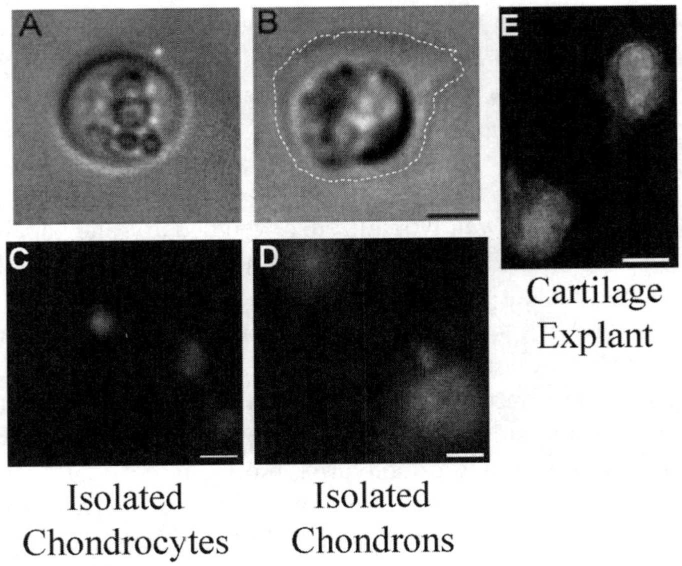

Fig. 2. Chondrocytes (A, C) and chondrons (B, D) seeded in agarose hydrogel at day 0. (A, B) DIC phase contrast imaging (scale bar represents 5 μm). (C, D) Type VI immunofluorescence, with propidium iodide counterstaining (scale bar represents 10 μm). (E) Type VI immunofluorescence, with propidium iodide counterstaining (scale bar represents 10 μm), of chondrons from the middle zone of an articular cartilage explant.

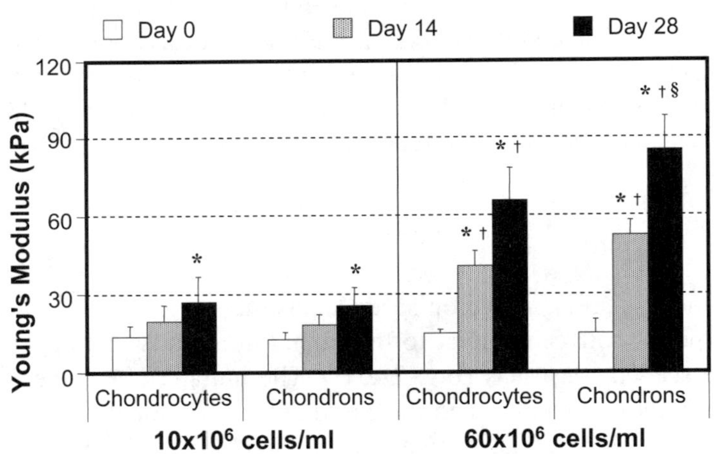

Fig. 3. Temporal development of Young's modulus of chondrocytes and chondrons seeded in agarose at 10 ($n = 9$) and 60 ($n = 6$) million/ml (Study 1). * Represents significant difference compared to day 0 ($p < 0.005$); † represents significant difference compared to 10 million/ml samples ($p < 0.005$); § represents significant difference compared to chondrocytes at same time point and seeding density ($p < 0.01$).

The biochemical content of samples is presented in Table 1. The thickness of the samples remained relatively constant throughout the study. All samples showed significant increases in wet weight, dry weight, and GAG and collagen contents over the 28 day culture period ($p < 0.05$). At both seeding densities, chondrocyte- and chondron-seeded samples showed similar increases in wet and dry weights. Furthermore, the GAG and collagen contents were significantly higher for the 60 million/ml samples compared to the 10 million/ml samples ($p < 0.05$).

Table 1
Biochemical composition (Study 1)

Group:	Chondrocytes			
Seeding density	Day 0		Day 28	
$\times 10^6$ cells/ml:	10	60	10	60
Thickness (mm)	2.3 ± 0.1	2.3 ± 0.1	2.5 ± 0.1	2.5 ± 0.0
Wet weight (mg)	29.1 ± 5.3	33.0 ± 5.5	$55.2 \pm 6.4^*$	$51.6 \pm 5.4^*$
Dry weight (mg)	1.1 ± 0.5	1.0 ± 0.3	$2.6 \pm 0.9^*$	$3.3 \pm 0.8^*$
Water (%)	96.2 ± 1.5	96.5 ± 1.2	95.3 ± 1.3	94.8 ± 2.0
GAG (%ww)	0.1 ± 0.0	0.1 ± 0.1	$0.8 \pm 0.2^*$	$1.1 \pm 0.3^{*\dagger}$
Collagen (%ww)	0.0 ± 0.0	0.1 ± 0.1	$0.8 \pm 0.1^*$	$1.2 \pm 0.5^{*\dagger}$

Group:	Chondrons			
Seeding density	Day 0		Day 28	
$\times 10^6$ cells/ml:	10	60	10	60
Thickness (mm)	2.3 ± 0.0	2.3 ± 0.0	2.4 ± 0.0	2.6 ± 0.0
Wet weight (mg)	24.0 ± 4.5	32.6 ± 4.9	$51.1 \pm 6.7^*$	$53.7 \pm 3.3^*$
Dry weight (mg)	1.3 ± 0.5	1.1 ± 0.3	$2.5 \pm 0.7^*$	$3.6 \pm 1.2^*$
Water (%)	95.7 ± 1.1	96.7 ± 0.7	95.6 ± 1.0	94.1 ± 2.9
GAG (%ww)	0.1 ± 0.0	0.1 ± 0.0	$0.7 \pm 0.2^*$	$1.3 \pm 0.2^{*\dagger}$
Collagen (%ww)	0.1 ± 0.1	0.1 ± 0.0	$0.9 \pm 0.1^*$	$1.4 \pm 0.1^{*\dagger}$

10×10^6/ml ($n = 9$); 60×10^6/ml ($n = 6$).
*Represents significant changes compared to day 0 ($p < 0.05$); † represents significant differences compared to 10 million/ml samples ($p < 0.05$).

3.2. *Study 2: Dynamic loading of chondrocyte- and chondron-seeded cultures*

In this study, the E_Y (Fig. 4A, B) of the free-swelling samples was comparable to those obtained in Study 1 at day 28. At 10 million/ml, the E_Y of dynamically loaded chondrocyte-seeded samples were significantly greater than free-swelling controls by day 42 ($p < 0.005$). However, in the chondron-seeded samples at this seeding density, dynamic loading significantly decreased E_Y by day 42 ($p < 0.05$). At 60 million/ml, the E_Y of dynamically loaded chondrocyte-seeded constructs was significantly greater than free-swelling controls by day 28 ($p < 0.005$). However, in chondron-seeded constructs, significant increases in E_Y, in response to dynamic loading, were not observed until day 42 ($p < 0.0005$). Similar results were obtained for the G^* (Fig. 4B) of these samples. However, the G^* of the dynamically loaded chondron constructs seeded at 60 million/ml was significantly greater than the free-swelling controls, by day 28 ($p < 0.005$). For all constructs, the G^* was proportional to the applied testing frequency and yielded similar variations with time and culture conditions, therefore, only data obtained at 0.5 Hz are reported.

The relative effect of deformational loading is more clearly seen by normalizing the material properties of the dynamically loaded groups by their respective free-swelling controls (Table 2). In 10 million/ml chondrocyte-seeded samples, E_Y increased by 49% and G^* increased by 65%. However, in the corresponding chondron-seeded samples, E_Y and G^* decreased by 14% and 18%, respectively. At 60 million/ml, E_Y increased by 61% and G^* increased by 92% in chondrocyte-seeded constructs by day 42. In contrast, chondron-seeded constructs exhibited a 31% increase in E_Y and a 67% increase in the G^*.

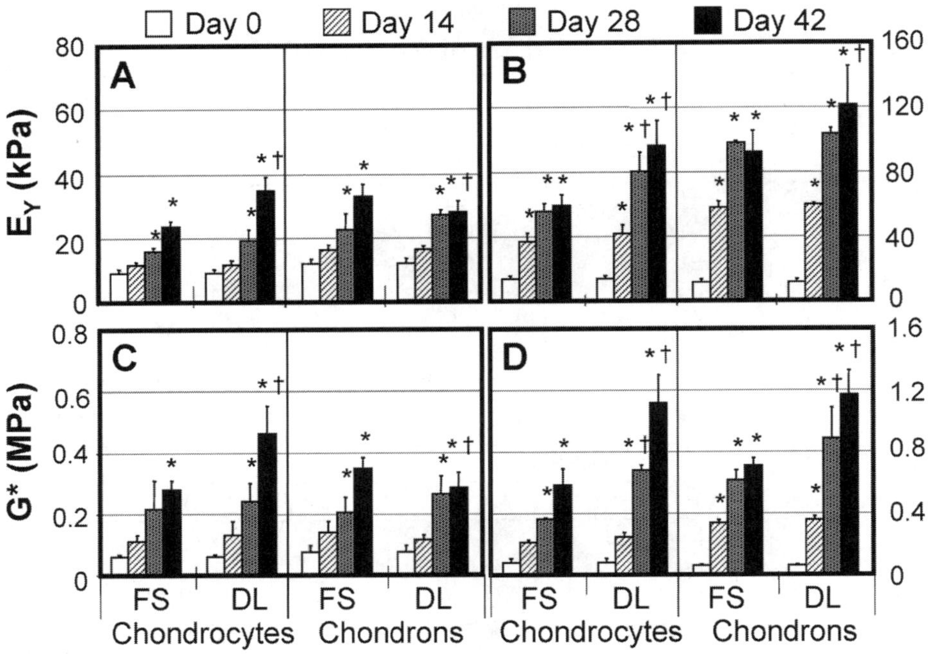

Fig. 4. Temporal development of Young's modulus (A, B) and dynamic modulus (C, D) of chondrocytes and chondrons seeded in agarose at 10 (A, C) and 60 (B, D) million/ml (Study 2; $n = 3$). Note that the scale for 60 million/ml groups is twice that for 10 million/ml groups. FS represents free-swelling; DL represents dynamically loaded; * represents significant difference compared to day 0 ($p < 0.05$); † represents significant difference compared to free-swelling controls ($p < 0.05$).

Table 2

Relative increases in Young's and dynamic moduli in response to dynamic deformational loading (Study 2)

	Chondrocytes				Chondrons			
	10 million/ml		60 million/ml		10 million/ml		60 million/ml	
	E_Y	G^*	E_Y	G^*	E_Y	G^*	E_Y	G^*
Day 0	1.00	1.00	1.00	1.00	1.00	1.00	1.00	1.00
Day 14	1.03	1.18	1.14	1.16	1.00	0.84	1.03	1.05
Day 28	1.23	1.12	1.44†	1.89†	1.19	1.29	1.06	1.45†
Day 42	1.49†	1.65†	1.61†	1.92†	0.86	0.82	1.31†	1.67†

E_Y represents Young's modulus; G^* represents dynamic modulus; † represents significant changes compared to day 0 ($p < 0.05$); $n = 3$.

Variations in the biochemical properties of these specimens are presented in Table 3. Under free-swelling culture conditions, the sample thickness increased significantly over the 6-week culture period in all groups ($p < 0.05$) except for the 10 million/ml chondron group. No significant increases in sample thickness were observed for the dynamically loaded samples. The wet and dry weights increased significantly over the 6 week study ($p < 0.05$). All groups exhibited a significant increase in GAG content over time in culture ($p < 0.05$). By day 42, dynamic loading produced a significant increase (versus free-swelling control) in the GAG content of the 60 million/ml chondrocyte constructs ($p < 0.005$). However, this response was absent in the chondron-seeded groups. Total collagen content also showed significant temporal increases in all groups ($p < 0.05$), however, the collagen content was not significantly altered by the application of dynamic loading.

Table 3
Biochemical composition (Study 2)

		Thickness (mm)	Wet weight (mg)	Dry weight (mg)	Water (%)	GAG (%ww)	Collagen (%ww)
Chondrocytes							
10×10^6/ml							
Day 0		2.29 ± 0.1	32.4 ± 2.2	1.5 ± 0.4	95.3 ± 1.3	0.08 ± 0.0	0.0 ± 0.0
Day 42	FS	$2.52 \pm 0.1^*$	$55.6 \pm 11.1^*$	$2.9 \pm 0.1^*$	94.7 ± 0.8	$1.21 \pm 0.4^*$	$1.43 \pm 0.4^*$
Day 42	DL	2.30 ± 0.0	$55.3 \pm 6.4^*$	$2.8 \pm 0.2^*$	94.9 ± 0.4	$1.27 \pm 0.2^*$	$1.56 \pm 0.2^*$
60×10^6/ml							
Day 0		2.34 ± 0.1	31.1 ± 7.0	1.3 ± 0.6	96.2 ± 1.0	0.06 ± 0.2	0.14 ± 0.0
Day 42	FS	$2.63 \pm 0.1^*$	$55.1 \pm 4.0^*$	$4.9 \pm 0.3^*$	95.0 ± 0.5	$1.23 \pm 0.2^*$	$1.58 \pm 0.2^*$
Day 42	DL	2.42 ± 0.1	$48.7 \pm 1.3^*$	$4.9 \pm 0.1^*$	93.5 ± 0.1	$1.91 \pm 0.4^{*\dagger}$	$1.98 \pm 0.3^*$
Chondrons							
10×10^6/ml							
Day 0		2.33 ± 0.0	26.7 ± 4.2	1.6 ± 0.0	93.9 ± 1.0	0.09 ± 0.0	0.08 ± 0.8
Day 42	FS	2.46 ± 0.0	$54.6 \pm 1.6^*$	$3.2 \pm 0.4^*$	94.1 ± 0.7	$1.39 \pm 0.5^*$	$1.42 \pm 0.1^*$
Day 42	DL	2.36 ± 0.0	$49.5 \pm 4.4^*$	$2.7 \pm 0.1^*$	94.6 ± 0.8	$1.08 \pm 0.1^*$	$1.33 \pm 0.2^*$
60×10^6/ml							
Day 0		2.33 ± 0.0	36.0 ± 4.9	1.3 ± 0.2	96.3 ± 0.6	0.09 ± 0.0	0.13 ± 0.0
Day 42	FS	$2.65 \pm 0.0^*$	$63.2 \pm 3.8^*$	$5.8 \pm 0.3^*$	$90.8 \pm 0.4^*$	$1.85 \pm 0.3^*$	$1.36 \pm 0.3^*$
Day 42	DL	2.53 ± 0.0	$60.1 \pm 5.9^*$	$5.9 \pm 0.1^*$	$90.1 \pm 0.8^*$	$2.09 \pm 0.4^*$	$1.84 \pm 0.6^*$

FS represents free-swelling cultures; DL represents dynamically loaded cultures; * represents significant changes compared to day 0 ($p < 0.05$); † represents significant differences compared to free-swelling controls ($p < 0.05$); $n = 3$.

Immunofluorescence labeling showed that all groups produced type II collagen by day 42 (Fig. 5), with qualitatively more intense staining in dynamically loaded constructs (Fig. 5B, D, F, H) compared to free-swelling controls (Fig. 6A, C, E, G). Furthermore, at 60 million/ml, type II collagen labeling appeared most intense in the dynamically loaded chondrocyte samples (Fig. 5F) compared to other samples at this cell density. At 60 million/ml, positive labeling for link protein (Fig. 6A–D) and type IX collagen (Fig. 6E–H) was observed in all groups at day 42. Dynamic loading also increased type IX collagen labeling in chondrocyte-seeded samples (Fig. 6F), and increased link protein (Fig. 6B) and type IX collagen (Fig. 6D) labeling in chondron-seeded samples, compared to respective free-swelling controls.

3.3. *Study 3: Dynamic loading of chondrocytes with pre-elaborated matrix*

All groups exhibited significant increases in E_Y (Fig. 7A) and G^* (Fig. 7B) by day 28 ($p < 0.05$). E_Y of the dynamically loaded samples was significant stiffer than free-swelling controls by day 28 for the 0 Day Delay constructs and by day 42 for the 14 Day Delay constructs. Similar results were obtained for G^* of these constructs. The best absolute material properties obtained on day 42 of total culture time was observed for the 0 Day Delay group ($p < 0.05$).

Normalizing E_Y and G^* of the free-swelling samples to their day 0 value showed that both parameters grew exponentially over the 42 days examined in this study (Fig. 8A). When the E_Y and G^* of the dynamically loaded samples were normalized to their values at the onset of loading (Fig. 8B), similar increases were observed in E_Y and G^* regardless of when dynamic loading was initiated.

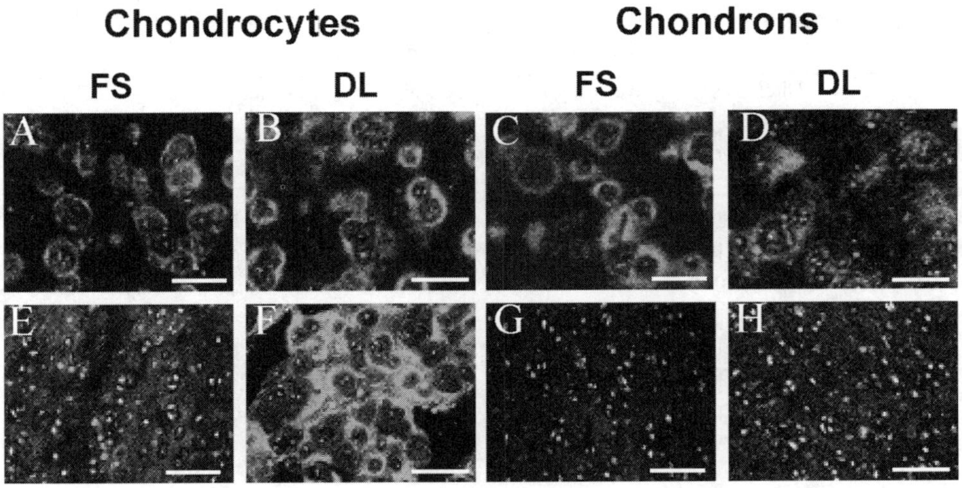

Fig. 5. Type II collagen immunofluorescence at day 42 for chondrocytes (A, B, E, F) and chondrons (C, D, G, H), seeded in agarose hydrogels at 10 million/ml (A–D) or 60 million/ml (E–H), and grown under free-swelling (A, C, E, G) or dynamic loading (B, D, F, H) culture conditions (Study 2). FS represents free-swelling cultures; DL represents dynamically loaded cultures. Scale bars are 100 μm.

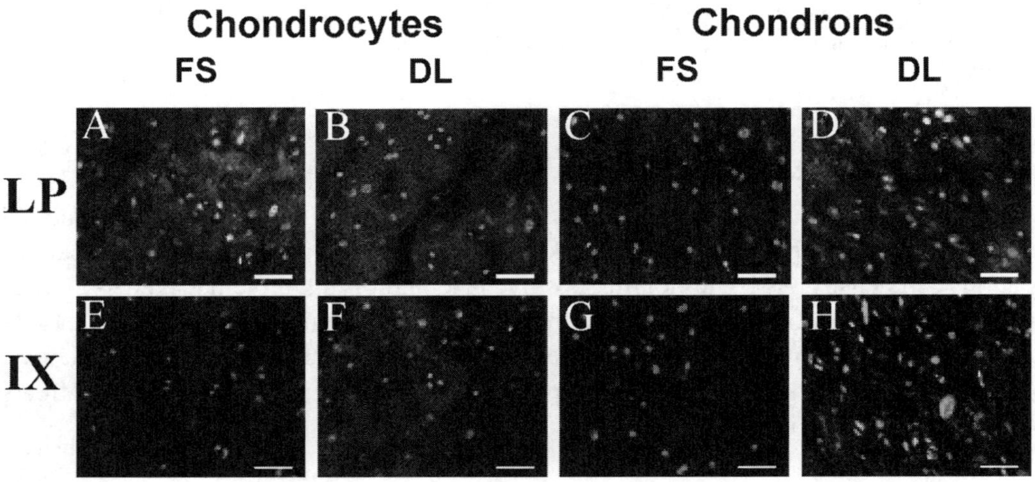

Fig. 6. Link protein (A–D) and type IX collagen (E–H) immunofluorescence at day 42 for chondrocytes (A, B, E, F) and chondrons (C, D, G, H), seeded in agarose hydrogels at 60 million/ml, and grown under free-swelling (A, C, E, G) or dynamic loading (B, D, F, H) culture conditions (Study 2). FS represents free-swelling cultures; DL represents dynamically loaded cultures; LP represents link protein; IX represents type IX collagen. Scale bars are 5 μm.

The biochemical content of these constructs is presented in Table 4. The thickness, except in the 0 day delay dynamically loaded, and wet and dry, weights increased significantly for all groups over the 42-day culture period. However, free-swelling controls showed the greatest increase in these parameters compared to the dynamically loaded samples. GAG and OHP exhibited significant increases with culture time for both groups ($p < 0.05$) by day 42, but were similar for all groups.

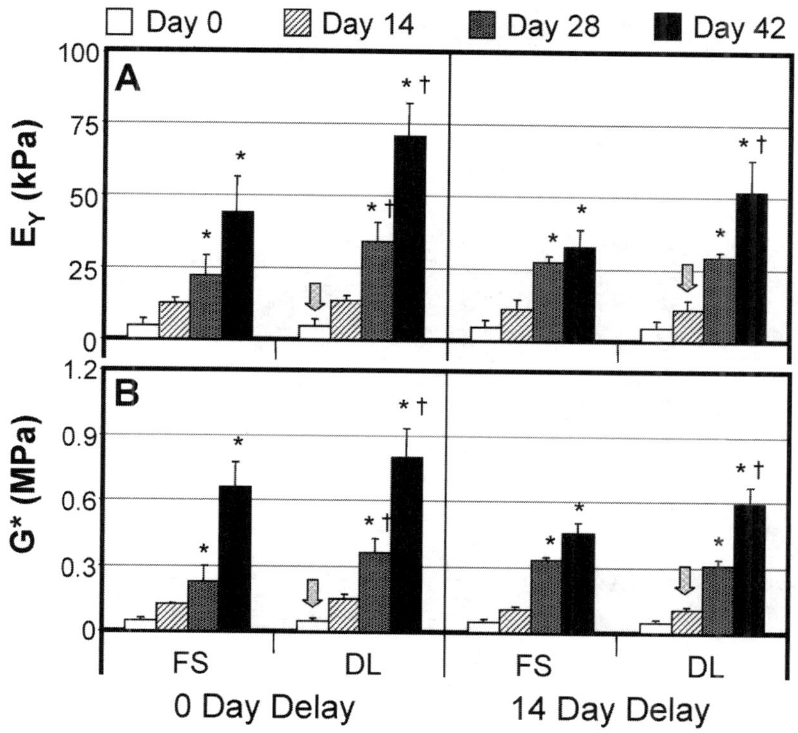

Fig. 7. Temporal development of Young's modulus (A) and dynamic modulus (B) of chondrocytes seeded at 60 million/ml ($n = 3$) with dynamic loading beginning at day 0 or day 14 (Study 3). FS represents free-swelling cultures; DL represents dynamically loaded cultures; arrow indicates when dynamic loading was initiated; * represents significant difference compared to day 0 ($p < 0.005$); † represents significant difference compared to FS controls ($p < 0.01$).

4. Discussion

We have previously demonstrated that deformational loading of chondrocyte-seeded agarose constructs enhances matrix elaboration and development of functional properties [24,25]. A temporal delay, dependent on the cell-seeding density, was observed before the beneficial effects of deformational loading became apparent (e.g., measurable change to biochemical and material properties verses free-swelling controls). The observed effects of deformational loading may parallel the development of a local ECM surrounding the cells, providing the cell–matrix interactions necessary for the chondrocyte to transduce the physical stimuli associated with applied deformational loading [14,22]. In the present study, we hypothesized that the effects of deformational loading could be expedited by loading agarose constructs seeded with chondrocytes surrounded by a pre-existing pericellular matrix. To test this hypothesis, we adopted enzymatic techniques for isolating chondrons [21], comprised of the chondrocyte and its cell-associated network of type VI collagen, and subjected these constructs to applied deformational loading. A number of investigators have studied chondrons in agarose hydrogel cultures [6,8,15, 17] as well as pellet cultures [8,13].

In the current study, free-swelling cultures of chondrocytes and chondrons appeared to develop at similar rates for the lower seeding density, whereas chondrons at 60 million/ml appeared to grow faster than chondrocytes seeded at the same density [24,26]. The accelerated growth by the higher seeding density cultures is similar to a study by Larson and co-workers [18] who observed that pellet cultures

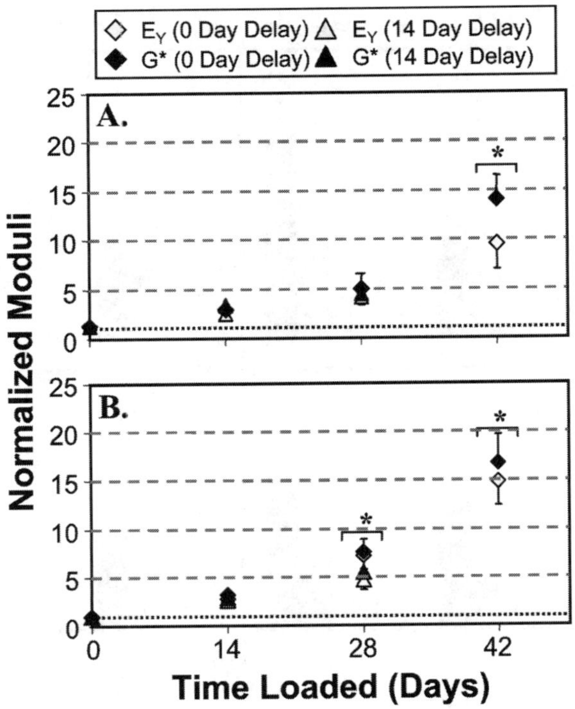

Fig. 8. Young's and dynamic moduli of (A) free-swelling and (B) dynamically loaded constructs normalized to the value at the onset of loading (Study 3). Dynamically loaded constructs were subjected to deformational loading after 0 and 14 days of free-swelling culture. * Represents significant difference compared to day 0 post-loading ($n = 3$; $p < 0.01$).

Table 4

Biochemical composition (Study 3)

			Thickness (mm)	Wet weight (mg)	Dry weight (mg)	Water (%)	GAG (%ww)	Collagen (%ww)
Day 0			2.33 ± 0.02	33.4 ± 3.0	1.1 ± 0.7	96.9 ± 1.7	0.03 ± 0.01	0.20 ± 0.12
Day 42	0 Day Delay	FS	$3.04 \pm 0.11^*$	$75.3 \pm 2.5^*$	$5.7 \pm 0.3^*$	92.4 ± 0.2	$1.55 \pm 0.35^*$	$2.46 \pm 0.28^*$
		DL	$2.57 \pm 0.03^\dagger$	$58.5 \pm 4.1^{*\dagger}$	$4.2 \pm 0.2^*$	92.8 ± 0.3	$1.63 \pm 0.22^*$	$2.06 \pm 0.18^*$
	14 Day Delay	FS	$3.08 \pm 0.12^*$	$71.4 \pm 5.1^*$	$5.2 \pm 0.2^*$	92.7 ± 0.4	$1.55 \pm 0.22^*$	$2.16 \pm 0.13^*$
		DL	$2.61 \pm 0.04^{*\dagger}$	$64.9 \pm 5.0^{*\dagger}$	$4.5 \pm 0.3^*$	93.0 ± 0.1	$1.63 \pm 0.13^*$	$1.85 \pm 0.36^*$

FS represents free-swelling cultures; DL represents dynamically loaded cultures; * represents significant changes compared to day 0 ($p < 0.05$); † represents significant differences compared to free-swelling controls ($p < 0.05$); $n = 3$–4.

of chondrons grew more quickly than chondrocyte cultures [18]. Studies by Knight and co-workers suggest differences between chondrons that have been mechanically extracted versus enzymatically extracted chondrons [16]. For the scope of the current study, the use of enzymatically-digested chondrons provided a convenient means to attain high densities of chondrocytes retaining their native pericellular matrix, and perhaps provide a more direct comparison to enzymatically-digested chondrocytes. Although isolated chondrocytes seeded in agarose produce a matrix rich in type II and type VI collagens [6], this pericellular matrix may develop differently (e.g., type and distribution) than that extracted of chondrons from the native tissue [17]. These potential differences would give rise to dissimilar local physical milieu (e.g., electrokinetic phenomena, osmotic pressure) [29,30].

In this respect, a more direct comparison can be drawn from the constructs in which a two-week delay of applied deformational loading was imposed (Study 3). Interestingly, chondrocytes with *de novo* cell-associated matrix appeared to respond similarly to those isolated as chondrons. Together, these findings suggest that the observed 2–3 week lag in significant material property development of chondrocyte-seeded agarose constructs with dynamic loading is not attributed to the absence of a cell-associated matrix at initial encapsulation. Moreover, there may be a period of preconditioning to physical stimuli (e.g., fluid flow, hydrostatic pressure, strain and stress fields [23]), arising from applied deformational loading that is necessary prior to appreciable matrix development by the chondrocytes.

Similar to our previous study [24], we observed a chondrocyte density-dependent matrix elaboration response, with 60 million chondrocytes/ml cultures growing more quickly and attaining greater functional properties compared to 10 million chondrocytes/ml in response to deformational loading. While 60 million chondrons/ml showed some loading-induced increases of material properties, constructs of 10 million chondrons/ml showed no apparent response to loading. Interestingly, total GAG and collagen content increased with culture time and were generally similar for all cultures, being insensitive to applied loading. This finding is similar to our previous observations that in some cases bulk content of these ECM constituents, particularly for the higher seeding density, does not correlate with the development of construct material properties [24,26]. Differences in GAG and collagen type and arrangement, may account for this disparity [26]. Our histological analyses also suggest that disparate levels of other ECM elements (i.e., type IX collagen [9]) may contribute to this finding. Increased type IX collagen levels with dynamic loading may improve the structural stability of the type II collagen network in dynamically loaded constructs by increasing crosslinking [44].

5. Conclusion

Under the conditions of the present study, the deformational loading response of chondrocytes with and without an initial cell-associated matrix were generally similar with respect to temporal development of material properties, thus disproving our hypothesis. From the perspective of utilizing this information for functional tissue engineering applications [5], it does not appear that encapsulation of cells with initial pericellular matrix (native or *de novo*) expedites tissue development with applied deformational loading.

Acknowledgements

The authors would like to thank Greta Lee for her assistance with the chondron isolation protocol. We also acknowledge Matthew Fisher and Mandy Ho for their invaluable contributions to the preparation of this manuscript. This work was supported by the NIH [R01 AR46568 (CTH) and a graduate research supplement (TNK)] and the Whitaker Foundation [pre-doctoral fellowship (RLM)].

References

[1] P.D. Benya and J.D. Shaffer, Dedifferentiated chondrocytes reexpress the differentiated collagen phenotype when cultured in agarose gels, *Cell* **30** (1982), 215–224.

[2] J.A. Buckwalter and H.J. Mankin, Articular cartilage: tissue design and chondrocyte–matrix interactions, *AAOS International Course Lectures* **47** (1998), 477–486.

[3] M.D. Buschmann, Y.A. Gluzband, A.J. Grodzinsky and E.B. Hunziker, Mechanical compression modulates matrix biosynthesis in chondrocyte/agarose cultures, *J. Cell Sci.* **108** (1995), 1497–1508.
[4] M.D. Buschmann, Y.A. Gluzband, A.J. Grodzinsky, J.H. Kimura and E.B. Hunziker, Chondrocytes in agarose culture synthesize a mechanically functional extracellular matrix, *J. Orthop. Res.* **10** (1992), 745–758.
[5] D.L. Butler, S.A. Goldstein and F. Guilak, Functional tissue engineering: the role of biomechanics, *J. Biomech. Eng.* **122** (2000), 570–575.
[6] J. Chang and C.A. Poole, Confocal analysis of the molecular heterogeneity in the pericellular microenvironment produced by adult canine chondrocytes cultured in agarose gel, *Histochem. J.* **29** (1997), 515–528.
[7] J.L. Cook, N. Williams, J.M. Kreeger, J.T. Peacock and J.L. Tomlinson, Biocompatibility of three-dimensional chondrocyte grafts in large tibial defects of rabbits, *Am. J. Vet. Res.* **64** (2003), 12–20.
[8] M.K. Elfervig, R.D. Graff, G.M. Lee, S.S. Kelley, A. Sood and A.J. Banes, ATP induces Ca^{2+} signaling in human chondrons cultured in three-dimensional agarose films, *Osteoarthritis Cart.* **9** (2001), 518–526.
[9] D.R. Eyre, S. Apone, J.J. Wu, L.H. Ericsson and K.A. Walsh, Collagen type IX: evidence for covalent linkages to type II collagen in cartilage, *FEBS Lett.* **220** (1987), 337–341.
[10] R.W. Farndale, D.J. Buttle and A.J. Barrett, Improved quantitation and discrimination of sulphated glycosaminoglycans by use of dimethylmethylene blue, *Biochim. Biophys. Acta* **883** (1986), 173–177.
[11] M.M. Fedewa, T.R. Oegema, M.H. Schwartz, A. MacLeod and J.L. Lewis, Chondrocytes in culture produce a mechanically functional tissue, *J. Orthop. Res.* **16** (1998), 227–236.
[12] R.D. Graff, S.S. Kelley and G.M. Lee, Role of pericellular matrix in development of a mechanically functional neocartilage, *Biotechnol. Bioeng.* **82** (2003), 457–464.
[13] R.D. Graff, E.R. Lazarowski, A.J. Banes and G.M. Lee, ATP release by mechanically loaded porcine chondrons in pellet culture, *Arthritis Rheum.* **43** (2000), 1571–1579.
[14] F. Guilak, R.L. Sah and L.A. Setton, Physical regulation of cartilage metabolism, in: *Basic Orthopaedic Biomechanics*, V.C. Mow and W.C. Hayes, eds, Lippincott-Raven, Philadelphia, 1997, pp. 179–207.
[15] W.A. Hing, A.F. Sherwin, J.M. Ross and C.A. Poole, The influence of the pericellular microenvironment on the chondrocyte response to osmotic challenge, *Osteoarthritis Cart.* **10** (2002), 297–307.
[16] M.M. Knight, D.A. Lee and D.L. Bader, The influence of elaborated pericellular matrix on the deformation of isolated articular chondrocytes cultured in agarose, *Biochem. Biophys. Acta* **1405** (1998), 67–77.
[17] M.M. Knight, J.M. Ross, A.F. Sherwin, D.A. Lee, D.L. Bader and C.A. Poole, Chondrocyte deformation within mechanically and enzymatically extracted chondrons compressed in agarose, *Biochim. Biophys. Acta* **1526** (2001), 141–146.
[18] C.M. Larson, S.S. Kelley, A.D. Blackwood, A.J. Banes and G.M. Lee, Retention of the native chondrocyte pericellular matrix results in significantly improved matrix production, *Matrix Biol.* **21** (2002), 349–359.
[19] D.A. Lee and D.L. Bader, The development and characterization of an in vitro system to study strain-induced cell deformation in isolated chondrocytes, *In Vitro Cell Dev. Biol. Anim.* **31** (1995), 828–835.
[20] D.A. Lee and D.L. Bader, Compressive strains at physiological frequencies influence the metabolism of chondroctyes seeded in agarose, *J. Orthop. Res.* **15** (1997), 181–188.
[21] G.M. Lee, C.A. Poole, S.S. Kelley, J. Chang and B. Caterson, Isolated chondrons: a viable alternative for studies of chondrocyte metabolism in vitro, *Osteoarthritis Cart.* **5** (1997), 261–274.
[22] R. Loeser, Integrin-mediated attachment of articular chondrocyte to extracellular matrix proteins, *Arthritis Rheum.* **36** (1993), 1103–1110.
[23] R.L. Mauck, C.T. Hung and G.A. Ateshian, Modeling of neutral solute transport in a dynamically loaded porous permeable gel, *J. Biomech. Eng.* **125** (2003), 602–614.
[24] R.L. Mauck, S.L. Seyhan, G.A. Ateshian and C.T. Hung, The influence of seeding density and dynamic deformational loading on the developing structure/function relationships of chondrocyte-seeded agarose hydrogels, *Ann. Biomed. Eng.* **30** (2002), 1046–1056.
[25] R.L. Mauck, M.A. Soltz, C.C.B. Wang, D.D. Wong, P.H. Chao, W.B. Valhmu, C.T. Hung and G.A. Ateshian, Functional tissue engineering of articular cartilage through dynamic loading of chondrocyte-seeded agarose gels, *J. Biomech. Eng.* **122** (2000), 252–260.
[26] R.L. Mauck, C.C.B. Wang, F.H. Chen, H.H. Lu, G.A. Ateshian and C.T. Hung, Dynamic deformational loading of chondrocyte-seeded agarose hydrogels modulates deposition and structural organization of matrix constituents, in: *Proceedings of 2003 Summer Bioengineering Conference*, Key Biscayne, FL, pp. 531–532.
[27] R.L. Mauck, C.C.B. Wang, E.S. Oswald, F.H. Chen, G.A. Ateshian and C.T. Hung, The role of seeding density and nutrient supply for articular cartilage tissue engineering with deformational loading, *Osteoarthritis Cart.* **11** (2003), 879–890.
[28] K.B. McGowan, M.S. Kurtis, L.M. Lottman, D. Watson and R.L. Sah, Biochemical quantification of DNA in human articular and septal cartilage using PicoGreen and Hoechst 33258, *Osteoarthritis Cart.* **10** (2002), 580–587.
[29] V.C. Mow, N.M. Bachrach, L.A. Setton and F. Guilak, Stress, strain, pressure, and flow fields in articular cartilage and chondrocytes, in: *Cell Mechanics and Cellular Engineering*, V.C. Mow, F. Guilak, R. Tran-Son-Tay and R.M. Hochmuth, eds, Springer-Verlag, New York, 1994, pp. 345–379.

[30] V.C. Mow, C.C.B. Wang and C.T. Hung, The extracellular matrix, interstitial fluid and ions as a mechanical signal transducter in articular cartilage, *Osteoarthritis Cart.* **7** (1999), 41–58.
[31] C.A. Poole, Articular cartilage chondrons: form, function and failure, *J. Anat.* **191** (1997), 1–13.
[32] C.A. Poole, S. Ayad and R.T. Gilbert, Chondrons from articular cartilage. V. Immunohistochemical evaluation of type VI collagen organisation in isolated chondrons by light, confocal and electron microscopy, *J. Cell Sci.* **103** (1992), 1101–1110.
[33] C.A. Poole, S. Ayad and J.R. Schofield, Chondrons from articular cartilage: I. Immunolocalization of type VI collagen in the pericellular capsule of isolated canine tibial chondrons, *J. Cell Sci.* **90** (1988), 635–643.
[34] C.A. Poole, M.H. Flint and B.W. Beaumont, Chondrons in cartilage: ultrastructural analysis of the pericellular microenvironment in adult human articular cartilages, *J. Orthop. Res.* **5** (1987), 509–522.
[35] C.A. Poole, M.H. Flint and B.W. Beaumont, Chondrons extracted from canine tibial cartilage: preliminary report on their isolation and structure, *J. Orthop. Res.* **6** (1988), 408–419.
[36] B. Rahfoth, J. Weisser, F. Sternkopf, T. Aigner, K. von der Mark and R. Brauer, Transplantation of allograft chondroctyes in agarose gel into cartilage defects in rabbits, *Osteoarthritis Cart.* **6** (1998), 50–65.
[37] M.A. Soltz, A. Stankiewicz, R.L. Mauck, C.T. Hung and G.A. Ateshian, Direct hydraulic permeability measurements of agarose hydrogels used as cell scaffolds, *Adv. Bioengng.* **43** (1999), 229–230.
[38] H. Stegeman and K. Stalder, Determination of hydroxyproline, *Clin. Chim. Acta* **19** (1967), 267–273.
[39] G.J.V.M. van Osch, W.B. van den Berg, E.B. Hunziker and H.J. Hauselmann, Differential effects of IGF-1 and TGF beta-2 on the assembly of proteoglycans in pericellular and territorial matrix by cultured bovine articular chondroctyes, *Osteoarthritis Cart.* **6** (1998), 187–195.
[40] G.J.V.M. van Osch, S.W. van der Veen, P. Buma and H.L. Verwoerd-Verhoef, Effect of transforming growth factor-beta on proteoglycan synthesis by chondrocytes in relation to differentiation stage and the presence of pericellular matrix, *Matrix Biol.* **17** (1998), 413–424.
[41] G. Vunjak-Novakovic, I. Martin, B. Obradovic, S. Treppo, A.J. Grodzinsky, R. Langer and L.E. Freed, Bioreactor cultivation conditions modulate the composition and mechanical properties of tissue-engineered cartilage, *J. Orthop. Res.* **17** (1999), 130–138.
[42] F.M. Watt and J. Dudhia, Prolonged expression of differentiated phenotype by chondrocytes cultured at low density on a composite substrate of collagen and agarose that restricts cell spreading, *Differentiation* **38** (1988), 140–147.
[43] M. Wong, M. Siegrist, X. Wang and E. Hunziker, Development of mechanically stable alginate/chondrocyte constructs: effects of guluronic acid content and matrix synthesis, *J. Orthop. Res.* **19** (2001), 493–499.
[44] J.J. Wu, P.E. Woods and D.R. Eyre, Identification of cross-linking sites in bovine cartilage type IX collagen reveals an antiparallel type II–type IX molecular relationship and type IX to type IX bonding, *J. Biol. Chem.* **267** (1992), 23007–23014.

Biorheology 41 (2004) 239–246
IOS Press

Flexibility of type I collagen and mechanical property of connective tissue

Kai-Nan An [a,*], Yu-Long Sun [a] and Zong-Ping Luo [b]
[a] *Mayo Clinic Rochester, Division of Orthopedic Research, Biomechanics Laboratory, Rochester, MN 55905, USA*
[b] *Department of Orthopedic Surgery, Baylor College of Medicine, Houston, TX 77030, USA*

Abstract. The hierarchical organization of the connective tissue, more specifically, the musculoskeletal soft tissue, has been extensively studied. With advancements in experimental methodology, investigation of the structure–function relationship has provided more insight into how the mechanical integrity of the tissue is created. Such information is essential in the linking the macroscopic loading environment of the tissue to the microscopic level of the tissue to be experienced by the cell. The flexibility and elastic modulus of gross connective tissue, the fascicle, the fiber and then the collagen molecule are compared based on the data available in the literature.

Keywords: Flexibility, elastic modulus, type I collagen, tendon, fascicle, connective tissue

1. Introduction

The hierarchical organization of the connective tissue, more specifically, the musculoskeletal soft tissue, has been extensively studied for half a century. The structural units [20] consist of the single protein macromolecule, the tropocollagen triple helix, the microfibril of five tropocollagen units, the tetragonal lattice of microfibrils, subfibrils and the collagen fibril. At this macroscopic level, with collagen fibrils embedded in the proteoglycan–water gel, fascicles and functional connective tissue are formed. With advancements in experimental methodology, investigation of the structure–function relationship has provided more insight into how the mechanical integrity of the tissue is created. Such information is essential in the linking the macroscopic loading environment of the tissue to the microscopic level of the tissue to be experienced by the cell.

2. Mechanical properties

The mechanical properties of tendon, ligament and other connective tissues were studied extensively in the eighties and nineties using bulk tissue. Soft tissues included the ligaments and tendons retrieved from a human or animal model and tested on a conventional material testing machine. Several technical challenges were encountered, and some solutions were provided to a certain extent of satisfaction. These technical difficulties should be kept in mind when analyzing the data of the material properties reported

*Address for correspondence: Kai-Nan An, Ph.D., Mayo Clinic Rochester, Division of Orthopedic Research, Biomechanics Laboratory, Rochester, MN 55905, USA. Tel.: +1 507 538 1717; Fax: +1 507 284 5392; E-mail: an@mayo.edu.

Table 1
Modulus of elasticity of a selected tendon and ligaments

	Modulus (MPa)	References
Human tendon		
Achiles	375 ± 102	[25]
Achilles	822 ± 211	[52]
Biceps	421 ± 212	[30]
Hamstring	145 ± 58	[51]
Medial gastrocnemius	280	[23]
Palmaris longus	1162∼1673	[7]
Patellar	660 ± 266	[19]
Patellar	225 ± 129	[51]
Supraspinatus	50∼150	[18]
Tibialis anterior	1200	[26]
Tibialis anterior	157∼530	[17]
Tibialis anterior	1200	[27]
Tricep surae	1048∼1474	[28]
Human ligament		
Thumb ulnar collateral	11.4 ± 1.2	[11]
Medial collateral ligament	38.6 ± 4.8	[38]
Meniscofemoral	355 ± 234	[24]
Interosseous membrane	608 ± 160	[36]
Rabbit tendon		
Patellar fascicle (300 μ)	216 ± 68	[53]
Patellar	1390 ± 53	[54]
Patellar fiber (1 μ)	54 ± 25	[32]
Rat tail tendon		
Fiber	1200∼1900	[46]
Fiber (50∼100 μ)	478 ± 130	[21]
Fascicle	250∼423	[9]
Fascicle	197 ± 38	[42,43]

in the literature. The grip of specimens has been, difficult on one hand, to eliminate the potential slippage of the specimen, and on the other hand, not to create stress concentrations, which introduce premature failure. Treatment of the specimens at the grip site by dehydration, freezing, and glue enhancement seem to be effective for overcoming the problem. The measurements of deformation and strain during testing of the soft tissue has been accomplished by using an extensometer or optical tracking. However, the measurement of the deformation based on the surface markers on the testing specimens when using the optical method might not be the true deformation of the substrate. Finally, the definition of the gauge lengths and cross-sectional areas of the soft tissue were sometimes equivocal. The data collected in the literature on the modulus of elasticity range from hundreds to thousands of MPa (Table 1).

More recently the focus has extended to the investigations of the properties of sub-units, including the fascicles and fibers performed primarily in rat tail tendon [9,15,21] or rabbit patellar tendon [32,53]. Micro-tensile testers were developed to determine the mechanical properties of the collagen fascicle with the diameter in the 100∼300 μm range. The collagen fiber, is the smallest unit observable with an optical microscope, having a diameter ranging from 1∼10 μm. The micro tensile test system utilized piezoelectric micromanipulators with the specimen gripped in microtubes and consisting of a bundle of fibrils. A systematic assessment of the material properties of the rabbit patellar tendon and its subunits was achieved [32,53,54]. Interestingly, it was noted that the stiffness, or moduli, increased with the level of structure, from fibers (54 ± 25 MPa), fascicle (216 ± 68 MPa) to tendon (1390 ± 53 MPa). The failure strain decreased with the increase of the lever at 21.6 ± 3%, 10.9 ± 1.6%, 14.1 ± 0.6%, respectively. There was no initial toe region observed for the collagen fibril, which is different from those of the gross tendon and collagen fascicle.

A novel approach to determine the strains, and thus, the material properties of the tendon fascicle using the cell nuclei was developed [44]. By monitoring the relative movement and deformation of fluorescently labelled tenocytes nuclei, the local strain fields of the fascicle were determined. It was found that as the fascicles were subjected to gross strains of up to 10%, the local strains of the fascicles were significantly less than the applied strain. The mean maximum modulus for a single fascicle of the rat tail tendon was thus measured to be 197 ± 38 MPa [43].

More recently, the physical properties of the lowest unit, type I procollagen, were studied using optical tweezers [45]. The force-extension relationship was measured and analyzed by fitting the data into a worm-like chain elasticity model. The persistence and contour lengths of this collagen molecule were found to be about 15 and 300 nm, respectively. The persistence length is less than 5% of the molecule length, suggesting that the molecule is relatively flexible. With a given assumed diameter of the molecule, the modulus was estimated to be much smaller than those of the higher unit, such as fiber, fascicle and tendon structures. Although there are potential arguments in terms of experimental methods and measurements, the observed changes of the mechanical properties as the collagen molecules organized in the hierarchical structure suggested the importance of the interfibrillar bonds and also the packaging.

As a basic structural element of connective tissues in animals, collagen molecules play a critical role in endowing tissues with their specific mechanical properties. The mechanical properties and flexibility of collagen molecules have been studied using hydrodynamic methods and electron microscopy [16,33,39,48]. Due to the intrinsic limitations of various techniques used to study flexibility of collagen molecules, and because of the restrictions imposed by models proposed, the conflicting conclusions about collagen flexibility were reached. With advances in nanoscale manipulation techniques, it is possible to stretch single molecules to measure their force–extension relationship and flexibility [31]. Since fewer assumptions are required [5,29], such a direct experimental approach enables a more precise determination of the flexibility of molecules.

Sun et al. [45] employed optical tweezers to stretch the collagen I molecules and determine the elastic properties of the collagen by fitting data to the worm-like chain elasticity models. The force–extension curves for collagen I monomers were quite non-linear as those of the typical soft tissue. In the equivalent toe region, a small force (<3 pN) stretched type I collagen monomer to nearly its entire contour length. The force increased sharply when the monomer was stretched close to its contour length (Fig. 1). To further characterize the flexibility of the molecules, the persistence length and contour length used in an entropic worm-like chain model were considered [13]. The persistence length and the contour length were simultaneously derived by fitting the force–extension measurements to the Marko–Siggia entropic elasticity model [5]. With 22 samples tests, the persistence length of 14.5 ± 7.3 nm was obtained for collagen I monomer. The contour length of collagen I was 309 ± 41 nm. The persistence length of collagen monomer is much shorter than previously described, such as 130 nm for the collagen isolated from lathyritic rat skin by using the hydrodynamic theories of worm-coil [48], and 160 to 180 nm for lugworm cuticle collagen which was calculated from the sedimentation constant and the intrinsic viscosity [39]. A similar result of 170 nm was derived from viscoelastic measurements carried out in acetate buffer at

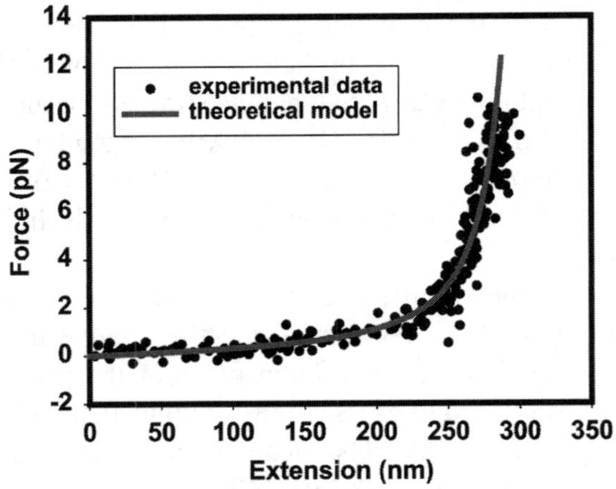

Fig. 1. The force–extension curve for stretching a single collagen I monomer. The data were fitted to the Marko–Siggia entropic elasticity model. The molecule length and persistence length of this sample are 320 nm and 8.0 nm, respectively. (From Sun, et al. [45]; used with permission.)

pH 4.0 [33]. The apparent difference between persistence length values could be attributed to: (1) the uncertainties of parameters of the theories in hydrodynamic measurements, and (2) the assumptions made on the uniform diameter of collagen molecule [16]. The persistence length of collagen also could be obtained from electron micrographs to be 57 nm for bovine skin collagen I [16].

The persistence length is related to the flexibility or stiffness of the molecule. The shorter the persistent length the more flexible. It is also directly proportional to the intrinsic elastic constant of polymers [2]. The elastic modulus (E) of collagen is related to its persistence length (L_p) as:

$$E = 4k_\mathrm{B}TL_\mathrm{p}/\pi r^4,$$

where k_B is Boltzmann's constant and T is absolute temperature. The determination of the elastic modulus is very sensitive to the value of the radius considered. Based on the data available in the literature, the elastic modulus of collagen I has been estimated to be 12.2 GPa by choosing a radius (r) of 0.28 nm [3], 350 MPa with a radius of 0.68 nm [4], and 32.7 MPa with r radius of 1.23 nm [40]. It compares with the values of 9.0 GPa estimated directly by Brillouin scattering [8,14] and 2.9 GPa by X-ray diffraction techniques [40].

The collagen triple helix has been assumed to be a rigid, rod-like structure, due to its constraints conferred by the imino groups of the prolines and hydroxyprolines, and supported in part by early physical studies [4,34]. On the other hand, some studies have indicated that collagen cannot simply be described as a stiff rod [12,47]. Since collagen is not uniform throughout the whole molecule the collagen triple helix undergoes micro-unfolding with the presence of interruptions between the Gly–Pro–Hyp triplets, caused by the Gly–X–Y-triplets that do not contain prolines or hydroxyprolines at the –X- and –Y-positions. It is expected that such local structural differences could possibly affect local flexibility of the collagen molecule.

By putting the data from Miyazaki [32] and that of Sun [45] with the assumption of the diameter of collagen I molecule at 1.23 nm, the material properties of the collagen connective tissue in hierarchical organization have specific characteristics (Fig. 2). Intuitively, it seems that the material properties of the fundamental building block should be stiffer than those of higher hierarchy. The possible explanation

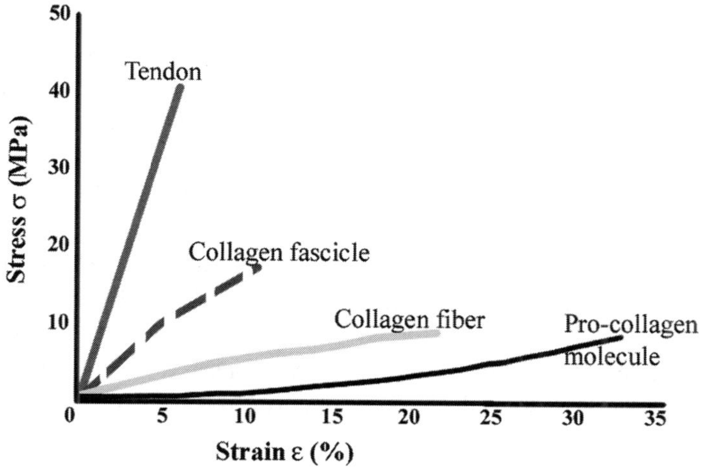

Fig. 2. By putting the data from Miyazaki [32] and that of Sun [45] with the assumption of the diameter of collagen I molecule at 1.23 nm, the material properties of the collagen connective tissue in hierarchical organization have specific characteristic.

for such observations would be the contribution of the "interface" reinforcement between the building blocks during the development of the tissue structures.

3. Cross-link and interfibrillar bond

Collagens are a large family of extracellular structural proteins made up of three α chains. The fibril forming collagens contain molecules with one large uninterrupted triple helical domain. These collagens form lateral aggregates in the fibrillogenesis. The other members of the collagen family have one or more non-triple helical domains. Most of those collagens do not form lateral aggregates, rather form complex aggregates and filamentous beaded structures [50]. Decorin interacts with fibrillar collagens and thus intervenes the collagen fibrillogenesis and the fibrils diameters [41,50]. The functional structure of collagen fibrils has intrigued many investigators in the past years and numerous models have been proposed to explain the experimental observation [37].

The strength of the collagen fibers depends upon the formation of cross-link between the telopeptide and adjacent helical domains of the collagen molecules. The material properties of the collagen can be modified by crosslinking through either chemical or physical processes [22,35].

All the collagen molecule consisted of three chains of polyamino acids or polypeptides arranged in a trihelical configuration ending in non-helical carboxyl and amino terminals. In the natural state, these collagen trihelical configurations are held in place by direct chemical bonds, hydrogen bonds, and water bridged cross-links. In addition, physical and chemical processes could also be used to introduce intra and inter molecular chemical bonds between collagen molecules [22]. In one study, for example, the glucose derived crosslinking of collagen film was introduced with combined exposure to ultraviolet light where the free-radicals were generated and expedited the process. The stiffness of such treated film was significantly greater than those of uncross-linked and crosslinked alone by ultraviolet or glucose [35].

Associated with the collagen in the connective tissue are the elastin and proteoglycans, which modulate collagen fibrillogenesis. In the study of structure–function relationship, it was observed that the decorin and biglycan influence the maturation of the collagen fibril and thus the size and mechanical properties of the fascicle in the rat rail tendon [10].

The structure–function relationship was investigated using rat tail tendon fascicle [9]. The structural and material properties were carefully obtained. Along with the measurements of the fibril diameter, it was found that in normal control, the stiffness and the modulus seem to be well correlated with the fibril size. However, by comparing the control with that of the Mov 13 transgenic mice where the fibril sizes were different, the stiffness and modulus were more or less equal. This further suggested that factors other than the fibril size, such as the cross-link and noncollageneous matrix may influence the material properties.

In the measurement of strain fields in the tendon fascicles under tensile loading it was noted that the fiber rotated and aligned to the direction of the strain applied and the fiber translated within the fascicle matrix [42].

Sliding of the reversible interfibrillar bond has been hypothesized as a mechanism for change in ligament and tendon length during growth. Inhibition of decorin/fibronectin binding by pentapetide NKISK results in lengthening of the tendon but not the modulus [6].

Furthermore, the fascicles are enveloped by sheath or reticular membrane. In recent studies, the difference in the maximum load of a tendon was found to be larger than the sum of the enclosed fascicles resected, which might be attributed to mechanical interaction between fibers and binding with the sheath [55].

4. Loading on cells

Mechanical loading has been found to induce matrix remodeling by the tenocytes [1], and potentially the mediator involved in tissue inflammation. The effects on the gene expression were studied using the cells isolated from the surface and internal tendon of chickens subjected to elongation of 5% at 1 Hz [1]. In another study, cyclic stretch at 4, 8 and 12% of the fibroblasts derived from patellar tendon in the microgroove culture proportionally induced the production of the prostaglandin E_2 [49]. The magnitude of the strains applied on the cell *in vitro* has been challenged as to its similarity *in vivo*. By staining the cell nuclei and using a confocal microscope, the strains of intra-fiber where the cell attached were in the range of 2–3% when the fascicles were loaded up to 8–10%. However, larger inter-fiber displacements were observed, representing the slide between adjacent fibers [42,43]. This implies that stain applied on the connective tissue observed at the macro-level may not transmit completely to the level of the cell. Further investigation of the relationship between the macro and micro levels is essential to the investigation and application of mechano-transduction of cell.

References

[1] A.J. Banes, G. Horesovsky, C. Larson, M. Tsuzaki, S. Judex, J. Archambault, R. Zernicke, W. Herzog, S. Kelley and L. Miller, Mechanical load stimulates expression of novel genes in vivo and in vitro in avian flexor tendon cells, *Osteoarthritis Cartilage* **7** (1999), 141–153.

[2] C.G. Baumann, S.B. Smith, V.A. Bloomfield and C. Bustamante, Ionic effects on the elasticity of single DNA molecules, *Proc. Natl. Acad. Sci. USA* **94** (1997), 6185–6190.

[3] K. Beck and B. Brodsky, Supercoiled protein motifs: the collagen triple-helix and the alpha-helical coiled coil, *J. Struc. Biol.* **122** (1998), 17–29.

[4] H. Boedtker and P. Doty, The native and denatured states of soluble collagen, *J. Am. Chem. Soc.* **78** (1956), 4267–4280.

[5] C. Bustamante, J.F. Marko, E.D. Siggia and S. Smith, Entropic elasticity of lambda-phage DNA, *Science* **265** (1994), 1599–1600.

[6] P.A. Caprise, G.E. Lester, P. Weinhold, J. Hill and L.E. Dahners, The effect of NKISK on tendon in an in vivo model, *J. Orthop. Res.* **19** (2001), 858–861.

[7] G.D. Carlson, M.J. Botte, M.S. Josephs, P.O. Newton, J.L. Davis and S.L. Woo, Morphologic and biomechanical comparison of tendons used as free grafts, *J. Hand Surg. (Am.)* **18** (1993), 76–82.
[8] S. Cusack and A. Miller, Determination of the elastic constants of collagen by Brillouin light scattering, *J. Molec. Biol.* **135** (1979), 39–51.
[9] K.A. Derwin and L.J. Soslowsky, A quantitative investigation of structure–function relationships in a tendon fascicle model, *J. Biomech. Eng.* **121** (1999), 598–604.
[10] K.A. Derwin, L.J. Soslowsky, J.H. Kimura and A.H. Plaas, Proteoglycans and glycosaminoglycan fine structure in the mouse tail tendon fascicle, *J. Orthop. Res.* **19** (2001), 269-277.
[11] K. Firoozbakhsh, I.S. Yi, M.S. Moneim and Y. Umada, A study of ulnar collateral ligament of the thumb metacarpophalangeal joint, *Clin. Orthop.* **403** (2002), 240–247.
[12] A. George and A. Veis, FTIRS in H_2O demonstrates that collagen monomers undergo a conformational transition prior to thermal self-assembly in vitro, *Biochemistry* **30** (1991), 2372–2377.
[13] P.J. Hagerman, Flexibility of DNA, *Ann. Rev. Biophysics Biophysical Chem.* **17** (1988), 265–286.
[14] R. Harley, D. James, A. Miller and J.W. White, Phonons and the elastic moduli of collagen and muscle, *Nature* **267** (1977), 285–287.
[15] R.C. Haut, The influence of specimen length on the tensile failure properties of tendon collagen, *J. Biomech.* **19** (1986), 951–955.
[16] H. Hofmann, T. Voss, K. Kuhn and J. Engel, Localization of flexible sites in thread-like molecules from electron micrographs. Comparison of interstitial, basement membrane and intima collagens, *J. Molec. Biol.* **172** (1984), 325–343.
[17] M. Ito, Y. Kawakami, Y. Ichinose, S. Fukashiro and T. Fukunaga, Nonisometric behavior of fascicles during isometric contractions of a human muscle, *J. Appl. Physiol.* **85** (1998), 1230–1235.
[18] E. Itoi, L.J. Berglund, J.J. Grabowski, F.M. Schultz, E.S. Growney, B.F. Morrey and K.N. An, Tensile properties of the supraspinatus tendon, *J. Orthop. Res.* **13** (1995), 578–584.
[19] G.A. Johnson, D.M. Tramaglini, R.E. Levine, K. Ohno, N.Y. Choi and S.L. Woo, Tensile and viscoelastic properties of human patellar tendon, *J. Orthop. Res.* **12** (1994), 796–803.
[20] J. Kastelic, A. Galeski and E. Baer, The multicomposite structure of tendon, *Connect. Tiss. Res.* **6** (1978), 11–23.
[21] Y.P. Kato, D.L. Christiansen, R.A. Hahn, S.J. Shieh, J.D. Goldstein and F.H. Silver, Mechanical properties of collagen fibres: a comparison of reconstituted and rat tail tendon fibres, *Biomaterials* **10** (1989), 38–42.
[22] E. Khor, Methods for the treatment of collagenous tissues for bioprostheses, *Biomaterials* **18** (1997), 95–105.
[23] K. Kubo, Y. Kawakami, H. Kanehisa and T. Fukunaga, Measurement of viscoelastic properties of tendon structures in vivo [comment], *Scand. J. Med. Sci. Sports* **12** (2002), 3–8.
[24] T. Kusayama, C.D. Harner, G.J. Carlin, J.W. Xerogeanes and B.A. Smith, Anatomical and biomechanical characteristics of human meniscofemoral ligaments, *Knee Surg. Sports Traumatol. Arthrosc.* **2** (1994), 234–237.
[25] G. Lewis and K.M. Shaw, Tensile properties of human tendo Achillis: effect of donor age and strain rate, *J. Foot Ankle Surg.* **36** (1997), 435–445.
[26] C.N. Maganaris and J.P. Paul, In vivo human tendon mechanical properties, *J. Physiol.* **521** (1999), 307–313.
[27] C.N. Maganaris and J.P. Paul, Tensile properties of the in vivo human gastrocnemius tendon, *J. Biomech.* **35** (2002), 1639–1646.
[28] S.P. Magnusson, P. Aagaard, P. Dyhre-Poulsen and M. Kjaer, Load-displacement properties of the human triceps surae aponeurosis in vivo [erratum appears in *J. Physiol.* **532**(Pt 3) (2001 May 1), 894], *J. Physiol.* **531** (2001), 277–288.
[29] J.F. Marko and E.D. Siggia, Stretching DNA, *Macromolecules* **28** (1995), 8759–8770.
[30] R.L. McGough, R.E. Debski, E. Taskiran, F.H. Fu and S.L. Woo, Mechanical properties of the long head of the biceps tendon, *Knee Surg. Sports Traumatol. Arthrosc.* **3** (1996), 226–229.
[31] A.D. Mehta, M. Rief, J.A. Spudich, D.A. Smith and R.M. Simmons, Single-molecule biomechanics with optical methods, *Science* **283** (1999), 1689–1695.
[32] H. Miyazaki and K. Hayashi, Tensile tests of collagen fibers obtained from the rabbit patellar tendon, *Biomed. Microdev.* **2** (1999), 151–157.
[33] F.H. Nestler, S. Hvidt, J.D. Ferry and A. Veis, Flexibility of collagen determined from dilute solution viscoelastic measurements, *Biopolymers* **22** (1983), 1747–1758.
[34] B. Obrink, Non-aggregated tropocollagen at physiological pH and ionic strength. A chemical and physico-chemical characterization of tropocollagen isolated from the skin of lathyritic rats, *Eur. J. Biochem.* **25** (1972), 563–572.
[35] M.P. Ohan, K.S. Weadock and M.G. Dunn, Synergistic effects of glucose and ultraviolet irradiation on the physical properties of collagen, *J. Biomed. Mat. Res.* **60** (2002), 384–391.
[36] H.J. Pfaeffle, M.M. Tomaino, R. Grewal, J. Xu, N.D. Boardman, S.L. Woo and J.H. Herndon, Tensile properties of the interosseous membrane of the human forearm, *J. Orthop. Res.* **14** (1996), 842–845.
[37] D.J. Prockop and A. Fertala, The collagen fibril: the almost crystalline structure, *J. Struc. Biol.* **122** (1998), 111–118.
[38] K.M. Quapp and J.A. Weiss, Material characterization of human medial collateral ligament, *J. Biomech. Eng.* **120** (1998), 757–763.

[39] T. Saito, N. Iso, H. Mizuno, N. Onda, H. Yamato and H. Odashima, Semiflexibility of collagens in solution, *Biopolymers* **21** (1982), 715–728.
[40] N. Sasaki and S. Odajima, Stress–strain curve and Young's modulus of a collagen molecule as determined by the X-ray diffraction technique, *J. Biomech.* **29** (1996), 655–658.
[41] E. Schonherr, H. Hausser, L. Beavan and H. Kresse, Decorin-type I collagen interaction. Presence of separate core protein-binding domains, *J. Biolog. Chem.* **270** (1995), 8877–8883.
[42] H. Screen, D. Lee, D. Bader and J. Shelton, Analysis of strain fields in tendon fascicles under tensile loading, in: *Proceedings of World Congress of Biomechanics*, Calgary, Alberta, Canada, 2002.
[43] H. Screen, D. Lee, D. Bader and J. Shelton, Structure–function relationship in tendon fascicles under tensile loading conditions, in: *Proceedings of World Congress of Biomechanics*, Calgary, Alberta, Canada, 2002.
[44] H.R. Screen, D.A. Lee, D.L. Bader and J.C. Shelton, Development of a technique to determine strains in tendons using the cell nuclei, *Biorheology* **40** (2003), 361–368.
[45] Y.L. Sun, Z.P. Luo, A. Fertala and K.N. An, Direct quantification of the flexibility of type I collagen monomer, *Biochem. Biophys. Res. Commun.* **295** (2002), 382–386.
[46] K.H. Svendsen and G. Thomson, A new clamping and stretching procedure for determination of collagen fiber stiffness and strength relations upon maturation, *J. Biomech.* **17** (1984), 225–229.
[47] J.C. Thomas and G.C. Fletcher, Dynamic light scattering from collagen solutions. II. Photon correlation study of the depolarized light, *Biopolymers* **18** (1979), 1333–1352.
[48] H. Utiyama, K. Sakato, K. Ikehara, T. Setsuiye and M. Kurata, Flexibility of tropocollagen from sedimentation and viscosity, *Biopolymers* **12** (1973), 53–64.
[49] J. Wang, D. Stone, F. Jia and S. Woo, in: *Orthopaedic Research Society*, San Francisco, CA, 2001, pp. 24.
[50] C. Wiberg, E. Hedbom, A. Khairullina, S.R. Lamande, A. Oldberg, R. Timpl, M. Morgelin and D. Heinegard, Biglycan and decorin bind close to the n-terminal region of the collagen VI triple helix, *J. Biolog. Chem.* **276** (2001), 18947–18952.
[51] T.W. Wilson, M.P. Zafuta and M. Zobitz, A biomechanical analysis of matched bone-patellar tendon-bone and double-looped semitendinosus and gracilis tendon grafts, *Am. J. Sports Med.* **27** (1999), 202–207.
[52] T.A. Wren, S.A. Yerby, G.S. Beaupre and D.R. Carter, Mechanical properties of the human achilles tendon, *Clin. Biomech.* **16** (2001), 245–251.
[53] E. Yamamoto, K. Hayashi and N. Yamamoto, Mechanical properties of collagen fascicles from the rabbit patellar tendon, *J. Biomech. Eng.* **121** (1999), 124–131.
[54] N. Yamamoto, K. Hayashi, H. Kuriyama, K. Ohno, K. Yasuda and K. Kaneda, Mechanical properties of the rabbit patellar tendon, *J. Biomech. Eng.* **114** (1992), 332–337.
[55] N. Yamamoto and E. Ikawa, in: *2002 World Congress of Biomechanics*, Calgary, Alberta, Canada, 2002.

Biorheology 41 (2004) 247–254
IOS Press

Single molecule mechanical properties of type II collagen and hyaluronan measured by optical tweezers

Zong-Ping Luo [a,*], Yu-Long Sun [b], Tadashi Fujii [b] and Kai-Nan An [b]
[a] *Department of Orthopedic Surgery, Baylor College of Medicine, Houston, TX 77030, USA*
[b] *Orthopedic Biomechanics Laboratory, Mayo Clinic/Mayo Foundation, Rochester, MN 55905, USA*

Abstract. Type II collagen and hyaluronan are the two major components of extracellular molecules in cartilage and play an important role in mechanical functions of extracellular matrix. Currently, their mechanical properties have been investigated only at the gross-level. In this study, the mechanical properties of single type II collagen and hyaluronan molecules were directly measured using optical tweezers technique. The persistence length was found to be 11.2 ± 8.4 nm in type II collagen and 4.5 ± 1.2 nm in hyaluronan. This result suggested that type II collagen is stiffer than hyaluronan at the individual molecule level, which supports the general concept that collagen is responsible for resisting tensile force. The experimental system developed here also provides a powerful tool for quantifying mechanical properties of extracellular matrix at the single molecule level.

Keywords: Type II collagen, hyaluronan, optical tweezers, persistence length, mechanical properties

1. Introduction

The major mechanical function of articular cartilage is to resist compressive loading. The three major components, proteoglycan aggregates, water and type II collagen, work closely to accomplish this function. Proteoglycan aggregates are mainly responsible for resisting compressive loading by trapping water molecules because of repelling of negative charges from each other. Type II collagen is responsible for holding proteoglycan aggregates in place under compression.

Alterations in any of these components are found in many joint diseases including osteoarthritis. Previous biomechanical studies suggest that the mechanical properties of cartilage at the gross level are related to the contents of molecules and assembly of the lower level structures. To date, little has been known regarding the mechanical properties of cartilage below the gross level.

With the development of nanoscale manipulation techniques, it is now possible to directly measure the force–extension relationship and flexibility of single molecules [9,12]. Mechanical experiments on single molecules such as DNA, titin, kinesin, and tenascin have been conducted. The objective of this study was to experimentally measure the mechanical properties of single type II collagen and hyaluronan (HA) molecules which are the backbone molecules of proteoglycan aggregates. We hypothesized that mechanical properties of collagen and hyaluronan are different at this fundamental level.

*Address for correspondence: Z.P. Luo, Department of Orthopedic Surgery, Baylor College of Medicine, One Baylor Plaza, Houston, TX 77030, USA. Tel.: +1 713 986 5522; Fax: +1 713 986 5521; E-mail: luo@bcm.tmc.edu.

2. Materials and methods

2.1. Loading apparatus

The experiment was performed on a single molecule mechanical testing system, consisting of an optical tweezers (Cell robotics, Inc.) and an interferometry system built in an inverted microscope (Axiovert 135, Carl Zeiss, Inc.) [10]. The optical tweezers were used to trap a bead and the interferometry system was constructed to monitor the bead displacement with an accuracy of nanometer on the basis of the concept described in literature [15–17]. The laser used in the optical tweezers had an 800 mW power output, resulting in a 124 mW power beam at the sample plane when 100× fluar objective lens was used. A piezo-stage with a resolution of 1 nm (P-731.20 high resolution XY positioning stage, Physik Instrumente) was mounted on the motorized x–y stage. A LabVIEW® program (National Instruments) was created to control /monitor the piezo-stage motion and measure the trapped bead displacement.

2.2. Testing procedure

In the HA experiment, a two-bead system was developed in which the molecule was linked to a large bead and a small bead, one at each side (Fig. 1). The bead/molecule samples were sandwiched between two coverglasses which were placed on the piezo-stage for the experiment. The large bead connecting one end of the molecule was immobilized between the coverglasses. The small bead was free in the solution except for connecting the opposite end of the HA. Under the microscope, the samples were carefully searched to identify possible linkages. After such a linkage was identified, the optical tweezers were turned on to trap the small bead. By slowly moving the piezo-stage at a speed of 45 nm/sec, one then stretched the molecule through the large bead, and moved the small bead away from the trapping (or laser beam) center (Fig. 2). The deformation of the HA was measured as the displacement difference between the large bead (or the piezo-stage) and the trapped small bead (monitored by the interferometry). The force applied on the HA was equal to the bead trapping force, which was calculated as the trapping stiffness times the departure distance of the small bead from the laser beam center [15–17]. The trapping stiffness was calibrated as $2.27 \pm 0.11 \times 10^{-2}$ pN/nm. The maximal trapping force on the small bead was measured as ~21 pN. Each molecule was stretched at least three times to ensure repeatability.

In the collagen study, one technical concern of using the above-described two-bead system was the optical interference of the large bead during the trapped bead measurement when the beads were near

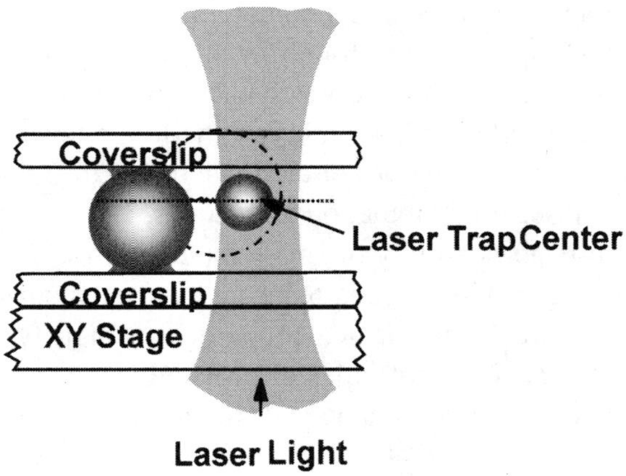

Fig. 1. Schematic illustration of mechanical testing system for single hyaluronan molecules.

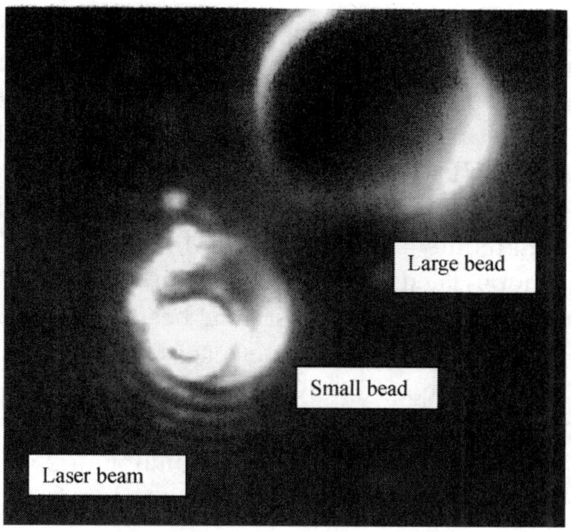

Fig. 2. A typical image from the experiment which shows the moment when the hyaluronan molecule was stretched.

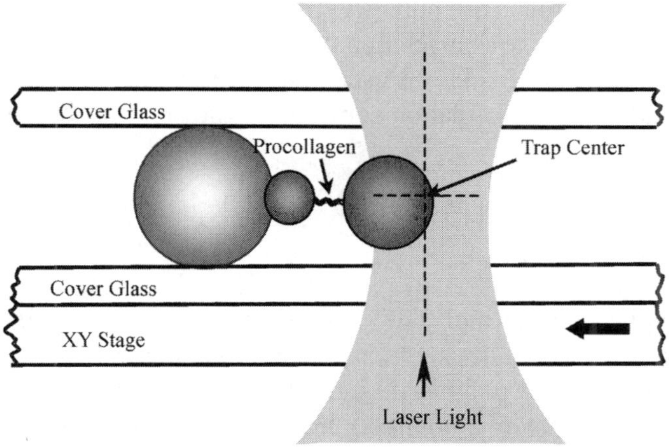

Fig. 3. Schematic illustration of mechanical testing system of single type II collagen molecules.

each other. This would introduce significant measurement errors when studying a short molecule (like collagen which is ~300 nm in length). In order to solve this problem, a third small bead fixed on the large bead was introduced [14]. The molecule was then linked between the small bead and the trapped bead (Fig. 3). The small bead served as an optical spacer to avoid the effect of diffraction interference on the trapped bead measurement.

2.3. Sample preparation

The typical HA molecule was attached onto two amino-modified polystyrene beads (Bangs Laboratories, Inc.) by covalent cross-links of amino groups on the surface of beads with the carboxyl groups of HA. First, HA molecules with a purity of 95% (Sigma) were reacted with 1-ethyl-3-(3-dimethylaminopropyl) carbodiimide hydrochloride and N-hydroxysuccinimide (Pierce) in 150 mM NaCl, phosphate sodium buffer pH 5.5. Then the small beads (5.20 μm in diameter) were added and pH was adjusted to 8. After 10 minutes, the beads were separated, mixed with the large beads (9.10 μm in diameter), and kept at room temperature for two hours. Finally, hydroxylamine was added to stop the reaction. The sample was ready for the test with a final concentration of 0.02% in H_2O.

Recombinant procollagen II molecules which were a gift from Fertala [4] were used for the experiments. Attachment of procollagen II to beads was achieved by the use of abundant disulfide bonds at propeptides. Type II procollagen was reduced with 1 mM dithiothreitol (Sigma) at pH 8.0 at room temperature for 15 minutes, then biotinylated using 10 mM EZ-Link™ PEO-maleimide actived biotin (Pierce) at pH 7 for 2 hours. Biotinylated type II procollagen was purified through a desalted column (Pierce) and coated on the streptavidin-coated polystyrene beads (diameter 2.17 μm, Spherotech Inc.), which had been bound to the biotin-coated polystyrene beads (diameter 6.7 μm). Under the microscope, the third bead coated with streptavidin-coated polystyrene beads (diameter 4.1 μm) was trapped by optical tweezers and tapped against the bead covered with the biotinylated procollagen molecules until connection was established. The samples were tested in a solution of 25 mM KCl, 1 mM EGTA, 0.2% Tween 20, 0.1% casein, and 25 mM HEPES pH 7.4.

2.4. Theoretical models for single molecules

Since both collagen and hyaluronan are polymer structures, polymer theory may be used to interpret the experimental measurement and to characterize their mechanical properties. One of the models, the Marko–Siggia wormlike chain model [11] was used in this study. The model assumes the polymer is an inexpansible structure, and all the deformation comes from the bending flexibility. Under this assumption, one has

$$F = \left(\frac{kT}{p}\right)\left[\frac{1}{4(1 - x/L)^2} - \frac{1}{4} + \frac{x}{L}\right],$$

where F is the force applied on the molecule, x is the deformation (the end-to-end distance of the molecule), L is the contour length under zero tension, and p is the persistence length which measures the tendency of a uniform, flexible polymer to persist in a given direction [2,7,19]. The model was fitted into the experimental data using a nonlinear least square method (SigmaPlot®, SPSS inc.). Two parameters, p and L, were then determined.

In this experiment, it was very critical to ensure that most measured molecules were single molecules. This was achieved based on the physical characteristics of molecules and measurement of a large number of samples, a technique used previously for other molecule studies [6,8,19]. First, using purified samples would ensure that the chance of testing other molecules was extremely small. In other words, even though some other molecules were tested, they would not show a peak in the histogram of the mechanical properties. Second, single molecules would be significantly different from the combined molecules in their mechanical properties, such as persistence length. Therefore, one would be able to distinguish peaks in the histogram of the persistence length [8], or to find only one peak when the possibility of forming combined molecules with identical persistence length was low.

3. Results

The result of HA molecules showed a typical nonlinear relationship between the loading and deformation, including the laxity region where the molecule was relaxed, the toe region when the molecule

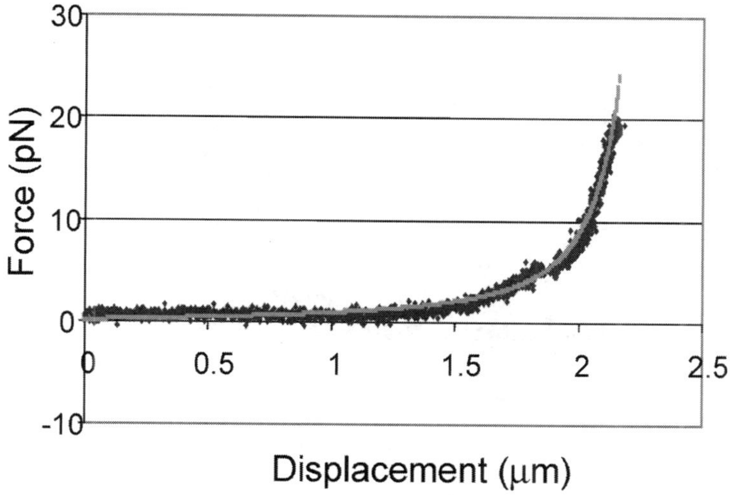

Fig. 4. A typical result of the force–displacement relationship from the single hyaluronan molecule measurement (dot: experimental data, and dash: Marko–Siggia model fitting).

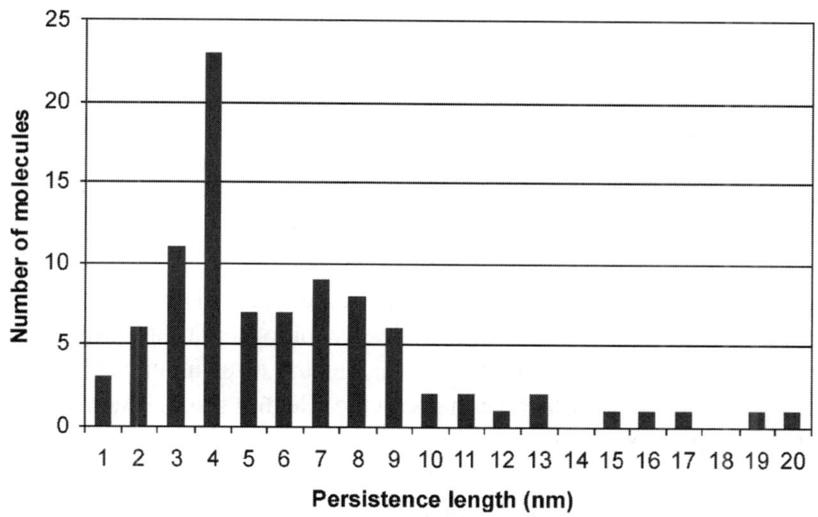

Fig. 5. Histogram of the persistence length distribution of 54 hyaluronan molecules.

started to resist tension, and the linear region where the molecule became significantly stiff (Fig. 4). The histogram of the persistence length distribution of all the specimens was determined (Fig. 5) ranging from 1.0 to 20.0 nm. The histogram results showed only one peak at ~4 nm persistence length (23 molecules). This result implied that in this experiment, most molecules tested were single molecules. By using the peak as the reference, we calculated the average persistence length from samples with persistence lengths ranging from 2.0 to 6.0 nm (54 molecules). The persistence length was 4.5 ± 1.2 nm (mean $\pm$ standard deviation). The corresponding molecule contour length was determined to be 2.65 ± 1.44 μm.

The highly nonlinear force–extension curves for type II collagen molecules were obtained (Fig. 6). The molecules were stretched by small forces (<3 pN) to near their contour length. Then the force increased sharply when the molecules were stretched close to their contour length. A total of 45 molecules were tested. The averaged persistence length was 11.2 ± 8.4 nm. The contour length was 296 ± 32 nm.

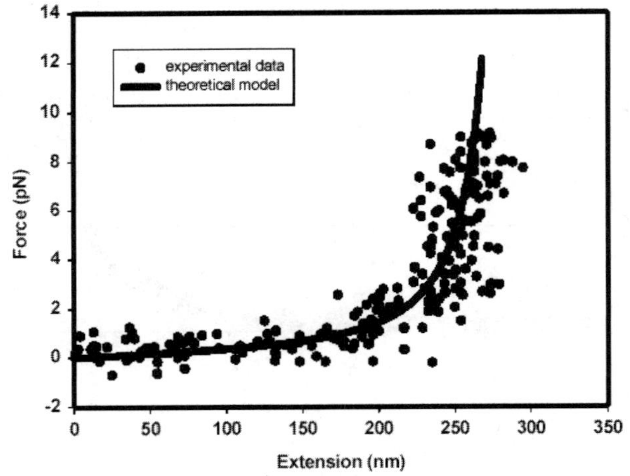

Fig. 6. A typical result of the force–displacement relationship from the type II collagen molecule measurement (dot: experimental data, and dash: Marko–Siggia model fitting).

4. Discussion

The present investigation measured mechanical properties of type II collagen and hyaluronan, two important macromolecules in cartilage extracellular matrix. To our knowledge, this was the first experiment of its type directly performed on these two molecules at the individual molecule level. The measurement was represented as a force–displacement relationship. Because the molecule is a macromolecule filament, the mechanical characteristics of the results could be interpreted by polymer theory. Specifically, the persistence length of the two molecules could be used to quantitatively describe the elastic stiffness and resiliency [1,13,19].

The results from this study were compared with the molecular structure data in the literature. Two other studies estimated the flexibility of HA molecules by the observation of the non-loaded molecule contours. The persistence length was about 20 nm using electric birefringence [18] or 4 nm using small-angle X-ray scattering [3]. Our results correspond with the Cleland study. The higher persistence length from the Trimm study might have been caused by allowing the molecules to contact the substrate resulting in artificial stiffening of the molecules. Our results of molecule length also showed agreement with a previous study (2.4 μm) based on direct measurement of HA molecules from electron microscope images [5]. In addition, the contour length of type II collagen was also consistent with that of 300 nm directly measured from electron micrographs (Fertala, personal communication).

Our results suggest that the type II collagen molecule is stiffer than the hyaluronan molecule. This rigidity allows the collagen to withstand tensile force, which is consistent with its speculated function in extracellular matrix. In articular cartilage, HA forms the backbone of the proteoglycan aggregates to absorb compression loading. Type II collagen is responsible for keeping those aggregates in place under compressive loading and therefore, experiences significant tensile loading.

Two concerns regarding the methodology used in the experiment had to be addressed. First, the attachment between the beads and the molecules had to be rigid. This was achieved by the covalent cross-links between the amino-modified beads and the carboxyl group on the HA chain or the biotin–streptavidin interaction in the collagen-bead linkage. Based on recent studies [6,7], the breaking load for a covalent bond was at least 1,400 pN and the breaking load for the biotin–streptavidin interaction was over 100 pN. Therefore, it was reasonable to assume that the attachment of the collagen and HA molecules to the beads was completely rigid. Second, the molecules were stretched up to $\sim$21 pN under the current setup. This loading range covered the laxity region where loading was small, and the toe region, where the molecule stiffness started to increase significantly (roughly within 5% variation from the molecule contour length). It also covered a part of the linear region, where molecule elastic deformation may take place. The covered range represents the spectrum of most interest since, under physiological conditions, most deformation occurs within this range. Unfortunately, this loading range together with errors in experiment (mostly due to fluctuation from Brownian motion of the trapped bead) made it difficult to accurately determine the elastic stiffness of the molecules in the linear region. The Marko–Siggia model, however, showed reasonable fitting to the experimental data, implying that within this loading range, both bending and extension of the molecule might contribute to the deformation. In the future, it would be interesting to measure the elastic stiffness more accurately in the higher linear region. This might be done by the use of a high power laser or an atomic force microscope. Higher laser power would also increase the trapping force and suppress the Brownian motion of the trapped bead, and the system sensitivity might also be enhanced.

In summary, this study measured the elastic properties of single type II collagen and hyaluronan molecules. The persistence length was found to be 11.2 ± 8.4 nm in type II collagen and 4.5 ± 1.2 nm in hyaluronan. This result suggested that type II collagen is stronger than hyaluronan at the individual molecule level, which supports the general concept that collagen is responsible for resisting tensile force. The experimental system developed here also provides a powerful tool for quantifying mechanical properties of extracellular matrix at the single molecule level.

Acknowledgements

This work was supported by grants from Arthritis Foundation and National Institutes of Health (R01AR44497). We are grateful to A. Fertala for his generous gift of recombinant procollagen II molecules.

References

[1] C. Bouchiat, M.D. Wang, J. Allemand, T. Strick, S.M. Block and V. Croquette, Estimating the persistence length of a worm-like chain molecule from force–extension measurements, *Biophys. J.* **76** (1999), 409–413.
[2] C. Bustamante, J.F. Marko, E.D. Siggia and S. Smith, Entropic elasticity of lambda-phage DNA, *Science* **265** (1994), 1599–1600.
[3] R.L. Cleland, The persistence length of hyaluronic acid: an estimate from small-angle X-ray scattering and intrinsic viscosity, *Arch. Biochem. Biophys.* **180** (1977), 57–68.
[4] A. Fertala, A.L. Sieron, A. Ganguly, S.W. Li, L. Ala-Kokko, K.R. Anumula and D.J. Prockop, Synthesis of recombinant human procollagen II in a stably transfected tumour cell line (HT1080), *Biochem. J.* **298** (1994), 31–37.
[5] J.H. Fessler and L.I. Fessler, Electron microscopic visualization of the polysaccharide hyaluronic acid, *Proc. Natl. Acad. Sci. USA* **56** (1966), 141–147.
[6] E.L. Florin, V.T. Moy and H.E. Gaub, Adhesion forces between individual ligand–receptor pairs, *Science* **264** (1994), 415–417.
[7] M. Grandbois, M. Beyer, M. Rief, H. Clausen-Schaumann and H.E. Gaub, How strong is a covalent bond?, *Science* **283** (1999), 1727–1730.
[8] M.S. Kellermayer, S.B. Smith, H.L. Granzier and C. Bustamante, Folding-unfolding transitions in single titin molecules characterized with laser tweezers, *Science* **276** (1997), 1112–1116.
[9] D. Leckband, Measuring the forces that control protein interactions, *Annu. Rev. Biophys. Biomol. Struct.* **29** (2000), 1–26.
[10] Z.P. Luo and K.N. An, Development and validation of a nanometer manipulation and measurement system for biomechanical testing of single macro-molecules, *J. Biomech.* **31** (1998), 1075–1079.

[11] J.F. Marko and E.D. Siggia, Stretching DNA, *Macromolecules* **28** (1995), 8759–8770.
[12] A.D. Mehta, M. Rief, J.A. Spudich, D.A. Smith and R.M. Simmons, Single-molecule biomechanics with optical methods, *Science* **283** (1999), 1689–1695.
[13] C. Rivetti, C. Walker and C. Bustamante, Polymer chain statistics and conformational analysis of DNA molecules with bends or sections of different flexibility, *J. Mol. Biol.* **280** (1998), 41–59.
[14] Y.L. Sun, Z.P. Luo and K.N. An, Stretching short biopolymers using optical tweezers, *Biochem. Biophys. Res. Commun.* **286** (2001), 826–830.
[15] K. Svoboda and S.M. Block, Biological applications of optical forces, *Annu. Rev. Biophys. Biomol. Struct.* **23** (1994), 247–285.
[16] K. Svoboda and S.M. Block, Force and velocity measured for single kinesin molecules, *Cell* **77** (1994), 773–784.
[17] K. Svoboda, C.F. Schmidt, B.J. Schnapp and S.M. Block, Direct observation of kinesin stepping by optical trapping interferometry, *Nature* **365** (1993), 721–727.
[18] H.H. Trimm and B.R. Jennings, Study of hyaluronic acid flexibility by electric birefringence, *Biochem. J.* **213** (1983), 671–677.
[19] M.D. Wang, H. Yin, R. Landick, J. Gelles and S.M. Block, Stretching DNA with optical tweezers, *Biophys. J.* **72** (1997), 1335–1346.

Biorheology 41 (2004) 255–261
IOS Press

Uniaxial tensile testing of canine annulus fibrosus tissue under changing salt concentrations

J.M. Huyghe [a,*] and M.R. Drost [b]
[a] *Department of Biomedical Engineering, Engineering Mechanics Institute, Eindhoven University of Technology, Eindhoven, The Netherlands*
[b] *University of Maastricht, Department of Movement Sciences, Maastricht, The Netherlands*

Abstract. Recent modelling efforts in the field of mechanics of the intervertebral disc, demonstrate that the deformation properties of intervertebral disc tissue are intimately linked to compositional changes. This paper presents uniaxial tensile relaxation experiments of canine annulus fibrosus tissue under stepwise changes of external salt concentration.

1. Introduction

Intervertebral disc tissue is a very deformable molecular mixture of an ionized solid and an ionised fluid. It is composed of a gelly-like core, the nucleus pulposus, enclosed within a fibrous ring, the annulus fibrosus. Different modelling strategies have been proposed to grasp its coupled electro-chemo-mechanical properties [5,7,8,10]. The material properties of intervertebral disc tissue are subdivided into two categories. One set relates to the relationship between on the one hand, the free energy of the mixture and on the other hand the composition of the mixture and its deformation [6]. The other are the frictional coefficients between its components. All of these parameters are generally found to be dependent on orientation and position within the disc [4], position of the disc within the spine and dependent on age and species. This paper concentrates on a experimental strategy to quantify coupling parameters associated with canine annulus fibrosus tissue. We present tensile relaxation experiments of canine annulus fibrosus tissue under stepwise changes of external salt concentration. Given the complexity of this undertaking, we restrict our attention to uni-axial experiments.

2. Methods

Lower lumbar motion segments were cut from spines of two mongrel dogs and were stored at −65 degrees centigrade. One of the vertebral bodies of a motion segments was sawed away almost completely, so that at most a thin layer of bone remained on top of the intervertebral discs. The posterior half of

*Address for correspondence: J.M. Huyghe, Department of Biomedical Engineering, Eindhoven University of Technology, P.O. Box 513, 5600MB Eindhoven, The Netherlands. Tel.: +31 40 2473137; Fax: +31 40 2447355; E-mail: j.m.r.huyghe@tue.nl.

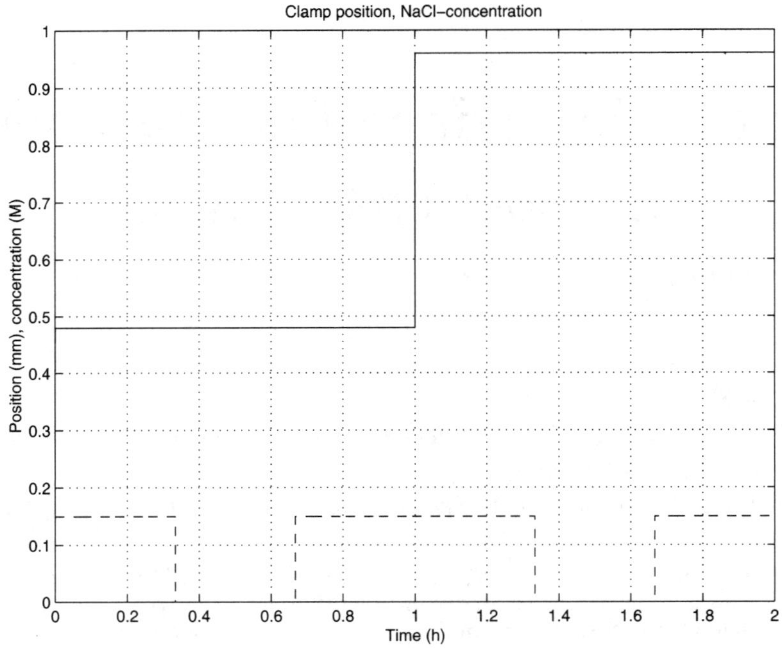

Fig. 1. Clamp displacement and external salt concentration as a function of time in all experiments.

the segment (arch) was sawed away. The anterior part of the motion segment was allowed to thaw. The surrounding muscle and fat tissues were removed with a scalpel. The nucleus pulposus was carefully removed. During this process the disc was kept moist with physiological saline. Lamellae at the anterior part of the annulus fibrosus and of at least 10 mm length and a minimal number of branches were selected under the stereo microscope. Lamellae were separated by blunt dissection. Finally a sample of circumferentially 10 mm by axially 2 to 8 mm was cut. The thickness of the lamellae was inhomogeneous and variable and was about 1 mm. The anterior side of each sample was glued upon two polystyrene lips using cyanoacrylate glue (Hyloglue M-100), such that traction exerted on the lips resulted in extension of the samples along their long axis, which corresponded with the circumferential direction in the disc. The long axis of the sample had an angle of approximately 30 degrees with the collagen fiber direction. The thickness of the polystyrene lips was 0.5 mm. The samples (6 in total??) were stored at −65 degrees centigrade until use.

A sample was taken out of the freezer, allowed to thaw and after about 15 minutes mounted into a horizontal mechanical drawing bench by clamping the polystyrene lips. The procedure was based upon the work of Myers et al. [9] and Akizuki et al. [1]. The displacement of the clamps of the drawing bench were computer controlled. The tensile force was measured by means of a force transducer. Above the samples two dripping devices were mounted, one filled with distilled water, and one filled with a physiological salt solution. The external salt concentration and the clamp position were changed in a stepwise fashion (Fig. 1). The clamp force was measured.

3. Results

The force tracing of the first experiment is shown in Fig. 2. The force tracings of the other experiments are illustrated in figures (Figs 3–7). At $t = 0$, the stepwise displacement of the clamp results in an

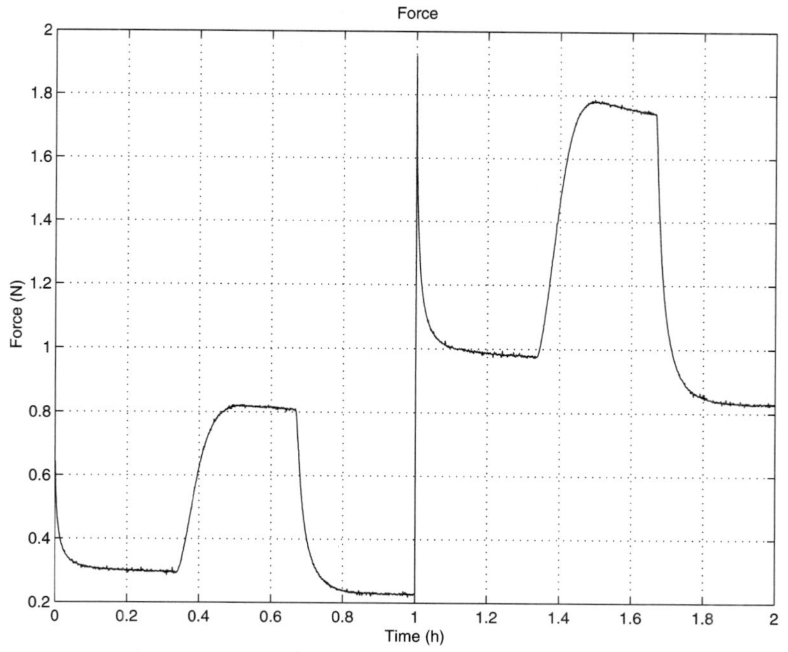

Fig. 2. Clamp force recording as a function of time in experiment 1. The thickness of the sample is less than 1 mm. The width is about 4 mm. The sample is taken halfway the outer edge of the annulus fibrosus and the nucleus-annulus interface. The clamp speed at which the step in the strain is applied is 0.1 mm/s.

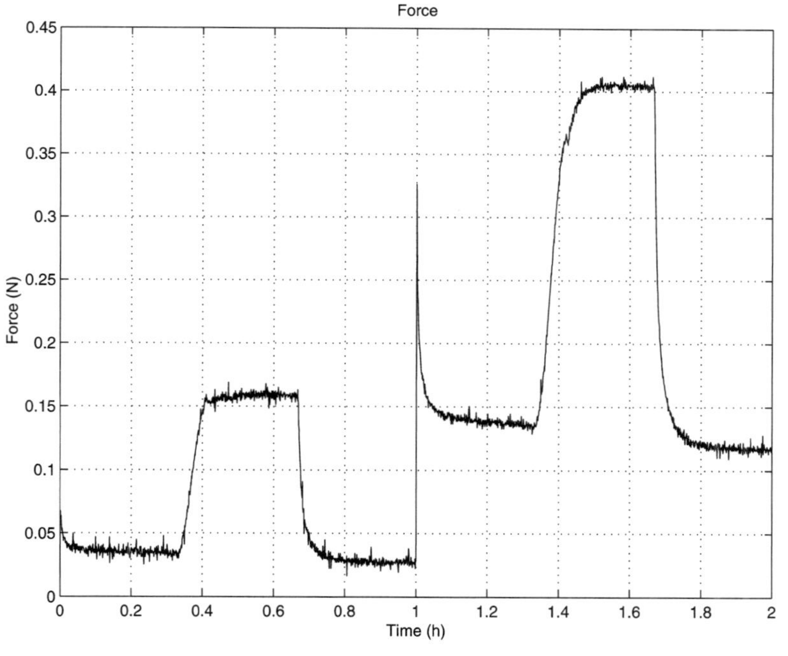

Fig. 3. Clamp force recording as a function of time in experiment 2. The thickness of the sample is less than 1 mm. The width is about 2 mm. The sample is taken from the annulus region contiguous to the nucleus. The clamp speed at which the step in the strain is applied is 0.1 mm/s.

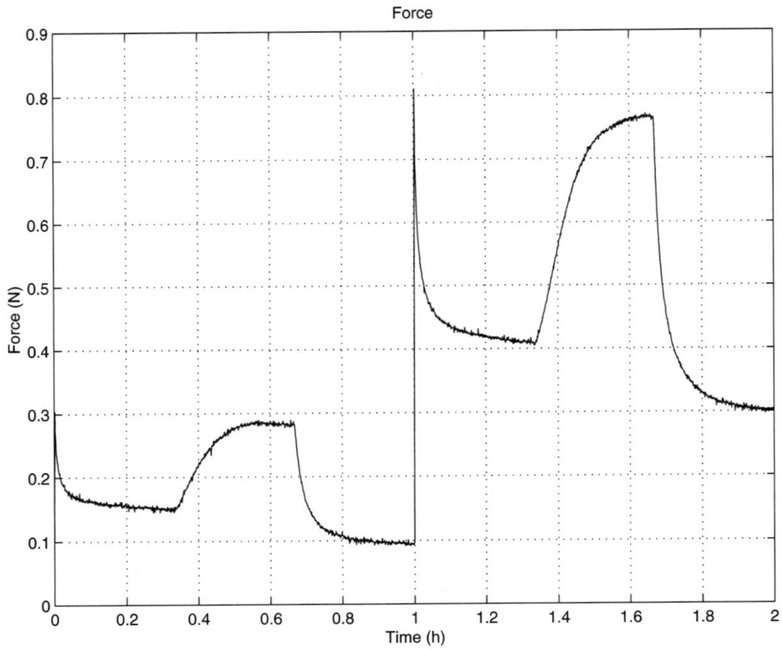

Fig. 4. Clamp force recording as a function of time in experiment 3. The thickness of the sample is about 1 mm. The width is 3 mm. The sample is taken halfway the outer edge of the annulus fibrosus and the nucleus–annulus interface. The clamp speed at which the step in the strain is applied is 0.05 mm/s.

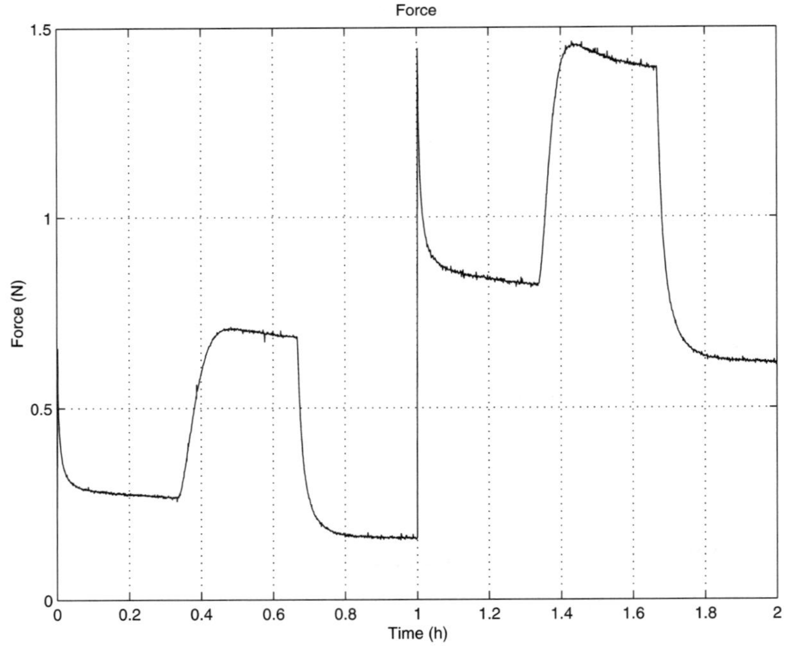

Fig. 5. Clamp force recording as a function of time in experiment 4. The thickness of the sample is less than 1 mm. The width is about 2 mm at one end and 3 mm at the other end. The sample is taken from the annulus region contiguous to the nucleus. The clamp speed at which the step in the strain is applied is 0.05 mm/s.

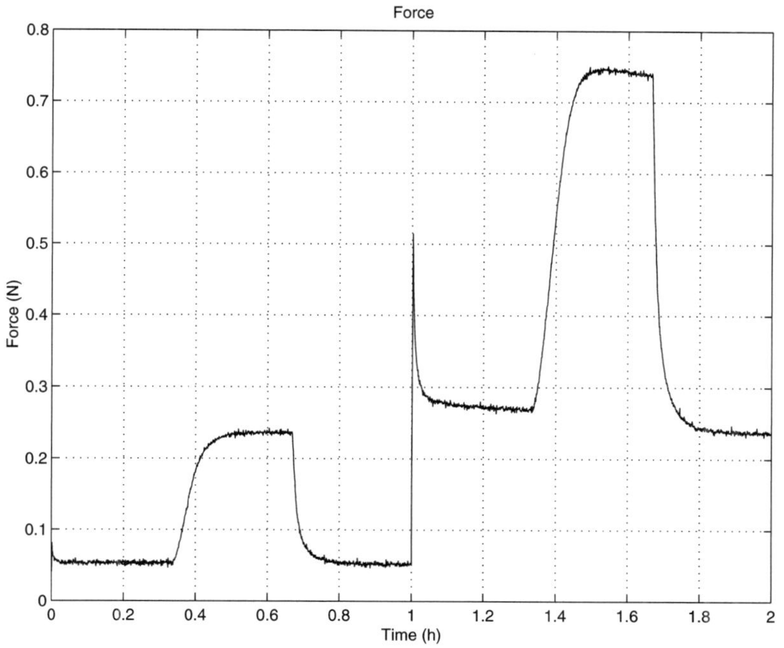

Fig. 6. Clamp force recording as a function of time in experiment 5. The thickness of the sample is less than 1 mm. The width is about 3 mm at one end and 4 mm at the other end. The sample is taken from the same disc as the sample of experiment 4. The annulus region from which it originates is further away from the nucleus than the region of the sample of experiment 4. The clamp speed at which the step in the strain is applied is 0.05 mm/s.

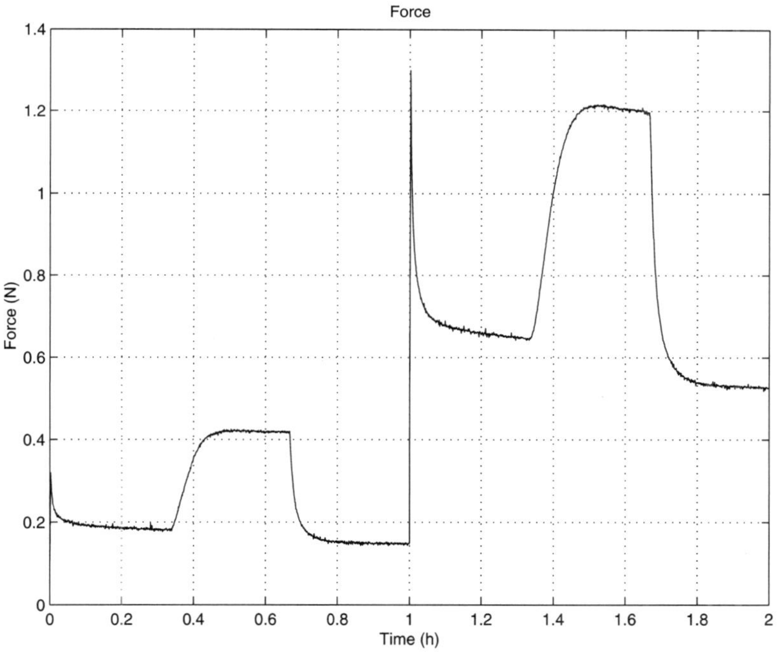

Fig. 7. Clamp force recording as a function of time in experiment 6. The thickness of the sample is about 1 mm. The width is about 4 mm. The sample is taken from the same disc as the sample of experiments 4 and 5. The annulus region from which it originates is close to the outer edge of the annulus. The clamp speed at which the step in the strain is applied is 0.05 mm/s.

increase of the force, which relaxes down to about half its value. At $t = 20$ min, the decrease in salt concentration from $c = 0.15$ M to $c = 0$ M induces a progressive increase of the force to a plateau level. At $t = 40$ min the external salt concentration is returned to its original level, causing the force to reduce to a value somewhat lower than its value in the first 20 min. The increase in strain at $t = 1$ h results in an increase in force proportionally larger than the increase in force observed in the first step in the strain. The relaxation results again in a reduction of the clamp force by half of its value. A similar pattern is repeated in the second hour of the experiment. All of the samples exhibit a creep behaviour which continues long after changes in external bath concentrations and strain.

4. Discussion

The force tracings show strong time dependent stress–strain relationships, which obviously are very sensitive to the external salt concentration. This sensitivity can be explained in the context of chemo-electromechanical theory [7]. When the external concentration c_{ext} of NaCl is reduced from 0.15 M to 0 M, the result is an immediate decrease in external osmotic pressure $\pi = 2RTc_{\text{ext}}$ and hence a steep gradient of the water chemical potential and electrochemical potentials of the ions between the tissue and the solution. These gradients induce an influx of water into the tissue and an outflux of ions from the tissue. Volumewise the outflux of ions is negligible compared to the influx of water. The influx of water results in an increase in clamp force with time. The time constants associated with it should therefore be the consolidation time of the sample:

$$\Delta t = \frac{(\Delta x)^2}{KE} \tag{1}$$

and the diffusion times of the ionic species

$$\Delta t = \frac{(\Delta x)^2}{D^\alpha}, \qquad \alpha = +,- \tag{2}$$

in which Δx is the typical distance traveled by the water and the ions, K is the hydraulic permeability of the tissue, E a measure of the elastic stiffness of the tissue and D^+ and D^- the ionic diffusion coefficients. Choosing $\Delta x = 0.5\ 10^{-3}$ m, $K = 10^{-16}$ m^4/NS [2,4] $E = 2\ 10^6$ Pa [2,4] and $D^+ \approx D^- \approx 0.5\ \ 10^{-9}$ m^2/s, we find a consolidation time of 1.2 10^3 s and a diffusion time of 0.5 10^3 s. These values are so close to each other that the time constants cannot be separated in the response of intervertebral disc tissue to changes in external salt concentrations [4]. The present experiments confirm this trend. The quantitative values of the time constants are consistent with all 6 experiments. A stepwise increase in the strain results in an immediate increase in the clamp forces. This increase in clamp force is distributed over (1) an increase of the strain dependent effective stress and (2) a decrease of the hydrostatic pressure. The latter induces an influx of water into the sample. This influx leads to a new equilibrium at a lower force level. The time constant 1 and – to a lesser extent – the time constant 2 are involved in this transition. All of the samples exhibit a sustained creep behaviour long after changes in external bath concentrations and strain. A similar phenomenon was observed in the confined swelling and compression experiments [4,6] on canine annulus fibrosus tissue. A leakage of proteoglycans from the samples is a possible explanation of this phenomenon.

Elastic stiffness as computed from these experiments is rather inaccurate, but lays in the order of one or a few MPa, which is consistent with earlier values [3,4,6].

Acknowledgements

This research is part of the project EURODISC (QLRT-2001-02852), funded by the European Union in the context of the key-action program "The Aging Population". The authors thank W. Meertens and M. van Hout for preparing the samples and performing the experiments.

References

[1] S. Akizuki, V.C. Mow, F. Mueller, J.C. Pita, D.S. Howell and D.H. Manicourt, Tensile properties of human knee joint cartilage: I. Influense of ionic conditions, weight bearing, and fibrillation on the tensile modulus, *J. Orthoped. Res.* **4** (1986), 379–392.

[2] A.J.H. Frijns, J.M. Huyghe and J.D. Janssen, A validation of the quadriphasic mixture theory for intervertebral disc tissue, *Int. J. Eng. Sci.* **35** (1997), 1419–1429.

[3] J.O. Galanta, Tensile properties of the human lumbar annulus fibrosus, *Acta Orthop. Scan. Suppl.* **100** (1967), 5–91.

[4] G.B. Houben, M.R. Drost, J.M. Huyghe, J.D. Janssen and A. Huson, Non-homogeneous permeability of canine anulus fibrosus, *Spine* **22** (1997), 7–16.

[5] J.M. Huyghe, Intra-extrafibrillar mixture formulation of soft charged hydrated tissues, *Journal of Theoretical and Applied Mechanics* **37** (1999), 519–536.

[6] J.M. Huyghe, G.B. Houben, M.R. Drost and C.C. van Donkelaar, An ionised/non-ionised dual porosity model of intervertebral disc tissue: experimental quantification of parameters, *Biomechanics and Modelling in Mechanobiology* **2** (2003), 3–19.

[7] J.M. Huyghe and J.D. Janssen, Quadriphasic mechanics of swelling incompressible porous media, *Int. J. Eng. Sci.* **35** (1997), 793–802.

[8] J.C. Iatridis, J.P. Laible and M.H. Krag, Influence of fixed charge density magnitude and distribution on the intervertebral disc: applications of a poroelastic and chemical ellectric (peace) model, *J. Biomech. Engng.* **125** (2003), 12–24.

[9] E.R. Myers, W.M. Lai and V.C. Mow, A continuum theory and an experiment for the ion-induced swelling behavior of articular cartilage, *J. Biomech. Eng.* **106** (1984), 151–158.

[10] R. van Loon, J.M. Huyghe, M.W. Wijlaars and F.P.T. Baaijens, 3d fe implementation of an incompressible quadriphasic mixture model, *Int. J. Numer. Meth. Engng.* **57** (2003), 1243–1258.

Biorheology 41 (2004) 263–269
IOS Press

Mathematical framework for modeling tissue growth

K.Y. Volokh
Faculty of Civil and Environmental Engineering, Technion, Haifa 32000, Israel
E-mail: cvolokh@tx.technion.ac.il

Abstract. A phenomenological continuum mechanics framework for modeling growth of living tissues is proposed. Tissue is considered as an open system where mass is not conserved. The momentum balance is completed with the full-scale mass balance. Constitutive equations define *simple growing materials*. 'Thermoelastic' formulation of a simple growing material is specified. Within this framework traction free growth of a cylinder is considered. It is shown that the theory accommodates the case where stresses are not generated in uniform volumetric growth. It is also found that surface growth corresponds to a *boundary layer* solution of the governing equations.

Keywords: Living tissue, volumetric growth, surface growth, mass balance

1. Introduction

Understanding growth of living tissues is of fundamental theoretical and practical interest. Analytical models of growth of both plant and animal tissues can predict the evolution of the tissue, which may improve treatment of pathological conditions and offer new prospects in tissue engineering. Though the early works on growth were kinematical [19], using continuum mechanics is now widely accepted for describing growth, remodeling, and morphogenesis [17]. Since the work of Hsu [8], there appeared various models of growth [2–4,6,7,9–12,14,15,18]. Multiplicative decompositions of the displacement gradient underlie many publications on analytical modeling of growth. In this case an intermediate configuration is introduced where the free deformation-growth is considered in the vicinity of every material point. Geometrical compatibility of the grown material particles is ensured by additional deformation and stress. Another line of the growth modeling is presented in [10] where tissue is considered as a classical mixture of solid and fluid. These authors show that several well-known models of adaptive growth can be embedded in their general theory. It should be noted that all cited works are devoted to volumetric growth while surface growth is still out of the scope of continuum mechanics [16,17]. Mathematical description of existing approaches is rather sophisticated and it includes variables that may be difficult to interpret in simple terms and to assess in measurements.

In the present work a novel continuum mechanics framework for modeling growth of living tissues is considered. It is assumed that deformation and mass flow can describe both volumetric and surface growth of living bodies. This assumption leads to the violation of mass conservation and to the introduction of the full-scale mass balance. A possible structure of constitutive equations is discussed with reference to *simple growing materials*. 'Thermoelastic' formulation of the simple growing material is specified. This formulation leads to the uncoupled mass-flow/deformation problem, which is analogous to the classical quasi-static and small strain thermoelasticity. Within this framework traction free growth

0006-355X/04/$17.00

of a cylindrical body is considered. It is shown that the theory accommodates the case where stresses are not generated in uniform volumetric growth. It is also found that surface growth corresponds to a *boundary layer* solution of the governing equations. This finding proves the ability of continuum mechanics to describe surface growth. This is in contrast to the widespread point of view that only purely kinematical theories are suitable for modeling surface growth.

2. Governing equations

The assumption that the continuous deformation and mass flow can describe growth of living bodies is central for further development. Sharp distinction between the real physical material and the mathematical concept of *material particle* should be kept in mind. The mapping $\mathbf{x} = \chi(\mathbf{X}, t)$ is considered in the case of deformation of a non-growing body. $\mathbf{X}$ designates an initial (reference) place of the material particle and $\mathbf{x}$ is its current place in space. $\mathbf{X}$ can be also interpreted as a label attached to the considered material particle. Since this mapping is one-to-one, we tacitly assume that the 'number' of material particles remains the same after deformation. For example, a rubber ball can be exposed to a significant outer pressure and its radius can decrease while preserving the spherical shape of the ball. Although the ball occupies less space after deformation, nobody doubts that the number of material particles remains the same or that the continuum mechanics mapping is applicable. The concept of the material particle is purely mathematical. It is physically meaningless. Particles-points do not exist: they are mathematical abstractions. Material always occupies some volume. One means a very small (infinitesimal) material volume saying 'material particle'. *Growth is considered as the deformation and mass change in this volume, i.e. in the vicinity of the given material particle.* The 'number' of the particles, however, is not changing during growth. If the connectivity of the living body is preserved during its growth and a continuous deformation of the grown body into its initial configuration can be imagined, then it is possible to claim that the number of particles remains the same, by analogy with deformation of non-growing materials, and the continuum mechanics mapping $\mathbf{x} = \chi(\mathbf{X}, t)$ is applicable to growing bodies too. To illustrate this statement, consider a sphere made up of a living material. Let this material experience negative growth (atrophy) and the radius decrease while preserving the shape. Assume now that an observer can follow both deformation of the sphere under pressure and negative growth of the living material simultaneously. If the (invisible gas) pressure increases slowly enough and the sphere gets smaller at the same rate, it is impossible to make a geometrical distinction between growth and deformation. A physical distinction, however, exists. Growth manifests itself in the change of mass, which becomes crucial for modeling growth. Mass balance should be considered in its completeness

$$\frac{\partial \rho_0}{\partial t} = \mathrm{Div}\boldsymbol{\psi}_0 + \xi_0, \tag{1}$$

where $\rho_0 \neq \mathrm{const}$ is the Lagrangian (referential) mass density; $\boldsymbol{\psi}_0$ is the Lagrangian mass flux per unit surface; ξ_0 is the Lagrangian mass supply per unit volume.

The classical mass conservation law is obtained when $\boldsymbol{\psi}_0 = \mathbf{0}$ and $\xi_0 = 0$.

The balance of linear and angular momentum takes the following form

$$\frac{\partial}{\partial t}(\rho_0 \mathbf{v}) = \mathrm{Div}(\mathbf{FS}) + \rho_0 \mathbf{b}, \tag{2}$$

$$\mathbf{S} = \mathbf{S}^{\mathrm{T}}. \tag{3}$$

where $\mathbf{v} = \partial\chi/\partial t$ is the velocity vector; $\mathbf{b}$ is the body force per unit mass; $\mathbf{S}$ is the 2nd Piola–Kirchhoff stress tensor. Growth and deformation are coupled in the deformation gradient $\mathbf{F} = \partial\chi/\partial\mathbf{X}$. It is worth emphasizing that no distinction is made between growth and purely mechanical deformation. This is in contrast to the works where a multiplicative decomposition of the deformation gradient is used. The latter presupposes existence of intermediate incompatible configurations of pure growth, which cannot be readily interpreted and measured in physical/biological terms.

Constitutive equations must be added to the mass and momentum balance. A possible form of these equations is as follows

$$\psi_0(t) = \hat{\psi}_0^{(t)}(\rho_0, \mathrm{Grad}\rho_0, \mathbf{F}, \mathbf{X}), \tag{4}$$

$$\mathbf{S}(t) = \hat{\mathbf{S}}^{(t)}(\rho_0, \mathrm{Grad}\rho_0, \mathbf{F}, \mathbf{X}), \tag{5}$$

$$\xi_0(t) = \hat{\xi}_0^{(t)}(\rho_0, \mathrm{Grad}\rho_0, \mathbf{F}, \mathbf{X}), \tag{6}$$

where caps designate constitutive functionals for the materials *with memory* up to the time t.

It is worth noting that these equations provide coupling between mass and momentum balance. Extending terminology of Truesdell and Noll [21], these equations define *simple growing materials*. It is essential that the mass density gradient *must* be included in the constitutive law. Indeed, after substituting (4)–(6) in (1), (2), (3), the system of governing equations is of the second order in spatial derivatives of ρ_0. The latter allows for imposing two boundary conditions on ρ_0 on opposite sides of the considered body. Assume, for example, that the constitutive relations do not include the mass density gradients: $\psi_0 = \hat{\psi}_0(\rho_0, \mathbf{F}, \mathbf{X})$, $\mathbf{S} = \hat{\mathbf{S}}(\rho_0, \mathbf{F}, \mathbf{X})$, and $\xi_0 = \hat{\xi}_0(\rho_0, \mathbf{F}, \mathbf{X})$. Substituting these relations in the balance equations, we obtain a system of governing equations of the first order in spatial derivatives of ρ_0. The first order differential equations require only one boundary condition and, generally, it is impossible to satisfy two boundary conditions on opposite sides of the body. It is also hardly possible to give an acceptable physical interpretation to such inconsistency between the number of reasonable boundary conditions and the order of the differential equations. The solution of the differential equations can be called over-determined in this case. In contrast to the over-determinacy, the use of the higher-grade materials where higher order gradients are presented in the constitutive equations can lead to the under-determinacy of differential equations. The latter happens if no additional boundary conditions are imposed. An example of the inconsistency of this kind can be found in [22] within the context of metal plasticity. The requirement of the strict correspondence between the number and character of boundary conditions and the structure of balance constitutive laws can be called the requirement of *mathematical consistency*.

In order to complete the formulation of the initial boundary value problem (IBVP) for simple growing materials it is necessary to formulate the initial and boundary conditions. The classical boundary and initial conditions take the form

$$\chi = \chi^* \in \partial\Omega_\chi, \qquad \mathbf{t}_0 = \mathbf{F}\mathbf{S}\mathbf{n}_0 = \mathbf{t}_0^* \in \partial\Omega_\mathrm{t}, \qquad \chi(t=0) = \chi^{**} \in \Omega, \qquad \mathbf{v}(t=0) = \mathbf{v}^{**} \in \Omega, \tag{7}$$

where $\mathbf{n}_0$ is a unit normal to the referential boundary surface; $\partial\Omega_\chi \cup \partial\Omega_\mathrm{t} = \partial\Omega$; $\partial\Omega_\chi \cap \partial\Omega_\mathrm{t} = 0$. The additional 'growth' conditions can be written as follows

$$\rho_0 = \rho_0^* \in \partial\Omega_\rho, \qquad \phi_0 = \psi_0 \cdot \mathbf{n}_0 = \phi_0^* \in \partial\Omega_\phi, \qquad \rho_0(t=0) = \rho_0^{**} \in \Omega, \tag{8}$$

where $\partial\Omega_\rho \cup \partial\Omega_\phi = \partial\Omega$; $\partial\Omega_\rho \cap \partial\Omega_\phi = 0$. All quantities with the asterisk are given.

Constitutive relations (4)–(6) are not specified yet. It is possible to specify them relying on the Skalak's [14] qualitative idea that growth is analogous to thermal expansion. The physical or biological basis for the analogy between growth and thermal expansion is simple: the increasing number of cells or the newly produced extracellular matrix material tends to expand the occupying volume as the increasing temperature does in structural materials.

It is assumed that (a) the process is quasi-static, i.e. transient behavior and inertia effects are ignored, and (b) deformations are small and body forces are ignored. The first restriction is reasonable because of the very slow growth process. The second restriction suppresses the difference between the Lagrangian and Eulerian descriptions of growth-deformation. Simplifying the notation ($\mathbf{S} = \boldsymbol{\sigma}$, Div = div, $\boldsymbol{\psi}_0 = \boldsymbol{\psi}$, $\xi_0 = \xi$, $\rho_0 = \rho$, Grad = ∇) we have for isotropic materials:

$$\mathrm{div}\boldsymbol{\psi} + \xi = 0, \tag{9}$$

$$\mathrm{div}\boldsymbol{\sigma} = \mathbf{0}, \tag{10}$$

$$\boldsymbol{\sigma} = \lambda \mathrm{tr}(\boldsymbol{\varepsilon})\mathbf{1} + 2\mu\boldsymbol{\varepsilon} - (3\lambda + 2\mu)\alpha\rho\mathbf{1}, \tag{11}$$

$$\boldsymbol{\psi} = \beta\nabla\rho, \tag{12}$$

$$\xi = \omega - \gamma\rho, \tag{13}$$

where $\rho := \rho(\omega) - \rho(0)$; $\boldsymbol{\varepsilon} = (\nabla\mathbf{u} + \nabla\mathbf{u}^{\mathrm{T}})/2$; $\mathbf{u} = \boldsymbol{\chi}(\mathbf{X}) - \mathbf{X}$; λ and μ are the Lame coefficients; $\mathbf{1}$ is the second-order identity tensor.

Increment of *mass supply* $\omega > 0$ is analogous to a quasi-static mechanical load. In contrast to the latter, however, ω is controlled by the tissue itself and its proper determination requires experiments. The dimension of ω is a unit of mass per volume and time. Time is not involved directly in quasi-static problems and can be replaced by some conditional units.

Coefficient of growth expansion $\alpha > 0$ determines how much the relative volume changes for the given increment of mass density. Its dimension is an inverse of a unit of mass.

Mass conductivity of solid $\beta > 0$ determines how much the mass supply changes for a given increment of the gradient of mass density. Its dimension is a unit of mass supply times length per mass density.

Coefficient of tissue resistance $\gamma > 0$ reflects the resistance of the tissue to accommodate new mass for increasing mass density. The second term on the right-hand side of Eq. (13) 'brakes' mass supply when the density increases. Roughly speaking, the more cells appear the less room remains for the new cells. The dimension of γ is the dimension of ω per a unit of mass density.

The similarity between the two first constitutive laws of growth and thermoelasticity is obvious after replacing the mass density increment by the temperature increment; the vector of mass flux by the vector of heat flux; the coefficient of growth expansion by the coefficient of thermal expansion; and the mass conductivity of solid by the thermal conductivity of solid. In this case Eq. (11) is nothing but the thermoelastic generalization of the Hooke's law, and Eq. (12) is just the Fourier law of heat conduction [1]. The constitutive law analogous to Eq. (13), however, is usually absent in thermoelasticity because of the lack of volumetric heat sources. The thermoelastic analogy allows for better understanding parameters of the growth model. The vector of mass flux is analogous to the vector of heat flux. We feel the heat flow by the changing temperature without directly defining what the heat is. The same is true for the mass flow. We 'feel' it by the changing mass density without directly defining what it is.

It is worth noting that the volumetric mass supply should include stresses or strains on the right-hand side of Eq. (13) when the *adaptive growth* is considered. However, we do not consider this issue in the present work. This allows for the uncoupling of the mass and momentum balance equations. Indeed, substituting Eqs (5) and (6) in Eq. (1) and assuming $\beta =$ constant we have

$$\beta\nabla^2\rho - \gamma\rho + \omega = 0. \tag{14}$$

Substituting solution of Eq. (14) in Eqs (11) and (9) it is possible to find the deformation characteristics and the corresponding stress field.

3. Results

We examine the proposed theory by considering traction-free radial growth of an infinite cylinder. The inner radius of the cylinder is a and the outer radius is $3a$.

In the case of volumetric growth we assume that material is supplied uniformly $\omega =$ constant, and $\gamma =$ constant, and boundary conditions take the form $\rho(r = a, 3a) = \rho^* = \omega/\gamma$. Solution of Eq. (14) is evident $\rho = \omega/\gamma$. This is the case of the uniform growth. Substituting this solution in Eqs (10)–(11) we get for the displacements and stresses $u = \alpha(1+\nu)r\omega/\gamma$; $\sigma_{rr} = \sigma_{\theta\theta} = 0$ (ν is the Poisson ratio). Thus uniform and traction-free growth does not produce stresses.

In the case of surface growth we assume that material is supplied on the outer surface only: $\omega = 0$; $\rho(r = a) = 0$; $\rho(r = 3a) = \rho^*$. Radial distributions of normalized mass densities, displacements, radial and circumferential stresses: $\bar{\rho} = \rho/\rho^*$; $\bar{u} = u/(\alpha\rho^* a)$; $\bar{\sigma}_{rr} = \sigma_{rr}/(\alpha\rho^* E)$; $\bar{\sigma}_{\theta\theta} = \sigma_{\theta\theta}/(\alpha\rho^* E)$ (E is the elasticity modulus) were computed for the different values of the normalized coefficient of tissue resistance $\tau = \sqrt{\gamma/\beta}$ (Fig. 1). All numerical results were calculated for $\nu = 1/4$. It is important to emphasize that mass densities, displacements, and circumferential stresses localize in a *boundary layer* at $r = 3a$ with increasing τ, while their magnitudes outside the boundary layer tend to zero. This is the surface growth.

4. Conclusions

The problem of establishing a simple analytical framework for modeling growth of living tissues has been addressed. General model of *simple growing materials* is presented where growth is considered as a mass-flow-deformation process. A novel theory of tissue growth is specified. This theory is analogous to thermoelasticity where temperature is replaced by mass density. In order to solve the growth problem for the given living body, it is necessary first to find the distribution of mass density from the mass balance equation. The thermoelastic counterpart of this equation is equation of heat conduction. When the mass density distribution is known, it is possible to find deformation from the momentum balance accounting for the generalized Hooke's law. The latter indicates close resemblance between growth and thermal expansion. Examples of volumetric growth of the living cylinder reveal the capacity of the theory to accommodate materials that can grow freely and uniformly without generating stresses.

An important feature of the proposed theory is its ability to reproduce surface growth. The latter appears as the *boundary layer solution* of governing equations. Thus surface growth may be interpreted as localization of growth in the vicinity of the surface. This finding seems to be of the principle matter

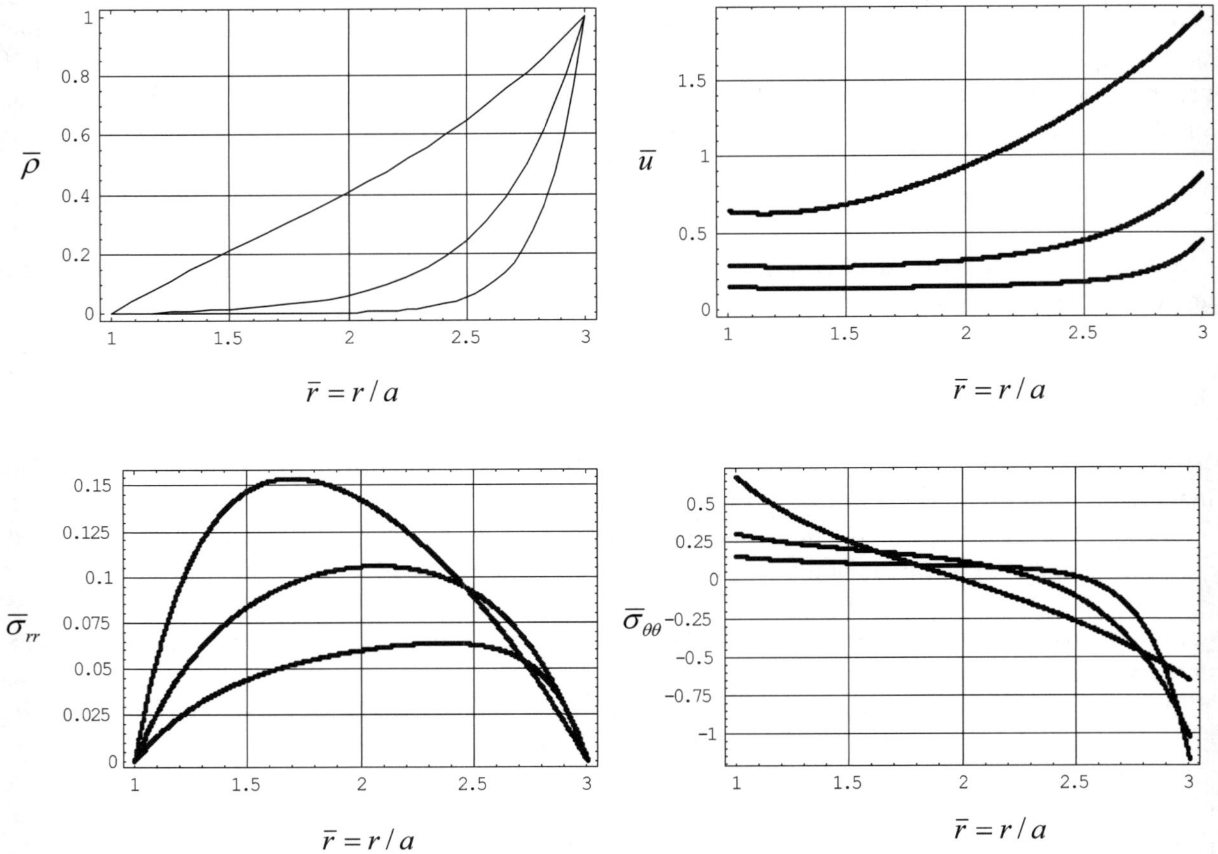

Fig. 1. Surface growth of the cylinder. Normalized density, displacements, radial stresses, and circumferential stresses along normalized radius for different values of the normalized coefficient of tissue resistance $\tau = \sqrt{\gamma/\beta}$. Graphs approach zero with increasing parameter $\tau = 1, 3, 6$.

proving the power of continuum mechanics to treat the phenomenon of surface growth. Traditionally, the description of surface growth is restricted by purely kinematical theories.

It is worth noting that the proposed approach is not restricted by a specific biological mechanism of growth. Indeed, the analytical model of growth is based on macroscopic variables: displacements and mass densities. It does not matter, in principle, what are the possible biological scenarios of the cells' evolution. Such evolution can occur volumetrically or at the surface or following some more complicated scenario. Information about the biological processes underlying macroscopic growth can be useful in creating phenomenological theories. Unfortunately, the cross-link between macro- and micro-scales remains an open challenging problem. The latter is true not only for living materials but even for much simpler structural materials [5].

Finally, it should be kept in mind that *only physically observable quantities – displacements and mass densities – are used as variables in the mathematical formulation*. Only these quantities should be measured in order to calibrate the theory. Recent developments in computer vision techniques (MRI, PET) [11,20] combined with the noninvasive densitometry (based, for example, on X-ray techniques) will hopefully allow for the calibration of the proposed analytical model.

References

[1] B. Boley and J.H. Weiner, *Theory of Thermal Stresses*, Dover, Mineola, 1997.

[2] Y.C. Chen and A. Hoger, Constitutive functions of elastic materials in finite growth and deformation, *Journal of Elasticity* **59** (2000), 175–193.

[3] S.C. Cowin and D.H. Hegedus, Bone remodeling I: Theory of adaptive elasticity, *Journal of Elasticity* **6** (1976), 313–326.

[4] A.D. Drozdov, *Viscoelastic Structures: Mechanics of Growth and Aging*, Academic Press, San Diego, 1998.

[5] G.J. Dvorak, Research trends in solid mechanics: executive summary, *International Journal of Solids and Structures* **37** (2000), 1–11.

[6] M. Epstein and G. Maugin, Thermomechanics of volumetric growth in uniform bodies, *International Journal of Plasticity* **16** (2000), 951–978.

[7] Y.C. Fung, *Biomechanics: Motion, Flow, Stress, and Growth*, Springer, New York, 1990.

[8] F.H. Hsu, The influence of mechanical loads on the form of a growing elastic body, *Journal of Biomechanics* **1** (1968), 303–311.

[9] S.M. Klisch, T.J. Van Dyke and A. Hoger, A theory of volumetric growth for compressible elastic biological materials, *Mathematics and Mechanics of Solids* **6** (2001), 551–575.

[10] S. Kuhn and W. Hauger, A theory of adaptive growth of biological materials, *Archive of Applied Mechanics* **70** (2000), 183–192.

[11] X. Papademetris, P. Shi, D.P. Dione, A.J. Sinusas, R.T. Constable and J.S. Duncan, Recovery of soft tissue object deformation from 3D image sequences using biomechanical models, *LNCS* **1613** (1999), 352–357.

[12] A. Rachev, A model of arterial adaptation to alterations in blood flow, *Journal of Elasticity* **61** (2000), 83–111.

[13] E.K. Rodriguez, A. Hoger and A.D. McCulloch, Stress-dependent finite growth in soft elastic tissues, *Journal of Biomechanics* **27** (1994), 455–467.

[14] R. Skalak, Growth as a finite displacement field, in: *Proceedings of the IUTAM Symposium on Finite Elasticity*, D.E. Carlson and R.T. Shield, eds, Martinus Nijhoff Publishers, 1982, pp. 347–355.

[15] R. Skalak, G. Dasgupta, M. Moss, E. Otten, P. Dullimeijer and H. Villeman, Analytical description of growth, *Journal of Theoretical Biology* **94** (1982), 555–577.

[16] R. Skalak, D.A. Farrow and A. Hoger, Kinematics of surface growth, *Journal of Mathematical Biology* **35** (1997), 869–907.

[17] L.A. Taber, Biomechanics of growth, remodeling, and morphogenesis, *Applied Mechanics Reviews* **48** (1995), 487–545.

[18] L.A. Taber and R. Perucchio, Modeling heart development, *Journal of Elasticity* **61** (2000), 165–197.

[19] D.W. Thompson, *On Growth and Form*, Cambridge University Press, Cambridge, 1948.

[20] P.M. Thompson, J.N. Giedd, R.P. Woods, D. MacDonald, A.C. Evans and A.W. Toga, Growth patterns in the developing brain detected by using continuum mechanical tensor maps, *Nature* **404** (2000), 190–193.

[21] C. Truesdell and W. Noll, *The Non-Linear Field Theories of Mechanics*, Springer-Verlag, Berlin, 1992.

[22] K.Y. Volokh and J.W. Hutchinson, Are lower order gradient theories of plasticity really lower order?, *Journal of Applied Mechanics* **69** (2002), 862–864.

Part II

Mechanobiology and chondrocytes

Biorheology 41 (2004) 273–281
IOS Press

NMDA receptor expression and roles in human articular chondrocyte mechanotransduction

D.M. Salter *, M.O. Wright and S.J. Millward-Sadler
Pathology, School of Molecular and Clinical Medicine, College of Medicine and Veterinary Medicine, Edinburgh University, Teviot Place, Edinburgh, EH8 9AG, UK

Abstract. Mechanical forces influence articular cartilage structure by regulating chondrocyte activity. Mechanical stimulation results in activation of an $\alpha5\beta1$ integrin dependent intracellular signal cascade involving focal adhesion kinase and protein kinase C, triggering the release of interleukin-4 from the cell. In normal HAC the response to physiological mechanical stimulation is characterised by increased levels of aggrecan mRNA and a decrease in levels of mRNA for matrix metalloproteinase 3 (MMP-3), the net result of which would be to maintain and optimise cartilage structure and function. This protective/anabolic response is not seen when chondrocytes from osteoarthritic cartilage are subjected to an identical mechanical stimulation regime.

Following the observation that the neurotransmitter substance P is involved in chondrocyte mechanotransduction the present study was undertaken to establish potential roles for glutamate receptors in the control of chondrocyte mechanical responses.

Using immunohistochemistry and RTPCR normal and OA chondrocytes are shown to express NR1 and NR2a subunits of the NMDA receptor. Addition of NMDA receptor agonists to chondrocytes in primary culture resulted in changes in membrane potential consistent with expression of functional receptors. NMDA receptor antagonists inhibited the hyperpolarisation response of normal chondrocytes to mechanical stimulation but had no effect on the depolarisation response of osteoarthritic chondrocytes to mechanical stimulation.

These studies indicate that at least one subset of the NMDA receptor family of molecules is expressed in cartilage and may have important modulatory effects on mechanotransduction and cellular responses following mechanical stimulation. Indeed the results suggest that there is an alteration of NMDA receptor signalling in OA chondrocytes, which may be critical in the abnormal response of OA chondrocytes to mechanical stimulation. Thus NMDA receptors appear to be involved in the regulation of human articular chondrocyte responses to mechanical stimulation, and in OA, mechanotransduction pathways may be modified as a result of altered activation and function of these receptors.

1. Introduction

Mechanical forces influence articular cartilage structure by regulating chondrocyte activity. Optimal mechanical stimulation maintains normal cartilage in a healthy state but mechanical forces, either abnormal loading of normal joints or physiological loading of diseased joints, lead to cartilage breakdown and the development of osteoarthritis (OA). We have been studying how human articular chondrocytes (HAC) recognise and respond to mechanical stimulation. In these studies we have shown, for the first time, that HAC use integrins to recognise mechanical forces [26]. Mechanotransduction via $\alpha5\beta1$ integrin leads to activation of an intracellular signalling cascade involving the actin cytoskeleton

*Address for correpondence: D.M. Salter, MD, Pathology, Edinburgh University Medical School, Teviot Place, Edinburgh, UK, EH8 9AG. Tel.: +44 01316502946; Fax: +44 01316506528; E-mail: Donald.Salter@ed.ac.uk.

0006-355X/04/$17.00

and interleukin-4 that results in the regulation of transcription of cartilage structural and remodelling genes [13,14,16]. In normal HAC the response to physiological mechanical stimulation is characterised by increased mRNA levels for aggrecan and a decrease in mRNA for matrix metalloproteinase 3 (MMP-3), the net result of which would be to maintain and optimise cartilage structure and function.

Studies in our laboratory have, however, shown that this protective/anabolic response is not seen when chondrocytes from osteoarthritic cartilage are mechanically stimulated [13,15]. We are in the process of identifying the mechanisms behind this abnormal response. To date we have shown that OA chondrocytes continue to use $\alpha 5\beta 1$ integrin as the prime mechanoreceptor, but downstream intracellular signalling pathways and activated autocrine cytokine loops differ significantly from these seen in normal chondrocytes [15].

As part of our ongoing studies into chondrocyte mechanotransduction in health and disease, following the observation that preprotachykinin gene is expressed in the cartilage of transgenic mice and substance P is involved in the mechanotransduction of human articular chondrocytes [15], we have been studying potential roles for neural associated molecules in control of chondrocyte function.

Substance P expression and function is closely associated with glutamate, a major excitory neurotransmitter in the central and peripheral nervous system. Glutamate acts on cells through interaction with members of the glutamate receptor family that includes N-methyl-D-aspartate (NMDA), amino-3-hydroxy-5-methylisoxazole-4-propionic acid (AMPA) and metabotropic glutamate receptors. These receptors are known to be widely expressed in the mammalian central nervous system but have also been shown to be expressed by non-neuronal tissues including bone, pancreas and skin [20,24]. Recent work has suggested possible role for NMDA receptors in responses to mechanical stimulation in neuronal and bone cells [19,20].

NMDA receptors are a subset of the glutamate receptor family, which also includes amino-3-hydroxy-5-methylisoxazole-4-propionic acid (AMPA) and metabotropic glutamate receptors. NMDA receptors are multimeric complexes made up of NR1 subunits, and one of four NR2 subunits (NR2A-D) [17]. The NR1 subunit is present in all NMDA receptors, whilst the NR2 subunits confer distinct properties on to the receptor. An NR3 subunit, of which two have so far been identified, can coassemble with the NR1/2 complex, and modify the receptor properties [7]. For the NR1 subunit, several splice variants have been identified [25], and these can affect the membrane localisation and clustering of the receptors, the phosphorylation state of the receptor subunit and intracellular protein interactions [8,18]. Splice variants of the NR2 subunits have also been identified, although the function of these is not clear [6].

The present study was undertaken to establish whether NMDA receptor components are expressed by human articular chondrocytes and to investigate potential roles for these molecules in chondrocyte mechanotransduction.

2. Materials and methods

2.1. Isolation and culture of chondrocytes

Chondrocytes were isolated by sequential enzymatic digestion from cartilage obtained from human adult knee joints following arthroplasty, resection for peripheral vascular disease with patient's consent. In all cases articular cartilage was assessed and graded macroscopically for the presence or absence of osteoarthritis using the Collins/McElligott system and samples processed for histological examination. Articular cartilage from different anatomical regions of the knee joint including the patella, tibial plateau, and femoral condyle were kept separate as were normal and osteoarthritic cartilage. Isolated cells were seeded in Iscoves modified Dulbeccos medium supplemented with 10% fetal calf serum (FCS), 2 mM L-glutamine, 100 IU/ml penicillin and 100 μg/ml streptomycin, at a concentration of 5×10^4 cells/ml into 58 mm diameter tissue culture dishes (Nunc). Primary, non-confluent cultures of chondrocytes were used in all experiments. Morphologically the cells from both normal and osteoarthritic cartilage were typically flattened with a polygonal cell shape. RT-PCR and immunohistochemistry showed expression of cartilage specific molecules including type II collagen and aggrecan.

2.2. Immunohistochemistry

Representative samples of normal articular cartilage (Collins grade 0) and osteoarthritic cartilage (Collins grade 2–3) were snap frozen in liquid nitrogen. 4 μm sections were cut with a Brights cryostat, mounted on superfrost glass slides (VNR), allowed to come to room temperature and fixed with acetone for 10 min. Sections were stained with a mouse monoclonal anti-human NR1 antibody (TCS Biologicals). Negative controls were obtained by using non-specific immunoglobulin instead of the specific primary antibody. All slides were immunostained using a Shandon Sequenza (Shandon).

2.3. RTPCR

Total RNA was extracted from cultured chondrocytes as described in the Micro RNA Isolation Kit (Stratagene), using a denaturing buffer of 4 M guanidine thiocyanate, 0.75 M sodium citrate, 10% (w/v) lauroyl sarcosine and 7.2 μl/ml β-mercaptoethanol. The quantity of RNA isolated was determined by the absorbance reading at 260 nm on a spectrophotometer (Genequant; Pharmacia). Template cDNA was synthesised using 0.5 μg RNA, Superscript II (Invitrogen) and oligo dT (12–18) (Amersham Pharmacia Biosciences) according to the manufacturers instructions. The primers used for the PCR reactions were:

GAPDH	5′-TCT AGA CGG CAG GTC AGG TCC ACC-3′
	5′-CCA CCC ATG GCA AAT TCC ATG GCA-3′
NR2a	5′-GCT TCA GGA GAC AGG TAA CCC AGC C-3′
	5′-CTG TCC CTG GAA CAG TAC GAT GCC G-3′

A typical 20 μl PCR reaction contained 20 mM ammonium sulphate, 75 mM Tris-HCl, pH 8.8, 0.01% (v/v) Tween-20, 1 μM each primer, 2 μl cDNA, 100 μM dNTPs, 0.1% (w/v) bovine serum albumin, 0.25 units Taq polymerase (Biogene, Kimbolton, UK). The magnesium chloride concentrations for each primer pair were: GAPDH – 2.5 mM; NMDA receptor 2A (NR2A) – 2 mM. The following programme was used for all reactions: 94°C for 3 minutes; 35 cycles of 94°C for 1 minute, 60°C for 1 minute, 72°C for 1 minute 30 seconds; 1 cycle of 72°C for 10 minutes. PCR products were analysed by electrophoresis using a 1% (w/v) agarose gel stained with ethidium bromide and visualised under UV.

2.4. Electrophysiological recording

Membrane potentials of chondrocytes were recorded using a single electrode bridge circuit and calibrator as previously described in detail [26]. Microelectrodes having tip resistances of between 40 to 60 mega-ohms and tip potentials of approximately 3 mV were used to impale the cells. Results from a cell were accepted if on impalement there was a rapid change in voltage to the membrane potential level of the cell and this voltage remained constant.

2.5. Experimental protocol

The day before experiments were carried out media containing FCS was replaced by FCS free media. In each set of experiments, a control dish of chondrocytes was examined whose culture medium did not contain any agent to be tested. For activation of glutamate receptors 50 μM glutamate, 50 μM N-methyl-D-aspartate, 20 μM glycine and the NMDA receptor agonist 20 μM MDL26630 (RBI-Sigma, UK) were used. For glutamate receptor blockade the following reagents, (obtained from RBI-Sigma, UK) were used: 30 μM carboxypiperazinyl propylphosphonic acid (CPP) a competitive NMDA receptor antagonist; 1 μM MK801, a non-competitive NMDA receptor antagonist; 25 μM 6-cyano-7-nitroquinoxaline-2,3-dione (CNQX), an AMPA/Kainate receptor antagonist; 500 μM methyl-(4-carboxyphenyl)glycine (MCPG), a competitive metabotropic receptor antagonist. Prior to addition of the chemical reagent to be tested or mechanical stimulation, membrane potentials of 5–10 cells were measured. Membrane potentials of a further 5–10 cells were measured immediately after addition of glutamate or NMDA and glycine, or 10 minutes after the application of the NMDA receptor agonist or chemical antagonists and thereafter following cyclical strain for 20 minutes.

2.6. Mechanical stimulation of chondrocytes in monolayer culture

The technique used for mechanical stimulation of primary monolayer cultures of human articular chondrocytes has been described in detail previously [13,14,16,26]. In brief, flexible, plastic tissue culture dishes are placed in a sealed pressure chamber with inlet and outlet ports. The chamber is pressurised with an inert gas from a cylinder, the frequency of pressurisation being dictated by an electronic timer controlling the inlet and outlet valves. The standard regime of cyclical pressurisation used consists of pressure pulses of 50 kPa above atmospheric pressure at a frequency of 0.33 Hz (2 seconds on, 1 second off) for 20 minutes. Cyclical pressurisation in this system has been shown to induce deformation and strain on the base of the plastic tissue culture dish and its adherent cells [22]. A pressure pulse of 50 kPa results in 3700 microstrain on the base of the dish with a maximum strain rate of 10,000 microstrain/sec.

2.7. Statistics

The mean, standard deviation and standard error of the mean were determined in each experiment. For statistical comparisons, when the F ratio of the 2 variances reached significance, the non-parametric Mann–Whitney test was used. When the ratio did not reach significance the Student t-test was used.

3. Results

3.1. Expression of NR1 and NR2a components of the NMDA receptor in adult human articular chondrocytes

Immunohistochemical staining of cryostat sections of snap frozen femoral and tibial knee joint articular cartilage showed extensive positivity for NR1 in greater than 90% of chondrocytes in all zones of all samples from normal and osteoarthritic cartilage (Fig. 1). RT-PCR with primers specific for NR2a cDNA on RNA extracted from normal and osteoarthritic articular cartilage chondrocytes (Fig. 2) demonstrated products of 490 bp for NR2a that were shown to be identical to the published sequences of this gene when sequenced (unpublished observations). There was a suggestion that expression of NR2a was greater in chondrocytes from normal cartilage in comparison to chondrocytes from osteoarthritic cartilage.

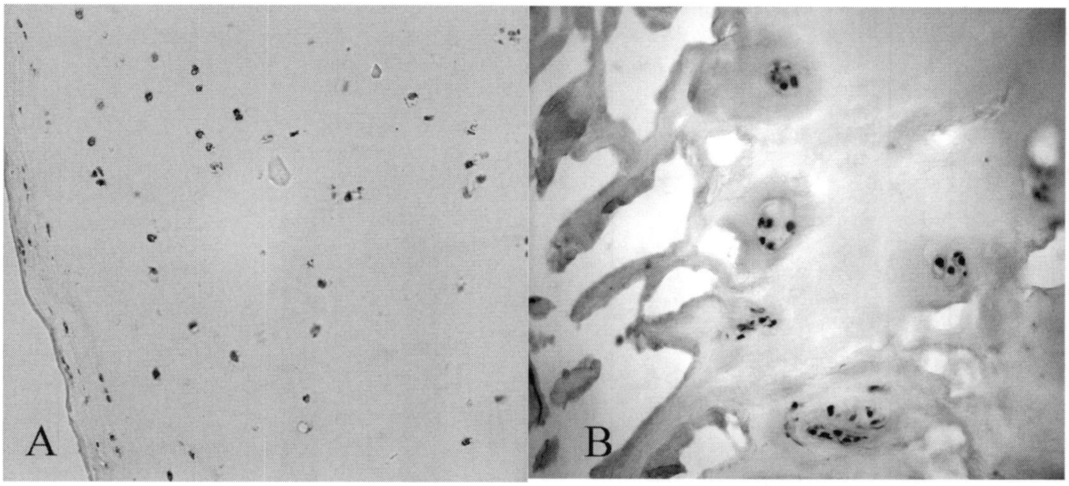

Fig. 1. Expression of NR1 in human articular cartilage. Cryostat sections were stained with antibodies against the NR1 subunit of the NMDA receptor. (A) Normal cartilage. (B) Osteoarthritic cartilage.

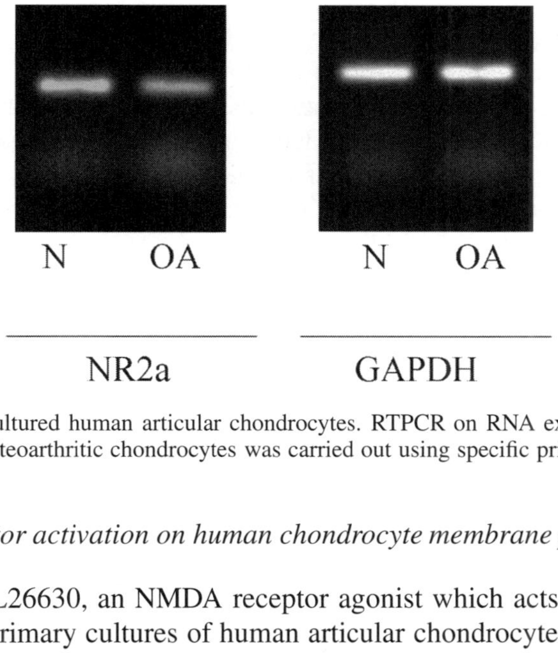

Fig. 2. Expression of NR2a in cultured human articular chondrocytes. RTPCR on RNA extracted from primary cultures of chondrocytes from normal and osteoarthritic chondrocytes was carried out using specific primers for the NR2a subunit of the NMDA receptor.

3.2. Effect of NMDA receptor activation on human chondrocyte membrane potential

Addition of 20 μM MDL26630, an NMDA receptor agonist which acts at the spermine site of the receptor, for 5 minutes to primary cultures of human articular chondrocytes causes a membrane depolarisation response in both normal and OA chondrocytes consistent with the expression of functional NMDA receptors (Fig. 3). Addition of both NMDA and glycine and glutamate with glycine to primary cultures of normal articular chondrocytes resulted in membrane hyperpolarisation whereas addition of these substances to osteoarthritic chondrocytes induced membrane depolarisation (Fig. 3).

3.3. Effect of glutamate receptor blockade on the electrophysiological response of normal chondrocytes to mechanical stimulation

The results are shown in Table 1. Chondrocytes from normal articular cartilage show a membrane hyperpolarisation response following 20 min 0.33 Hz mechanical stimulation. This was inhibited by

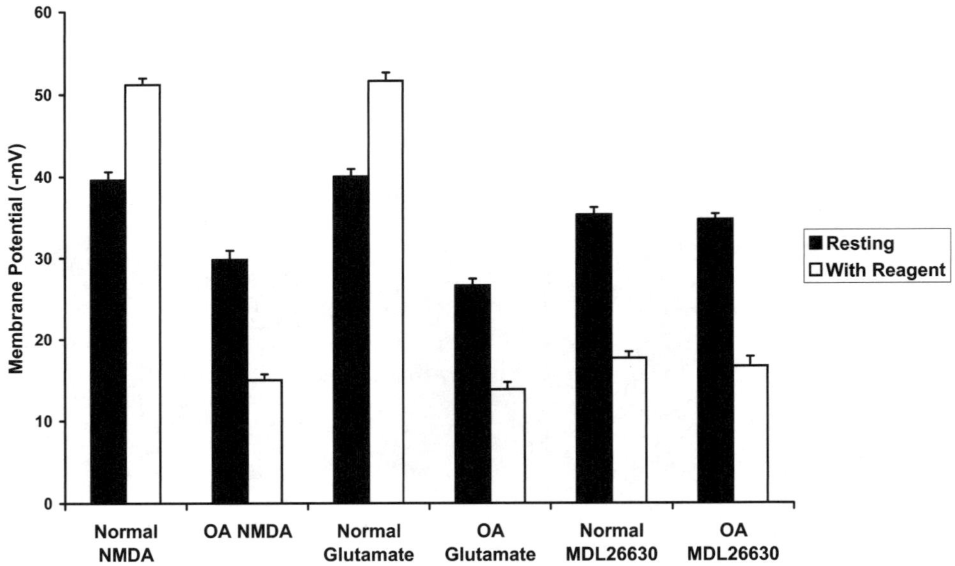

Fig. 3. Effect of NMDA receptor activation on membrane potential of normal and osteoarthritic chondrocytes in primary culture.

preincubation of the cells with either 30 μM CPP a competitive NMDA receptor antagonist or 1 μM MK801, a non-competitive NMDA receptor antagonist. However these NMDA receptor antagonists had no effect on the mechanical stimulation induced membrane depolarisation response of chondrocytes from osteoarthritic cartilage under similar conditions. The AMPA/Kainate receptor antagonist CNQX and MCPG, a competitive metabotropic receptor antagonist had no effect on the electrophysiological response of chondrocytes from either normal or osteoarthritic cartilage to mechanical stimulation.

4. Discussion

The present study demonstrates that the NR1 component of the NMDA receptor is expressed by human articular chondrocytes *in vivo*. Demonstration of NR2a mRNA expression *in vitro* and changes in membrane potential following application of glutamate and glycine, NMDA and glycine or the NMDA receptor agonist MDL26630, which acts at the spermine site of the receptor, to cultures of human chondrocytes suggest expression of at least one functional member of the NMDA receptor family. Inhibition of the electrophysiological response of normal human articular chondrocytes to mechanical stimulation by competitive and non-competitive NMDA receptor antagonists would be consistent with roles for these receptors in the stimulated integrin and IL4 dependent mechanotransduction pathway. Furthermore the differences in electrophysiological response following the addition of NMDA and glutamate, in the presence of glycine, of osteoarthritic chondrocytes in comparison to normal chondrocytes and the failure of NMDA receptor antagonists to influence the electrophysiological response of these cells to mechanical stimulation raise the possibility that there is an abnormality of NMDA receptor signalling in osteoarthritis.

The cytoplasmic domains of NMDA receptor subunits form part of a multiprotein complex termed the post-synaptic density (PSD), which interacts with the actin cytoskeleton. This complex consists of a number of proteins involved in the regulation of the receptor and subsequent intracellular signalling,

Table 1

Effect of glutamate receptor antagonists on the membrane potential response of normal and osteoarthritic chondrocytes to 0.33 Hz mechanical stimulation

Reagent (conc.)	Cartilage type	Membrane potential (−mV) (mean ± SEM)			% Change	P-value*
		Resting	Reagent alone	Reagent post-0.33 Hz	post-0.33 Hz	
No treatment	Normal	21.8 ± 0.80		43.2 ± 1.1	+98	<0.0001
	Osteoarthritic	36.8 ± 0.92		17.0 ± 0.45	−54	<0.0001
CPP (5 μM)	Normal	30.2 ± 0.86	30.8 ± 1.36	30.2 ± 0.66	0	ns
	Osteoarthritic	35.6 ± 0.51	35.6 ± 0.51	13.6 ± 0.81	−62	<0.0001
MK801 (1 μM)	Normal	28.4 ± 0.81	28.6 ± 0.68	28.6 ± 0.93	0	ns
	Osteoarthritic	34.0 ± 0.84	35.0 ± 0.45	16.4 ± 0.75	−53	<0.0001
CNQX (25 μM)	Normal	33.0 ± 1.05	33.2 ± 0.92	52.4 ± 0.93	+58	<0.0001
	Osteoarthritic	34.0 ± 1.10	35.2 ± 0.73	16.8 ± 1.1	−52	<0.0001
MCPG (500 μM)	Normal	32.6 ± 1.25	33.0 ± 0.63	53.4 ± 1.29	+62	<0.0001
	Osteoarthritic	34.4 ± 1.5	34.8 ± 0.97	15.8 ± 0.86	−55	<0.0001

CPP: carboxypiperazinyl propylphosphonic acid (CPP) a competitive NMDA receptor antagonist; MK801, a non-competitive NMDA receptor antagonist; CNQX: 6-cyano-7-nitroquinoxaline-2,3-dione, an AMPA/Kainate receptor antagonist; MCGP: methyl-(4-carboxyphenyl)glycine (MCPG), a competitive metabotropic receptor antagonist; ns = not significant.
*Compared with reagent alone.

including post-synaptic density protein 95 (PSD-95), calcium/calmodulin kinase II (CamKII), Src kinase and neuronal nitric oxide synthetase (nNOS). NMDA receptor activity and its intracelluar interactions with the cytoskeleton and proteins of the PSD are modulated by phosphorylation of the receptor subunits. Several protein kinases have been shown to phosphorylate the NMDA receptor, including protein kinase C (PKC) and phosphotidylinositol 3-kinase (PI3K) [9,21], molecules which we have previously shown to be activated following mechanical stimulation of human articular chondrocytes [12,26]. Activation of the NMDA receptor leads to an influx of calcium into the post-synaptic neuron, resulting in an increase in NOS activity, together with an increase in kinase activity including Src kinase, and CaMKII.

The present study supports the idea that NMDA receptors are involved in normal chondrocyte mechanotransduction and would be consistent with similar suggested roles for involvement of this family of molecules in mechanical signalling of bone and neural cells [19,20]. The specific roles of NMDA receptors in chondrocyte mechanotransduction remains to be ascertained. The mechanotransduction pathway activated in normal human articular chondrocytes in the current *in vitro* system involves $\alpha 5\beta 1$ integrin, substance P and interleukin 4. Activity of each molecule is necessary for the induced anabolic response represented by an increase in relative levels of aggrecan mRNA [13,16]. A number of studies in different fields of biological research have demonstrated relationships between these molecules and NMDA/NMDA receptors that would support a role for NMDA in chondrocyte mechanotransduction. Integrin signalling is known to influence NMDA receptor activity, directing the pattern of expression of splice variants, modulating subunit composition [4], and affecting the clustering and cytoskeletal interactions of receptors in hippocampal neurons [2]. Neurokinin (substance P, neurokinin A and B) secretion has been shown to be induced by NMDA [5] and studies by other workers investigating mechanisms of neuronal damage in brain ischaemia suggest that NMDA receptor-mediated signaling can activate cytokine gene expression [11]. Thus in normal human articular chondrocyte NMDA receptor activation may be a result of mechanical stimulation induced release of glutamate, the effects of mechanical stimulation directly on the NMDA receptor [3] or as a result of an integrin-dependent kinase-linked regulation of the NMDA receptor [1].

We have previously shown the chondrocytes from osteoarthritic cartilage show a number of differences in their response to mechanical stimulation in comparison to chondrocytes from normal cartilage [13, 15]. These differences include membrane depolarisation of osteoarthritic chondrocytes in contrast to membrane hyperpolarisation of normal chondrocytes and failure of osteoarthritic chondrocytes to show the increase of relative levels of aggrecan subsequent to the mechanical stimulus seen in normal chondrocytes. The reasons for these differences are not yet clear but may be a result of pro-inflammatory cytokine activity as interleukin 1β appears to be involved in the membrane depolarisation response [23]. Interleukin 1 can influence NMDA receptor signalling through effects on glutamate uptake, by a mechanism that may involve NO [10]. It is of interest that NMDA and glycine induce membrane depolarisation in osteoarthritic chondrocytes but a membrane hyperpolarisation response in normal chondrocytes raising the intriguing possibility that in osteoarthritis chondrocyte mechanotransduction pathways are modified as a result of altered activation and function of NMDA receptors.

Acknowledgements

We would like to acknowledge the help of orthopaedic surgeons at Edinburgh Royal Infirmary in obtaining tissue. The work was funded by grants from Arthritis Research Campaign, Action Medical Research and The Chief Scientist Office of the Scottish Executive Health Department.

References

[1] D.C. Ali and M.W. Salter, NMDA receptor regulation by Src kinase signalling in excitory synaptic transmission and plasticity, *Curr. Opin. Neurobiol.* **11** (2001), 336–42.

[2] B.A. Bahr, Integrin-type signalling has a distinct influence on NMDA-induced cytoskeletal disassembly, *J. Neurosci. Res.* **59** (2000), 827–832.

[3] M. Cassado and P. Ascher, Opposite modulation of NMDA receptors by phospholipids and arachidonic acid: common features with mechanosensitivity, *J. Physiol.* **513**(2) (1998), 317–330.

[4] P. Chavis and G. Westbrook, Integrins mediate functional pre- and postsynaptic maturation at a hippocampal synapse, *Nature* (2001), 317–321.

[5] I. Colin, C. Blondeau and A. Baude, Neurokinin release in the rat nucleus of the solitary tract via NMDA and AMPA receptors, *Neuroscience* **115** (2002), 1023–33.

[6] S. Cull-Candy, S. Brickley and M. Farrant, NMDA receptor subunits: diversity, development and disease, *Curr. Op. Neuro.* **11** (2001), 327–335.

[7] S. Das, Y.F. Sasaki, T. Rothe, L.S. Premkumar, M. Takasu, J.E. Crandall, P. Dikkes, D.A. Conner, P.V. Rayudu, W. Cheung, H.S.V. Chen, S.A. Lipton and N. Nakanishi, Increased NMDA current and spine density in mice lacking the NMDA receptor subunit NR3A, *Nature* **393** (1998), 377–381.

[8] M.D. Ehlers, E.T. Fung, R.J. O'Brien and R.L. Huganir, Splice variant-specific interaction of the NMDA receptor subunit NR1 with neuronal intermediate filaments, *J. Neuroscience* **18** (1998), 720–730.

[9] C. Hisatsune, H. Umemori, M. Mishina and T. Yamamoto, Phosphorylation-dependent interaction of the N-methyl-D-aspartate receptor $\varepsilon 2$ subunit with phosphatidylinositol 3-kinase, *Genes to Cells* **4** (1999), 657–666.

[10] S. Hu, W.S. Sheng, L.C. Ehrlich, P.K. Peterson and C.C. Chao, Cytokine effects on glutamate uptake by human astrocytes, *Neuroimmunomodulation* **7** (2000), 153–159.

[11] S. Jander, M. Schroeter and G. Stoll, Role of NMDA receptor signaling in the regulation of inflammatory gene expression after focal brain ischemia, *J. Neuroimmunol.* **109** (2000), 181–7.

[12] H.S. Lee, S.J. Millward-Sadler, M.O. Wright, G. Nuki, R. Al-Jamal and D.M. Salter, Activation of Integrin-RACK1/PKCalpha signalling in human articular chondrocyte mechanotransduction, *Osteoarthritis Cartilage* **10** (2002), 890–7.

[13] S.J. Millward-Sadler, M.O. Wright, L.W. Davies, G. Nuki and D.M. Salter, Mechanotransduction via integrins and IL4 results in altered aggrecan and MMP3 gene expression in normal but not osteoarthritic human articular chondrocytes, *Arthritis and Rheumatism* **43**, 2091–9.

[14] S.J. Millward-Sadler, M.O. Wright, H.-S. Lee, K. Nishida, H. Caldwell, G. Nuki and D.M. Salter, Integrin-regulated secretion of interleukin-4: a novel pathway of mechanotransduction in human articular chondrocytes, *J. Cell Biol.* **145** (1999), 183–9.

[15] S.J. Millward-Sadler, M.O. Wright, H.-S. Lee, H. Caldwell, G. Nuki and D.M. Salter, Altered electrophysiological responses to mechanical stimulation and abnormal signalling through $\alpha 5\beta 1$ integrin in chondrocytes from osteoarthritic cartilage, *Osteoarthritis and Cartilage* **8** (2000), 272–8.

[16] S.J. Millward-Sadler, A. Mackenzie, H. Monaghan, M.O. Wright, C.E. Fiskerstrand, D.M. Salter and J.P. Quinn, Tachykinin expression in cartilage and function in human articular chondrocyte mechanotransduction, *Arthritis Rheum.* **48** (2003), 146–56.

[17] H. Monyer, R. Sprengel, R. Schoepfer, A. Herb, M. Higuchi, H. Lomeli, N. Burnashev, B. Sakmann and P.H. Seeburg, Heteromeric NMDA receptors: molecular and functional distinction of subtypes, *Science* **256** (1992), 1217–21.

[18] S. Okabe, A. Miwa and H. Okado, Alternative splicing of the C-terminal domain regulates cell surface expression of the NMDA receptor NR1 subunit, *J. Neuroscience* **19** (1999), 7781–7792.

[19] P. Paoletti and P. Ascher, Mechanosensitivity of NMDA receptors in cultured mouse central neurons, *Neuron* **13** (1994), 645–655.

[20] A.J. Patton, P.G. Genever, M.A. Birch, L.J. Suva and T.M. Skerry, Expression of an N-methyl-D-aspartate-type receptor by human and rat osteoblasts and osteoclasts suggests a novel glutamate signaling pathway in bone, *Bone* **22**(6) (1998), 645–649.

[21] K.W. Roche, W.G. Tingley and R.L. Huganir, Glutamate receptor phosphorylation and synaptic plasticity, *Current Opinion in Neurobiology* **4** (1994), 383–388.

[22] D.M. Salter, J.E. Robb and M.O. Wright, Electrophysiological responses of human bone cells to mechanical stimulation: evidence for specific integrin function in mechanotransduction, *J. Bone Miner. Res.* **12** (1997), 1133–1141.

[23] D.M. Salter, S.J. Millward-Sadler, G. Nuki and M.O. Wright, Differential responses of chondrocytes from normal and osteoarthritic human articular cartilage to mechanical stimulation, *Biorheology* **39** (2002), 97–108.

[24] T.M. Skerry and P.G. Genever, Glutamate signalling in non-neuronal tissues, *TIPS* **22** (2001), 174–181.

[25] H. Sugihara, K. Moriyoshi, T. Ishii, M. Masu and S. Nakanishi, Structures and properties of seven isoforms of the NMDA receptor generated by alternative splicing, *Biochem. Biophys. Res. Comm.* **185** (1992), 826–832.

[26] M.O. Wright, J.L. Godolphin, E. Dunne, C. Bavington, P. Jopanbutra, G. Nuki and D.M. Salter, Hyperpolarisation of cultured human chondrocytes following cyclical pressurisation involves $\alpha 5\beta 1$ integrin and integrin-associated intracellular pathways, *J. Orth. Res.* **15** (1997), 742–747.

Biorheology 41 (2004) 283–298
IOS Press

Passage in monolayer influences the response of chondrocytes to dynamic compression

Mike Wiseman [a], Dan L. Bader [b], Tom Reisler [a] and David A. Lee [b,*]
[a] *Institute of Orthopaedics, University College London Medical School, Brockley Hill, Stanmore, Middlesex HA7 4LP, UK*
[b] *IRC in Biomedical Materials and Medical Engineering Division, Department of Engineering, Queen Mary, University of London, Mile End Road, London E1 4NS, UK*

Abstract. This study tests the hypothesis that expansion by passage in monolayer influences the response of isolated articular chondrocytes to dynamic compression. Chondrocytes, isolated from bovine articular cartilage, were seeded in monolayer and passaged 4 times (P1–4). For assessment of chondrocytic and fibroblastic phenotype, freshly isolated and passaged cells were seeded on glass coverslips or in 2% alginate beads and cultured for 7 days in DMEM + 10% FCS. Samples were assayed for DNA and GAG content and stained for collagen types I and II. In separate experiments, freshly isolated or passaged chondrocytes were seeded at 10×10^6 cells.ml^{-1} in 4% cylindrical agarose constructs and subjected to 15% dynamic compressive strain at 1 Hz for 24 hours. [^{3}H]-thymidine incorporation, SO_4 incorporation and nitrite release were analysed. Immediately following isolation (P0), chondrocytes seeded in alginate expressed high levels of type II collagen, but did not stain for type I collagen. Following repeat passage the cells expressed enhanced levels of type I collagen, with an associated reduction in type II collagen staining. These data indicate a modulation to a fibroblastic phenotype during monolayer expansion which was not rapidly reversed by culture in a 3D hydrogel. Dynamic compression down-regulated SO_4 incorporation at P0, but did not affect [^{3}H]-thymidine incorporation. By contrast the incorporation of both SO_4 and [^{3}H]-thymidine was enhanced by dynamic compression at both P1 and to a lesser extent P2. SO_4 and [^{3}H]-thymidine incorporation were inhibited at P3 and P4. Nitrite release was down-regulated by dynamic compression at all passages. These data demonstrate a clear modulation in the response of bovine articular chondrocytes to dynamic compression following passage in monolayer.

Keywords: Chondrocyte, dynamic compression, tissue engineering, cell phenotype, cell expansion, monolayer culture

1. Introduction

There is growing interest in the development of tissue engineered cartilage repair systems which involve the implantation of autologous chondrocytes into defect sites for the treatment of localised lesions in the young adult patient group [2,9,17,25,39,44]. Mechanical compression is known to modulate the metabolic activity and the biomechanical properties of articular cartilage. *In vivo* studies have shown that joint immobilisation or disuse can lead to articular cartilage atrophy, characterised by significant changes in the proteoglycanaceous matrix and loss of mechanical function [27,41]. *In vitro* studies, utilising cartilage explants or isolated cells maintained in 3D constructs typically demonstrate that static strain inhibits cellular activity but dynamic strain may enhance activity in a manner dependent on the

*Address for correspondence: Dr David A. Lee, Department of Engineering, Queen Mary, University of London, Mile End Road, London E1 4NS, UK. Tel.: +44 20 7882 5433; Fax: +44 20 8983 5032; E-mail: D.A.Lee@qmul.ac.uk.

dynamic frequency [11,18,26,28,35,42]. Accordingly, it has been proposed that mechanical conditioning, applied using bioreactor systems or by defined exercise regimes, may enhance the efficacy of tissue engineered repair systems [1].

The cells used for tissue engineered cartilage repair systems are typically isolated from a small tissue biopsy, removed from a low load-bearing site at the periphery of the cartilaginous joint surface [9]. Using a small biopsy minimises potentially detrimental effects at the donor site, but limits the number of available cells. The isolated cells must, therefore, be cultured *in vitro* in order to increase cell number to a level sufficient for implantation [9]. Monolayer culture techniques are routinely used to expand the chondrocytes prior to implantation. While these methods are extremely effective at inducing cell proliferation, chondrocytes are known to dedifferentiate and adopt a fibroblastic phenotype when cultured in monolayer and this process is only slowly reversible [3,4,14,36]. The dedifferentiation process is believed to be mediated by the formation of actin stress fibres which occurs when the cells spread on an adhesion-permitting substrate [4,36]. Culture systems which maintain the chondrocyte in a rounded morphology, such as agarose and alginate prevent the formation of stress fibres and therefore induce retention of the chondrocytic phenotype or enhance the re-expression of chondrocytic phenotype by monolayer expanded chondrocytes [5,7,8,10,15,21–23,33,38,43]. More recent studies have demonstrated that the provision of specified differentiation factors, such as FGF and TGF-β, either during expansion or subsequent culture in 3D, can influence the re-expression of the chondrogenic phenotype [14,24,32,46].

The phenotypic modulation of chondrocytes associated with passage in monolayer may influence specific mechanotransduction processes and thereby alter the response of the cells to the application of mechanical conditioning regimes. The majority of studies investigating the influence of loading on chondrocytes within 3D constructs have used freshly isolated cells. A few studies have used passaged cells but the presence of compounding variables such as tissue age, scaffold material and load regime renders definitive conclusions regarding the effect of passage on mechanotransduction processes difficult to sustain. Accordingly this study uses a systematic approach to test the hypothesis that expansion by passage in monolayer influences the response of isolated articular chondrocytes to dynamic compression.

2. Materials and methods

2.1. Chondrocyte isolation

Full depth slices of cartilage were removed from the metacarpalphalangeal joints of 18 month-old steers. The cartilage slices were diced finely, and incubated at 37°C for 1 hr in Dulbecco's minimal essential medium supplemented with 10% (v/v) foetal calf serum, 2 mM L-glutamine, 20 mM HEPES, 100 unit.ml^{-1} penicillin, 100 μg.ml^{-1} streptomycin and 150 μg.ml^{-1} L-ascorbic acid (DMEM + 10% FCS, all Gibco, Paisley, UK) + 700 unit.ml^{-1} pronase (BDH Ltd., Poole, UK) and for 16 hr at 37°C in DMEM + 10% FCS + 100 unit.ml^{-1} collagenase type XI (Sigma, Poole, UK). The supernatant containing released chondrocytes was passed through a 70 μm pore size sieve (Falcon, Oxford, UK), washed twice in DMEM + 10% FCS and finally resuspended in DMEM + 10% FCS.

2.2. Preparation and passage in monolayer culture

For monolayer cultures, the cell suspension was adjusted to 8.5×10^4 cells.ml^{-1} and 35 ml was added to tissue culture flasks (175 cm^2 growth area, Falcon, Oxford, UK). On initial seeding each flask contained approximately 3×10^6 cells at an initial seeding density of 1×10^4 cells.cm^{-2}. The cells were maintained at 37°C/5% CO_2 in DMEM + 10% FCS and the culture media were changed every two days. After 7 days in culture, the chondrocytes were rinsed twice in phosphate buffered saline pH 7.4 (PBS, Sigma, Poole, UK) and recovered by incubation for 15 min in 0.25% (w/v) trypsin (Sigma, Poole, UK) in PBS. A single cell suspensions was produced by incubation in DMEM + 10% FCS + 700 unit.ml^{-1} pronase for 30 minutes. A proportion of the recovered cells were re-seeded into flasks as above for further expansion. At each passage, the number of cells recovered and cell viability was determined using a haemacytometer and the trypan blue exclusion test. Expansion of cells was calculated using the following equation:

$$\text{Cell expansion}_{(p_n - p_{n+1})} = \frac{\text{Viable cell number recovered}_{(p_{n+1})}}{\text{Cell number seeded}_{(p_n)}}, \tag{1}$$

where p_n = passage number.

A total of four passages, at 7 day intervals, were performed. The initial 7 day culture period was termed passage 0 and subsequent 7 day culture periods were termed passages 1–4, as indicated schematically in Fig. 1.

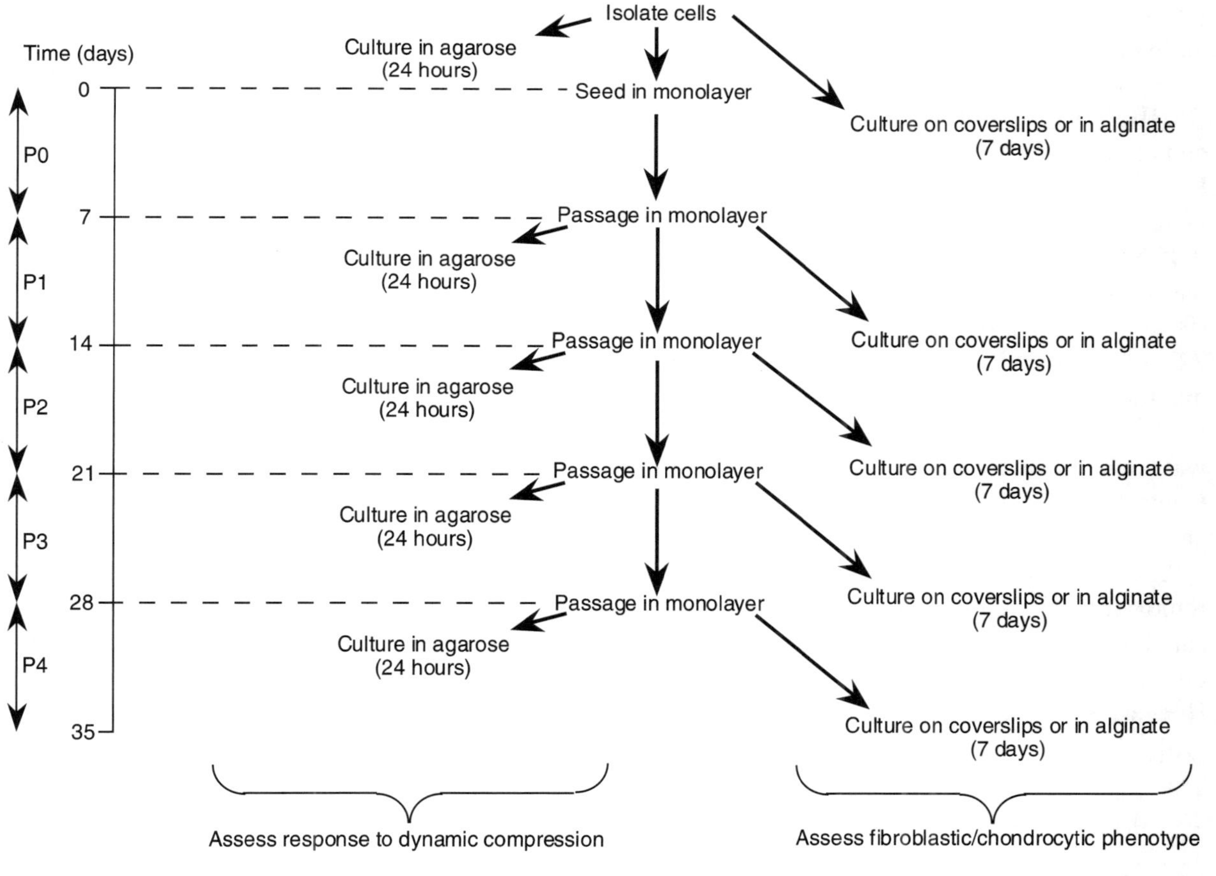

Fig. 1. Schematic representation of culture methodologies used to examine the chondrocytic and fibroblastic phenotypes and the response of chondrocytes to dynamic compression following expansion by repeat passage in monolayer.

2.3. Assessment of chondrocytic and fibroblastic phenotypic potential

For assessment of chondrocytic potential, freshly isolated cells and cells recovered after passage in monolayer were seeded in alginate beads. The cell suspension was adjusted to 20×10^6 cells.ml^{-1} and added to an equal volume of 4% (w/v) alginate (Keltone LV, Kelco Nutrasweet, Poole, UK) in Earle's Balanced Salt Solution (EBSS, Gibco, Paisley, UK) to give a final concentration of 10×10^6 cells.ml^{-1} in 2% (w/v) alginate. The cell/alginate suspension was added drop-wise through a 23-gauge needle into a solution containing 100 mM $CaCl_2$ in DMEM + 10% FCS and were incubated for 10 min at room temperature to induce crosslinking of the alginate gel. The resultant alginate beads were cultured for 7 days in DMEM + 10% FCS. For assessment of fibroblastic potential freshly isolated cells and cells recovered after passage in monolayer were seeded on glass coverslips (13 mm diameter, Agar Scientific, Stansted, UK) at 2×10^4 cells.cm^{-2} and cultured for 7 days in DMEM + 10% FCS. Culture media were replaced every two days.

At the end of the culture period representative alginate bead and monolayer samples were fixed by incubation in 1 ml of 4% paraformaldehyde in 0.1 M sodium cacodylate buffer + 10 mM $CaCl_2$, pH 7.4 for 4 hr at room temperature. All samples were washed exhaustively in 0.1 M sodium cacodylate buffer + 10 mM $CaCl_2$, pH 7.4 and stored in 0.1 M sodium cacodylate buffer + 50 mM $BaCl_2$, pH 7.4 at 4°C. The alginate beads were dehydrated and embedded in paraffin wax. Sections, approximately 9 μm in thickness were cut from the specimens and rehydrated prior to immunolocalisation. The streptavidin biotin–immunoperoxidase method [45] was used for immunolocalisation using primary antibodies raised against collagen type I (COL-1, Sigma, Poole, UK, used at 1 : 100) and collagen type II (CIICI, DSHB, Iowa, USA, used at 1 : 2). Sections and coverslips were incubated for 15 minutes in a solution containing hydrogen peroxide and methanol in a 50 : 1 ratio to block endogenous peroxidase activity. Enzymatic pre-treatment with 0.1% (w/v) trypsin for 10 minutes at 37°C was used to unmask the antigenic sites on the collagen molecules. The samples were incubated with primary antibody, diluted in 0.05 M Tris, 0.15 M NaCl, pH 7.6, for 30 minutes at room temperature, washed in 0.05 M Tris, 0.15 M NaCl, pH 7.6 and incubated at room temperature for 30 minutes with biotinylated rabbit anti-mouse Ig (Dako, Ely, UK), diluted 1 : 200 in TBS. The samples were washed in 0.05 M Tris, 0.15 M NaCl, pH 7.6 and incubated with streptavidin–biotin complex (Dako, Ely, UK) for 30 minutes, followed by 3′,3′-diaminobenzidine tetrahydrochloride liquid substrate solution (Sigma, Poole, UK). The samples were washed in tap water and counterstained with Harris's haematoxylin, dehydrated and mounted. Staining was ranked using a scale from −, indicating minimal staining to + + + indicating intense staining. Control samples were incubated in 0.05 M Tris, 0.15 M NaCl, pH 7.6, without addition of the primary antibodies.

At time 0 and at the end of the culture period, further alginate bead and glass coverslip samples were removed from culture and solubilised by incubation in 50 mM sodium citrate, 150 mM NaCl, pH 7.4 supplemented with 10 mM cysteine HCl and 10 mM EDTA and 5 μl.ml^{-1} papain suspension (391 unit.ml^{-1}, Sigma, Poole, UK) for 24 hr at 60°C. The digests were assayed for total DNA, using the Hoechst 33258 method [40] and for glycosaminoglycan using the method of Enobakhare and others [16].

2.4. Application of dynamic compression

Freshly isolated cells and cells recovered after passage in monolayer were resuspended in DMEM + 10% FCS to give a cell concentration of 20×10^6 cells.ml^{-1}. The chondrocyte suspension was added to an equal volume of molten 8% (w/v) agarose type IXa (Sigma, Poole, UK) in EBSS (Sigma, Poole, UK) to yield a final cell concentration of 10×10^6 cells.ml^{-1} in 4% (w/v) agarose. The chondrocyte/agarose suspension was transferred into a sterile stainless steel mould, containing holes measuring 5 mm in diameter and 5 mm in height. The chondrocyte/agarose suspension was allowed to gel at 4°C for 20 minutes to yield cylindrical constructs.

A well characterised cell-straining system (Zwick Testing Machines Ltd, Leominster, UK) was used to apply dynamic compression to chondrocyte/agarose constructs, as described previously [13,28–31]. Representative constructs were transferred into a 24-well culture plate (Costar, High Wycombe, UK) and mounted within the apparatus. One ml of DMEM + 10% FCS + 1 μCi.ml^{-1} [^{3}H]-thymidine and 10 μCi.ml^{-1} $^{35}SO_4$ was introduced into each well. One group of constructs were subjected to 15% amplitude dynamic strain (0–15%) applied in a sinusoidal waveform at 1 Hz for 24 hours. A second group of constructs remained unstrained, but were maintained within the incubator of the cell-straining system.

At the end of the culture period, the medium was removed and stored at −20°C for subsequent analysis. The constructs were incubated at 70°C in 1 ml PBS (Gibco, Paisley, UK) supplemented with 10 mM EDTA and 10 mM cysteine HCl until molten. The suspension was cooled to 40°C, 5 μl papain suspension (Sigma, Poole, UK containing 3 IU papain) and 10 IU agarase (Sigma, Poole, UK) were added and the solution was incubated for 24 hr at 40°C.

Incorporation of $^{35}SO_4$ into newly synthesized proteoglycans was determined in both medium and agarase/papain digests using the alcian blue precipitation method [34]. Aliquots (75 μl) of 50 mM sodium acetate, 0.5% (v/v) triton X-100, pH 5.8, 50 μl of alcian blue solution consisting of 0.2% (w/v) alcian blue 8GX (BDH, Poole, UK) in 50 mM sodium acetate + 85 mM magnesium chloride and 25 μl aliquots of culture medium or agarase/papain digest were added to individual wells of a multiscreen plate (0.45 μm pore filter, Millipore, Watford, UK). The plate was agitated for 1 hr at room temperature, vacuum aspirated and washed three times with 50 mM sodium acetate + 100 mM sodium sulphate + 50 mM magnesium chloride, pH 5.8. The filters were punched out into scintillation vials, agitated for 1 hr in 0.5 ml of 4 M guanidine HCl in 4.3 M propan-2-ol and counted in 4 ml scintillation fluid (Emulsifier Safe, Packard, Pangbourne, UK) using a Tricarb 4000 series counter (Packard, Pangbourne, UK). Ten μl aliquots of culture medium were added to scintillation vials, mixed with 4 ml scintillation fluid and counted to determine total $^{35}SO_4$.

Incorporation of [^{3}H]-thymidine was quantified using a trichloroacetic acid (TCA) precipitation technique. One hundred μl of papain/agarase digest was added to individual wells of a multiscreen plate (0.45 μm pore filter, Millipore, Watford, UK). One hundred μl of 20% (w/v) TCA was added to each well and incubated at 4°C for 30 min. The multiscreen plate was vacuum aspirated and 100 μl of 10% (w/v) TCA was added to each well. The plate was vacuum aspirated, dried and the filters were punched out into scintillation vials. The filters were agitated for 30 min in 0.5 μl 0.05 M KOH to release bound tritiated thymidine into solution and counted in 4 ml scintillation fluid (Emulsifier plus, Packard, Pangbourne, UK) using the Tricarb 4000 series counter.

Total nitrite was measured in the culture medium using the Griess assay [19,20]. A volume of 100 μl of sample was added in a 1 : 1 ratio to a solution containing 0.5% (w/v) sulphanilamide (Sigma, Poole, UK) and 0.1% naphthylethylediamide (Sigma, Poole, UK) in 5% (v/v) phosphoric acid (BDH, Poole, UK). Absorbance at 550 nm was measured immediately on a microplate reader. Absolute concentrations of nitrite in culture medium and agarase/papain digests were determined by comparison with standard solutions of sodium nitrite (Sigma, Poole, UK) in freshly prepared culture medium.

DNA concentrations in the constructs were determined using the bisbenzamide Hoechst 33258 method [40]. DNA analysis was used to normalise values of [^{3}H]-thymidine incorporation, $^{35}SO_4$ incorporation and nitrite release were expressed per μg DNA.

2.5. *Statistical analysis*

Where appropriate, two tailed Student's unpaired t-tests were used to compare results.

3. Results

3.1. *Passage in monolayer culture*

On initial seeding in monolayer (P0), the chondrocytes rapidly attached to the growth surface but typically retained a rounded morphology for approximately 3 days. Between days 3 and 7 the cells began to spread on the growth surface. After the first passage in monolayer, however, the cells attached and spread more rapidly, such that after 2 days in culture the majority of cells had adopted a fibroblastic morphology with a small proportion retaining a rounded morphology. The cells proliferated rapidly and had formed a confluent monolayer by day 7. A similar pattern was observed following subsequent passages in monolayer. Cell expansion, assessed as an increase in cell number, was slightly lower during P0 compared to later passages, suggesting an initial lag phase (Fig. 2). Thereafter the increase in cell number was exponential, indicated by the linearity of the log/linear plot presented in Fig. 2 ($r^2 = 0.997$).

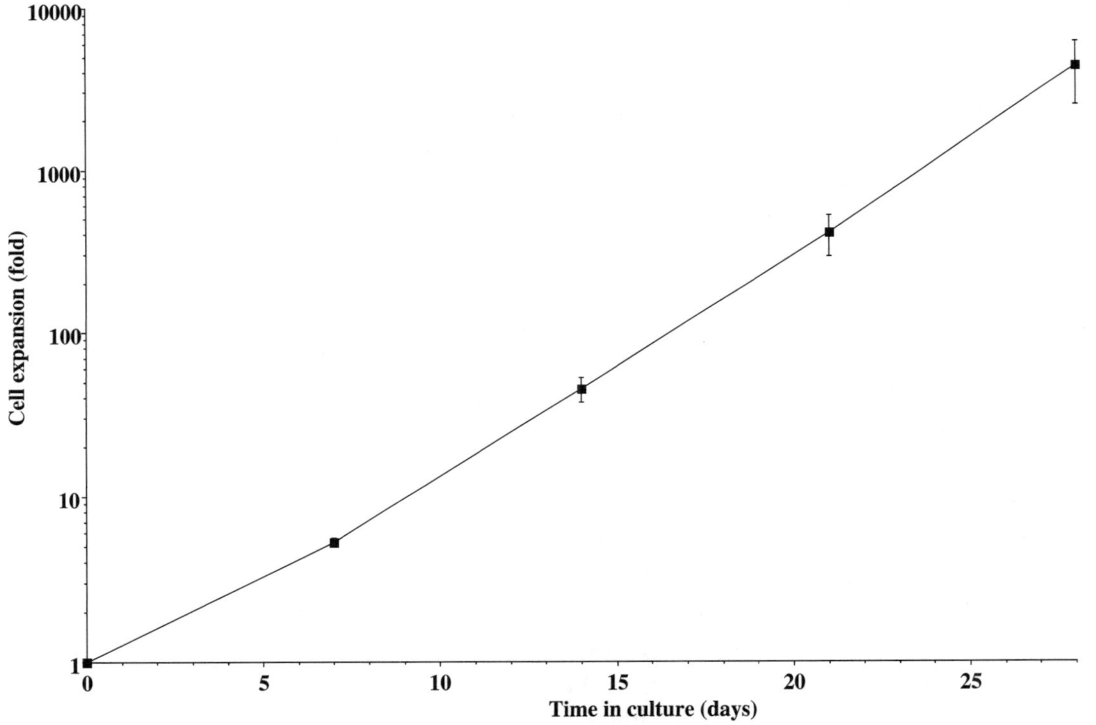

Fig. 2. Cumulative cell expansion for chondrocytes cultured in monolayer for up to 28 days and passaged at 7 day intervals. Each point represents the mean ± standard deviation of 6 measurements from two separate experiments.

These data suggest a doubling time of approximately 2.5 days during the exponential growth phase. The viability of cells recovered following in monolayer ranged from 92.0% to 96.7%.

3.2. Assessment of chondrocytic and fibroblastic phenotypic potential

In order to assess fibroblastic phenotype freshly isolated and passaged chondrocytes were cultured on glass coverslips for 7 days. Collagen type I and II immunolocalisation was performed. Representative images for P0 and P4 cells are presented in Fig. 3 and a summary of the intensity of straining is provided in Table 1. Staining for both collagens I and II staining was detected for P0 cells cultured on coverslips for 7 days (Fig. 3A,B). Intense staining for type I collagen was also evident for P1–4 cells following culture on coverslips (Fig. 3C and Table 1). By contrast, the intensity of staining for type II collagen was markedly reduced for P1 cells and absent at P2–4 (Fig. 3D and Table 1).

Culture within alginate beads was used to determine the ability of cells passaged in monolayer to re-express a chondrocytic phenotype. Representative images for collagen type I and II immunolocalisation performed at P0 and P4 are presented in Fig. 4 and a summary of the intensity of straining is provided in Table 2. Morphological assessment of immunolocalisation micrographs indicated that alginate beads prepared using freshly isolated cells contained large clusters of cells, characterised by intense pericellular staining for collagen type II (Fig. 4B). By contrast collagen type I staining was absent (Fig. 4A). With increasing passage number the intensity of staining for type II collagen was reduced and type I collagen

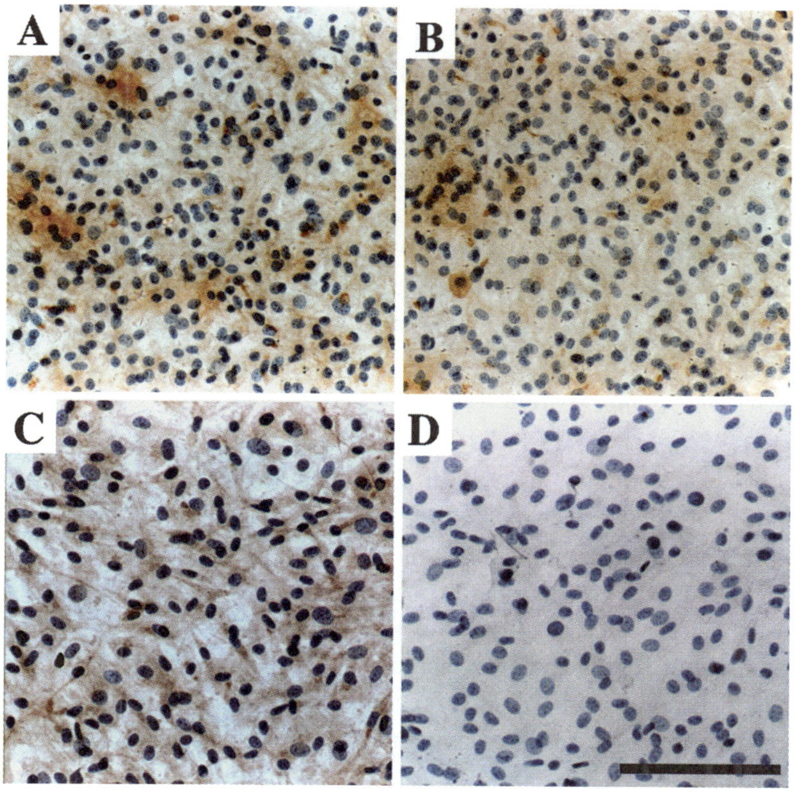

Fig. 3. Micrographs representing immunolocalisation for type I collagen (A, B) or type II collagen (C, D) for freshly isolated chondrocytes (A, C) or cells passaged four times in monolayer (B, D). The cells were cultured on glass coverslips for 7 days prior to immunolocalisation. All micrographs are at the same magnification, scale bar = 100 μm.

Table 1

Intensity of immunostaining for monolayer cultures prepared from freshly isolated chondrocytes (P0) and chondrocytes passaged between one and four times in monolayer and subsequently cultured on glass coverslips for 7 days

	P0	P1	P2	P3	P4
Collagen I	++	++	+++	+++	+++
Collagen II	+++	+	–	–	–

Scale ranges from –, indicating minimal staining to +++ indicating intense staining.

Table 2

Intensity of immunostaining within alginate bead cultures prepared from freshly isolated chondrocytes (P0) and chondrocytes passaged between one and four times in monolayer and subsequently cultured in alginate beads for 7 days

	P0	P1	P2	P3	P4
Collagen I	–	+	+	++	++
Collagen II	+++	++	–	–	–

Scale ranges from –, indicating minimal staining to +++ indicating intense staining.

staining increased (Table 2). Accordingly at P3 and P4 collagen II staining was not observed (Fig. 4C,D). Moreover, the size of cell clusters formed within alginate beads prepared using P4 cells was markedly smaller compared to those observed at P0.

DNA values, used to indicate an increase in cell number during the 7 day culture period in alginate, are presented in Fig. 5A for freshly isolated cells and cells passaged between 1 and 4 times in monolayer. DNA values were significantly greater after 7 days of culture in alginate for P1 cells compared to freshly isolated cells (P0). Thereafter a clear inverse relationship between passage number and DNA content was observed for cells from P1–P4. The glycosaminoglycan content of alginate beads prepared from freshly isolated cells and cells passaged in monolayer is presented in Fig. 5B. An increase in glycosaminoglycan content was observed within beads prepared with P1 cells compared to freshly isolated cells. Thereafter, however, there was a clear inverse relationship between passage number and glycosaminoglycan content within beads containing cells passaged in monolayer between 1 and 4 times.

3.3. *Application of dynamic compression*

Absolute values for [^{3}H]-thymidine, incorporation, SO_4 incorporation and nitrite release are presented in Table 3, for chondrocytes passaged between one and four times in monolayer and subsequently cultured in agarose constructs which remained unstrained within the cell-straining system. Levels of both [^{3}H]-thymidine, and SO_4 incorporation increased significantly from P0 to P1 and declined thereafter. By contrast nitrite release values were lowest for freshly isolated cells and highest for P4 cells (Table 3). Dynamic compression significantly down-regulated SO_4 incorporation at P0 ($p < 0.05$), but did not affect [^{3}H]-thymidine incorporation. By contrast the incorporation of both SO_4 and [^{3}H]-thymidine was significantly enhanced by the application of dynamic compression at P1 ($p < 0.05$ for both SO_4 and [^{3}H]-thymidine incorporation). However both SO_4 and [^{3}H]-thymidine incorporation were inhibited by the application of dynamic compression for cells which had been passaged 3 or 4 times in monolayer prior to compression. Nitrite release was down-regulated by dynamic compression at all passages with the differences being significant at P0, P1 and P4.

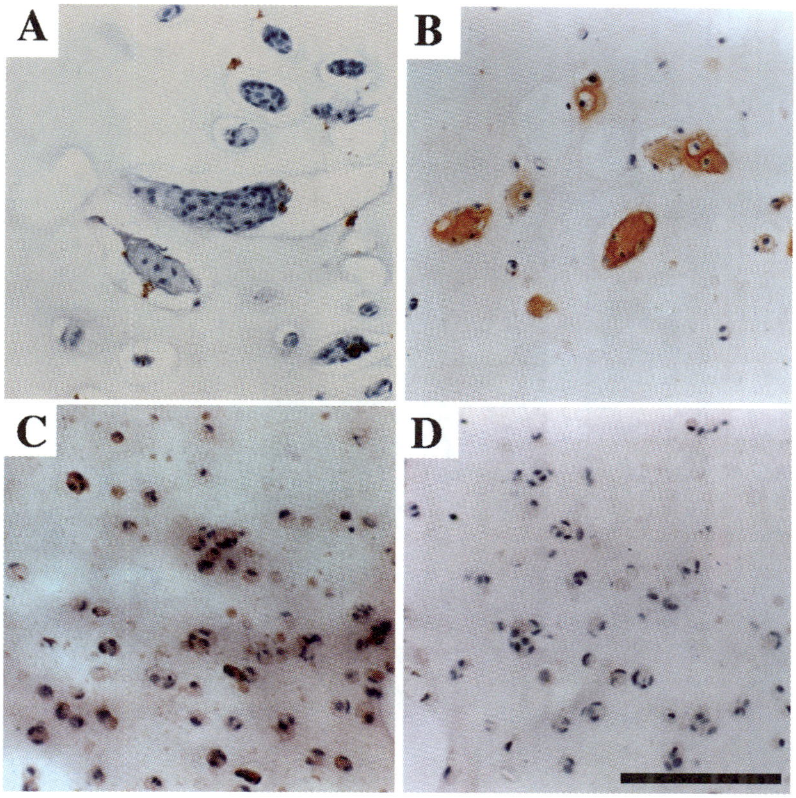

Fig. 4. Micrographs representing immunolocalisation for type I collagen (A, B) or type II collagen (C, D) for freshly isolated chondrocytes (A, C) or cells passaged four times in monolayer (B, D). The cells were cultured in alginate beads for 7 days prior to immunolocalisation. All micrographs are at the same magnification, scale bar = 100 μm.

Table 3

Absolute values for [^{3}H]-thymidine, incorporation, SO_4 incorporation and nitrite release for freshly isolated chondrocytes (P0) and chondrocytes passaged between one and four times in monolayer. The cells were cultured, unstrained in agarose constructs for 24 hours in DMEM + 10% FCS. Values are presented as the mean ± standard error of at least 12 replicates

	P0	P1	P2	P3	P4
[^{3}H]-thymidine (cpm.μg^{-1} DNA)	296.0 ± 11.9	745.0 ± 18.7	469.4 ± 11.9	163.3 ± 5.1	96.9 ± 2.7
SO_4 incorporation (μmol.μg^{-1} DNA.hr^{-1})	0.0411 ± 0.0012	0.1152 ± 0.0043	0.0505 ± 0.0023	0.0270 ± 0.0004	0.0035 ± 0.0001
Nitrite release (μM.μg^{-1} DNA)	1.047 ± 0.096	2.433 ± 0.332	1.115 ± 0.086	3.550 ± 0.182	7.199 ± 0.299

4. Discussion

The aim of this study was to test the hypothesis that expansion by passage in monolayer influences the response of isolated articular chondrocytes to dynamic compression. Accordingly bovine chondrocytes were passaged up to four times in monolayer culture prior to analysis of cell phenotype and response to mechanical loading. Cell expansion data revealed that the increase in cell number was lowest during the initial 7 days in monolayer culture. During further 7 day passage periods the increase in cell number

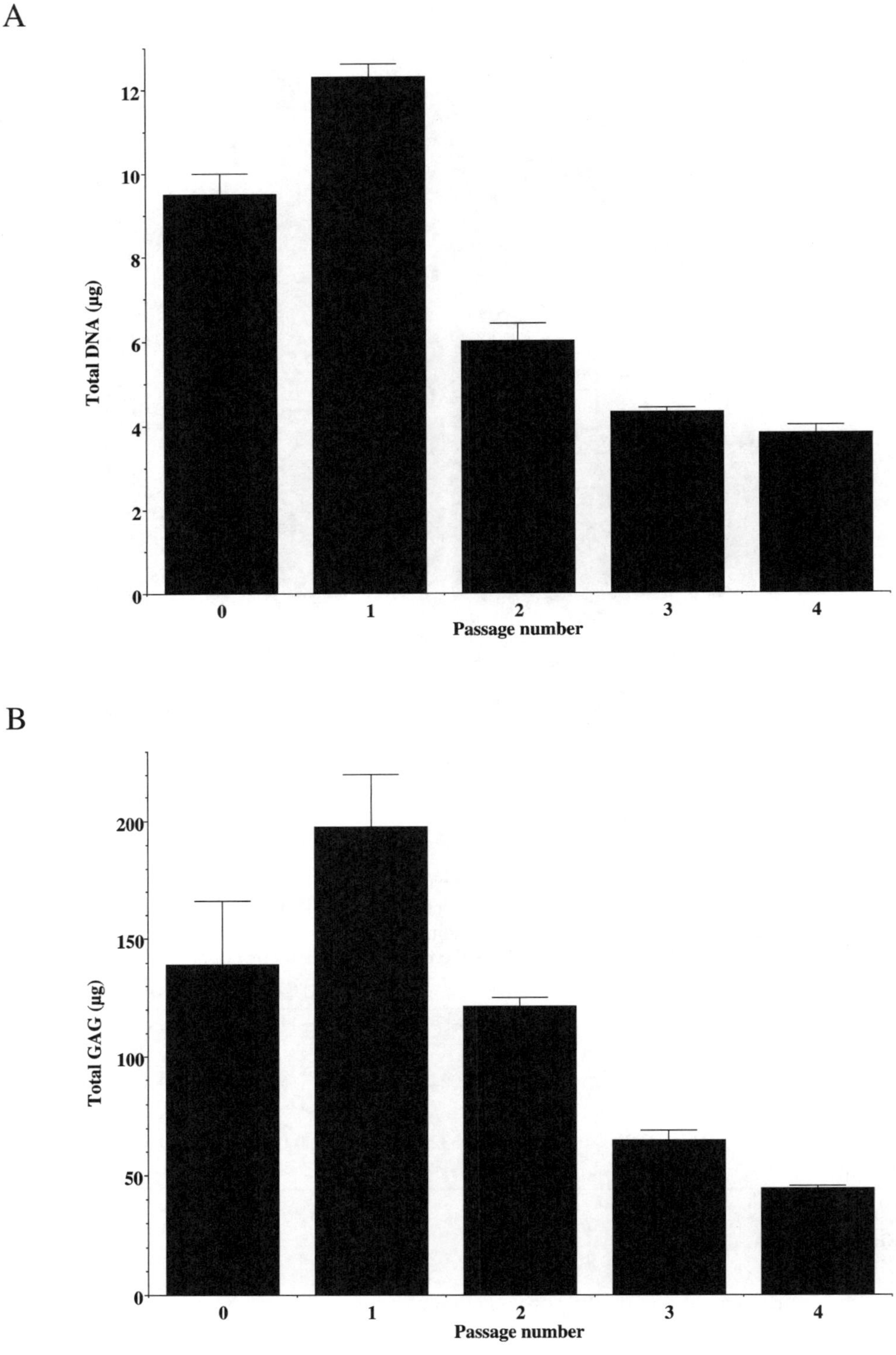

Fig. 5. Total DNA (A) and glycosaminoglycan (B) content of alginate bead cultures prepared from chondrocytes passaged between one and four times in monolayer and subsequently cultured in alginate beads for 7 days. Each value represents the mean and standard deviation of 3 replicates.

remained relatively constant. After initial seeding in monolayer, freshly isolated chondrocytes undergo a process of attachment and spreading, associated with actin stress fibre formation and modulation to a fibroblastic phenotype [4,5,38]. While cell attachment typically occurs rapidly, full spreading, may take up to 5 days, thereby precluding rapid proliferation over the initial 7 day culture period. Once chondrocytes have adapted to a fibroblastic morphology in monolayer, the rate of cell spreading post-passage is more rapid, leading to the enhanced proliferation rate observed at P1–3 (Fig. 2).

Freshly isolated cells expressed the chondrocytic marker collagen type II when cultured in monolayer or alginate. Freshly isolated cells cultured in alginate beads did not express type I collagen indicating a retention of the chondrocytic phenotype. By contrast cells cultured in monolayer expressed both collagens types I and II. Type I collagen is a marker for fibrous or fibrocartilaginous tissue but is not present in bovine articular cartilage. These findings suggest that a process of phenotypic modulation occurs during the initial 7 days of monolayer culture but that this process is not complete at this stage. Immunolocalisation further revealed a complete absence of collagen type II staining and intense staining for collagen type I for cells cultured on coverslips after repeat passage in monolayer (Table 1). Both collagen types I and II were expressed within alginate beads prepared from P1 cells, but cells passaged in monolayer more than once failed to express the hyaline cartilage-specific marker when cultured in alginate beads, suggesting that chondrocytic phenotypic, once lost, is not rapidly re-expressed. These data are in agreement with previous studies which have reported limited re-expression of chondrogenic markers and maintenance of fibroblastic markers during the first few days of culture in alginate following monolayer expansion [7,14].

Biochemical analysis of alginate beads or agarose constructs prepared from freshly isolated or passaged chondrocytes revealed an up-regulation of both cell proliferation and glycosaminoglycan synthesis for P1 cells when compared to freshly isolated cells. The magnitude of up-regulation was greater when measured over the initial 24 hour period in gel culture, using radioisotope incorporation (Table 3), compared to values obtained from total biochemical analysis at day 7 (Fig. 5). Previous studies have suggested that metabolic activity tends to be relatively low directly following isolation, but increases rapidly after 24–48 hours in culture [12]. The modulation in metabolic activity post-isolation is believed to be related to recovery following enzyme treatment, potentially associated with the re-expression of key cell surface markers. Alternatively this process may represent an adaptive response to culture conditions. Biochemical analysis further indicated that P2–4 cells exhibited lower rates of proliferation and glycosaminoglycan synthesis when cultured in alginate beads compared to P1 or freshly isolated expanded cells (Fig. 3A). Examination of micrographs confirmed these findings. Large cell clusters, surrounded by extensive pericellular matrix, were observed in beads prepared from freshly isolated cells compared beads prepared from P4 cells which were characterised by the presence of single cells, couplets or small groups (Fig. 4). Accordingly, modulation to a three-dimensional culture system appears to impede both cell cycle activity and the ability to synthesise or accumulate glycosaminoglycan-rich matrix by monolayer-expanded cells, a phenomenon which may be associated with adaptation to a fibroblastic phenotype.

The data presented in this study suggest that dynamic compression does not significantly enhance either cell proliferation or glycosaminoglycan synthesis by freshly isolated chondrocytes, in contrast with previous studies using the chondrocyte/agarose system [13,28–31]. It should be noted, however, that compression was applied to the agarose constructs for 24 hours directly after formation in the current study, in order to limit phenotypic modulation. This protocol contrasts with previous studies where an initial 24 hour unstrained culture period was included prior to the application of compression. As discussed above, the re-expression of key cell surface markers, such as integrins, is believed to occur

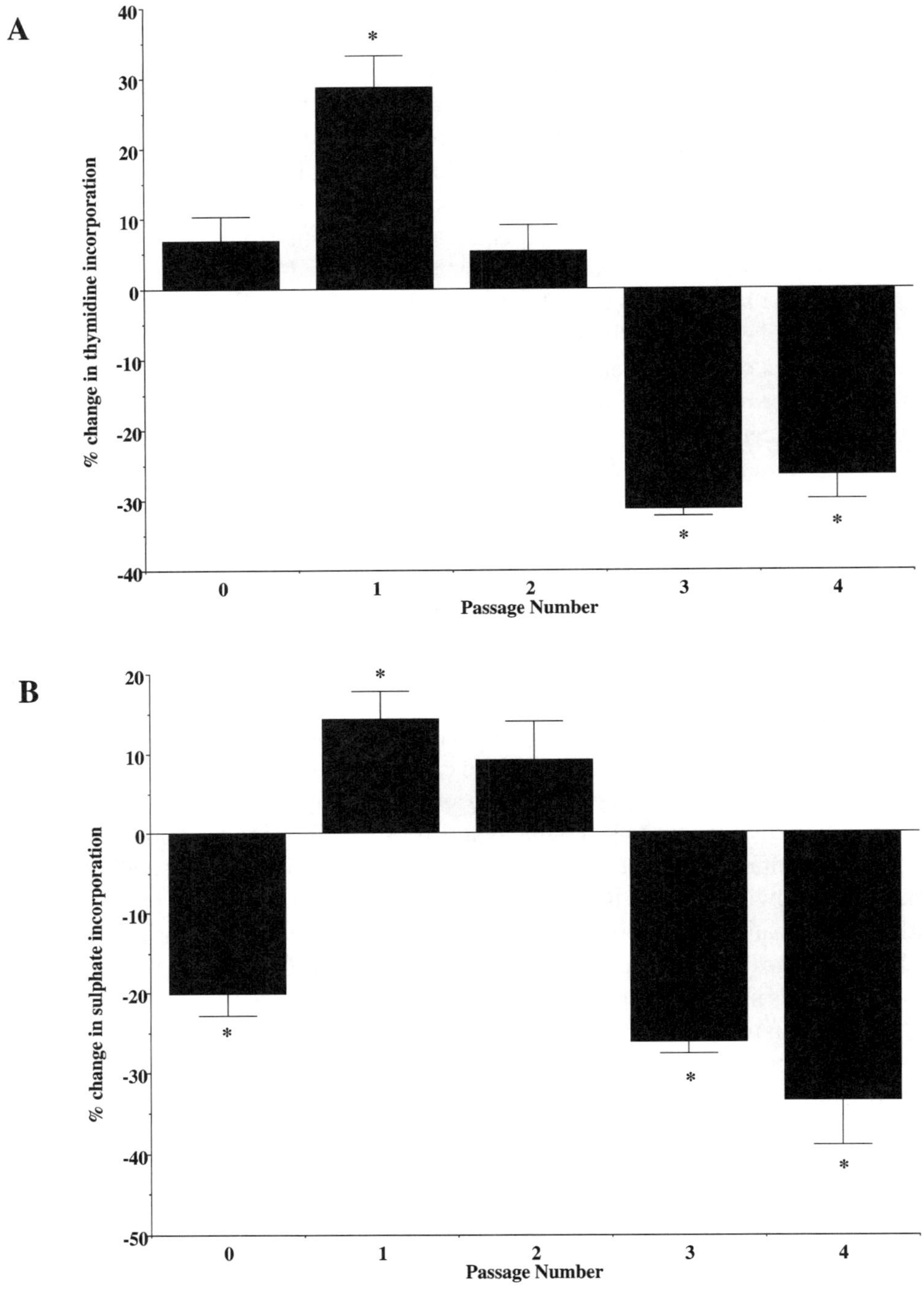

Fig. 6. The influence of dynamic compression on [^{3}H]-thymidine incorporation (A) sulphate incorporation (B) and nitrite release (C) by freshly isolated chondrocytes (designates passage 0) and chondrocyte passaged in monolayer between one and four times. The cells were subsequently embedded in agarose constructs and subjected to 15% dynamic compressive strain amplitude at 1 Hz for 24 hours. The values are presented as % change from unstrained control levels. Each value represents the mean and standard error of at least 12 replicates. Unpaired Student's t-test results indicate differences from control values as follows: $^*p < 0.05$.

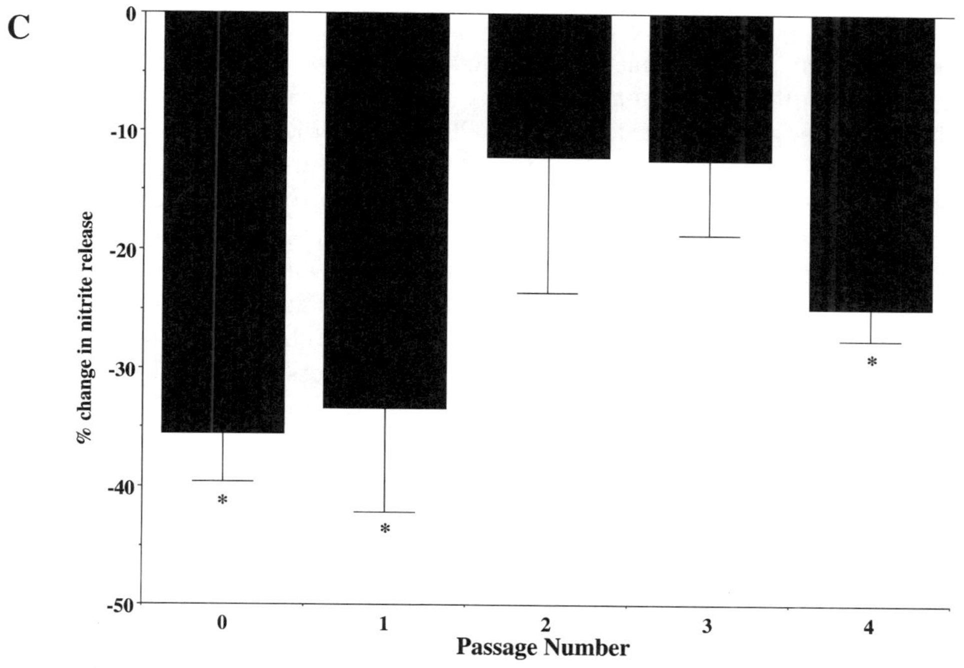

Fig. 6. (Continued.)

immediately post-isolation. As integrins have been reported to act as mechanoreceptors in chondrocytes [37], these data suggest that a rapid alteration in chondrocyte mechanosensitivity may occur during initial culture post-isolation.

Passage in monolayer clearly induced an alteration in the response of the chondrocytes to the application of dynamic compression (Fig. 6). The response of P1 cells to dynamic compression was broadly similar to that reported for freshly isolated cells in previous studies, characterised by a stimulation of both cell proliferation or glycosaminoglycan synthesis [13,28–31] (Fig. 6A,B). As discussed above, P1 cells exhibit a mixed chondrocytic and fibroblastic phenotype, characterised by staining for both type I and type II collagen (Table 2). At later passages, once collagen type II staining was lost, the response of the cells was completely reversed, resulting in an inhibition of both cell proliferation or glycosaminoglycan synthesis in response to the application of dynamic compression (Fig. 6A,B). Accordingly, it appears that passage in monolayer induces an alteration in mechanotransduction pathways associated with these processes. Interestingly, nitrite release was inhibited by dynamic compression in all cases, suggesting that the mechanotransduction processes involved are not affected by monolayer expansion (Fig. 6C). Previous studies have reported that nitric oxide mediates compression-induced alterations in cell proliferation by chondrocytes in agarose construct [12,13,29,31]. These current findings suggest that this relationship may not be maintained following passage in monolayer.

It is tempting to speculate that a relationship exists between the phenotypic status of the cells and the modulation in response to dynamic compression. Although a clear association between these processes has been demonstrated in the current study, the nature of the methodology used cannot confirm a causal link and other explanations are possible. For example, previous studies have demonstrated that cellular age can influence the response of cells to mechanical conditioning [6]. Expansion in monolayer will induce an alteration in the telomeric age of the cells associated with an increase in population doubling. It may be predicted from the data presented in Fig. 2 that P4 cells will have undergone mitotic cell division approximately 11 times more than freshly isolated cells. Accordingly further studies should alter the phenotype of the cells without an associated induction of cell aging, in order to elucidate the mechanisms underlying the modulation of response to dynamic compression. This may be achieved by using defined chondrogenic media containing TGF-β or BMP-2, which have been demonstrated to enhance the re-expression of chondrocytic phenotype by monolayer-expanded cells [14,24,32,46].

In summary this study has demonstrated a clear alteration in the response of chondrocytes to dynamic compression following passage in monolayer. At early passage, dynamic compression up-regulated both cell proliferation and glycosaminoglycan synthesis, processes which are fundamental to the success of cell-based cartilage repair systems. However at later passage numbers (P3–P4), which correspond to those used during the expansion phase of existing autologous chondrocyte repair systems, the cells exhibited a compression-induced down-regulation of these processes. Accordingly, the application of dynamic compression may be detrimental to the development of a functional neo-tissue within a clinical setting. These findings are of major significance, therefore, to researchers proposing that mechanical conditioning, applied using bioreactor systems or by defined exercise regimes, may enhance the efficacy of tissue engineered repair systems.

Acknowledgements

Support for this work was provided by the Engineering and Physical Sciences Research Council and the Higher Education Funding Council for England.

References

[1] D.L. Bader and D.A. Lee, Structure-properties of soft tissues: Articular cartilage, in: *Structural Biological Materials: Design and Structure Property Relationships*, M. Elices, ed., Pergamon, 2000, pp. 73–104.

[2] G. Bentley and R.B. Greer, Homotransplantation of isolated epiphyseal and articular cartilage chondrocytes into joint surfaces of rabbits, *Nature* **230** (1971), 184–197.

[3] P.D. Benya and J.D. Shaffer, Dedifferentiated chondrocytes re-express the differentiated collagen phenotype when cultured in agarose gels, *Cell* **30** (1982), 215–224.

[4] P.D. Benya, S. Padilla and M.E. Nimni, Independent regulation of collagen types of chondrocytes during the loss of differentiated function in culture, *Cell* **15** (1978), 1313–1321.

[5] P.D. Benya, P.D. Brown and S.R. Padilla, Microfilament modification by dihydrocytochalasin B causes retinoic acid-modified chondrocytes to reexpress the differentiated collagen phenotype without a change in shape, *J. Biol. Chem.* **106** (1988), 161–170.

[6] C.C. Berry, C. Cacou, D.A. Lee, D.L. Bader and J.C. Shelton, Dermal fibroblasts respond to mechanical conditioning in a strain profile dependent manner, *Biorheology* **40** (2003), 337–345.

[7] F. Binette, M.P. McQuaid, D.R. Haudenschild, P.C. Yaeger, J.M. McPherson and R. Tubo, Expression of a stable articular cartilage phenotype without evidence of hypertrophy by human articular chondrocytes *in vitro*, *J. Orthop. Res.* **16** (1998), 207–216.

[8] K.M. Bonaventure, N. Kadhom, L. Cohen-Solan, K.H. Ng, J. Bouguisgnon, C. Lasselin and P. Freisinger, Re-expression of cartilage-specific genes by de-differentiated human articular chondrocytes cultured in alginate beads, *Exp. Cell Res.* **212** (1994), 97–104.

[9] M. Brittberg, A. Lindahl, A. Nilsson, C. Ohlsson, O. Isaksson and L. Peterson, Treatment of deep cartilage defects in the knee with autologous chondrocyte transplantation, *New Eng. J. Med.* **331** (1994), 889–895.

[10] P.D. Brown and P.D. Benya, Alterations in chondrocyte cytoskeletal architecture during phenotypic modulation by retinoic acid and dihydrocytochalasin B-induced re-expression, *J. Cell Biol.* **106** (1988), 171–179.

[11] M.D. Buschmann, Y.A. Gluzband, A.J. Grodzinsky and E.B. Hunziker, Mechanical compression modulates matrix biosynthesis in chondrocyte/agarose culture, *J. Cell Sci.* **108** (1995), 1497–1508.

[12] T.T. Chowdhury, Role of nitric oxide and PGE_2 in chondrocyte mechanotransduction, PhD Dissertation, University of London, 2002.

[13] T.T. Chowdhury, D.L. Bader and D.A. Lee, Dynamic compression inhibits the synthesis of nitric oxide and PGE_2 by IL-1β stimulated chondrocytes cultured in agarose constructs, *Biochem. Biophys. Res. Comm.* **285** (2001), 1168–1174.

[14] M. deHaart, W.J.C.M. Marijnissen, G.J.V.N. van Osch and J.A.N. Verhaar, Optimization of chondrocyte expansion in culture: Effects of TGF-β, bFGF and L-ascorbic acid on bovine articular chondrocytes, *Acta Orthop. Scand.* **70** (1999), 55–61.

[15] B.O. Enobakhare, In vitro evaluation of chondrocyte/alginate constructs for tissue engineered articular cartilage, PhD Dissertation, University of London, 1998.

[16] B.O. Enobakhare, D.L. Bader and D.A. Lee, Quantification of sulphated glycosaminoglycans in chondrocyte/alginate cultures, by use of 1,9-dimethylmethylene blue, *Analyt. Biochem.* **243** (1996), 189–191.

[17] D.A. Grande, M.I. Pitman, L. Peterson, D. Menche and M. Klein, The repair of experimentally produced defects in articular cartilage by autologous chondrocyte transplantation, *J. Orthop. Res.* **7** (1989), 208–218.

[18] M.L. Gray, A.M. Pizzanelli, R.C. Lee, A.J. Grodzinsky and D.A. Swann, Kinetics of the chondrocyte biosynthetic response to compressive load and release, *Biochim. Biophys. Acta* **991** (1989), 415–425.

[19] L.C. Green, D.A. Wagner, J. Glogowski, P.L. Skipper, J.S. Wishnok and S.R. Tannenbaum, Analysis of nitrate, nitrite, and [15N]nitrate in biological fluids, *Analyt. Biochem.* **126** (1982), 131–138.

[20] S.S. Gross, A.E. Jaffee, R. Levi and R.G. Kilbourn, Cytokine-activated endothelial cells express an isotope of nitric oxide synthase which is tetrahydrobiopterin-dependent, calmodulin-independent and inhibited by arginine analogs with a rank-order of potency characteristic of activated macrophages, *Biochem. Biophys. Res. Comm.* **178** (1991), 823–829.

[21] J. Guo, G.W. Jourdan and D.K. MacCallum, Culture and growth characteristics of chondrocytes encapsulated in alginate beads, *Conn. Tiss. Res.* **19** (1989), 277–297.

[22] H.J. Hauselmann, M.B. Aydelotte, B.L. Schumacher, K.E. Kuettner, S.H. Gillelis and E.J.M.A. Thonar, Synthesis and turnover of proteoglycan by human and bovine articular chondrocytes cultured in alginate beads, *Matrix* **12** (1992), 116–129.

[23] H.J. Hauselmann, R.J. Fernandes, S. Mok, T.M. Schmid, J.A. Block, M.B. Aydelotte, K.E. Kuettner and E.J.M.A. Thonar, Phenotypic stability of bovine articular chondrocytes after long term culture in alginate, *J. Cell Sci.* **104** (1994), 17–27.

[24] M. Jakob, O. Demarteau, D. Schafer, B. Hintermann, W. Dick, M. Heberer and I. Martin, Specific growth factors during the expansion and redifferentiation of adult human articular chondrocytes enhance chondrogenesis and cartilaginous tissue formation in vitro, *J. Cell Biochem.* **81** (2001), 368–377.

[25] S. Kawamura, S. Wakitani, T. Kimura, A. Maeda, A.I. Caplan, K. Shin and T. Ochi, Articular cartilage repair. Rabbit experiments with a collagen gel-biomatrix and chondrocytes cultured in it, *Acta Orthop. Scand.* **69** (1998), 56–62.

[26] J. Kisiday, M. Jin and A.J. Grodzinsky, Effects of dynamic compressive loading duty cycle on in vitro conditioning of chondrocyte seeded peptide and agarose scaffolds, *Trans. Orthop. Res. Soc.* **27** (2002), 216.

[27] I. Kiviranta, M. Tammi, J. Jurvelin, A.-M. Saamanen and H.J. Heliminen, Moderate running exercise augments glycosaminoglycans and thickness of articular cartilage in knee joint of young beagle dogs, *J. Orthop. Res.* **6** (1988), 188–195.

[28] D.A. Lee and D.L. Bader, Compressive strains at physiological frequencies influence the metabolism of chondrocytes seeded in agarose, *J. Orthop. Res.* **15** (1997), 181–188.

[29] D.A. Lee, S.P. Frean, P. Lees and D.L. Bader, Dynamic compressive strain influences nitric oxide production by articular chondrocytes seeded in agarose, *Biochem. Biophys. Res. Comm.* **251** (1998), 580–585.

[30] D.A. Lee, T. Noguchi, M.M. Knight, L. O'Donell, G. Bentley and D.L. Bader, Response of chondrocyte sub-populations cultured with unstrained and strained agarose, *J. Orthop. Res.* **16** (1998), 726–733.

[31] D.A. Lee, T. Noguchi, S.P. Frean, P. Lees and D.L. Bader, The influence of mechanical loading on isolated chondrocytes seeded in agarose constructs, *Biorheology* **37** (2000), 149–161.

[32] F. Lemare, N. Steinberg, C. Le Griel, S. Demignot and M. Adolphe, Dedifferentiated chondrocytes cultured in alginate beads: restoration of the differentiated phenotype and of the metabolic response to interleukin-1 beta, *J. Cell Physiol.* **176** (1998), 303–313.

[33] S. Loty, N. Forest, H. Boulekbache and J. Sautier, Cytochalasin D induces changes in cell shape and promotes in vitro chondrogenesis: A morphological study, *Biol. Cell* **83** (1995), 149–161.

[34] K. Masuda, H. Shirota and J.M.-A. Thonar, Quantification of ^{35}S-labeled proteoglycans complexed to alcian blue by rapid filtration in multiwell plates, *Analyt. Biochem.* **217** (1994), 167–175.

[35] R.L. Mauck, M.A. Soltz, C.C. Wang, D.D. Wong, P.H. Chao, W.B. Valhmu, C.T. Hung and G.A. Ateshian, Functional tissue engineering of articular cartilage through dynamic loading of chondrocyte-seeded agarose gels, *J. Biomech. Eng.* **122** (2000), 252–260.

[36] R. Mayne, M.S. Vail, P.M. Mayne and E.J. Miller, Changes in the type of collagen synthesised as clones of chick chondrocytes grow and eventually lose division capacity, *Proc. Nat. Acad. Sci. USA* **73** (1976), 1674–1678.

[37] S.J. Millward-Sadler, M.O. Wright, H. Lee, K. Nishida, H. Caldwell, G. Nuki and D.M. Salter, Integrin-regulated secretion of interleukin 4: A novel pathway of mechanotransduction in human articular chondrocytes, *J. Cell Biol.* **145** (1999), 183–189.

[38] P. Newman and F.M. Watt, Influence of cytochalasin D-induced changes in cell shape on proteoglycan synthesis by cultured articular chondrocytes, *Exp. Cell Res.* **178** (1988), 192–210.
[39] K.T. Paige, L.G. Cima, M.J. Yaremchuk, B. Schloo, J.P. Vacanti and C.A. Vacanti, De novo cartilage generation using calcium alginate-chondrocyte constructs, *Plastic Reconstruct. Surg.* **97** (1996), 168–178.
[40] J. Rao and W.R. Otto, Fluorimetric assay for cell growth estimation, *Analyt. Biochem.* **207** (1992), 186–192.
[41] A.-M. Saamanen, M. Tammi, J. Jurvelin, I. Kiviranta and H.J. Helminen, Proteoglycan alteration following immobilisation and remobilisation in the articular cartilage of young canine joint, *J. Orthop. Res.* **8** (1990), 863–873.
[42] R.L.Y. Sah, Y.J. Kim, J.Y. Doong, A.J. Grodzinsky, A.H. Plaas and J.D. Sandy, Biosynthetic response of cartilage explants to dynamic compression, *J. Orthop. Res.* **7** (1989), 619–636.
[43] M. Takigawa, T. Takano, E. Shirai and F. Suzuki, Cytoskeleton and differentiation: effects of cytochalasin B and colchicine on expression of the differentiated phenotype of rabbit costal chondrocytes in culture, *Cell Diff.* **14** (1984), 197–204.
[44] C.A. Vacanti, R. Langer, B. Schloo and J.P. Vacanti, Synthetic biodegradable polymers seeded with chondrocytes provide a template for new cartilage formation *in vivo*, *J. Plast. Reconstruct. Surg.* **87** (1991), 753–759.
[45] G.S. Woods and R. Warnke, Suppression of endogenous avidin-binding activity and its relevance to biotin–avidin detection systems, *J. Histochem.* **29** (1981), 1196–1204.
[46] P.C. Yaeger, T.L. Masi, J.L. deOrtiz, F. Binette, R. Tubo and J.M. McPherson, Synergistic action of transforming growth factor-beta and insulin-like growth factor-I induces expression of type II collagen and aggrecan genes in adult human articular chondrocytes, *Exp. Cell Res.* **237** (1997), 318–325.

Biorheology 41 (2004) 299–308
IOS Press

The influence and interactions of hydrostatic and osmotic pressures on the intracellular milieu of chondrocytes

J.A. Browning, K. Saunders, J.P.G. Urban * and R.J. Wilkins
University Laboratory of Physiology, Oxford University, Parks Rd, Oxford OX1 3PT, UK

Abstract. The intracellular milieu of chondroctyes is regulated by an array of proteins in the cell membrane which operate as transport pathways, allowing ions and nutrients such as glucose and amino acids and metabolites such as lactate to cross the plasma membrane. Here we investigated the influence of hydrostatic pressure on intracellular calcium concentrations ($[Ca^{2+}]_i$) in isolated bovine articular chondrocytes. We found that short applications of high hydrostatic pressures led to a significant increase in $[Ca^{2+}]_i$. The pressure-induced rise was abolished for long (240 sec) but not short (30 sec) pressure applications by removal of extracellular Ca^{2+}. The rise in pressure was also blocked by the inhibitors neomycin and thapsigargin confirming that pressure, by generating IP_3, led to an increase in $[Ca^{2+}]_i$ by mobilising the pool of Ca^{2+} ions contained within intracellular stores. We also found that intracellular $[Na^+]$ was affected by a rise in osmotic pressure and further affected by application of hydrostatic pressure. The effect of hydrostatic pressure on sulphate incorporation depended strongly on extracellular osmolality. Since significant gradients in extracellular osmolality exist across intact cartilage, the results imply that responses of chondrocytes to the same pressure will vary depending on location in the joint. The results also indicate that hydrostatic pressures can affect several different transporter systems thus influencing the intracellular milieu and chondrocyte metabolism.

Keywords: Calcium, sodium pump, mechanotransduction, ion transport, osmotic pressure

1. Introduction

Chondrocytes, like all cells, expend a large proportion of their energy on "housekeeping". This housekeeping involves general maintenance of a constant intracellular milieu, supply of nutrients and removal of metabolic wastes in order to sustain viable cellular function. These housekeeping functions are regulated by an array of proteins in the cell membrane which operate as transport pathways, allowing ions and nutrients such as glucose and amino acids and metabolites such as lactate to cross the plasma membrane [29].

Mechanical stress has been shown to influence these transport pathways in a number of different ways [22]. Fluid expression alters the ionic environment by concentrating extracellular cations such as Ca^{2+}, H^+ and Na^+ [27] and hence alters the electrochemical gradient across the cell membrane. The resulting increase in extracellular osmolality leads to a rapid loss of cell volume [4,7], concentrating intracellular ionic and other solutes which can upregulate expression and activity of transporters [12,32]. Application of very short bursts of hydrostatic pressure can affect activity of Na^+ and H^+ transporters [2,

*Address for correspondence: Dr J. Urban, Physiology Laboratory, Oxford University, Parks Rd, Oxford OX1 3PT, UK. Tel.: +44 18965 272509; Fax: +44 1865 272469; E-mail: jpgu@physiol.ox.ac.uk.

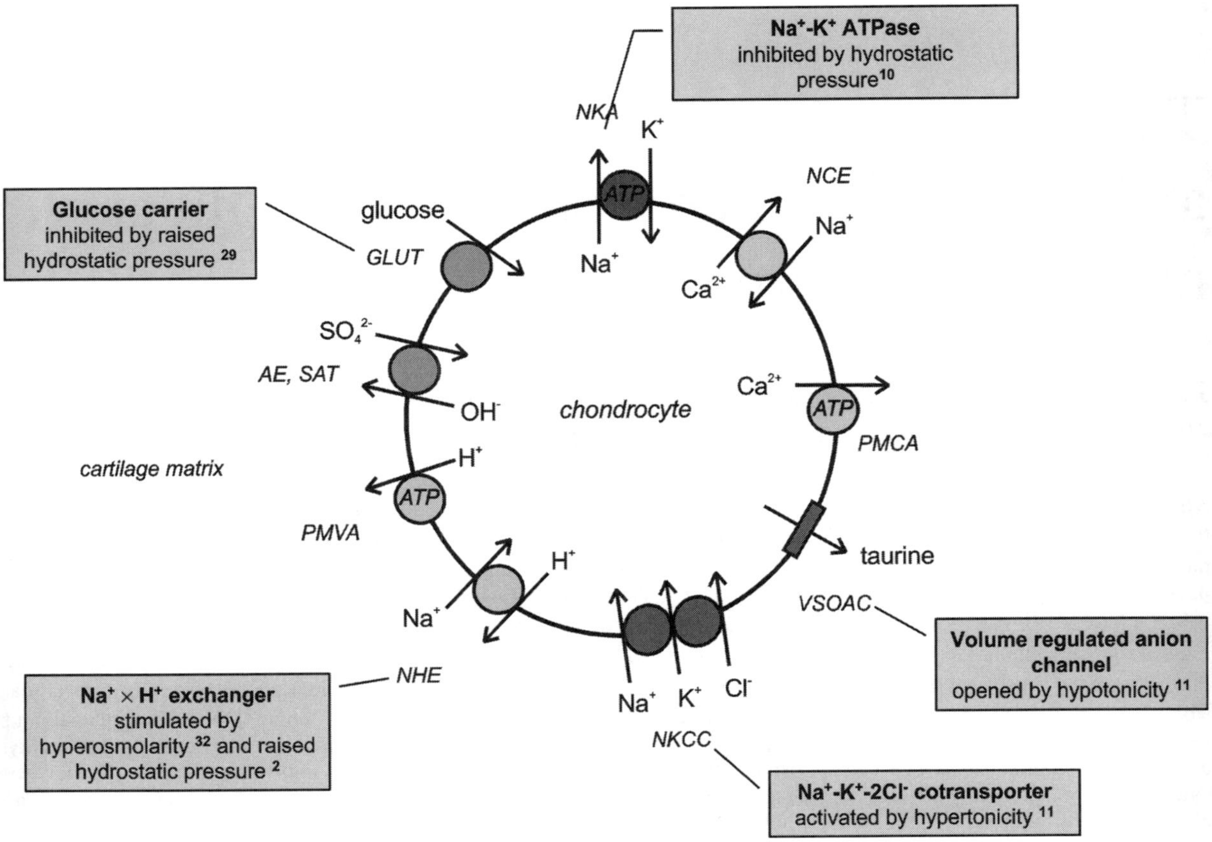

Fig. 1. Schematic view of some of the membrane transporters identified in chondrocytes indicating those which have been shown to be affected by mechanical signals.

10]. Stretch and flow can open channels allowing ingress of Ca^{2+} and other ions [31,34]. Figure 1 summarises some of the transporters known to be sensitive to mechanical signals in chondrocytes.

Concentrations of intracellular solutes depend both on transmembrane electrochemical gradients and on activity of transporters. Mechanical stress may thus, through its effects on transporter activity and gradients, alter the intracellular milieu, further activating "housekeeping" processes in an attempt to maintain homeostasis. Such changes in the intracellular concentrations of signalling molecules such as Ca^{2+} and of ions such as Na^+ and K^+ which affect enzyme activity [17] could have a rapid influence on cellular metabolism. Here we examine the effects of hydrostatic pressure on intracellular Ca^{2+} concentration ($[Ca^{2+}]_i$). We also examine how changes in hydrostatic and osmotic pressure affect intracellular concentration of Na^+ ($[Na^+]_i$) and the consequent effects on rates of matrix synthesis.

2. Materials and methods

2.1. Cell isolation

For some tests, we used bovine metacarpal phalangeal articular cartilage from 18–24 month steers. Articular cartilage was shaved from the opened joint as described previously [24] and then placed in a 25 ml tissue culture flasks containing Dulbecco's Modified Eagles Medium (DMEM) (1 g l^{-1} glucose, 25 mM HEPES, no $NaHCO_3$; Life Technologies, Paisley, Scotland) supplemented with antibiotics/antimycotics (500 units ml^{-1} penicillin G, 500 $\mu g\,ml^{-1}$ streptomycin sulphate: Life Technologies, Paisley, Scotland). The dissected tissues were then enzymatically digested as previously described [27] at 37°C in DMEM containing 1% (vol/vol) antibiotic/antimycotic solutions with 1 mg ml^{-1} collagenase (collagenase 1, Sigma Chemical Co., Poole, Dorset, England) under 95% air/5% CO_2 for 18–20 h. After incubation the digested tissue suspension was filtered to isolate the cells which were then washed three times by repeated centrifugation (1000g for 5 min) and resuspended in DMEM. The cell suspension was assessed microscopically for cell viability using trypan blue exclusion and cell number counted. Only cell preparations with viabilities >95% were then used.

2.2. *Culture of C-20/A4 cells*

In other experiments, a transformed human chondrocyte cell line was employed. C-20/A4 cells (the kind gift of Professor Mary Goldring, Harvard Medical School, USA) were cultured to confluency at 37°C in a humidified 5% CO_2 incubator in vented 75 cm^2 culture flasks containing 20 ml Dulbecco's modified Eagle's Medium/Ham's F-12 nutrient mixture (DMEH) supplemented with 10% fetal calf serum (FCS) and 1% penicillin (1250 U ml^{-1})–streptomycin (5 mg ml^{-1}) antibiotic. The medium was changed twice-weekly and, once confluency had been reached (after approximately 3–5 days of culture), chondrocytes were removed from the culture flask by digestion in 5 ml trypsin-EDTA (ethylenediaminetetraacetic acid) solution for 5 min at 37°C and passaged.

Cells for use in experiments were removed from the culture flask when confluent by digestion in 5 ml trypsin-EDTA solution for 5 min at 37°C. 15 ml DMEH containing FCS and penicillin–streptomycin (as above) was added after this time to prevent further digestion by trypsin. The resulting suspension was centrifuged and washed twice in HEPES buffered solution (HBS) before the cells were counted and resuspended at a final concentration of $1\text{–}2 \times 10^6$ cells ml^{-1}.

2.3. *Hydrostatic pressure application*

For the measurement of transport or metabolism, pressure was applied in a thermostatically controlled fluid-filled pressure vessel as described previously [9,13]. Briefly, the chondrocytes were suspended in the required medium at a known density in vials filled with medium and containing no air. The vials were sealed with a plunger or rubber seal to allow pressure equilibration between the fluid in the pressure chamber and the fluid in the vial. A similar procedure was used for measurement of intracellular Ca^{2+}, however the pressure vessel in this case contained quartz windows and was incorporated into a fluorescence spectrometer [29].

2.4. *Measurement of intracellular ions*

2.4.1. *Intracellular Na^+ and K^+*

Briefly, intracellular Na^+ and K^+ were measured in isolated bovine chondrocytes using solutions held at different osmolalities containing tracer ^{22}Na; NaCl was added to DMEM to alter the osmolality in the range 280–550 mOsm [27] and the osmolality determined using a freezing-point osmometer (Wescor, UK). In some cases, the cells were subjected to hydrostatic pressure. After equilibration at atmospheric pressure, cell volumes were determined as described previously [5]. The cell suspensions were centrifuged and the cell pellet resuspended, washed once rapidly in ice-cold $MgCl_2$ solution at an equivalent osmolality, recentrifuged and the supernatant decanted rapidly. The pellet was weighed and digested in 0.1 ml Na^+-free buffer [28]. The activity of ^{22}Na and the concentration of Na^+ and K^+ in the pellet and supernatant was determined, respectively, using liquid scintillation counting and flame-photometry [12] and intracellular ion concentrations estimated.

2.4.2. *Intracellular Ca^{2+}*

The effect of pressure on $[Ca^{2+}]_i$ in a human cell line was measured using fluorimetric methods previously described [3]. Chondrocytes were suspended in solutions containing the calcium fluoroprobe Fura-2 (5 μM, 30 min at 21°C, 15 min at 37°C) and subjected to pulses (30 s or 240 s) of hydrostatic pressure as described above. The solutions contained either various concentrations of Ca^{2+} or else were Ca^{2+} free and contained the calcium chelator, EGTA (1 mM). Effect of pressure on release of Ca^{2+} from intracellular stores was determined by measuring Ca^{2+} and cytoplasmic IP_3 after pressure application (0.1–30 MPa) with and without the presence of thapsigargin (1 μM), a specific inhibitor of the Ca^{2+}-ATPase which restocks intracellular stores.

2.5. *IP_3 assay*

A D-*myo*-inositol assay system ("Biotrak", Amersham Pharmacia Biotech, Amersham, UK) was employed. C-20/A4 cells were lysed and protein removed by precipitation. The resulting solution was incubated with IP_3 binding protein, and free IP_3 was separated from that which had bound by centrifugation. The level of radioactivity in the pellet was measured using scintillation counting. The amount of IP_3 in the original sample was determined by interpolation from a standard curve.

2.6. *^{35}S-sulphate incorporation*

Synthesis of sulphated glycosaminoglycans (GAG) was determined from ^{35}S-sulphate incorporation [21] determined as described previously [16]. Briefly the cells were suspended in the desired medium, supplemented with 5 $\mu Ci\,ml^{-1}$ of ^{35}S-sulphate (Amersham Pharmacia Biotech, Little Chalfont, UK), incubated at 37°C and 5% CO_2 and subjected to the desired pressure protocol. At the end of the labeling period, a sample of medium was withdrawn and the cells and medium frozen at −20°C until processed. The bound and free ^{35}S-sulphate radiolabel were separated by dialysis [16]. When the dialysate was at background levels, the tube contents and the tubing were placed into a scintillation vial with 4 ml of scintillant and the activity present in the GAGs synthesised by the cells and in the culture medium was measured on a beta scintillation counter. Rates of proteoglycan synthesis were calculated assuming that the specific activity of ^{35}S-sulphate was the same in the medium and newly synthesised proteoglycans [21].

3. Results

3.1. *Effect of hydrostatic pressure on intracellular $[Ca^{2+}]$*

In the presence of extracellular Ca^{2+} and at atmospheric pressure, intracellular was approximately 80–100 nM (Figs 2 and 3); even short applications of pressure had a marked effect on $[Ca^{2+}]_i$ in C-20/A4 cells; $[Ca^{2+}]_i$ increased approximately three-fold after only 30 s of high (30 MPa) pressure application (Fig 2). No further change in Ca^{2+} was observed if pressure was maintained for 240 s rather than 30 s.

When extracellular Ca^{2+} was removed, $[Ca^{2+}]_i$ was reduced at 0.1 MPa and the effects of high hydrostatic pressure were significantly diminished after 30 s pressure application; after a longer pressure

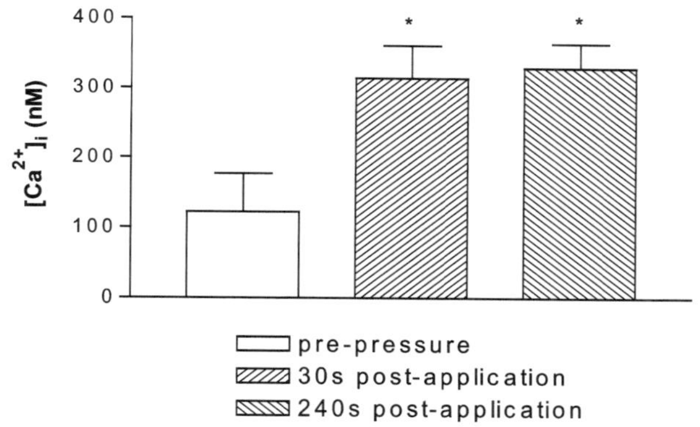

Fig. 2. Effect of pressure application on intracellular calcium concentrations. Pressure (30 MPa) was applied to C-20/A4 cells for 30 s and 240 s and intracellular calcium measured. Results were compared to control cells at atmospheric pressure. Results are mean + SE ($n = 3$).

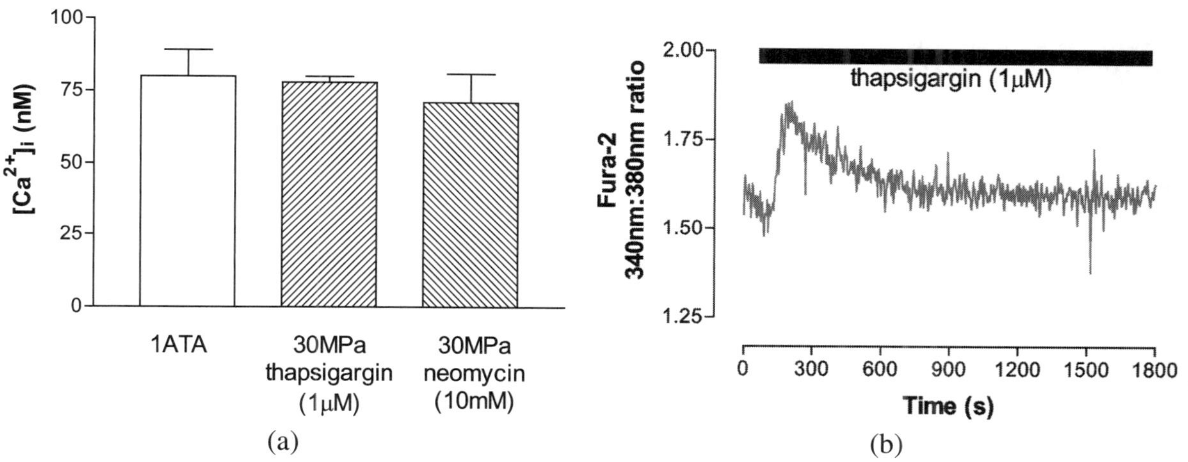

Fig. 3. (a) Effect of pressure application on intracellular calcium concentrations in cells pretreated with thapsigargin. Pressure (30 MPa) was applied to C-20/A4 cells for 30 s and 240 s and intracellular calcium measured. Results were compared to control cells at atmospheric pressure. Results are mean + SE ($n = 3$). (b) Measurements of $[Ca^{2+}]_i$ in thapsigargin-treated cells. C-20/A4 cells were suspended in solutions containing EGTA. A typical trace is shown.

application (240 s), intracellular $[Ca^{2+}]$ was no different to that at 0.1 MPa ($[Ca^{2+}]_i = 48 \pm 4$ nM in control conditions, 67 ± 8 nM at 30 s, 44 ± 2 nM at 240 s). These results suggest that release from intracellular stores – which are partially depleted by suspension in $0Ca^{2+}$ solutions – underly the initial pressure induced-increase in $[Ca^{2+}]_i$, and that a subsequent capacitative influx of Ca^{2+} ions is responsible for sustaining the elevated levels of Ca^{2+} under pressure.

3.2. The role of intracellular stores on pressure-induced increase in intracellular $[Ca^{2+}]$

The requirement for IP_3 generation in the response to hydrostatic pressure was confirmed by experiments employing the phospholipase C inhibitor neomycin. This compound abolished the increase in Ca^{2+} levels associated with pressure, such that $[Ca^{2+}]_i$ after 240 s was not significantly different from control levels (Fig. 2). Moreover, radioassay measurements of cytoplasmic $[IP_3]$ after exposure to high pressure showed a significant elevation in the levels of this messenger. IP_3 levels were increased by ap-

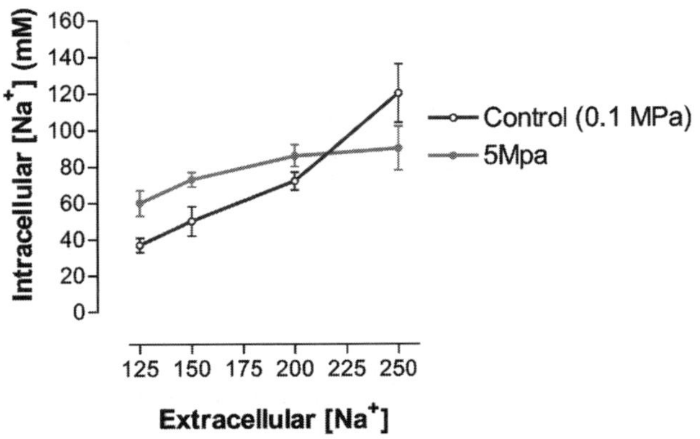

Fig. 4. Effect of extracellular osmolality and hydrostatic pressures on intracellular Na^+ in bovine articular chondrocytes. Chondrocytes were isolated enzymatically, cultured in suspension overnight in solutions whose osmolality was altered by Na^+ addition and hydrostatic pressure applied for 2 hrs. Results for 5 MPa are shown as mean and SE ($n = 3$ separate experiments).

proximately 20-fold following application of 300 ATA hydrostatic pressure for 30 s (39 ± 12 vs 2 ± 1 pmol $(10^6 \text{ cells})^{-1}$, $n = 3$, $P < 0.05$). Smaller increases were also observed at lower (10 MPa) pressure (data not shown), suggesting that the response we observe also occurs within the physiological range.

To confirm that the liberated IP_3 induces Ca^{2+} release from intracellular stores, cells were treated with thapsigargin. This compound is a specific inhibitor of the Ca^{2+}-ATPase which restocks intracellular stores [23]. For cells exposed to pressure, pretreatment with thapsigargin (1 μm) abolished the increase in $[Ca^{2+}]_i$, (Fig. 3a), confirming that pressure, by generating IP_3, mobilises the pool of Ca^{2+} ions contained within intracellular stores. Measurements of $[Ca^{2+}]_i$ in thapsigargin-treated cells which had been suspended in solutions containing the Ca^{2+} chelator EGTA showed a rapid rise, followed by a slow return to resting levels (Fig. 3b). This represents the uncompensated leak of Ca^{2+} ions from intracellular stores, and slower extrusion across the cell membrane, where these ions become chelated, and results in depletion of the intracellular stores.

3.3. The effect of pressure on intracellular Na^+ concentrations

Intracellular sodium concentration ($[Na^+]_i$) was found to increase as extracellular $[Na^+]$ or osmolality was raised (Fig. 4). Application of pressure influenced the change in $[Na^+]_i$; in the range 250–450 mOsm, $[Na^+]_i$ was higher in pressurised than unpressurised samples and less sensitive to changes in extracellular osmolality or $[Na^+]$.

3.4. The effect of pressure on sulphate incorporation rates

Measurement of the rates of ^{35}S-sulphate incorporation found that rates were higher in pressurised samples than in samples held at atmospheric pressure under all conditions. However the prevailing osmolality had a strong effect on pressure-induced stimulation with rates for the same pressure protocol increasing approximately 10–20% at 300 mOsm and 500 mOsm but by approximately 60% at 450 mOsm.

4. Discussion

These results confirm previous reports which show the intracellular ionic milieu of chondrocytes is affected by mechanical signals; fluid shear, compression and osmotic shock have all been found to af-

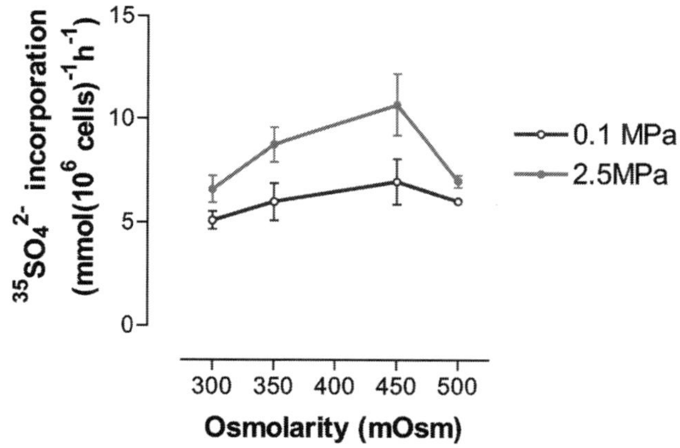

Fig. 5. Effect of extracellular osmolality and hydrostatic pressure on sulphate incorporation rates for bovine articular chondrocytes. Chondrocytes were isolated enzymatically, cultured in suspension overnight in solutions whose osmolality was altered by Na^+ addition. Radioactive tracer (^{35}S-sulphate) was added and hydrostatic pressure applied for 2 hrs and rates of incorporation were measured during this period. Results for 2.5 MPa are shown as mean and SE ($n = 3$ separate experiments).

fect intracellular calcium concentrations [6,8,25,33,34]. Here we show that application of short bursts of hydrostatic pressure are also able to affect chondrocyte $[Ca^{2+}]_i$ (Fig. 2). Concentrations of other ions are also affected by hydrostatic pressure; intracellular $[Na^+]_i$ increased on application of hydrostatic pressure (Fig. 4) with the extent of pressure-induced change depending on extracellular osmolality. Extracellular osmolality itself influenced $[Na^+]_i$, (Fig. 4) in part through its effect on cell volume and on transmembrane ion gradients [11]. Intracellular ion concentrations have been shown to affect rates of protein synthesis [17] and thus the osmotically sensitive increase in sulphate incorporation induced by hydrostatic pressure (Fig. 5) could result from such changes in intracellular ionic composition.

Changes in intracellular composition could arise from alterations to transporter activity. Short pressure applications are know to affect sodium transport, inhibiting the sodium pump and thus allowing Na^+ influx [9] with the degree of inhibition depending on $[Na^+]_i$ hence possibly explaining in part the interaction between osmotic and hydrostatic pressures on $[Na^+]_i$. Pressure and osmolality are known to affect other K^+ and Na^+ pathways (Fig. 1) [10,12]. For example, both application of hydrostatic pressure and hyperosmotic shock stimulates sodium–hydrogen exchange [30]. Activation of these transporters, resulting in changes in intracellular Na^+, K^+ and H^+ will modulate other transporters. At present there is not sufficient information on transporter kinetics in chondrocytes to predict how the effects of manoeuvres such as changing transporter activity by increasing pressure will influence intracellular composition, although models exist in better studied systems such as the red cell [19]. It is worth emphasising that some changes in intracellular ions result from complex and interacting changes in transmembrane gradients, transporter action and, in the case of calcium, from release from intracellular stores as shown in Fig. 6.

The interactions seen between pressure and osmolality (Figs 4 and 5) indicate that in order to understand the responses of chondrocytes to mechanical stress, gradients across the tissue need to be considered. In intact tissue, osmolality varies from ca 350 mOsm at the surface of cartilage to around 450 mOsm in the deep zone [15,20]; the response of chondrocytes to the same applied pressure could thus vary across the joint depending on ambient osmolality. Gradients in other factors, such as extracellular oxygen or pH could also influence responses to mechanical stress. Hence responses to mechanical signals measured under standard *in vitro* conditions (300 mOsm, 21% oxygen, pH 7.4) could be very different from responses *in vivo* where oxygen varies from ca 5–10% at the synovial surface to ca 1% in the deep zone [26] and extracellular pH is ca pH 6.8–7.0 [1]. It should also be noted that normally no

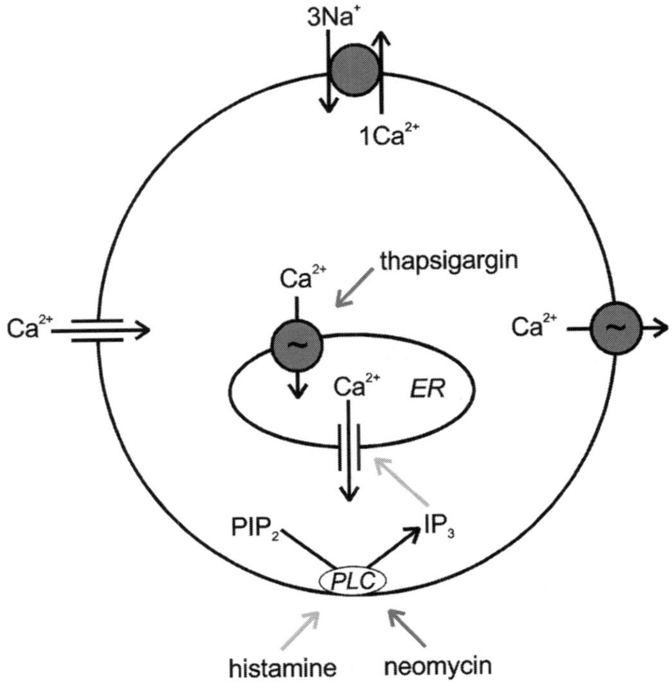

Fig. 6. Schematic view of the pathways influencing intracellular Ca^{2+} in chondrocytes after application of hydrostatic pressure. The increase in Ca^{2+} after pressure application results from the increase in generation of IP_3 which mobilises Ca^{2+} from intracellular stores. Capacitative influx of Ca^{2+} ions is responsible for sustaining the elevated levels of Ca^{2+} under pressure. Histamine stimulates, neomycin inhibits phospholipase C (PLC).

attempt is made here to separate cells from different zones although they respond differently to a variety of extracellular signals [14,18]. At present, little is known about the influence of environmental gradients such as those of oxygen tension, pH or osmolality or their interactions with chondrocyte phenotype, on responses of cartilage to mechanical signals.

Acknowledgements

We thank the Arthritis Research Campaign (U0511, W0616), Halley Stewart Trust and EPA Cephalosporin Fund for support and Dr A.C. Hall for his help with the measurements of the effects of pressure on $[Na^+]_i$ and synthesis. We thank Professor Mary Goldring, Harvard Medical School, USA, for the gift of C-20/A4 cells.

References

[1] N. Boustany, M.L. Gray, A.C. Black and E.B. Hunziker, Correlation between synthetic activity and glycosaminoglycan concentration in epiphyseal cartilage raises questions about the regulatory role of interstitial pH, *J. Orthop. Res.* **13** (1995), 733–739.

[2] J.A. Browning, R.E. Walker, A.C. Hall and R.J. Wilkins, Modulation of $Na^+ \times H^+$ exchange by hydrostatic pressure in isolated bovine articular chondrocytes, *Acta Physiol. Scand.* **166** (1999), 39–45.

[3] J.A. Browning and R.J. Wilkins, The effect of intracellular alkalinisation on intracellular Ca^{2+} homeostasis in a human chondrocyte cell line, *Pflugers Arch.* **444** (2002), 744–751.

[4] P.G. Bush and A.C. Hall, Regulatory volume decrease (RVD) by isolated and in situ bovine articular chondrocytes, *J. Cell Physiol.* **187** (2001), 304–314.
[5] C.M.P. Deshayes, A.C. Hall and J.P.G. Urban, Effect of extracellular osmolytes on porcine articular chondrocyte volume, *J. Physiol.* **467** (1993), 214P.
[6] G.R. Erickson, L.G. Alexopoulos and F. Guilak, Hyper-osmotic stress induces volume change and calcium transients in chondrocytes by transmembrane, phospholipid, and G-protein pathways, *Journal of Biomechanics* **34** (2001), 1527–1535.
[7] R.J. Errington, M.D. Fricker, J.L. Wood, A.C. Hall and N.S. White, Four-dimensional imaging of living chondrocytes in cartilage using confocal microscopy: a pragmatic approach, *American Journal of Physiology – Cell Physiology* **41** (1997), C1040–C1051.
[8] M. Grandolfo, A. Calabrese and P. D'Andrea, Mechanism of mechanically induced intercellular calcium waves in rabbit articular chondrocytes and in HIG-82 synovial cells, *Journal of Bone and Mineral Research* **13** (1998), 443–453.
[9] A.C. Hall, Differential effects of hydrostatic pressure on cation transport pathways of isolated articular chondrocytes, *Journal of Cellular Physiology* **178** (1999), 197–204.
[10] A.C. Hall, Hydrostatic pressure directly affects the activity of articular chondrocyte membrane transporters, *Trans. ORS* **22** (1997), 177.
[11] A.C. Hall and I. Starks, Changes to the activity of volume-sensitive transporters of articular chondrocytes following osmotic perturbation, *Trans. Am. Orthop. Res. Soc.* **20** (1995), 395.
[12] A.C. Hall, I. Starks, C.L. Shoults and S. Rashidbigi, Pathways for K^+ transport across the bovine articular chondrocyte membrane and their sensitivity to cell volume, *Am. J. Physiol.* **270** (1996), C1300–C1310.
[13] A.C. Hall, J.P.G. Urban and K.A. Gehl, The effects of hydrostatic pressure on matrix synthesis in articular cartilage, *J. Orthop. Res.* **9** (1991), 1–10.
[14] H.J. Hauselmann, J. Flechtenmacher, L. Michal, E.J. Thonar, M. Shinmei, K.E. Kuettner and M.B. Aydelotte, The superficial layer of human articular cartilage is more susceptible to interleukin-1-induced damage than the deeper layers, *Arthritis Rheum.* **39** (1996), 478–488.
[15] B. Hopewell and J.P. Urban, Adaptation of articular chondrocytes to changes in osmolality, *Biorheology* **40** (2003), 73–77.
[16] H.A. Horner, S. Roberts, R.C. Bielby, J. Menage, H. Evans and J.P. Urban, Cells from different regions of the intervertebral disc: effect of culture system on matrix expression and cell phenotype, *Spine* **27** (2002), 1018–1028.
[17] S. Horowitz and Y.-T. Lau, A function that relates protein synthetic rates to potassium activity in vivo, *J. Cell Physiol.* **135** (1988), 425–434.
[18] D.A. Lee, T. Noguchi, M.M. Knight, L. O'Donnell, G. Bentley and D.L. Bader, Response of chondrocyte subpopulations cultured within unloaded and loaded agarose, *J. Orthop. Res.* **16** (1998), 726–733.
[19] V.L. Lew and A.R. Hockaday, The effects of transport perturbations on the homeostasis of erythrocytes, *Novartis Found. Symp.* **226** (1999), 37–50.
[20] A. Maroudas and H. Evans, A study of ionic equilibria in cartilage, *Conn. Tiss. Res.* **1** (1974), 69–79.
[21] A. Maroudas, Metabolism of cartilaginous tissues, a quantitative approach, in: *Studies in Joint Disease 1*, A. Maroudas and J. Holborow, eds, Pitman Medical, London, 1981.
[22] A. Mobasheri, S.D. Carter, P. Martin-Vasallo and M. Shakibaei, Integrins and stretch activated ion channels; putative components of functional cell surface mechanoreceptors in articular chondrocytes, *Cell Biol. Int.* **26** (2002), 1–18.
[23] T. Pozzan, R. Rizzuto, P. Volpe and J. Meldolesi, Molecular and cellular physiology of intracellular calcium stores, *Physiol. Rev.* **74** (1994), 595–636.
[24] S. Razaq, J.P. Urban and R.J. Wilkins, Regulation of intracellular pH by bovine intervertebral disc cells, *Cell Physiol. Biochem.* **10** (2000), 109–115.
[25] S.R. Roberts, M.M. Knight, D.A. Lee and D.L. Bader, Intracellular calcium signalling in compressed chondrocyte agarose constructs, *Journal of Physiology London* **523** (2000), 12P–12P.
[26] I.A. Silver, Measurement of pH and ionic composition of pericellular sites, *Philos. Trans. R. Soc. Lond. B Biol. Sci.* **271** (1975), 261–272.
[27] J.P.G. Urban, A.C. Hall and K.A. Gehl, Regulation of matrix synthesis rates by the ionic and osmotic environment of articular chondrocytes, *J. Cell Physiol.* **154** (1993), 262–270.
[28] M.R. Urban, B. Fermor, R.B. Lee and J.P. Urban, Measurement of DNA in intervertebral disc and other autofluorescent cartilages using the dye hoechst 33258, *Anal. Biochem.* **262** (1998), 85–88.
[29] R.J. Wilkins, J.A. Browning and J.C. Ellory, Surviving in a matrix: membrane transport in articular chondrocytes, *J. Membr. Biol.* **177** (2000), 95–108.
[30] R.J. Wilkins, J.A. Browning and J.P. Urban, Chondrocyte regulation by mechanical load, *Biorheology* **37** (2000), 67–74.
[31] M. Wright, P. Jobanputra, C. Bavington, D.M. Salter and G. Nuki, Effects of intermittent pressure-induced strain on the electrophysiology of cultured human chondrocytes: evidence for the presence of stretch-activated membrane ion channels, *Clinical Science* **90** (1996), 61–71.

[32] N. Yamazaki, J.A. Browning and R.J. Wilkins, Modulation of $Na^+ \times H^+$ exchange by osmotic shock in isolated bovinearticular chondrocytes, *Acta Physiol. Scand.* **169** (2000), 221–228.
[33] C.E. Yellowley, J.C. Hancox and H.J. Donahue, Effects of cell swelling on intracellular calcium and membrane currents in bovine articular chondrocytes, *Journal of Cellular Biochemistry* **86** (2002), 290–301.
[34] C.E. Yellowley, C.R. Jacobs, Z.Y. Li, Z.Y. Zhou and H.J. Donahue, Effects of fluid flow on intracellular calcium in bovine articular chondrocytes, *American Journal of Physiology – Cell Physiology* **42** (1997), C30–C36.

Biorheology 41 (2004) 309–313
IOS Press

Hydrostatic pressure-induced changes in cellular protein synthesis

Mikko J. Lammi *, Mika A. Elo, Reijo K. Sironen, Hannu M. Karjalainen, Kai Kaarniranta and Heikki J. Helminen

Department of Anatomy, University of Kuopio, P.O. Box 1627, 70211 Kuopio, Finland

Abstract. Hydrostatic pressure is a well-known effector of cellular protein synthesis. High continuous hydrostatic pressure inhibits protein synthesis in general. It has been known for a long time that 30S ribosomal subunit is associated with the effects of pressure on protein synthesis in prokaryotes, however, the mechanisms of action are still not completely understood. Our new data suggest that synthesis of eukaryotic elongation factor-2 (eEF-2) is decreased under 30 MPa continuous hydrostatic pressure. Thus, eEF-2 may have a role in the synthesis of pressure-regulated proteins in eukaryotic cells. The presence of pressure-sensitive proteins indicate that hydrostatic pressure can induce very specific responses in stressed cells. Accumulation of heat shock protein 70 and 90 beta occurs under high pressure, independent of the general inhibition of protein synthesis, although this response appears clearly weaker than during heat stress.

Keywords: Hydrostatic pressure, translation, chondrocyte

1. Introduction

Cells of our body have to sense and withstand various mechanical stresses, such as shearing, compression, tension and osmotic pressure [1]. Hydrostatic pressure is present in various tissues and organs of our body, the highest levels of the pressure occurring in articular cartilage during locomotion. In chondrocytic cells, numerous cellular events arise in response to cyclic or continuous hydrostatic pressure, including effects on extracellular matrix production [2–9], modulation of Na^+/H^+ exchange [10], transmembrane potentials [11] and cation transport [12], inhibition of shear stress-induced NO release [13] and cAMP accumulation [14], alterations in actin cytoskeleton [15] and Golgi apparatus [16], and cellular proliferation [17]. This review will summarize our present knowledge of the effects of hydrostatic pressure on cellular protein synthesis.

2. General effects of hydrostatic pressure on protein synthesis

Protein synthesis is a complicated event that involves a large number of molecules acting together. Control of mRNA translation is a key regulator of gene expression in eukaryotes [18], and it appears to be exerted mainly at the stage of initiation. During this event, ribosomal 40S subunit binds to the 5′-end of mRNA, the start codon is located and ribosomal 60S subunit joins the complex forming a functional

*Address for correspondence: Mikko Lammi, Department of Anatomy, University of Kuopio, P.O. Box 1627, 70211 Kuopio, Finland. Tel.: +358 17 163 027; Telefax: +358 17 163 032; E-mail: mikko.lammi@uku.fi.

ribosome that enters the elongation stage of translation [19]. In eukaryotes, more than 10 soluble initiation factors participate in the formation of the mRNA recognition and preinitiation complexes [20], which enable unwinding and scanning of mRNA secondary structure to find a correct initiator AUG codon [21]. When the translation initiator codon is found, eukaryotic initiation factor-2A (eIF-2A) hydrolyses its bound GTP, and the preinitiation factors will dissociate from the small subunit of ribosome allowing the elongation of the polypeptide chain to begin.

Phosphorylation events of various eukaryotic initiation factors (eIFs) are involved in the initiation stage. Phosphorylation of eIF-2A frequently occurs under stressful conditions reducing the translation rates [22]. Activity of eIF-2B is altered in response to viral infection, hormones, nutrients, growth factors and certain stresses, and it may be regulated directly by its own phosphorylation [23]. eIF-4E also plays an important role by binding the 5′-cap structure, and it serves to recruit other initiation complex components to the 5′-end of mRNA [24]. Phosphorylation of eIF-4E is increased by variety of conditions, including exposures of insulin and growth factors [25]. On the other hand, eukaryotic elongation factor (eEF-2) is inactivated by its phosphorylation during stimuli that, e.g., increase cellular energy demand or reduce its supply [26]. This protein functions in protein translation facilitating the movement of peptidyl-tRNA from ribosome A site to the P site [26].

The effects of hydrostatic pressure on translational step of protein synthesis were investigated already in 1960's, mainly in bacterial cells [27–29] and later in bacterial cell-free systems [30]. In cell-free *E. coli* preparations, a progressive inhibition of protein synthesis was observed starting at approximately 20 MPa that was instantaneously reversible after release of the pressure [31]. Then, the protein synthesis proceeded at a rate similar to that in the non-pressurised control [31]. Studies on whole-cell preparations of *E. coli* has shown that 67 MPa hydrostatic pressure totally inhibited protein synthesis at elongation stage [32]. The response of reticulocyte system was basically similar to prokaryotic systems, although minor differences existed [33]. The response of bacterial protein synthesis to hydrostatic pressure has been associated with the structure of 30S ribosomal subunit [34,35], while it did not affect amino acid permeability, aminoacyl-tRNA formation [32], transpeptidation, initiation [31] or ribosome stability [36].

In our own recent studies, cell cultures of immortalized human chondrocytes [37] and HeLa cells were labelled for 12 hours with ^{35}S-methionine and ^{35}S-cysteine during 30 MPa hydrostatic pressurisation, and the cellular proteins were separated with two-dimensional gel electrophoresis (manuscript submitted). The pressure inhibited total protein synthesis by approximately 30–40%, suggesting that the translation machinery was markedly affected by the application of continuous high hydrostatic pressure. Radiography of two-dimensional gels showed a number of proteins whose relative synthesis level was altered. The protein spots that changed consistently in both cell types were chosen for mass spectrometric identification. One of the down-regulated proteins could be identified as eEF-2. In Western blot, a decrease in its amount was observed, while no change in the mRNA level of eEF-2 was evident after pressurisation. It is known that phosphorylation of eEF-2 prevents its participation in the translocation of ribosome along mRNA, however, we could not observe increased phosphorylation. It is possible that the main regulatory mechanism of protein synthesis in pressurised cells acts on inititation factors similarly to heat stress [38], however, elongation stage may be an additional regulator.

3. Hydrostatic pressure-sensitive proteins

Pressure treatment can result in certain specific responses in the protein expression of chondrocytic cells. Aggrecan, biglycan, decorin and type II collagen are the most studied molecules in chondrocytic cells. The investigations have mainly shown that intermittent loading at physiological level and frequencies stimulate the synthesis, while overloading has an inhibitory effect [2–4,9,39,40]. Mechanical compression and hydrostatic loading had similar frequency-dependent stimulation of proteoglycan synthesis in tissue explants, while monolayer cultures required longer loading period to be stimulated [39,41].

High continuous hydrostatic pressure increases especially the accumulation of heat shock protein 70 in the pressurised cells [42–45]. The synthesis of heat shock protein 90 beta was also induced in immortalised human chondrocytes and HeLa cells, while the alpha form remained at the control level [44, 46]. It is noteworthy that in bovine primary chondrocytes no heat shock protein response was detected, while, e.g., primary fibroblast and synovial cells responded strongly under pressure [47]. Although the mechanism behind this barotolerance is not known [48], the result may be interpreted to indicate that the loading history of the cells may affect their barotolerance. Interestingly, barotolerance of bacteria differs in various strains, and appears to be associated with the structure of 30S ribosomal subunit [34].

The effects of hydrostatic pressure on various proteins have been mainly investigated at mRNA level. High pressure induces the expression of interleukin-6, tumor necrosis factor-alpha and decreases transforming growth factor-beta mRNA levels [42,49]. In intervertebral disc, a physiological level of hydrostatic pressure had an anabolic influence on the tissue, while pressure levels of 3 MPa or higher caused a catabolic effect [50]. Recently, a cDNA array data showed that intermittent and static 5 MPa pressure had differential effects in chondrosarcoma cells [51]. High hydrostatic pressure revealed other specific cDNA array changes, of which the decreased mRNA and protein level of Id-1 and -3 proteins are interesting [52], since they are capable of directing cellular differentiation.

4. Messenger RNA stabilisation by high continuous hydrostatic pressure

High continuous pressure promotes an interesting cell biological phenomenon by increasing hsp70 mRNA stability, without any detectable induction of the actual gene activity [43]. High pressure may have selective effect on mRNA stability. It has been previously reported that histone mRNA levels were reduced partly because of suppression of transcription (one third) and partly due to loss of mRNA stability (two thirds) [53]. A recent cDNA study confirmed that sensitivity of mRNA stabilisation varies among the different transcripts [54]. Thus, high hydrostatic pressure may turn out an useful tool when the mechanisms of mRNA stabilisation are investigated.

References

[1] M.J. Lammi, R.K. Sironen, M.A. Elo, N. Oksala, K. Kaarniranta, H.M. Karjalainen and H.J. Helminen, Responses of mammalian cells to mechanical forces, in: *Recent Research Developments in Biophysics and Biochemistry*, Vol. 1. Research Signpost, Trivandrum, India, 2001, pp. 77–98.

[2] G.P. van Kampen, J.P. Veldhuijzen, R. Kuijer, R.J. van de Stadt and C.A. Schipper, Cartilage response to mechanical force in high-density chondrocyte cultures, *Arthritis Rheum.* **28** (1985), 419–424.

[3] L. Lippiello, C. Kaye, T. Neumata and H.J. Mankin, In vitro metabolic response of articular cartilage segments to low levels of hydrostatic pressure, *Connect. Tiss. Res.* **13** (1985), 99–107.

[4] A.C. Hall, J.P. Urban and K.A. Gehl, The effects of hydrostatic pressure on matrix synthesis in articular cartilage, *J. Orthop. Res.* **9** (1991), 1–10.

[5] F. Lafeber, J.P. Veldhuijzen, J.L. Vanroy, O. Huber-Bruning and J.W. Bijlsma, Intermittent hydrostatic compressive force stimulates exclusively the proteoglycan synthesis of osteoarthritic human cartilage, *Br. J. Rheumatol.* **31** (1992), 437–442.

[6] M.J. Lammi, R. Inkinen, J.J. Parkkinen, T. Häkkinen, M. Jortikka, L.O. Nelimarkka, H.T. Järveläinen and M.I. Tammi, Expression of reduced amounts of structurally altered aggrecan in articular cartilage chondrocytes exposed to high hydrostatic pressure, *Biochem. J.* **304** (1994), 723–730.
[7] R.L. Smith, S.F. Rusk, B.E. Ellison, P. Wessells, K. Tsuchiya, D.R. Carter, W.E. Caler, L.J. Sandell and D.J. Schurman, In vitro stimulation of articular chondrocyte mRNA and extracellular matrix synthesis by hydrostatic pressure, *J. Orthop. Res.* **14** (1996), 53–60.
[8] M.O. Jortikka, J.J. Parkkinen, R.I. Inkinen, J. Kärner, H.T. Järveläinen, L.O. Nelimarkka, M.I. Tammi and M.J. Lammi, The role of microtubules in the regulation of proteoglycan synthesis in chondrocytes under hydrostatic pressure, *Arch. Biochem. Biophys.* **374** (2000), 172–180.
[9] T. Ikenoue, M.C. Trindade, M.S. Lee, E.Y. Lin, D.J. Schurman, S.B. Goodman and R.L. Smith, Mechanoregulation of human articular chondrocyte aggrecan and type II collagen expression by intermittent hydrostatic pressure in vitro, *J. Orthop. Res.* **21** (2003), 110–116.
[10] J.A. Browning, R.E. Walker, A.C. Hall and R.J. Wilkins, Modulation of Na^+ x H^+ exchange by hydrostatic pressure in isolated bovine articular chondrocytes, *Acta Physiol. Scand.* **166** (1999), 39–45.
[11] M.O. Wright, R.A. Stockwell and G. Nuki, Response of plasma membrane to applied hydrostatic pressure in chondrocytes and fibroblasts, *Connect. Tiss. Res.* **28** (1992), 49–70.
[12] A.C. Hall, Differential effects of hydrostatic pressure on cation transport pathways of isolated articular chondrocytes, *J. Cell. Physiol.* **178** (1999), 197–204.
[13] M.S. Lee, M.C. Trindade, T. Ikenoue, D.J. Schurman, S.B. Goodman and R.L. Smith, Intermittent hydrostatic pressure inhibits shear stress-induced nitric oxide release in human osteoarthritic chondrocytes in vitro, *J. Rheumatol.* **30** (2003), 326–328.
[14] L.A. Bourret and G.A. Rodan, Inhibition of cAMP accumulation in epiphyseal cartilage cells exposed to physiological pressure, *Calcif. Tiss. Res.* **21**(Suppl.) (1976), 431–436.
[15] J.J. Parkkinen, M.J. Lammi, R. Inkinen, M. Jortikka, M. Tammi, I. Virtanen and H.J. Helminen, Influence of short-term hydrostatic pressure on organization of stress fibers in cultured chondrocytes, *J. Orthop. Res.* **13** (1995), 495–502.
[16] J.J. Parkkinen, M.J. Lammi, A. Pelttari, H.J. Helminen, M. Tammi and I. Virtanen, Altered Golgi apparatus in hydrostatically loaded articular cartilage chondrocytes, *Ann. Rheum. Dis.* **52** (1993), 192–198.
[17] D.B. Saris, A. Sanyal, K.N. An, J.S. Fitzsimmons and S.W. O'Driscoll, Periosteum responds to dynamic fluid pressure by proliferating in vitro, *J. Orthop. Res.* **17** (1999), 668–677.
[18] T.E. Dever, Gene-specific regulation by general translation factors, *Cell* **108** (2002), 545–556.
[19] G.S. Wilkie, K.S. Dickson and N.K. Gray, Regulation of mRNA translation by 5′- and 3′-UTR-binding factors, *Trends Biochem. Sci.* **28** (2003), 182–188.
[20] T.V. Pestova, V.G. Kolupaeva, I.B. Lomakin, E.V. Pilipenko, I.N. Shatsky, V.I. Agol and C.U. Hellen, Molecular mechanisms of translation initiation in eukaryotes, *Proc. Natl. Acad. Sci. USA* **98** (2001), 7029–7036.
[21] M. Kozak, Pushing the limits of the scanning mechanism for initiation of translation, *Gene* **299** (2002), 1–34.
[22] M.J. Clemens, Initiation factor eIF2 alpha phosphorylation in stress responses and apoptosis, *Prog. Mol. Subcell. Biol.* **27** (2001), 57–89.
[23] B.L. Webb and C.G. Proud, Eukaryotic initiation factor 2B (eIF2B), *Int. J. Biochem. Cell Biol.* **29** (1997), 1127–1131.
[24] G.C. Scheper and C.G. Proud, Does phosphorylation of the cap-binding protein eIF4E play a role in translation initiation?, *Eur. J. Biochem.* **269** (2002), 5350–5359.
[25] A. Flynn and C.G. Proud, Insulin and phorbol ester stimulate initiation factor eIF-4E phosphorylation by distinct pathways in Chinese hamster ovary cells overexpressing the insulin receptor, *Eur. J. Biochem.* **236** (1996), 40–47.
[26] G.J. Browne and C.G. Proud, Regulation of peptide-chain elongation in mammalian cells, *Eur. J. Biochem.* **269** (2002), 5360–5368.
[27] E.C. Pollard and P.K. Weller, Effect of hydrostatic pressure on the synthetic processes in bacteria, *Bibl. Laeger* **112** (1966), 573–580.
[28] J.V. Landau, Protein and nucleic acid synthesis in Escherichia coli: pressure and temperature effects, *Science* **153** (1966), 1273–1274.
[29] J.V. Landau, Induction, transcription and translation in Escherichia coli: a hydrostatic pressure study, *Biochim. Biophys. Acta* **149** (1967), 506–512.
[30] R.M. Arnold and L.J. Albright, Hydrostatic pressure effects on the translation stages of protein synthesis in a cell-free system from Escherichia coli, *Biochim. Biophys. Acta* **238** (1971), 347–354.
[31] J.R. Schwarz and J.V. Landau, Inhibition of cell-free protein synthesis by hydrostatic pressure, *J. Bacteriol.* **112** (1972), 1222–1227.
[32] J.R. Schwarz and J.V. Landau, Hydrostatic pressure effects on Escherichia coli: site of inhibition of protein synthesis, *J. Bacteriol.* **109** (1972), 945–948.
[33] A.C. Scheck and J.V. Landau, The stability of rabbit reticulocyte polysomes at high hydrostatic pressure, *Biochim. Biophys. Acta* **699** (1982), 226–231.

[34] W. Smith, D. Pope and J.V. Landau, Role of bacterial ribosome subunits in barotolerance, *J. Bacteriol.* **124** (1975), 582–584.

[35] J.V. Landau, W.P. Smith and D.H. Pope, Role of the 30S ribosomal subunit, initiation factors, and specific ion concentration in barotolerant protein synthesis in Pseudomonas bathycetes, *J. Bacteriol.* **130** (1977), 154–159.

[36] D.H. Pope, N.T. Connors and J.V. Landau, Stability of Escherichia coli polysomes at high hydrostatic pressure, *J. Bacteriol.* **121** (1975), 753–758.

[37] M.B. Goldring, J.R. Birkhead, L.F. Suen, R. Yamin, S. Mizuno, J. Glowacki, J.L. Arbiser and J.F. Apperley, Interleukin-1 beta-modulated gene expression in immortalized human chondrocytes, *J. Clin. Invest.* **94** (1994), 2307–2316.

[38] R. Panniers, Translational control during heat shock, *Biochimie* **76** (1994), 737–747.

[39] J.J. Parkkinen, J. Ikonen, M.J. Lammi, J. Laakkonen, M. Tammi and H.J. Helminen, Effects of cyclic hydrostatic pressure on proteoglycan synthesis in cultured chondrocytes and articular cartilage explants, *Arch. Biochem. Biophys.* **300** (1993), 458–465.

[40] R.L. Smith, J. Lin, M.C. Trindade, J. Shida, G. Kajiyama, T. Vu, A.R. Hoffman, M.C. van der Meulen, S.B. Goodman, D.J. Schurman and D.R. Carter, Time-dependent effects of intermittent hydrostatic pressure on articular chondrocyte type II collagen and aggrecan mRNA expression, *J. Rehabil. Res. Dev.* **37** (2000), 153–161.

[41] J.J. Parkkinen, M.J. Lammi, H.J. Helminen and M. Tammi, Local stimulation of proteoglycan synthesis in articular cartilage explants by dynamic compression in vitro, *J. Orthop. Res.* **10** (1992), 610–620.

[42] K. Takahashi, T. Kubo, K. Kobayashi, J. Imanishi, M. Takigawa, Y. Arai and Y. Hirasawa, Hydrostatic pressure influences mRNA expression of transforming growth factor-beta 1 and heat shock protein 70 in chondrocyte-like cell line, *J. Orthop. Res.* **15** (1997), 150–158.

[43] K. Kaarniranta, M. Elo, R. Sironen, M.J. Lammi, M.B. Goldring, J.E. Eriksson, L. Sistonen and H.J. Helminen, Hsp70 accumulation in chondrocytic cells exposed to high continuous hydrostatic pressure coincides with mRNA stabilization rather than transcriptional activation, *Proc. Natl. Acad. Sci. USA* **95** (1998), 2319–2324.

[44] M.A. Elo, R.K. Sironen, K. Kaarniranta, S. Auriola, H.J. Helminen and M.J. Lammi, Differential regulation of stress proteins by high hydrostatic pressure, heat shock, and unbalanced calcium homeostasis in chondrocytic cells, *J. Cell. Biochem.* **79** (2000), 610–619.

[45] R. Sironen, M. Elo, K. Kaarniranta, H.J. Helminen and M.J. Lammi, Transcriptional activation in chondrocytes submitted to hydrostatic pressure, *Biorheology* **37** (2000), 85–93.

[46] M.A. Elo, R.K. Sironen, H.M. Karjalainen, K. Kaarniranta, H.J. Helminen and M.J. Lammi, Specific induction of heat shock protein 90beta by high hydrostatic pressure, *Biorheology* **40** (2003), 141–146.

[47] K. Kaarniranta, C.I. Holmberg, M.J. Lammi, J.E. Eriksson, L. Sistonen and H.J. Helminen, Primary chondrocytes resist hydrostatic pressure-induced stress while primary synovial cells and fibroblasts show modified Hsp70 response, *Osteoarthritis Cartilage* **9** (2001), 7–13.

[48] K. Kaarniranta, C.I. Holmberg, H.J. Helminen, J.E. Eriksson, L. Sistonen and M.J. Lammi, Protein synthesis is required for stabilization of hsp70 mRNA upon exposure to both hydrostatic pressurization and elevated temperature, *FEBS Lett.* **475** (2000), 283–286.

[49] K. Takahashi, T. Kubo, Y. Arai, I. Kitajima, M. Takigawa, J. Imanishi and Y. Hirasawa, Hydrostatic pressure induces expression of interleukin 6 and tumour necrosis factor alpha mRNAs in a chondrocyte-like cell line, *Ann. Rheum. Dis.* **57** (1998), 231–236.

[50] T. Handa, H. Ishihara, H. Ohshima, R. Osada, H. Tsuji and K. Obata, Effects of hydrostatic pressure on matrix synthesis and matrix metalloproteinase production in the human lumbar intervertebral disc, *Spine* **22** (1997), 1085–1091.

[51] H.M. Karjalainen, R.K. Sironen, M.A. Elo, K. Kaarniranta, M. Takigawa, H.J. Helminen and M.J. Lammi, Gene expression profiles in chondrosarcoma cells subjected to cyclic stretching and hydrostatic pressure. A cDNA array study, *Biorheology* **40** (2003), 93–100.

[52] R.K. Sironen, H.M. Karjalainen, M.A. Elo, K. Kaarniranta, K. Törrönen, M. Takigawa, H.J. Helminen and M.J. Lammi, cDNA array reveals mechanosensitive genes in chondrocytic cells under hydrostatic pressure, *Biochim. Biophys. Acta* **1591** (2002), 45–54.

[53] A.L. Symington, S. Zimmerman, J. Stein, G. Stein and A.M. Zimmerman, Hydrostatic pressure influences histone mRNA, *J. Cell Sci.* **98** (1991), 123–129.

[54] R.K. Sironen, H.M. Karjalainen, K. Törrönen, M.A. Elo, K. Kaarniranta, M. Takigawa, H.J. Helminen and M.J. Lammi, High pressure effects on cellular mRNA stability. A cDNA array analysis, *Biorheology* **39** (2002), 111–117.

Biorheology 41 (2004) 315–322
IOS Press

Cycle number and waveform of fluid flow affect bovine articular chondrocytes [1]

Maximilian Edlich [a,*], Clare E. Yellowley [b], Christopher R. Jacobs [b] and Henry J. Donahue [b]
[a] *Department of Orthopaedics, University Hospital, University of Freiburg, Hugstetter Strasse 55, 79106 Freiburg, Germany*
[b] *Musculoskeletal Research Laboratory, Departments of Orthopaedics and Rehabilitation and Cellular and Molecular Physiology, The Pennsylvania State University College of Medicine, 500 University Dr, Hershey, PA 17033, USA*

Abstract. Many studies have shown that a loading-induced (bio)physical signal regulates chondrocyte behavior. In a recent study our group has demonstrated the shear stress level- and frequency-dependent effect of sinusoidal oscillatory fluid flow on bovine articular chondrocyte (BAC) cytosolic calcium concentration ($[Ca^{2+}]_i$), neglecting the fact that chondrocytes are not likely to see these ideal waveform *in vivo* or *in vitro*. Furthermore, possible overload of articular cartilage or excessive shear stress in chondrocyte cultures are more likely to be of a short nature. Therefore, in this study we choose to investigate a saw-tooth waveform oscillating fluid flow at varying exposure times in comparison to the established sinusoidal oscillatory waveform. $[Ca^{2+}]_i$, as an early signaling molecule, was quantified using the fluorescent dye fura-2. BAC were exposed to 1 Hz sinusoidal or saw-tooth waveform oscillating fluid flow at 2.2 Pa flow rates in a parallel plate flow chamber for 8 different loading times. As little as 5 cycles of oscillatory fluid flow were sufficient to increase $[Ca^{2+}]_i$ significantly over baseline. The number of responding cells could not be increased any further after a sufficient number of cycles (11), regardless of the waveform. Furthermore, a saw-tooth waveform appeared to be more stimulatory than regular sinusoidal oscillating flow at higher cycle numbers. BAC appear to be able to respond to these biophysical stimuli in a differentiated manner. This ability might give every single chondrocyte the capability to maintain its territory autonomously, since chondrocytes distributed in articular cartilage without the possibility to interact, e.g., via cell processes.

Keywords: Chondrocytes, *in vitro*, mechanotransduction, fluid flow, shear stress, calcium

1. Introduction

Normal joint function depends on an intact, "healthy", articular cartilage. The extracellular matrix (ECM) and the bound water are the main determents of articular cartilage's properties. Chondrocytes are randomly distributed within the cartilage and maintain its homeostasis [8,17]. Whereas physiological mechanical load is necessary for the chondrocytes' normal function, abnormal loads such as overload due to non-physiologic joint axis, may lead to diseases such as degenerative osteoarthritis [2,21,23]. This knowledge has led to different options for treatment of cartilage diseases. Besides the orthopaedic standard of correcting joint axis, more recently tissue engineered cartilage implants

[1]Presented in part at the 45th Annual Meeting of the Orthopedic Research Society in Orlando, FL, 2000 and the 12th Conference of the European Society of biomechanics in Dublin, Ireland, 2000.

*Address for correspondence: Maximilian Edlich, Department of Orthopaedics, University Hospital, University of Freiburg, Hugstetter Strasse 55, 79106 Freiburg, Germany. Tel.: +49 717 531 4819; Fax: +49 717 531 7583; E-mail: edlich@uni-freiburg.de.

have been investigated. Still, the cellular level reaction of chondrocytes to biophysical stimuli remains poorly understood.

Physical loading of articular cartilage may lead to direct cell deformation/strain [9,16] bioelectrical phenomena [3], convective mass transport [20], tissue pressurization [15] and fluid flow/shear stress [4]. Fluid flow has been shown to influence chondrocyte biosynthetic activity [4,10,25]. In a recent study our group has demonstrated the shear stress level and frequency dependent effect of oscillatory fluid flow on bovine articular chondrocytes intracellular calcium concentration [6]. This study investigated the effect of sinusoidal oscillating fluid flow, neglecting the fact that chondrocytes are not likely to see these ideal waveforms *in vivo* or *in vitro* (tissue engineering). Additionally, previous investigations focused on a long-term fluid flow exposure from several minutes to some hours. Possible overload of articular cartilage or excessive shear stress in chondrocyte cultures are more likely to be of a short nature. Therefore, in this study we chose to investigate a saw-tooth waveform oscillating fluid flow at varying exposure times in comparison to the established sinusoidal oscillating waveform.

2. Materials and methods

2.1. Cell culture

BAC were isolated and cultured as described previously [25]. Briefly, articular cartilage from bovine hock joints was chopped into small pieces and digested for 2 hours at 37°C in a mixture of 0.15 mg/ml DNAse, 2 mg/ml collagenase and 0.1 mg/ml hyaluronidase in a spinner flask. Cells were cultured in media containing RPMI 1640 with HEPES, 20% FBS and 2% penicillin + streptomycin. Cells were then subcultured onto microscope quartz glass slides (19 cm^2) (Fridrich & Dimmock, Miville, NJ) and grown to 80% confluency. We have previously demonstrated that cells isolated in this manner display characteristics of the chondrocyte phenotype including expression of type II collagen [25].

2.2. Fluid flow

Cells were exposed to fluid flow as previously described [6,12]. Briefly, the slides with cells were mounted in a parallel plate flow chamber. The flow chamber was then attached to a custom designed fluid pump via rigid wall tubing. The oscillating flow was provided by a 500 μl syringe mounted in a computer controlled servopneumatic material testing machine (Quick Test; Endura-Tec, Eden Prairie, MN). A sinusoidal oscillating flow profile was achieved via programming appropriate piston displacement in the computer control software. Additionally, a saw-tooth waveform was created (Fig. 1). The resulting flow rate (300 μm/sec) was monitored with a flow probe (T106, Transsonic Systems Inc., Ithaca, NY). From the given flow channel size, of 38 $\times$ 20 $\times$ 0.28 mm, the resulting shear stress on the cells (2.2 Pa) can be estimated by the equation:

$$\tau = \frac{6\mu Q}{\beta h^2},$$

where Q is the flow rate, τ is the shear stress, μ is the viscosity, β is the chamber width, and h the chamber height. The cells were exposed to flow regimes utilizing Tyrode's solution with 2% FBS. No-flow controls were obtained by following the flow protocol in all aspects without activating the pump.

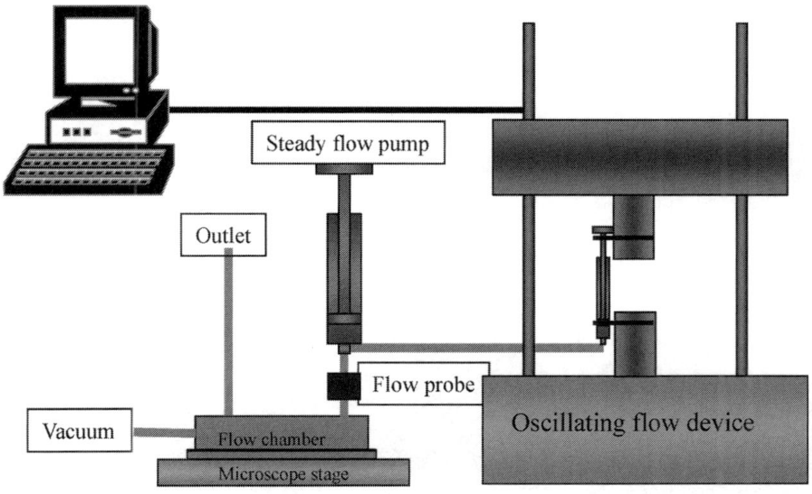

Fig. 1. Experimental setup to apply steady and oscillating flow *in vitro*. The slide with cells facing the flow channel is held in place via vacuum sealing. The flow chamber is mounted upside-down on the stage of an inverted microscope. Flow is applied by either a steady flow pump or a custom made, computer controlled, oscillating flow device. Flow velocity is measured by an ultrasound flow probe. Control experiments were done with the same setup without applying flow.

2.3. Calcium imaging

$[Ca^{2+}]_i$ was quantified as previously described [25], using the fluorescent Ca^{2+} indicator fura-2. Calibration ratios were determined with calcium standards supplied by the manufacturer (Molecular Probes, Eugene, OR). Cells were loaded with 1 μM fura-2-am for 30 minutes, and extracellular dye was removed. The cells were then mounted in a parallel plate flow chamber and placed on to the microscope stage (Nikon Diaphot 300, Nikon, Melville, NY City). Cells were alternately illuminated at 340 nm and 380 nm and the emitted light was detected by an ICCD camera at a rate of 1 image every 2.5 seconds. Data acquisition and analysis software were used to capture and calculate fluorescent signal intensity which was converted to $[Ca^{2+}]_i$ values (Metafluor; Universal Imaging, West Chester, PA). Each experiment consisted of 1-min of imaging without applied flow (baseline), followed by a flow period of 180 cycles (3 min), 90, 45, 22, 11, 5, 2 and 1 cycle. In all cases imaging was done for 3 minutes. Each cell in the field was manually outlined, utilizing analysis software and a time course of $[Ca^{2+}]_i$ changes was computed for each cell independently (Fig. 2). All intracellular Ca^{2+} data was transferred to a Microsoft Excel spread sheet for further analysis.

2.4. Data analysis

Transients in $[Ca^{2+}]_i$ were examined by utilizing an algorithm, which allows one to separate smaller transients, superimposed over larger alterations in $[Ca^{2+}]_i$ [13]. The maximum $[Ca^{2+}]_i$ transient was determined for both the baseline and the flow periods for each outlined cell. Maximum $[Ca^{2+}]_i$ transients for the baseline period for the cells in each slide were averaged and twice this value used as a cutoff to eliminate background noise. Cells exhibiting a $[Ca^{2+}]_i$ transient larger than the cutoff during flow, but not during baseline, were considered responsive. The number of responding cells was expressed as a percentage of total number of cells analyzed. Data are shown as percentage of cells responding,

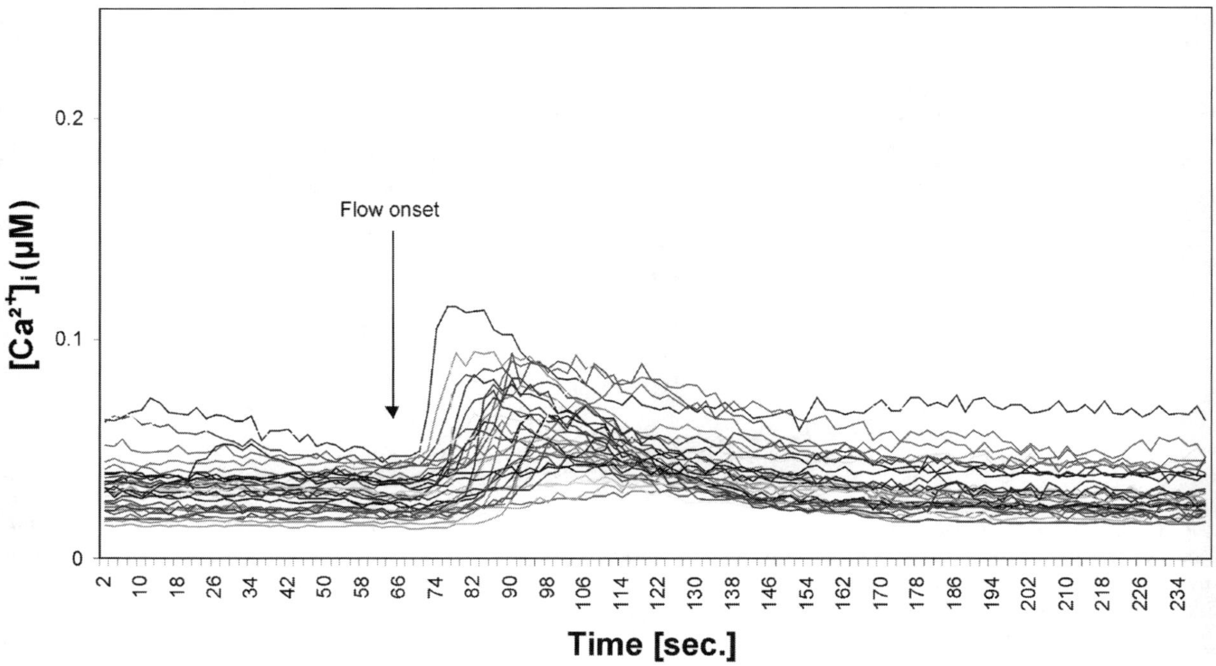

Fig. 2. Timecourse of changes in $[Ca^{2+}]_i$ of BAC. Each line represents 1 individual cell. The arrow indicates the onset of a 1 Hz 300 μl/s peak amplitude flow profile with 2% FBS. Note the larger $[Ca^{2+}]_i$ transients, indicating a response, with smaller superimposed transients, after flow onset.

±standard error of proportion. The percentage of responding cells for each group was compared utilizing analysis of variance. The difference between waveforms at 180 cycles was compared utilizing a Chi-square test with orthogonal contrast.

3. Results

As few as 5 cycles of oscillatory fluid flow producing shear stress (2.2 Pa) were sufficient to create an increase of $[Ca^{2+}]_i$ for sinusoidal and saw-tooth waveforms. For sinusoidal oscillating flow regimes 61–109 individual cells were examined and for saw-tooth oscillating flow 62–87 cells were examined on a minimum of 3 slides each. Under control no-flow conditions 3 of 32 cells displayed spontaneous $[Ca^{2+}]_i$ oscillations (data not shown). The number of cells responding to sinusoidal oscillating flow was maximal at 22 cycles (63% of cells responding) (Fig. 3A) and at 180 cycles (74%) for saw-tooth oscillating flow (Fig. 3B). There was a statistically significant difference between 1, 2 or 5 cycle(s) and 11, 22, 45, 90 and 180 cycles for sinusoidal and saw-tooth oscillating flow ($p < 0.05$). 1 and 2 cycle(s) of both waveforms showed no statistically significant difference from no flow controls. These results show three important trends. First, the number of cycles has an influence on the response of BAC to fluid flow such that 1 and 2 cycles do not stimulate BAC, whereas higher cycle numbers do. Second, after a sufficient number of cycles, 11 cycles in this experiment, no further stimulation can be achieved with additional cycles. Third, a saw-tooth waveform appears to be more stimulatory than a regular sinusoidal waveform.

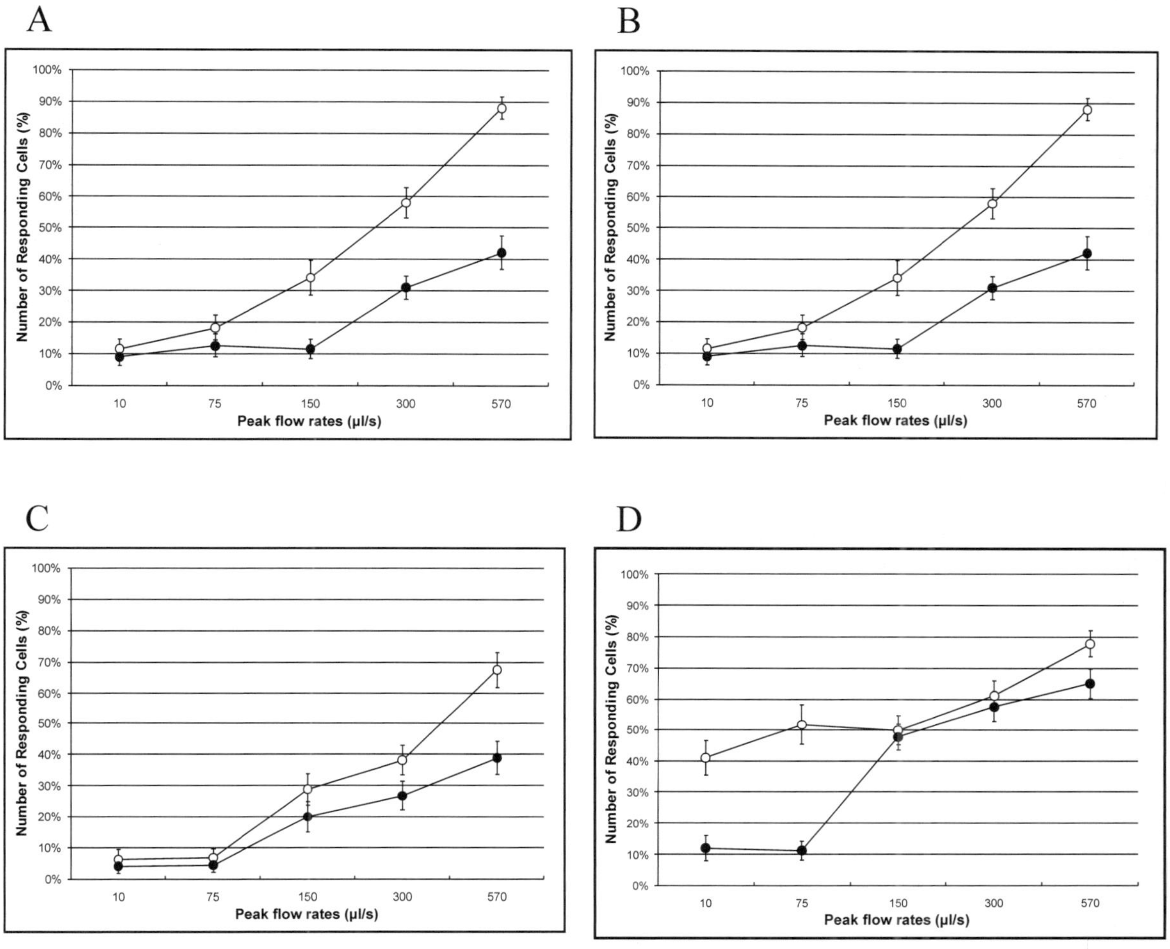

Fig. 3. Flow rate-dependent effect of fluid flow on $[Ca^{2+}]_i$ in BAC. The y axis shows the percentage of cells responding, ±standard error of proportion. The x-axis shows all tested flow rates, with (empty circles) and without 2% FBS (full circles). Panel A shows 0.5 Hz oscillating flow, panel B shows 1 Hz oscillating flow panel, C shows 5 Hz oscillating flow and panel D shows steady flow. $N = 89$–148 individual cells in each condition.

4. Discussion

Articular cartilage is known to react to tissue-level mechanical load with synthesis or degradation of parts of the ECM. This connection has been shown in many different studies. The exact biophysical signal chondrocytes perceive and react to remains poorly understood. Data from our group has shown the influence of oscillatory fluid flow on $[Ca^{2+}]_i$ of bone cells [12] and BAC [6]. To our knowledge we have shown for the first time a dose and frequency depending effect of fluid flow on BAC. The aim of this study was to investigate the possible effect of cycle number (duration of fluid flow) and waveform on BAC. As little as 5 cycles of sinusoidal or saw-tooth waveform at 1 Hz oscillatory fluid flow were sufficient to increase $[Ca^{2+}]_i$ significantly over baseline. The number of responding cells could not be increased any further after a sufficient number of cycles (11) regardless of the waveform. Furthermore, saw-tooth oscillating flow appeared to be more stimulatory than regular sinusoidal oscillating flow at higher cycle numbers.

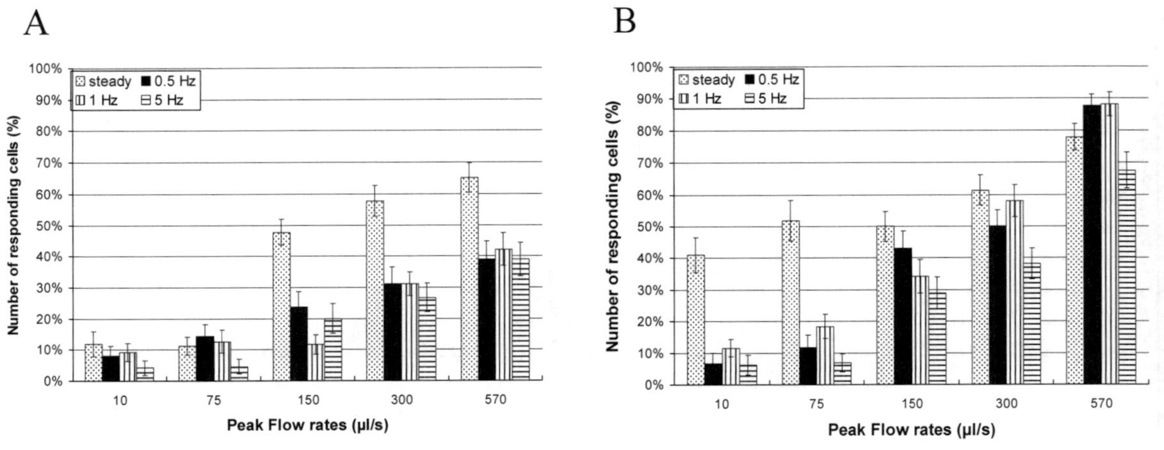

Fig. 4. Effect of flow rate on the number of cells responding ($n = 89$–148). The y-axis shows the percentage of cells responding, ±standard error of proportion. Panel A shows conditions without and Panel B shows conditions with 2% FBS. Note the increasing number of cells reponding at increasing flow rates ($p < 0.0001$). Steady flow was more stimulatory than oscillating fluid flow over all conditions tested ($p < 0.0001$). However, in the absence of FBS there was no difference between 10 mm^3/s and 75 μl/s peak flow amplitudes between oscillating and steady flow. In contrast, in the presence of 2% FBS was no significant difference between 300 μl/s and 570 mm^3/s peak flow rate. $N = 89$–148 individual cells in each condition.

Table 1

Tested flow rates in μl/s and resulting shear stresses

Flow rate [μl/s]	Shear stress [Pa]
10	0.05
75	0.5
150	1.1
300	2.2
570	4.4

Fluid flow has been shown to affect many different cell types. Although numerous investigators have demonstrated the effect of fluid flow, and therefore possibly shear stress on articular chondrocytes, predicted fluid flow and shear stress values *in vivo* might be lower than those applied in *in vitro* studies including our own [18]. Therefore other fluid flow associated phenomena might play a role in mechanotransduction in BAC *in vitro*. Streaming potentials are one possible candidate. There are several studies demonstrating that streaming potentials are not crucial in mechanotransduction in bone cells *in vitro* [11, 14,22]. Another possible way fluid flow might influence mechanotransduction in chondrocytes might be the transport of biochemical agents including nutrients, (ant)agonists or para/autocrine components. Studies from our lab show the influence of FBS on BAC's $[Ca^{2+}]_i$ [6]. Similar results were reported by Allen et al in bone cells [1]. Whether electrical phenomena or biochemical transport is crucial or just sustaining mechanotransduction in chondrocytes *in vitro/vivo* remains unclear and opens interesting perspectives for further studies. At the current stage of research fluid flow/shear stress is a well defined biophysical stimulus for *in vitro* studies of mechanotransduction of articular chondrocytes, helping to understand diseases such as osteoarthritis and possibly develop ideal environments for tissue engineering requirements.

Various waveforms are known to effect trabecular bone remodeling *in vivo* [7]. Endothelial cells *in vitro* are able to differentiate between different forms of intermittent loading, such as various fluid flow regimes [5] or pulsatile and square waveforms of fluid flow [19]. We have found more BACs responding to oscillatory fluid flow with a saw-tooth waveform, compared to a regular sinusoidal waveform. This effect might be due to the higher rate of change (slope) in flow rate in the saw-tooth waveform compared to the sinusoidal waveform and/or to the larger fluid volume moved, indicating a possible chemotransport-dependant effect.

Interestingly we have found BAC to respond to oscillatory fluid flow in a cycle-number dependent manner, not in an all-or-nothing manner. In other cell types time dependency to cellular responses is well described, e.g., for bovine aortic endothelial cells [26]. For cartilage, similar results are found by Steinmeyer et al. [24]. The effect of fluid flow on chondrocytes *in vitro* is time dependent as well [10]. Although, other authors found the time-dependent cellular response to mechanical load on a much larger scale (hours to days) than we did, they did not investigate early messenger, like $[Ca^{2+}]_i$ as we did.

In conclusion, our results indicate that chondrocytes have the ability to sense different waveforms and cycle numbers of fluid flow, as wells as fluid flow level and rate [6]. Furthermore they appear to be able to respond to these biophysical stimuli in a differentiated manner. This ability might give every single chondrocyte the capability to maintain his territory autonomously.

Acknowledgements

This work was supported by the Whitaker foundation and the US Army Medical Research Acquisition Activity Grant 17-98-85-09 (CRJ) and the National Institutions of Health Grant AG 15107 (HJD).

References

[1] F.D. Allen et al., Serum modulates the intracellular calcium response of primary cultured bone cells to shear flow, *J. Biomech.* **33**(12) (2000), 1585–1591.
[2] R.D. Altman et al., Preliminary observations of chondral abrasion in a canine model, *Ann. Rheum. Dis.* **51**(9) (1992), 1056–62.
[3] C.T. Brighton et al., Proliferative and synthetic response of bovine growth plate chondrocytes to various capacitively coupled electrical fields, *J. Orthop. Res.* **7**(5) (1989), 759–65.
[4] M.D. Buschmann et al., Stimulation of aggrecan synthesis in cartilage explants by cyclic loading is localized to regions of high interstitial fluid flow, *Arch. Biochem. Biophys.* **366**(1) (1999), 1–7.
[5] D.C. Chappell et al., Oscillatory shear stress stimulates adhesion molecule expression in cultured human endothelium, *Circ. Res.* **82**(5) (1998), 532–9.
[6] M.J. Edlich et al., Oscillating fluid flow regulates cytosolic calcium concentration in bovine articular chondrocytes, *J. Biomech.* **31**(1) (2001), 59–64.
[7] S.A. Goldstein et al., Trabecular bone remodeling: an experimental model, *J. Biomech.* **24** (Suppl. 1) (1991), 135–50.
[8] D. Heinegard et al., Structure and biology of cartilage and bone matrix noncollagenous macromolecules, *FASEB J.* **3**(9) (1989), 2042–51.
[9] K. Honda et al., The effects of high magnitude cyclic tensile load on cartilage matrix metabolism in cultured chondrocytes, *Eur. J. Cell. Biol.* **79**(9) (2000), 601–9.
[10] C.T. Hung et al., Mitogen-activated protein kinase signaling in bovine articular chondrocytes in response to fluid flow does not require calcium mobilization, *J. Biomech.* **33**(1) (2000), 73–80.
[11] C.T. Hung et al., Real-time calcium response of cultured bone cells to fluid flow, *Clin. Orthop.* (313) (1995), 256–69.
[12] C.R. Jacobs et al., Differential effect of steady versus oscillating flow on bone cells, *J. Biomechan.* **31**(11) (1998), 969–976.
[13] C.R. Jacobs et al., A novel application of rainflow counting to time-varying biophysical data, *Medical and Biological Engineering and Computing* **3** (2000), 31–40.
[14] J. Klein-Nulend et al., Sensitivity of osteocytes to biomechanical stress *in vitro*, *FASEB J.* **9**(5) (1995), 441–5.
[15] J. Klein-Nulend et al., Influence of intermittent compressive force on proteoglycan content in calcifying growth plate cartilage *in vitro*, *J. Biol. Chem.* **262**(32) (1987), 15490–5.

[16] D.A. Lee et al., Dynamic mechanical compression influences nitric oxide production by articular chondrocytes seeded in agarose, *Biochem. Biophys. Res. Commun.* **251**(2) (1998), 580–5.
[17] V.C. Mow et al., Stress, strain pressure and flow fields in articular cartilage and chondrocytes, in: *Cell Mechanics and Cellular Engineering*, V.C. Mow, F. Guilak, R. Tran-Son-Tay and R.M. Hochmuth, eds, Springer-Verlag, New York, 1994, pp. 345–379.
[18] V.C. Mow et al., The extracellular matrix, interstitial fluid and ions as a mechanical signal transducer in articular cartilage, *Osteoarthritis Cartilage* **7**(1) (1999), 41–58.
[19] M. Noris et al., Nitric oxide synthesis by cultured endothelial cells is modulated by flow conditions, *Circ. Res.* **76**(4) (1995), 536–43.
[20] B.P. O'Hara et al., Influence of cyclic loading on the nutrition of articular cartilage, *Ann. Rheum. Dis.* **49**(7) (1990), 536–9.
[21] M. Palmoski et al., Development and reversal of a proteoglycan aggregation defect in normal canine knee cartilage after immobilization, *Arthritis Rheum.* **22**(5) (1979), 508–17.
[22] K.M. Reich et al., Fluid shear stress as a mediator of osteoblast cyclic adenosine monophosphate production, *J. Cell. Physiol.* **143**(1) (1990), 100–4.
[23] L.A. Setton et al., Altered mechanics of cartilage with osteoarthritis: human osteoarthritis and an experimental model of joint degeneration, *Osteoarthritis Cartilage* **7**(1) (1999), 2–14.
[24] J. Steinmeyer et al., Effects of intermittently applied cyclic loading on proteoglycan metabolism and swelling behaviour of articular cartilage explants, *Osteoarthritis Cartilage* **7**(2) (1999), 155–64.
[25] C.E. Yellowley et al., Effects of fluid flow on intracellular calcium in bovine articular chondrocytes, *American Journal of Physiology – Cell Physiology* **273**(1) (Pt 1) (1997), C30–C36.
[26] T. Ziegler et al., Influence of oscillatory and unidirectional flow environments on the expression of endothelin and nitric oxide synthase in cultured endothelial cells, *Arterioscler. Thromb. Vasc. Biol.* **18**(5) (1998), 686–92.

Biorheology 41 (2004) 323–333
IOS Press

The influence of oxygen and hydrostatic pressure on articular chondrocytes and adherent bone marrow cells *in vitro*

Katrin Scherer [a], Michael Schünke [a], Roland Sellckau [b], Joachim Hassenpflug [c] and Bodo Kurz [a,*]

[a] *Anatomisches Institut, CAU zu Kiel, Olshausenstr. 40, D-24098 Kiel, Germany*
[b] *Endo-Klinik, Hamburg, Germany*
[c] *Klinik für Orthopädie des Universitätsklinikums Schleswig-Holstein, Campus Kiel, Kiel, Germany*

Abstract. Tissue engineering of articular cartilage from chondrocytes or stem cells is considered to be a potential aspect in the treatment of cartilage defects. In order to optimize culture conditions the influence of low oxygen tension (5%) – single or in combination with intermittent hydrostatic pressure (HP: 30/2 min on/off loading; 0.2 MPa) – on the biosynthetic activity (sulfate and proline incorporation) of human osteoarthritic chondrocytes cultured on collagen I/III membranes was investigated. Additionally, chondrogenesis from high density or monolayer cultures of bovine adherent bone marrow cells (aBMC) with and without chondrogenic medium supplements (CM) was analyzed by RT-PCR (mRNA expression of aggrecan and collagen type II). We could show that low oxygen tension increases significantly the biosynthesis of collagen I/III membrane-associated chondrocytes and even higher under co-stimulation with HP. While there is no chondrogenesis in monolayer cultures, CM induces expression of cartilage matrix molecules in high density cultures of aBMC which is even increased under the influence of low oxygen tension. Both, low oxygen tension and HP without CM are alone not sufficient stimuli for chondrogenesis. It can be concluded that low oxygen tension and HP might be useful tools in cartilage tissue engineering and that these physico-chemical factors promote but do not induce chondrogenesis under the given conditions.

Keywords: Chondrogenesis, biosynthesis, high density culture, collagen I/III membrane

1. Introduction

Since articular cartilage is a tissue with a low regenerative capacity, cartilage defects appear to be a major problem. Without treatment they usually lead to degenerative diseases like osteoarthritis. Tissue engineering of articular cartilage is one potential aspect in the treatment of cartilage defects. However, the source of cells and culture conditions still have to be optimized. It might be speculated, that factors that influence the microenvironment of the cells *in vivo* (like oxygen and hydrostatic pressure (HP)) also influence chondrocyte proliferation, matrix production, or phenotype stability as well as chondrogenesis *in vitro*.

1.1. Articular chondrocytes and oxygen tension

Articular cartilage is an avascular tissue, nutrified via the synovial fluid only, and thus it has to be adapted to a low oxygen tension environment [34]. The oxygen tension *in vivo* has been described to

*Address for correspondence: Tel.: +49 431 880 3080; Fax: +49 431 880 2699; E-mail: bkurz@anat.uni-kiel.de.

be lower than 8% in cartilage in contrast to 21% in common cell culture incubators [5]. Low oxygen tension (but not anoxia) stimulates proliferation, increases matrix production and stabilizes the chondrogenic phenotype of primary and dedifferentiated chondrocytes in the monolayer and alginate culture system [9,14,24]. Additionally, chondrocytes exhibit properties typical for juvenile cartilage under these conditions [26].

So far, it is unknown how oxygen regulates articular chondrocyte signaling pathways. There is evidence that articular chondrocytes express hypoxia-inducible factor-1α (HIF-1α) [33], a factor which has recently been demonstrated to control matrix synthesis of epiphyseal chondrocytes [32]. It seems to be reasonable, that chondrocytes have the same oxygen-sensing pathways as other mammalian cells, where HIF-1α is hydroxylated by an oxygen-dependent mechanism [35]. This hydroxylation allows binding of HIF-1α to the von Hippel–Lindau tumor suppressor protein and subsequent enzymatic degradation. Without oxygen HIF-1α cannot be degraded, translocates into the nucleus, forms a heterodimer with the constitutively expressed HIF-1β, and activates hypoxia-inducible target genes by binding to HIF-1 binding sites in hypoxia response elements. Nevertheless, although all this is still pure speculation for articular chondrocytes, we and others have suggested that oxygen can be considered as an important parameter in cartilage tissue engineering.

1.2. Articular chondrocytes and hydrostatic pressure

Articular cartilage has to bear mechanical loading intermittently, indicating that the tissue must be adapted to mechanical stimulation. HP is one component of the biomechanical cascade of physico-chemical events during compression of the tissue. It can be applied even to single cells, monolayer cultures or cell suspensions which might be of interest in the context of tissue engineering. Several *in vitro* models have been designed in order to study the influence of HP on chondrocyte metabolism [6,13,14, 17,20,22,23,28,31,38,41,42]. They demonstrate a significant influence of HP on chondrocyte biosynthetic activities depending on frequency and amplitude of HP (Table 1). Recent studies suggest, that also sequence and duration of loading periods regulate the metabolic activity of the cells [16]. Little is known about how HP is sensored by the cells and transduced into cell signaling. Fluid is incompressible and the same has been suggested for the solid cartilage matrix under physiological levels of HP [2]. Hall [12] and Browning et al. [6] demonstrated the involvement of cation transport pathways, but it is still unknown how HP is transduced into these signals. So it is still a challenge to encode the complexity of chondrocyte regulation by HP.

1.3. Articular chondrocytes and scaffolds

All data from the literature suggest that HP and oxygen might be supportive parameters in cartilage tissue engineering. Since chondrocytes seem to need some kind of 3D-scaffold in order to maintain their phenotype we investigate whether oxygen alone or in combination with HP might influence the biosynthetic activity of human osteoarthritic chondrocytes cultured on collagen I/III membranes. A variety of different scaffolds for cartilage tissue engineering has been described [4,7,8,11,18,25,27,31,36]. Many of these scaffolds maintain differentiated chondrocytes and accumulate extracellular matrix *in vitro*, but so far most of them are not available for clinical application. One commercially available scaffold is the acellular porcine collagen I/III membrane (Chondro-Gide®) which has been used clinically for the treatment of articular cartilage defects already [4]. Since little is known about the effect of oxygen and HP on scaffold-associated chondrocytes we now demonstrate that human OA chondrocytes cultured on collagen I/III membranes are able to respond to these physico-chemical stimuli.

Table 1

Data from the literature about the influence of static or intermittent hydrostatic pressure (HP, applied by compression of the medium = fluid or the gas phase of culture) on sulfate incorporation or protein and mRNA synthesis in articular chondrocyte or cartilage cultures in comparison to control cultures without HP (Con)

HP type	Time of incorporation/biosynthesis	Amplitude (frequency)	Type of culture	Sulfate incorporation (HP/Con)	Matrix production (HP/Con)	Reference
Gas phase	90 min during end of 24 hrs HP	0.013 MPa (0.3 Hz)	Chick embryo high density	1.4-fold	–	[17]
Gas phase	24 hrs during HP	0.2 MPa (30/2 min on/off loading)	Bovine monolayer	–	Collagen-like proteins 1.88-fold	[14]
Fluid	On day 5 (sulfate) or 15 (matrix) during HP	2.8 MPa (0.015 Hz)	Bovine 3D-collagen sponges	1.4-fold	Proteoglycan 2.7-fold	[23]
Fluid	0–2 hrs after HP	10 MPa (static) 50 MPa (static)	HCS-2/8 monolayer	0.88-fold 0.71-fold	– –	[41]
Fluid	4 hrs during HP	10 MPa (static) 10 MPa (1 Hz)	Bovine high density	1.32-fold 1.65-fold	– mRNA Coll II ↑ mRNA Agg ↑	[38,39]
Fluid	4 hrs HP/day	10 MPa (1 Hz)	Human high density	–	mRNA Coll II ↑ mRNA Agg ↑	[16]
Fluid	4 hrs during HP	5 MPa (static)	Bovine agarose	–	mRNA Coll II ↑ mRNA Agg ↑	[42]
Fluid	20 hrs during HP	5 MPa (0.5 Hz) 30 MPa (static)	Bovine monolayer	1.11-fold 0.63-fold	– –	[20]
Fluid	20 hrs during HP	5 MPa (0.25 Hz) 5 MPa (∼0.02 Hz)	Bovine monolayer	∼1.2-fold ∼0.75-fold	– –	[28]
Fluid	2 hrs during HP	0.5 MPa (static) 30 MPa (static)	Bovine tissue	∼1.36-fold ∼0.5-fold	∼1.25-fold ∼0.58-fold	[13]

1.4. Oxygen, hydrostatic pressure and chondrogenesis

So far most oxygen or HP studies are dealing with the influence of these parameters on primary or dedifferentiated chondrocytes. But nowadays additionally the potential influence of these factors on chondrogenesis from stem cells is getting into the focus of interest. About 40 years ago it was suggested that oxygen or HP might have a promoting effect on chondrogenesis from mesenchymal stem cells [3,29,30], but since then this topic had been ignored almost completely. Since it still remains unknown whether oxygen or HP might induce cartilage formation directly we looked for the effect of these parameters on adherent bone marrow cells. Our data suggest that low oxygen tension alone or in combination with HP might support but does not induce chondrogenesis under the given circumstances.

2. Materials and methods

2.1. Isolation and cultivation of human osteoarthritic articular chondrocytes on collagen I/III membranes

Articular cartilage samples were harvested from human femur and tibia condyles which were isolated during implantation of knee endoprotheses (Endo-Klinik, Hamburg, Germany). The samples were washed in Hank's buffered salt solution (HBSS, containing 100 U/ml penicillin/streptomycin and 2.5 μg/ml amphotericin; Biochrom, Berlin, Germany), diced aseptically into pieces, and treated with a mixture of pronase (32 U/ml; Boehringer Mannheim, Germany) and hyaluronidase (2.3 KU/ml; Sigma, Deisenhofen, Germany) in HBSS for one hour. After washing with HBSS (antibiotics see above) extracellular collagen was digested with collagenase [Sigma, 918 U/ml Ham's F-12 medium (Biochrom) with vitamin C (50 μg/ml; Sigma) and E (50 μM; Sigma T-3251)] for 6–8 hours. The cell-suspension was centrifuged (10 min at $400 \times g$), the pellet resuspended in Ham's F-12 medium with 10% fetal calf serum (FCS, Biochrom), antibiotics and vitamins (see above), and filtered through a gauze filter (mesh size 20 μm, Hydrobios, Kiel, Germany). Viable cells were counted using a Neubauer chamber and the trypan blue exclusion method and seeded as primary cultures on collagen I/III membranes with a density of 100,000 cells/cm^2. Medium was exchanged twice a week. The chondrocytes were seeded on the rough side of the collagen I/III membranes (Chondro-Gide®) in a density of 100,000 cells/cm^2. Before use the collagen I/III membranes were washed several times with PBS.

2.2. Isolation and culturing of bovine adherent bone marrow cells

Spongious bone tissue with adherent bone marrow tissue was isolated from the sternum of cattle during the slaughter process. The tissue samples were kept in DMEM with antibiotics (see above) until a 15 min treatment with trypsin. The resulting cell suspension was harvested without bone remnants and centrifuged (10 min, $400 \times g$). Then, the pellet was resuspended in 20 ml medium (DMEM, 10% FCS, antibiotics).The nucleated single cells were isolated by a Ficoll gradient centrifugation (20 min, $2000 \times g$). Finally, the cells were seeded in culture flasks in a density of 10^5 cells/cm^2 and cultured under common culture conditions until the cells were confluent. The cells were multiplied during 2 passages and then used for further monolayer or high density culture experiments. In monolayer cultures the cells were seeded with a density of 100,000 cells/cm^2. In high density cultures 7.5×10^5 cells were cultured as a pellet at the bottom of Eppendorf tubes. The lids of the tubes had a little hole in order to allow exchange of gases. The monolayer and high density cultures were cultured in medium with or without chondrogenic supplements (10 ng/ml TGF-β1, 0.1 μg/ml dexamethasone, 40 μg/ml ascorbic acid; [27,37]) under different oxygen tensions or under the influence of intermittent HP.

2.3. Cultivation under different oxygen tension and intermittent hydrostatic pressure (HP)

Cell cultures were kept under an atmosphere of 5% O_2, 5% CO_2, and 90% N_2 in an O_2-/CO_2-regulated incubator (BB 6220 O_2, Heraeus Instruments, Hanau, Germany) or under 21% O_2, 5% CO_2, and 74% N_2 in a CO_2-regulated incubator (3862, Forma Scientific, Marietta, Ohio, USA). In order to apply intermittent HP the cell cultures were kept in a combined intermittent HP/oxygen cell culture chamber which has been introduced in detail previously [14]. In brief, it consists of two connected work and storage chambers. Cells are cultured in the work chamber at 37°C and 95% humidity. The chambers are filled with a defined gas pressure and mixture of 5% O_2, 5% CO_2, rest N_2 from a gas cylinder. The gas pressure induces HP in the culture medium and is transferred to the cells via the fluid. In the present study the cell cultures were incubated under pressure amplitudes of 0.2 MPa with an intermittent HP cycle of 30/2 min (on/off loading) from culture day 2 on. Chondrocyte cultures were incubated every other day and adherent bone marrow cells on culture day 2 and 5 for 24 hrs each. In between, cells were cultured without loading in common cell incubators with corresponding oxygen tension (see above).

2.4. [^{3}H]-proline and [^{35}S]-sulfate incorporation/DNA measurement

Collagen I/III membrane cultures were incubated overnight (on culture day 10) in culture medium (see above) containing [^{3}H]-proline or [^{35}S]-sulfate (10 μCi/ml each; Amersham Pharmacia) and washed in PBS afterwards. Each sample was digested separately at 65°C overnight in 1 ml papain-solution [0.125 mg/ml (2.125 units/ml, Sigma), 0.1 M Na_2HPO_4, 0.01 M EDTA, pH 6.5]. 200 μl of each sample were added to 2 ml of scintillation fluid (Hydroluma, Baker Chemicals) and measured using a scintillation counter (Wallac 1904, Turku, Finland). 500 μl of each sample were mixed with 500 μl PBS and used to determine the DNA-content. Samples and blanks (containing 1 ml PBS) were treated for 15 sec with an ultrasonic beam (sonoplus gm70, Bandelin, Berlin, Germany). 0.5 ml of RNAse [5 KU/ml RNAse/pronase-buffer (PBS diluted 1 : 5 with buffer (50 μM $MgCl_2{\cdot}6H_2O$ and 90 μM $CaCl_2{\cdot}6H_2O$))] and 0.5 ml of pronase (0.7 U/ml RNAse/pronase-buffer) were added and incubated for 30 min at 37°C. 0.5 ml of ethidium bromide (25 μg/ml) were added, samples incubated for 30 min, and measured with a fluorometer (Hitachi F2000; $\lambda_{ex} = 365$ nm; $\lambda_{em} = 590$ nm). DNA-content was computed using a standard curve. Radiolabelling was normalized to DNA content. Significance of differences was calculated using the two-tailed Student's t-test (with significance at $p < 0.05$).

2.5. Reverse transcription-polymerase chain reaction (RT-PCR)

RNA was isolated from the cell cultures on culture day 10 using the Qiagen Rneasy Mini-Kit according to the manufacturer's instructions. Lysed samples were homogenized using the QIAshredder spin columns (Quiagen). Isolated RNA was determined for quantity and quality spectrophotometrically (260 and 280 nm). cDNA was generated using the QUIAGEN OneStep RT-PCR Kit according to the manufacturer's instructions. For PCR the following bovine primers were used (0.5g μM): collagen type II (sense: 5′-AAG AAA CAC ATC TGG TTT GGA GAA ACC-3′; anti-sense: 5′-ATG GGT GCA ATG TCA ATG ATG GG-3′; product size 342 bp; annealing temp. 59°C), G3PDH (S: 5′-TCA CTG GCA TGG CCT TCC GTG TCC-3′; AS: 5′-AAA TGA GCT TGA CAA AGT GGT CGT TGA AGG-3′; product size 252 bp; annealing temp. 64°C); collagen type I (S: 5′-GAT GGA GAC TTC TAC AGG GCT GAC-3′; AS: 5′-CGA TGT CCA AAG GTG CAA TAT CAA GG-3′; product size 712 bp; annealing temp. 62°C); aggrecan core protein (S: 5′-TCC ACT GAC ACC AAA GAG T-3′; AS: 5′-TCT GGA TTT AGT GGT GAG TAT T-3′; product size 252 bp; annealing temp. 54°C). RT-PCR reaction conditions were: reverse transcription 30 min, 50°C; initial activation step 15 min, 95°C; 3-step cycles: melting 30 sec 94°C, annealing 60 sec primer specific temp. (59–64°C, see above); a final extension step 10 min 72°C. Amplified products were separated on a 2% agarose gel, visualized fluorometrically and digitalized.

3. Results

3.1. Biosynthetic activity of human OA articular chondrocytes on collagen I/III membranes

We isolated human articular chondrocytes from cartilage which was harvested during the implantation of knee endoprotheses because of osteoarthritis. The cells were seeded successfully on the rough side of collagen I/III membranes and cultured under the influence of low oxygen tension and additionally under the influence of intermittent HP (30/2 min on/off loading). We found a significant increase of both, sulfate and proline incorporation, under the influence of low oxygen tension in comparison to common culture conditions with 21% oxygen (Figs 1 and 2). Sulfate incorporation was increased 2.2-fold and proline incorporation 1.24-fold. This increase was even higher during co-stimulation with low oxygen tension and intermittent HP (sulfate: 4.6-fold; proline: 1.42-fold).

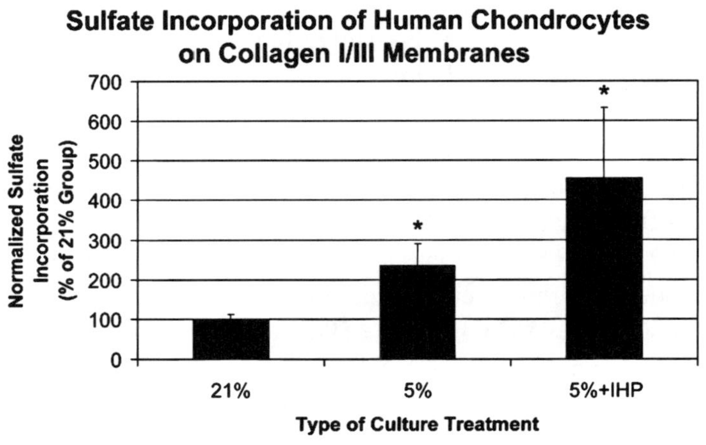

Fig. 1. Influence of low oxygen conditions (5%) without or with intermittent hydrostatic pressure (IHP) on overnight sulfate incorporation of human chondrocytes after 10 days of culture on collagen I/III membranes. Mean values ± SEM; $n = 10$; $^{*}p < 0.05$ vs 21% group.

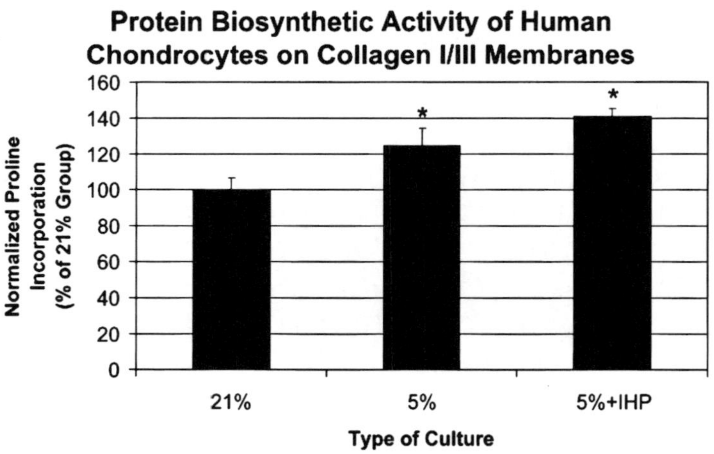

Fig. 2. Influence of low oxygen conditions (5%) without or with intermittent hydrostatic pressure (IHP) on overnight proline incorporation of human chondrocytes after 10 days of culture on collagen I/III membranes. Mean values ± SEM; $n = 10$; $^{*}p < 0.05$ vs 21% group.

3.2. Chondrogenesis under the influence of oxygen and intermittent hydrostatic pressure

For chondrogenesis experiments we isolated adherent bone marrow cells from cattle sternum and cultured the cells in monolayers or in high density cultures under the influence of chondrogenic medium (CM: containing TGF-β, dexamethasone, ascorbic acid), oxygen, and intermittent HP. Chondrogenesis was monitored by the detection of aggrecan and collagen type II mRNA.

In monolayer culture the adherent bone marrow cells multiplied and got confluent, but even under all different culture conditions no chondrogenesis could be detected (data not shown). This was different in high density cultures. Here, chondrogenesis could be observed especially under the influence of CM. Clear bands for aggrecan or collagen type II Rt-PCR products were visible in the agarose gel (Fig. 3). In contrast, low oxygen tension without CM and with or without addition of intermittent HP was no sufficient stimulus for chondrogenesis (Figs 3 and 4). In these samples RT-PCR showed very weak or no bands for aggrecan or collagen type II. However, CM-induced chondrogenesis was influenced by oxygen. Under the influence of low oxygen tension aggrecan and collagen type II mRNA-expression of the cells was increased in comparison to cells cultured under 21% oxygen.

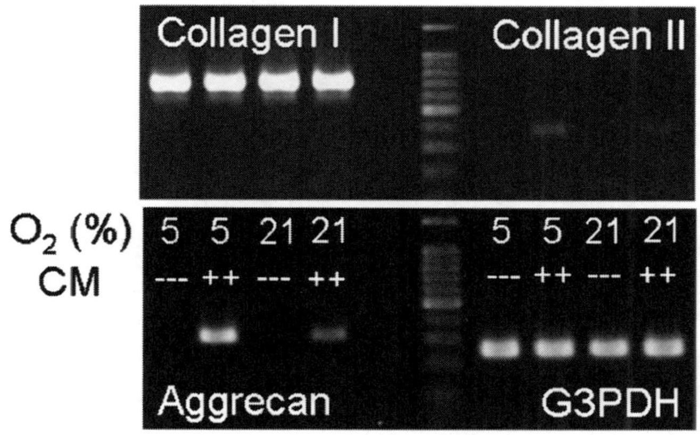

Fig. 3. Influence of chondrogenic medium (with ++ or without --- CM) and oxygen (%) on mRNA expression (RT-PCR products in agarose gels) of adherent bone marrow cells after 10 days in high density culture.

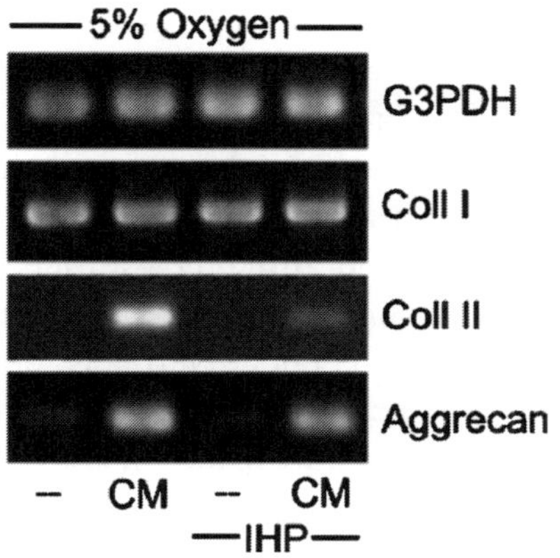

Fig. 4. Influence of chondrogenic medium (CM) and intermittent hydrostatic pressure (IHP) under low oxygen conditions on mRNA expression (RT-PCR products in agarose gels) of adherent bone marrow cells after 10 days in high density culture.

In all types of adherent bone marrow cell cultures collagen type I mRNA was expressed constitutively under all experimental conditions (monolayers: data not shown, high density cultures (Figs 3 and 4)). The same could be observed for G3PDH.

4. Discussion

4.1. Biosynthetic activity of human OA articular chondrocytes on collagen I/III membranes

In the present study we have investigated the ability of articular chondrocytes from osteoarthritic human knee joints to respond to oxygen and HP stimuli. This ability seems to be an important issue since so far there is still no way to recover bone areas which lost their cartilage surface in osteoarthritic joints. We try to answer the question whether these cells still might be able to produce a sufficient matrix under certain conditions. The biosynthesis stimulating effect of low oxygen tension that we found is of interest since most oxygen studies have been published with bovine, rabbit, or avian cartilage so far [5,9,10,14,24,26,34]. The results of an increased biosynthetic activity under low oxygen tension support our data from the studies with animal cells [9,14] and suggest that human OA chondrocytes might be responsive to oxygen, too.

We added an intermittent HP (30/2 min on/off loading, 0.2 MPa) to the low oxygen tension since we found an increased biosynthetic activity in bovine monolayer cultures under these combined conditions in earlier studies [14]. Our data with human OA cells on collagen I/III membranes support our previous findings and suggest that a combination of HP and low oxygen tension might be a supportive tool for human cartilage engineering, too. Like in the oxygen studies little is known about the influence of HP on human primary articular chondrocyte biosynthesis so far. Ikenoue et al. [16] demonstrate an increase in collagen type II and aggrecan mRNA in human high density monolayer cultures which also suggests that the cells increase their biosynthetic activity under these conditions. Human osteoarthritic cartilage tissue might be even more sensitive to intermittent HP stimulation ($1/2$ sec on/off loading, 0.013 MPa) than normal cartilage which was demonstrated by an increased sulfate incorporation [19]. Our biosynthesis data also go well with certain HP studies using avian or bovine cells (Table 1). The authors describe in general, that static short-term or low-frequency loading increases biosynthetic activities of chondrocytes (GAG synthesis, protein or collagen synthesis) if applied with low amplitudes only ($\sim$0.01–2.8 MPa). Loading with higher frequencies ($\geqslant$0.1 Hz) increase biosynthesis even under higher amplitudes (up to 10 MPa). However, amplitudes higher than 10 MPa and long-term static loads seem to inhibit the biosynthetic activity of the cells. Taken together it can be concluded, that human (osteoarthritic) chondrocytes can be stimulated under the influence of specific loading regimes of HP.

Another issue in our study is the problem that it is still unknown how chondrocytes might be fixed in cartilage lesions. Here some kind of scaffold seems to be the key although it still remains unknown how this scaffold should really look like. A variety of different scaffolds for cartilage tissue engineering has been described [4,7,8,11,18,25,27,31,36]. Namely, these are polyglycolic acid and polylactic acid, fibrin, alginate, polyethylene oxide, self-assembling peptide hydrogels, or collagen type I gels, and collagen type I and II sponges. In the present study we have used the biodegradable and commercially available acellular porcine collagen I/III membrane (Chondro-Gide®), since this has been used clinically for the treatment of articular cartilage defects already [4]. We see that human OA chondrocytes can successfully be seeded onto this membrane and that these scaffold-associated cells are able to respond to the physico-chemical stimuli low oxygen tension and intermittent HP with an increase in the biosynthetic activity, just like bovine cells cultured in 3D-collagen sponges or agarose [23,42].

4.2. Chondrogenesis under the influence of oxygen and intermittent hydrostatic pressure

We still do not know what source of cells might be adequate in order to treat cartilage defects. Next to autologous chondrocytes – which have been used clinically for about a decade already – stem cells are considered to be very useful in this context. This raises the question how chondrogenesis might be induced in these cells. Certain medium supplements, like growth factors, have successfully been used in order to induce expression of cartilage-specific matrix molecules (like collagen type II or aggrecan) in mesenchymal stem cell cultures [27,37]. But it is also suggested that physico-chemical factors might induce or support chondrogenesis. In 1999, Tagil and Aspenberg [40] found in a bone defect rat model that HP induces and maintains cartilage instead of bone. Recently, Angele et al. [1] showed that cartilage constructs which were generated from mesenchymal progenitor cells under the influence of chondrogenic medium components, are able to respond to intermittent HP stimulation with an increased matrix production. On the other hand, low oxygen tension was shown to promote proteoglycan synthesis in chick limb bud chondrogenesis *in vitro* [15]. Additionally, low oxygen tension supported osteogenesis from rat mesenchymal stem cells *in vitro* [21]. We now demonstrate in our preliminary study that low oxygen tension alone or in combination with intermittent HP supports but not induces chondrogenesis in adherent bone marrow cell high density cultures. It can be concluded that physico-chemical factors that stimulate the biosynthetic activity of adult chondrocytes must not necessarily induce chondrogenesis. However, these factors might support the chondrogenic function once the cells have been differentiated into chondrocytes. Hydrostatic pressure is a very complicated factor and it is not recommended to draw conclusions too early. It still might be that other frequencies and amplitudes or even durations and sequences of HP loading have to be found that are able to induce chondrogenesis without co-stimulation with other chondrogenic factors. For that reason, the investigation of the influence of physico-chemical factors like oxygen and HP on chondrogenesis is still a challenge and needs further investigation.

5. Conclusion

Our study demonstrates, that human osteoarthritic chondrocytes cultured on collagen I/III membranes are able to respond to low oxygen tension with an increase of the biosynthetic activity. Intermittent HP supports this increase and it might be concluded that these physico-chemical parameters alone or in combination are useful tools in human cartilage tissue engineering. Next to the chondrocyte biosynthetic activity chondrogenesis from adherent bone marrow cells seems to be supported but not induced by these factors in our study. However, since our bone marrow cells were not specifically isolated stem cells and especially HP seems to be a very complex parameter further studies will be necessary, in order to understand the influence of oxygen and HP on chondrogenesis and cartilage tissue engineering completely.

Acknowledgements

The authors would like to thank Claudia Kremling, Rita Kirsch, and Frank Lichte for their excellent technical support. Collagen I/III membranes (Chondro-Gide®) were a gift from Geistlich Biomaterials, Wolhusen, Switzerland. The study was funded by the "Gemeinnütziger Verein der Endo-Klinik, Hamburg", Hamburg, Germany.

References

[1] P. Angele, J.U. Yoo, C. Smith, J. Mansour, K.J. Jepsen, M. Nerlich and B. Johnstone, Cyclic hydrostatic pressure enhances the chondrogenic phenotype of human mesenchymal progenitor cells differentiated in vitro, *J. Orthop. Res.* **21** (2003), 451–457.

[2] N.M. Bachrach, V.C. Mow and F. Guilak, Incompressibility of the solid matrix of articular cartilage under high hydrostatic pressures, *J. Biomech.* **31** (1998), 445–451.

[3] C.A.L. Bassett and I. Herrmann, Influence of oxygen concentration and mechanical factors on differentiation of connective tissues in vitro, *Nature* **190** (1961), 460.

[4] P. Behrens, E.M. Ehlers, K.U. Kochermann, J. Rohwedel, M. Russlies and W. Plötz, New therapy procedure for localized cartilage defects. Encouraging results with autologous chondrocyte implantation, *MMW Fortschr. Med.* **141** (1999), 49–51.

[5] C.T. Brighton and R.B. Heppenstall, Oxygen tension in zones of the epiphyseal plate, the metaphysis and diaphysis. An in vitro and in vivo study in rats and rabbits, *J. Bone Joint Surg. Am.* **53** (1971), 719–728.

[6] J.A. Browning, R.E. Walker, A.C. Hall and R.J. Wilkins, Modulation of $Na^+ \times H^+$ exchange by hydrostatic pressure in isolated bovine articular chondrocytes, *Acta Physiol. Scand.* **166** (1999), 39–45.

[7] S.J. Bryant and K.S. Anseth, The effects of scaffold thickness on tissue engineered cartilage in photocrosslinked poly(ethylene oxide) hydrogels, *Biomaterials* **22** (2001), 619–626.

[8] L.E. Freed, J.C. Marquis, A. Nohria, J. Emmanual, A.G. Mikos and R. Langer, Neocartilage formation in vitro and in vivo using cells cultured on synthetic biodegradable polymers, *J. Biomed. Mater. Res.* **27** (1993), 11–23.

[9] C. Domm, M. Schünke, K. Christesen and B. Kurz, Redifferentiation of dedifferentiated bovine articular chondrocytes in alginate culture under low oxygen tension, *Osteoarthritis Cartilage* **10** (2002), 13–22.

[10] M.J. Grimshaw and R.M. Mason, Bovine articular chondrocyte function in vitro depends upon oxygen tension, *Osteoarthitis Cartilage* **8** (2000), 386–392.

[11] H.J. Häuselmann, R.J. Fernandes, S.S. Mok, T.M. Schmid, J.A. Block, M.B. Aydelotte, K.E. Kuettner and E.J. Thonar, Phenotypic stability of bovine articular chondrocytes after long-term culture in alginate beads, *J. Cell Sci.* **107** (1994), 17–27.

[12] A.C. Hall, Differential effects of hydrostatic pressure on cation transport pathways of isolated articular chondrocytes, *J. Cell Physiol.* **178** (1999), 197–204.

[13] A.C. Hall, J.P.G. Urban and K.A. Gehl, The effects of hydrostatic pressure on matrix synthesis in articular cartilage, *J. Orthop. Res.* **9** (1991), 1–10.

[14] U. Hansen, M. Schünke, C. Domm, N. Ioannidis, J. Hassenpflug, T. Gehrke and B. Kurz, Combination of reduced oxygen tension and intermittent hydrostatic pressure: a useful tool in articular cartilage tissue engineering, *J. Biomech.* **34** (2001), 941–949.

[15] P.M. Hwang, D.H. Byrne and P.A. Kitos, Effects of molecular oxygen on chick limb bud chondrogenesis, *Differentiation* **37** (1988), 14–19.

[16] T. Ikenoue, M.C. Trindade, M.S. Lee, E.Y. Lin, D.J. Schurman, S.B. Goodman and R.L. Smith, Mechanoregulation of human articular chondrocyte aggrecan and type II collagen expression by intermittent hydrostatic pressure in vitro, *J. Orthop. Res.* **21** (2003), 110–116.

[17] G.P. van Kampen, J.P. Veldhuijzen, R. Kuijer, R.J. van de Stadt and C.A. Schipper, Cartilage response to mechanical force in high-density chondrocyte cultures, *Arthr. Rheum.* **28** (1985), 419–424.

[18] J. Kisiday, M. Jin, B. Kurz, H. Hung, C. Semino, S. Zhang and A.J. Grodzinsky, Self-assembling peptide hydrogel fosters chondrocyte extracellular matrix production and cell division: implications for cartilage tissue repair, *Proc. Natl. Acad. Sci. USA* **99** (2002), 9996–10001.

[19] F. Lafeber, J.P. Veldhuijzen, J.L. Vanroy, O. Huber-Bruning and J.W. Bijlsma, Intermittent hydrostatic compressive force stimulates exclusively the proteoglycan synthesis of osteoarthritic human cartilage, *Br. J. Rheumatol.* **31** (1992), 437–442.

[20] M.J. Lammi, R. Inkinen, J.J. Parkkinen, T. Häkkinen, M. Jortikka, L.O. Nelimarkka, H.T. Järveläinen and M.I. Tammi, Expression of reduced amounts of structurally altererd aggrecan in articular cartilage chondrocytes exposed to high hydrostatic pressure, *Biochem. J.* **304** (1994), 723–730.

[21] D.P. Lennon, J.M. Edmison and A.I. Caplan, Cultivation of rat marrow-derived mesenchymal stem cells in reduced oxygen tension: effects on in vitro and in vivo osteochondrogenesis, *J. Cell Physiol.* **187** (2001), 345–355.

[22] L. Lippiello, C. Kaye, T. Neumata and H.J. Mankin, In vitro metabolic response of articular cartilage segments to low levels of hydrostatic pressure, *Connect. Tiss. Res.* **13** (1985), 99–107.

[23] S. Mizuno, T. Tateishi, T. Ushida and J. Glowacki, Hydrostatic fluid pressure enhances martrix synthesis and accumulation by bovine chondrocytes in three-dimensional culture, *J. Cell Physiol.* **193** (2002), 319–327.

[24] C.L. Murphy and A. Sambanis, Effect of oxygen tension and alginate encapsulation on restoration of the differentiated phenotype of passaged chondrocytes, *Tiss. Eng.* **7** (2001), 791–803.

[25] S. Nehrer, H.A. Breinan, A. Ramappa, S. Shortkroff, G. Young, T. Minas, C.B. Sledge, I.V. Yannas and M. Spector, Canine chondrocytes seeded in type I and type II collagen implants investigated in vitro, *J. Biomed. Mater. Res.* **38** (1997), 95–104.

[26] Z. Nevo, A. Beit-Or and Y. Eilam, Slowing down ageing of cultured embryonal chick chondrocytes by maintenance under low oxygen tension, *Mech. Ageing Dev.* **45** (1988), 157–165.

[27] U. Nöth, R. Tuli, A.M. Osyczka, K.G. Danielson and R.S. Tuan, In vitro engineered cartilage constructs produced by press-coating biodegradable polymer with human mesenchymal stem cells, *Tiss. Eng.* **8** (2002), 131–144.

[28] P.P. Parkkinen, J. Ikonen, M.J. Lammi, J. Laakkonen, M. Tammi and H.J. Helminen, Effects of cyclic hydrostatic pressure on proteoglycan synthesis in cultured chondrocytes and articular cartilage explants, *Arch. Biochem. Biophys.* **300** (1993), 458–465.

[29] F. Pauwels, Eine neue Theorie über den Einfluss mechanischer Reize auf die Differenzierung der Stützgewebe, *Z. Anat. Entwicklungsgesch.* **121** (1960), 478–515.

[30] J.M. Pawelek, Effects of thyroxine and low oxygen tension on chondrogenic expression in cell culture, *Dev. Biol.* **19** (1969), 52–72.

[31] C. Perka, R.S. Spitzer, K. Lindenhayn, M. Sittinger and O. Schultz, Matrix-mixed culture: new methodology for chondrocyte culture and preparation of cartilage transplants, *J. Biomed. Mater. Res.* **49** (2000), 305–311.

[32] D. Pfander, T. Cramer, E. Schipani and R.S. Johnson, HIF-1α controls extracellular matrix synthesis by epiphyseal chondrocytes, *J. Cell Sci.* **116** (2003), 1819–1826.

[33] T. Pufe, A. Lemke, B. Kurz, W. Petersen, B. Tillmann, A.J. Grodzinsky and R. Mentlein, Mechanical overload induces VEGF in cartilage discs via hypoxia-inducible factor, *Am. J. Pathol.* **164** (2004), 185–192.

[34] R. Rajpurohit, C.J. Koch, Z. Tao, C.M. Teixeira and I.M. Shapiro, Adaptation of chondrocytes to low oxygen tension: relationship between hypoxia and cellular metabolism, *J. Cell Physiol.* **168** (1996), 424–432.

[35] M. Safran and W.G. Kaelin Jr., HIF hydroxylation and the mammalian oxygen-sensing pathway, *J. Clin. Invest.* **111** (2003), 779–783.

[36] L. Schuman, P. Buma, D. Versleyen, B. de Man, P.M. van der Kraan, W.B. van den Berg and G.N. Homminga, Chondrocyte behaviour within different types of collagen gel in vitro, *Biomaterials* **16** (1995), 809–814.

[37] I. Sekiya, J.T. Vuoristo, B.L. Larson and D.J. Prockop, In vitro cartilage formation by human adult stem cells from bone marrow stroma defines the sequence of cellular and molecular events during chondrogenesis, *Proc. Nalt. Acad. Sci.* **99** (2002), 4397–4402.

[38] R.L. Smith, J. Lin, M.C. Trindade, J. Shida, G. Kajiyama, T. Vu, A.R. Hoffman, M.C. Meulen, S.B. Goodman, D.J. Schurman and D.R. Carter, Time-dependent effects of intermittent hydrostatic pressure on articular chondrocyte type II collagen and aggrecan mRNA expression, *J. Rehabil. Res. Dev.* **37** (2000), 153–161.

[39] R.L. Smith, S.F. Rusk, B.E. Ellison, P. Wessels, K. Tsuchiya and D.R. Carter, In vitro stimulation of articular chondrocyte mRNA and extracellular matrix synthesis by hydrostatic pressure, *J. Orthop. Res.* **14** (1996), 53–60.

[40] M. Tagil and P. Aspenberg, Cartilage induction by controlled mechanical stimulation in vivo, *J. Orthop. Res.* **17** (1999), 200–204.

[41] K. Takahashi, T. Kubo, K. Kobayashi, J. Imanishi, M. Takigawa, Y. Arai and Y. Hirasawa, Hydrostatic pressure influences mRNA expression of transforming growth factor-beta 1 and heat shock protein 70 in chondrocyte-like cell line, *J. Orthop. Res.* **15** (1997), 150–158.

[42] T. Toyoda, B.B. Seedhom, J. Kirkham and W.A. Bonass, Upregulation of aggrecan and type II collagen mRNA expression in bovine chondrocytes by the application of hydrostatic pressure, *Biorheology* **40** (2003), 79–85.

[43] J.P. Veldhuijzen, L.A. Bourret and G.A. Rodan, In vitro studies of the effect of intermittent compressive forces on cartilage cell proliferation, *J. Cell Physiol.* **98** (1979), 299–306.

Biorheology 41 (2004) 335–346
IOS Press

Cyclic, mechanical compression enhances chondrogenesis of mesenchymal progenitor cells in tissue engineering scaffolds

Peter Angele [a,*], Detlef Schumann [a], Martin Angele [b], Bernd Kinner [a], Carsten Englert [a], Reiner Hente [a], Bernd Füchtmeier [a], Michael Nerlich [a], Carsten Neumann [a] and Richard Kujat [a]

[a] *Department of Trauma Surgery, University Hospital of Regensburg, 93053 Regensburg, Germany*
[b] *Department of Surgery, Ludwig Maximilians University, Klinikum Grosshadern, 81377 Munich, Germany*

Abstract. The effects of cyclic, mechanical compression on human bone marrow-derived mesenchymal progenitor cells undergoing chondrogenic differentiation were examined in this study. Mesenchymal progenitor cells were injected into cylindrical biodegradable scaffolds (hyaluronan–gelatin composites), cultured in a defined, serum-free chondrogenic medium and subjected to cyclic, mechanical compression. Scaffolds were loaded for 4 hours daily in the first 7 days of culture. At 1, 7, 14 and 21 days of culture, scaffolds were harvested for reverse transcriptase Polymerase Chain Reaction (RT-PCR), histology, quantitative DNA, proteoglycan and collagen analysis. Scaffolds loaded for 7 days showed a significant upregulation especially of chondrogenic markers (type II collagen, aggrecan; $p < 0.0001$). No significant difference could be found for DNA content between loaded samples and unloaded controls. At day 1 in culture no significant differences in proteoglycan- and collagen contents could be detected between unloaded and loaded samples. After 21 days the proteoglycan ($p < 0.001$) and collagen contents ($p < 0.0001$) were significantly higher in the loaded samples compared to unloaded controls. By histological analysis (toluidine blue) a higher amount of proteoglycan-rich, extracellular matrix production throughout the matrix could be detected for loaded samples compared to unloaded controls. This study indicates that cyclic, mechanical compression enhances the expression of chondrogenic markers in mesenchymal progenitor cells differentiated *in vitro* resulting in an increased cartilaginous matrix formation, and suggests that mechanical forces may play an important role in cartilage repair.

Keywords: Chondrogenesis, stem cell, bone marrow, mechanobiology, compression

1. Introduction

During the *in vivo* repair process of skeletal tissue, the tissue is subject to a variety of mechanical forces that may play an important role in the rate and/or quality of tissue repair. Mesenchymal progenitor cells (MPC) are identified as possible repair cells of various connective tissues, because of their potential to differentiate into a variety of cell types of the mesenchymal lineage [1,22,24–26,35,41,44,45,63]. Incorporation of MPC in suitable tissue engineering scaffolds and culture in chondrogenic medium are described in the literature [1,21,26,42,50,59]. MPCs can be harvested from bone marrow by a small puncture of the iliac crest of patients. In contrast to cartilage based repair this small procedure creates no additional harvest defect in the knee joint of the patients.

*Corresponding author: Dr. Peter Angele, Department of Trauma Surgery, University Hospital of Regensburg, Franz Josef Strauss Allee 11, 93053 Regensburg, Germany. Tel.: +49 9419446805; Fax: +49 0419446975; E-mail: angelepeter@aol.com.

Numerous publications showed the beneficial influence of compression (e.g., mechanical, hydrostatic, osmotic, shear, ultrasonic) on the chondrogenic differentiation of dedifferentiated chondrocytes [6,11, 14,20,33,34,37,46,56,61]. It has been shown that compressive stress is one of the most prominent extracellular factors that regulate biosynthetic and catabolic activities of chondrocytes [62] and that isolated chondrocytes respond to mechanical forces by upregulation of RNA levels for collagen, aggrecan and other matrix molecules [4]. Dynamic loading, which causes intra-tissue fluid flow and dynamic tissue deformation, stimulates the synthesis of matrix molecules at certain amplitudes and frequencies [16,28, 29,32,47].

In the case of mesenchymal progenitor cell mediated repair of skeletal tissue that involves a chondrogenic phase, such as fracture repair or articular cartilage regeneration, it is likely that mesenchymal progenitor cells are subjected to mechanical factors that will influence the quality of repair. Theoretical models have predicted the influence of biomechanical loading on tissue differentiation during skeletal growth and morphogenesis [11–13,53] suggesting that different loading conditions should result in different types of tissue. Hydrostatic pressure may encourage cartilage formation [8–10,52]. *In vivo* application of 2 MPa of hydrostatic pressure to mesenchymal cells in a conduction chamber, implanted in the tibiae of Sprague–Dawley rats was shown to promote cartilage formation [53]. Some forces enhance collagen production (direct compression) whereas others better stimulate proteoglycan deposition (hydrostatic pressure). Mass transfer for constructs under direct compression is expected to be better than for those cultured under hydrostatic pressure or static compression. Dynamic compression helps to alleviate diffusion limitations through a pressure gradient within the scaffold as well as by a secondary mixing effect on the surrounding medium. The compression of the scaffold creates a higher hydrostatic pressure at the center of the constructs than at the surface, which produces different fluid velocities within the construct as the applied load changes [15].

Although much attention has been given to the influences of bioactive factors on mesenchymal progenitor cell differentiation and proliferation, there are few studies that examine the effect of mechanical factors on these cells. A previous study showed the enhancement of chondrogenic differentiation of mesenchymal progenitor cells in aggregate culture by hydrostatic compression [2]. The influence of biomechanical loading on the expression of chondrogenic markers of mesenchymal progenitor cells from post-natal tissues undergoing differentiation in tissue engineering scaffolds is still unclear and should be analysed in this study.

2. Material and methods

Human bone marrow was obtained from the iliac crests of patients undergoing spine fusion. After Percoll gradient fractionation, Dulbecco's modified Eagle's Medium with 10% fetal bovine serum was added to the aspirate and 10×10^6 nucleated cells/100 mm dish were plated and grown at 37°C with 5% CO_2 until the cells reached 80% confluency. Adherent cell colonies were trypsinized, counted, and 2×10^6 synchronized mesenchymal progenitor cells were injected into cylindrical biodegradable scaffolds (hyaluronan–gelatin composites) with defined pore size (250 μm) and volume (70 mm^3) and cultured at 37°C/5% CO_2 in a defined medium, previously shown to induce chondrogenic differentiation of mesenchymal progenitor cells [1,24,63].

To apply cyclic, mechanical compression, scaffolds were transferred under sterile conditions in load definition units (for a total compression of 40%) at the bottom of 15 ml polypropylene tubes (sterile compression chambers) (NeoLab, Germany), which were filled with 4 ml fresh chondrogenic medium

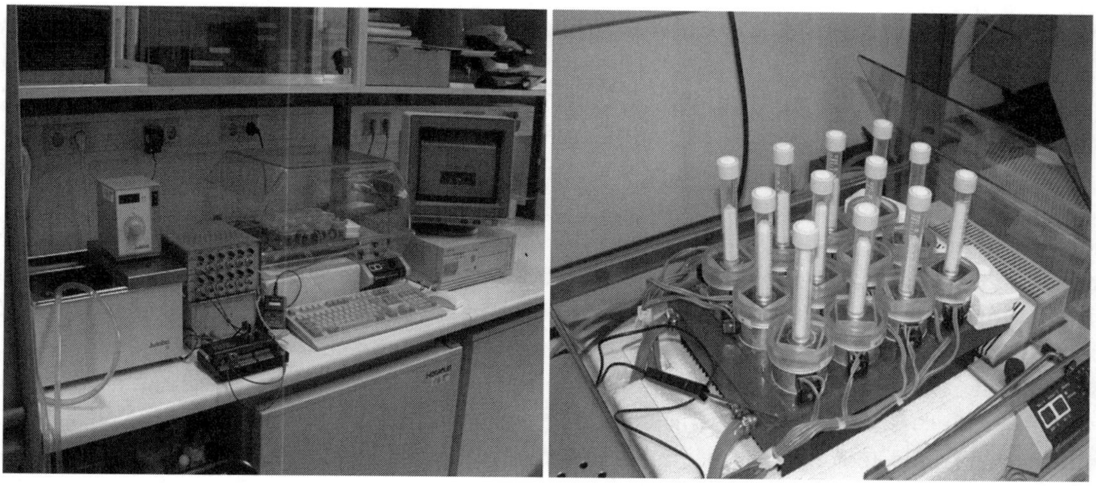

Fig. 1. *In vitro* loading system for the application of cyclic mechanical compression in sterile culture conditions (left half). Sterile compression chambers with sterile bar magnets (right half). Note the electro-magnets below the sterile compression chamber.

(Fig. 1). A sterile bar magnet (NeoLab, Germany) (Fig. 1) with defined weight was inserted under sterile conditions into the tubes. The tubes were placed in a custom-built mechanical compression apparatus (Fig. 1) on electromagnets (RS Components, Germany), located within a temperature-controlled incubator (37°C). Cyclic, mechanical compression was applied to the scaffolds using a computer-controlled interface SIOS and the software Do-it (AK Modul-Bus GmbH Saerbeck, Germany). The samples were subjected to a continuous, uniaxial and unconfined compressive load using a sinusoidal waveform of 0.33 Hz and a peak stress of 7994 Pa over a period of 4 hours/day for the first 7 days. After each daily loading period, the scaffolds were cultured in chondrogenic medium without load application [1,24,63].

The scaffolds were kept in culture for a total of either 1, 7, 14 or 21 days, allowing time for chondrogenic differentiation, matrix formation and biosynthetic response [1,24,63], and then harvested for RT-PCR, biochemical and histological analysis. Each group of cell-scaffold composites subjected to cyclic biomechanical compression was compared with two sets of matched control scaffolds. To investigate the influence of magnetic effects the first set of control scaffolds was manipulated in the same manner as the experimental group, i.e. they were transferred into containers and placed in the loading chamber incubator for the same time period, but was not subjected to cyclic biomechanical compression. The second set of control scaffolds was left in standard culture conditions [1,24,63] without any manipulation. All experiments were performed with three sets of samples.

For RT-PCR RNA was extracted using RNA-Bee (Biozol, Eching, Germany) according to the single-step acid-phenol guanidinium method. cDNA synthesis was performed by using Superscript RNase H^- reverse transcriptase (Invitrogen, Karlsruhe, Germany) in the presence of oligo-dt primers. We performed and monitored RT-PCR by Light Cycler analysis with Light Cycler Kits (Roche, Mannheim, Germany). The level of each target gene was normalized to the reference gene glyceraldehyde phosphate dehydrogenase (GAPDH). The primers were as follows:

GAPDH *Forward* 5′-gAA ggT gAA ggT Cgg AgT C Reverse 5′-gAA gAT ggT gAT ggg ATT TC
Aggrecan *Forward* 5′-ACT TCC gCT ggT CAg ATg gA *Reverse* 5′-TCT CgT gCC AgA TCA TCA CC
Col 1 *Forward* 5′-Agg gCC AAg ACg AAg ACA TC *Reverse* 5′-AgA TCA CgT CAT CgC ACA ACA
Col 2 *Forward* 5′-TTC AgC TAT ggA gAT gAC AAT C *Reverse* 5′-AgA gTC CTA gAg TgA CTg Ag
Col 10 *Forward* 5′-gAg gAA gCT TCA gAA AgC Tg *Reverse* 5′-CTg gTT TCC CTA CAg CTg A.

DNA contents of the scaffolds were assayed using a two-step fluorometric assay. Scaffolds were prepared by individual digestion with Soluene 350 (Packard) for 1.5 hours at 60°C. Aliquots of the digest (100 μl) were mixed with 1 ml of diluted Hoechst 33258 Dye Solution (1 μg/ml). The fluorescence emission of the samples was measured with Cytofluor at $A_{360\text{ nm}}$ excitation, $A_{460\text{ nm}}$ emission with a gain of 70. A standard curve was produced with calf thymus DNA (Sigma) and used to determine the DNA content of the experimental samples.

The glycosaminoglycan content of the scaffolds, used as an indicator of proteoglycan production, was quantified with a spectrophotometric assay as described elsewhere [2]. Scaffolds were digested in papain solution overnight at 60°C. After adding 100% ethanol the mixture was incubated at −20°C for 3 hours. After centrifugation (14000 rpm, 10 minutes) the supernatant was discarded and the pellets resuspended in 200 μl distilled water. Samples were allowed to precipitate in Safranin O reagent on nitrocellulose filter (BioRad, USA), which covered wells in a dot-blot apparatus (BioRad, USA). The precipitates were collected on the nitrocellulose filter and then dissolved in 10% cetylpyridinium chloride at 37°C for 20 minutes. The absorbance of the liquid was spectrophotometrically read at 536 nm. For the experimental groups and the unloaded controls the proteoglycan-content of the negative controls (cell-free matrix) was substracted to get a minimal estimation of the produced extracellular proteoglycan even when a complete degradation of the matrices would influence the measurement.

The hydroxyproline content of the scaffolds was measured as described elsewhere [2]. Scaffolds were hydrolyzed with 6 M HCl at 108°C for 24 hours. The hydrolysates were repeatedly dried with resuspension in distilled water in between each drying cycle to wash out the acid. The neutralized samples were resuspended in hydroxyproline buffer. Freshly prepared, chloramine T solution was added and the vortexed suspension incubated for 20 minutes at room temperature. Perchloric acid and freshly prepared P-dimethylaminobenzaldehyde solution was then added and the suspension incubated at 60°C for 20 minutes with intermittent shaking. The absorbance was spectrophotometrically read at 560 nm. Cis-4-hydroxy-d-proline was used to construct a standard curve. The collagen content was calculated by multiplying the total hydroxyproline content by assay with a conversion factor (×7.5). For the experimental groups and the unloaded controls the content of the negative controls (cell-free matrix) was substracted to get a minimal estimation of the produced extracellular collagen even when a complete degradation of the matrices would influence the measurement.

Histological (toluidine blue) and immunohistochemical (type II collagen) analysis was performed on frozen sections of 10% formaldehyde fixed materials as described elsewhere [2]. Serial sections (12 μm) of all samples were cut and sections stained with toluidine blue for histological evaluation. For immunohistochemical analysis, unstained sections were washed in phosphate buffered saline (PBS). Blocking of nonspecific antibody binding was achieved by incubating the slides with bovine serum albumin (BSA) (15 minutes). The sections were digested for 30 minutes with 1% pronase in 1% BSA, PBS and then for 30 minutes with chondroitinase ABC to facilitate the collagen antibody access to the extracellular matrix. The sections were probed with anti-collagen antibody raised against type II collagen (II-II6B3, obtained from the Developmental Studies Hybridoma Bank). Visualization was achieved with a monoclonal fluorescein-conjugated secondary antibody.

The expression of molecular markers, DNA, proteoglycan and collagen contents in the scaffolds with and without application of mechanical compression were compared between the groups at each time point using two way ANOVA with Fisher's post hoc test for multiple comparisons. The tests were performed using Sigma Stat™ Software for Windows Version 2.03, SPSS Inc. Significance was accepted at a level of $p < 0.05$.

3. Results

There were no differences in expression of chondrogenic markers (aggrecan, type II collagen, type X collagen and type I collagen) between the non-loaded control scaffolds and the control scaffolds cultured without any manipulation (data not shown). The non-loaded controls showed a statistically significant increase in chondrogenic marker expression, especially type II collagen, after 21 days in culture compared to unloaded controls after 1 day in culture ($p < 0.0001$) (Fig. 2).

The aggrecan expression (major component of hyaline cartilage) of the 7 days loaded scaffolds showed a 98% increase versus non-loaded control at day 7 ($p < 0.001$), an even higher increase of aggrecan expression (loaded versus non-loaded group) was found at day 14 of culture (+386%; $p < 0.0001$). At day 21 the aggrecan expression of loaded scaffolds showed an increase of 92% over unloaded controls (Fig. 3; $p < 0.001$).

The type II collagen expression (major component of hyaline cartilage) of 7 days loaded versus non loaded scaffolds showed a significant increase at day 14 (687%; $p < 0.0001$) and day 21 in culture (90%; $p < 0.0001$) compared to unloaded samples (Fig. 3).

The type I collagen expression of 7 days loaded scaffolds showed a significant increase at day 14 (221%) and at day 21 in culture (76%) over non-loaded controls (Fig. 2; $p < 0.001$), but did not reach the increased expression levels of hyaline cartilage specific markers.

No statistical significant differences in type X collagen expression were found between any loaded group and its controls at either day 7, 14 or day 21 (Fig. 3; $p = 0.45$).

No statistical significant differences in DNA content were found in unloaded and loaded groups during chondrogenesis (day 1 versus day 21). Also, no differences were found between the non-loaded control scaffolds (those manipulated in the same way as the loaded scaffolds but kept non-loaded for 4 hours per day in the compression chamber incubator) and the control scaffolds cultured without any manipulation (data not shown). There were no significant differences in the DNA contents between the loaded and non-loaded scaffolds (Fig. 4).

The proteoglycan content was significantly higher after 21 days in culture compared to the corresponding samples on day 1 ($p < 0.0001$) (Fig. 4). At day 1 no significant differences in proteoglycan content could be detected between unloaded and loaded samples ($p = 0.23$), however after 21 days in culture a significant increase for loaded samples could be seen ($p < 0.001$) (Fig. 4).

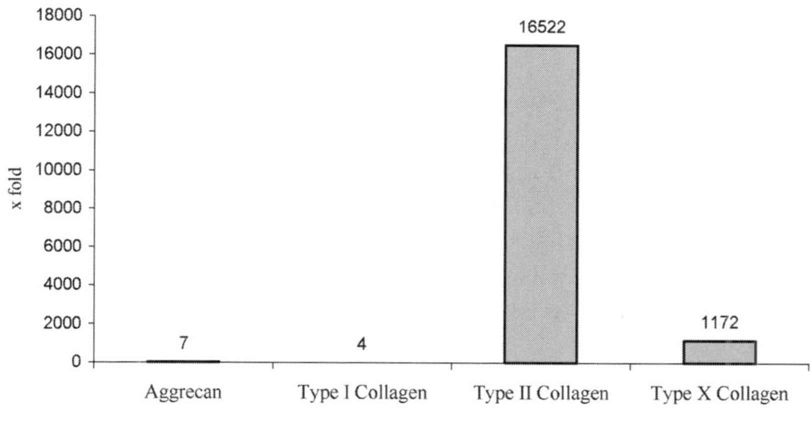

Fig. 2. Increase of relative chondrogenic marker expression (x fold) of unloaded group (day 21) versus unloaded group (day 1).

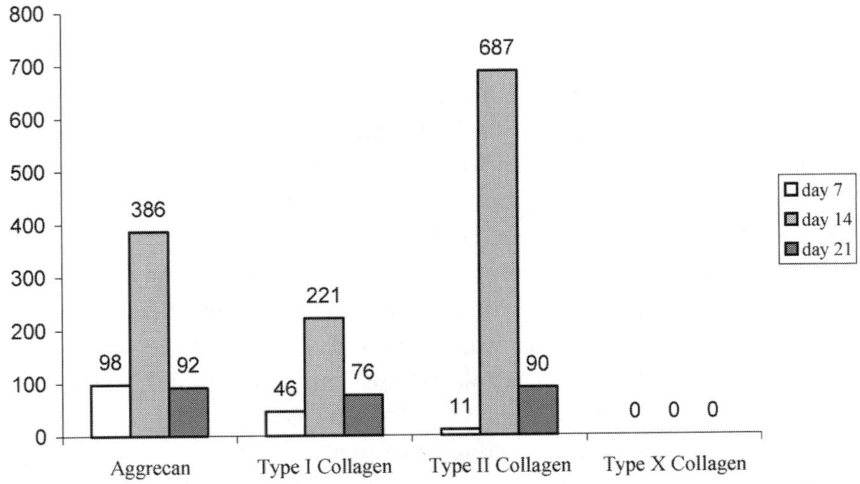

Fig. 3. Increase of relative chondrogenic marker expression of loaded versus unloaded group at days 7/14/21 in percent (%).

The collagen content was significantly higher after 21 days in culture compared to the corresponding samples on day 1 ($p < 0.0001$) (Fig. 4). At day 1 no significant differences in collagen content could be detected between unloaded and loaded samples ($p = 0.12$), however after 21 days in culture a significant increase for loaded samples could be observed ($p < 0.0001$) (Fig. 4).

The histological and immunohistochemical analysis confirmed chondrogenic differentiation after 21 days by extracellular, metachromatic matrix deposition (toluidine blue stained sections) and positive staining for collagen type II in the cell-scaffold constructs (Fig. 5). A comparison of toluidine blue stained sections indicated no qualitative microscopic differences between 14 (not shown) or 21 days in culture. The 7 days loaded scaffolds had a quantitative increase in extracellular matrix production (increased metachromatic, extracellular matrix deposition) throughout the scaffolds, especially in the center of the scaffolds, in comparison with non-loaded controls after 14 days and 21 days in culture (Fig. 5). Immunohistochemical staining of the scaffolds indicated positive expression of collagen type II in all control and experimental groups, confirming successful chondrogenic differentiation of mesenchymal progenitor cells (hyaline cartilage-like appearance) (not shown). No qualitative difference in collagen distribution was found between the controls and the loaded scaffolds by immunohistochemistry.

4. Discussion

This study was performed using a custom-built system (Fig. 1), which allowed application of cyclic mechanical compression to cell – scaffold constructs under sterile culture conditions. The applied mechanical compression was completely transferred to mesenchymal progenitor cells in tissue engineered scaffolds over a pressure range between 0 and 8 kPa. Mesenchymal progenitor cells injected into biodegradable hyaluronan–gelatin-composite scaffolds, which had proven ability to allow the differentiation of mesenchymal progenitor cells to cartilage [1], were used to study the effects of mechanical loading during chondrogenesis. For the first 7 days no extracellular deposition of cartilage-like matrix was detected in the cell-matrix constructs [1]. Therefore this timeframe was used in this study to selectively load mesenchymal progenitor cells and not chondrocytic cells.

In an *in vitro* aggregate chondrogenesis model, the progenitor cells are densely packed as they aggregate and condense prior to cartilaginous matrix production [24,63]. The condensation of progenitor

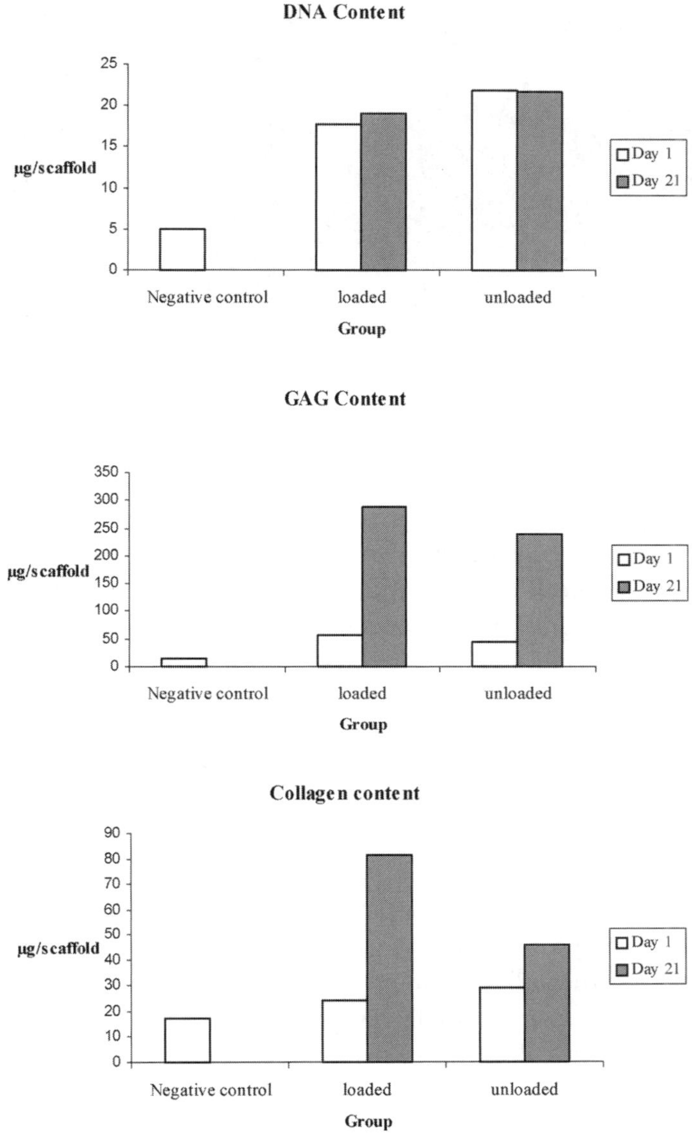

Fig. 4. DNA, proteoglycan and collagen content of loaded and unloaded samples (mean of 3 samples). Negative control represents cell free scaffolds.

cells also occurred when the progenitor cells were injected in biodegradable hyaluronan–gelatin composite matrices as seen in the present study. Densely packed cells were seen early in articular cartilage regeneration, when the subchondral bone is penetrated and marrow cells migrate into the defect [49], during the development of limbs and also in fracture repair [5,23]. The influence of mechanical forces on progenitor cells during cell condensation and chondrogenesis is not well understood. This is partly due to the complex distribution of stresses and strains in each of these situations. For example, the form of mechanical stress in a joint *in vivo* is more complex than simply mechanical compression. According to Mow et al. [38], the typical total stress on the surface of human joints ranges from 2.96 to 9.86 MPa. A complex state of pressure in the interstitial fluid and spatially varying normal and shear stresses in

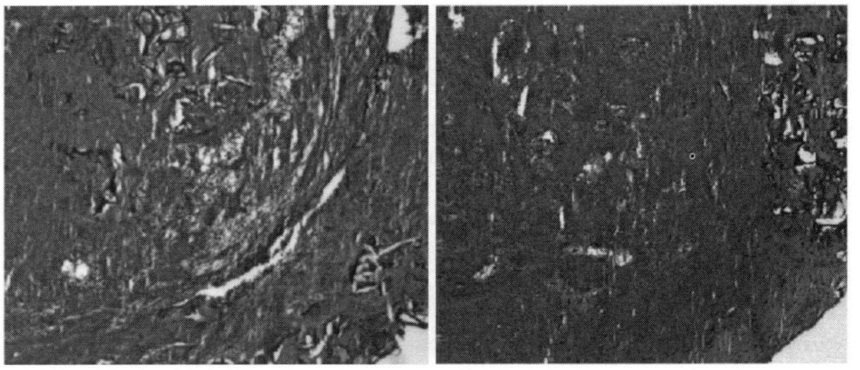

Fig. 5. Toluidine blue stained sections of unloaded (left half) and loaded scaffolds (right half) after 21 days in culture.

the solid matrix are developed in response to the surface stress. According to the mechanical tissue differentiation theory [11,12] an undifferentiated mesenchymal tissue will differentiate into fibrous tissue, cartilage or bone depending on the mechanical load stimulation. Although mechanical compression does not represent the mechanical forces seen by cartilage *in vivo*, compression and its effects on chondrogenic differentiation of mesenchymal progenitor cells were examined selectively in the present study.

More information on the effect of mechanical load exists for chondrocytes and cartilage tissue. Loads below the normal physiological levels appear to stimulate catabolic effects in chondrocytes [3]. In contrast, a load within the physiologic range stimulates production of cartilaginous matrix [36]. Supraphysiologic loads appear to result in tissue damage [19]. An upregulation of aggrecan and type II collagen mRNA expression in bovine chondrocytes by the application of hydrostatic pressure has been described [58]. Convective transport is a critical component of nutrient transfer, especially in the early stages of culture when the constructs are still very porous. In developing soft tissues, the extracellular matrix is soft, so the total hydrostatic stress consists mainly of pressure in the fluid phase [11].

The highly porous tissue engineering scaffolds used in the present study cannot bear physiologic load levels in its initial state. Therefore a lower compression of 8 kPa was applied. A cyclic load of 0.33 Hz for 4 hours per day was used because this rate was previously shown to stimulate extracellular matrix synthesis of articular chondrocytes [48,51]. The results of the present study indicate that with low levels of mechanical compression a similar effect occurred with differentiating cells.

Mechanical influences are also found on the precursor cells of forming limbs [54,55]. Takahashi et al. [54] demonstrated that significant differences exist between undifferentiated mesenchymal cells and differentiated chondrocytes in terms of their responses to compressive forces. They reported enhanced chondrogenesis of mouse embryonic limb bud mesenchymal cells following static compressive force. In contrast, static compressive force was shown to down-regulate cartilage matrix deposition [32] and aggrecan expression [31] in mature articular cartilage. In a separate study of chick limb bud cells, Elder et al. [17,18] found that cyclic compressive loading enhanced chondrogenic differentiation, but they did not find an influence of static loading. They noted an increase in proteoglycan production when differentiating chick limb bud cells were loaded with cyclic compressive displacement. However, they also noted an increase in the total number of cartilaginous nodules in their cultures of cells entrapped in agarose, thus the increase may have been due to increased recruitment of cells into the chondrogenic lineage, rather than to an increase in per cell production. In the present study no significant differences could be seen for DNA-content between loaded samples and unloaded controls, consistent with the literature for chondrocyte loading, which postulated that mechanical stressed cells may direct their anabolic efforts towards production and maintenance of matrix rather than cell proliferation [40,60].

Hydrostatic pressure in the physiologic range (5.03 MPa) was chosen in a previous study to evaluate its effect on chondrogenesis of mesenchymal progenitor cells [2]. A significant increase in proteoglycan- and collagen-content without change in DNA-content was detected in the loaded groups compared to unloaded controls. In the present study similar results were seen. The mechanical compression applied on multiple days for 4 hours per day significantly increased the chondrogenic marker expression and matrix deposition. The increase in type II collagen and aggrecan expression were noted in the loaded groups after 14 and 21 days in culture. This finding is similar to the effects of cyclical loading on articular cartilage explants or chondrocytes in culture: increased proteoglycan synthesis and overall matrix production [3,7,57]. However, the effect of loading on mature chondrocytes may be more selective. Carver and Heath [14] reported that the matrix produced by mature equine chondrocytes subjected to 6.90 MPa of intermittent loading increased collagen content but there was no significant difference in the glycosaminoglycan content. At a lower level of pressure (3.45 MPa), the collagen concentration did not change but a small increase in glycosaminoglycan concentration was noted. In contrast, the present study had shown that low level compression (8 kPa) resulted in a significant enhancement of chondrogenic differentiation of human mesenchymal progenitor cells, seen by an increase in type II collagen and aggrecan expression and followed by cartilage specific protein production.

In the present study, both the loaded and non-loaded scaffolds were cultured in conditions that promote the chondrogenic differentiation of mesenchymal progenitor cells. Therefore, the influence of mechanical compression on the initiation of chondrogenic differentiation was not directly explored. Nevertheless, the results of this study point to the importance of load during skeletal tissue regeneration, such as is found during the chondrogenic differentiation of mesenchymal cells in fracture repair and articular cartilage regeneration. The study suggested that an appropriate load applied to mesenchymal progenitor cells in the differentiation process enhance the chondrogenic marker expression. The increase in chondrogenic marker expression may facilitate the formation of an appropriate matrix, improving the final quality of either fracture callus or repair cartilage of osteochondral defects.

Direct compression enhances the access of mesenchymal progenitor cells to TGF-β1 (mass transfer) and accelerates the biosynthetic response of the mesenchymal progenitor cells to the growth factor. Because soluble low molecular weight polypeptides such as TGF-β1 tend to be cleared rapidly after intra-articular administration, this ability of dynamic compression to accelerate transport into the tissue is of potential clinical significance [6,30,39,43]. In an *in vitro* system mechanical load might also increase the access of exogenous added TGF-β1 and other supplements to the center of the cell-matrix constructs. This could explain the increase in extracellular matrix deposition especially in the center of the scaffolds seen in the present study.

Mechanical compression was the only component of loading considered in this study. Although mechanical compression is an important component of load seen by cartilage in a joint, additional studies will be needed to evaluate the effects of shear and complex mechanical deformation on progenitor cells undergoing chondrogenic differentiation. The mechanism by which the mechanical loading is sensed by the differentiating cells is presently unknown, as is the mechanism by which the load is transduced into an increase in chondrogenic marker expression and extracellular matrix deposition.

The present study described a system that allows the study of the effects of cyclic mechanical compression on differentiating mesenchymal cells in tissue engineering scaffolds. This system can be used to elicit mechanoreceptors and signal transduction pathways influenced by mechanical forces during chondrogenesis of mesenchymal cells. Furthermore the system can be used to optimize tissue engineered implants for cartilage repair.

Acknowledgements

The authors thank Tanja Weinfurtner and Daniela Drenkard for their excellent technical assistance.

References

[1] P. Angele, R. Kujat, M. Nerlich, J. Yoo, V. Goldberg and B. Johnstone, Engineering of osteochondral tissue with bone marrow mesenchymal progenitor cells in a derivatized hyaluronan–gelatin composite sponge, *Tissue Eng.* **5** (1999), 545–554.
[2] P. Angele, J.U. Yoo, C. Smith, J. Mansour, K.J. Jepsen, M. Nerlich and B. Johnstone, Cyclic hydrostatic pressure enhances the chondrogenic phenotype of human mesenchymal progenitor cells differentiated in vitro, *J. Orthop. Res.* **21** (2003), 451–457.
[3] N.M. Bachrach, W.B. Valhmu, E. Stazzone, A. Ratcliffe, W.M. Lai and V.C. Mow, Changes in proteoglycan synthesis of chondrocytes in articular cartilage are associated with the time-dependent changes in their mechanical environment, *J. Biomech.* **28** (1995), 1561–1569.
[4] E.J. Blain, S.J. Gilbert, R.J. Wardale, S.J. Capper, D.J. Mason and V.C. Duance, Up-regulation of matrix metalloproteinase expression and activation following cyclical compressive loading of articular cartilage in vitro, *Arch. Biochem. Biophys.* **396** (2001), 49–55.
[5] M.E. Bolander, Regulation of fracture repair by growth factors, *Proc. Soc. Exp. Biol. Med.* **200** (1992), 165–170.
[6] L.J. Bonassar, A.J. Grodzinsky, E.H. Frank, S.G. Davila, N.R. Bhaktav and S.B. Trippel, The effect of dynamic compression on the response of articular cartilage to insulin-like growth factor-I, *J. Orthop. Res.* **19** (2001), 11–17.
[7] N. Burton-Wurster, M. Vernier-Singer, T. Farquhar and G. Lust, Effect of compressive loading and unloading on the synthesis of total protein, proteoglycan, and fibronectin by canine cartilage explants, *J. Orthop. Res.* **11** (1993), 717–729.
[8] M.D. Buschmann, Y.A. Gluzband, A.J. Grodzinsky and E.B. Hunziker, Mechanical compression modulates matrix biosynthesis in chondrocyte/agarose culture, *J. Cell. Sci.* **108** (1995), 1497–1508.
[9] M.D. Buschmann, E.B. Hunziker, Y.J. Kim and A.J. Grodzinsky, Altered aggrecan synthesis correlates with cell and nucleus structure in statically compressed cartilage, *J. Cell. Sci.* **109** (1996), 499–508.
[10] M.D. Buschmann, Y.J. Kim, M. Wong, E. Frank, E.B. Hunziker and A.J. Grodzinsky, Stimulation of aggrecan synthesis in cartilage explants by cyclic loading is localized to regions of high interstitial fluid flow, *Arch. Biochem. Biophys.* **366** (1999), 1–7.
[11] D.R. Carter, G.S. Beaupre, N.J. Giori and J.A. Helms, Mechanobiology of skeletal regeneration, *Clin. Orthop.* (1998), S41–55.
[12] D.R. Carter, P.R. Blenman and G.S. Beaupre, Correlations between mechanical stress history and tissue differentiation in initial fracture healing, *J. Orthop. Res.* **6** (1988), 736–748.
[13] D.R. Carter and M. Wong, Mechanical stresses and endochondral ossification in the chondroepiphysis, *J. Orthop. Res.* **6** (1988), 148–154.
[14] S.E. Carver and C.A. Heath, Increasing extracellular matrix production in regenerating cartilage with intermittent physiological pressure, *Biotechnol. Bioeng.* **62** (1999), 166–174.
[15] E.M. Darling and K.A. Athanasiou, Articular cartilage bioreactors and bioprocesses, *Tissue Eng.* **9** (2003), 9–26.
[16] T. Davisson, S. Kunig, A. Chen, R. Sah and A. Ratcliffe, Static and dynamic compression modulate matrix metabolism in tissue engineered cartilage, *J. Orthop. Res.* **20** (2002), 842–848.
[17] S.H. Elder, S.A. Goldstein, J.H. Kimura, L.J. Soslowsky and D.M. Spengler, Chondrocyte differentiation is modulated by frequency and duration of cyclic compressive loading, *Ann. Biomed. Eng.* **29** (2001), 476–482.
[18] S.H. Elder, J.H. Kimura, L.J. Soslowsky, M. Lavagnino and S.A. Goldstein, Effect of compressive loading on chondrocyte differentiation in agarose cultures of chick limb-bud cells, *J. Orthop. Res.* **18** (2000), 78–86.
[19] T. Farquhar, Y. Xia, K. Mann, J. Bertram, N. Burton-Wurster, L. Jelinski and G. Lust, Swelling and fibronectin accumulation in articular cartilage explants after cyclical impact, *J. Orthop. Res.* **14** (1996), 417–423.
[20] M.J. Grimshaw and R.M. Mason, Modulation of bovine articular chondrocyte gene expression in vitro by oxygen tension, *Osteoarthritis Cartilage* **9** (2001), 357–364.
[21] N. Jaiswal, S.E. Haynesworth, A.I. Caplan and S.P. Bruder, Osteogenic differentiation of purified, culture-expanded human mesenchymal stem cells in vitro, *J. Cell. Biochem.* **64** (1997), 295–312.

[22] Y. Jiang, B.N. Jahagirdar, R.L. Reinhardt, R.E. Schwartz, C.D. Keene, X.R. Ortiz-Gonzalez, M. Reyes, T. Lenvik, T. Lund, M. Blackstad, J. Du, S. Aldrich, A. Lisberg, W.C. Low, D.A. Largaespada and C.M. Verfaillie, Pluripotency of mesenchymal stem cells derived from adult marrow, *Nature* **418** (2002), 41–49.
[23] S. Jingushi, M.E. Joyce and M.E. Bolander, Genetic expression of extracellular matrix proteins correlates with histologic changes during fracture repair, *J. Bone Miner. Res.* **7** (1992), 1045–1055.
[24] B. Johnstone, T.M. Hering, A.I. Caplan, V.M. Goldberg and J.U. Yoo, In vitro chondrogenesis of bone marrow-derived mesenchymal progenitor cells, *Exp. Cell. Res.* **238** (1998), 265–272.
[25] B. Johnstone and J. Yoo, Bone marrow-derived mesenchymal progenitor cells, *Methods Mol. Biol.* **137** (2000), 313–315.
[26] B. Johnstone and J.U. Yoo, Autologous mesenchymal progenitor cells in articular cartilage repair, *Clin. Orthop.* (1999), S156–162.
[27] K. Kaarniranta, M. Elo, R. Sironen, M.J. Lammi, M.B. Goldring, J.E. Eriksson, L. Sistonen and H.J. Helminen, Hsp70 accumulation in chondrocytic cells exposed to high continuous hydrostatic pressure coincides with mRNA stabilization rather than transcriptional activation, *Proc. Natl. Acad. Sci. USA* **95** (1998), 2319–2324.
[28] K. Kaarniranta, M.A. Elo, R.K. Sironen, H.M. Karjalainen, H.J. Helminen and M.J. Lammi, Stress responses of mammalian cells to high hydrostatic pressure, *Biorheology* **40** (2003), 87–92.
[29] K. Kaarniranta, C.I. Holmberg, M.J. Lammi, J.E. Eriksson, L. Sistonen and H.J. Helminen, Primary chondrocytes resist hydrostatic pressure-induced stress while primary synovial cells and fibroblasts show modified Hsp70 response, *Osteoarthritis Cartilage* **9** (2001), 7–13.
[30] H.M. Karjalainen, R.K. Sironen, M.A. Elo, K. Kaarniranta, M. Takigawa, H.J. Helminen and M.J. Lammi, Gene expression profiles in chondrosarcoma cells subjected to cyclic stretching and hydrostatic pressure. A cDNA array study, *Biorheology* **40** (2003), 93–100.
[31] Y.J. Kim, A.J. Grodzinsky and A.H. Plaas, Compression of cartilage results in differential effects on biosynthetic pathways for aggrecan, link protein, and hyaluronan, *Arch. Biochem. Biophys.* **328** (1996), 331–340.
[32] Y.J. Kim, R.L. Sah, A.J. Grodzinsky, A.H. Plaas and J.D. Sandy, Mechanical regulation of cartilage biosynthetic behavior: physical stimuli, *Arch. Biochem. Biophys.* **311** (1994), 1–12.
[33] J. Klein-Nulend, J.P. Veldhuijzen, R.J. van de Stadt, G.P. van Kampen, R. Kuijer and E.H. Burger, Influence of intermittent compressive force on proteoglycan content in calcifying growth plate cartilage in vitro, *J. Biol. Chem.* **262** (1987), 15490–15495.
[34] H.S. Lee, S.J. Millward-Sadler, M.O. Wright, G. Nuki, R. Al-Jamal and D.M. Salter, Activation of Integrin-RACK1/PKCalpha signalling in human articular chondrocyte mechanotransduction, *Osteoarthritis Cartilage* **10** (2002), 890–897.
[35] D.P. Lennon, S.E. Haynesworth, D.M. Arm, M.A. Baber and A.I. Caplan, Dilution of human mesenchymal stem cells with dermal fibroblasts and the effects on in vitro and in vivo osteochondrogenesis, *Dev. Dyn.* **219** (2000), 50–62.
[36] L. Lippiello, C. Kaye, T. Neumata and H.J. Mankin, In vitro metabolic response of articular cartilage segments to low levels of hydrostatic pressure, *Connect. Tissue Res.* **13** (1985), 99–107.
[37] S.J. Millward-Sadler, M.O. Wright, H. Lee, H. Caldwell, G. Nuki and D.M. Salter, Altered electrophysiological responses to mechanical stimulation and abnormal signalling through alpha5beta1 integrin in chondrocytes from osteoarthritic cartilage, *Osteoarthritis Cartilage* **8** (2000), 272–278.
[38] V.C. Mow, A. Ratcliffe and A.R. Poole, Cartilage and diarthrodial joints as paradigms for hierarchical materials and structures, *Biomaterials* **13** (1992), 67–97.
[39] M. Ohno, J.P. Cooke, V.J. Dzau and G.H. Gibbons, Fluid shear stress induces endothelial transforming growth factor beta-1 transcription and production. Modulation by potassium channel blockade, *J. Clin. Invest.* **95** (1995), 1363–1369.
[40] M.J. Palmoski and K.D. Brandt, Effects of static and cyclic compressive loading on articular cartilage plugs in vitro, *Arthritis Rheum.* **27** (1984), 675–681.
[41] M. Pittenger, A. Mackay, S. Beck, R. Jaiswal, R. Douglas, J. Mosca, M. Moorman, D. Simonetti, S. Craig and D. Marshak, Multilineage potential of adult human mesenchymal stem cells, *Science* **284** (1999), 143–148.
[42] M.S. Ponticiello, R.M. Schinagl, S. Kadiyala and F.P. Barry, Gelatin-based resorbable sponge as a carrier matrix for human mesenchymal stem cells in cartilage regeneration therapy [In Process Citation], *J. Biomed. Mater. Res.* **52** (2000), 246–255.
[43] J.P. Pujol, P. Galera, S. Pronost, K. Boumediene, D. Vivien, M. Macro, W. Min, F. Redini, H. Penfornis, M. Daireaux et al., Transforming growth factor-beta (TGF-beta) and articular chondrocytes, *Ann. Endocrinol.* **55** (1994), 109–120.
[44] M. Reyes and C.M. Verfaillie, Characterization of multipotent adult progenitor cells, a subpopulation of mesenchymal stem cells, *Ann. N. Y. Acad. Sci.* **938** (2001), 231–233; discussion 233–235.
[45] J.J. Rogers, H.E. Young, L.R. Adkison, P.A. Lucas and A.C. Black Jr., Differentiation factors induce expression of muscle, fat, cartilage, and bone in a clone of mouse pluripotent mesenchymal stem cells, *Am. Surg.* **61** (1995), 231–236.
[46] R.L. Sah, J.Y. Doong, A.J. Grodzinsky, A.H. Plaas and J.D. Sandy, Effects of compression on the loss of newly synthesized proteoglycans and proteins from cartilage explants, *Arch. Biochem. Biophys.* **286** (1991), 20–29.

[47] R.L. Sah, Y.J. Kim, J.Y. Doong, A.J. Grodzinsky, A.H. Plaas and J.D. Sandy, Biosynthetic response of cartilage explants to dynamic compression, *J. Orthop. Res.* **7** (1989), 619–636.
[48] D.M. Salter, S.J. Millward-Sadler, G. Nuki and M.O. Wright, Differential responses of chondrocytes from normal and osteoarthritic human articular cartilage to mechanical stimulation, *Biorheology* **39** (2002), 97–108.
[49] F. Shapiro, S. Koide and M.J. Glimcher, Cell origin and differentiation in the repair of full-thickness defects of articular cartilage, *J. Bone Joint Surg. Am.* **75** (1993), 532–553.
[50] M. Sittinger, C. Perka, O. Schultz, T. Haupl and G.R. Burmester, Joint cartilage regeneration by tissue engineering, *Z. Rheumatol.* **58** (1999), 130–135.
[51] R.L. Smith, S.F. Rusk, B.E. Ellison, P. Wessells, K. Tsuchiya, D.R. Carter, W.E. Caler, L.J. Sandell and D.J. Schurman, In vitro stimulation of articular chondrocyte mRNA and extracellular matrix synthesis by hydrostatic pressure, *J. Orthop. Res.* **14** (1996), 53–60.
[52] J.K. Suh, G.H. Baek, A. Aroen, C.M. Malin, C. Niyibizi, C.H. Evans and A. Westerhausen-Larson, Intermittent sub-ambient interstitial hydrostatic pressure as a potential mechanical stimulator for chondrocyte metabolism, *Osteoarthritis Cartilage* **7** (1999), 71–80.
[53] M. Tagil and P. Aspenberg, Cartilage induction by controlled mechanical stimulation in vivo, *J. Orthop. Res.* **17** (1999), 200–204.
[54] I. Takahashi, G.H. Nuckolls, K. Takahashi, O. Tanaka, I. Semba, R. Dashner, L. Shum and H.C. Slavkin, Compressive force promotes sox9, type II collagen and aggrecan and inhibits IL-1beta expression resulting in chondrogenesis in mouse embryonic limb bud mesenchymal cells, *J. Cell Sci.* **111** (1998), 2067–2076.
[55] K. Takahashi, T. Kubo, Y. Arai, I. Kitajima, M. Takigawa, J. Imanishi and Y. Hirasawa, Hydrostatic pressure induces expression of interleukin 6 and tumour necrosis factor alpha mRNAs in a chondrocyte-like cell line, *Ann. Rheum. Dis.* **57** (1998), 231–236.
[56] K. Takahashi, T. Kubo, K. Kobayashi, J. Imanishi, M. Takigawa, Y. Arai and Y. Hirasawa, Hydrostatic pressure influences mRNA expression of transforming growth factor-beta 1 and heat shock protein 70 in chondrocyte-like cell line, *J. Orthop. Res.* **15** (1997), 150–158.
[57] P.A. Torzilli, R. Grigiene, C. Huang, S.M. Friedman, S.B. Doty, A.L. Boskey and G. Lust, Characterization of cartilage metabolic response to static and dynamic stress using a mechanical explant test system, *J. Biomech.* **30** (1997), 1–9.
[58] T. Toyoda, B.B. Seedhom, J. Kirkham and W.A. Bonass, Upregulation of aggrecan and type II collagen mRNA expression in bovine chondrocytes by the application of hydrostatic pressure, *Biorheology* **40** (2003), 79–85.
[59] M.R. Urist, Y. Terashima, M. Nakagawa and C. Stamos, Cartilage tissue differentiation from mesenchymal cells derived from mature muscle in tissue culture, *In Vitro* **14** (1978), 697–706.
[60] G.P. van Kampen, J.P. Veldhuijzen, R. Kuijer, R.J. van de Stadt and C.A. Schipper, Cartilage response to mechanical force in high-density chondrocyte cultures, *Arthritis Rheum.* **28** (1985), 419–424.
[61] M.O. Wright, K. Nishida, C. Bavington, J.L. Godolphin, E. Dunne, S. Walmsley, P. Jobanputra, G. Nuki and D.M. Salter, Hyperpolarisation of cultured human chondrocytes following cyclical pressure-induced strain: evidence of a role for alpha 5 beta 1 integrin as a chondrocyte mechanoreceptor, *J. Orthop. Res.* **15** (1997), 742–747.
[62] Q.Q. Wu and Q. Chen, Mechanoregulation of chondrocyte proliferation, maturation, and hypertrophy: ion-channel dependent transduction of matrix deformation signals, *Exp. Cell. Res.* **256** (2000), 383–391.
[63] J.U. Yoo, T.S. Barthel, K. Nishimura, L. Solchaga, A.I. Caplan, V.M. Goldberg and B. Johnstone, The chondrogenic potential of human bone-marrow-derived mesenchymal progenitor cells, *J. Bone Joint Surg. Am.* **80** (1998), 1745–1757.

Part III

Subchondral bone and osteoarthritis

Biorheology 41 (2004) 349–358
IOS Press

Biochemical and mechanical properties of subchondral bone in osteoarthritis

Allen J. Bailey [a,*], Jason P. Mansell [b], Trevor J. Sims [c] and Xavier Banse [d]
[a] *Collagen Research Group, University of Bristol, Bristol, UK*
[b] *Oral and Dental Science, University of Bristol, Bristol, UK*
[c] *Academic Rheumatology Unit, Southmead Hospital, Bristol, UK*
[d] *Orthopaedic Research Laboratory, University of Louvain, Brussels, Belgium*

Abstract. The subchondral bone has long been known to thicken in osteoarthritis. However, recent evidence has demonstrated that the turnover of the bone is increased several fold, and further suggests that the thickening occurs prior to degradation of the articular cartilage, indicating that it plays a role in the pathogenesis of osteoarthritis. The mechanical and biochemical properties of the subchondral bone are therefore of particular interest in any attempt to determine the nature of the factors initiating osteoarthritis.

We have shown that the subchondral bone collagen of the femoral head possessed a 20-fold increase in turnover, as assessed by procollagen rate of synthesis and metalloproteinase degradation, and a 25% decrease in mineralisation. This increased metabolism and high lysyl hydroxylation leads to narrower and weaker fibres. Additionally the phenotypic expression of the osteoblasts is modified to produce increasing proportions of type I homotrimer in addition to the normal type I heterotrimer, which further reduces the mechanical strength of the bone. Overall, the narrow immature collagen fibres, the reduction in pyrrole cross-linking, decreased mineralisation, and increased amounts of type I homotrimer, all contribute to a weakening of the mechanical properties of the subchondral bone.

Keywords: Subchondral bone, osteoarthritis, collagen, cross-linking, metabolism, biomechanics

The majority of the basic research into the pathogenesis of osteoarthritis (OA) has concentrated on the mechanisms involved in the characteristic focal destruction of the articular cartilage. The initiating events of this destruction are believed to be changes in the proteoglycans followed by the collagenous framework, where after the disease is irreversible.

The well recognised additional characteristics of bony osteophytes and subchondral bone thickening have to a large extent been considered secondary and unimportant in the pathogenesis of osteoarthritis. However, some workers have suggested the involvement of bone. Over four decades ago Johnson [29] postulated that subtle changes in bone remodelling might precipitate irregularities in the articular cartilage. Radin and Rose [48] later proposed that the increased bone mass and sclerosis of the subchondral bone plate would cause stiffening of the bone and result in cartilage destruction on repetitive loading. More recent studies have re-emphasised the importance of the bone changes. Newer radiographic techniques [11] have demonstrated progressive thickening of the subchondral bone with joint narrowing and technetium-labelled bisphosphonate demonstrated that the heightened bone activity of the knee was related to the progression to severe OA, as monitored radiographically by joint space narrowing [20].

The question then arises as to whether the bone thickening was a result of cartilage degradation, occurred at the same time, or preceded cartilage degradation [4,12,52]. If the latter, then the subchondral

*Address for correspondence: Professor Allen J. Bailey, Collagen Research Group, University of Bristol, Bristol BS40 5DS, UK. E-mail: A.J.Bailey@bristol.ac.uk.

bone thickening is indeed important in the pathogenesis of OA, either as an initiating factor or, at the very least, important in the progression of OA. To follow the progression of OA recourse has to be made to animal models. Thickening of the subchondral bone was shown to occur prior to cartilage destruction in elderly cynomolgus macaques and that the thickness of the subchondral bone was related to the onset of cartilage fibrillation [14]. We have demonstrated a similar finding during spontaneous OA in the guinea pig knee [3] and have reported that remodelling of the cruciate ligament also occurred prior to cartilage destruction [2,47]. The latter finding may be related to the changes in the subchondral bone because of a change in stress on the bone due to the remodelling of the ligament. In a mechanically induced model it was reported that sub-fracture impacts applied to the patellofemoral joint of rabbits resulted in subchondral bone thickening after 6 months whilst cartilage stiffness only decreased after 12 months [43]. In an attempt to detect earlier changes than the histological techniques revealed ultrasonic echography was employed [51] in a rodent inflammatory joint model and reported changes in both bone and cartilage before any histological evidence of cartilage damage.

However, regardless of whether thickening of the subchondral bone occurs at a late stage or precedes cartilage destruction it is important to know the biochemical and mechanical properties of bone, since they may be important in the pathogenesis of OA. In an analysis of human femoral head bone collagen we found that the collagen was turning over at a rate several fold greater than non-OA bone [39] and is not a slow fibrosis due to remodelling of micro-fractures as suggested by Radin [48]. The increased rate of synthesis (20-fold) was determined from the levels of the C-terminal pro-peptide of type I collagen (PICP) and the increased degradation by the levels of MMPs (20-fold) (Fig. 1). The elevated type I collagen synthesis was accurately corroborated by the increase in alkaline phosphatase activity. The relative contribution of cathepsin K, which has been shown to cleave the collagen triple helix at several locations [23] and the MMPs remains to be determined. Transforming growth factor-beta (TGF-β) levels were increased 4-fold. TGF-β plays a major role in remodelling and recruitment of osteoblast precursors [46], stimulating collagen synthesis, inhibiting MMPs and increasing TIMPs thus leading to a net gain in collagen deposition. In contrast the mineral content of the new bone was significantly reduced (30%) (Fig. 2) which is consistent with previous studies reporting hypomineralisation of the

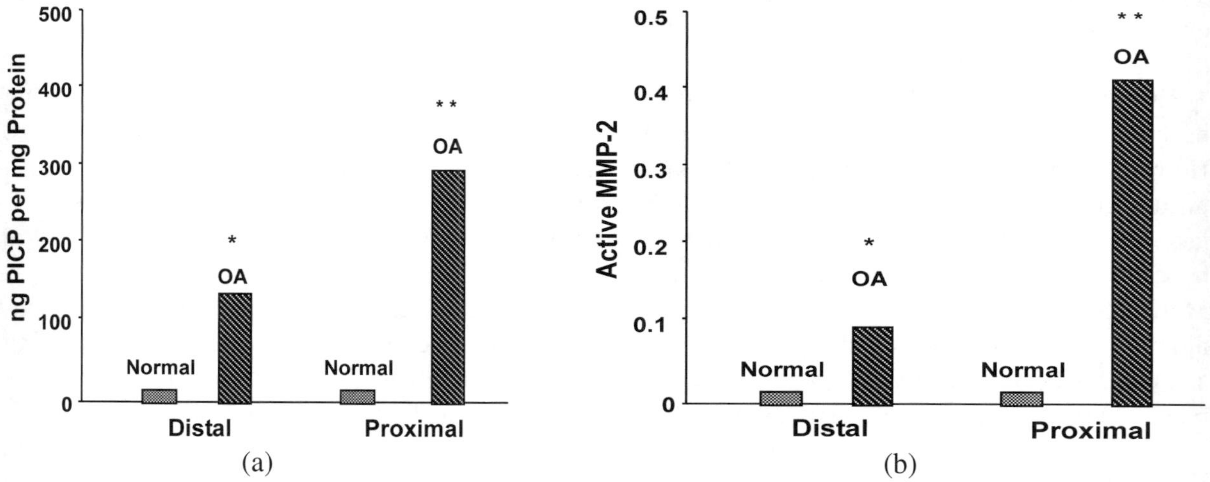

Fig. 1. Comparison of the metabolism of subchondral and normal bone from the femoral head; (a) relative rate of synthesis of type I collagen determined by type I propeptide (PICP) antibody in both distal and proximal regions of the femoral head. (b) Relative levels of MMP-2 determined by zymography against a standard MMP-2.

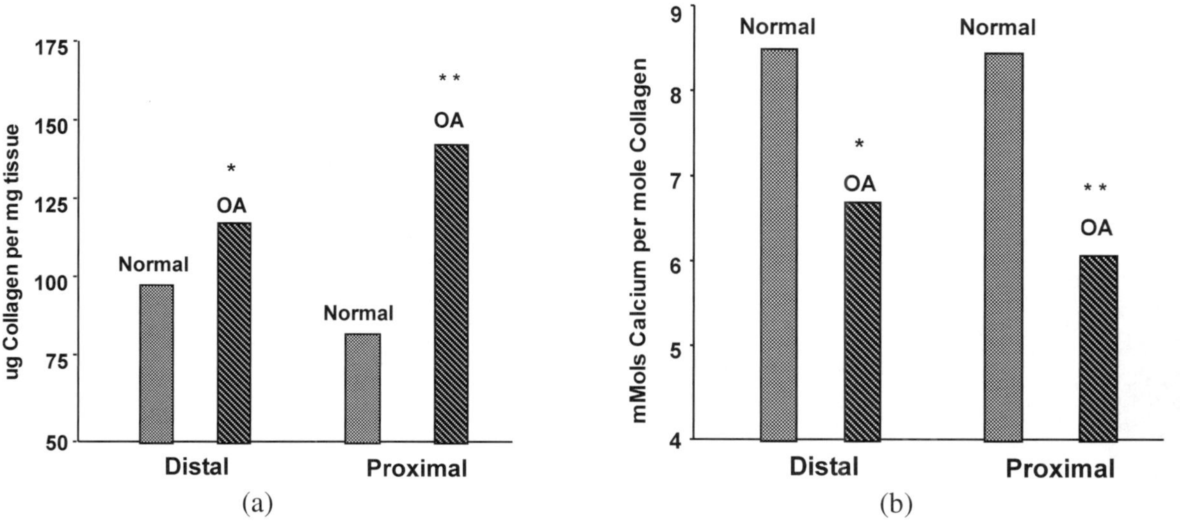

Fig. 2. Comparison of the quality of the subchondral bone (a) collagen content as μg per mg of bone (b) mineralisation in terms of mM calcium per mole of collagen, in distal and proximal domains of the femoral head. ($^{*}p < 0.005$ and $^{**}p < 0.001$.)

bone [25] and the increased presence of osteoid [50]. The greatest increase in the rate of turnover was in the subchondral bone plate but high levels were maintained in the distal regions of the femoral head. The higher turnover rate would lead to a higher proportion of immature bone suggesting a reduced mechanical strength. Indeed, a detailed biomechanical analysis reported that cancellous bone from OA femoral head was significantly weaker than age-matched controls [36].

The increased metabolism of bone collagen may be a feature of other bones in OA subjects. Interestingly, increased bone density and increased levels of TGF-β have been reported in the iliac crest bone of OA subjects [18] and they suggest that the increased bone density of OA subjects protects them against osteoporosis [19]. Certainly spontaneous fracture of femoral heads tends not to occur in OA subjects.

Of additional importance is the fact that the increased turnover is not a function of age. We have investigated bone collagen turnover in the iliac crest bone collagen over a wide age range of mature non-OA subjects. These individuals did not show any changes in rate of turnover or the nature of the stabilising cross-links through into old age. There was, as expected, a steady loss of bone collagen, which correlated with the loss of density and mechanical strength [5]. Any changes found with increasing age are, therefore, likely to be due to disease processes, for example, in the case of osteoporosis the increased hydroxylation leads to narrow fibres, increased pyridinoline cross-links and loss of strength [40].

Simplistically bone can be considered to be a two phase system of a collagenous framework providing tensile strength and hydroxyapatite providing rigidity to that framework. The collagen fibres themselves are formed by a quarter-staggered end-overlap parallel alignment of the collagen molecules. This very precise organization is crucial to both the stabilisation of bone by intermolecular cross-linking and the nucleation of mineralisation within the gap region of the fibre. The crystals of apatite are initially formed in the gap region and have been shown to be plate-like crystals penetrating the fibre prior to covering the fibre surface [34]. The remodelling of bone therefore requires several stages, removal of the mineral, degradation of the collagen, re-synthesis of the collagen and finally re-mineralisation.

The tensile strength of the collagen framework is provided by the intermolecular cross-linking of the collagen molecules making up the fibres of the framework. The initial head to tail cross-linking occurs enzymically through lysyl or hydroxylysyl aldehydes formed in the non-helical telopeptide region by the

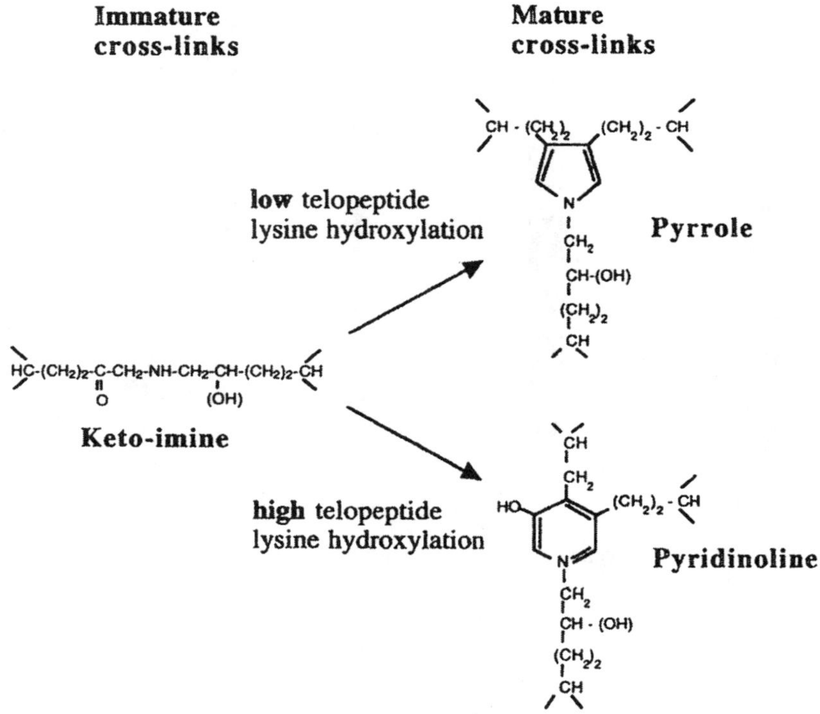

Fig. 3. Alternate pathways for the maturation of the divalent keto-imine to the mature trivalent hydroxylysyl and lysyl-pyrrole or pyridinoline cross-links.

action of lysyl oxidase. In the case of bone collagen these lysyl aldehydes then react with the ε-amino group of specific lysine residues within the triple helical domain exactly opposite the telopeptide to form the keto-imine intermolecular cross-link. These divalent cross-links are capable of providing tensile strength to young bone collagen but spontaneously react with another lysyl or hydroxylysyl aldehyde to form a trivalent mature cross-link, the pyrroles and the pyridinolines respectively (Fig. 3). The proportion of these mature cross-links increases as turnover decreases with maturity. The relative amount of each particular cross-link depends on the relative levels of hydroxylation of the triple helical and telopeptide lysines, each site being controlled by a different lysyl hydroxylase [7]. In addition there is no oxidizable lysine in the C-telopeptide of the $\alpha 2$(I) chain thus restricting the type of cross-link at the C-terminal. The pyrrole cross-link appears to be concentrated at the N-terminal whereas the pyridinolines are present at both termini [26], although there is more lysyl pyridinoline than hydroxylysyl-pyridinoline at the N-terminus. This finding suggests a distinct role for each cross-link and in view of the apparent correlation of the pyrroles with bone strength we have proposed that the latter form interfibrillar cross-links [31]. The hydroxylation state of the telopeptide lysine residues is clearly crucial to the type of immature and subsequent mature cross-link, and hydroxylation of the specific triple helix lysine determines whether it is the lysyl or hydroxylysyl derivative of the cross-link.

The bone matrix varies with skeletal site reflecting the different functions of bone. For example, a study of avian bone [32] revealed a higher turnover of the collagen in the tibiotarsus and increased lysyl hydroxylation compared with the humerus. Variable turnover rates must also occur in different human bones and will also influence the lysine hydroxylation and the nature of the cross-links and therefore their properties. This suggestion is currently being investigated.

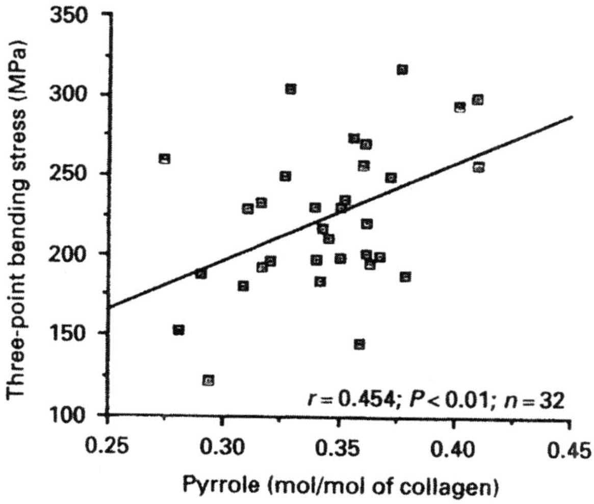

Fig. 4. Correlation of the pyrrole cross-links and the strength of bone.

A major clinical problem in bone biology is the prediction of bone strength. There is no doubt that the amount of bone per unit volume (BV/TV), or apparent density, is a major parameter influencing the stiffness and strength of bone but a substantial part of the variance of these mechanical properties, as much as 30%, is unexplained [33,42]. Several studies have suggested that the unexplained variation is due to the different trabecular architecture, for example, the thickness, spacing and connectivity of the trabeculae [15], but this still does not permit a total prediction of the properties [17]. However, despite its role as the fundamental framework of bone few attempts have been made to include the nature of the collagen cross-links in the prediction of the mechanical properties of bone. As discussed above the stability of the collagen may vary depending on the nature and extent of the collagen cross-links. It has long been known that inhibition of the lysyl oxidase results in bone of little or no tensile strength. Even a partial reduction in the extent of cross-linking produces a significant loss of mechanical strength in the bone [30,35,44].

Initial attempts to correlate mechanical strength and intermolecular cross-linking revealed a surprising relationship between the tensile strength and the pyrrole cross-link rather than the pyridinoline cross-link (Fig. 4). These studies were carried out on avian tibiotarsus bone employing three point bending to determine the tensile strength [30]. Further studies on avian bone, during an assessment of genetic selection for resistance to osteoporosis, again revealed a positive correlation between humerus and tibiotarsal pyrrole content and bone strength [54].

Encouraged by these relationships we carried out a more detailed study of the mechanical properties of human vertebral cancellous bone and the nature of the intermolecular cross-linking of the collagen. We determined the density with peripheral quantitative computer tomography (pQCT) and mechanically tested the bone by compression [8,9]. The amount of collagen and the nature of the cross-links were determined on these samples and on applying multiple regression we found that the prediction of mechanical properties was improved by combining the cross-link assessment with density.

Further, the hydroxylysyl-pyridinoline/lysyl-pyridinoline ratio appeared to be a significant predictor of strength ($p = 0.001$) and stiffness ($p = 0.001$), samples with a high ratio being stronger and stiffer. The ultimate strain correlated with the hydroxylysyl-pyridinoline or lysyl-pyridinoline concentration. However, there was a wide variation between the different subjects analysed. This variation is caused by the different degrees of lysyl hydroxylation, which is controlled by different isomers of lysyl hydroxylases acting on either the specific lysines in the triple helix or in the non-helical telopeptide. These variations lead to different cross-links. The variation of hydroxylysyl- and lysyl-pyridinolines between subjects indicates that the activity of the lysyl hydroxylase acting on the triple helical lysines also varies significantly from one subject to another.

In a continuation of these studies on the vertebral cancellous bone we compared the nature of the collagen with the structural organisation of the trabecular network [9]. This bone has a low bone volume fraction (BV/YV); only 7–14% of the total bone is occupied by bone tissue, the rest being marrow. We determined bone volume fraction, BV/TV; trabecular thickness, TbTh; trabecular number, TbN; trabecular spacing, TbSp; strut length TSL; number of nodes, Nd; number of free ends, Fe; and amount of osteoid bone. Again we observed a relationship between the nature of the bone matrix and the concentration of the cross-links (the pyrrole and pyridinolines) and the structural organisation of the trabeculae. The concentration of pyrrole and pyridinoline cross-links correlated with the structural organisation, but in opposite directions. Thus, the pyrrole/pyridinoline ratio was a good predictor of the above parameters. Subjects with high pyrrole and low pyridinoline in their bone collagen had a thicker and simpler structure, whereas those with a low pyrrole and a high pyridinoline content possessed thin trabeculae that were more numerous. This indicates a relationship between the structure and the stability of the collagen fibre controlling the micro-architecture of the bone. Again high pyrrole appears to be important and this indicates that the regulation may be in the rate of collagen turnover and the activity of the lysyl hydroxylase.

We have not, as yet, carried out a similar study of the cancellous subchondral bone of the OA femoral head. However, based on the above studies on the vertebral bone the high pyridinoline low pyrrole ratio observed in OA subchondral bone should result in weaker and narrow numerous struts, which is consistent with our findings to date. Indeed, morphological changes, which are different from those associated with ageing, increased trabeculae number and reduced separation between trabeculae have been reported for OA bone [22].

The extremely rapid turnover of the collagen in OA subchondral bone means that the environment of the osteoblasts is very different from normal bone, and the consequent change in cytokines and growth factors could lead to differences in their phenotypic expression. Indeed, osteoblasts isolated from the bone of OA subjects were found to be capable of degrading cartilage proteoglycans in contrast to osteoblasts from normal bone [55] and possessed an altered response of OA osteoblasts to various cytokines and growth factors [27]. We therefore investigated the phenotypic expression of the osteoblasts *in situ* and demonstrated the presence of collagen, type I homotrimer $[\alpha 1(\mathrm{I})]_3$, in OA femoral heads in addition to the normal type I heterotrimer $[(\alpha 1)_2\alpha 2]$. The amount of the homotrimer, demonstrated by the ratio of the $\alpha 1$ to $\alpha 2$ chains varied considerably from subject to subject, from $4:1$ to $17:1$ compared with $2:1$ for normal bone [6] (Fig. 5).

The amount of the excess $\alpha 1$ chain presumably depends on the rate of turnover and hence the severity of the OA, but further studies are required on clinically classified samples to confirm this suggestion. Type I homotrimer has previously been reported in tumours and in embryonic tissue, again tissues similarly involved in a high turnover of collagen, but were reported to be genetically distinct collagens. However, cyanogen bromide cleavage of the homotrimer derived from OA subchondral bone revealed an identical peptide profile to the $\alpha 1$ chain of the heterotrimer isolated from the type I collagen of non-OA bone [6]. The presence of the homotrimer is therefore due to a loss of the regulation controlling the $\alpha 1/\alpha 2$ ratio to produce an excess of $\alpha 1$, rather than a genetically distinct type I collagen $\alpha 1$-chain.

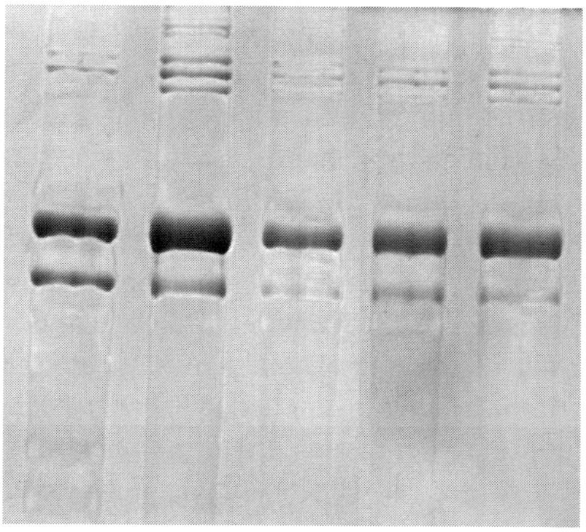

Fig. 5. SDS gel electrophoretogram demonstration of increasing ratio of type I $\alpha1/\alpha2$ from 2.4 : 1 for normal bone ranging up to 17 : 1 for the subchondral bone, indicating the presence additional $\alpha1$(I) chains of type I homotrimer.

The identification of type I homotrimer prompted us to search for a possible association of Col1A1Sp1 polymorphism in OA [37] by employing a modification of the technique for a Col1A1 association in osteoporosis [24] we observed moderate evidence for increased type I homotrimer in OA females. Although not statistically significant the result was consistent with increased synthesis of $\alpha1$(I) collagen and hence the presence of the homotrimer.

To determine the effect of increasing amounts of type I homotrimer on the mechanical properties of the OA subchondral bone we studied the mouse model expressing a variant of osteogenesis imperfecta (*oim*) in which the $\alpha2$ chain had been deleted [15]. This *oim* mutation is a single nucleotide deletion in the ColA2 gene that alters the terminal 50 amino acid residues of the pro-$\alpha2$(I) chain and prevents its association with the pro-$\alpha1$(I) chains to form the type I heterotrimer. The mechanical strength of the bones of these animals is about half that of the wild type mouse [13]. X-ray diffraction studies of the tail tendon indicated an altered collagen structure [38], but the striated pattern of the fibre in the electron microscope appeared normal suggesting that the alteration is small. However, the covalent intermolecular cross-links are the major factor in governing the mechanical strength of the fibre and we have demonstrated a significant decrease in the cross-linking compared with the wild type mouse, which would certainly account for the loss of mechanical strength. Quite why the loss of the $\alpha2$ chain should have such an effect on the cross-linking is not immediately obvious, particularly since the deletion did not appear to affect the molecular alignment of the molecules in the fibre as observed in the electron microscope. In addition the mutation would certainly not have affected the activity of the lysyl oxidase. To account for the decrease in cross-linking we carried out a detailed study of the integrity of the triple helix of the homotrimer by differential scanning calorimetry. The denaturation temperature (Td) of the *oim* molecule was 2.6°C higher than the wild type and the Tm for the *oim* fibre was 1.4°C higher than the wild type fibre. Previous studies had reported identical denaturation temperatures using the less accurate circular dichroism [38]. The increase in denaturation temperature is presumably due to the increase in hydroxyproline content on replacement of the $\alpha2$ chain by the $\alpha1$ which contains a higher hydroxyproline content, with 348 hydroxyprolines compared with 327 for the heterotrimer triple helix.

The mechanism of stabilisation of the triple helix by hydroxyproline is still controversial. It may form stabilising hydrogen-bonded water-bridges as originally proposed by Ramachandran [49] a proposal currently supported by recent X-ray studies defining the location of the water molecules on the triple helix [10]. Alternatively other workers [28] have proposed that the hydroxyproline stabilises the *trans* configuration of the prolyl peptide bond by the inductive effect of its hydroxyl group suggesting that water hydrogen bonding plays no role in the stabilisation of the triple helix. This is supported by earlier work on the effects of alcohol on collagen triple helical stability which demonstrated that water was not an essential component in the stabilisation of the triple helix [21]. In contrast, our recent NMR/DSC studies indicate that H-bonded water does play a role in stabilizing the collagen molecule [53].

We found only a small decrease in the enthalpy of the *oim* fibre suggesting that the amount of disorder of the triple helical molecules was small. However, the difference in denaturation temperature between the molecules in solution and that of the aggregated fibres was 19.9°C against 23.1°C for the wild type, indicating reduced molecular interactions and hence looser packing of the molecules in the fibre. Calculation of the volume fraction of water revealed that the interaxial separation of the molecules in the *oim* fibre was increased by 1.4 Angstrom. This is equivalent to the length of a C–C bond and would therefore certainly decrease the ability of the telopeptide lysine-aldehyde to interact with the ε-amino group from an adjacent molecule to form an intermolecular cross-link. We therefore proposed that the decrease in cross-linking and hence in mechanical properties, was due to the increased water content of the fibre rather than a distortion of the molecular structure. An increasing proportion of the weaker type I homotrimer fibres in the subchondral bone of OA subjects would clearly have a significant effect on decreasing its mechanical properties.

The rapid turnover and fibrosis of bone collagen together with the ability of the ‘modified’ osteoblasts to degrade articular cartilage [55] need to be taken into account by tissue engineers attempting to replace, or regenerate, the articular cartilage of osteoarthritic femoral head and knee bones.

It is clear that osteoarthritis is a disease encompassing articular cartilage degradation and fibrosis of the subchondral bone and therapies need to consider both aspects of the disease rather than concentrating solely on the articular cartilage. Partial regression of the symptoms of OA of the knee have been achieved by high tibial osteotomy [1] which reduced the subchondral thickening. More recent studies [45] have proposed an application of licofelone to control abnormal bone remodelling in OA since licofelone was able to inhibit OA osteoblast prostaglandin E2 and leukotriene B4 *in vitro*.

In summary, the ability of bone collagen to provide a strong framework and to fully mineralise depends on the very precise alignment of the type I collagen heterotrimer. However, in osteoarthritis the increased narrowing of the fibres diameters, the reduced level of the pyrrole cross-link, the decreased mineralization, together with the increasing proportion of the homotrimer, all contribute to a weakening of the mechanical properties of subchondral bone.

References

[1] Y. Akamatsu, T. Koshino, T. Saito and J. Wada, Changes in osteosclerosis of the osteoarthritic knee after high tibial osteotomy, *Clin. Orthop. Rel. Res.* **334** (1997), 207–214.

[2] J.M. Anderson-MacKenzie, M.E.J. Billingham and A.J Bailey, Collagen remodelling in the anterior cruciate ligament associated with developing spontaneous murine osteoarthritis, *Biochem. Biophys. Res. Commun.* **258** (1999), 763–767.

[3] J.M. Anderson-MacKenzie, H.L. Quasnichka, R.L. Starr, E.J. Lewis, M.E.J. Billingham and A.J. Bailey, Comparison of osteoarthritic and control guinea pigs reveals fundamental and early bone changes in osteoarthritis aetiology, *submitted.*

[4] A.J. Bailey and J.P. Mansell, Do subchondral bone changes exacerbate or precede articular cartilage destruction in osteoarthritis of the elderly?, *Gerontology* **43** (1997), 296–304.

[5] A.J. Bailey, T.J. Sims, E.N. Ebbesen, J.P. Mansell, J.S. Thomsen and L. Moskilde, Age-related changes in the biochemical properties of human cancellous bone collagen: Relationships to bone strength, *Calcif. Tissue Int.* **65** (1999), 203–210.
[6] A.J. Bailey, T.J. Sims and L. Knott, Phenotypic expression of osteoblast collagen in osteoarthritic bone. Production of type I homotrimer, *Intern. J. Biochem. & Cell Biology* **34** (2002), 176–182.
[7] R.A. Bank, S.P. Robins, C. Wijmenga, L.J. Breslau-Siderius, A.F. Bardoel, H.A. van der Sliujs, H.E. Pruijs and J.M. TeKoppele, Defective collagen cross-linking in bone, but not in ligament or cartilage, in Buck syndrome: Indications for a bone specific telopeptide lysyl hydroxylase on chromosome 17, *Proc. Natl. Acad. Sci. USA* **96** (1999), 1054–1058.
[8] X. Banse, T.J. Sims and A.J. Bailey, Mechanical properties of adult vertebral cancellous bone: Correlation with collagen intermolecular cross-links, *J. Bone Miner. Res.* **17** (2002), 1621–162.
[9] X. Banse, J.P. Devogelaer, A. Lafosse, T.J. Sims, M. Grynpass and A.J. Bailey, The cross-link profile of bone collagen correlates with the structural organization of the trabeculae, *Bone* **31** (2002), 70–76.
[10] J. Bella, B. Brodsky and H.M. Berman, Hydration structure of a collagen peptide, *Structure* **3** (1995), 893–906.
[11] J.C. Buckland-Wright, D.G. Macfarlane and J.A. Lynch, Quantitiative microfocal radiographic assessment of osteoarthritis of the knee from weight bearing tunnel and semi-flexed standing views, *J. Rhematol.* **21** (1994), 1734–1741.
[12] D.B. Burr, The importance of subchondral bone in osteoarthrosis, *Curr. Opinions Rheum.* **10** (1998), 256–262.
[13] N.P. Camacho, L. Hou, T.R. Toledano, W.A. Ilg, C.F. Brayton, C.L. Raggio, L. Root, K. Misof, W.J. Landes, K. Klaushofer and P. Fratzl, Collagen from osteogenesis imperfecta mouse model (oim) shows reduced resistance against tensile stress, *J. Clin. Invest.* **100** (1997), 40–45.
[14] C.S. Carlson, R.E. Loeser, C.B. Purser, J.F. Gardin and C.P. Jerome, Osteoarthritis in cynomolgus macques. II Effects of age, gender and subchondral bone thickness on the severity of the disease, *J. Bone Miner. Res.* **1** (1996), 1209–1217.
[15] S.D. Chipman, H.O. Sweet, D.J. McBride, M.T. Davisson, S.C. Marks, A.R. Shuldiner, R.J. Wenstrup, D.C.W. Rowe and J.R. Shapiro, Defective pro-α2(I) collagen synthesised in a recessive mutation in mice. A model for human osteogenesis imperfecta, *Proc. Natl. Acad. Sci. USA* **90** (1993), 1701–1705.
[16] T.E. Ciarelli, D.P. Fybrie, M.B. Schaffler and S.A. Goldstein, Variations in three-dimensional cancellous bone architecture of the proximal femur in female hip fractures and in controls, *J. Bone Miner. Res.* **15** (2000), 32–40.
[17] D.W. Dempster, The contribution of trabecular architecture to cancellous bone quality, *J. Bone Miner. Res.* **15** (2000), 20–23.
[18] J. Dequeker, S. Mohan, R.D. Finkelman, J. Aerssens and D.J. Baylink, Generalised OA associated with increased insulin-like growth factor I and II and transforming growth factor-β in cortical bone from the iliac crest, *Arth. Rheum.* **36** (1993), 1702–1708.
[19] J. Dequecker and O. Johnell, Osteoarthritis protects against femoral neck fracture: the MEDOS study experience, *Bone (NY)* **14** (1993), S51–S56.
[20] P.A. Dieppe, J. Cushnaghan, P. Young and J. Kirwan, Prediction of progression of joint space narrowing in osteoarthritis of the knee, *Ann. Rheum. Dis.* **52** (1993), 557–563.
[21] J. Engel, H.-T. Chen, D.J. Prockop and J.T. Klump, The triple helix-coil conversion of collagen-like polypeptides in aqueous and nonaqueous solvents. Comparison of thermodynamic parameters and the binding of water to (L-Pro-L-Pro-Gly)n and (L-Pro-L-Hyp-Gly)n, *Biopolymers* **16** (1977), 601–622.
[22] N.L. Fazzalari and I.H. Parkinson, Fractal properties of subchondral bone in severe osteoarthrhritis of the hip, *J. Bone Miner. Res.* **12** (1997), 632–640.
[23] P. Garnero, O. Borel, I. Byralsen, M. Ferreras, F.H. Drake, M.S. McQueney, N.T. Fogel, P.D. Delmas and J.M. Delaisse, The collagenolytic activity of cathepsin K is unique among mammalian proteinases, *J. Biol. Chem.* **273** (1998), 32347–52.
[24] S.F.A. Grant, D.M. Reid, G. Blake, R. Herd, I. Fogelman and S.H. Ralston, Reduced bone density and osteoporosis associated with a polymorphic Sp1 binding site in the collagen type I α1 gene, *Nature Genet.* **14** (1996), 203–205.
[25] M.D. Grynpas, B. Alpert, I. Katz, J. Lieberman and K.P.H. Pritzkeer, Subchondral bone in osteoarthritis, *Calcif. Tissue Intl.* **49** (1991), 20–26.
[26] D.A. Hansen and D.R. Eyre, Molecular site specificity of pyridinoline and pyrrole cross-links in type I collagen of human bone, *J. Biol. Chem.* **271** (1996), 26508–26516.
[27] G. Hilal, J.F. Martel-Pelletier, P. Ranger and D. Lajeunesse, Osteoblastic-like cells from human subchondral osteoarthritic bone demonstrate an altered phenotype in vitro: possible role of subchondral bone sclerosis, *Arth. Rheum.* **41** (1998), 981–999.
[28] S.K. Holmgren, K.M. Taylor, L.E. Bretscher and R.T. Raines, Code for collagen stability deciphered, *Nature* **392** (1998), 666–667.
[29] L.C. Johnson, Joint remodelling as a basis for osteoarthritis, *J. Amer. Vet. Med. Assoc.* **141** (1962), 1237–1241.
[30] L. Knott, C.C. Whitehead, R.H. Fleming and A.J. Bailey, The biochemistry of the collagenous matrix of osteoporotic avian bone, *Biochem. J.* **310** (1995), 1045–1051.
[31] L. Knott and A.J. Bailey, Collagen cross-links in mineralising tissue: a review of their chemistry, function and clinical relevance, *Bone* **22** (1998), 181–187.

[32] L. Knott and A.J. Bailey, The collagen biochemistry of avian bone. A comparison of bone type and skeletal site, *Brit. J. Poultry Sci.* **40** (1999), 371–379.
[33] D.L. Kopperdahl and T.M. Keaveny, Yield strain behaviour of trabecular bone, *J. Biomech.* **31** (1998), 601–608.
[34] W.J. Landis, K.J. Hodgens and M.J. Song, Mineralization of collagen may occur on fibril surfaces. Evidence from conventional and high-voltage electron microscopy and three-dimensional imaging, *J. Struct. Biol.* **117** (1996), 24–35.
[35] S. Lees, D.R. Eyre and S.M. Barnard, BAPN dose dependence of mature crosslinking in bone matrix collagen of rabbit compact bone: Corresponding variation of sonic velocity and equatorial diffraction spacing, *Connect. Tissue Res.* **24** (1990), 95–105.
[36] B. Li and R.M. Aspden, Composition and mechanical properties of cancellous bone from the femoral head of patients with osteoporosis and osteoarthritis, *J. Bone Min. Res.* **12** (1997), 641–651.
[37] J. Loughlin, J.S. Sinsheimer, Z. Mustafa, A.J. Carr, K. Clipsham, V.A. Bloomfield, J. Chitnavis, A.J. Bailey, B. Sykes and K. Chapman, Association analyses of the vitamin receptor gene, the type I collagen gene and the oestrogen receptor gene in idiopathic osteoarthritis, *J. Rheum.* **27** (1999), 779–784.
[38] D.J. McBride, V. Choe, J.R. Shapiro and B. Brodsky, Altered collagen structure in mouse tail tendon lacking the $\alpha 2$(I) chain, *J. Mol. Biol.* **270** (1997), 275–284.
[39] J.P. Mansell and A.J. Bailey, Abnormal cancellous bone collagen metabolism in osteoarthritis, *J. Clin. Invest.* **101** (1998), 1596–1603.
[40] J.P. Mansell and A.J. Bailey, Increased metabolism of collagen in post-menopausal osteoporotic female femoral heads, *Intl. J. Biochem. Cell Biol.* **35** (2003), 522–529.
[41] C.A. Miles, T.J. Sims, N.P. Camacho and A.J. Bailey, The role of the $\alpha 2$ chain in the stabilization of the collagen type I heterotrimer: a study of the type I homotrimer in oim mouse tissue, *J. Mol. Biol.* **321** (2002), 797–805.
[42] L. Moskilde and C.C. Danielsen, Biomechanical competence of vertebral trabecular bone in relation to ash, density and age in normal individuals, *Bone* **8** (1987), 79–85.
[43] W.N. Newberry, D.K. Zukosky and R.C. Haut, Subfracture insult to a knee joint causes alterations in the bone and in the functional stiffness of overlying cartilage, *J. Orthop. Res.* **15** (1997), 450–455.
[44] H. Oxlund, M. Barckman, G. Ortoft and T.T. Andreassen, Reduced concentration of collagen cross-links are associated with reduced strength of bone, *Bone* **17** (Suppl. 4) (1995), 365S–371S.
[45] Y. Paredes, F. Massicotte, J.P. Pelletier, J. Martel-Pelletieer, S. Laufer and D. Lajeunesse, *Arth. Rheum.* **46** (2002), 1804–1812.
[46] J. Pfeilschifter, O. Wolf, A. Nautman, H.W. Minne, G.R. Munday and R. Ziegler, Chemotactic response of osteoblastic cells to TGF-β, *J. Bone Miner. Res.* **5** (1990), 852–830.
[47] H.L. Quasnichka, J.M. Anderson-MacKenzie, M.E.J. Billingham and A.J. Bailey, Biochemical and ultrastructural changes in the cruciate ligament of osteoarthritic knees, *Osteoarthritis and Cartilage* **10A** (2000), S 54 (Abstract).
[48] E.L. Radin and R.M. Rose, Role of subchondral bone in the initiation and progression of cartilage damage, *Clin. Orthop.* **213** (1986), 34–40.
[49] G.N. Ramachandran, M. Bansal and R.S. Bhatnagar, A hypothesis on the role of hydroxyproline in stabilising the collagen structure, *Biochim. Biophys. Acta* **323** (1973), 166–171.
[50] I. Reinmann, H.J. Mankin and C. Trahan, Quantitative histologic analyses of articular cartilage and subchondral bone from osteoarthritic and normal human hips, *Acta Ortho. Scand.* **48** (1997), 63–67.
[51] A. Saied, E. Cherin, H. Gaucher, P. Lasugier, P. Gillet, J. Floquet, P. Netter and G. Berger, Assessment of articular cartilage and subchondral bone: subtle and progressive changes in experimental osteoarthritis using 50 MHz echography *in vitro*, *J. Bone Miner. Res.* **12** (1997), 1378–1386.
[52] K. Senior, Osteoarthritis research: on the veerge of a revolution, *Lancet* **355** (2000) 208 (Editorial).
[53] D.A. Slatter, C.A. Miles and A.J. Bailey, Asymmetry of the triple helix of collagen-like heterodimers confirm that external bonds stabilize collagen structure, *J. Mol. Biol.* **329** (2003) 175–183.
[54] A.J. Sparke, T.J. Sims, N.C. Avery, A.J. Bailey, R.H. Fleming and C.C. Whitehead, Difference in composition of avian bone following genetic selection for resistance to osteoporosis, *Brit. J. Poultry Sci.* **43** (2002), 127–14.
[55] C. Westacott, G.R. Webb, M.G. Warnock, J.V. Sims and C.J. Elson, Alteration of cartilage metabolism by cells from osteoarthritic bone, *Arthritis and Rheum.* **40** (1997), 1282–1291.

Biorheology 41 (2004) 359–368
IOS Press

Adaptation of subchondral bone in osteoarthritis

J.S. Day [a], J.C. van der Linden [a], R.A. Bank [b], M. Ding [c], I. Hvid [c], D.R. Sumner [d] and H. Weinans [a,*]

[a] *Erasmus Orthopaedic Research Lab, University Rotterdam, Rotterdam, The Netherlands*
[b] *TNO Prevention and Health, Leiden, The Netherlands*
[c] *Orthopaedic Research Laboratory, Aarhus University Hospital, Aarhus, Denmark*
[d] *Department of Anatomy and Cell Biology, Rush Medical College, Chicago, IL, USA*

Abstract. Osteoarthritis is a chronic joint disease with pathological changes in the articulating cartilage and all other tissues that occupy the joint. Radin and coworkers have suggested the involvement of subchondral bone in the disease process. However, evidence for an essential role in the etiology has never been proven. Recent studies showing reduced chemical and mechanical properties of subchondral bone in various stages of the disease have invigorated interest in the role of subchondral bone in the development and progression of the disease. The current study showed that the concept of bone adaptation might explain subchondral stiffening, a process where subchondral bone becomes typically sclerotic in osteoarthritis. In addition, we report reduced mechanical matrix tissue properties as well as an increase in denatured collagen content. In conclusion, although osteoarthritic bone tissue contains increased denatured collagen and has reduced matrix mechanical properties, the widely accepted concept of subchondral stiffening is compatible with the process of normal bone adaptation.

1. Introduction – The role of bone in osteoarthritis

Osteoarthritis has been described as an "organ failure" where pathological changes in the cartilage, bone, synovium and other soft tissues interact resulting in failure of the joint. Indications of disease progression include destruction of the cartilage in combination with abnormal growth of the subchondral bone resulting in thickened subchondral bone plate, osteophytes and deformation of the affected joint (Fig. 1). Although little is known about the relation between bone and cartilage in the etiology of osteoarthritis, clinical management has focused mainly on chondroprotective strategies.

The role of bone in disease pathogenesis requires examination, as the subchondral bone plate provides a mechanical base for the cartilage while supplying a large proportion of its nutrient requirements [8, 35]. Sclerosis of the subchondral plate can be observed on X-rays of osteoarthritic patients and increased stiffness and bone mineral density (BMD) of the subchondral plate is considered a hallmark of osteoarthritis. Subchondral bone is altered in the earliest stages of osteoarthritis [9,10,18,20,45] as the subchondral plate thickens and bone volume fraction is increased in the weight bearing areas of trabecular bone [23,25,31,39]. The structure of the trabecular bone is also altered to consist of thick, widely spaced trabeculae [24,36]. Although osteoarthritis has been repeatedly associated with increased BMD,

[*] Address for correspondence: Harrie Weinans, Erasmus University Rotterdam, Erasmus Orthopaedic Research Lab, EE1614, P.O. Box 1738, 3000 DR Rotterdam, The Netherlands. Tel.: +31 (0) 10 40 87367/87384; Fax: +31 (0) 10 40 89415; E-mail: weinans@ortho.fgg.eur.nl.

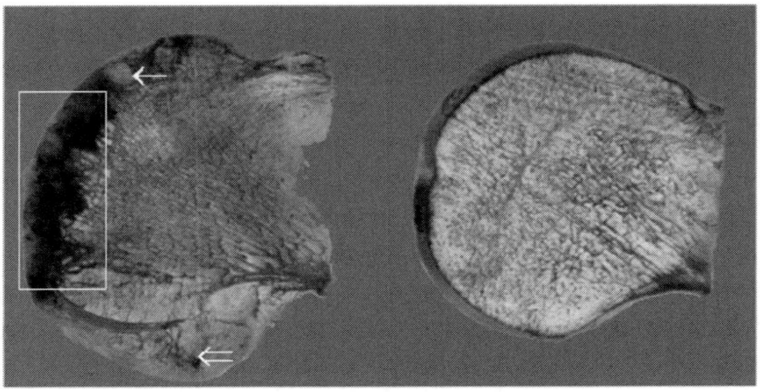

Fig. 1. Pathological adaptation of an osteoarthritic femur head. A section through an osteoarthritic (left) and normal femur head (right). Prominent osteophytosis (⇐), cyst formation (←), and subchondral sclerosis (box) are indicated. Note the flattening of the osteoarthritic femur head and widening of the femoral neck.

both in the affected joint and appendicular skeleton [7,17,18,27,29,38,45,56,60], the relation between BMD and disease progression is not clear [6,29,34].

Radin et al. [53,54] were the first to propose a link between subchondral bone mechanics and disease progression. Based on observations of increased stifness [51] and decreased energy absorbing capacity [52] of osteoarthritic bone they proposed that stiffening of the subchondral plate was an initiating factor in osteoarthritis. According to their hypothesis, trabecular microfracture due to impulsive loading initiates bone remodeling in the subchondral plate. This leads to localized stiffening that in turn produces increased shear stress in the cartilage, culminating in cartilage breakdown.

2. Mechanical testing of arthritic bone: Scale effects

In spite of evidence of the involvement of mechanical factors in the development of osteoarthritis, there is little data available regarding the mechanical properties of osteoarthritic bone. Upon examination of the available literature (summarized in Table 1), the concept of subchondral stiffening seems to present a paradox since arthritic bone has been reported to be both stiffer and more compliant than normal bone. However many of these studies fail to recognize that bone's mechanical properties can be altered at a minimum of two distinct levels of organization. Because bone is a cellular (i.e. foam-like) material, its mechanical properties are influenced both by the material properties of the calcified matrix and the porosity of the structure as a whole. Matrix properties are measured at the micron scale and are commonly referred to as 'bone matrix quality'. Apparent or functional level measurements are made at the millimeter to centimeter scale and reflect the matrix level properties combined with the effects of bone mass and trabecular architecture (Fig. 2).

At this time, only two studies have measured both the apparent and matrix level properties of arthritic bone. Day et al. [16] used a combination of standard mechanical testing and micro finite element modeling to determine both the apparent and matrix Young's modulus of cadaveric specimens. Tibiae with mild cartilage damage in the medial compartment and intact cartilage on the lateral side were compared to intact controls. In these specimens, the matrix modulus was reduced by approximately 50% in the arthritic group with a larger reduction on the medial side. The apparent modulus was not significantly different from controls, but the ultimate stress was reduced on the medial side in the arthritic specimens [23]. Li and Aspen [39–41] used a combination of standard compression and ultrasonic testing to measure the

Table 1

A review of previously published mechanical testing results for osteoarthritic specimens. Although there have been multiple tests of osteoarthritic bone, many have used test methods that do not discriminate between matrix and apparent properties

Author	Type of bone	Location	Disease progression	Test method	Finding	Comments
Radin et al. (1970)	subchondral plate	tibial plateau	early and late stage OA	drop test	65% reduction in impact energy in early OA	apparent properties
Lereim et al. (1974)	subchondral plate	tibial plateau	late stage OA	Brinell hardness (5 mm indentor)	50% reduction in hardness in OA	mixture of apparent and tissue properties
Pugh et al. (1974)	trabecular bone	femoral condyle	mild cartilage damage	dynamic apparent stiffness	40% increase in apparent modulus	apparent properties
Finlay et al. (1988)	trabecular bone	tibial plateau	late stage OA	4 mm indentor	result varied by location and disease progression	apparent properties
Hvid et al. (1988)	trabecular bone	tibial plateau	late stage OA	osteopenetrometer (2.5 mm indentor)	result varied by location and knee alignment	mixture of apparent and tissue properties
Li & Aspden (1997)	subchondral plate and trabecular bone	femoral head	late stage OA	standard compression testing and ultrasound	15% increase in apparent modulus	apparent properties
					15% decrease in matrix modulus	tissue properties
Ding et al. (2001) Day et al. (2001)	trabecular bone	tibial plateau	mild cartilage damage	standard compression testing and finite element modelling	30% decrease in apparent modulus	apparent properties
					45% decrease in matrix modulus	tissue properties
Brown et al. (2002)	trabecular bone	femoral head	late stage OA	standard compression testing	Arthritic apparent modulus was decreased compared to normal bone with equal bone mass	apparent properties (also with correction for BMD)

mechanical properties of bone taken from patients undergoing hip arthroplasty for osteoarthritis. They found a 15% increase in the apparent modulus of the trabecular bone and a 15% decrease in the matrix modulus of the subchondral bone plate compared to a control group. In common to both of these studies, one on early and the other on late-stage osteoarthritis, was a finding of increased bone volume fraction and decreased matrix properties in the arthritic groups.

3. Bone mechanosensors, remodelling and adaptation

It is important to note that bone remodeling and adaptation is driven by the cells and thus occurs at a scale that is closer to the tissue level than the apparent level. The cells that generate and resorb the bone matrix are the osteoblasts and osteoclasts, respectively. Osteoclasts make resorption cavities on the surfaces of the trabeculae in the range of 20 to 60 microns deep. During normal turnover osteoblasts fill this cavity with collagen that slowly mineralizes thereafter (Fig. 3). It has long been known that the osteoclastic resorption and osteoblastic apposition are a coupled mechanism, often referred to as a Bone

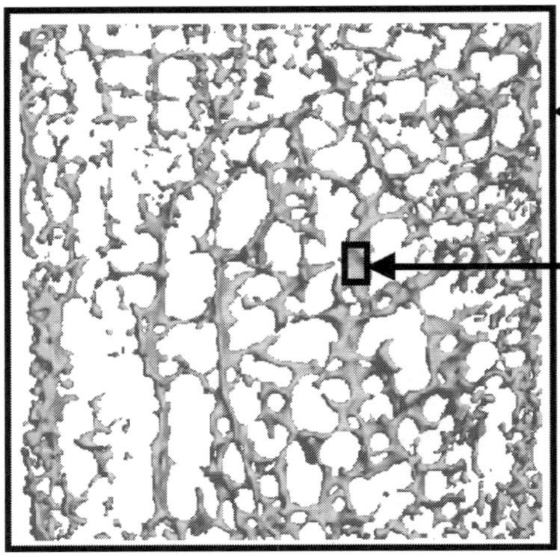

Fig. 2. Definitions of the apparent level and the matrix level. The matrix modulus was defined as the modulus effective at the scale of tens to hundreds of microns. The apparent modulus was effective at the scale of millimeters to centimeters and includes both the influence of the matrix modulus and the influence of trabecular architecture.

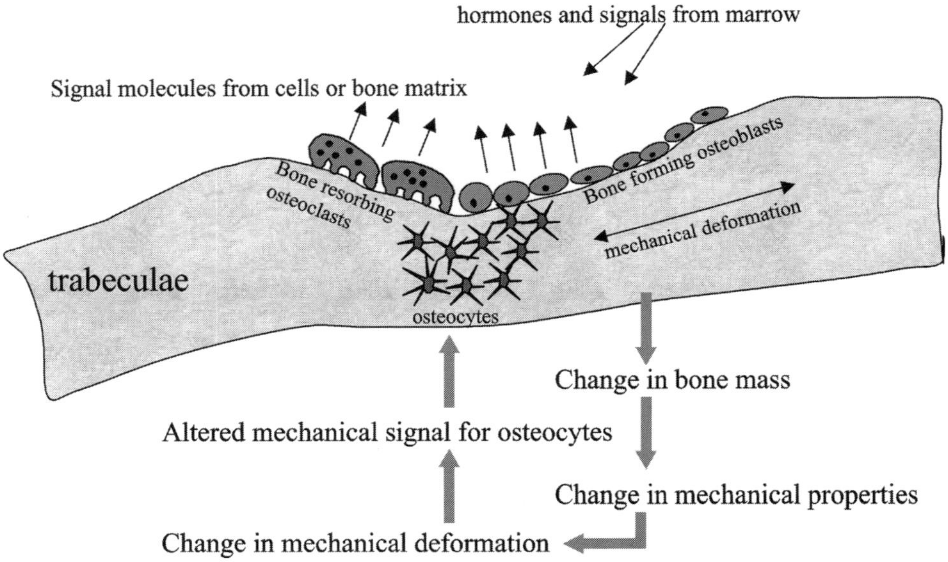

Fig. 3. Illustration of a bone remodeling unit. Osteocytes and/or osteoblasts act to modulate the resorptive activity of osteoclasts. These cells are thought to be sensitive to both systemic signals and local mechanical stimulation.

Multicellular Unit or a BMU. It has been recently discovered that an important coupling mechanism between osteblasts and osteoclasts occurs due to cell binding molecules expressed on their cell membranes. Osteoblasts express a molecule called RANKL on the cell membrane and by physically making contact with the osteoclast they initiate osteoclast activity by binding to RANK at the osteoclast membrane [11, 48,58,66]. However many details on where osteoclasts start their activity, why they stop resorbing bone and what triggers the osteoblasts to express RANKL are unknown.

It is generally believed that mechanical loading or deformation of the bone matrix is an important aspect in bone remodeling regulation. During normal bone turnover there is often a small deficit in refilling the resorption cavity by the osteoblasts. Therefore each BMU cycle results in a small amount of net bone loss. Disuse (unloading the bone) stimulates this net bone loss while mechanical loading counteracts it. Although several hypotheses have been proposed in the literature, it is unclear how the hormonal, paracrine and autocrine factors interact with mechanical loading [13,14,48,57,61].

An important consequence of turnover is the renewal of bone and repair of microdamage within the matrix. However, abnormally high turnover leads to bone with a relatively lower mean age. Since the mineralization of bone is a slow process, there is also a resultant reduction in mean mineralization and a concomitant reduction of bone stiffness [49,50].

4. Simulation of bone adaptation in osteoarthritis

Previously in this paper we have demonstrated that the calcified matrix stiffness is reduced and the bone volume fraction increased in osteoarthritis. We have also demonstrated that bone adaptation is most likely driven by either osteoblasts or the osteocyte network acting as mechanoreceptors at a local (matrix) level. It therefore seems plausible that the increase in bone volume fraction observed in osteoarthritis is an adaptive response to the increased bone strain that is experienced as the matrix becomes more compliant. In order to test whether this sequence of events is compatible with the concept of subchondral stiffening we have constructed a simulation of bone adaptation driven by matrix level strain. We have created models of the subchondral trabecular bone from high-resolution microCT scans and simulated bone apposition in response to degeneration of the matrix. By analyzing the resulting apparent Young's modulus we could investigate the compatibility of the concept of subchondral stiffening with that of a reduced matrix modulus.

Cylindrical bone specimens of epiphyseal trabecular bone were obtained post mortem from the medial compartment of 4 donors (ages: 58, 61, 62, 80) [21]. The donors had no record of musculoskeletal disease and the donor tibiae did not exhibit any macroscopic pathology. The samples were taken 1 mm distal to the subchondral bone plate, were 7.5 mm in both diameter and length and were orientated such that the axes of the cylinders were parallel to the longitudinal axes of the tibiae. Specimens were scanned with a 3D voxel size of 20 μm using a microtomographic system (μCT20, Scanco Medical AG, Zurich, Switzerland).

Cubic data sets were extracted from the resulting image stacks. Each cube was taken from the centre of the sample and had an edge length of 4.3 mm. The image data were converted into finite element models with $20 \times 20 \times 20$ μm 8 node brick elements and isotropic material properties. A linear elastic modulus of 5 GPa and Poisson's ratio of 0.3 were assigned to all bone elements and confined compression tests were simulated [16,32,37,64]. The models were solved using in-house code that implemented the element-by-element method on a desktop computer [64]. The apparent stiffness (quotient of applied stress and the applied strain at the scale of the entire sample) of each sample was then calculated from the output of the finite element models.

Degeneration of the calcified matrix and the subsequent adaptation process was simulated using an inverse approach. Instead of reducing the elastic modulus of the elements and then incrementally adding bone voxels, the adaptive response was first simulated by adding bone voxels to all surfaces using a dilation algorithm (Morph3D, R.A. Peters, Vanderbilt University). These bone voxels were assigned the same material properties as the existing matrix resulting in homogenous matrix properties. Next, the finite element models were solved and the maximal principal strain was calculated for each element. Because of the linear nature of the finite element models, the matrix modulus could be scaled to a value

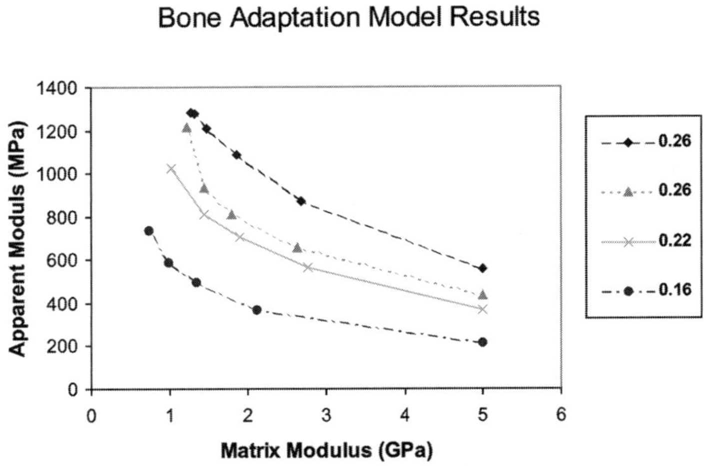

Fig. 4. Results of the adaptive bone model. The matrix modulus (x axis) of each sample was initially assigned a value of 5 GPa and the apparent moduli of the 4 samples were calculated (y axis). The matrix modulus was then reduced for each specimen and the adaptive response was simulated. The new apparent modulus was then plotted. This was continued until the sample was a solid block. The initial bone volume fraction of each sample is indicated in the legend.

that would normalize the median strain for the model. By repeating this process for a number of steps a tissue response curve could be generated.

As expected, an increase in the bone volume fraction was necessary to compensate for the decreased matrix modulus. In spite of the decline in the local matrix modulus, the increase in volume fraction resulted in an overall increase in the apparent modulus of the models (Fig. 4). When the matrix modulus was reduced to approximately 25% of its original value the adaptive response led to a solid block of bone with an apparent Young's modulus that was 4 times the original. A more moderate(and realistic) reduction of 20% for the matrix modulus resulted in a 25% increase in the apparent Young's modulus.

5. More on bone matrix properties in osteoarthritis

The reason for a reduced bone matrix modulus in osteoarthritis is unclear. Possible explanations include: lower mineralization levels due to increased bone remodeling, abnormal mineralization, or defects in the collagen matrix.

While there is some debate on whether systemic breakdown of mature collagen is elevated in arthritic patients [1,30,42,59,62], there is considerable evidence that remodeling activity is increased in the subchondral bone of arthritic joints [2,20]. Increased remodeling activity results in a lower mean age of the bone matrix by reducing the time available for passive mineralization of the matrix [50]. Thus increased remodeling activity results in a lower mean level of mineralization, and thus a reduced modulus in osteoarthritic bone [12,31,43,45,55,56].

In addition to lower mineralization levels, there have also been observations of abnormalities in the collagenous matrix of osteoarthritic bone. Bailey et al. [3] have reported both increased collagen hydroxylation and the presence of an unusual type I homotrimer [$(\alpha 1)_3$] as opposed to the normal heterotrimer [$(\alpha 1)_2(\alpha 2)_1$] in osteoarthritic subchondral bone. Both increased hydroxylation and the formation of a type I homotrimer have been linked to a reduction in bone mechanical properties [5,46].

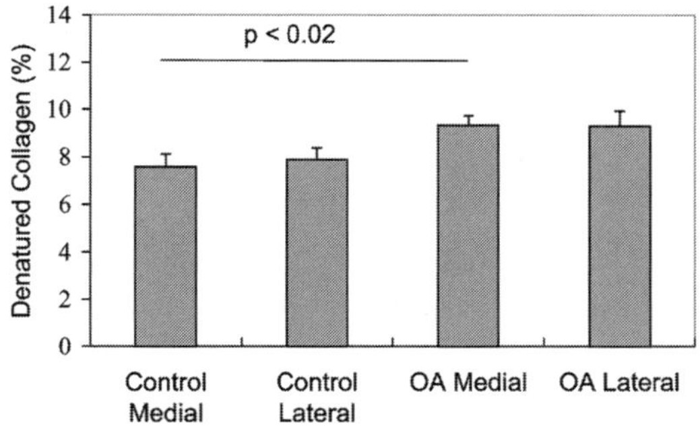

Fig. 5. Denatured collagen content of subchondral trabecular bone in donors with early osteoarthritis (mean + SEM). The amount of denatured collagen was measured for each sample using a selective digestion technique. Denatured collagen content was increased by approximately 22% in the arthritic medial condyle.

We have reported an increase in the proportion of denatured collagen in the epiphyseal trabecular bone of donors with mild cartilage damage (i.e. early osteoarthritis) [15]. In brief, trabecular bone samples were harvested from donors with early-stage osteoarthritis and normal controls. These samples were decalcified in EDTA before selectively digesting denatured collagen using α-chymotrypsin. By comparing the amount of hydroxyproline in the supernatant to that in the remaining insoluble matrix, the proportion of denatured collagen with respect to intact collagen could be measured in each specimen [4]. We found a 22% increase in denatured collagen in the medial condyle of the arthritic donors with respect to controls (Fig. 5). It has previously been demonstrated that mechanical toughness is inversely correlated to the denatured collagen content of bone [65]. The presence of elevated levels of denatured collagen may be the result of increased levels of MMPs in osteoarthritic bone [43,44].

6. Discussion

Recent advances have made it possible to examine the mechanical properties of bone matrix at multiple levels of organization. New methods such as high-resolution ultrasound and nanoindentation can be used to resolve the mechanical properties of bone osteons at the scale of single lamellae. Comparatively, the finite element derived matrix modulus is in fact the average effective matrix stiffness throughout the whole sample and includes effects such as trabecular architecture, matrix porosity, microdamage, etc. In this paper we have described a reduction in the matrix Young's modulus in osteoarthritic bone as measured by multiple methods [16,39–41]. At this time, we are only beginning to understand the underlying chemical and/or physical explanation of this phenomenon. Reduced matrix properties may be, in part, due to hypomineralization of the bone matrix; a normal consequence of the increased remodeling activity in osteoarthritic joints. However, it is also likely that there is also a contribution due to degradation of the collagen network as evidenced by the presence of increased collagen hydroxylation, an unusual collagen homotrimer and an increase in denatured collagen content.

We have applied a simulation model to investigate the apparent paradox of subchondral stiffening in the presence of reduced bone quality (matrix stiffness). We employed an equilibrium model constructed assuming homogenous tissue properties and neglecting the temporal aspects of the bone remodeling cycle [33,63]. The model demonstrates that, in an equilibrium state, the concept of subchondral stiffening at the apparent level is not incompatible with reduced calcified matrix stiffness when adaptation is driven by strain at the cell level. The model does not attempt to model the full remodeling cycle (i.e.

resorption and formation phases) and does not examine the possibility of phase lags between tissue level degeneration and the adaptive response. The presence of a phase lag between degeneration of the matrix and the adaptive response may explain why an increased apparent modulus is reported in end-stage osteoarthritis [26,39–41] and a decreased apparent modulus is reported in less severe osteoarthritis [16, 26].

In vivo it is likely that the driving factors behind the formation of pathological bone are more complex than the adaptive model proposed here. It is unlikely that such a model can explain the formation of subchondral cysts or osteophytes. However, a great deal of insight into the disease progress can be gained by applying such models. For example, we can apply this model to the original hypothesis proposed by Radin who suggested that subchondral sclerosis in osteoarthritis created stiffness gradients in the subchondral bone resulting in shearing of the overlying cartilage. From the model results we can see that it is possible for the reduced matrix modulus to result in subchondral stiffening but from the experimental data it is likely that this is preceded by a period of increased compliance of the subchondral bone [16, 22,26]. It is of interest to note that the presence of a compliant region of subchondral bone would still, however, create stiffness gradients in the subchondral bone and thus shearing of the overlying cartilage.

Subchondral bone and overlying cartilage perform as a functional unit. Presently little is known about the interaction between the cartilage and bone in the etiology of osteoarthritis. Recent research indicates that bone is intimately involved in disease progression. While there is currently little evidence of a direct cause and effect relation between subchondral bone sclerosis and degeneration of the overlying cartilage, recent studies using bisphosphonates (bone antiresorptive compounds) have yielded interesting results with regard to both pain and cartilage degeneration [19,28,47]. Thus, true understanding of the etiology of osteoarthritis will require study of all of the tissues of the joint.

References

[1] C. Astbury, H.A. Bird, A.M. McLaren and S.P. Robins, Urinary excretion of pyridinium crosslinks of collagen correlated with joint damage in arthritis, *Br. J. Rheumatol.* **33** (1994), 11–15.

[2] A.J. Bailey and J.P. Mansell, Do subchondral bone changes exacerbate or precede articular cartilage destruction in osteoarthritis of the elderly?, *Gerontology* **43** (1997), 296–304.

[3] A.J. Bailey, T.J. Sims and L. Knott, Phenotypic expression of osteoblast collagen in osteoarthritic bone: production of type I homotrimer, *Int. J. Biochem. Cell. Biol.* **34** (2002), 176–182.

[4] R.A. Bank, M. Krikken, B. Beekman, R. Stoop, A. Maroudas, F.P. Lafeber and J.M. te Koppele, A simplified measurement of degraded collagen in tissues: application in healthy, fibrillated and osteoarthritic cartilage, *Matrix Biol.* **16** (1997), 233–243.

[5] X. Banse, T.J. Sims and A.J. Bailey, Mechanical properties of adult vertebral cancellous bone: correlation with collagen intermolecular cross-links, *J. Bone Miner. Res.* **17** (2002), 1621–1628.

[6] R.J. Bruno, P.A. Sauer, A.G. Rosenberg, J. Block and D.R. Sumner, The pattern of bone mineral density in the proximal femur and radiographic signs of early joint degeneration, *J. Rheumatol.* **26** (1999), 636–640.

[7] H. Burger, P.L. van Daele, E. Odding, H.A. Valkenburg, A. Hofman, D.E. Grobbee, H.E. Schutte, J.C. Birkenhager and H.A. Pols, Association of radiographically evident osteoarthritis with higher bone mineral density and increased bone loss with age. The Rotterdam Study, *Arthritis Rheum.* **39** (1996), 81–86.

[8] D. Burstein, J. Velyvis, K.T. Scott, K.W. Stock, Y.J. Kim, D. Jaramillo, R.D. Boutin and M.L. Gray, Protocol issues for delayed Gd(DTPA)(2-)-enhanced MRI (dGEMRIC) for clinical evaluation of articular cartilage, *Magn. Reson. Med.* **45** (2001), 36–41.

[9] C.S. Carlson, R.F. Loeser, M.J. Jayo, D.S. Weaver, M.R. Adams and C.P. Jerome, Osteoarthritis in cynomolgus macaques: a primate model of naturally occurring disease, *J. Orthop. Res.* **12** (1994), 331–339.

[10] C.S. Carlson, R.F. Loeser, C.B. Purser, J.F. Gardin and C.P. Jerome, Osteoarthritis in cynomolgus macaques. III: Effects of age, gender, and subchondral bone thickness on the severity of disease, *J. Bone Miner. Res.* **11** (1996), 1209–1217.

[11] D.R. Carter, Mechanical loading history and skeletal biology, *J. Biomech.* **20** (1987), 1095–1109.

[12] P. Christensen, J. Kjaer, F. Melsen, H.E. Nielsen, O. Sneppen and P.S. Vang, The subchondral bone of the proximal tibial epiphysis in osteoarthritis of the knee, *Acta Orthop. Scand.* **53** (1982), 889–895.

[13] S.C. Cowin, L. Moss-Salentijn and M.L. Moss, Candidates for the mechanosensory system in bone, *J. Biomech. Eng.* **113** (1991), 191–197.

[14] S.C. Cowin, Wolff's law of trabecular architecture at remodeling equilibrium, *J. Biomech. Eng.* **108** (1986), 83–88.

[15] J.D. Day, O. Manintveld, R.A. Bank, M. Ding, D.R. Sumner, I. Hvid, G.J.V.M. van Osch and H. Weinans, Denatured collagen in osteoarthritic bone, in: *49th Annual Meeting of the Orthopaedic Research Society*, New Orleans, LA, 2003, p. 746.

[16] J.S. Day, M. Ding, J.C. van der Linden, I. Hvid, D.R. Sumner and H. Weinans, A decreased subchondral trabecular bone tissue elastic modulus is associated with pre-arthritic cartilage damage, *J. Orthop. Res.* **19** (2001), 914–918.

[17] J. Dequeker, L. Mokassa and J. Aerssens, Bone density and osteoarthritis, *J. Rheumatol. Suppl.* **43** (1995), 98–100.

[18] J. Dequeker, L. Mokassa, J. Aerssens and S. Boonen, Bone density and local growth factors in generalized osteoarthritis, *Microsc. Res. Tech.* **37** (1997), 358–371.

[19] J.P. Devogelaer, Treatment of bone diseases with bisphosphonates, excluding osteoporosis, *Curr. Opin. Rheumatol.* **12** (2000), 331–335.

[20] P. Dieppe, J. Cushnaghan, P. Young and J. Kirwan, Prediction of the progression of joint space narrowing in osteoarthritis of the knee by bone scintigraphy, *Ann. Rheum. Dis.* **52** (1993), 557–563.

[21] M. Ding, Age variations in the properties of human tibial trabecular bone and cartilage, *Acta Orthop. Scand. Suppl.* **292** (2000), 1–45.

[22] M. Ding, M. Dalstra, F. Linde and I. Hvid, Changes in the stiffness of the human tibial cartilage–bone complex in early-stage osteoarthrosis, *Acta Orthop. Scand.* **69** (1998), 358–362.

[23] M. Ding, C.C. Danielsen and I. Hvid, Bone density does not reflect mechanical properties in early-stage arthrosis, *Acta Orthop. Scand.* **72** (2001), 181–185.

[24] N.L. Fazzalari and I.H. Parkinson, Femoral trabecular bone of osteoarthritic and normal subjects in an age and sex matched group, *Osteoarthritis Cartilage* **6** (1998), 377–382.

[25] N.L. Fazzalari and I.H. Parkinson, Fractal properties of subchondral cancellous bone in severe osteoarthritis of the hip, *J. Bone Miner. Res.* **12** (1997), 632–640.

[26] J.B. Finlay, R.B. Bourne, W.J. Kraemer, T.K. Moroz and C.H. Rorabeck, Stiffness of bone underlying the tibial plateaus of osteoarthritic and normal knees, *Clin. Orthop.* (1989), 193–201.

[27] M.V. Foss and P.D. Byers, Bone density, osteoarthrosis of the hip, and fracture of the upper end of the femur, *Ann. Rheum. Dis.* **31** (1972), 259–264.

[28] T. Fujita, Y. Fujii, S.F. Okada, A. Miyauchi and Y. Takagi, Analgesic effect of etidronate on degenerative joint disease, *J. Bone Miner. Metab.* **19** (2001), 251–256.

[29] B. Goker, D.R. Sumner, D.E. Hurwitz and J.A. Block, Bone mineral density varies as a function of the rate of joint space narrowing in the hip, *J. Rheumatol.* **27** (2000), 735–738.

[30] M.P. Graverand, A.M. Tron, M. Ichou, M.C. Dallard, M. Richard, D. Uebelhart and E. Vignon, Assessment of urinary hydroxypyridinium cross-links measurement in osteoarthritis, *Br. J. Rheumatol.* **35** (1996), 1091–1095.

[31] M.D. Grynpas, B. Alpert, I. Katz, I. Lieberman and K.P. Pritzker, Subchondral bone in osteoarthritis, *Calcif. Tissue Int.* **49** (1991), 20–26.

[32] F.J. Hou, S.M. Lang, S.J. Hoshaw, D.A. Reimann and D.P. Fyhrie, Human vertebral body apparent and hard tissue stiffness, *J. Biomech.* **31** (1998), 1009–1015.

[33] R. Huiskes, R. Ruimerman, G.H. van Lenthe and J.D. Janssen, Effects of mechanical forces on maintenance and adaptation of form in trabecular bone, *Nature* **405** (2000), 704–706.

[34] D.E. Hurwitz, D.R. Sumner and J.A. Block, Bone density, dynamic joint loading and joint degeneration. A review, *Cells Tissues Organs* **169** (2001), 201–209.

[35] H. Imhof, I. Sulzbacher, S. Grampp, C. Czerny, S. Youssefzadeh and F. Kainberger, Subchondral bone and cartilage disease: a rediscovered functional unit, *Invest. Radiol.* **35** (2000), 581–588.

[36] L. Kamibayashi, U.P. Wyss, T.D. Cooke and B. Zee, Trabecular microstructure in the medial condyle of the proximal tibia of patients with knee osteoarthritis, *Bone* **17** (1995), 27–35.

[37] A.J. Ladd, J.H. Kinney, D.L. Haupt and S.A. Goldstein, Finite-element modeling of trabecular bone: comparison with mechanical testing and determination of tissue modulus, *J. Orthop. Res.* **16** (1998), 622–628.

[38] M.W. Layton, S.A. Goldstein, R.W. Goulet, L.A. Feldkamp, D.J. Kubinski and G.G. Bole, Examination of subchondral bone architecture in experimental osteoarthritis by microscopic computed axial tomography, *Arthritis Rheum.* **31** (1988), 1400–1405.

[39] B. Li and R.M. Aspden, Composition and mechanical properties of cancellous bone from the femoral head of patients with osteoporosis or osteoarthritis, *J. Bone Miner. Res.* **12** (1997), 641–651.

[40] B. Li and R.M. Aspden, Material properties of bone from the femoral neck and calcar femorale of patients with osteoporosis or osteoarthritis, *Osteoporos. Int.* **7** (1997), 450–456.

[41] B. Li and R.M. Aspden, Mechanical and material properties of the subchondral bone plate from the femoral head of patients with osteoarthritis or osteoporosis, *Ann. Rheum. Dis.* **56** (1997), 247–254.

[42] A.G. MacDonald, P. McHenry, S.P. Robins and D.M. Reid, Relationship of urinary pyridinium crosslinks to disease extent and activity in osteoarthritis, *Br. J. Rheumatol.* **33** (1994), 16–19.
[43] J.P. Mansell and A.J. Bailey, Abnormal cancellous bone collagen metabolism in osteoarthritis, *J. Clin. Invest.* **101** (1998), 1596–1603.
[44] J.P. Mansell, J.F. Tarlton and A.J. Bailey, Biochemical evidence for altered subchondral bone collagen metabolism in osteoarthritis of the hip, *Br. J. Rheumatol.* **36** (1997), 16–19.
[45] H. Matsui, M. Shimizu and H. Tsuji, Cartilage and subchondral bone interaction in osteoarthrosis of human knee joint: a histological and histomorphometric study, *Microsc. Res. Tech.* **37** (1997), 333–342.
[46] D.J. McBride, Jr., J.R. Shapiro and M.G. Dunn, Bone geometry and strength measurements in aging mice with the oim mutation, *Calcif. Tissue Int.* **62** (1998), 172–176.
[47] C. Muehleman, J. Green, J.M. Williams, K.E. Kuettner, E.J. Thonar and D.R. Sumner, The effect of bone remodeling inhibition by zoledronic acid in an animal model of cartilage matrix damage, *Osteoarthritis Cartilage* **10** (2002), 226–233.
[48] B.S. Noble, N. Peet, H.Y. Stevens, A. Brabbs, J.R. Mosley, G.C. Reilly, J. Reeve, T.M. Skerry and L.E. Lanyon, Mechanical loading: biphasic osteocyte survival and targeting of osteoclasts for bone destruction in rat cortical bone, *Am. J. Physiol. Cell. Physiol.* **284** (2003), C934–943.
[49] A.M. Parfitt, Bone age, mineral density, and fatigue damage, *Calcif. Tissue Int.* **53**(Suppl. 1) (1993), S82–85; discussion S85–86.
[50] A.M. Parfitt, in: *Bone Mineral Measurement by Photon Absorptiometry*, J. Dequeker, P. Geusens and H.W. Wahner, eds, Leuven University Press, Leuven, 1988.
[51] J.W. Pugh, E.L. Radin and R.M. Rose, Quantitative studies of human subchondral cancellous bone. Its relationship to the state of its overlying cartilage, *J. Bone Joint Surg. Am.* **56** (1974), 313–321.
[52] E.L. Radin, I.L. Paul and M. Lowy, A comparison of the dynamic force transmitting properties of subchondral bone and articular cartilage, *J. Bone Joint Surg. Am.* **52** (1970), 444–456.
[53] E.L. Radin, I.L. Paul and R.M. Rose, Role of mechanical factors in pathogenesis of primary osteoarthritis, *Lancet* **1** (1972), 519–522.
[54] E.L. Radin and R.M. Rose, Role of subchondral bone in the initiation and progression of cartilage damage, *Clin. Orthop.* (1986), 34–40.
[55] G. Raymaekers, J. Aerssens, R. Van den Eynde, J. Peeters, P. Geusens, P. Devos and J. Dequeker, Alterations of the mineralization profile and osteocalcin concentrations in osteoarthritic cortical iliac crest bone, *Calcif. Tissue Int.* **51** (1992), 269–275.
[56] I. Reimann, H.J. Mankin and C. Trahan, Quantitative histologic analyses of articular cartilage and subchondral bone from osteoarthritic and normal human hips, *Acta Orthop. Scand.* **48** (1977), 63–73.
[57] C. Rubin, A.S. Turner, S. Bain, C. Mallinckrodt and K. McLeod, Anabolism. Low mechanical signals strengthen long bones, *Nature* **412** (2001), 603–604.
[58] J. Rubin, T.C. Murphy, L. Zhu, E. Roy, M.S. Nanes and X. Fan, Mechanical strain differentially regulates endothelial nitric-oxide synthase and receptor activator of nuclear kappa B ligand expression via ERK1/2 MAPK, *J. Biol. Chem.* **278** (2003), 34018–34025.
[59] L. Sinigaglia, M. Varenna, L. Binelli, F. Bartucci, M. Arrigoni, R. Ferrara and G. Abbiati, Urinary and synovial pyridinium crosslink concentrations in patients with rheumatoid arthritis and osteoarthritis, *Ann. Rheum. Dis.* **54** (1995), 144–147.
[60] A. Stewart, A. Black, S.P. Robins and D.M. Reid, Bone density and bone turnover in patients with osteoarthritis and osteoporosis, *J. Rheumatol.* **26** (1999), 622–626.
[61] M.M. Thi, T. Kojima, S.C. Cowin, S. Weinbaum and D.C. Spray, Fluid shear stress remodels expression and function of junctional proteins in cultured bone cells, *Am. J. Physiol. Cell. Physiol.* **284** (2003), C389–403.
[62] P.W. Thompson, T.D. Spector, I.T. James, E. Henderson and D.J. Hart, Urinary collagen crosslinks reflect the radiographic severity of knee osteoarthritis, *Br. J. Rheumatol.* **31** (1992), 759–761.
[63] J.C. Van Der Linden, J.A. Verhaar and H. Weinans, A three-dimensional simulation of age-related remodeling in trabecular bone, *J. Bone Miner. Res.* **16** (2001), 688–696.
[64] B. van Rietbergen, H. Weinans, R. Huiskes and A. Odgaard, A new method to determine trabecular bone elastic properties and loading using micromechanical finite-element models, *J. Biomech.* **28** (1995), 69–81.
[65] X. Wang, R.A. Bank, J.M. TeKoppele, G.B. Hubbard, K.A. Athanasiou and C.M. Agrawal, Effect of collagen denaturation on the toughness of bone, *Clin. Orthop.* (2000), 228–239.
[66] H. Yasuda, N. Shima, N. Nakagawa, K. Yamaguchi, M. Kinosaki, S. Mochizuki, A. Tomoyasu, K. Yano, M. Goto, A. Murakami, E. Tsuda, T. Morinaga, K. Higashio, N. Udagawa, N. Takahashi and T. Suda, Osteoclast differentiation factor is a ligand for osteoprotegerin/osteoclastogenesis-inhibitory factor and is identical to TRANCE/RANKL, *Proc. Natl. Acad. Sci.* **31** (1998), 3597–3602.

Part IV

Cartilage remodeling and engineering

Biorheology 41 (2004) 371–375
IOS Press

Organ printing: Fiction or science

Karoly Jakab [a], Adrian Neagu [a,b], Vladimir Mironov [c] and Gabor Forgacs [a,d,*]
[a] *Department of Physics, University of Missouri, Columbia, MO 65211, USA*
[b] *Department of Biophysics and Medical Informatics, Victor Babes University of Medicine and Pharmacy Timisoara, 1900 Timisoara, Romania*
[c] *Department of Cell Biology and Anatomy, Medical University of South Carolina, SC 29425, USA*
[d] *Department of Biology, University of Missouri, Columbia, MO 65211, USA*

Abstract. Aggregates of living cells (i.e. model tissue fragments) under appropriate conditions fuse like liquid drops. According to Steinberg's differential adhesion hypothesis (DAH), this may be understood by assuming that cells are motile and tissues made of such cells possess an effective surface tension. Here we show that based on these properties three-dimensional cellular structures of prescribed shape can be constructed by a novel method: cell aggregate printing. Spherical aggregates of similar size made of cells with known adhesive properties were prepared. Aggregates were embedded into biocompatible gels. When the cellular and gel properties, as well as the symmetry of the initial configuration were appropriately adjusted the contiguous aggregates fused into ring-like organ structures. To elucidate the driving force and optimal conditions for this pattern formation, Monte Carlo simulations based on a DAH motivated model were performed. The simulations reproduced the experimentally observed cellular arrangements and revealed that the control parameter of pattern evolution is the gel-tissue interfacial tension, an experimentally accessible parameter.

Keywords: Tissue engineering, differential adhesion, fusion, spherical aggregate, tissue liquidity

1. Introduction

It has long been suggested that under appropriate conditions morphogenetic tissue motions can be interpreted by assuming that tissues possess liquid-like properties [10]. Aggregates made of motile and adhesive cells round up when incubated in gyratory shakers. When aggregates are prepared from cells of different types randomly intermixed, the morphogenetic phenomenon of sorting takes place and the final equilibrium pattern of the system corresponds to a configuration in which cells of one type form a sphere surrounded by the cells of the other type. Such behaviors are characteristic for liquids. A liquid drop in the absence of external forces assumes a spherical shape to minimize its surface energy. Two immiscible liquids (e.g., water and oil) when intermixed, phase separate and the more cohesive liquid (water) becomes engulfed (surrounded) by the less cohesive one (oil). Such configuration again minimizes the total surface or interfacial energy of the system.

The molecular basis of tissue liquidity has been established by Steinberg's Differential Adhesion Hypothesis (DAH) [10]. DAH stipulates that tissues possess measurable surface or interfacial tension whose origin is in the adhesive properties of their constituent cells. Different tissues have different surface tensions due to the differential adhesion between the cells that compose them. Foty and co-workers have measured the tensions of a number of embryonic tissues [2]. Their values where consistent

*Address for correspondence: Gabor Forgacs, Department of Physics, University of Missouri, Columbia, MO 65211, USA. Tel.: +1 573 882 3036; Fax: +1 573 882 4195; E-mail: forgacsg@missouri.edu.

with the mutual engulfment behavior of these tissues, that is the more cohesive tissue with the higher surface tension always sorted out of the less cohesive tissue with lower surface tension [2]. Computer simulations [3] and *in vivo* experiments have provided further support for DAH [4,5].

The analogy between tissues and liquids makes it possible to apply quantitative modeling methods to morphogenetic phenomena. However, it is important to keep in mind that despite analogies, fundamental differences remain. Liquids equilibrate by molecular motions driven by thermal fluctuations, whereas the motion of cells is ameboid and is fueled by energy producing chemical reactions (i.e. ATP hydrolysis).

In the present work we demonstrate that tissue liquidity can be used to construct organ structures in a novel way: by layer-by-layer cell aggregate printing. To print one needs ink, paper and printer. Here we describe some details of the bioink and biopaper and present results obtained by manual printing. Details of the true printer have been described elsewhere [8].

2. Bioink and biopaper

Traditional organ or tissue engineering uses biocompatible gels as scaffolds to accommodate cells to be grown in number until they can be harvested for specific applications [1,6]. This is a slow process, which has also limitations as far as the shape of the tissue to be engineered is concerned. We propose instead, to use spherical cell aggregates consisting of many thousands of cells. These can be delivered (e.g., printed) into gels in specified geometry. Due to the liquid properties of these tissue fragments, provided the gel properties are appropriately adjusted, they will fuse into organ structures (see below). Thus our bioink consists of spherical cellular aggregates and the biopaper is a biocompatible gel. Figure 1 shows unicolor and multicolor bioinks. In the former case the aggregate contains cells of one type, whereas in the latter case several cell types may be included, depending on the specific requirements. For example if blood vessels are to be constructed, the multicolor aggregate will contain both endothelial and smooth muscle cells. In this case, the self-organizing properties of cells and tissues will guarantee the "right" (physiological) arrangement of cells in the final structure. In Fig. 1 this is illustrated by the sorted cellular pattern.

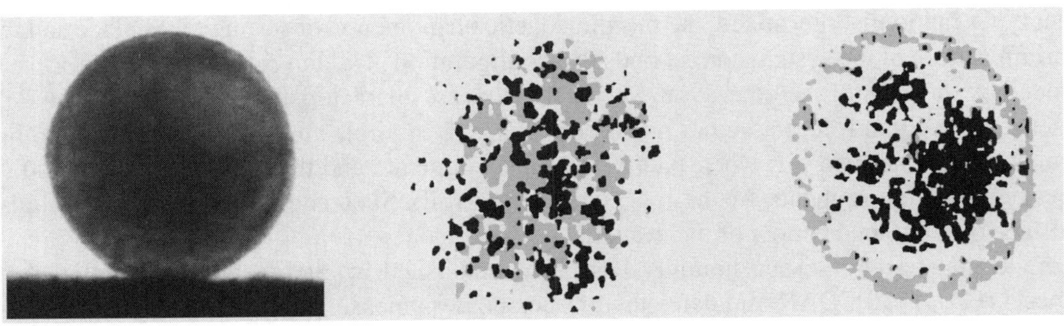

Fig. 1. Left panel: spherical aggregate (of 500 micron diameter) of chicken embryonic heart cells containing approximately 40,000 living cells. Such aggregates, made of cells of one type constitute the unicolor bioink. Right panel: an initially random mixture of two cell types when equilibrated (i.e. sorted) assumes a configuration with one tissue surrounded by the other. Such mixed aggregates comprise the multicolor bioink. The aggregates shown in the figure contain two types of Chinese hamster ovary cells (CHO) transfected with E-cadherin, one containing 25% more of the adhesion molecule than the other. The diameter of these aggregates is 450 microns. (Image prepared and kindly provided by R. Foty.)

3. Printing tubular organ structures

To demonstrate the feasibility of organ printing, we have manually printed a circle of aggregates using unicolor bioink (Fig. 2) into gels of varying composition. Here we report results with collagen gels of two different concentrations. Once the aggregates have been placed into the gel, we let the system evolve in time. Depending on the gel properties the system reached various configurations. As can be seen in Fig. 2, for a permissive gel, with collagen concentration of 1.7 mg/ml, the system evolves towards the minimum energy configuration, corresponding to a single sphere. Permissive in the present context means a gel, which favors cell movement. It is well known that cells are able to crawl along collagen fibers [9] and the more concentrated the gel the more intensive is this movement. Thus a gel with 1.0 mg/ml is much less permissive. At the end of the observation (168 h after preparation) in this case the aggregates fuse into a toroid. Printing the circular pattern layer-by-layer and using an appropriate biopaper thus can lead to a tubular organ structure with lumen. When the bioink has the right composition such a process may result in a blood vessel or gut fragment.

Figure 2 demonstrates the importance of the biopaper; its properties are crucial for structure formation. Note also the shrinking of the gel, especially in the 1.7 mg/ml case. In order to gain further insight into controlling the biopaper properties, we have carried out Monte Carlo simulations of the described experimental process. We have placed cells and gel particles into a three-dimensional discrete lattice and assumed specific adhesive molecular interactions between them (gel–gel, cell–cell and gel–cell). We then prepared the experimental initial arrangement (Fig. 2) and let this system evolve towards its lowest energy configuration according to the standard Metropolis Monte Carlo algorithm [7]. A more detailed analysis of such a model reveals that the only control parameter of this pattern evolution is the interfacial tension γ_{cg} between the gel and the cellular material, which is the combination of the three molecular interaction parameters. Thus, in the simulations it is this parameter that determines how permissive is the biopaper. As Fig. 2 demonstrates, a relatively low value of γ_{cg} is analogous to the

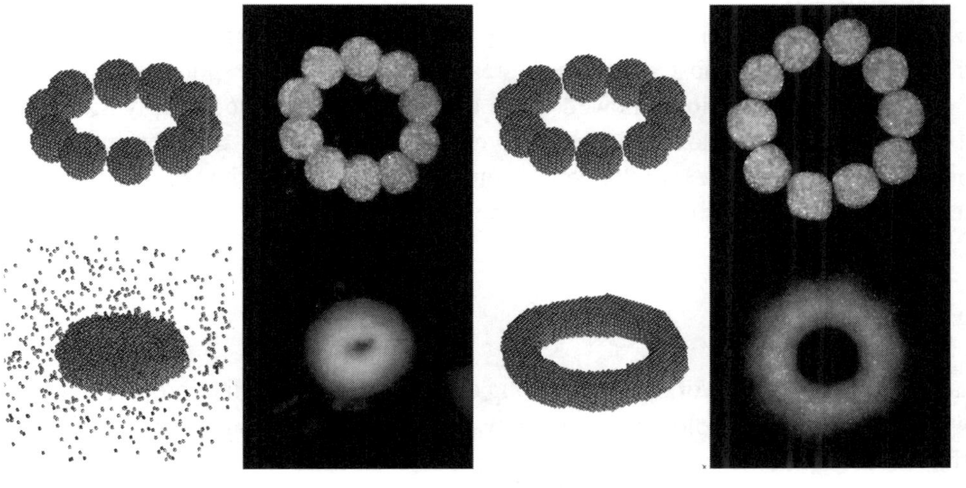

Fig. 2. Experimental and computational realization of a simple organ structure. Upper panels show initial configurations, lower panels show final configurations. First panel from left: simulation of the fusion of 10 spherical aggregates (each containing 925 cells) for a permissive gel with small gel–tissue interfacial tension $\gamma_{cg} = 0.25$. Second panel: manual printing of 10 spherical aggregates of CHO cells with diameter of 500 microns into a permissive collagen gel of 1.7 mg/ml concentration. The final configuration shown has been reached in 168 h. Third panel: model simulations for a non-permissive gel with $\gamma_{cg} = 0.9$. Fourth panel: manual printing of 10 spherical aggregates of CHO cells with diameter of 500 microns into a non-permissive collagen gel of 1.0 mg/ml concentration. The final configuration shown has been reached in 168 h.

Fig. 3. Initial (*left panel*) and final (*right panel*) configuration of 15 rings, 10 aggregate each, each aggregate containing 257 cells in the simulations with $\gamma_{cg} = 1.0$.

concentrated collagen solution, whereas a larger interfacial tension corresponds to a less permissive gel. The almost perfect match between experiments and simulations suggests that the parameter γ_{cg} largely determines the final geometry after aggregate fusion takes place. Clearly γ_{cg} depends in a complicated way on the chemical composition of the gel, but once this dependence is established, the outcome of pattern evolution can be predicted.

In Fig. 3 we present simulation results for the case when a large number (15) of layers (of 10 aggregates each) are printed. As the figure indicates, for the right interfacial tension the desired tubular structure can be achieved.

It has to be emphasized that rings and tubes are not final equilibrium states of the cellular systems and thus do not represent the most stable structures. These would correspond to spheres, as discussed earlier. However, as the simulations showed, rings and tubes can be made deeply metastable, meaning they could persist for a long time, long enough to eliminate the gel and thus conserve these structures. Elimination of the gel can, for example, be accomplished by a slight change in temperature (in some cases a one degree change is sufficient to induce the sol–gel transition).

4. Conclusion

The results presented in this work provide strong evidence that liquid-like properties of tissues, established many years ago can be exploited in a novel way to construct three-dimensional organ structures of specified shape. Moreover, the same liquid properties allow the introduction of the concept of bioink and the delivery by printing of cells in large numbers into scaffold gels, the biopaper. We have demonstrated the feasibility of such an ambitious program with printing very simple albeit nontrivial organ structures using unicolor bioink. We have shown how computer simulations can aid in the optimization of this technology. Although many technical difficulties will have to be overcome before noble organs, such as liver will be possible to construct, we hope the results of this work are strongly suggestive of organ printing being much more than fiction.

Acknowledgement

This work was partially supported by NASA.

References

[1] L.J. Bonassar and C.A. Vacanti, Tissue engineering: the first decade and beyond, *J. Cell. Biochem. Suppl.* **30–31** (1998), 297–303.
[2] R.A. Foty, C.M. Pfleger, G. Forgacs and M.S. Steinberg, Surface tensions of embryonic tissues predict their mutual development behavior, *Development* **122** (1996), 1611–1620.
[3] J.A. Glazier and F. Graner, Simulation of the differential adhesion driven rearrangement of biological cells, *Phys. Rev. E* **47** (1993), 2128–2154.
[4] D. Godt and U. Tepass, Drosophila oocyte localization is mediated by differential cadherin-based adhesion, *Nature* **395** (1998), 387–391.
[5] A. González-Reyes and D. St Johnston, Patterning of the follicle cell epithelium along the anterior–posterior axis during Drosophila oogenesis, *Development* **125** (1998), 2837–2846.
[6] J. Marler, J. Upton, R. Langer and J. Vacanti, Transplantation of cells in matrices for tissue regeneration, *Advanced Drug Delivery Reviews* **33** (1998), 165–182.
[7] N. Metropolis, A.W. Rosenbluth, M.N. Rosenbluth, A.H. Teller and E. Teller, Equation of state calculations by fast computing machines, *J. Chem. Phys.* **21** (1953), 1087–1092.
[8] V. Mironov, T. Boland, T. Trusk, G. Forgacs and R.R. Markwald, Organ printing: Computer-aided jet-based 3D tissue engineering, *Trends in Biotechnology* **21** (2003), 157–161.
[9] R.K. Shawny and J. Howard, Slow local movements of collagen fibers by fibroblasts drive the rapid global self-organization of collagen gels, *J. Cell Biol.* **157** (2002), 1083–1091.
[10] M.S. Steinberg, Mechanism of tissue reconstruction by dissociated cells. II. Time course of events, *Science* **137** (1962), 762–763.

Biorheology 41 (2004) 377–387
IOS Press

The influence of cyclic tension amplitude on chondrocyte matrix synthesis: Experimental and finite element analyses

John T. Connelly, Eric J. Vanderploeg and Marc E. Levenston *
George W. Woodruff School of Mechanical Engineering and Parker H. Petit Institute for Bioengineering and Bioscience, Georgia Institute of Technology, Atlanta, GA 30332-0405, USA

Abstract. While not generally viewed as physiologically significant in articular cartilage, substantial tension can develop in fibrocartilage structures and in articular cartilage injuries. This study examined how different amplitudes of cyclic tension influence chondrocyte matrix synthesis. Bovine articular chondrocytes seeded in fibrin gels were loaded continuously for 48 hours at 1.0 Hz with displacements of 5%, 10%, or 20%. Protein and proteoglycan synthesis were measured by ^{3}H-proline and ^{35}S-sulfate incorporation, respectively. A poroelastic finite element model of the fibrin gel was developed to determine the strain distributions, hydrostatic pressures, and fluid velocities within the constructs at the various levels of displacement. Compared to unloaded controls, 10% and 20% displacements inhibited proteoglycan synthesis to the same extent, while 5% displacement had no effect. Tensile loading did not significantly affect protein synthesis. The finite element model predicted a wide range of strains and fluid velocities within the region of the gel analyzed for matrix synthesis, and the ranges overlapped for the different levels of displacement. These results indicate that the cyclic tension amplitude influences chondrocyte proteoglycan synthesis and that there may be a threshold in the response.

Keywords: Cartilage, poroelastic, fibrin, tissue engineering

1. Introduction

During normal physiologic loading, articular cartilage primarily bear compressive forces. Based on this function, it has been hypothesized that compressive loading is a key stimulus for chondrocytes, and numerous studies have investigated how both dynamic and static compression influence matrix synthesis [9,10,13,15,19,20]. In general, dynamic compression in explant and 3D culture increases chondrocyte synthesis of the major extracellular molecules, type II collagen and aggregating proteoglycans. Furthermore, loading parameters such as magnitude and frequency influence the extent of matrix synthesis [2, 13,20].

Although much work has been done with compression, the effects of tensile loading on chondrocyte biosynthesis have received little attention. While it is not generally considered a major component of normal physiologic loading, significant tensile strains can develop near cracks or tears in the tissue as a result of injury or arthritic degradation. Unlike articular cartilage, fibrocartilage tissues normally experience substantial levels of tensile loading. For example, the loading conditions and geometry of the

*Address for correspondence: Georgia Institute of Technology, 315 Ferst Dr, IBB rm 2312, Atlanta, GA 30332-0405, USA. Tel.: +1 404 894 4219; Fax: +1 404 385 1397; E-mail: marc.levenston@me.gatech.edu.

meniscus produce tensile hoop stresses aligned with the circumferentially oriented collagen fibers [16]. Studying the role of tensile loading on chondrocyte and fibrochondrocyte matrix synthesis is therefore important for understanding both articular cartilage injury and normal fibrocartilage function. In addition to investigating the fundamental cellular responses, oscillatory tension may also be useful for developing tissue engineered constructs by providing a means to control matrix synthesis and organization.

A loading system was previously developed for applying cyclic tension to cells within a three-dimensional fibrin construct [24]. With this system, it was shown that cyclic loading at 10% overall displacement and 1.0 Hz inhibited both chondrocyte and fibrochondrocyte synthesis of protein and proteoglycan as measured by ^{3}H-proline and ^{35}S-sulfate incorporation, respectively. Although these results indicated an inhibition of matrix synthesis due to cyclic tension, only one set of loading parameters has been investigated. As seen in compressive loading, the cellular response may depend on the strain magnitude and loading frequency. The decreases in matrix synthesis previously observed at 10% tensile displacement may reflect an adaptive response to non-physiologic loading conditions. Alternatively, it could represent a response to excessive tensile strains. Examining the response to various amplitudes of cyclic tension would further our fundamental understanding of how tensile loading affects matrix synthesis and possibly lead to the development of an optimal loading regime for tissue engineered cartilages.

One complicating factor in studying different amplitudes of tension with the current system is determining the mechanical environment of the cells during dynamic loading. While different amplitudes of tension can be applied to the fibrin constructs, the precise stress and strain distributions have not been determined, making it difficult to correlate cellular responses with specific mechanical stimuli. In addition, the fibrin construct is a hydrated gel that allows culture media to flow through its porous structure during loading. As a result, different tension protocols may alter the hydrostatic pressures, fluid velocities, and convective transport throughout the construct.

The goal of this study was to investigate the influence of oscillatory tension amplitude on chondrocyte matrix synthesis. Experimental studies examined the biosynthetic response to a range of tension amplitudes at a fixed oscillatory frequency. A poroelastic finite element model of the fibrin construct was also created to determine how the different amplitudes of overall displacement affect the mechanical environment of cells within the construct.

2. Methods

2.1. Materials

Bovine stifle joints were from Research 87 (Marlborough, MA). Dimethyl sulfoxide (DMSO), calcium chloride, and Triton X-100 were from Fisher Scientific (Pittsburgh, PA). Fetal bovine serum (FBS) was from Mediatech (Herndon, VA) and fibrinogen and thrombin were from ICN (Costa Mesa, CA). Ascorbate, ε-aminocaproic acid, ammonium acetate, and Hoechst 33258 dye were from Sigma (St. Louis, MO). Proteinase-K and formaldehyde were from EM Science (Gibbstown, NJ) and ^{3}H-proline and ^{35}S-sodium sulfate were from American Radiolabeled Chemicals (St. Louis, MO). Alexa Fluor® 546 conjugated phalloidin (A-22283) was from Molecular Probes (Eugene, OR). Collagen type II antibody (ab300) was from Abcam (Cambridgeshire, UK), aggrecan antibody was donated by Dr. John Sandy (Shriner's Hospital, Tampa, FL), and the FITC conjugated rabbit IgG antibody was from Sigma (St. Louis, MO). All other cell isolation and culture media reagents were from Invitrogen (Carlsbad, CA).

2.2. Cell isolation and construct seeding

Articular cartilage was aseptically harvested from the femoropatellar grooves of a pair of immature bovine stifle joints from a single animal, avoiding the deep vascularized and calcified cartilage layers. Tissue was then minced, and articular chondrocytes were isolated enzymatically using 0.2% collagenase in antibiotic/antimycotic supplemented Dulbecco's Modified Eagle's Medium (DMEM plus 50 μg/ml gentamycin sulfate, 100 μg/ml kanamycin sulfate, 100 U/ml penicillin, 100 μg/ml streptomycin, and 0.25 μg/ml Fungizone) for approximately 40 hours at 37°C with gentle agitation. Cell solutions were filtered through a 74 μm mesh, centrifuged and washed multiple times, counted using a Coulter Multisizer II, frozen at 1°C per minute in DMEM with 20% FBS and 10% DMSO, and finally stored in liquid nitrogen.

Chondrocytes were later thawed, re-counted, and seeded into 1 ml oval shaped fibrin constructs at a density of 5×10^6 cells/ml using custom polycarbonate molds. In addition to the cells, the fibrinogen solution was comprised of 50 mg/ml bovine fibrinogen, 2 mg/ml ε-aminocaproic acid, and 10% FBS (final concentrations), with 50 U/ml of the enzyme thrombin added to catalyze the gelation of the fibrinogen/cell mixture. The constructs were incubated for approximately 90 minutes and transferred from the custom molds to 8-well tissue culture plates with supplemented media (DMEM, 10% FBS, 2 mg/ml ε-aminocaproic acid, 10 mM non-essential amino acids, 10 mM HEPES buffer, 50 mg/ml ascorbate, 0.4 mM L-proline, 50 μg/ml gentamycin, and 0.25 μg/ml Fungizone). Culture media were changed every two to three days throughout the duration of the study.

2.3. Cyclic tensile loading

After four days of unloaded culture, the final construct geometry was created by punching two 6 mm diameter holes in the constructs using a dermal biopsy punch and a custom template (Fig. 1A). All constructs were cultured for an additional 3 days before being transferred to either the unloaded chambers or tensile loading chambers. Cyclic tension was applied to the fibrin gels using a custom loading device, which consisted of both stationary and moveable pegs inserted into the holes of the constructs. The moveable pegs connected to a DC motor via a linear sliding mechanism, and the magnitude and frequency of tension was controlled by the position of the slide and input voltage, respectively (Fig. 1C).

Two experiments were conducted to examine the effects of various amplitudes of cyclic tension on matrix synthesis. For the first experiment, constructs were subjected to 48 hours of continuous tensile loading at a frequency of 1.0 Hz and displacement amplitudes of 5% or 10% (peak-to-peak). In the second experiment, constructs were again loaded continuously for 48 hours at 1.0 Hz but with displacement amplitudes of 10% or 20%. For both experiments, unloaded controls were cultured in chambers with fixed pegs to impose diffusive limitations similar to those seen by the loaded groups. During the final 20 hours of culture, the medium for six constructs in each group was further supplemented with 7.4 μCi/ml ^{3}H-proline and 5 μCi/ml ^{35}S-sodium sulfate. Two additional samples from each loading group were later used for confocal imaging.

2.4. Biochemical and morphological analyses

Constructs for biochemical analysis were washed 4 times for 30 minutes in phosphate buffered saline (PBS) supplemented with 0.8 mM sulfate and 1.0 mM L-proline to remove unincorporated radiolabeled precursors. Constructs were then sliced such that only the central region was used in the analyses, avoiding the areas outside the pegs (Fig. 1A). These pieces were weighed, frozen, lyophilized, weighed dry,

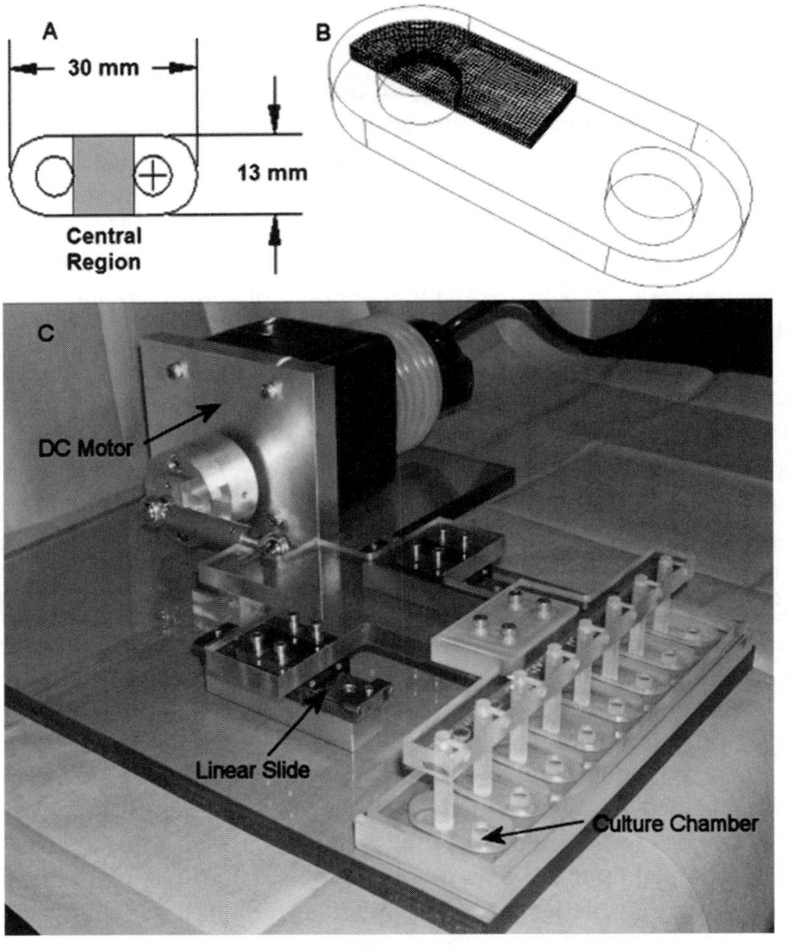

Fig. 1. (A) Geometry of the oval fibrin constructs indicating the central region used for biochemical and morphological analyses. (B) The finite element mesh of the gel and peg modeled one-eighth of the construct based on the three symmetry planes. (C) The custom tensile loading system dynamically loaded the constructs within the culture chambers via moveable pegs attached to a linear sliding mechanism and driven by a DC motor.

and solubilized using Proteinase-K (80 mg per 1 mg tissue) in 100 mM ammonium acetate for approximately 24 hours at 60°C. Digest solutions were analyzed for ^{3}H and ^{35}S incorporation as indicators of protein and proteoglycan synthesis, respectively, using a liquid scintillation counter. DNA and total sGAG content were also measured using the Hoechst 33258 dye [12] and 1,9-DMMB [5] dye assays, respectively. All data were normalized to the wet mass of the section analyzed.

Two constructs from each group were analyzed for cell morphology and extracellular matrix deposition using confocal microscopy. Constructs were fixed in 10% neutral buffered formalin for 20 minutes at 4°C, washed twice for 5 minutes in PBS, permeabilized for 20 minutes using 1% Triton X-100, and again washed twice for 5 minutes in PBS. Blocking buffer (5% FBS in PBS) was added to the constructs for at least 10 minutes to minimize nonspecific antibody binding and reduce background staining. A primary solution containing either a 1 : 100 dilution of rabbit IgG anti-type II collagen or 1 : 50 dilution of rabbit IgG anti-aggrecan core protein antibodies was added to the constructs and incubated for 60 to 90 minutes with gentle agitation. Constructs were again washed with blocking buffer for at least 10 minutes and then a secondary solution containing 0.1 μg/ml Hoechst 33258 dye, a 1 : 50 dilution Alexa Fluor® 546 conjugated phalloidin, and a 1 : 150 dilution of FITC conjugated goat anti-rabbit IgG was added. Constructs were then incubated for 60 to 90 minutes with gentle agitation before finally being washed twice for 5 minutes in PBS. All images were obtained using laser scanning confocal microscopy and oil emersion objectives at a magnification of either 40× or 63× (Zeiss LSM 510).

Biochemical data were analyzed using ANOVA with significance at $p \leqslant 0.05$ and Tukey's test for pairwise comparisons. All data are presented as mean ± standard error of the means (SEM).

2.5. Finite element model

A poroelastic finite element (FE) model of the fibrin construct was developed with Abaqus 6.3-3 (HKS, Pawtucket, RI) to analyze the strain distributions during tensile loading (Fig. 1B). The fibrin gel and infiltrating medium were modeled using 6972 linear brick elements coupled with pore pressure (C3D8P), and the peg was modeled using 1536 linear elastic elements (C3D8). Contact surfaces with the small sliding formulation modeled the interaction between the two objects. The gel was assigned an isotropic modulus of 22 kPa [21], Poisson's ratio of 0.3, and permeability of 5×10^{-10} m^4/N-s [1], and the peg was assigned a modulus substantially higher than that of the gel (220 MPa). Based on the symmetry, only one-eighth of the construct was required for the FE model, and displacements were constrained across the planes of symmetry. Zero pore pressure was fixed at the construct's free surfaces and there was no fluid flow between the contact surfaces of the gel and the peg. The initial pore pressure throughout the gel was also assumed to be zero. The loading experiments were simulated by sinusoidally displacing the peg 0.5 mm, 1 mm, or 2 mm, corresponding to the 5%, 10%, and 20% loading experiments, respectively. The displacements were applied at a frequency of 1.0 Hz for five cycles using 100 uniform time steps and the non-linear geometry option for large deformations. Within the region of interest, principal strains, pore pressures, and fluid velocities at the maximum displacement were calculated for post-processing analysis.

3. Results

3.1. Radiolabel and biochemical analyses

Protein synthesis in cyclically stretched gels was not significantly different from that of the unloaded controls for any amplitude of tension (Fig. 2A). In the first experiment, however, 10% displacement significantly decreased proteoglycan (PG) synthesis ($p < 0.05$), while 5% displacement had no effect on PG synthesis (Fig. 2B). In the second loading experiment, both 10% and 20% displacements inhibited PG synthesis compared to the controls ($p < 0.05$), but there were no differences between the two loaded groups (Fig. 2B). There were no significant differences between any of the loading groups in the DNA or total sGAG contents accumulated during the entire nine day culture.

3.2. Confocal imaging

Immuno-histochemical staining revealed that the chondrocytes synthesized and assembled extracellular type II collagen (Fig. 3A) and aggrecan core protein (Fig. 3B). Most of the cells also maintained a rounded morphology as indicated by the F-actin cytoskeletal staining. Some of the cells at the surface of the gel had a more spread, fibroblastic morphology. In addition, these cells did not stain positively for

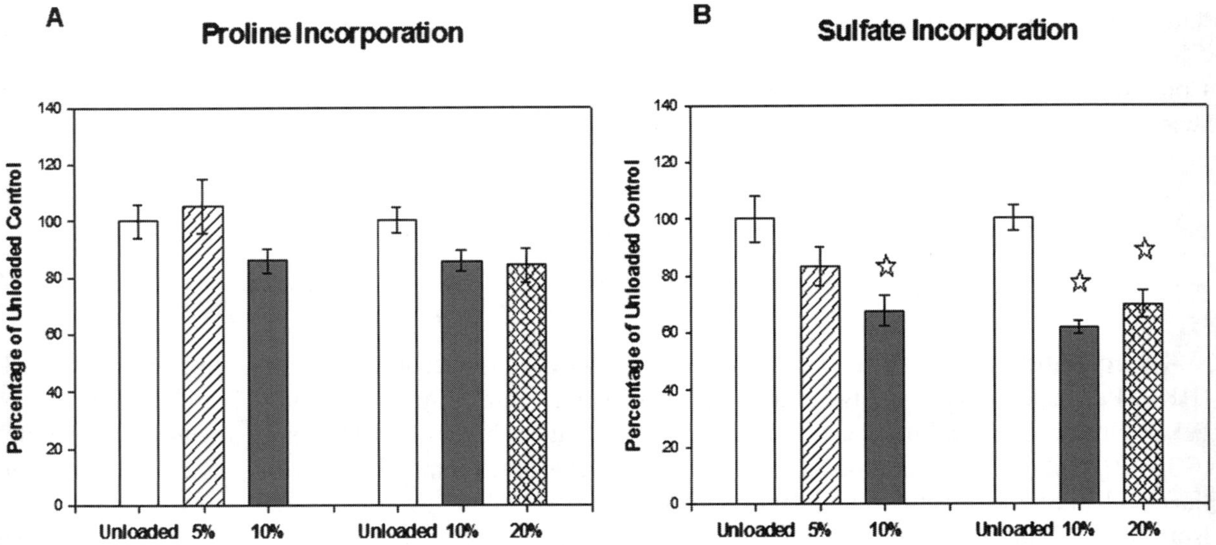

Fig. 2. (A) Protein synthesis measured by ^{3}H-proline incorporation was not significantly affected by any amplitude of dynamic tension. (B) Proteoglycan synthesis measured by ^{35}S-sulfate incorporation was significantly inhibited by 10% and 20% tension, but not by 5% tension. $n = 6$ per loading group. $p < 0.05$.

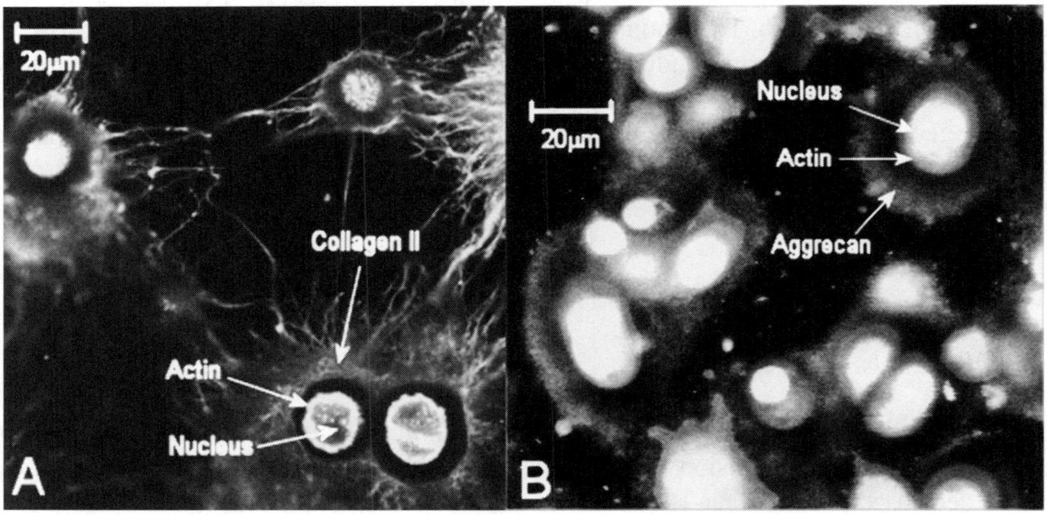

Fig. 3. For all loading groups, chondrocytes in the construct interiors stained positively for extra-cellular (A) type II collagen and (B) aggrecan core protein and maintained a rounded morphology as seen by F-actin staining with phalloidin.

extracellular type II collagen or aggrecan. There were no observable differences in morphology between any of the loading groups.

3.3. Finite element analysis

Contour plots of the maximum and minimum principal strains show the relative distribution of strains through the middle cross-section of the gel (Fig. 4). At the maximum displacement during cyclic loading, the peak predicted strains are located around the holes in the construct where the peg makes contact.

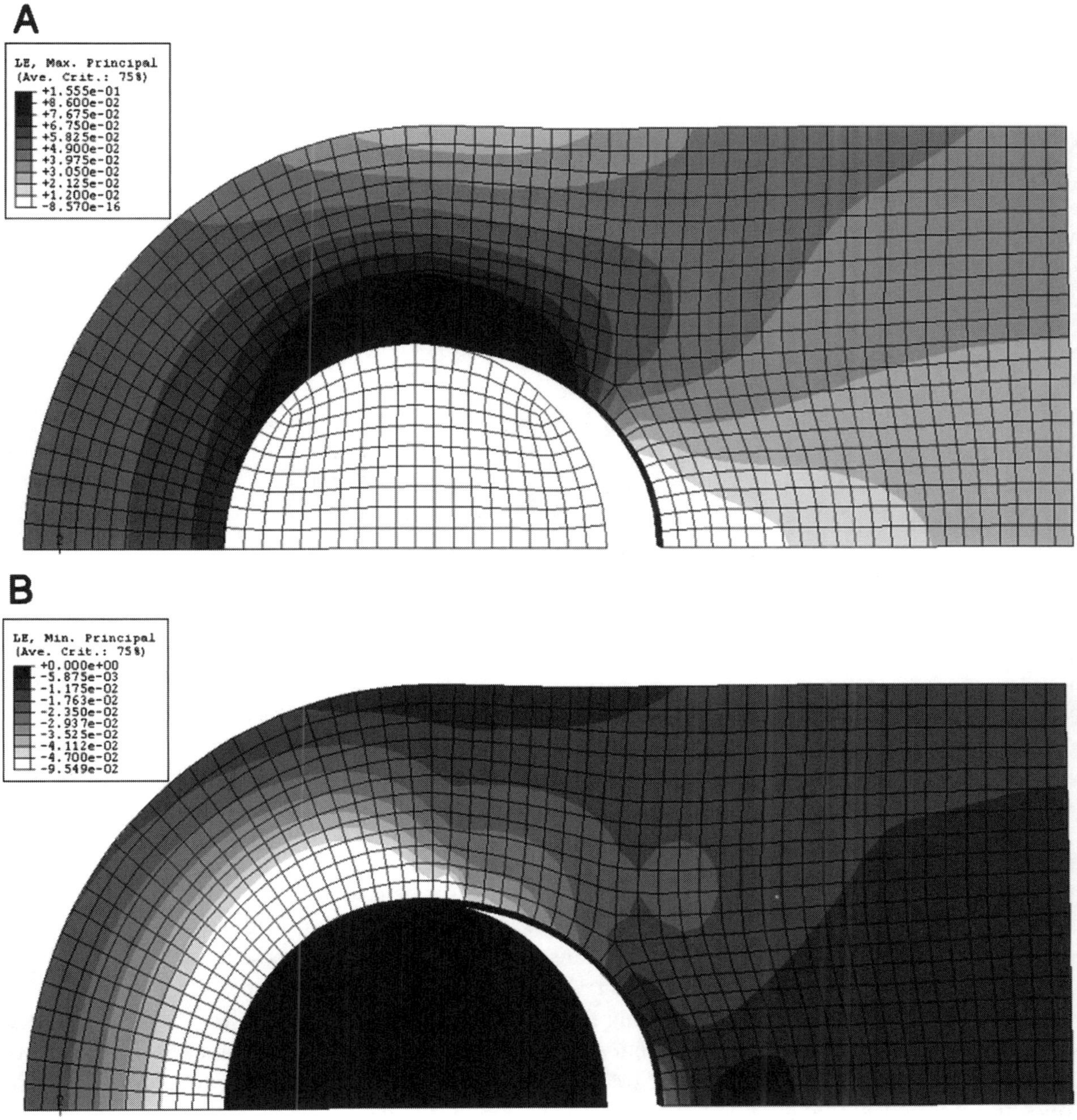

Fig. 4. (A) Maximum principal strains predicted by the finite element model varied across the central region of the construct and were greatest around the hole. (B) Minimum principal strains also varied across the central region and were greatest at the outer edge of the hole.

Within the central region (isolated for biochemical analyses), the maximum principal strains were greatest along the outer edge of the construct and varied across the section (XY plane). However, there was little variation in predicted strains through the thickness of the construct (along the Z-direction).

For 10% total displacement, the maximum principal strains in the region of interest ranged from 0.93–5.2% (Fig. 5). The magnitude of strains was approximately proportional to the overall level of

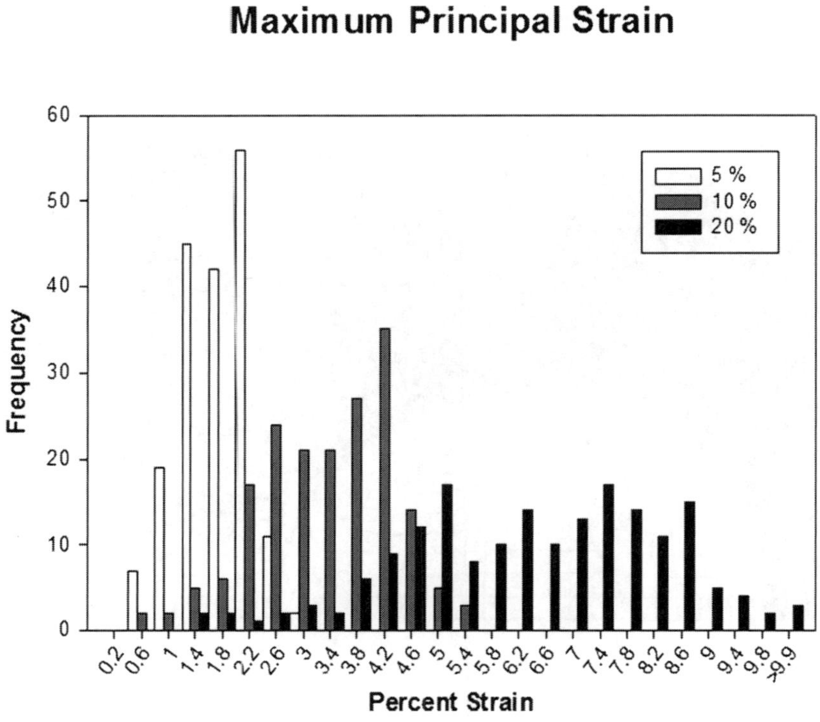

Fig. 5. The histogram of predicted maximum principal strains for elements within the central region of the construct indicates both a large variation in strain magnitude for each displacement amplitude and large overlaps in strains between the different displacement amplitudes. The strains were approximately proportional to the level of displacement.

displacement such that the range of maximum principal strains was 0.51–2.6% for 5% total displacement and 1.8–10% for 20% total displacement. These results indicate that the principal strains varied greatly within the central region of the construct and there was a large degree of overlap in strains among the different levels of displacement. It should be noted that the strain scaling was not perfectly proportional due to non-linear contact, non-linear geometry, and poroelastic behavior.

The poroelastic behavior of the fibrin gel was examined under tensile loading. A contour plot of the interstitial fluid pressure at the maximum displacement demonstrates how the three-dimensional gradients developed throughout the construct. At this point in the cycle, the pressures were negative (~ -40 Pa) with respect to the surrounding fluid (Fig. 6A). The negative pressure gradients resulted in fluid flow into the construct in all directions (Fig. 6B,C,D) and area averaged relative fluid velocity magnitudes ranged from 0 to 20 μm/s at 10% total displacement. Within the central region of interest, velocities were greatest at the free surfaces, zero at the center of the construct, and varied non-linearly along each axis. In addition, fluid flow was greatest along the Z-axis due to the steep pressure gradient through the thickness of the gel. Similar trends in fluid pressure and velocity were seen for all displacement amplitudes, with magnitudes roughly proportional to the overall displacement.

4. Discussion

The goal of this study was to examine how various amplitudes of dynamic tension influence chondrocyte matrix synthesis. In both loading experiments, displacement amplitudes of 10% consistently

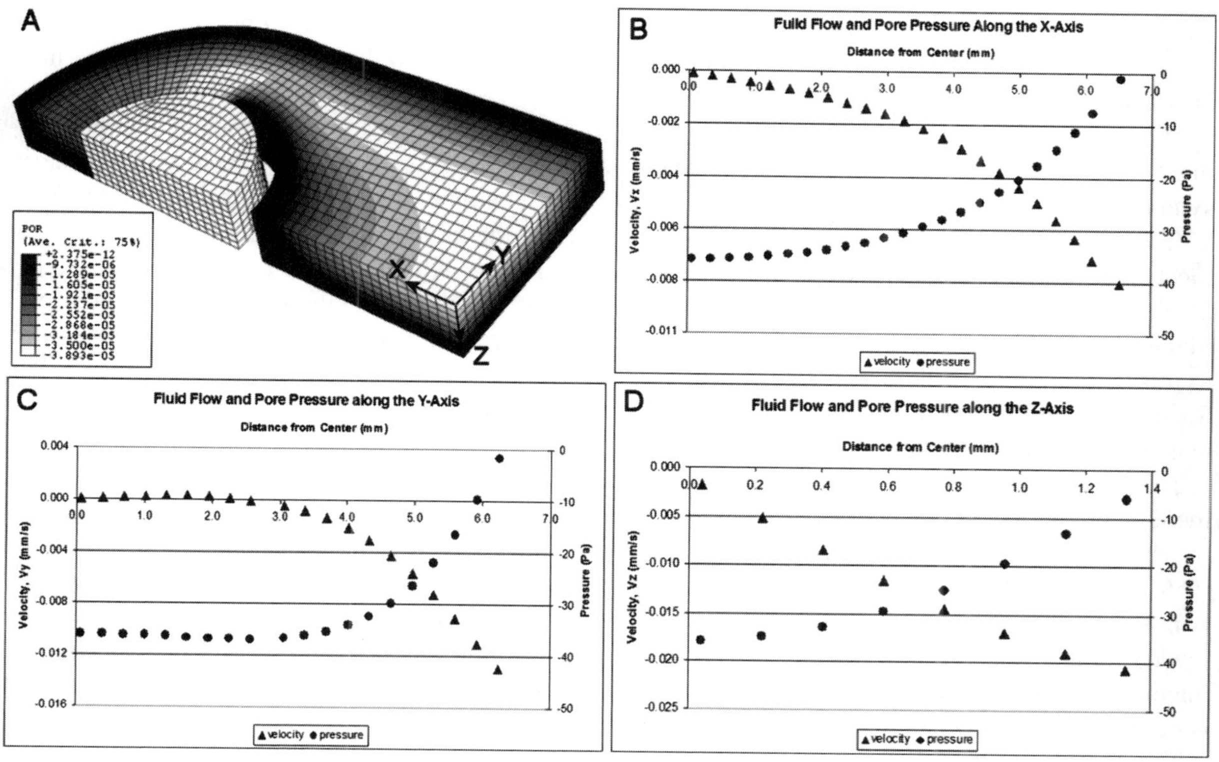

Fig. 6. (A) The predicted hydrostatic pressure (MPa) within the gel was greatest at the center and zero at the outer surface. (B,C,D) The predicted fluid flow varied non-linearly along each of the central axes of the gel and was greatest in the Z-direction due to the high pressure gradients across the thickness. All results correspond to the point of maximal peg displacement.

inhibited proteoglycan (PG) synthesis but not protein synthesis relative to the unloaded controls. The 20% displacement loading also inhibited PG synthesis to the same extent as 10% and did not affect protein synthesis, while the 5% displacement did not significantly decrease PG or protein synthesis. These results suggest that there may be a threshold amplitude of cyclic tension that inhibits chondrocyte production of proteoglycans. This response is also consistent with previous experiments where 10% displacement at 1.0 Hz decreased PG synthesis in chondrocytes and fibrochondrocytes [24]. However, the significant decreases in protein synthesis seen in the previous study were not seen here. This inconsistency may be due to variability in cellular activity between animals or to a relatively small inhibition of protein synthesis, making any effect difficult to consistently detect at statistically significant levels.

A small number of previous studies have investigated the effects of cyclic tension on chondrocyte matrix synthesis in monolayer culture. In general, higher magnitudes of tensile strain ($\sim$17%) inhibited PG synthesis while smaller strains ($\sim$5%) increased synthesis [6,7,23]. Although, the local environment and cellular deformations in 3D and monolayer culture are significantly different, these results are generally consistent with the magnitude dependent inhibition of PG synthesis seen in the current study.

The immunostaining and confocal microscopy images demonstrated that chondrocytes within the fibrin gels maintained their phenotype in all loading groups. These cells stained positively for extracellular type II collagen and aggrecan core protein indicating synthesis of the major articular cartilage matrix components. In addition, cells within the construct maintained rounded chondrocyte morphologies. Based on the spread morphology and lack of matrix staining, cells at the surface of the construct appeared to dedifferentiate similarly to chondrocytes in monolayer culture [25]. In previous studies, cyclic tension induced a significant number of chondrocytes within the gel to take on a three-dimensional stellate morphology [24]. However this phenomenon was not observed in the current study and may have been due to inter-animal variability.

The results of the finite element analyses revealed important insights into the mechanical environment of the fibrin construct during cyclic tensile loading. Within the central region (analyzed for matrix synthesis), there was a fairly wide range of strains. For example, the maximum principal strains at 10% displacement were approximately 1–5%. As a result, the chondrocytes within the central region most likely experienced a wide range of cellular deformations, which could have produced a variety of biosynthetic responses. Previous poroelastic simulations of the micro-environment in articular cartilage predicted large variability in the local cellular strains even for uniform macroscopic strains due to mismatches in the cell–matrix mechanical properties and complexity of the extracellular matrix [8]. Such a non-uniform strain field therefore makes it difficult to isolate how specific amplitudes of dynamic tension affect matrix synthesis.

This issue is further complicated by an overlap in the range of principal strains at different displacement magnitudes. As seen in Fig. 5, the maximum principal strains for the 10% and 20% loading groups overlapped between 1.4% and 5.4% strains, and the 5% and 10% loading groups overlapped between 0.6% and 3.0%. With a large number of elements in each group experiencing similar deformations, it is not surprising that no significant differences in matrix synthesis were detected between any of the loaded groups. Therefore, it is possible that cyclic tension inhibits PG synthesis in a dose dependent manner rather than as a threshold response, but an experimental system with a more uniform strain field is necessary to distinguish the two hypotheses.

The poroelastic model of the fibrin gel also provided interesting information about the hydrostatic pressure and fluid flow distributions, which have been theorized as stimuli of chondrocyte matrix synthesis [3,11,14,18,22]. Due to the high permeability of the fibrin gel, the media easily flowed through the construct, and only minimal hydrostatic pressures developed (~ -40 Pa). As a result, the total stress within the gel was primarily carried by the solid component, unlike articular cartilage tissue where the low permeability causes the total stress to include larger hydrostatic pressures [17]. It should be noted that *in vivo* fluid velocities in articular cartilage have also been estimated within the same order of magnitude as those calculated in this study [3,4]. As seen with the calculated strain distributions, identifying the extent to which pressure, fluid flow or other aspects of the local environment regulate matrix synthesis is difficult due to the variability throughout the construct.

The combination of loading experiments and finite element analysis allowed for detailed examination of how the amplitude of cyclic tension influences chondrocyte matrix synthesis, and several conclusions can be drawn from this study. (1) Dynamic tension consistently inhibits proteoglycan production. (2) This response depends on the magnitude of tension in either a threshold or dose dependent manner. (3) Mechanical parameters vary greatly within the central region of the fibrin construct, possibly inducing a variety of cellular responses. Clearly, future studies using an experimental system with a more uniform mechanical environment are necessary to isolate the effects of various amplitudes. In addition, other loading parameters such as strain rate and duration of loading should be investigated to fully characterize the effects of oscillatory tension on chondrocyte matrix synthesis.

References

[1] R.A. Ariens, H. Philippou, C. Nagaswami, J.W. Weisel, D.A. Lane and P.J. Grant, The factor XIII V34L polymorphism accelerates thrombin activation of factor XIII and affects cross-linked fibrin structure, *Blood* **96** (2000), 988–995.

[2] M.D. Buschmann, Y.A. Gluzband, A.J. Grodzinsky and E.B. Hunziker, Mechanical compression modulates matrix biosynthesis in chondrocyte/agarose culture, *J. Cell Sci.* **108** (Pt 4) (1995), 1497–1508.
[3] M.D. Buschmann, Y.J. Kim, M. Wong, E. Frank, E.B. Hunziker and A.J. Grodzinsky, Stimulation of aggrecan synthesis in cartilage explants by cyclic loading is localized to regions of high interstitial fluid flow, *Arch. Biochem. Biophys.* **366** (1999), 1–7.
[4] F. Eckstein, M. Tieschky, S. Faber, K.H. Englmeier and M. Reiser, Functional analysis of articular cartilage deformation, recovery, and fluid flow following dynamic exercise in vivo, *Anat. Embryol. (Berl.)* **200** (1999), 419–424.
[5] R.W. Farndale, C.A. Sayers and A.J. Barrett, A direct spectrophotometric microassay for sulfated glycosaminoglycans in cartilage cultures, *Connect. Tissue Res.* **9** (1982), 247–248.
[6] T. Fujisawa, T. Hattori, K. Takahashi, T. Kuboki, A. Yamashita and M. Takigawa, Cyclic mechanical stress induces extracellular matrix degradation in cultured chondrocytes via gene expression of matrix metalloproteinases and interleukin-1, *J. Biochem. (Tokyo)* **125** (1999), 966–975.
[7] K. Fukuda, S. Asada, F. Kumano, M. Saitoh, K. Otani and S. Tanaka, Cyclic tensile stretch on bovine articular chondrocytes inhibits protein kinase C activity, *J. Lab. Clin. Med.* **130** (1997), 209–215.
[8] F. Guilak and V.C. Mow, The mechanical environment of the chondrocyte: a biphasic finite element model of cell–matrix interactions in articular cartilage, *J. Biomech.* **33** (2000), 1663–1673.
[9] C.J. Hunter, S.M. Imler, P. Malaviya, R.M. Nerem and M.E. Levenston, Mechanical compression alters gene expression and extracellular matrix synthesis by chondrocytes cultured in collagen I gels, *Biomaterials* **23** (2002), 1249–1259.
[10] C.J. Hunter and M.E. Levenston, The influence of repair tissue maturation on the response to oscillatory compression in a cartilage defect repair model, *Biorheology* **39** (2002), 79–88.
[11] Y.J. Kim, L.J. Bonassar and A.J. Grodzinsky, The role of cartilage streaming potential, fluid flow and pressure in the stimulation of chondrocyte biosynthesis during dynamic compression, *J. Biomech.* **28** (1995), 1055–1066.
[12] Y.J. Kim, R.L. Sah, J.Y. Doong and A.J. Grodzinsky, Fluorometric assay of DNA in cartilage explants using Hoechst 33258, *Anal. Biochem.* **174** (1988), 168–176.
[13] Y.J. Kim, R.L. Sah, A.J. Grodzinsky, A.H. Plaas and J.D. Sandy, Mechanical regulation of cartilage biosynthetic behavior: physical stimuli, *Arch. Biochem. Biophys.* **311** (1994), 1–12.
[14] P. Malaviya and R.M. Nerem, Fluid-induced shear stress stimulates chondrocyte proliferation partially mediated via TGF-beta1, *Tissue Eng.* **8** (2002), 581–590.
[15] R.L. Mauck, M.A. Soltz, C.C. Wang, D.D. Wong, P.H. Chao, W.B. Valhmu, C.T. Hung and G.A. Ateshian, Functional tissue engineering of articular cartilage through dynamic loading of chondrocyte-seeded agarose gels, *J. Biomech. Eng.* **122** (2000), 252–260.
[16] K. Messner and J. Gao, The menisci of the knee joint. Anatomical and functional characteristics, and a rationale for clinical treatment, *J. Anat.* **193** (Pt 2) (1998), 161–178.
[17] N. Mukherjee and J.S. Wayne, Load sharing between solid and fluid phases in articular cartilage: II – Comparison of experimental results and u-p finite element predictions, *J. Biomech. Eng.* **120** (1998), 620–624.
[18] J.J. Parkkinen, J. Ikonen, M.J. Lammi, J. Laakkonen, M. Tammi and H.J. Helminen, Effects of cyclic hydrostatic pressure on proteoglycan synthesis in cultured chondrocytes and articular cartilage explants, *Arch. Biochem. Biophys.* **300** (1993), 458–465.
[19] J.J. Parkkinen, M.J. Lammi, H.J. Helminen and M. Tammi, Local stimulation of proteoglycan synthesis in articular cartilage explants by dynamic compression in vitro, *J. Orthop. Res.* **10** (1992), 610–620.
[20] R.L. Sah, Y.J. Kim, J.Y. Doong, A.J. Grodzinsky, A.H. Plaas and J.D. Sandy, Biosynthetic response of cartilage explants to dynamic compression, *J. Orthop. Res.* **7** (1989), 619–636.
[21] D.H. Sierra, A.W. Eberhardt and J.E. Lemons, Failure characteristics of multiple-component fibrin-based adhesives, *J. Biomed. Mater. Res.* **59** (2002), 1–11.
[22] R.L. Smith, B.S. Donlon, M.K. Gupta, M. Mohtai, P. Das, D.R. Carter, J. Cooke, G. Gibbons, N. Hutchinson and D.J. Schurman, Effects of fluid-induced shear on articular chondrocyte morphology and metabolism in vitro, *J. Orthop. Res.* **13** (1995), 824–831.
[23] T. Toyoda, S. Saito, S. Inokuchi and Y. Yabe, The effects of tensile load on the metabolism of cultured chondrocytes, *Clin. Orthop.* (1999), 221–228.
[24] E.J. Vanderploeg, S.M. Imler, K.R. Brodkin, A.J. García and M.E. Levenston, Oscillatory tension differentially modulates matrix metabolism and cytoskeletal organization in chondrocytes and fibrochondrocytes, *J. Biomech.* (in press).
[25] K. von der Mark, V. Gauss, H. von der Mark and P. Muller, Relationship between cell shape and type of collagen synthesised as chondrocytes lose their cartilage phenotype in culture, *Nature* **267** (1977), 531–532.

Biorheology 41 (2004) 389–399
IOS Press

Adipose-derived adult stem cells for cartilage tissue engineering

Farshid Guilak [a,*], Hani A. Awad [a], Beverley Fermor [a], Holly A. Leddy [a] and Jeffrey M. Gimble [b]
[a] *Departments of Surgery and Biomedical Engineering, Duke University Medical Center, Durham, NC 27710, USA*
[b] *Pennington Biomedical Research Center, Louisiana State University System, Baton Rouge, LA 70808, USA*

Abstract. Tissue engineering is a promising therapeutic approach that uses combinations of implanted cells, biomaterial scaffolds, and biologically active molecules to repair or regenerate damaged or diseased tissues. Many diverse and increasingly complex approaches are being developed to repair articular cartilage, with the underlying premise that cells introduced exogenously play a necessary role in the success of engineered tissue replacements. A major consideration that remains in this field is the identification and characterization of appropriate sources of cells for tissue-engineered repair of cartilage. In particular, there has been significant emphasis on the use of undifferentiated progenitor cells, or "stem" cells that can be expanded in culture and differentiated into a variety of different cell types. Recent studies have identified the presence of an abundant source of stem cells in subcutaneous adipose tissue. These cells, termed adipose-derived adult stem (ADAS) cells, show characteristics of multipotent adult stem cells, similar to those of bone marrow derived mesenchymal stem cells (MSCs), and under appropriate culture conditions, synthesize cartilage-specific matrix proteins that are assembled in a cartilaginous extracellular matrix. The growth and chondrogenic differentiation of ADAS cells is strongly influenced by factors in the biochemical as well as biophysical environment of the cells. Furthermore, there is strong evidence that the interaction between the cells, the extracellular biomaterial substrate, and growth factors regulate ADAS cell differentiation and tissue growth. Overall, ADAS cells show significant promise for the development of functional tissue replacements for various tissues of the musculoskeletal system.

Keywords: Articular cartilage, chondrocyte, pre-adipocyte, stromal cell, osteoarthritis, collagen, proteoglycan

1. Introduction

Under normal physiologic circumstances, articular cartilage may function for decades as a nearly frictionless articulating surface in diarthrodial joints, while exposed to loads of several times body weight. This remarkable function is attributed to the unique structure and composition that determine the mechanical properties of the cartilage extracellular matrix [51]. The cartilage extracellular matrix is maintained by the metabolic activity of a sparse population of cells (chondrocytes) embedded within the tissue. Due to its lack of vascularity and the low metabolic activity of chondrocytes, articular cartilage exhibits a limited capacity for intrinsic repair. Isolated chondral or osteochondral lesions may be a significant source of pain and loss of function, and will rarely, if ever, heal spontaneously. Even minor lesions or injuries may lead to progressive damage and joint degeneration [34,35].

*Address for correspondence: Farshid Guilak, PhD, Orthopaedic Research Laboratories, Department of Surgery, Division of Orthopaedic Surgery, 375 MSRB, Box 3093, Duke University Medical Center, Durham, North Carolina 27710, USA. Tel.: +1 919 684 2521; Fax: +1 919 681 8490; E-mail: guilak@duke.edu.

One of the potential explanations for the poor repair response of articular cartilage is the lack of a blood supply or a source of undifferentiated cells that can promote repair. To overcome these biological limitations, many surgeons have used drilling, abrasion, and microfracture of the subchondral bone to induce bleeding in a repair site in the cartilage [2,8,22,24,35]. This approach generally promotes the formation of a fibrocartilaginous repair tissue, suggesting that the cells responsible for cartilage repair do not differentiate into a true chondrocytic lineage. While the repair tissue is in some cases satisfactory and can decrease pain and morbidity in the short term, fibrocartilaginous repair tissue differs in its mechanical properties in comparison with native articular cartilage [28] and therefore may not function effectively as a long term replacement for normal tissue [55]. Other techniques for cartilage repair have included the transplantation of allograft or autograft cartilage in an effort to restore tissue function. One of the main challenges with these techniques is promoting the functional integration between the implanted graft and the host cartilage. Furthermore, the long-term efficacy of such methods remains unproven, and there is now evidence that significant morbidity may be associated with the donor site of autograft cartilage in the joint [43].

2. Tissue engineering of articular cartilage

To address the clinical need for the repair or regeneration of articular cartilage, significant attention has turned to tissue engineering approaches [3,5,13,15,20,21,23,26,33,47–49,59,65,68]. Despite rapid and exciting advances in the field, few clinical applications have been developed for cartilage tissue engineering, and tissue engineers continue to face significant challenges in repairing or replacing tissues that serve a predominantly biomechanical function. At this writing, one cell-based therapy is available clinically for cartilage repair. The Carticel™ procedure (Genzyme Biosurgery, Cambridge, MA, USA) involves the isolation and amplification of autologous chondrocytes from articular cartilage of an unaffected region of the joint, followed by surgical implantation of cells into the cartilage defect [11]. A flap of autologous periosteal tissue is then used to cover the defect. Clinical outcomes of this procedure are good to excellent [10,50], although animal studies have not shown long-term success [9]. Furthermore, it is now apparent that the harvest procedure required to provide autologous cells may initiate joint degeneration [43]. Therefore there still exists a continuing and unmet need for a readily available and abundant source of chondrocyte progenitor cells for cartilage tissue engineering.

3. Adult stem cells in tissue engineering

The adult stem cell is defined as an "undifferentiated (unspecialized) cell that is found in a differentiated (specialized) tissue; it can renew itself and become specialized to yield all of the specialized cell types of the tissue from which it originated" [39]. In other words, the adult stem cell retains the capacity for self-renewal as well as the potential for differentiation into one or more specialized cell types. It is now apparent that many adult tissues harbor cells with the potential to differentiate into multiple cell types. These cells have been described most commonly as mesenchymal stem cells (or MSCs), but also as multipotential adult stem cells, human marrow stromal cells, or mesenchymal progenitors [14,16,17, 37,52,58,61–63,67,72].

Due to their accessibility, adult stem cells have been used extensively in a variety of tissue engineering applications. Numerous sources of cells with multipotent or pluripotent differentiation capabilities have been found in various musculoskeletal tissues, including bone marrow [14,61,63], adipose tissue [20,29,

74,75], trabecular bone [58], periosteum [53,59], synovial tissue [18,57], muscle [36,44,73], and several other tissues. These cells have formed the basis for tissue-engineered repair of different musculoskeletal tissues [19,23,65]. However, despite several common characteristics among these cells, significant differences exist in their proliferation and differentiation capabilities, their expression of various cell surface markers, their abundance and ease of harvest, and therefore, their potential utility in tissue engineering applications.

Specific to articular cartilage, a number of tissue engineering approaches have explored embryonic or adult stem cells as a source of progenitor cells. For example, murine embryonic stem cells stimulated with bone morphogenetic protein 4 (BMP-4) have been shown to differentiate along the mesodermal lineages [40,54]. Subpopulations of the totipotent embryonic stem cells expressing the VEGF receptor (flk-1) and/or PDGFR-alpha differentiated into the chondrogenic lineage when cultured in 3D pellets and exposed to TGF-β3. The chondrogenic lineage was determined by mRNA expression of cartilage-specific genes (proteoglycans and type II collagen) and synthesis of type II collagen protein. The chondrogenic differentiation was synergistically enhanced in the presence of platelet-derived growth factor (PDGF) and the subsequent culture with BMP-4 [54]. The primary source of adult stem cells for cartilage tissue engineering has been from the bone marrow. Bone marrow-derived MSCs undergo chondrogenesis *in vitro* when maintained in pellet culture in the presence of TGF-β, based on the expression and synthesis of proteoglycans and collagen type II [7,38,46,61,72]. Other tissues have also been shown as sources of chondrocyte progenitors, including muscle [1], periosteum [60], synovium [18,57], skin/foreskin [56], and trabecular bone [58,66].

4. Adipose tissue as a source of adult stem cells

Adipose tissue harbors a population of multipotent progenitor cells that can be induced to differentiate along multiple mesodermal and ectodermal lineages under controlled *in vitro* culture conditions (Fig. 1)

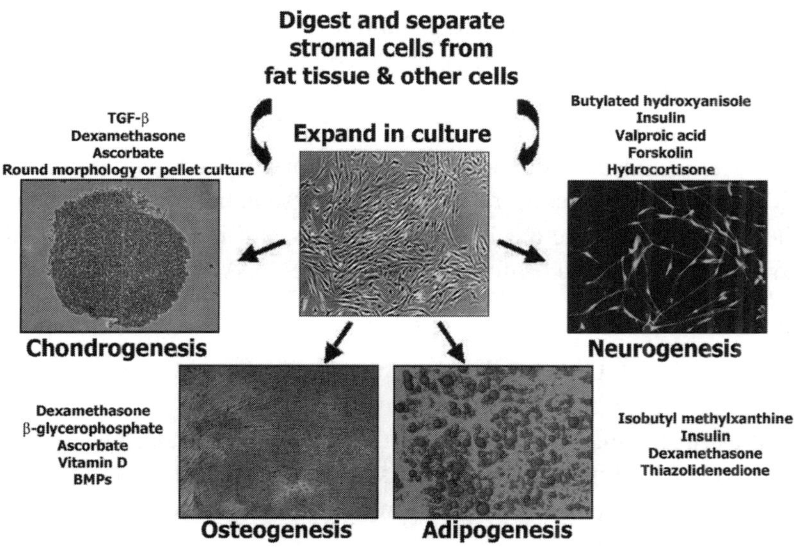

Fig. 1. Multipotent capabilities of adipose-derived adult stem (ADAS) cells. Under specific and controlled culture conditions, ADAS cells can be induced to express the phenotypic characteristics of chondrocytes, osteoblasts, adipocytes, or neurons.

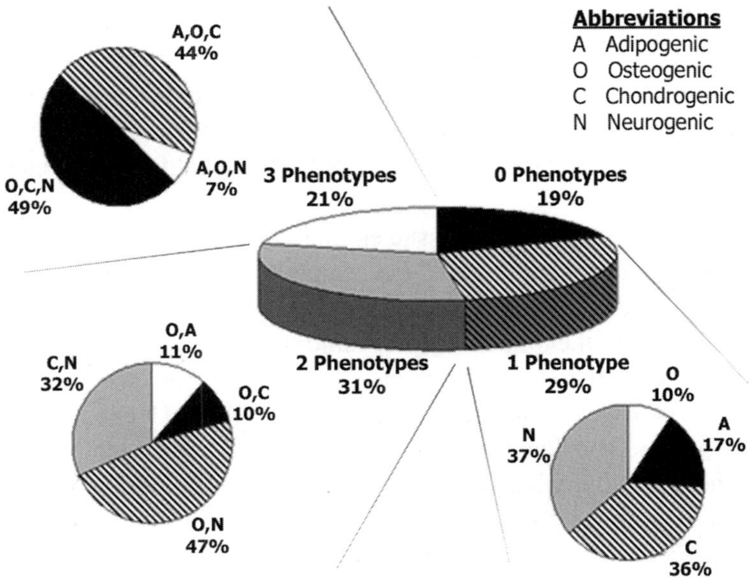

Fig. 2. Clonal analysis of the multipotent differentiation capabilities of ADAS cells [45]. The cell clones were assayed for adipogenic, osteogenic, chondrogenic, and neurogenic differentiation. 52% of the cell clones demonstrated stem cell characteristics by displaying bi- and tri-potent differentiation potential.

[4,5,20,30,31,64,70]. Furthermore, the cells seem to maintain the differentiated phenotype for long periods of time when implanted *in vivo* in a subcutaneous pouch in the mouse [20,32]. Human ADAS cells display a heterogeneous immunophenotype, expressing a panel of cell surface proteins associated with adhesion and signal transduction pathways, and displaying a remarkably similar immunophenotype to marrow derived MSCs [25,70]. Further, ADAS cells have been induced to differentiate along a number of lineage pathways relevant to musculoskeletal repair, including chondrogenic, osteogenic, myogenic, neurogenic, and adipogenic cell types (Fig. 1). Importantly, recent studies indicate that significant percentages of individual clones (Fig. 2), selected at random from primary ADAS cell populations using ring-cloning techniques, can differentiate into multiple cell types. This finding suggests that ADAS cell populations contain "true" stem cells rather than simply a mixed, heterogeneous population of unipotent progenitors [45]. The differentiation process can be controlled by biochemical factors, including the presence and exposure time of exogenous growth factors, hormones, vitamins, and other soluble mediators [4,20,31,70], as well as factors related to the physicochemical environment such as oxygen tension [69]. Furthermore, the interaction between the cells and the biomaterial substrate affects cell shape, which in turn appears to have a profound effect on the ADAS cells' phenotype [6].

5. Chondrogenic potential of *h*ADAS cells

When cultured with transforming growth factor (TGF-β1), ascorbate, and dexamethasone, human ADAS cells express biochemical markers characteristic of articular chondrocytes [4,20,70,75]. An important requirement for chondrogenic differentiation appears to be the maintenance of a rounded cell shape with a 3-dimensional culture system such as a micromass pellet [7,45] or as individual cells suspended in gel matrix such as alginate or agarose [5,20]. When maintained in pellet culture or encapsulated in alginate beads and cultured with 10 ng/ml of TGF-β1, ADAS cells express a chondrocyte-like phenotype and synthesize collagen type II, aggrecan, link protein, and chondroitin sulfate in a time-dependent manner based on mRNA analysis, immunohistochemistry, and radiolabel incorporation [20]. The chondrogenic gene profile expressed by the ADAS cells appears similar to the chondrogenic phenotype of bone marrow derived stromal cells characterized using microarray analysis [71]. Following this *in vitro* differentiation process, human ADAS cells can retain the chondrocyte phenotype for up to 12 weeks when implanted subcutaneously *in vivo* in immunodeficient mice [20].

The influence of various growth factors and mediators on ADAS cell chondrogenesis may be additive or synergistic depending on concentration and time of exposure [4]. For example, the combination of the serum substitute ITS+ and TGF-β1 can increase ADAS cell proliferation in an additive manner in alginate beads relative to control cultures. Similarly, protein synthesis rates are increased by TGF-β1 and dexamethasone in the presence of ITS+ or fetal bovine serum (FBS). TGF-β1 significantly increased proteoglycan synthesis and accumulation by 1.5 to 2 fold, although this effect appears to be suppressed by dexamethasone (Fig. 3).

Furthermore, the choice of the biomaterial scaffold also influences the differentiation of ADAS cells and the functional properties of the tissue engineered construct when grown under *in vitro* chondrogenic culture conditions consisting of basal medium with TGF-β1, ascorbate, and dexamethasone [5, 41]. Tissue engineered constructs constructed from alginate and agarose hydrogels, and porous gelatin scaffolds (Surgifoam®) seeded with ADAS cells respond to these chondrogenic culture conditions with increased rates of protein and proteoglycan synthesis, cell proliferation, and sulfated glycosaminoglycans and hydroxyproline accumulation in the extracellular matrix compared to control culture conditions of basal medium [5]. Furthermore, chondrogenic culture conditions resulted in 86%, and 160% increases ($p < 0.05$) in the equilibrium compressive and shear moduli of the gelatin scaffolds, although they did not affect the mechanical properties of the agarose or alginate constructs over 28 days in culture [5]. Under these conditions, the equilibrium compressive and shear moduli of the gelatin scaffolds were comparable to agarose by day 28 and both were significantly greater than alginate.

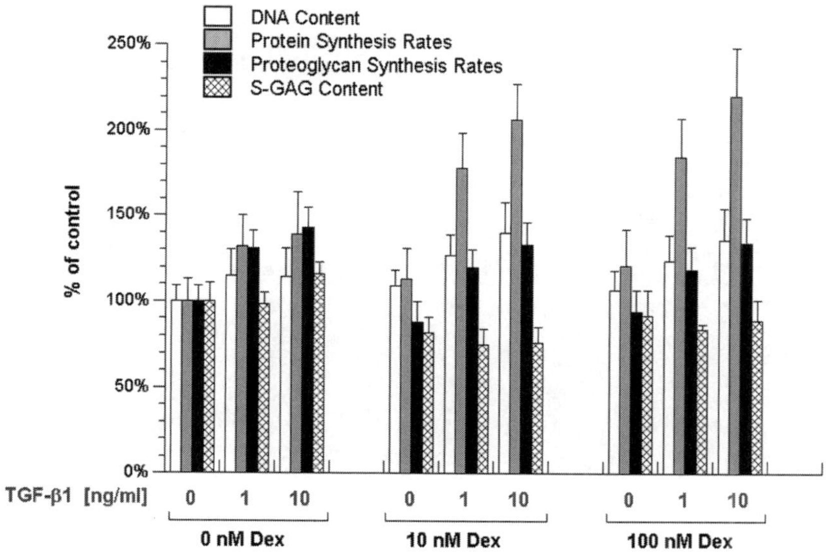

Fig. 3. Effects of Transforming Growth Factor (TGF-β1) and the glucocorticoid Dexamethasone on the growth and chondrogenic differentiation of ADAS cells seeded in alginate beads after 9 days in culture [4]. Bars represent means ($\pm$ standard error of the mean) for each culture combination with FBS and ITS+.

Interestingly, cells encapsulated in alginate or agarose exhibited a spherical cellular morphology, while cells in the gelatin scaffolds showed a more elongated shape as they exerted significant compaction on the scaffold. This elongated fibroblastic morphology did not appear to hinder the chondrogenic differentiation of the cells in the gelatin scaffold. However, immunohistochemical analysis of the constructs demonstrated that while collagen type II and chondroitin sulfate were detected in all the biomaterial scaffolds grown in chondrogenic conditions, collagen type I was also detected in the newly synthesized matrix in the gelatin constructs only (Fig. 4), suggesting that gelatin might promote the differentiation of ADAS cells into a fibrocartilaginous phenotype [41].

Although the choice of the biomaterial scaffold is important, changes in the biochemical composition of the tissue engineered cartilage constructs over time can also influence their biomechanical properties. For example, increases in the shear moduli were significantly associated with increases in S-GAG content ($R^2 = 0.36, p < 0.05$) and with the interaction between S-GAG and hydroxyproline ($R^2 = 0.34, p < 0.05$). These observations suggest that the choice of the biomaterial scaffold influences the chondrogenic differentiation of ADAS cells, and that manipulating the composition of these tissue engineered constructs may have significant effects on their mechanical and functional properties.

The biomaterial scaffold also affects the diffusion of various molecules in cartilage engineered from ADAS cells. Diffusion is likely to be the primary mechanism for macromolecular transport in tissue-engineered cartilage, and providing an adequate nutrient supply via diffusion may be necessary for cell proliferation and extracellular matrix production. The diffusion coefficients of four different sized (ranging from 3 to 500 kDa), fluorescent, uncharged dextrans were measured using fluorescence recovery after photobleaching (FRAP) in tissue engineered cartilage constructs [41]. The constructs comprised

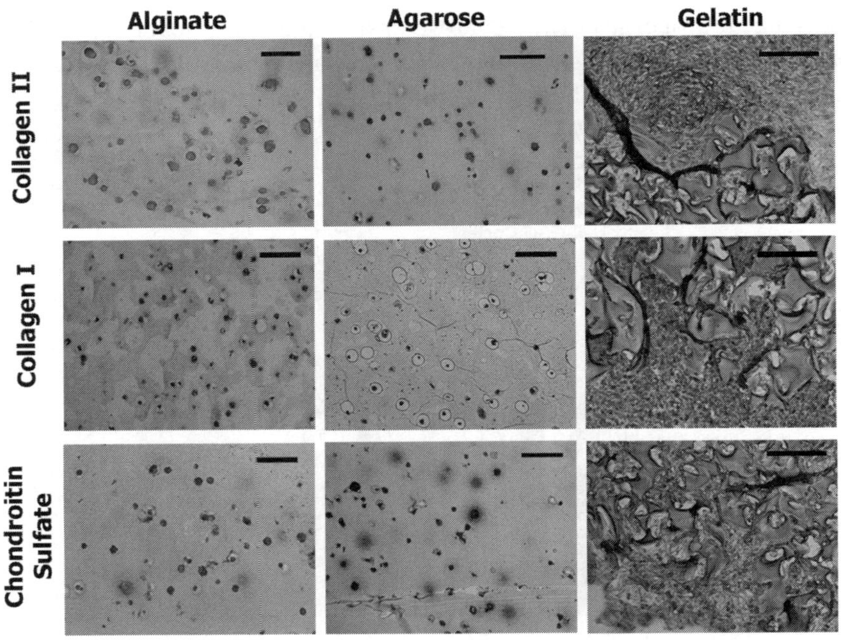

Fig. 4. Immunohistochemistry of cartilage constructs engineered with human adipose-derived adult stem cells [6]. Constructs were grown in chondrogenic media for a period of 28 days, and subsequently fixed, sectioned, and labeled against collagens type I and II and the 2B6 epitope of chondroitin sulfate. All biomaterial constructs showed synthesis of the cartilage matrix proteins collagens type II and chondroitin sulfate. However, cells within gelatin synthesized collagen type I as well, suggesting perhaps that gelatin supports more of a fibrocartilagenous differentiation of ADAS cells. Scale bar represents 100 μm.

alginate, agarose, gelatin, or fibrin scaffolds that were seeded with human adipose derived stem cells and cultured for 1 or 28 days in either chondrogenic or control conditions. Irrespective of the size of the molecule, the diffusion of uncharged dextran molecules in engineered tissue constructs depended on the biomaterial composition of the construct, the presence of cells, the culture conditions, and the culture time. By 28 days in culture, the diffusivity of the constructs decreased 42% and 27% in chondrogenic and control cultures, respectively. This decrease in diffusivity likely reflects a combination of changes that include a net increase in the synthesis, assembly, and retention of neomatrix macromolecules, which may be enhanced with chondrogenic media, over degradative changes in the scaffold. Most importantly, diffusivity in all the constructs tested up to 28 days in culture were significantly greater than those of native cartilage [42], suggesting that the transport of nutrients and metabolites to cells within the constructs might not be hindered in the early stages of tissue generation (Fig. 5).

Moreover, it is widely accepted that laboratory cell culture conditions, under which most tissue engineered constructs are grown, do not mimic the environment that would be present in the normal synovial joint. Oxygen tension is one of the most important biophysical culture variables that seem to regulate engineered tissues' growth and differentiation [69]. ADAS cells were suspended in alginate beads and cultured in control or chondrogenic media in either low oxygen (5%) or ambient oxygen tension (20%) for up to 14 days. Low oxygen tension, which would more closely resemble the *in vivo* joint environment, significantly inhibited the proliferation of ADAS cells, while inducing a two-fold increase in the rate of protein synthesis and a three-fold increase in total collagen synthesis. Low oxygen tension also increased glycosaminoglycan synthesis, as well as lactate production. Immunohistochemical analysis showed significant production of cartilage-associated matrix molecules, including collagen type II, VI, and chondroitin-4-sulfate. These findings suggest that oxygen tension may play an important role in regulating the switch between proliferation and differentiation of ADAS cells during chondrogenesis, and may provide a means of controlling cell growth and biosynthetic activity *in vitro*.

In summary, a number of previous studies show that the biochemical environment such as growth factors and hormones and biophysical environmental variables such as oxygen tension may have significant effects on the growth and chondrogenic differentiation of ADAS cells. Furthermore, the data provide strong evidence that the interplay between the cells, the extracellular biomaterial substrate, and growth factors regulate ADAS cell differentiation and tissue growth. The associated intracellular signaling mechanisms leading to this regulation remain to be elucidated and offer exciting opportunities for future research and discovery.

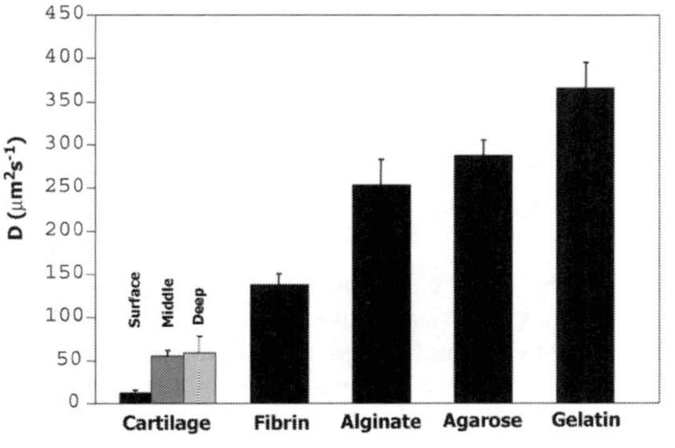

Fig. 5. Diffusion coefficients of 70 kDa dextran in the three zones of native cartilage are less than those in tissue-engineered cartilage constructs [41]. Bars represent means (± standard error of the mean) for each scaffold type averaged across all times and conditions.

6. Challenges and future goals in cartilage tissue engineering using stem cells

Adipose tissue is abundant and readily accessible as compared to most sources of viable tissue, making it a rich potential source of adult stem cells for regenerative medical applications and for functional tissue engineering approaches to cartilage repair. A number of basic and applied issues still remain regarding the functionality, safety, and cost-effectiveness of these cells, as well as any other cell source for tissue engineering.

In addition, because of the predominately mechanical role that cartilage plays in diarthrodial joints, issues of biomechanical function have been addressed by a series of guidelines and recommendations for "functional tissue engineering" proposed to promote a more rational design approach for engineered tissue replacement [12,26]. These guidelines, summarized in a recent text [27], include: (1) Improved definitions of function and success in tissue engineering; (2) A more thorough understanding of the *in vivo* biomechanical environment and mechanical properties of native tissues, and the associated design criteria that meet the biomechanical and metabolic demands of the tissue to be replaced; (3) Design and functional assessment of the biomechanical and biophysical properties of scaffolds for tissue engineering; and (4) A more thorough understanding of the role of biophysical factors on cells and tissue-engineered constructs.

The application of adult stem cells in tissue engineering shows tremendous promise for the future. Despite rapid advances and many early successes, there are few clinical applications of tissue engineering based on adult stem cells. Early work has often focused on trial-and-error methods, and there is a need to balance such an approach with more conservative "engineering" based approaches for product design and manufacture. Furthermore, other rapidly evolving new technologies, especially in biomaterials chemistry and in bioreactor design, may have a significant impact on stem-cell based tissue engineering. It is important to consider the principles of functional tissue engineering when implementing such new technologies towards the development of novel stem cell-based therapeutic products.

Acknowledgements

Supported in part by grants from Artecel Sciences, Inc., the North Carolina Biotechnology Center, the Kenan Institute for Engineering, Technology, and Science, and NIH grants AR50245, AG15768, AR48182, AR49294, and GM08555. We would like to thank Geoffrey Erickson, Yuan-Di Halvorsen, Kristen Lott, Henry Rice, Lori Setton, Robert Storms, David Wang, Quinn Wickham, and the R&D staff at Artecel Sciences for their contributions to this work.

References

[1] N. Adachi, K. Sato, A. Usas, F.H. Fu, M. Ochi, C.W. Han, C. Niyibizi and J. Huard, Muscle derived, cell based ex vivo gene therapy for treatment of full thickness articular cartilage defects, *J. Rheumatol.* **29** (2002), 1920–1930.

[2] S. Akizuki, Y. Yasukawa and T. Takizawa, Does arthroscopic abrasion arthroplasty promote cartilage regeneration in osteoarthritic knees with eburnation? A prospective study of high tibial osteotomy with abrasion arthroplasty versus high tibial osteotomy alone, *Arthroscopy* **13** (1997), 9–17.

[3] E. Alsberg, K.W. Anderson, A. Albeiruti, J.A. Rowley and D.J. Mooney, Engineering growing tissues, *Proc. Natl. Acad. Sci. USA* **99** (2002), 12025–12030.

[4] H.A. Awad, Y.D. Halvorsen, J.M. Gimble and F. Guilak, Effects of transforming growth factor beta1 and dexamethasone on the growth and chondrogenic differentiation of adipose-derived stromal cells, *Tiss. Eng.* **9** (2003), 1301–1312.

[5] H.A. Awad, M.Q. Wickham, H.A. Leddy, J.M. Gimble and F. Guilak, Functional properties of biomaterials for cartilage tissue engineering using adipose derived adult stem cells, in: *Summer Bioengineering Conference*, 2003, pp. 273–274.

[6] H.A. Awad, M.Q. Wickham, H.A. Leddy, J.M. Gimble and F. Guilak, Chondrogenic differentiation of adipose-derived adult stem cells in agarose, alginate, and gelatin scaffolds, *Biomaterials* **25** (2004), 3211–3220.

[7] F. Barry, R.E. Boynton, B. Liu and J.M. Murphy, Chondrogenic differentiation of mesenchymal stem cells from bone marrow: differentiation-dependent gene expression of matrix components, *Exp. Cell Res.* **268** (2001), 189–200.

[8] F.T. Blevins, J.R. Steadman, J.J. Rodrigo and J. Silliman, Treatment of articular cartilage defects in athletes: an analysis of functional outcome and lesion appearance, *Orthopedics* **21** (1998), 761–767.

[9] H.A. Breinan, T. Minas, H.P. Hsu, S. Nehrer, C.B. Sledge and M. Spector, Effect of cultured autologous chondrocytes on repair of chondral defects in a canine model, *J. Bone Joint Surg. Am.* **79** (1997), 1439–1451.

[10] M. Brittberg, Autologous chondrocyte transplantation, *Clin. Orthop.* **367** (1999), S147–155.

[11] M. Brittberg, A. Lindahl, A. Nilsson, C. Ohlsson, O. Isaksson and L. Peterson, Treatment of deep cartilage defects in the knee with autologous chondrocyte transplantation, *N. Engl. J. Med.* **331** (1994), 889–895.

[12] D.L. Butler, S.A. Goldstein and F. Guilak, Functional tissue engineering: the role of biomechanics, *J. Biomech. Engng.* **122** (2000), 570–575.

[13] Y. Cao, A. Rodriguez, M. Vacanti, C. Ibarra, C. Arevalo and C.A. Vacanti, Comparative study of the use of poly(glycolic acid), calcium alginate and pluronics in the engineering of autologous porcine cartilage, *J. Biomater. Sci. Polym. Ed.* **9** (1998), 475–487.

[14] A.I. Caplan and S.P. Bruder, Mesenchymal stem cells: building blocks for molecular medicine in the 21st century, *Trends Mol. Med.* **7** (2001), 259–264.

[15] E.J. Caterson, W.J. Li, L.J. Nesti, T. Albert, K. Danielson and R.S. Tuan, Polymer/alginate amalgam for cartilage-tissue engineering, *Ann. NY Acad. Sci.* **961** (2002), 134–138.

[16] E.J. Caterson, L.J. Nesti, K.G. Danielson and R.S. Tuan, Human marrow-derived mesenchymal progenitor cells: isolation, culture expansion, and analysis of differentiation, *Mol. Biotechnol.* **20** (2002), 245–256.

[17] D.C. Colter, I. Sekiya and D.J. Prockop, Identification of a subpopulation of rapidly self-renewing and multipotential adult stem cells in colonies of human marrow stromal cells, *Proc. Natl. Acad. Sci. USA* **98** (2001), 7841–7845.

[18] C. De Bari, F. Dell'Accio, P. Tylzanowski and F.P. Luyten, Multipotent mesenchymal stem cells from adult human synovial membrane, *Arthritis Rheum.* **44** (2001), 1928–1942.

[19] D.R. Diduch, L.C. Jordan, C.M. Mierisch and G. Balian, Marrow stromal cells embedded in alginate for repair of osteochondral defects, *Arthroscopy* **16** (2000), 571–577.

[20] G.R. Erickson, J.M. Gimble, D.M. Franklin, H.E. Rice, H. Awad and F. Guilak, Chondrogenic potential of adipose tissue-derived stromal cells in vitro and in vivo, *Biochem. Biophys. Res. Commun.* **290** (2002), 763–769.

[21] L.E. Freed, G. Vunjak-Novakovic, R.J. Biron, D.B. Eagles, D.C. Lesnoy, S.K. Barlow and R. Langer, Biodegradable polymer scaffolds for tissue engineering, *Biotechnology (NY)* **12** (1994), 689–693.

[22] D.D. Frisbie, G.W. Trotter, B.E. Powers, W.G. Rodkey, J.R. Steadman, R.D. Howard, R.D. Park and C.W. McIlwraith, Arthroscopic subchondral bone plate microfracture technique augments healing of large chondral defects in the radial carpal bone and medial femoral condyle of horses, *Vet. Surg.* **28** (1999), 242–255.

[23] J. Gao, J.E. Dennis, L.A. Solchaga, A.S. Awadallah, V.M. Goldberg and A.I. Caplan, Tissue-engineered fabrication of an osteochondral composite graft using rat bone marrow-derived mesenchymal stem cells, *Tiss. Eng.* **7** (2001), 363–371.

[24] S.C. Ghivizzani, T.J. Oligino, P.D. Robbins and C.H. Evans, Cartilage injury and repair, *Phys. Med. Rehab. Clin. N. Am.* **11** (2000), 289–307.

[25] S. Gronthos, D.M. Franklin, H.A. Leddy, P.G. Robey, R.W. Storms and J.M. Gimble, Surface protein characterization of human adipose tissue-derived stromal cells, *J. Cell. Physiol.* **189** (2001), 54–63.

[26] F. Guilak, D.L. Butler and S.A. Goldstein, Functional tissue engineering: the role of biomechanics in articular cartilage repair, *Clin. Orthop.* **391** (2001), S295–305.

[27] F. Guilak, D.L. Butler, S.A. Golstein and D.J. Mooney, *Functional Tissue Engineering*, Springer-Verlag, New York, 2003.

[28] J.E. Hale, M.J. Rudert and T.D. Brown, Indentation assessment of biphasic mechanical property deficits in size-dependent osteochondral defect repair, *J. Biomech.* **26** (1993), 1319–1325.

[29] Y.C. Halvorsen, W.O. Wilkison and J.M. Gimble, Adipose-derived stromal cells – their utility and potential in bone formation, *Int. J. Obes. Relat. Metab. Disord.* **24** (2000), S41–44.

[30] Y.D. Halvorsen, A. Bond, A. Sen, D.M. Franklin, Y.R. Lea-Currie, D. Sujkowski, P.N. Ellis, W.O. Wilkison and J.M. Gimble, Thiazolidinediones and glucocorticoids synergistically induce differentiation of human adipose tissue stromal cells: biochemical, cellular, and molecular analysis, *Metabolism* **50** (2001), 407–413.

[31] Y.D. Halvorsen, D. Franklin, A.L. Bond, D.C. Hitt, C. Auchter, A.L. Boskey, E.P. Paschalis, W.O. Wilkison and J.M. Gimble, Extracellular matrix mineralization and osteoblast gene expression by human adipose tissue-derived stromal cells, *Tiss. Eng.* **7** (2001), 729–741.

[32] K.C. Hicok, T.V. duLaney, Y.S. Zhou, Y.D.C. Halvorsen, D.C. Hitt, L.F. Cooper and J.M. Gimble, Human adipose derived adult stem cells produce osteoid in vivo, *Tiss. Eng.* **10** (2004), 371–380.

[33] J.I. Huang, S.R. Beanes, M. Zhu, H.P. Lorenz, M.H. Hedrick and P. Benhaim, Rat extramedullary adipose tissue as a source of osteochondrogenic progenitor cells, *Plast. Reconstr. Surg.* **109** (2002), 1033–1041; discussion 1042–1033.
[34] E.B. Hunziker, Articular cartilage repair: are the intrinsic biological constraints undermining this process insuperable?, *Osteoarthritis Cartilage* **7** (1999), 15–28.
[35] E.B. Hunziker, Articular cartilage repair: basic science and clinical progress. A review of the current status and prospects, *Osteoarthritis Cartilage* **10** (2002), 432–463.
[36] R.J. Jankowski, B.M. Deasy and J. Huard, Muscle-derived stem cells, *Gene Therapy* **9** (2002), 642–647.
[37] Y. Jiang, B.N. Jahagirdar, R.L. Reinhardt, R.E. Schwartz, C.D. Keene, X.R. Ortiz-Gonzalez, M. Reyes, T. Lenvik, T. Lund, M. Blackstad, J. Du, S. Aldrich, A. Lisberg, W.C. Low, D.A. Largaespada and C.M. Verfaillie, Pluripotency of mesenchymal stem cells derived from adult marrow, *Nature* **418** (2002), 41–49.
[38] B. Johnstone, T.M. Hering, A.I. Caplan, V.M. Goldberg and J.U. Yoo, In vitro chondrogenesis of bone marrow-derived mesenchymal progenitor cells, *Exp. Cell Res.* **238** (1998), 265–272.
[39] R. Kirschstein and S.R. Skirboll, Stem cells: Scientific progress and future research directions, Department of Health and Human Services, 2001.
[40] J. Kramer, C. Hegert, K. Guan, A.M. Wobus, P.K. Muller and J. Rohwedel, Embryonic stem cell-derived chondrogenic differentiation in vitro: activation by BMP-2 and BMP-4, *Mech. Dev.* **92** (2000), 193–205.
[41] H.A. Leddy, H.A. Awad, M.Q. Wickham and F. Guilak, Molecular diffusion in articular cartilage and tissue engineered cartilage contructs, *Trans. Orthop. Res. Soc.* **28** (2003), 289.
[42] H.A. Leddy and F. Guilak, Site-specific molecular diffusion in articular cartilage measured using fluorescence recovery after photobleaching, *Ann. Biomed. Engng.* **31** (2003), 753–760.
[43] C.R. Lee, A.J. Grodzinsky, H.P. Hsu, S.D. Martin and M. Spector, Effects of harvest and selected cartilage repair procedures on the physical and biochemical properties of articular cartilage in the canine knee, *J. Orthop. Res.* **18** (2000), 790–799.
[44] J.Y. Lee, Z. Qu-Petersen, B. Cao, S. Kimura, R. Jankowski, J. Cummins, A. Usas, C. Gates, P. Robbins, A. Wernig and J. Huard, Clonal isolation of muscle-derived cells capable of enhancing muscle regeneration and bone healing, *J. Cell Biol.* **150** (2000), 1085–1100.
[45] K.E. Lott, H.A. Awad, J.M. Gimble and F. Guilak, Clonal analysis of multipotent differentiation of human adipose-derived adult stem sells, *Trans. Orthop. Res. Soc.* **29** (2004), 162.
[46] A.M. Mackay, S.C. Beck, J.M. Murphy, F.P. Barry, C.O. Chichester and M.F. Pittenger, Chondrogenic differentiation of cultured human mesenchymal stem cells from marrow, *Tiss. Eng.* **4** (1998), 415–428.
[47] I. Martin, G. Vunjak-Novakovic, J. Yang, R. Langer and L.E. Freed, Mammalian chondrocytes expanded in the presence of fibroblast growth factor 2 maintain the ability to differentiate and regenerate three-dimensional cartilaginous tissue, *Exp. Cell Res.* **253** (1999), 681–688.
[48] K. Masuda, R.L. Sah, M.J. Hejna and E.J. Thonar, A novel two-step method for the formation of tissue-engineered cartilage by mature bovine chondrocytes: the alginate-recovered-chondrocyte (ARC) method, *J. Orthop. Res.* **21** (2003), 139–148.
[49] R.L. Mauck, M.A. Soltz, C.C. Wang, D.D. Wong, P.H. Chao, W.B. Valhmu, C.T. Hung and G.A. Ateshian, Functional tissue engineering of articular cartilage through dynamic loading of chondrocyte-seeded agarose gels, *J. Biomech. Engng.* **122** (2000), 252–260.
[50] T. Minas and S. Nehrer, Current concepts in the treatment of articular cartilage defects, *Orthopedics* **20** (1997), 525–538.
[51] V.C. Mow, A. Ratcliffe and A.R. Poole, Cartilage and diarthrodial joints as paradigms for hierarchical materials and structures, *Biomaterials* **13** (1992), 67–97.
[52] H. Nakahara, S.P. Bruder, S.E. Haynesworth, J.J. Holecek, M.A. Baber, V.M. Goldberg and A.I. Caplan, Bone and cartilage formation in diffusion chambers by subcultured cells derived from the periosteum, *Bone* **11** (1990), 181–188.
[53] T. Nakase, H. Nakahara, M. Iwasaki, T. Kimura, K. Kimata, K. Watanabe, A.I. Caplan and K. Ono, Clonal analysis for developmental potential of chick periosteum-derived cells: agar gel culture system, *Biochem. Biophys. Res. Commun.* **195** (1993), 1422–1428.
[54] N. Nakayama, D. Duryea, R. Manoukian, G. Chow and C.Y. Han, Macroscopic cartilage formation with embryonic stem-cell-derived mesodermal progenitor cells, *J. Cell Sci.* **116** (2003), 2015–2028.
[55] S. Nehrer, M. Spector and T. Minas, Histologic analysis of tissue after failed cartilage repair procedures, *Clin. Orthop.* **365** (1999), 149–162.
[56] S.B. Nicoll, A. Wedrychowska, N.R. Smith and R.S. Bhatnagar, Modulation of proteoglycan and collagen profiles in human dermal fibroblasts by high density micromass culture and treatment with lactic acid suggests change to a chondrogenic phenotype, *Connect. Tiss. Res.* **42** (2001), 59–69.
[57] K. Nishimura, L.A. Solchaga, A.I. Caplan, J.U. Yoo, V.M. Goldberg and B. Johnstone, Chondroprogenitor cells of synovial tissue, *Arthritis Rheum.* **42** (1999), 2631–2637.
[58] U. Noth, A.M. Osyczka, R. Tuli, N.J. Hickok, K.G. Danielson and R.S. Tuan, Multilineage mesenchymal differentiation potential of human trabecular bone-derived cells, *J. Orthop. Res.* **20** (2002), 1060–1069.

[59] S.W. O'Driscoll and J.S. Fitzsimmons, The role of periosteum in cartilage repair, *Clin. Orthop.* (2001), S190–207.
[60] S.W. O'Driscoll, A.D. Recklies and A.R. Poole, Chondrogenesis in periosteal explants. An organ culture model for in vitro study, *J. Bone Joint Surg. Am.* **76** (1994), 1042–1051.
[61] M.F. Pittenger, A.M. Mackay, S.C. Beck, R.K. Jaiswal, R. Douglas, J.D. Mosca, M.A. Moorman, D.W. Simonetti, S. Craig and D.R. Marshak, Multilineage potential of adult human mesenchymal stem cells, *Science* **284** (1999), 143–147.
[62] D.J. Prockop, Adult stem cells gradually come of age, *Nature Biotechnology* **20** (2002), 791–792.
[63] D.J. Prockop, I. Sekiya and D.C. Colter, Isolation and characterization of rapidly self-renewing stem cells from cultures of human marrow stromal cells, *Cytotherapy* **3** (2001), 393–396.
[64] K.M. Safford, K.C. Hicok, S.D. Safford, Y.D. Halvorsen, W.O. Wilkison, J.M. Gimble and H.E. Rice, Neurogenic differentiation of murine and human adipose-derived stromal cells, *Biochem. Biophys. Res. Commun.* **294** (2002), 371–379.
[65] L.A. Solchaga, V.M. Goldberg and A.I. Caplan, Cartilage regeneration using principles of tissue engineering, *Clin. Orthop.* (2001), S161–170.
[66] V. Sottile, C. Halleux, F. Bassilana, H. Keller and K. Seuwen, Stem cell characteristics of human trabecular bone-derived cells, *Bone* **30** (2002), 699–704.
[67] C.M. Verfaillie, Adult stem cells: assessing the case for pluripotency, *Trends Cell Biol.* **12** (2002), 502–508.
[68] S.D. Waldman, C.G. Spiteri, M.D. Grynpas, R.M. Pilliar, J. Hong and R.A. Kandel, Effect of biomechanical conditioning on cartilaginous tissue formation in vitro, *J. Bone Joint Surg. Am.* **85-A** (2003), 101–105.
[69] D.W. Wang, B. Fermor, J.M. Gimble and F. Guilak, Effects of hypoxia on the proliferation and chondrogenic potential of human adipose-derived stromal cells, *Trans. Orthop. Res. Soc.* **28** (2003), 875.
[70] M.Q. Wickham, G.R. Erickson, J.M. Gimble, T.P. Vail and F. Guilak, Multipotent stromal cells derived from the infrapatellar fat pad of the knee, *Clin. Orthop.* **412** (2003), 196–212.
[71] A. Winter, S. Breit, D. Parsch, K. Benz, E. Steck, H. Hauner, R.M. Weber, V. Ewerbeck and W. Richter, Cartilage-like gene expression in differentiated human stem cell spheroids: a comparison of bone marrow-derived and adipose tissue-derived stromal cells, *Arthritis Rheum.* **48** (2003), 418–429.
[72] J.U. Yoo, T.S. Barthel, K. Nishimura, L. Solchaga, A.I. Caplan, V.M. Goldberg and B. Johnstone, The chondrogenic potential of human bone-marrow-derived mesenchymal progenitor cells, *J. Bone Joint Surg. Am.* **80** (1998), 1745–1757.
[73] H.E. Young, T.A. Steele, R.A. Bray, K. Detmer, L.W. Blake, P.W. Lucas and A.C. Black, Jr., Human pluripotent and progenitor cells display cell surface cluster differentiation markers CD10, CD13, CD56, and MHC class-I, *Proc. Soc. Exp. Biol.* **221** (1999), 63–71.
[74] P.A. Zuk, M. Zhu, P. Ashjian, D.A. De Ugarte, J.I. Huang, H. Mizuno, Z.C. Alfonso, J.K. Fraser, P. Benhaim and M.H. Hedrick, Human adipose tissue is a source of multipotent stem cells, *Mol. Biol. Cell* **13** (2002), 4279–4295.
[75] P.A. Zuk, M. Zhu, H. Mizuno, J. Huang, J.W. Futrell, A.J. Katz, P. Benhaim, H.P. Lorenz and M.H. Hedrick, Multilineage cells from human adipose tissue: implications for cell-based therapies, *Tiss. Eng.* **7** (2001), 211–228.

Biorheology 41 (2004) 401–410
IOS Press

The effect of media perfusion on three-dimensional cultures of human chondrocytes: Integration of experimental and computational approaches

Manuela T. Raimondi [a,*], Federica Boschetti [a], Leonarda Falcone [b], Francesco Migliavacca [a], Andrea Remuzzi [c] and Gabriele Dubini [a]

[a] *Laboratory of Biological Structure Mechanics, Politecnico di Milano, Milano, Italy*
[b] *Simple Structure of Tissue Therapy, Niguarda Cà Granda Hospital, Milano, Italy*
[c] *Biomedical Engineering Laboratory, Mario Negri Institute for Pharmacological Research, Bergamo, Italy*

Abstract. This work examines the effect of perfusion on human mature articular chondrocytes cultured on synthetic biodegradable scaffolds (DegraPol). Human chondrocytes were isolated, seeded on the scaffolds and subjected to perfused culture at a flow rate of 0.5 ml/min, corresponding to an average inlet fluid velocity of 44 μm/s, with flow inversion every 1 minute. The flow was imposed at the construct surface in some constructs, it was forced through the construct thickness in other constructs and was absent in the static controls. We compared cell viability and morphology and we evaluated material properties of the constructs at 1 month of culture. Thickness-perfused constructs showed significantly higher material properties and roughly a two-fold cell viability, when compared both to surface-perfused constructs and to static controls. Chondrocytes maintained a phenotypic morphology in all experiments, probably favoured by a limited cell–scaffold interaction. Biosynthetic activity could be demonstrated only in the bioreactor-cultured constructs. In this experimental model, a bi-directional flow of culture medium was applied to the cells at a macroscopic level and computational modelling was used to quantify the fluid-dynamic environment induced on the cells at a microscopic level. This method may be used to quantify the effects of fluid-dynamic shear on the growth modulation of tissue-engineered cartilage constructs, to potentially enhance tissue growth *in vitro*.

Keywords: Tissue engineering, articular cartilage, mechanobiology, chondrocyte, bioreactor, computational fluid dynamics

1. Introduction

The generation of autologous cartilaginous grafts requires the development of procedures not only to quickly expand human chondrocytes, but also to reproducibly promote their re-differentiation and to maintain their chondrogenic potential in three-dimensional (3D) culture systems [5]. The conditions which favor *in vitro* chondrogenesis include high cell density and maintenance of spherical cell shape by reduced cell/matrix interaction. This condition may be achieved if the cells are carried by a 3D scaffold, depending on its material, geometry and pore size [3]. Other conditions which are believed

*Address for correspondence: Manuela Teresa Raimondi, PhD, Dipartimento di Bioingegneria, Politecnico Di Milano, Piazza Leonardo da Vinci, 32, 20133 Milano, Italy. Tel.: +39 02 2399 3349; Fax: +39 02 2399 3360; E-mail: manuela.raimondi@polimi.it.

to favor *in vitro* chondrogenesis include efficient nutrient delivery and mechanical stimulation [4,6]. Perfusion systems may provide an effective strategy to overcome current diffusion limitations associated with culturing 3D tissues *in vitro*, by enhancing nutrient delivery and catabolites removal within the constructs and by providing flow-mediated mechanical stimuli [2]. However, in most of the systems currently available, the cells located inside the constructs are not exposed to a fluid flow. Here, fluid-induced shear is absent and mass transfer occurs by molecular diffusion, as in static culture dishes. A better control of the hydrodynamic environment is a condition required to quantify the response to flow-mediated mechanical stimuli of these 3D cell systems.

We have developed a perfused-column bioreactor in which the culture medium flows through the inner regions of porous cellular constructs, together with a computational fluid dynamic (CFD) model of the flow through the construct microstructure [8]. The experimental set-up, in conjunction with the CFD model, provides a technique in which a bi-directional flow of culture medium can be applied to the cells on a macroscopic level and CFD modelling can be used to quantify the shear stress level induced on the cells on a microscopic level. These predictions provide insight into the potential mechanical interactions between the cells and their surrounding environment and the model may be used to investigate possible mechanisms of mechanical signal transduction. Furthermore, new scaffold design criteria may be derived, such as indication on a microstructure which would optimise transport to and from the cells. The bioreactor has been employed in a preliminary set of experiments of human chondrocyte growth, under static versus dynamic culture conditions. The CFD model has been applied to numerically predict the shear stress acting on the cells in the experiments.

2. Materials and methods

2.1. Chondrocyte expansion

Biopsies of articular cartilage were harvested aseptically from two patients aged 31 and 28, scheduled for autologous chondrocyte implantation (ACI). Chondrocyte isolation and expansion were carried out as described by Brittberg [1]. The reagents used were from Euroclone (Devon, UK), unless specified differently. The complete medium contained Dulbecco's Modified Eagle Medium (DMEM), antibiotics, L-glutamine, L-ascorbic acid and 10% aliquot of foetal calf serum (FCS). Cells were either directly used for scaffold seeding, or stored in liquid nitrogen and subsequently seeded for a second expansion.

2.2. Construct preparation and cell culture

The scaffold used is DegraPol®, a biodegradable polyestherurethane foam developed at the Swiss Federal Institute of Technology (ETH) in Zurich [9]. The scaffolds employed in our study were sterile 16-mm diameter discs of 2 mm thickness and nominal pore size 100 μm. The discs were wetted in medium for 2 hours at 37°C then positioned in a culture cluster and seeded with the cells. Two independent experiments were conducted, at a cell seeding density of 1.5×10^6 cells/disk. The discs were incubated for 3 hours at 37°C to allow for cell adhesion and then complete medium was added to each well. Seeded constructs were placed in a closed-loop recirculation system, previously described [8]. In this bioreactor system, the culture medium may be conveyed through the porous structure of the construct, thus exposing the cells located in the entire scaffold thickness both to convective solute transport and to a flow-induced mechanical stimulus. The flow was either imposed at the construct surface or forced through the construct thickness (Fig. 1). The flow rate was regulated at 0.5 ml/min, the flow

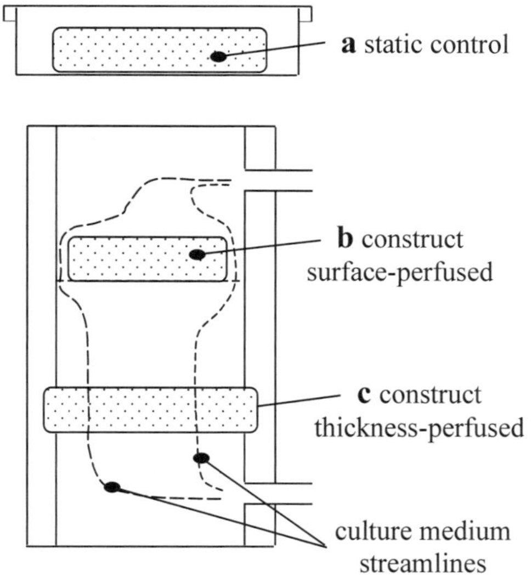

Fig. 1. Scheme of the experimental set-up. Constructs seeded with human chondrocytes were placed in: (a) static control dishes and (b) in a closed-loop recirculation system, shown in section, in which the culture medium flow was imposed either at the construct surface or (c) conveyed through the porous structure of the construct, thus exposing the cells located in the entire scaffold thickness both to convective solute transport and to a flow-induced mechanical stimulus.

direction was kept constant for 4 days and then cyclically inverted every 1 minute. Both the bioreactor system and the static controls were incubated for 4 weeks, with medium completely replaced every 2 days.

2.3. Biological assays

At 4 weeks of culture, constructs were prepared for colorimetric assays and for electron microscopy. Cell viability was determined using mitochondral dye inclusion (MTT) of thiazolyl blue (Sigma Chemical, St. Luis, MO, USA) following the product indications. Absorbance of the converted dye was measured with a DU-65 spectrophotometer (Beckman Instruments GmbH, München, Germany) at a wavelength of 570 nm, with background subtraction at 630 nm. Samples of the seeded constructs were prepared for scanning electron microscopy [8] and observed at 10 kV using a Stereoscan S260 electron microscope (Leica Cambridge Ltd., Cambridge, UK).

2.4. Biomechanical testing

Cylindrical samples of 5 mm in diameter were punched from the constructs and tested in dynamic shear using a Bohlin CVO 120 controlled stress rheometer (Bohlin Rheology AB, Lund, Sweden). Samples were tested by imposing a fixed shear stress level, falling in the range of linear material behaviour, at increasing frequencies, in the range $10^{-2}\div 5$ Hz (frequency sweep). The material behaviour was described using the storage modulus, G', calculated from the complex shear modulus in the frequency sweeps [10].

2.5. Computational model

This part of the study was designed to characterise the local fluid dynamics to which the chondrocytes were exposed inside the thickness-perfused constructs, during the experiments. To this purpose, a CFD

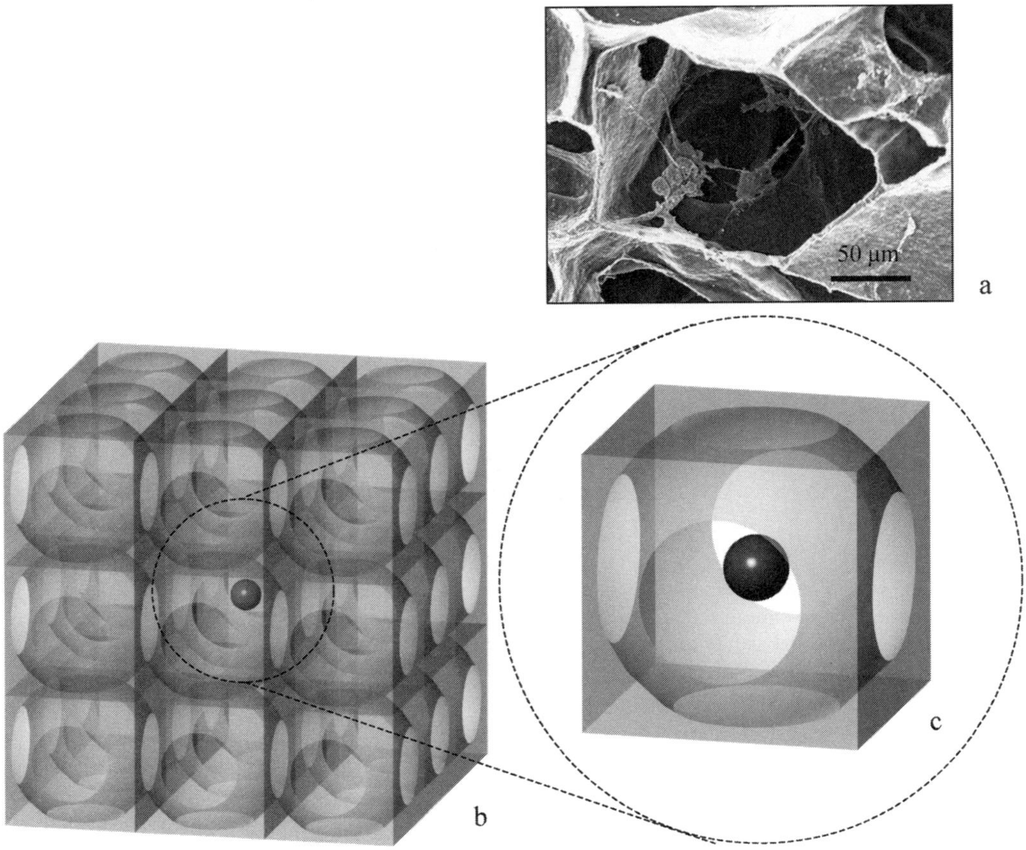

Fig. 2. Method employed to set up the CFD model. (a) SEM image of the construct microstructure at the single porosity level, (b) solid model of the honeycomb structure representing a cubic scaffold portion of 260 μm in side, (c) detail of the central cubic sub-unit.

model of the flow through the construct microstructure was developed, based on a simplification of the actual scaffold geometry. The simplified geometry consisted of a 3D cubic domain, 260 μm in side, in which a group of 27 cubic sub-units were arranged in a honeycomb-like pattern (Fig. 2b). The geometrical definition of a single sub-unit was obtained by subtracting a solid sphere, 100 μm in diameter, from a concentric solid cube, 86.2 μm in side. These dimensions were selected in order to mimic, in the modelled geometry, the actual porosity of the scaffolds used in the experiments, which was equal to 77%. In the central sub-unit of the honeycomb structure (Fig. 2c), a solid sphere of diameter 20 μm was subtracted to the model, in order to simulate a cluster of chondrocytes configured as the one represented in Fig. 2a. The ACIS-based, solid modeller Gambit (Fluent, Lebanon, NH, USA) was used to set-up the solid model and relevant mesh. Due to the existing symmetries, only one fourth of the model was studied (Fig. 3a). A 4 node-tetrahedral mesh was used, with a higher density in correspondence of the central sub-unit (Fig. 3b) and the solid internal sphere. The mesh consisted of 97090 elements with a corresponding number of nodes of 17393. The commercial finite-element code FIDAP (Fluent, Lebanon, NH, USA) was used to set up and solve the problem and to analyse the results. No-slip boundary conditions were applied to the solid surfaces, under the hypothesis of rigid, impermeable walls. The culture medium flows in from bottom of the domain and out from top. A flat velocity profile was applied at the inlet, corresponding to an average flow rate, Q, of 0.5 ml/min, while null total stress was

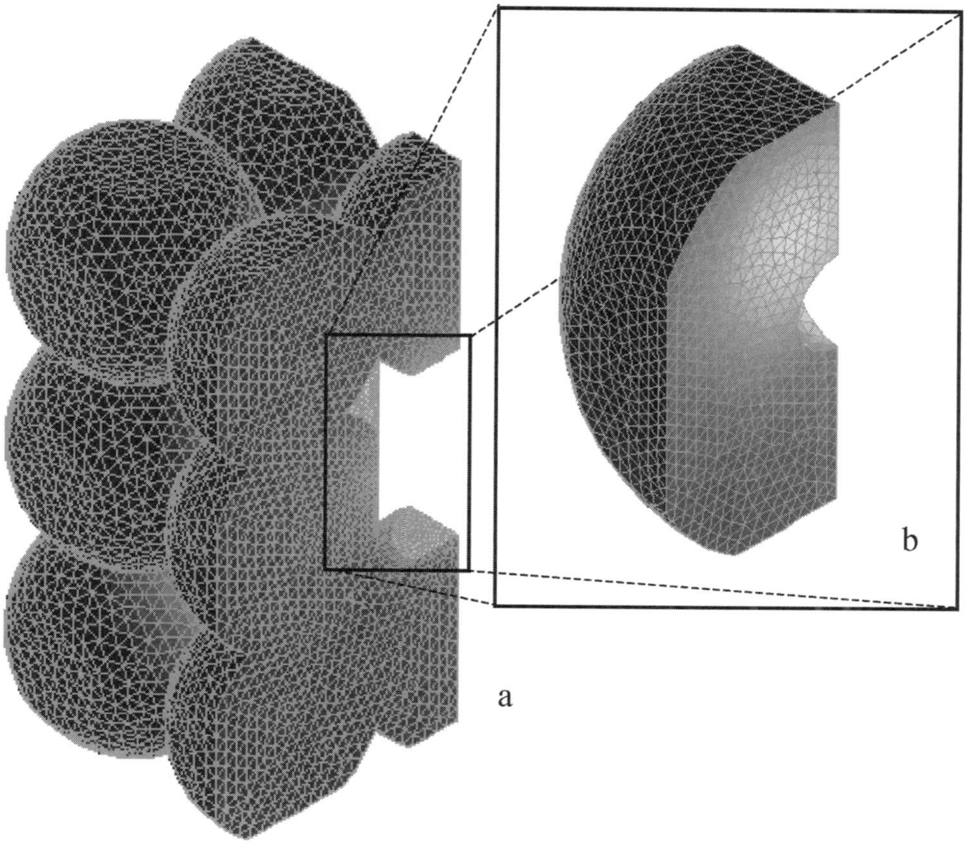

Fig. 3. (a) Mesh of the fluid domain, (b) detail of the mesh of the fluid domain relevant to the central sub-unit.

applied at the outlet. A symmetry boundary condition was used for the lateral sides of the fluid domain. The culture medium was modelled as an incompressible, homogeneous, Newtonian fluid with density, ρ, equal to 1000 $\mathrm{kg\,m^{-3}}$. The viscosity, μ, was set at 8.2×10^{-4} $\mathrm{kg\,m^{-1}\,s^{-1}}$. Reynolds number was equal to 1.77×10^{-2}. The computations were carried out on a Windows NT PC with 512 Mb RAM. The results of the computations were analysed in the central sub-unit of the honey-comb structure. The distribution of the shear stress, τ, were determined separately for the internal walls of the sub-unit and for the external surface of the central spherical element.

3. Results

3.1. Experimental results

In this work we compared cell viability and morphology and we evaluated material properties of chondrocyte-seeded constructs cultured for 1 month in three experimental conditions. The main experimental finding was that thickness-perfused constructs showed roughly a two-fold cell viability, when compared both to surface-perfused constructs and to static controls (Fig. 4). Chondrocytes maintained a phenotypic spherical morphology in all the experiments (Fig. 5), probably favoured by a limited cell–scaffold interaction. However, biosynthetic activity was observed in the bioreactor-cultured constructs

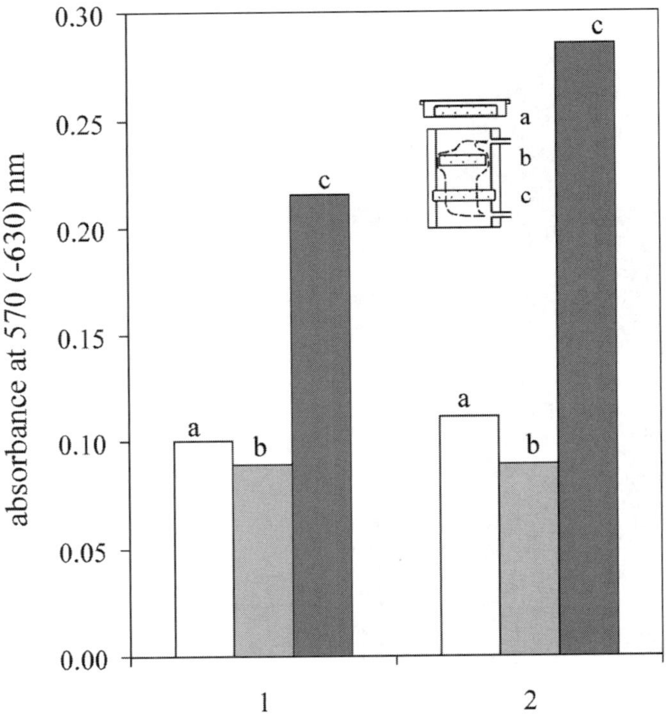

Fig. 4. Results of the MTT colorimetric assay for the chondrocyte-seeded scaffolds at 4 weeks of culture. (a) Static controls, (b) constructs surface-perfused and (c) constructs thickness-perfused. Values are reported for two experiments, as mean of two readings.

ant not in the static controls. At the SEM, surface-perfused constructs showed few attached cells (Fig. 5a) and a spherical cell morphology with membrane vesicles (Fig. 5b), consistent with ongoing apoptosis. In thickness-perfused constructs cells were organised in large aggregates of neo-formed matrix (Fig. 5c), by close intercellular contact (Fig. 5d). Thickness-perfused constructs also showed higher material properties at all testing frequencies, if compared to surface-perfused constructs (Fig. 6).

3.2. Computational results

Figure 7 shows the shear stress contour plot at the internal wall of the central sub-unit and at the wall of the central spherical element. The highest shear stress values are found in correspondence of the inlet and outlet flow regions (top and bottom of the domain, where the velocity is higher). Indeed, the spherical element behaves as a barrier to the fluid. The shear stress values range between 9 and 10.5 mPa, with a mean value of 10 mPa (0.1 dyn/cm^2) for the central sphere. At the inner walls of the central sub-unit the shear stresses range between 1.5 and 13.5 mPa, with a large central area subjected to 3 mPa (0.03 dyn/cm^2). As expected, the central element is the area subjected to the higher stresses because it behaves as a barrier to the fluid.

4. Discussion

Our main result from the cell culture experiments is that thickness-perfused constructs maintained a higher degree of cell viability, if compared to surface-perfused constructs and to static controls. We

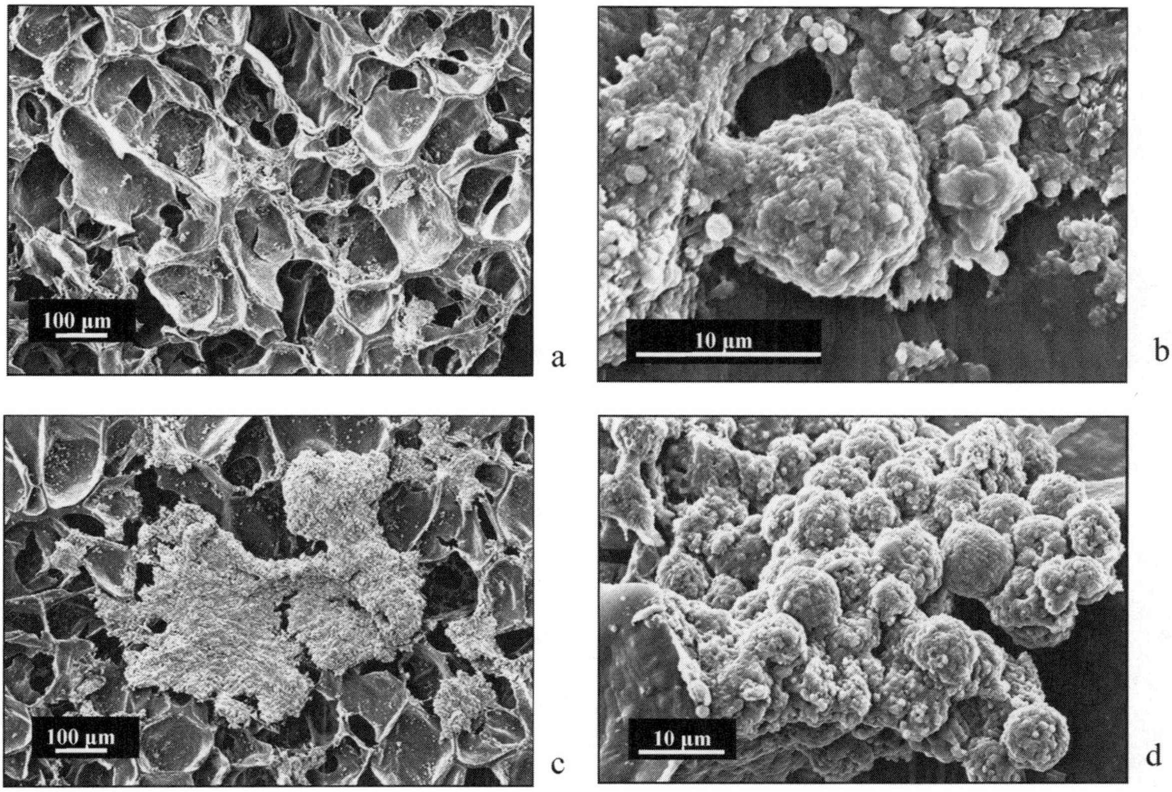

Fig. 5. SEM images of chondrocyte-seeded constructs at 4 weeks of culture. Surface-perfused constructs show (a) few attached cells and (b) a spherical cell morphology with sparse vesicles (white particles), originating from the cell membrane. In thickness-perfused constructs (c) cells are organised in large aggregates of neo-formed matrix, (d) by close intercellular contact and show a spherical morphology.

found signs of ongoing apoptosis both in the surface-perfused constructs and in the static controls, indicating that a slower apoptotic rate, rather than a faster proliferation rate, was responsible of the higher cell viability measured in the thickness-perfused constructs. The stimulatory effects of fluid flow on overall cell viability observed here are generally consistent with previous observations [2,6] but here the medium was perfused through the constructs and the observations were conducted on human cells, giving indications of potential use for tissue growth enhancement in clinical human application.

We observed a scaffold degradation rate, both in static and in dynamic culture conditions, significantly slower than what reported by previous studies conducted using other scaffolds, derived from biological polymers, such as hyaluronic-acid fibre pads or collagen sponges, suggesting that a foam of synthetic polymer may be more adequate in supporting cell survival and mechanical properties of tissue-engineered constructs during long term *in vitro* cultivation.

The target of the CFD model was to quantify the shear stresses acting on the cells in the bioreactor-cultured constructs. Our previous work [8] showed that in fibre scaffolds, the seeded cells adhere preferentially at the fibres' intersections, which were shown by CFD modelling to be regions of low flow, with low median shear stresses imposed to the cells, in the order of 0.003 Pa (0.03 dyn/cm^2), at a flow rate of 0.5 ml/min, corresponding to an inlet fluid velocity of 44 μm/s. These preliminary results indicated that a homogeneous open-pore structure, rather than a fibre structure, was likely to optimise mass transport

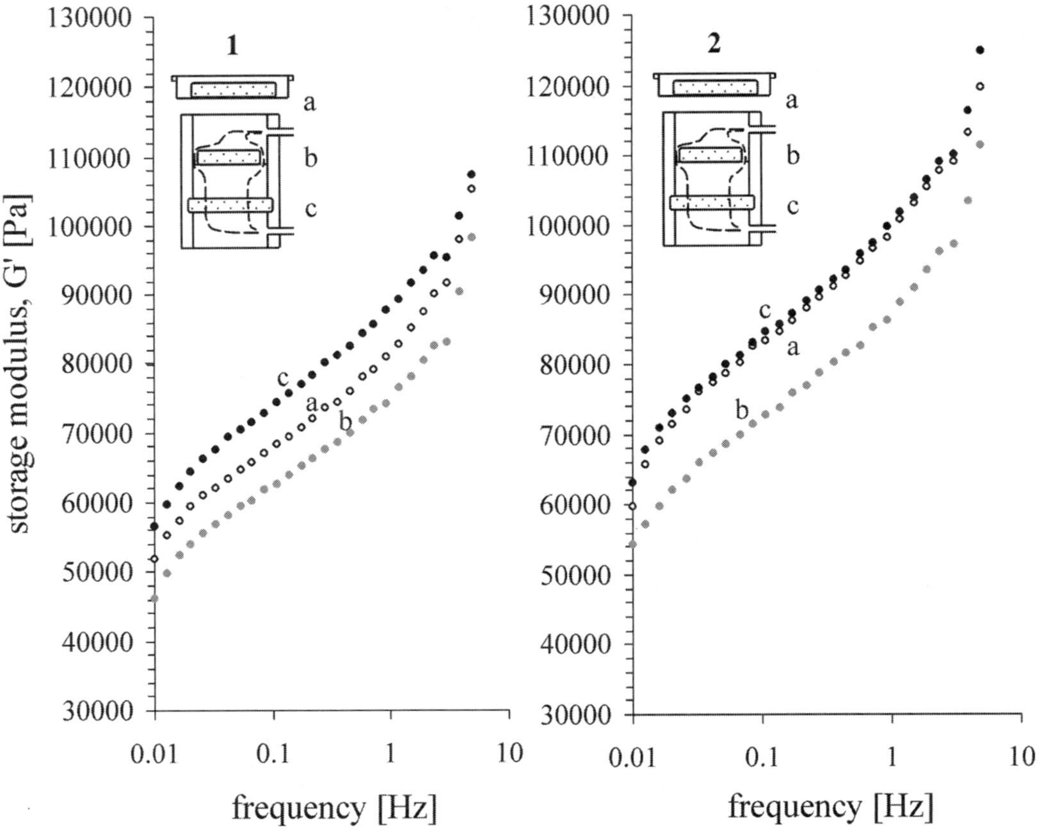

Fig. 6. Results of the dynamic shear testing at 4 weeks of culture. The material behaviour is described by the storage modulus, G', which measures the purely elastic response of viscoelastic materials. The values of G' are shown separately for the two experiments, as a function of the testing frequency, for (a) static controls, (b) constructs surface-perfused and (c) constructs thickness-perfused.

to and from the cells. In this study, a considerably different fluid-dynamic behaviour is estimated in the DegraPol® foams. The average shear level predicted for cells adhered to the scaffold walls is comparable to what calculated for fibre scaffolds, and acts on most of the surface. This level of shear is estimated to be of the same order of magnitude as the level of shear to which chondrocytes may be exposed within native cartilage [7].

The most severe simplification in our previous CFD study [8] was the two-dimensional flow modelling. In such a configuration, the fibre axes extended infinitely in the direction normal to the plane of the model, which was rather unrealistic. Here, the microstructure was modelled in 3D, hopefully giving more realistic predictions of the flow dynamics inside the perfused scaffold. Considerable simplifications were however introduced in order to model the actual scaffold microstructure. The magnitude of shear stress applied to chondrocytes in this study was calculated from a numerical analysis of flow through this 3D simplified domain, which is yet to be validated by experimental measurements of actual geometry and of flow characteristics. A second relevant problem in our previous 2D model arose with regard to the boundary conditions, especially at the lateral sides of the fluid domain. Here, the CFD results were extracted from the simulation only in the central sub-unit, that is in a neighbourhood significantly smaller with respect to the overall domain modelled, in order to avoid edge effects altering the results.

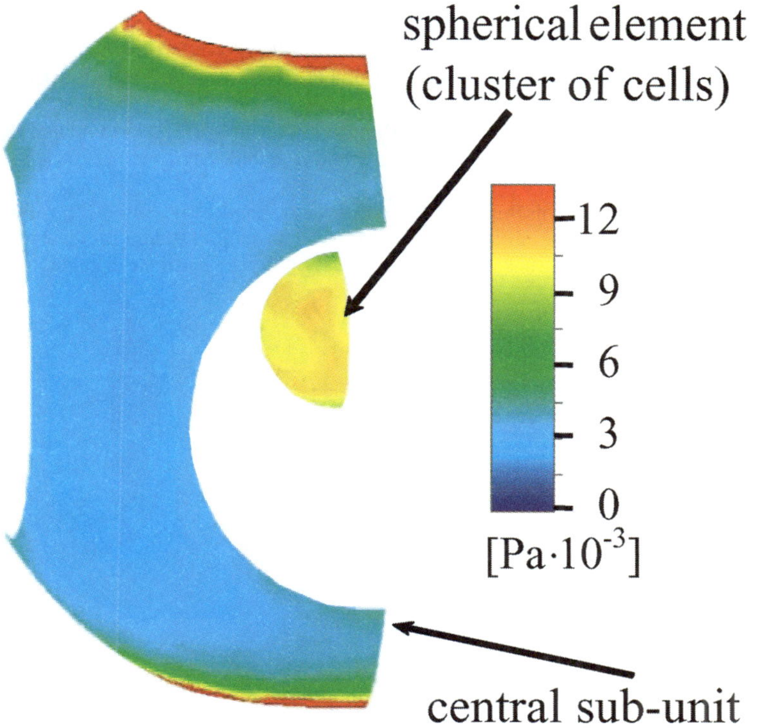

Fig. 7. Fluid-dynamic shear stresses inside the construct during bioreactor culture, as computed by the CFD analysis, mapped on the lateral wall of the central sub-unit, simulating the surface where the seeded cells adhere, and mapped on the spherical element, simulating a cell cluster partially attached to the scaffold, thus suspended in the flow.

In conclusion, our results provide a basis for the completion of more exhaustive quantitative studies to further assess the relationship between perfusion, at known micro-fluid dynamic conditions, and tissue growth *in vitro*.

Acknowledgements

This study was supported in part by the grant Young Researchers Project 2000 from Politecnico di Milano. The DegraPol® scaffolds were kindly provided by Dr. Peter Neuenschwander from ETH, Zurich. We thank Gianfranco B. Fiore and Maurizio Colombo for their skilled help.

References

[1] M. Brittberg, A. Lindahl, A. Nilsson, C. Ohlsson, O. Isaksson and L. Peterson, Treatment of deep cartilage defects in the knee with autologous chondrocyte transplantation, *N. Engl. J. Med.* **33** (1994), 889–895.

[2] L.E. Freed and G. Vunjak-Novakovic, Tissue engineering bioreactors, in: *Principles of Tissue Engineering*, R.P. Lanza, R. Langer and J. Vacanti, eds, 2nd edn, Academic Press, San Diego, 2000.

[3] D.W. Hutmacher, Scaffolds in tissue engineering bone and cartilage, *Biomaterials* **21** (2000), 2529–2543.

[4] D.E. Ingber, Mechanical and chemical determinants of tissue development, in: *Principles of Tissue Engineering*, R.P. Lanza, R. Langer and J. Vacanti, eds, 2nd edn, Academic Press, San Diego, 2000.

[5] M. Jakob, O. Demarteau, D. Schafer, B. Hintermann, W. Dick, M. Heberer and I. Martin, Specific growth factors during the expansion and redifferentiation of adult human articular chondrocytes enhance chondrogenesis and cartilaginous tissue formation in vitro, *J. Cell Biochem.* **81**(2) (2001), 368–377.

[6] I. Martin, B. Obradovic, S. Treppo, A.J. Grodzinsky, R. Langer, L.E. Freed and G. Vunjak-Novakovic, Modulation of the mechanical properties of tissue engineered cartilage, *Biorheology* **37** (2000), 141–147.
[7] V.C. Mow and C.C. Wang, Some bioengineering considerations for tissue engineering of articular cartilage, *Clin. Orthop.* **367**(Suppl.) (1999), S204–S223.
[8] M.T. Raimondi, F. Boschetti, L. Falcone, G.B. Fiore, A. Remuzzi, M. Marazzi, E. Marinoni and R. Pietrabissa, Mechanobiology of engineered cartilage cultured under a quantified fluid dynamic environment, *Biomechanics and Modelling in Mechanobiology* **1** (2002), 69–82.
[9] B. Saad, Y. Kuboki, M. Welti, G.K. Uhlschmid, P. Neuenschwander and U.W. Suter, DegraPol-foam: a degradable and highly porous polyesterurethane foam as a new substrate for bone formation, *Artif. Organs* **24**(12) (2000), 939–945.
[10] M. Stading and R. Langer, Mechanical shear properties of cell–polymer cartilage constructs, *Tissue Eng.* **5**(3) (1999), 241–250.

Biorheology 41 (2004) 411–421
IOS Press

Considerations on the use of ear chondrocytes as donor chondrocytes for cartilage tissue engineering

Gerjo J.V.M. van Osch [a,b,*], Erik W. Mandl [a], Holger Jahr [a,c], Wendy Koevoet [a,b], Gilbert Nolst-Trenité [d] and Jan A.N. Verhaar [a]

[a] *Department of Orthopaedics, Erasmus University Medical Center, Rotterdam, The Netherlands*
[b] *Department of Otorhinolaryngology, Erasmus University Medical Center Rotterdam, The Netherlands*
[c] *Department of Internal Medicine, Erasmus University Medical Center Rotterdam, The Netherlands*
[d] *Department of Otorhinolaryngology, Academic Medical Center, Amsterdam, The Netherlands*

Abstract. Articular cartilage is often used for research on cartilage tissue engineering. However, ear cartilage is easier to harvest, with less donor-site morbidity. The aim of this study was to evaluate whether adult human ear chondrocytes were capable of producing cartilage after expansion in monolayer culture.

Cell yield per gram of cartilage was twice as high for ear than for articular cartilage. Moreover, ear chondrocytes proliferated faster. Cell proliferation could be further stimulated by the use of serum-free medium with Fibroblast Growth Factor 2 (FGF2) in stead of medium with 10% serum. To evaluate chondrogenic capacity, multiplied chondrocytes were suspended in alginate and implanted subcutaneously in athymic mice. After 8 weeks the constructs demonstrated a proteoglycan-rich matrix that contained collagen type II. Constructs of ear chondrocytes showed a faint staining for elastin. Quantitative RT-PCR revealed that expression of collagen type II was 2-fold upregulated whereas expression of collagen type I was 2-fold down regulated in ear chondrocytes expanded in serum-free medium with FGF2 compared to serum-containing medium. Expression of alkaline phosphatase and collagen type X were low indicating the absence of terminal differentiation.

We conclude that ear chondrocytes can be used as donor chondrocytes for cartilage tissue engineering. Furthermore, it may proof to be a promising alternative cell source to engineer cartilage for articular repair.

Keywords: FGF2, dedifferentiation, elastin, articular

1. Introduction

Tissue engineering techniques might offer possibilities of restoring missing or damaged cartilage structures. Although significant progression has been made during the last decade, some major problems still have to be addressed before widespread clinical use can be expected. The chondrocytes, being the key factor of tissue engineering, will largely determine the quality of the engineered tissue. Thus the choice of donor cell source is of utmost importance. For clinical tissue engineering, an autologous cell source is a prerequisite, which will limit the risk of infection, disease transmission and potential immune

[*]Address for correspondence: Gerjo J.V.M. van Osch, PhD, Erasmus University Medical Center, Dept. Orthopaedics & Dept. Otorhinolaryngology, Room Ee1655, PO Box 1738, 3000 DR Rotterdam, The Netherlands. Tel.: +31 10 4087661; Fax: +31 10 4089441; E-mail: g.vanosch@erasmusmc.nl.

rejection. Consequently, treatment of cartilage defects with tissue engineering techniques will most often be performed with donor tissue of mature origin, as cartilage defects are most commonly found in adult patients [2]. However, compared to immature tissues, mature tissues are often characterised by lower cell yield and lower metabolism, which could have disadvantageous effects on the quality of the engineered repair tissue [1,7,13,19].

The most obvious choice of an autologous cell source is cartilage itself. From the various cartilage subtypes existing in the body, articular cartilage is most frequently used for research on cartilage tissue engineering, although some experiments have been published on the use of nasal septum cartilage and ear cartilage [3,4,8,20]. Articular cartilage as a donor cell source has some very obvious disadvantages: it is frequently harvested from a diseased joint and harvesting itself may induce further degenerative changes in the joint. In this respect, ear cartilage has the advantages that it is healthy cartilage, easy to harvest and little donor-site morbidity [5,12,16]. However, ear cartilage is elastic cartilage and amongst other things differs from hyaline cartilage by the presence of elastin. It is our goal to investigate the possibility of using ear chondrocytes as donor source for reconstruction of articular cartilage. The aim of this study was to evaluate whether adult human ear chondrocytes were capable of producing cartilage after expansion in monolayer culture. The performance of human ear chondrocytes was compared to human articular chondrocytes using data of various experiments previously performed in our laboratory. Also, porcine chondrocytes were used to compare ear and articular chondrocytes from the same individual cultured at similar conditions. Furthermore, addition of Fibroblast Growth Factor 2 (FGF2) during expansion culture has been demonstrated to support chondrogenic capacity of bone marrow stromal chondrocytes [11]. Therefore, in this study we evaluated the effect of addition of FGF2 during expansion culture in monolayer on chondrogenic capacity of expanded ear chondrocytes after subcutaneous implantation in mice.

2. Materials & methods

2.1. Cell sources

Adult human cartilage from the ear (three donors) obtained as leftover material from patients undergoing reconstructive surgery of the nose using ear cartilage, was used.

For comparative studies on ear and articular cartilage, adult human cartilage from healthy knees was obtained from road victims (aged 19–40 years) and osteoarthritic knee cartilage was obtained from donors (aged 35–70 years) undergoing surgery for total knee replacement. For additional data on adult human cartilage from the ear, material was used that was obtained as left-over material from donors (aged 26–52 years) undergoing meatoplasty or reconstructive surgery on the nasal septum using ear cartilage. In addition we have used cartilage from healthy knees and ears of porcine origin (approx. 4 months old).

2.2. Cell isolation

Slices of cartilage were washed with saline and subsequently incubated in Pronase E (Sigma, St Louis, MO, 2 mg/ml) for 2 hours at 37°C and in Collagenase B (Boehringer Mannheim, Germany; 1.5 mg/ml DMEM and 10% Fetal Calf Serum (FCS)) overnight. The next day cell suspensions were filtered through a 100 μm filter, centrifuged and washed with saline.

2.3. Monolayer culture

Adult human ear chondrocytes were seeded at 7500 chondrocytes/cm^2 or 15000 chondrocytes/cm^2 and cultured in DMEM (Life Technologies, Breda, The Netherlands) with 50 μg/ml gentamycin and 0.5 μg/ml fungizone with either 10% FCS (serum-containing medium: SCM) or in serum-free medium containing ITS+ (1 : 100, Becton Dickinson, Bedford, MA) and 100 ng/ml FGF2 (Recombinant Human, Instruchemie, Hilversum, The Netherlands) (SFM+FGF2). Porcine chondrocytes were seeded at 7500 chondrocytes/cm^2 and cultured in medium with 10% FCS. Human chondrocytes used for the additional data were seeded at a density of 40000 chondrocytes/cm^2 and cultured in medium with 10% FCS. When chondrocytes reached confluency they were trypsinized with trypsin-EDTA (Life Technologies).

2.4. Subcutaneous implantation in vivo

The expanded chondrocytes were suspended in 1.8% alginate (LV, Keltone) at a concentration of 20 million chondrocytes/ml [17]. Constructs of approximately 3 mm height and 8 mm diameter were prepared by polymerisation of alginate in 102 mM $CaCl_2$ and implanted subcutaneously in athymic mice. This experiment was performed with cartilage of two healthy adult human knee donors, three adult human ear donors, one immature human ear donor and both ear and knee cartilage of one pig.

Animals were housed under sterile conditions at the Center for Animal Research. After eight weeks the animals were killed by cervical dislocation and the constructs were harvested. Animal experiments were approved by the University Ethics Committee and carried out as outlined in the University guidelines for the care and use of laboratory animals, which in general follows the NIH "Guide for the care and use of laboratory animals".

2.5. Analyses

Constructs after *in vivo* implantation were digested in proteinase K solution (1 mg/ml Proteinase K; Boehringer Mannheim, Germany) overnight at 56°C. The amount of DNA in the beads was measured fluorometrically using Hoechst 33258 dye, using calf thymus DNA (Sigma, St Louis, MO) as standard. The amount of GAG was quantified spectrophotometrically by the absorption ratio A_{540}/A_{595} using a metachromatic reaction of GAG with dimethylmethylene blue, with chondroitin sulfate C (Shark; Sigma, St Louis, MO) as a standard.

2.6. Histology

Evaluation of tissue engineered constructs after subcutaneous implantation in athymic mice was performed using paraffin sections. Sections were stained with thionin and immunohistochemically stained for collagen II (using mab II-II6B3, Developmental Studies Hybridoma Bank, University of Iowa, under contract N01-HD-6-2915) and elastin (using mab BA4, Sigma, St Louis, MO). For procollagen type I (using mab M38, DSHB) and a fibroblast membrane marker (11-Fibrau, Imgen, distributed by ITK diagnostics, Uithoorn, The Netherlands) cryosections were prepared. For all immunohistochemical stainings the primary antibodies were coupled overnight to a secondary antibody directed against mouse, before incubation on the sections.

2.7. Gene expression of expanded adult human ear chondrocytes after subcutaneous implantation

Constructs were snap frozen in LiN2 and disrupted in a Microdismembrator-S (B. Braun Biotech International GmbH, Melsungen, Germany). Chondrocytes were then homogenized in RNABee, the RNA was isopropanol precipitated and purified using RNeasy Mini Kit (Qiagen GmbH, Hilden, Germany).

Quantities and purity of extracted total RNA were determined spectrophotometrically at 260/280 nm. Ratios between 1.7 and 2.0 indicated reasonable quality. 500 ng total RNA of each sample was reverse transcribed into cDNA using RevertAid™ First Strand cDNA Synthesis Kit (MBI Fermentas, Germany).

Primer and probes were designed using PrimerExpress1.5 software (Applied Biosystems, Foster City, CA, USA). Primer sets were first tested in reverse transcription PCR (RT-PCR). Real-time quantitative PCR analysis was performed using the ABI 7700 Sequence Detection System (Applied Biosystems) with the following probes:

aggrecan:	5′-GGCTACCCCGACCCCTC-3′ (hAGC1_F),
	5′-TCCACAAAGTCTTCACCTCTGTGT-3′ (hAGC1_R),
	5′-TCCCGCTACGACGCCATCTGC-3′ (hAGC1_FAM);
alkaline phosphatase:	5′-GACCCTTGACCCCCACAAT-3′ (hALPL_F),
	5′-GCTCGTACTGCATGTCCCCT-3′ (hALPL_R),
	5′-TGGACTACCTATTGGGTCTCTTCGAGCCA-3′ (hALPL_FAM);
elastin:	5′-AGCCGCTAAGGCAGCCA-3′ (hELN_F),
	5′-ACCAACTACTCCCGGGCC-3′ (hELN_R),
	5′-TATGGAGCTGCTGCAGGCTTAGTGCC-3′ (hELN_FAM).

The primer and probe nucleotide sequences, as well as assay concentrations, for collagen type I, collagen type II, collagen type X, versican, and GAPDH were adopted from [10]. All Taqman assays were performed using qPCR™ Core Kit (Eurogentec Nederland B.V., Maastricht, Nederland B.V.) according to user's manual. Per reaction, 300 pg of gene-specific cDNA (assuming 3% mRNA) was amplified in 40 cycles (15 sec at 95°C, 1 min at 60°C) after initially activating *Taq* DNA polymerase 10 min at 95°C.

The Ct value for each sample was defined as the cycle number in PCR at which the fluorescence intensity reached a value significantly above the background. The Ct value of the genes of interest was subtracted from the Ct value of GAPDH. The level of expression for each gene was then calculated as $2^{\text{Ct`gapdh'-Ct`gene of interest'}}$. The efficiency of amplification of each set of primers and probes was always higher than 90% [10].

3. Results

3.1. Cell expansion in monolayer

Adult human ear chondrocytes (four donors) cultured for 3–4 passages at relatively high seeding densities (40000 chondrocytes/cm^2) in medium with 10% FCS, showed 3.2 ± 1.6 times multiplication in 32 ± 10 days. Chondrocytes of healthy knees (three donors), expanded 1.3 ± 0.4 times in 42 ± 2 days, whereas chondrocytes of osteoarthritic knees (5 donors) expanded 5.1 ± 4.8 times in 52 ± 13 days. In general, articular chondrocytes more often showed problems with cell attachment and sometimes suddenly ceased to grow at various time points in the monolayer culture.

Using animal experiments we can compare the capacities of ear and knee chondrocytes of the same animal to proliferate under exactly similar conditions. Seeded at 7500 chondrocytes/cm^2 and cultured for 28 days in medium with 10% FCS, porcine ear chondrocytes were passaged three times and yielded 47 times multiplication, whereas knee chondrocytes were passaged only two times in 28 days and cell number increased only 13 times.

Human ear chondrocyte seeded at high densities (40000 chondrocytes/cm^2) and cultured in serum-free medium with FGF2 (SFM+FGF2), expanded similar to serum containing conditions [9]. When lower seeding densities were used (either 7500 chondrocytes/cm^2 or 15000 chondrocytes/cm^2), expansion of human ear chondrocytes during four weeks was significantly increased. Furthermore, with lower seeding densities, the use of SFM+FGF2 further increased cell expansion on average four times compared to SCM. When cultured in SFM+FGF2, initial cell number was expanded 150 times when chondrocytes seeded at a density of 7500 chondrocytes/cm^2 whereas seeded at a density of 15000 chondrocytes/cm^2, cell number expanded 60 times.

3.2. Comparison of cell yields from donor tissue

Besides a better capacity to proliferate, the number of vital chondrocytes isolated per gram of healthy mature ear cartilage was higher than per gram of healthy mature articular cartilage. Average cell yield of human adult healthy ear cartilage (three donors) was 9.7 ± 0.8 million chondrocytes/gram of tissue, but from human adult healthy knee cartilage (four donors) only 4.9 ± 1.4 million chondrocytes/gram of tissue were obtained. We propose that 0.5 gram of ear cartilage can be easily harvested from a human ear with minimal cosmetic problems, from which 5 million donor chondrocytes can be routinely obtained for tissue engineering purposes. Previously we have calculated that a minimum of 50 million chondrocytes (20 million/chondrocytes per cm^3 is considered to be a minimum) is needed to repair an average defect of 5 cm^2 and thickness of 5 mm. This implies that the amount of donor chondrocytes has to be expanded approximately 10 times. Expansion from a seeding density of 7500 chondrocytes/cm^2 in medium with 10% FCS would take approximately 3 weeks to reach 10 times multiplication, whereas it would take approximately 2 weeks using SFM+FGF2, considering cell expansion rates we found in this study.

3.3. Cartilage matrix production in constructs

At harvesting after eight weeks subcutaneous implantation in athymic mice, the constructs with human ear chondrocytes had a cartilage-like macroscopic appearance. On average, the constructs contained 20 μg GAG/mg wet weight, being half the amount present in native articular or nasal septum cartilage [18]. To get a fair comparison, we compared chondrocytes that had been through similar expansion. Chondrocytes expanded in monolayer at a seeding density of 7500 chondrocytes/cm^2 in SCM had similar multiplication as chondrocytes expanded from 15000 chondrocytes/cm^2 in SFM+FGF2. Chondrocytes that had been expanded in SFM+FGF2 produced 1.5 times more GAG per milligram tissue than constructs with chondrocytes expanded in SCM. On histology, abundant GAG was visible by thionin staining. Collagen type II was expressed in all grafts. In constructs of chondrocytes expanded in SCM the expression of collagen type II was lower and appeared to be mainly associated with chondrocytes embedded in alginate (Fig. 1A). In the constructs of chondrocytes expanded in SFM+FGF2, expression occurred though out the construct (Fig. 1B) with exception of the capsule surrounding the construct. Procollagen type I staining with antibody M38 was found in the capsule and in the dense, more fibrous-appearing tissue in the construct, but was absent in chondrocytes associated with alginate. Immunohistochemical staining for elastin was only faintly positive (Fig. 2). Differences could be found within and between the constructs. In general, dense more fibrous-appearing tissue between alginate remnants showed

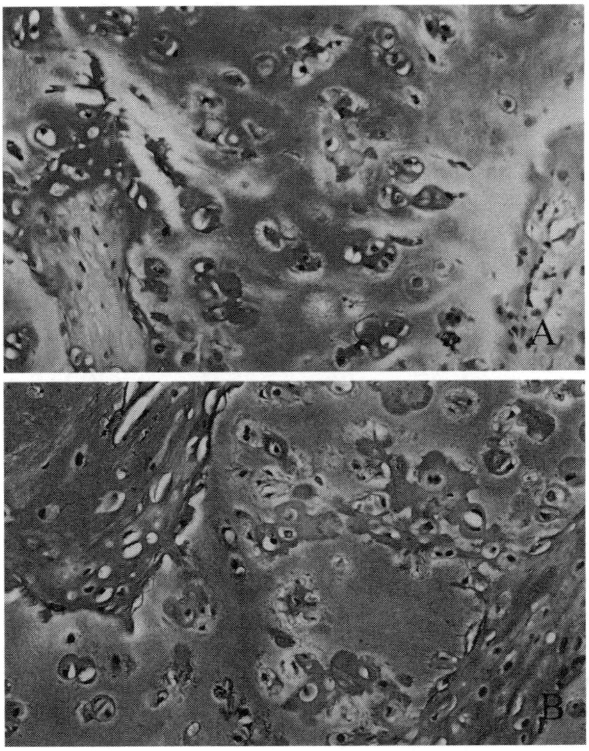

Fig. 1. Immunohistochemical staining for collagen type II of constructs generated from human adult ear chondrocytes expanded in monolayer from seeding density of 7500 cells/cm^2 in medium with 10% FCS (A) and from seeding density of 15000 cells/cm^2 in serum-free medium with 100 ng/ml FGF2 (B). Red colour indicates a positive signal; original magnification 200×.

the most intense staining for elastin (Fig. 2A). On locations were alginate was still present (showing by amorphous substance between the chondrocytes), hardly any staining was found when chondrocytes were expanded with SFM+FGF2 (Fig. 2B), whereas chondrocytes expanded in SCM stained positive (Fig. 2C). This was only a faint staining, compared to native human ear cartilage (Fig. 2E). In contrast, construct of human articular chondrocytes were negative for elastin (Fig. 2D).

Gene expression was determined to evaluate cellular phenotype of the neo-cartilage harvested from the athymic mice in more detail. In human ear chondrocytes expanded in SCM, expression of collagen type I was always higher than expression of collagen type II, resulting in an expression ratio of collagen type II/collagen type I of 0.39. Expression ratio of collagen type II to collagen type I was 1.9 for chondrocytes expanded in SFM+FGF2. Chondrocytes expanded in SFM+FGF2 actually expressed two times more collagen II and two times less collagen type I than chondrocytes expanded in SCM. Expressions of collagen were 20–40 times as high as that of GAPDH. Elastin expression of chondrocytes expanded in SFM+FGF2 and SCM were similar. Expression of elastin was 5–10 times as high as that of GAPDH. Aggrecan and versican were not differently expressed in the different expansion media. Expression of chondrocyte hypertrophy markers, such as alkaline phosphatase and collagen type X, were low (Table 1).

With the porcine cartilage it was possible to harvest enough chondrocytes to study the behaviour of differentiated ear and knee chondrocytes in constructs. This enabled us to compare the performance of differentiated chondrocytes with that of chondrocytes, which were culture-expanded in monolayer

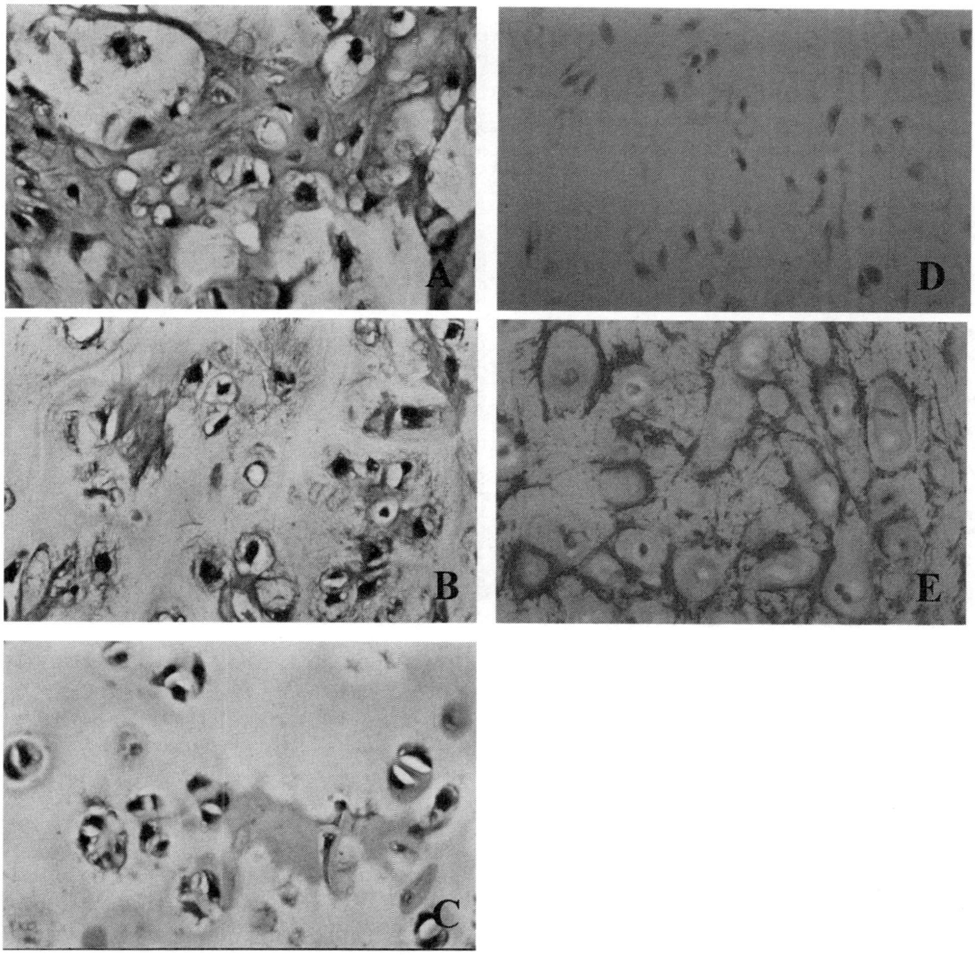

Fig. 2. Immunohistochemical staining for elastin. (A) Example of dense fibrous tissue from a construct generated from ear chondrocytes. (B) Ear chondrocytes still surrounded by alginate, that have been expanded in monolayer from seeding density of 7500 cells/cm^2 in medium with 10% FCS. (C) Ear chondrocytes still surrounded by alginate, that have been expanded in monolayer from seeding density of 15000 cells/cm^2 in serum-free medium with 100 ng/ml FGF2. (D) Construct of articular chondrocytes expanded in monolayer in medium with 10% FCS. (E) Native human ear cartilage. Red colour indicates a positive signal; original magnification 400×.

Table 1

Expression of genes normalized for GAPDH (mean ± sd of chondrocytes from three donors) in constructs of human ear chondrocytes expanded in monolayer from seeding density of 7500 cells/cm^2 in medium with 10% FCS and from seeding density of 7500 cells/cm^2 and 15000 cells/cm^2 in serum-free medium with 100 ng/ml FGF2. Data on chondrocytes expanded from 15000 cells/cm^2 in SCM are missing because not enough chondrocytes were obtained to generate a construct. AGC = aggrecan; VERS = versican; COL1 = collagen type I; COL2 = collagen type II; ELN = elastin. Collagen type X and alkaline phosphatase were not demonstrated because gene expression was at highest 0.001

	AGC	VERS	COL1	COL2	ELN
7500SCM	1.38 ± 0.10	0.05 ± 0.01	57.99 ± 19.12	22.83 ± 7.68	10.48 ± 4.94
7500SFM+FGF2	1.89 ± 0.30	0.06 ± 0.02	25.05 ± 8.13	35.63 ± 9.44	6.25 ± 3.37
15000SFM+FGF2	1.79 ± 0.16	0.05 ± 0.01	20.06 ± 5.33	36.58 ± 5.12	8.36 ± 2.33

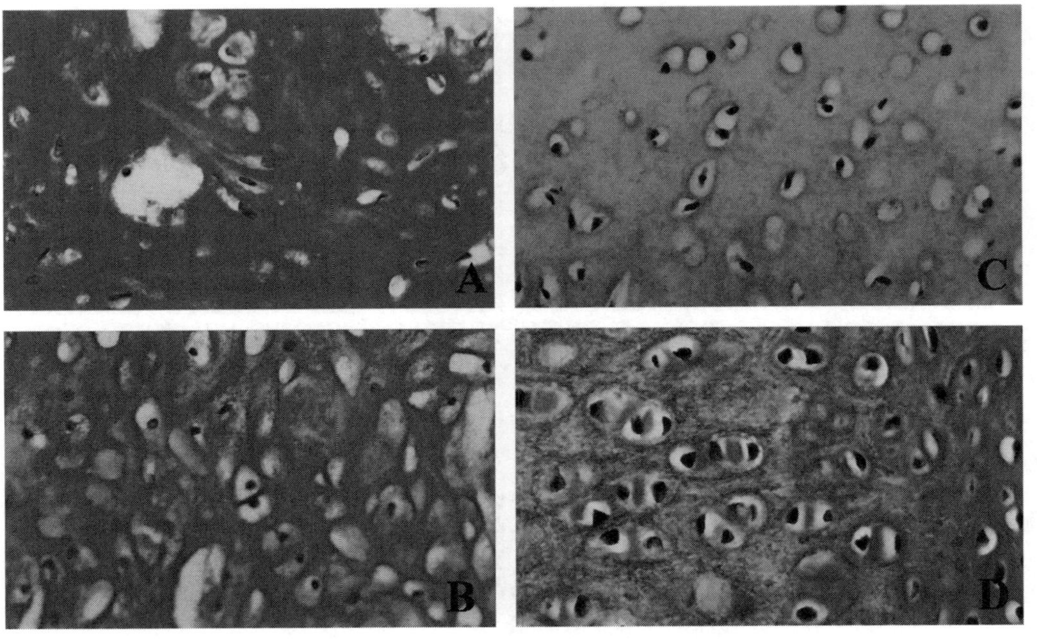

Fig. 3. Immunohistochemical staining for collagen type II. (A) A tissue engineered construct generated from differentiated porcine articular chondrocytes. (B) A tissue engineered construct generated from differentiated porcine ear chondrocytes. (C) The original porcine articular cartilage where the cells have been isolated from. (D) The original porcine ear cartilage where the cells have been isolated from. Red colour indicates a positive signal; original magnification 400×.

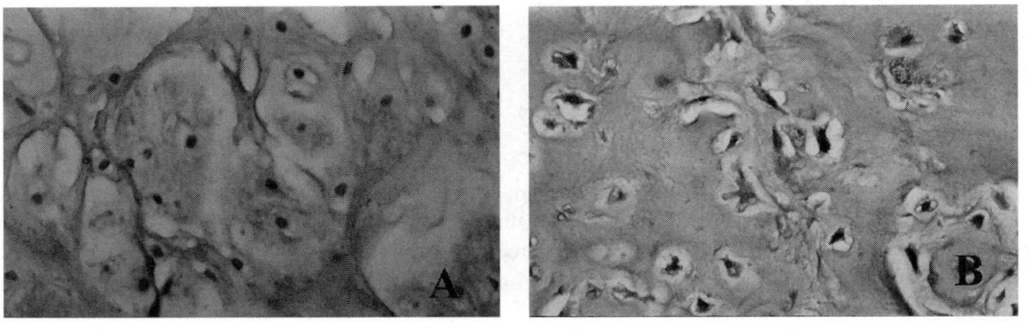

Fig. 4. Immunohistochemical staining for elastin in constructs generated from differentiated porcine ear chondrocytes (A) and monolayer expanded porcine ear chondrocytes (B). Red colour indicates a positive signal; original magnification 400×.

before seeding in a construct. It appeared that differentiated (seeded immediately after isolation) ear chondrocytes produced more GAG than differentiated articular chondrocytes. However, in similar experiments with ear and articular chondrocytes that had been expanded in monolayer culture for 28 days, this difference was absent. It should be noted that the ear chondrocytes were expanded more (47 times) than the knee chondrocytes (13 times) and were therefore probably more dedifferentiated.

On histology, constructs generated from ear cells and constructs generated from articular cells were both positive for collagen type II (Fig. 3). Constructs made from previously expanded cells appeared less positive than constructs from differentiated cells. Constructs from ear cells demonstrated some elastin whereas no stain for elastin was found in constructs from articular chondrocytes. Constructs from expanded ear chondrocytes showed less elastin than those initially seeded with differentiated ear chondrocytes (Fig. 4). Also these constructs with differentiated ear chondrocytes more often contained the typical "big" cells which are characteristic for the central part of ear cartilage (Fig. 3).

4. Discussion

It has been demonstrated before that ear chondrocytes can be used to generate cartilage. However, these studies used immature cartilage, mostly from porcine origin and only rarely from human tissues. Furthermore, most studies were performed with primary, differentiated chondrocytes. However, in a clinical setting, the donor tissue most often will be mature and, due to the limited amount of donor chondrocytes, the chondrocytes have to be culture-expanded. Our study showed that ear chondrocytes of adult human origin were able to multiply in monolayer culture. Evenmore, the cell yields after cultures were always higher compared to experiments with articular chondrocytes. A serious side effect of monolayer expansion culture is chondrocyte dedifferentiation. During expansion, chondrocytes change the expression pattern from the cartilage specific collagen type II and aggrecan towards less functional molecules like collagen type I and versican. The ability to re-express the differentiated phenotype is, besides the ability to multiply *in vitro*, a crucial prerequisite to be a suitable donor cell source for cartilage tissue engineering. To assess whether ear chondrocytes possessed this ability, we subcutaneously implanted culture-expanded human ear chondrocytes in nude mice. Subsequent histology revealed cartilage-like tissue that was highly positive for collagen type II and glycosaminoglycans.

Recently, Kafienah et al. [6] compared the capacity of chondrocytes from articular cartilage and nasal septum cartilage and concluded that human adult nasal chondrocytes proliferate approximately four times faster than human articular chondrocytes in monolayer culture. Evenmore, they possessed a markedly higher chondrogenic capacity, as assessed by the mRNA and protein analysis of *in vitro*-engineered constructs. Our study, together with the study by Kafienah [6], clearly show a growing interest, as well as the possibilities, for the use of alternative cartilage types as donor tissue for articular cartilage engineering. We hypothesise that chondrocytes exhibit certain plasticity in the expression of their phenotype, which means that ear chondrocytes could be persuaded to express a more hyaline-like phenotype. Local environmental factors within a joint might provide the necessary stimuli to accomplish this. Whether ear chondrocytes can respond to these stimuli is not known. Furthermore, as mechanical loading is not physiological for ear chondrocytes it remains questionable whether these chondrocytes and their produced extracellular matrix can prevail themselves within a joint. In pilot experiments we found that immature bovine ear chondrocytes seeded in agarose discs remained vital after four consecutive days of 5–10% cyclic compression. In future experiments we will investigate whether ear chondrocytes subjected to a mechanical stimulus will produce a more hyaline-like matrix. So far, no data on mechanical loading of ear cartilage or ear chondrocytes seeded in a biomaterial are present in literature.

Obviously, the ultimate test to assess the feasibility of using ear chondrocytes for articular cartilage repair would be the implantation of dedifferentiated ear chondrocytes in an articular cartilage defect. Again, no data are found in literature. However, in reconstructive surgery of the nose, ear cartilage transplants have been used with success for many years to repair nasal septum defects, which is in fact hyaline cartilage. In a study on the fate of the implanted ear cartilage within the nasal septum, Min et al. [14] showed that hyaline-like neo-cartilage was formed after 4 weeks from within the transplanted ear cartilage. In our opinion, this study might represent an example on how local environmental factors can alter cellular phenotype.

A repair tissue that exactly resembles the native tissue might not be obligatory for a functional cartilage repair. In other words, cartilage repair is not similar to cartilage regeneration. Therefore, a sustained expression of elastin might be of less importance. Mechanical characteristics similar to the native tissue and good integrative capacities could be more important. Data on the mechanical characteristics of cartilage generated by ear chondrocytes are lacking and even data on mechanical characteristics of native ear cartilage are scarce. In experiments comparing characteristics of the different cartilage subtypes in rabbits, Naumann et al. [15] found that the aggregate modulus of ear cartilage was higher and the permeability was lower than that of articular cartilage. In our experiments we did not assess the mechanical characteristics of the neo-cartilage engineered with adult human ear chondrocytes.

Summarising, in this paper we combined data from previous experiments in our laboratory with articular and ear chondrocytes as donor chondrocytes for cartilage tissue engineering. We found that both the cell yield of the donor tissue and monolayer expansion culture favoured the use of ear chondrocytes. Cell yield and performance could further be increased by the use of serum-free medium with 100 ng/ml FGF2 for the expansion culture in monolayer. Furthermore, we speculated on the use of ear chondrocytes to engineer repair tissue for articular cartilage defects. To make more solid statements on this issue, the mechanical properties of engineered cartilage with ear chondrocytes should be assessed in future. Furthermore, experiments in which culture expanded ear chondrocytes, seeded onto a scaffold, are implanted in articular cartilage defects will be performed.

Acknowledgements

The authors would like to thank Corrina de Ridder and Cindy Bolder for technical assistance with the animal experiments. This work was funded by the Dutch Ministry of Economic Affairs in a grant of Senter (nr. BTS00021).

References

[1] M. Adolphe, X. Ronot, P. Jaffray, C. Hecquet, J. Fontagne and P. Lechat, Effects of donor's age on growth kinetics of rabbit articular chondrocytes in culture, *Mech. Ageing Dev.* **23** (1983), 191–8.

[2] J.A. Buckwalter, Articular cartilage injuries, *Clin. Orthop.* (2002), 21–37.

[3] J. Bujia, P. Pitzke, E. Kastenbauer, E. Wilmes and C. Hammer, Effect of growth factors on matrix synthesis by human nasal chondrocytes cultured in monolayer and in agar, *Eur. Arch. Otorhinolaryngol.* **253** (1996), 336–40.

[4] J. Bujia, M. Sittinger, E. Wilmes and C. Hammer, Effect of growth factors on cell proliferation by human nasal septal chondrocytes cultured in monolayer, *Acta Otolaryngol.* **114** (1994), 539–43.

[5] B. Guyuron, Simplified harvesting of the ear cartilage graft, *Aesthetic Plast. Surg.* **10** (1986), 37–9.

[6] W. Kafienah, M. Jakob, O. Demarteau et al., Three-dimensional tissue engineering of hyaline cartilage: comparison of adult nasal and articular chondrocytes, *Tissue Eng.* **8** (2002), 817–26.

[7] H. Kamada, K. Masuda, A.L. D'souza et al., Age-related differences in the accumulation and size of hyaluronan in alginate culture, *Arch. Biochem. Biophys.* **408** (2002), 192–9.

[8] A. Lavezzi, M. Mantovani, L.G. Della Berta and L. Matturri, Cell kinetics of human nasal septal chondrocytes in vitro: importance for cartilage grafting in otolaryngology, *J. Otolaryngol.* **31** (2002), 366–70.

[9] E.W. Mandl, S.W. Van Der Veen, J.A. Verhaar and G.J. Van Osch, Serum-free medium supplemented with high-concentration FGF2 for cell expansion culture of human ear chondrocytes promotes redifferentiation capacity, *Tissue Eng.* **8** (2002), 573–80.

[10] I. Martin, M. Jakob, D. Schafer, W. Dick, G. Spagnoli and M. Heberer, Quantitative analysis of gene expression in human articular cartilage from normal and osteoarthritic joints, *Osteoarthritis Cartilage* **9** (2001), 112–8.

[11] M. Mastrogiacomo, R. Cancedda and R. Quarto, Effect of different growth factors on the chondrogenic potential of human bone marrow stromal cells, *Osteoarthritis Cartilage* **9** (Suppl. A) (2001), S36–40.

[12] C.A. Megerian, B.D. Weitzner, B. Dore and L.J. Bonassar, Minimally invasive technique of auricular cartilage harvest for tissue engineering, *Tissue Eng.* **6** (2000), 69–74.

[13] H. Messai, Y. Duchossoy, A.M. Khatib, A. Panasyuk and D.R. Mitrovic, Articular chondrocytes from aging rats respond poorly to insulin-like growth factor-1: an altered signaling pathway, *Mech. Ageing Dev.* **115** (2000), 21–37.

[14] Y.G. Min, Y.K. Kim and H.S. Kim, The histology of autologous auricular cartilage implanted into the nasal septum, *Clin. Otolaryngol.* **21** (1996), 339–42.

[15] A. Naumann, J.E. Dennis, A. Awadallah et al., Immunochemical and mechanical characterization of cartilage subtypes in rabbit, *J. Histochem. Cytochem.* **50** (2002), 1049–58.

[16] F.V. Nicolle and A.O. Grobbelaar, Technique for harvesting of conchal cartilage grafts, *Aesthetic Plast. Surg.* **21** (1997), 243–4.

[17] A. Panossian, S. Ashiku, C.H. Kirchhoff, M.A. Randolph and M.J. Yaremchuk, Effects of cell concentration and growth period on articular and ear chondrocyte transplants for tissue engineering, *Plast. Reconstr. Surg.* **108** (2001), 392–402.

[18] N. Rotter, G. Tobias, M. Lebl et al., Age-related changes in the composition and mechanical properties of human nasal cartilage, *Arch. Biochem. Biophys.* **403** (2002), 132–40.

[19] G.J. Van Osch, W.J. Marijnissen, S.W. Van Der Veen and H.L. Verwoerd-Verhoef, The potency of culture-expanded nasal septum chondrocytes for tissue engineering of cartilage, *Am. J. Rhinol.* **15** (2001), 187–92.

[20] G.J. Van Osch, S.W. Van Der Veen and H.L. Verwoerd-Verhoef, In vitro redifferentiation of culture-expanded rabbit and human auricular chondrocytes for cartilage reconstruction, *Plast. Reconstr. Surg.* **107** (2001), 433–40.

Biorheology 41 (2004) 423–431
IOS Press

Expression of adhesion molecules and collagen on rat chondrocyte seeded into alginate and hyaluronate based 3D biosystems. Influence of mechanical stresses

C. Gigant-Huselstein [a,*], P. Hubert [b], D. Dumas [a], E. Dellacherie [b], P. Netter [c], E. Payan [c] and J.F. Stoltz [a]

[a] *Mécanique et Ingénierie Cellulaire et Tissulaire, UMR-CNRS-UHP-INPL 7563 LEMTA et IFR 111, Faculté de médecine, 54505 Vandoeuvre-lès-Nancy, France*

[b] *Laboratoire de Chimie Physique Macromoléculaire, UMR CNRS 7568 et IFR 111, Groupe ENSIC, 54000 Nancy, France*

[c] *Physiopathologie et Pharmacologie Articulaires, UMR CNRS UHP 7561 et IFR 111, Faculté de médecine, 54505 Vandoeuvre-lès-Nancy, France*

Abstract. Chondrocytes use mechanical signals, via interactions with their environment, to synthesize an extracellular matrix capable to withstanding high loads. Most chondrocyte–matrix interactions are mediated via transmembrane receptors such as integrins or non-integrins receptors (i.e. annexin V and CD44).

The aim of this study was to analyze, by flow cytometry, the adhesion molecules ($\alpha 5/\beta 1$ integrins and CD44) on rat chondrocytes seeded into 3D biosystem made of alginate and hyaluronate. These biosystems were submitted to mechanical stress by knocking the biosystems between them for 48 hours. The expression of type I and type II collagen was also evaluated.

The results of the current study showed that mechanical stress induced an increase of type II collagen production and weak variations of $\alpha 5/\beta 1$ receptors expression no matter what biosystems. Moreover, our results indicated that hyaluronan receptor CD44 expression depends on extracellular matrix modifications. Thus, these receptors were activated by signals resulted from cell environment variations (HA addition and modifications owing to mechanical stress). It suggested that this kind of receptor play a crucial role in extracellular matrix homeostasis. Finally, on day 24, no dedifferentiation of chondrocytes was noted either in biosystems or under mechanical stress. For all biosystems, the neosynthesized matrix contained an important level of collagen, which was type II, whatever biosystems.

In conclusion, it appeared that the cells, under mechanical stress, maintained their phenotype. In addition, it seems that, on rat chondrocytes, $\alpha 5/\beta 1$ integrins did not act as the main mechanoreceptor (as described for human chondrocytes). In return, hyaluronan receptor CD44 seems to be in relation with matrix composition.

1. Introduction

The perception of mechanical signals and biological responses to such stimuli are fundamental properties of load bearing articular cartilage in diarthrodial joints. Under mechanical stimuli, chondrocytes

*Corresponding author: Céline Huselstein, Mécanique et Ingénierie Cellulaire et Tissulaire, LEMTA UMR CNRS 7563, Faculté de Médecine - B.P. 184, 54505 Vandoeuvre-lès-Nancy Cedex, France. Tel.: +33 (0) 3 83 68 34 52; Fax: (33) 03 83 68 34 59; E-mail: Celine.Huselstein@medecine.uhp-nancy.fr.

regulate their metabolic activity through complex biological and biophysical interactions with the extracellular matrix (ECM). This ECM is composed of proteoglycans (PGs); predominantly aggrecan aggregates, and collagens, predominantly type II, IX and XI, as well as other proteins.

At the cell surface, matrix receptors link the ECM to the cell interior through elements of the cytoskeleton and other component proteins of signal transduction pathways. Among these receptors, articular chondrocytes have been shown to express both members of the integrin family [11,20] as well as non-integrin receptor: cluster of differentiation (CD) 44 of the hyaluronan (HA) binding protein family, intercellular adhesion molecule (ICAM)-1 of the immunoglobulin superfamily, annexin V (anchorin CII) [9].

The route by which a particular mechanical signal is transduced into an intracellular response is now well established for a role for integrins [17,18]. Integrins are heterodimeric (α and β subunits) transmembrane glycoproteins. A variety of $\beta 1$ integrins, including $\alpha 1\beta 1, \alpha 2\beta 1, \alpha 3\beta 1, \alpha 5\beta 1, \alpha 6\beta 1, \alpha 10\beta 1$ and $\alpha 11\beta 1$, have been identified on normal human chondrocytes [11,20]. Under mechanical stimulus, integrin mediated adhesion to ECM protein (collagen II, fibronectin, laminin, vitronectin) can activate several cytoskeleton associated protein, such as paxillin, tensin and intracellular signalling protein such as FAK (focal adhesion kinase). All these interactions play a key role in the regulation of gene expression, proliferation, differentiation and matrix remodeling. Integrin signaling is coordinated with both the organization of cytoskeleton and with signaling by growth factor and cytokine receptor (Fig. 1) [12].

Another cell receptor expressed by chondrocytes is the single-pass transmembrane glycoprotein CD44. This receptor represents the principal receptor responsible for HA binding [10]. In cartilage, HA is seldom present as pure molecule. Often more than 50 aggrecan proteoglycan monomers together with an associated link protein become bound to a single filament of HA [2]. Thus, the binding of HA to CD44 in cartilage facilitates the retention of large, highly hydrated, HA/PG/link protein aggregates at the surface of chondrocyte. The CD44 receptor structure and more exactly the cytoplasmic domain suggest a role in signal transmission toward cell interior ("outside-in signaling"). In response to intracellular signals ("inside-out signaling"), ECM modifications may occur. Through interactions with HA, CD44 has been shown to have a role in assembly, organization and maintenance of chondrocyte pericellular matrix [1,3].

Both integrins and CD44 receptors have the capacity for signal transduction as well as modulated interactions with the actin cytoskeleton. So, in order to study the chondrocyte mechanobiology, in an established chondrocyte-alginate culture system supplemented or no with HA [4], we analyzed, the expression of adhesion molecules on chondrocytes and compared the synthesized ECM under mechanical stimulus. This investigation was carried out according to the main criteria: expression of type I and type II collagens and adhesion molecules like integrin $\alpha 5/\beta 1$ and CD44.

2. Materials and methods

2.1. Isolation of chondrocytes and cell culture

Chondrocytes were isolated enzymatically from femoral heads of male Wistar rats (10 weeks old) (Charles River, France) as described previously [13].

The cell culture was performed in DMEM F-12 medium, supplemented with glutamine (1%), gentamicin (0.7%), amphotericin B (0.1%) and fetal bovine serum (10%, FBS), purchased from Life Technologies. Then, 3 mM $CaCl_2$ (Merck, Germany) was added to the medium in order to maintain the structural

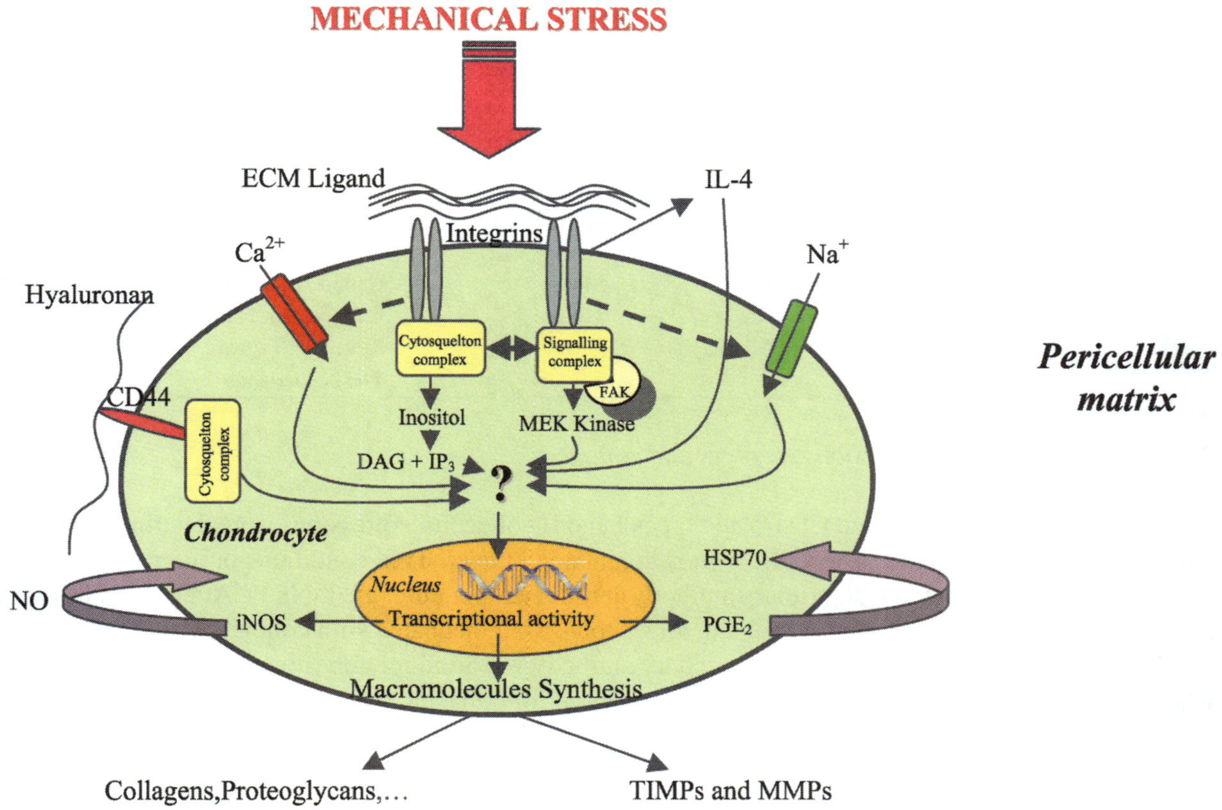

Fig. 1. Schematic representation of chondrocyte mechanobiology.

integrity of alginate-based materials. Throughout the following text, this mixture will be named "complete medium".

2.2. Cell seeding into three-dimensional constructs

Sodium alginate (medium viscosity) from *Macrocystis pyrifera* (Sigma Aldrich) and HA from bacterial origin (Acros Organics, Belgium) were used. An alginate (Alg) solution (16 g/l) or a mixture containing Alg (16 g/l) and HA (4 g/l) were prepared by dissolution of autoclaved polymers in sterile 0.9% NaCl. The cell pellet was suspended at 3×10^6 cells/ml in polymer solutions. Then, the mixtures were dripped in 100 mM $CaCl_2$, using the method described by Guo et al. [6]. An average number of 10^5 cells/bead was obtained and placed in complete medium. Cultures in beads were maintained in a humidified atmosphere of 5% CO_2 at 37°C for 10 and 24 days.

2.3. Mechanical testing

An agitator 10°/tridimensional Polymax 1040 (Fisher Bioblock Scientific, France) was used to stimulate the cells embedded in 3D constructs. Beads were deposed in tubes (Costar, France, 10 beads/tube) containing complete medium supplemented with Hepes 10 mM. Then, these tubes were all aligned parallel with the agitation plate and rotated at 30 cycles per minute.

Table 1

Characteristics of primary antibodies and conjugated-secondary antibody

Antibodies	Species specificity	Isotype	Commercial origin
$\alpha 5/\beta 1$	Rat	IgG1 (Hamster)	BD Pharmingen
CD44	Rat	IgG1 (Hamster)	BD Pharmingen
FITC conjugated antibody	Hamster	IgG2b (Mouse)	BD Pharmingen
Type I collagen	Rat	Mouse	Calbiochem, France
Type II collagen	Rat	Mouse	Calbiochem, France
Alexa 488 conjugated antibody	Mouse	Goat	Molecular Probes, The Netherlands

2.4. *Immunolabelling of adhesion receptors and collagens*

Adhesion receptor ($\alpha 5/\beta 1$ and CD44) and types I and II collagens expressions were determined by an indirect immunofluorescence assay. Biosytems, maintained under static conditions or after mechanical stimulation, were incubated with primary antibody at room temperature (cf. Table 1). After several washing with culture medium without phenol red, all samples were incubated with conjugated-secondary antibody. The fluorescein (FITC) labeled secondary antibody was used for adhesion molecules expression whereas Alexa 488 labeled secondary antibody was used for collagen expression (cf. Table 1).

2.5. *$\alpha 5/\beta 1$ integrins, CD44 and collagens analysis by flow cytometry*

2.5.1. *Semi-quantitative expression of types I et II collagens*

Embedded cells onto Alg or Alg/HA biosystems were fixed with 1% paraformaldehyde solution. Then, cell suspension was obtained by treatment with sodium citrate (55 mM) and EDTA (50 mM). Cells were washed with buffer solution and analyzed by using a FACScan flow cytometer (Becton Dickinson, France). Dead cells were excluded by forward and side scatter gating. At least, 5 000 cells were counted and list mode files were then analyzed using WinMDI software.

2.5.2. *Quantitative expression of $\alpha 5/\beta 1$ integrins and CD44*

The receptor analysis principle was the same as for collagens expressions ($n = 6$ for each time and each type of biosystem). The quantification of antigen expression performed by using Qifikit (Dako, France) in five populations of beads; each coated with a different amount of antibodies. Beads were stained and analyzed parallel to the samples. Immunolabelling was performed under saturating conditions; antibodies were considered to bind to surface antigens through monovalent interaction. The number of antigenic sites per bead is proportional to the mean intensity fluorescence (MIF) measured. Thus, for each sample tested, corrected MIF (i.e., specific MIF minus that of the negative control measured under the same conditions) could therefore be correlated with antibody molecules bound per cell [15].

2.6. *Statistical analysis*

Results are expressed as mean $\pm$ standard deviation (SD). Two-way analysis of variance followed by Student's comparison test was used for statistical analyses. Significant differences were taken into account only when a probability of less than or equal to 5% was obtained.

3. Results

3.1. Expression of types I and II collagens

The flow cytometry analysis showed that, under mechanical stimulation, chondrocyte embedded onto Alg matrixes significantly increased the collagen type I (+115%, $p < 0.001$; Fig. 2) whereas any significant variation of this type of collagen was observed in Alg/HA biosystems. The expression of collagen type II, whatever the bioconstruct, was significantly increased when biosystems were submitted to mechanical stimulation (+87%, $p < 0.025$ for Alg and +126%, $p < 0.001$ for Alg/HA).

Finally, it appeared that, on 24 days-ECMs, type II collagen was more expressed than type I collagen, whatever the biosystems. This increase was more important when matrix component as HA was added into Alg matrixes than in Alg matrixes alone ($p < 0.01$ whatever mechanical conditions).

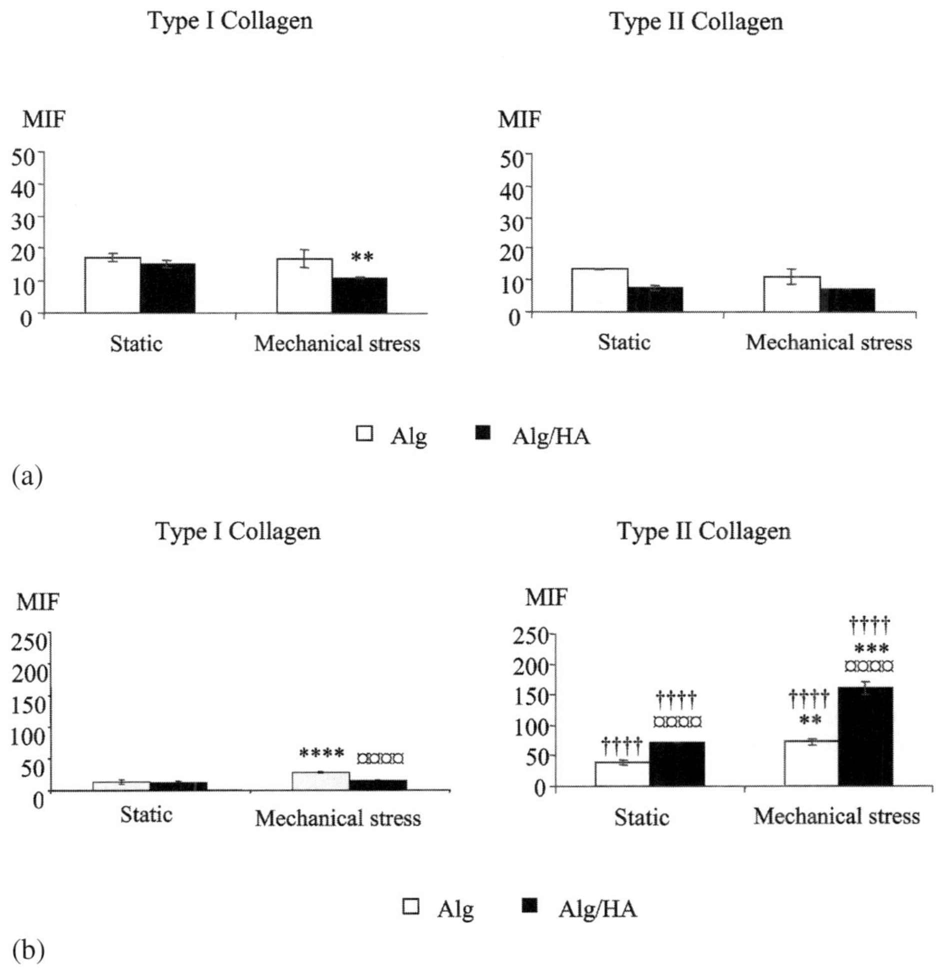

Fig. 2. Expression of types I and II collagens on neosynthesized matrix of 10 days (a) and 24 days (b) during static conditions and mechanical stimulus ($N = 3$). MIF = mean intensity fluorescence. $^{**}p < 0.025$, $^{***}p < 0.01$, $^{****}p < 0.001$ (knocking process versus static conditions, ¤¤¤¤ $p < 0.001$ (Alg/HA versus Alg).

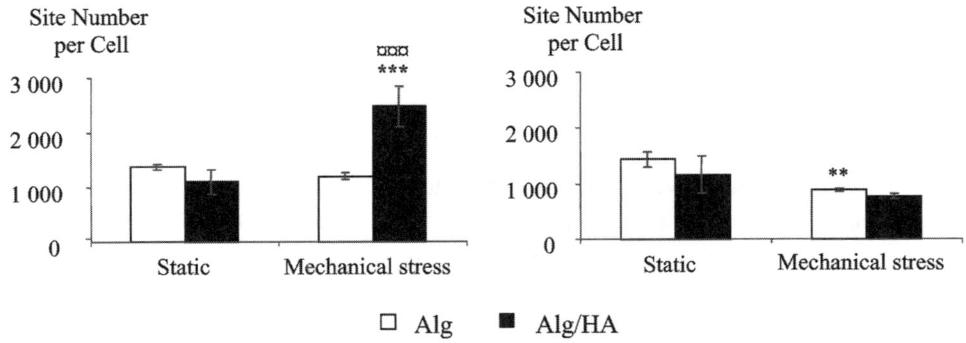

Fig. 3. Expression of $\alpha5/\beta1$ integrins on rat chondrocytes seeded onto neosynthesized matrix of 10 days (left) and 24 days (right) under static conditions and knocking process ($N = 6$). $^{**}p < 0.025$, $^{***}p < 0.01$ (knocking process versus static conditions), ¤¤¤$p < 0.01$ (Alg/HA versus Alg).

3.2. Expression of $\alpha5/\beta1$ integrins receptors

The expression of $\alpha5\beta1$ integrin on chondrocytes submitted to mechanical stimulation showed significant variations, especially Alg/HA biosystems ($p < 0.001$, Fig. 3(left)), as far as the extracellular matrix was not yet synthesized (day 10). In other hand, on day 24, the mechanical stress did not induce major variations of this expression in all biosystems seeded with rat articular chondrocytes (Fig. 3(right)).

3.3. Expression of CD44 receptors

The expression of CD44 receptors showed, whatever the conditions, a high frequency of positive cells in Alg/HA biosystems than in Alg biosystems ($p < 0.001$, Fig. 4). Moreover, it seems that mechanical stimulation increased the site number per cell only in Alg/HA matrixes ($p < 0.05$ on day 10 and $p < 0.01$ on day 24).

4. Discussion

In this study, we investigated the influence of mechanical stimulation on the expression of CD44 and integrin $\alpha5\beta1$ receptors on chondrocyte embedded onto polysaccharidic matrixes. The expression of matrix components characterizing chondrocyte phenotype (type I and II collagen) was also analyzed by flow cytometry. Mechanical stimulation was performed on biosystems containing 10 and 24 days-ECMs.

Articular chondrocytes normally produce type II collagen as the major constituent of cartilage, but do not produce type I collagen except for the dedifferentiated chondrocytes. Thus, type II collagen and proteoglycans assemble in the extracellular space to form an electromechanically and biochemically functional ECM [5] which is essential to the function of cartilage as a load-bearing, wear-resistant tissue. In contrast, type I collagen characterize an ECM whose mechanical stability is mediocre in time [8]. According to our results, whatever the mechanical conditions, on 24 days-ECMs, type II collagen was more synthesized than type I collagen ($\times2$ in Alg matrixes and $\times11$ in Alg/HA matrixes). These results obtained are agreed with previous study using alginate as culture scaffold. In fact, it was shown that under static conditions, chondrocytes cultivated during 28 days expressed a large part of type II collagen [14,16]. Moreover, Petit et al. also observed the presence of type I collagen into the synthesized 28 days-ECM [14]. Since type I collagen degradation time is very long [19], this protein synthesized at

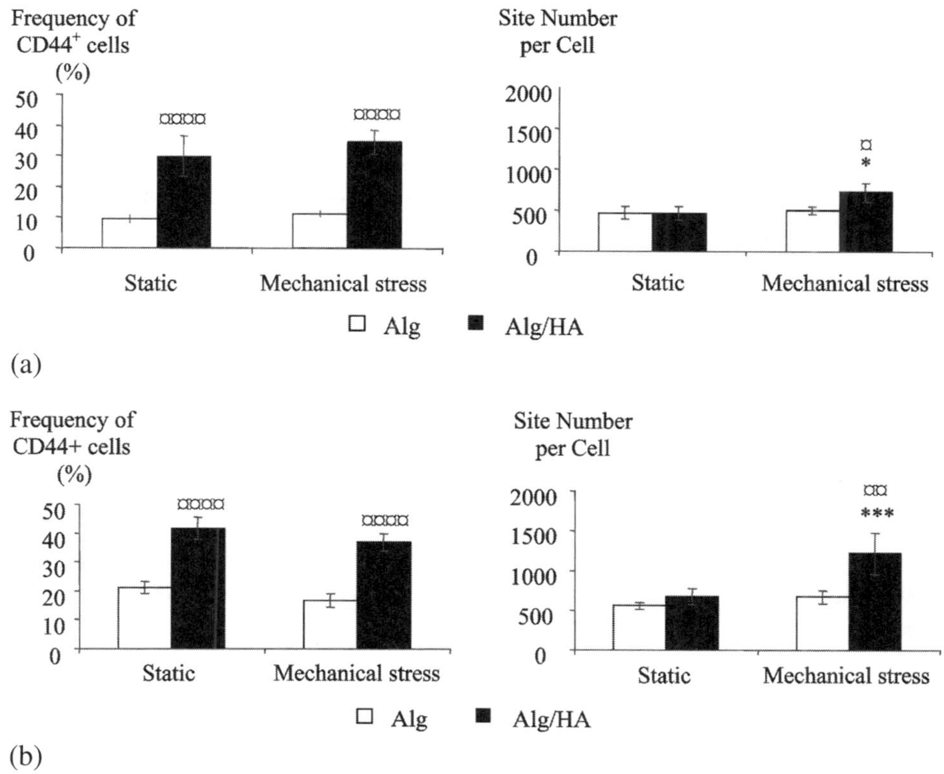

Fig. 4. Expression of CD44 receptors on neosynthesized matrix of 10 days (a) and 24 days (b) during static conditions and mechanical stimulus ($N = 4$). $^{*}p < 0.05$, $^{***}p < 0.01$ (knocking process versus static conditions), ¤$p < 0.05$, ¤¤$p < 0.025$, ¤¤¤¤$p < 0.001$ (Alg/HA versus Alg).

the beginning of the 3D culture was still observed in 24 days-ECMs. When cells were submitted to mechanical stress, the neosynthesized matrix contained an important level of collagen, which was in part of type II. It seems that, under dynamic mechanical stimulation, first, cells lose their fibroblastic phenotype acquired during monolayer expansion more quickly than under static condition, then, they are able to maintain their chondrocyte phenotype. Finally, the addition of HA into biosystems induces an increase of the expression of type II collagen as compared to Alg matrixes alone. These results associated with those obtained in previous study analyzing proteoglycans synthesis and chondrocyte proliferation during mechanical stimulation permitted to conclude that under knocking process, cells significantly increase their metabolic activities without losing their phenotype and the neosynthetized matrix obtained seems to be comparable to native cartilage [4].

The quantitative analysis of $\alpha5\beta1$ receptors showed, under mechanical stress, significant variations as far as the extracellular matrix was not yet synthesized (D10). No variation at all is observed as soon as the matrix is synthesized (D24). This is contrary to experiments performed on human, in which these receptors are considered as mechanoreceptor. From our results with the rat, it seems that $\alpha5\beta1$ receptors are not involved in the signal transduction during mechanical stress. Another hypothesis concerning the influence of the culture conditions could be proposed. In fact, it was shown that in chondrosarcoma cells, mechanical stress increased $\alpha5$ chain expression when cells were adherent to plastic while cells adherent to collagen type II increased expression of $\alpha2$ [7]. It could be suggested that 3D polysaccharidic scaffolds stimulate the expression of other integrins. For completing these preliminary results, it would be interesting to study the influence of mechanical stress on human cells in order to compare the variations between species but also to analyze the expression of other integrins described in the literature as being involved in cell–matrixes interactions.

In chondrocytes, CD44 represents the primary receptor responsible for hyaluronan binding. And the binding of HA to CD44 in cartilages facilitates the retention of large, highly hydrated aggregates at the surface of chondrocytes. From this study, it appeared that the addition of HA in biomaterial could favour the adhesion of chondrocytes to their extracellular matrix. Furthermore, in several studies, it was described that CD44 receptors have the capacity to respond to changes in the extracellular matrix [10]. Since the intracytoplasmic domain of CD44 receptors has binding sites for cytoskeleton and especially for actin filaments, the changes in matrix composition modulate interactions with actin and lead to changes in cell metabolism [10]. As previously described, mechanical stimulation induces variation of ECM composition (increase of proteoglycans and collagen synthesis). Therefore, it is possible that these modifications may affect the expression of CD44 receptors. An increase of CD44 site number at cell surface could then favour the interactions of CD44 receptors with other adjacent CD44 molecules. According to our results, when biosystems are supplemented with HA, the CD44+ cells number significantly increases as compared with plain Alg. So, it seems that the modifications by mechanical stimulation associated with the addition of HA in biomaterials could increase the site number per cell.

5. Conclusion

In conclusion, it is clear that under mechanical stimulation, cells maintained their chondrocyte phenotype. Moreover, the add of HA in biosystems increase the type II collagen production, which is in favor of neosynthetized ECM functional and comparable with native cartilage.

In return, if the function of CD44 adhesion molecule in ECM homeostasis is clearly defined, the implication of integrin $\alpha 5\beta 1$ expressed by rat chondrocyte in mechanotransduction pathways may be elucidated.

References

[1] D.J. Aguiar, W. Knudson and C.B. Knudson, Internalization of the hyaluronan receptor CD44 by chondrocytes, *Exp. Cell. Res.* **252** (1999), 292–302.

[2] J.A. Buckwalter and L.C. Rosenberg, Electron microscopic studies of cartilage proteoglycans. Direct evidence for the variable length of the chondroitin sulfate-rich region of proteoglycan subunit core protein, *J. Biol. Chem.* **257** (1982), 9830–9839.

[3] G. Chow, J.J. Nietfeld, C.B. Knudson and W. Knudson, Antisense inhibition of chondrocyte CD44 expression leading to cartilage chondrolysis, *Arthritis Rheum.* **41** (1998), 1411–1419.

[4] C. Gigant-Huselstein, D. Dumas, P. Hubert, D. Baptiste, E. Dellacherie, D. Mainard, P. Netter, E. Payan and J.F. Stoltz, Influence of mechanical stress on extracellular matrixes synthesized by chondrocytes seeded onto alginate and hyaluronate-based 3D biosystems, *JMMB* **3** (2003), 59–70.

[5] A.J. Grodzinsky, Electromechanical and physicochemical properties of connective tissue, *Crit. Rev. Biomed. Eng.* **9** (1983), 133–199.

[6] J. Guo, G.W. Jourdian and D.K. McCallum, Culture and growth characteristics of chondrocytes encapsulated in alginate beads, *Connect. Tissue Res.* **19** (1989), 277–297.

[7] K. Holmvall, L. Camper, S. Johansson, J.H. Kimura and E. Lundgren-Akerlund, Chondrocyte and chondrosarcoma cell integrins with affinity for collagen type II and their response to mechanical stress, *Exp. Cell. Res.* **221** (1995), 496–503.

[8] E.B. Hunziker, Articular cartilage repair: are the intrinsic biological constraints undermining this process insuperable?, *Osteoarthritis Cartilage* **7** (1999), 15–28.

[9] W. Knudson, D.J. Aguiar, Q. Hua and C.B. Knudson, CD44-anchored hyaluronan-rich pericellular matrices: an ultrastructural and biochemical analysis, *Exp. Cell. Res.* **228** (1996), 216–228.

[10] W. Knudson and R.F. Loeser, CD44 and integrin matrix receptors participate in cartilage homeostasis, *Cell. Mol. Life Sci.* **59** (2002), 36–44.

[11] R.F. Loeser, C.S. Carlson and M.P. McGee, Expression of beta 1 integrins by cultured articular chondrocytes and in osteoarthritic cartilage, *Exp. Cell. Res.* **217** (1995), 248–257.

[12] R.F. Loeser, Integrins and cell signaling in chondrocytes, *Biorheology* **39** (2002), 119–124.

[13] G. Miralles, R. Baudoin, D. Dumas, D. Baptiste, P. Hubert, J.F. Stoltz, E. Dellacherie, D. Mainard, P. Netter and E. Payan, Sodium alginate sponges with or without sodium hyaluronate: in vitro engineering of cartilage, *J. Biomed. Mater. Res.* **57** (2001), 268–278.

[14] B. Petit, K. Masuda, A.L. D'Souza, L. Otten, D. Pietryla, D.J. Hartmann, N.P. Morris, D. Uebelhart, T.M. Schmid and E.J. Thonar, Characterization of crosslinked collagens synthesized by mature articular chondrocytes cultured in alginate beads: comparison of two distinct matrix compartments, *Exp. Cell. Res.* **225** (1996), 151–161.

[15] P. Poncelet, F. George, S. Papa and F. Lanza, Quantitation of hemopoietic cell antigens in flow cytometry, *Eur. J. Histochem.* **40** (1996), 15–32.

[16] P.M. Ragan, V.I. Chin, H.H. Hung, K. Masuda, E.J. Thonar, E.C. Arner, A.J. Grodzinsky and J.D. Sandy, Chondrocyte extracellular matrix synthesis and turnover are influenced by static compression in a new alginate disk culture system, *Arch. Biochem. Biophys.* **383** (2000), 256–264.

[17] D.M. Salter, S.J. Millward-Sadler, G. Nuki and M.O. Wright, Integrin–interleukin-4 mechanotransduction pathways in human chondrocytes, *Clin. Orthop.* **391** (Suppl.) (2001), S49–60.

[18] D.M. Salter, J.E. Robb and M.O. Wright, Electrophysiological responses of human bone cells to mechanical stimulation: evidence for specific integrin function in mechanotransduction, *J. Bone Min. Res.* **12** (1997), 1133–1141.

[19] M. Shinmei, S. Miyauchi, A. Machida and K. Miyazaki, Quantitation of chondroitin 4-sulfate and chondroitin 6-sulfate in pathologic joint fluid, *Arthritis Rheum.* **35** (1992), 1304–1308.

[20] V.L. Woods, Jr, P.J. Schreck, D.S. Gesink, H.O. Pacheco, D. Amiel, W.H. Akeson and M. Lotz, Integrin expression by human articular chondrocytes, *Arthritis Rheum.* **37** (1994), 537–544.

Biorheology 41 (2004) 433–443
IOS Press

Cartilage repair: Surgical techniques and tissue engineering using polysaccharide- and collagen-based biomaterials

L. Galois [a,b,e,*] A.M. Freyria [b], L. Grossin [a,e], P. Hubert [c,e], D. Mainard [a,e], D. Herbage [b,e], J.F. Stoltz [d,e], P. Netter [a,e], E. Dellacherie [c,e] and E. Payan [a,e]

[a] *UMR 7561 CNRS-UHP Nancy I, Faculté de Médecine, 54505 Vandoeuvre les Nancy, France*
[b] *UMR 5086 CNRS-UCB Lyon I & IFR 128-Biosciences Lyon-Gerland, 7 passage du Vercors, 69367 Lyon, France*
[c] *UMR 7568 CNRS-INPL, ENSIC, 1 rue Grandville, 54000 Nancy, France*
[d] *UMR 7563 CNRS-INPL-UHP Nancy I, Faculté de Médecine, 54505 Vandoeuvre les Nancy, France*
[e] *IFR 111 "Bioingénierie", France*

Abstract. Lesions of articular cartilage have a large variety of causes among which traumatic damage, osteoarthritis and osteochondritis dissecans are the most frequent. Replacement of articular defects in joints has assumed greater importance in recent years. This interest results in large part because cartilage defects cannot adequately heal themselves. Many techniques have been suggested over the last 30 years, but none allows the regeneration of the damaged cartilage, i.e. its replacement by a strictly identical tissue. In the first generation of techniques, relief of pain was the main concern, which could be provided by techniques in which cartilage was replaced by fibrocartilage. Disappointing results led investigators to focus on more appropriate bioregenerative approaches using transplantation of autologous cells into the lesion.

Unfortunately, none of these approaches has provided a perfect final solution to the problem. The latest generation of techniques, currently in the developmental or preclinical stages, involve biomaterials for the repair of chondral or osteochondral lesions. Many of these scaffolds are designed to be seeded with chondrocytes or progenitor cells. Among natural and synthetic polymers, collagen- and polysaccharide-based biomaterials have been extensively used. For both these supports, studies have shown that chondrocytes maintain their phenotype when cultured in three dimensions. In both types of culture, a glycosaminoglycan-rich deposit is formed on the surface and in the inner region of the cultured cartilage, and type II collagen synthesis is also observed. Dynamic conditions can also improve the composition of such three-dimensional constructs. Many improvements are still required, however, in a number of key aspects that so far have received only scant attention. These aspects include: adhesion/integration of the graft with the adjacent native cartilage, cell-seeding with genetically-modified cell populations, biomaterials that can be implanted without open joint surgery and combined therapies, aimed at disease modification, pain relief and reduction of inflammation.

Keywords: Cartilage repair, surgical techniques, tissue engineering, collagen, polysaccharide-based biomaterials, chondrocyte

1. Introduction

Over the past decade, one of the most active areas of biomaterials research has involved cartilage repair. Replacement of articular defects in joints has assumed greater importance in recent years because

*Corresponding author: L. Galois, Laboratoire de Pharmacologie et UMR 7561 CNRS-UHP, Faculté de Médecine, Avenue de la Forêt de Haye, BP 184, 54505 Vandoeuvre les Nancy Cedex, France. Tel.: +33 3 83 68 39 81; Fax: +33 3 83 68 39 59; E-mail: lgalois@free.fr.

of two main factors. The first is the increasing longevity of the population which is leading to a higher incidence of osteoarthritis initiated in some cases by early injuries to the joint, and the second is the increased incidence of joint injuries as a result of sports-related and other accidents in the young population [28]. This interest results in large part because cartilage defects cannot adequately heal themselves. This fact has been known for some time, probably even before it was first noted in 1743 by Hunter – "From Hippocrates to the present age, it is universally allowed that ulcerated cartilage is a troublesome thing and that when destroyed, it is not recovered" [17]. Many techniques have been suggested over the last 30 years, but none allows the regeneration of the damaged cartilage, i.e. its replacement by a strictly identical tissue (in terms of composition, structure and resistance). Surgical treatments available for treatment include debridement [21,27,33], drilling [18,38,47], micro-fractures [52,53], abrasion arthroplasty [27], autogenous [3,15,19] and allogeneic osteochondral grafts [2,6,14] periostal and perichondral grafting [7,16,44,45,50] and autogenous chondrocyte transplantation [4]. Each of these methods has had a measure of success. Methods that rely on fibrocartilage replacement are limited by the inherent altered biomechanics of replacing hyaline cartilage with fibrocartilaginous tissue. Other methods have specific problems (fixation of the graft, maintenance of the phenotype, cost ...). These disappointing results have led investigators to focus on more appropriate bioregenerative approaches using tissue engineering techniques especially with collagen- and polysaccharide-based biomaterials.

2. Conventional surgical techniques (articular resurfacing techniques)

Following articular resurfacing techniques such as chondral shaving [21,30], abrasion arthroplasty [27], subchondral drilling [40] and microfracture [52,53], the repair tissue changes to fibrocartilage after about 6 months, thereby losing the biomechanical properties of normal articular cartilage. Chondral shaving of partial-thickness defects provides symptomatic relief for a brief period but does not stimulate cartilage regeneration [21,37]. The biological rationale behind this empirical type of intervention remains unclear. It may represent merely an intuitive move by the surgeon to do the patient "some good" by removing the fibrillated articular cartilage seen by arthroscopic inspection, and thus by "smoothing the surface" reduce possible friction. But from a biomechanical point of view, such an undertaking makes no sense at all.

Abrasion arthroplasty yields variable results in the repair tissue produced ranging from fibrous tissue to hyaline-like cartilage with type II collagen [5,21]. Nevertheless, abrasion arthroplasty seems to give worse long-term results with increasing breakdown of the repair compared to subchondral drilling [27].

Pridie and Insall described debridement along with drilling in the exposed subchondral bone to encourage fibrocartilaginous repair [18]. The newly formed fibrocartilaginous layer that develops over the lesion in osteoarthritis may be enough to decrease patient symptoms. This procedure appears, however, to confer only short-term benefits; in the long run (several years after surgery), these are lost [47,48]. The drilling operation remains useful however in osteoarthritis and osteochondrosis dissecans.

Spongialization is an extension of the concept of Pridie for resurfacing damaged joints. The diseased cartilage is excised and the subchondral plate completely removed exposing the cancellous bone or "spongiosa" from which a new fibrous tissue surface can grow [12]. This technique was developed initially for diseased patellae but has been progressively discontinued [35].

The microfracture technique was developed by Steadman [52,53] to enhance chondral resurfacing by providing a suitable environment for new tissue formation and taking advantage of the body's own healing potential. Specially designed awls are used to make multiple perforations, or microfractures, into the subchondral bone plate. Perforations are made as close together as possible, but not so close that one breaks into another. They are usually approximately 3 to 4 mm apart. The integrity of the subchondral bone plate must be maintained. The principle of repair by micro-fractures is based on the construction of a super-clot that is progressively invaded by multipotent cells from the marrow and platelets. The cells first form a fibrous tissue that progressively undergoes metaplasia leading to active chondrocytes. The repair tissue is habitually a mixture of fibrocartilage and efficient "hyaline-like" cartilage. Microfracture has been done in more than 1800 patients. The rehabilitation program is crucial to optimise the results of the surgery. The other factors determining the result are the quality of the technical procedure, the oxygen content, and the absence of constraints on the repair zone.

3. Biological techniques using transplantation

3.1. Soft tissue grafts

Among soft tissue grafts, perichondrial and periosteal grafts have had the best results. Autologous perichondrial and periosteal grafts were tested in the tarsocrural joint of horses. Periosteal grafts resulted in neochondrogenesis in 83% of the animals [56]. The chondrogenic potential of free autogenous periosteal grafts was also studied histologically in rabbits [34,44,45]. The results revealed that the defects were repaired and filled after 4 weeks with a hyaline-like cartilage that was histologically similar to the cartilage adjacent to the transplant.

Cartilage repair tissue derived from rib perichondrium could also mature into hyaline articular cartilage over time [10] in rabbits. Clinical trials in humans utilising free perichondrial grafts in hand surgery [50] or in chondral lesions of the knee [16] have had mixed results. Some of the failures of this type of graft were deemed secondary to inadequate fixation and graft instability. Despite these interesting results, few publications however using these techniques can be found in the literature.

3.2. Allografts

A frozen or fresh allograft has the advantage of being able to replace the destroyed articular zone exactly [24]. Osteochondral allografting also appears to be effective in treating larger osteochondral lesions, where there are few other attractive alternatives [6]. While antigenic effects due to the chondrocytes are avoided with this approach, there is a risk of viral transmission (HIV, hepatitis C) with fresh allografts. The transplanted chondrocytes remain alive for several years after transplantation, even in the event of failure. The graft must to be tailored perfectly to fit the loss of substance and fixation must be firm. Nevertheless, the anchorage of such a graft is not easy and is still problematic. The results obtained in the literature are good: 73% to 85% success at 10 years [2,9,14]. Bipolar replacements are possible, but the success rate falls to 50%. Frozen allografts have the advantage of a better detection of viral disorders, but at the cost of a less satisfying mechanical quality of the cartilage and a less viability of the transplanted cells.

3.3. Osteochondral autografts

This concerns a technique, which is long recommended and has been revived in recent years, as it offers advantages over allografts in presenting no risk of viral transmission and permitting one-stage surgery, with encouraging results. This always involves grafts of rather large size which risks morbidity at the donor site and incongruity at the receptor site hence the concept of packing extensive losses of substance with several small cylindrical autografts, removed at the knee and capable of restoring better congruence. These techniques have been especially developed since 1992 under the influence of Hangody [15], Bobic [3], Matsusue [26] and, more recently, Jakob [19].

In summary, autologous osteochondral mosaicplasty – as an easy, one-step procedure, providing a relatively quick rehabilitation – can be an alternative in the treatment of small and medium sized lesions. Excellent clinical outcome, low costs of the treatment and short rehabilitation time represent the main advantages of this method. Treatment of full-thickness cartilage damage to weight bearing surfaces depends not only on the method of cartilage repair but also on treatment of the underlying cause. Accordingly, effective treatment of full thickness defects on the weight bearing surfaces requires careful patient selection, complex operative planning and a well organised treatment course.

3.4. Chondrocyte transplantation

The first human implantation (autologous chondrocyte transplantation or ACT) was made by a Swedish team (Brittberg, Lindahl and Peterson) in 1994 in 23 cases [4]. The principle of the technique consists of filling the cartilaginous defect using the biological dynamic represented by the periosteal–chondrocytic complex. Three stages are necessary. The first consists of the removal of around 200 mg of healthy cartilage from a non-weight bearing zone of the knee. Then, chondrocytes are removed from the cartilage after an enzymatic digestion and cultured in a monolayer procedure over a period of 3 weeks. The third stage is implantation which is performed by arthrotomy. The slightly over-sized periosteal flap is raised from the medial surface of the tibia and is attached very carefully around the loss of substance with its bony aspect towards the defect. The cell culture is then injected under the periosteal patch. Active and passive mobilisation of the knee is possible but some weight relief by crutches for 8 weeks is necessary so as not to compromise the development of the periosteal flap. The results are based essentially on the Swedish studies, whose follow-up is remarkable. Peterson et al. [39] reported their experience with 110 implantations performed between October 1987 and December 1995 for lesions of 1.3 cm^2 to 12 cm^2 with a minimum follow-up of 2 years. Among these, 92% were found to be improved in isolated lesions of the femoral condyle, 89% in osteochondritis, and 68% in cases of associated lesions of the anterior cruciate ligament or patella. A hyaline-like phenotype (type II collagen) was found in 85% of the biopsies and was well correlated with the good clinical results. The problem is the interpretation of findings relating to the autologous chondrocyte transplantation approach. It represents theoretically not a single but potentially a triple chondrogenic system, in that repair responses may originate from the transplanted cells, the cambial layer of the periosteal flap, and the subchondral bone-tissue spaces. Finally, no prospective, double-blinded clinical studies have been undertaken to compare this methodology with alternative, well-established and less-expansive treatment protocols.

4. Tissue engineering and bioregeneration approaches

Tissue engineering can be defined as the art of reconstituting mammalian tissues both structurally and functionally. Such material could be used at sites where native tissue has been lost or compromised by trauma or pathological processes. Articular cartilage provides its own particular challenges for tissue engineering. Its structure appears simple, it only contains one cell type but it shows different phenotypes, with a complex, highly organised extracellular matrix (ECM). Cartilage has no blood vessels, it is not innervated and normal mechanisms of tissue repair, involving the recruitment of cells to the site of damage, do not occur. The challenge for cartilage tissue engineering is to produce cartilage tissue with suitable structure and properties *ex vivo*, which can be implanted into joints to provide a natural repair that with time will become integrated with the patient's tissues.

Cartilage regeneration continues to challenge state-of-the-art three dimensional scaffold technology. Three key constituents usually form the basis of a tissue engineering approach: a matrix scaffold, cells and signalling molecules (such as growth factors or genes). Many different types of matrix have been tested *in vitro*, as well as in experimental animals and in human patients. These matrices can be broadly categorised according to their chemical nature into protein-based polymers, carbohydrate-based polymers, artificial materials and combinations of these.

Interactions between cells and their environment play an important role in many biological processes. Most studies on the role of the extracellular matrix components have been carried out with cells cultured on matrix proteins deposited on plastic. This kind of system is far from that found *in vivo*. *In vivo*, cells are in a three-dimensional matrix and receive physical and chemical signals from their environment. Cells are capable of recognising collagen by specific membrane receptors called integrins. Integrins $\alpha 1\beta 1$, $\alpha 2\beta 1$ and $\alpha 3\beta 1$ are the principal integrins involved in interactions with collagen.

4.1. Polysaccharide-based matrices

Numerous polysaccharide polymers have been proposed to build three-dimensional scaffolds to enhance neosynthesis of extracellular matrix by cells placed inside them. Among these, agarose, alginate or chitosan are used either alone or associated with other polymers to act as biocompatible and biodegradable networks for cartilage repair. Since some years, another group of naturally derived components, the glycosaminoglycans, existing in the cartilaginous matrix, have also proposed to conceive new biosystems for cartilage engineering. Glycosaminoglycans may create an appropriate environment for the regeneration of cartilage-like tissue when associated with other polymers. The main glycosaminoglycans, hyaluronic acid (HA) and chondroitin sulphate (CS) are already commercialised for treatment of osteoarthritis by intra-articular injection [32] and oral [43,55] administration, respectively. Numerous studies in the literature have proposed a combination of CS with porous collagen matrix [46] or chitosan [49] and have found that the resulting scaffold supports the production of proteoglycans and type II collagen by seeded chondrocytes.

Among the HA-based devices, the benzylic esters of hyaluronan, referred to as HYAFF, are a promising new class of semi-synthetic biomaterials in tissue engineering because they support the growth of human chondrocytes and maintain their original phenotype [13]. Esterified HA has also been investigated in a gelatin-based matrix to facilitate the differentiation of bone marrow mesenchymal progenitor cells into chondral or osseous tissue [1]. In the same way, sponges, constituted of fibronectin coated with hyaluronan, provide a structural support that can be filled with progenitor cells. This composite material might be superior to others for osteochondral repair and appears to organise the natural response and facilitate the integration of the neocartilage with the adjacent tissue [51].

Our own experience on HA-based biomaterials consists of associating this polymer chemically modified or not with calcium alginate to obtain hydrogels or sponges usable for *in vitro* or *in vivo* cartilage neosynthesis. The *in vitro* investigations have been carried out by seeding rat chondrocytes inside hydrogels or sponges to test different parameters such as proliferation, synthesis activity of proteoglycans and production of types I and II collagens. The results obtained showed a higher proliferation and synthesis activity for rat cells in the presence of hyaluronate (Fig. 1).

For all the biosystems, the neosynthesised matrix contained a high level of collagen which was in part type II collagen [29]. The same result was obtained *in vitro* with a hyaluronate substituted by C18

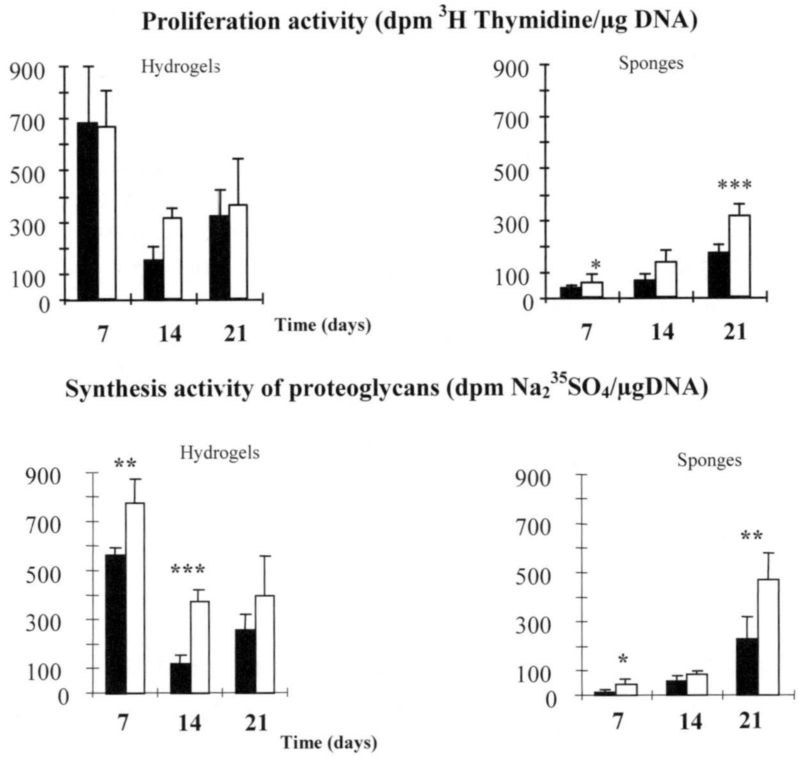

Fig. 1. Proliferation activity and synthesis activity of proteoglycans in hydrogels and sponges-based on alginate (black) or on alginate/hyaluronate (white). Results are the mean of $n = 3$ experiments $\pm$SD.

alkyl chains (HAC18). The advantage of this chemical modification is that it leads to viscoelastic properties mimicking perfectly those of matrix surrounding the chondrocytes. Moreover, this hydrophobic polysaccharide implanted in a rat model of an osteochondral lesion showed a good biotolerance and biocompatibility. The reconstructed tissue obtained is very similar to the native cartilage [11]. In the future, the concept of cell seeding will be associated with the design of this chemically modified HA, so as to develop an innovative injectable device affording optimal adhesion/integration of the graft with the tissue surrounding the defect.

4.2. Collagen-based matrices

Among biological scaffolds, collagen-based biomaterials have been extensively used in tissue engineering [54] as they have low immunogenicity and are easily configured into different shapes. They have been used for skin replacement, bone substitutes, artificial blood vessels and valves and more recently in cartilage repair. Chondrocyte seeded-3D scaffolds have an advantage over conventional autologous chondrocyte transplantation (ACT) as they help the growth of chondrocytes and enhance matrix production [22]. Previous *in vitro* studies have shown that differentiated bovine chondrocytes seeded in type I collagen sponges proliferated (2.3×10^6 cells/cm^3) and maintained their phenotype (2.3×10^7 cells/cm^3) [41]. Thus it did not seem necessary to use type II collagen matrices to produce neocartilage. Moreover the arthritogenic potential of type II collagen does not favour its clinical use in humans. The composition of the construct may be also modified using culture conditions involving mechanical and/or chemical stimuli. Indeed the intensity of the compression applied on the construct may or may not favour the

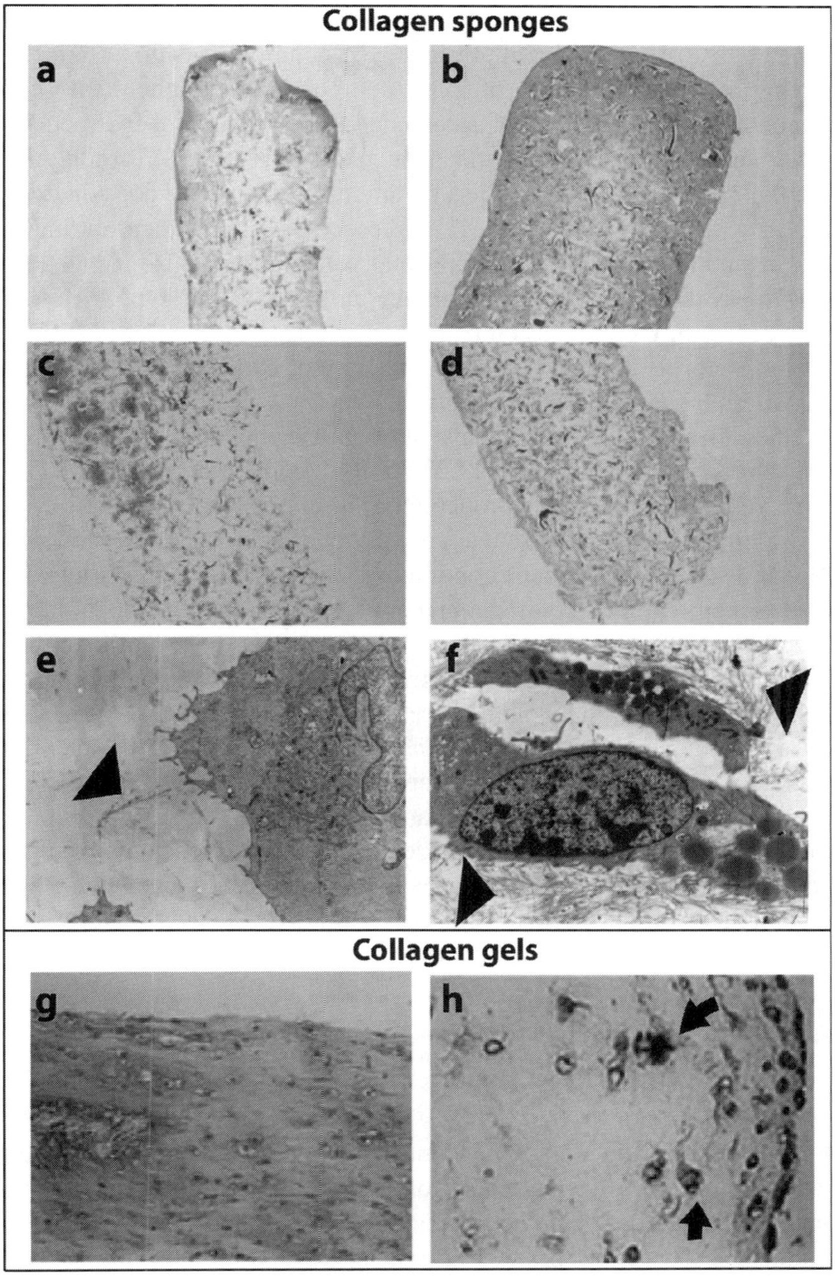

Fig. 2. Analysis of the neomatrix deposited by bovine chondrocytes grown in collagen-based biomaterials: collagen sponges (a–f) and collagen gels (g and h). Glycosaminoglycan staining and scanning electron micrographs of collagen sponges initially seeded with one million foetal bovine chondrocytes and grown for three months in serum-supplemented medium 10 % FCS (a, b and f) or serum-free medium supplemented with 1% ITS and cysteine (c, d and e). When static conditions (a and c) were applied the glycosaminoglycan deposits were at the border of the sponge or in clusters whereas stirred conditions favoured homogenous deposition of glycosaminoglycans (b and d). Collagen fibrils (arrowheads), abundant in FCS conditions, are fewer in ITS conditions. Glycosaminoglycan staining (g) and immunohistology of collagen type II (h) of a collagen gel seeded with two million veal chondrocytes and grown for 2 weeks in 10% FCS supplemented media. Deposits of glycosaminoglycan were present at the periphery of the gels and collagen type II had mostly a pericellular distribution. Bar: a, b, c, d: 640 μm; e: 4.3 μm; f: 2.7 μm; g: 160 μm and h: 40 μm.

deposition of ECM as the biomechanical environment is a key factor in the modulation of chondrocyte proliferation and metabolism [8,31,57]. Culture in stirred conditions using an orbital shaker (30 rpm) favoured a more homogenous distribution of ECM components throughout the scaffold (Fig. 2b–d) compared to static conditions (Fig. 2a–c) and accelerated the deposition of ECM components. The substitution of foetal calf serum (FCS; 10%) by media supplement (ITS; 1%) for long term cultures led to a neocartilage with the same cell number, the same amount of aggrecan deposited (Fig. 2c,d) but less collagen deposited than with FCS (Fig. 2e,f). The presence of ascorbic acid during the culture stimulated cell proliferation but did not increase cartilage matrix synthesis. It also induced a partial loss of the chondrocyte phenotype as the cells produced high amounts of collagen type I as well as a maturation of the chondrocytes based on the expression of type X collagen mRNA and MMP-13 [42]. Thus the presence or absence of ascorbic acid in the culture medium of cartilage construct has to be considered in the light of its effect on the chondrocyte phenotype.

Experimental studies using collagen gels have shown that chondrocytes maintain their phenotype when cultured in such 3D biomaterials [20,22,23,25,58]. A glycosaminoglycan-rich deposition was formed on the surface of cultured cartilage probably because more nutrition and oxygen were provided from the medium, which both help chondrocyte growth of enhanced matrix production [59]. Synthesis of collagen type II was also observed in such conditions. When maintained in culture in these conditions for 2–3 weeks, the collagen scaffold acquired a higher elastic resistance without alterations in chondrogenic phenotype. Our own *in vitro* study suggests that collagen gels support primary bovine chondrocyte proliferation with accumulation of an extracellular matrix (Fig. 2g,h), as well as maintenance of a round morphology and phenotype (assessed by gene expression, unpublished data) during 2-week culture periods. By allowing the cartilage matrix to accumulate while culturing the collagen–chondrocyte composite, partial protection against mechanical pressure or metabolic changes is provided. This hardened gel–chondrocyte composite is easy to manipulate and shape and could serve as an appropriate vehicle for chondrogenic cell-mediated repair of cartilage defects. The first clinical use of collagen gels for the treatment of full-thickness defects of cartilage in 28 knees has recently been published with promising results [36]. Transplantation eliminated locking of the knee and reduced pain and swelling in all patients. Chondrocytes in a newly formed collagen gel matrix can promote restoration of the articular cartilage of the knee.

5. Conclusion

The goal of cartilage engineering is to achieve mechanical properties identical to those of the native tissue. A simple and minimal invasive procedure to collect cells from the patient is mandatory. One of the most important problems to be solved is the fixation of the construct to the subchondral bone in the joint. Composite biomaterials seem to provide promising ways to organise the natural response and to facilitate the integration of the neocartilage with the adjacent tissue. Further studies are mandatory to optimise culture methodologies to ensure the successful function of engineered tissue.

Acknowledgements

This work was funded by a “Programme Thématique Prioritaire” (085113) from the Rhône-Alpes region, a CNRS-INSERM grant “Ingénierie Tissulaire” (2000-03), and one grant from the Fondation de l’Avenir pour la recherche appliquée (Paris, ET2-320).

References

[1] P. Angele, R. Kujat, M. Nerlich, J. Yoo, V. Goldberg and B. Johnstone, Engineering of osteochondral tissue with bone marrow mesenchymal progenitor cells in a derivatized hyaluronan-gelatin composite sponge, *Tissue Eng.* **5** (1999), 545–554.

[2] P.P. Aubin, H.K. Cheah, A.M. Davis and A.E. Gross, Long-term follow-up of fresh femoral osteochondral allografts for post-traumatic knee defects, *Clin. Orthop.* **391** (2001), S318–327.

[3] V. Bobic, Autologous osteo-chondral grafts in the management of articular cartilage lesions, *Orthopadedics* **28** (1999), 19–25.

[4] M. Brittberg, A. Lindahl, A. Nilsson, C. Ohlsson, O. Isaksson and L. Peterson, Treatment of deep cartilage defects in the knee with autologous chondrocyte transplantation, *N. Engl. J. Med.* **331** (1994), 889–895.

[5] J.A. Buckwalter and S. Lohmander, Operative treatment of osteoarthrosis. Current practice and future development, *J. Bone Joint Surg. [Am.]* **76** (1994), 1405–1418.

[6] W.D. Bugbee and F.R. Convery, Osteochondral allograft transplantation, *Clin. Sports Med.* **18** (1999), 67–75.

[7] S.K. Bulstra, G.N. Homminga, W.A. Buurman, E. Terwindt-Rouwenhorst and A.J. van der Linden, The potential of adult human perichondrium to form hyaline cartilage in vitro, *J. Orthop. Res.* **8** (1990), 328–335.

[8] S.E. Carver and C.A. Heath, Semi-continuous perfusion system for delivering intermittent physiological pressure to regenerating cartilage, *Tissue Engin.* **5** (1999), 1–11.

[9] F.R. Convery, M.H. Meyers and W.H. Akeson, Fresh osteochondral allografting of the femoral condyle, *Clin. Orthop.* **273** (1991), 139–145.

[10] R.D. Coutts, S.L. Woo, D. Amiel, H.P. von Schroeder and M.K. Kwan, Rib perichondrial autografts in full-thickness articular cartilage defects in rabbits, *Clin. Orthop.* **275** (1992), 263–273.

[11] Y. Dausse, L. Grossin, G. Miralles, S. Pelletier, D. Mainard, P. Hubert, D. Baptiste, P. Gillet, E. Dellacherie, P. Netter and E. Payan, Cartilage repair using new polysaccharidic biomaterials: macroscopic, histological and biochemical approaches in a rat model of cartilage defects, *Osteoarthritis and Cartilage* **11** (2003), 16–28.

[12] R.P. Ficat, C. Ficat, P. Gedeon and J.B. Toussaint, Spongialization: a new treatment for diseased patellae, *Clin. Orthop.* **144** (1979), 74–83.

[13] B. Grigolo, G. Lisignoli, A. Piacentini, M. Fiorini, P. Gobbi, G. Mazzotti, M. Duca, A. Pavesio and A. Facchini, Evidence for redifferentiation of human chondrocytes grown on a hyaluronan-based biomaterial (HYAff 11): molecular, immunohistochemical and ultrastructural analysis, *Biomaterials* **23** (2002), 1187–1195.

[14] A.E. Gross, P. Aubin, H.K. Cheah, A.M. Davis and M.T. Ghazavi, A fresh osteochondral allograft alternative, *J. Arthroplasty* **17** (Suppl.) (2002), 50–53.

[15] L. Hangody, G. Kish, Z. Karpati, I. Udvarhelyi, I. Szigeti and M. Bely, Mosaicplasty for the treatment of articular cartilage defects: application in clinical practice, *Orthopedics* **21** (1998), 751–756.

[16] G.N. Homminga, S.K. Bulstra, P.S. Bouwmeester and A.J. van der Linden, Perichondral grafting for cartilage lesions of the knee, *J. Bone Joint Surg. [Br.]* **72** (1990), 1003–1007.

[17] W. Hunter, On the structure and disease of articular cartilage, *Philos. Trans. R. Soc. London Biol.* **42** (1743), 514–521.

[18] J. Insall, The Pridie debridement operation for osteoarthritis of the knee, *Clin. Orthop.* **101** (1974), 61–67.

[19] R.P. Jakob, T. Franz, E. Gautier and P. Mainil-Varlet, Autologous osteochondral grafting in the knee: indication, results, and reflections, *Clin. Orthop.* **401** (2002), 170–184.

[20] K. Katsube, M. Ochi, Y. Uchio, S. Maniwa, M. Matsusaki, M. Tobita and J. Iwasa, Repair of articular cartilage defects with cultured chondrocytes in Atelocollagen gel. Comparison with cultured chondrocytes in suspension, *Arch. Orthop. Trauma. Surg.* **120** (2000), 121–127.

[21] H.K. Kim, M.E. Moran and R.B. Salter, The potential for regeneration of articular cartilage in defects created by chondral shaving and subchondral abrasion. An experimental investigation in rabbits, *J. Bone Joint Surg. [Am.]* **73** (1991), 1301–1315.

[22] T. Kimura, N. Yasui, S. Ohsawa and K. Ono, Chondrocytes embedded in collagen gels maintain cartilage phenotype during long-term cultures, *Clin. Orthop.* **186** (1987) 231–239.

[23] C.R. Lee, H.A. Breinan, S. Nehrer and M. Spector, Articular cartilage chondrocytes in type I and type II collagen-GAG matrices exibit contractile behaviour in vitro, *Tissue Eng.* **6** (2000), 555–565.

[24] E. Lexer, Substitution of whole or half joints from freshly amputated extremities by freeplastic operation, *Surg. Gynecol. Obstetr.* **6** (1908), 601–605.

[25] M. Matsusaki, M. Ochi, Y. Uchio, N. Shu, H. Kurioka, K. Kawasaki and N. Adachi, Effects of basic fibroblast growth factor on proliferation and phenotype expression of chondrocytes embedded in collagen gel, *Gen. Pharmacol.* **31** (1998), 759–764.

[26] Y. Matsusue, T. Kotake, Y. Nakagawa and T. Nakamura, Arthroscopic osteochondral autograft transplantation for chondral lesion of the tibial plateau of the knee, *Arthroscopy* **17** (2001), 653–659.

[27] D.S. Menche, S.R. Frenkel, B. Blair, N.F. Watnik, B.C. Toolan, R.S. Yaghoubian and M.I. Pitman, A comparison of abra-

sion burr arthroplasty and subchondral drilling in the treatment of full-thickness cartilage lesions in the rabbit, *Arthroscopy* **12** (1996), 280–286.
[28] K. Messner and W. Maletius, The long-term prognosis for severe damage to weight-bearing cartilage in the knee – a 14-year clinical and radiographic follow-up in 28 young athletes, *Acta Orthop. Scand.* **67** (1996), 165–168.
[29] G. Miralles, R. Baudoin, D. Dumas, D. Baptiste, P. Hubert, J.F. Stoltz, E. Dellacherie, D. Mainard, P. Netter and E. Payan, Sodium alginate sponges loaded or not with sodium hyaluronate: In vitro engineering of cartilage, *J. Biomed Mater. Res.* **57** (2001), 268–278.
[30] Mitchell and N. Shepard, Effect of patellar shaving in the rabbit, *J. Orthop. Res.* **5** (1987), 388–392.
[31] S. Mizuno, T. Tateishi, T. Ushida and J. Glowacki, Hydrostatic fluid pressure enhances matrix synthesis and accumulation by bovine chondrocytes in three-dimensional culture, *J. Cell. Physiol.* **193** (2002), 319–327.
[32] L.W. Moreland, Intra-articular hyaluronan (hyaluronic acid) and hylans for the treatment of osteoarthritis: mechanisms of action, *Arthritis Res. Ther.* **5** (2003), 54–67.
[33] J.B. Moseley, K. O'Malley, N.J. Petersen, T.J. Menke, B.A. Brody, D.H. Kuykendall, J.C. Hollingsworth, C.M. Ashton and N.P. Wray, A controlled trial of arthroscopic surgery for osteoarthritis of the knee, *N. Engl. J. Med.* **347** (2002), 81–88.
[34] S.W. O'Driscoll, F.W. Keeley and R.B. Salter, The chondrogenic potential of free autogenous periosteal grafts for biological resurfacing of major full-thickness defects in joint surfaces under the influence of continuous passive motion. An experimental investigation in the rabbit, *J. Bone Joint Surg. [Am.]* **68** (1986), 1017–1035.
[35] M. Oberlander, C. Baker and B. Morgan, Patellofemoral arthrosis: the treatment options. *Am. J. Orthop.* **27** (1998), 263–270.
[36] M. Ochi, Y. Uchio, K. Kawasaki, S. Wakitani and J. Iwasa, Transplantation of cartilage-like tissue made by tissue engineering in the treatment of cartilage defects of the knee, *J. Bone Joint Surg.* **84** (2002), 571–578.
[37] D.J. Ogilvie-Harris and R.W. Jackson, The arthroscopic treatment of chondromalacia patellae, *J. Bone Joint Surg. [Br.]* **66** (1984), 660–665.
[38] M. Pedersen, A. Moghaddam, K. Bak and J. Koch, The effect of bone drilling on pain in gonarthrosis, *Int. Orthop.* **19** (1995), 12–15.
[39] L. Peterson, T. Minas, M. Brittberg, A. Nilsson, E. Sjogren-Jansson and A. Lindahl, Two- to 9-year outcome after autologous chondrocyte transplantation of the knee, *Clin. Orthop.* **374** (2000), 212–234.
[40] K.H. Pridie, A method of resurfacing osteoarthritic knee joint, *J. Bone Joint Surg. [Br.]* **41** (1959), 618.
[41] S. Roche, M.C. Ronzière, D. Herbage and A.M. Freyria, Native and DPPA cross-linked collagen sponges seeded with fetal bovine epiphyseal chondrocytes used for cartilage tissue engineering, *Biomaterials* **22** (2001), 9–18.
[42] M.C. Ronzière, S. Roche, J. Gouttenoire, O. Demarteau, D. Herbage and A.M. Freyria, Ascorbate modulation of bovine chondrocyte growth, matrix protein gene expression and synthesis in three-dimensional collagen sponges, *Biomaterials* **24**, (2003), 851–861.
[43] G. Rovetta, P. Monteforte, G. Molfetta and V. Balestra, Chondroitin sulfate in erosive osteoarthritis of the hands, *Int. J. Tissue React.* **24** (2002), 29–32.
[44] J.M. Rubak, M. Poussa and V. Ritsila, Chondrogenesis in repair of articular cartilage defects by free periosteal grafts in rabbits, *Acta Orthop. Scand.* **53** (1982), 181–186.
[45] J.M. Rubak, Reconstruction of articular cartilage defects with free periosteal grafts. An experimental study, *Acta Orthop. Scand.* **53** (1982), 175–180.
[46] R.E. Samuel, C.R. Lee, S.C. Ghivizzani, C.H. Evans, I.V. Yannas, B.R. Olsen and M. Spector, Delivery of plasmid DNA to articular chondrocytes via novel collagen-glycosaminoglycan matrices, *Hum. Gene Ther.* **13** (2002), 791–802.
[47] H. Schmidt, K. Schulze and R. Cyffka, Results of treatment of cartilage damage by Pridie drilling of the knee joint, *Beitr. Orthop. Traumatol.* **35** (1988), 117–122.
[48] H. Schmidt and E. Hasse, Arthroscopic surgical treatment of circumscribed cartilage damage with spongiolization or Pridie drilling, *Beitr. Orthop. Traumatol.* **36** (1989), 35–37.
[49] V.F. Sechriest, Y.J. Miao, C. Niyibizi, A. Westerhausen-Larson, H.W. Matthew, C.H. Evans, F.H. Fu and J.K. Suh, GAG-augmented polysaccharide hydrogel: a novel biocompatible and biodegradable material to support chondrogenesis, *J. Biomed. Mater. Res.* **49** (2000), 534–541.
[50] T. Skoog and S.H. Johansson, The formation of articular cartilage from free perichondrial grafts, *Plast. Reconstr. Surg.* **57** (1976), 1–6.
[51] L.A. Sochalga, J. Gao, J.E. Dennis, A. Awadallah, M. Lundberg, A.I. Caplan and V.M. Goldberg, Treatment of osteochondral defects with autologous bone marrow in a hyaluronan-based delivery vehicle, *Tissue Eng.* **8** (2002), 333–347.
[52] J.R. Steadman, W.G. Rodkey and J.J. Rodrigo, Microfracture: surgical technique and rehabilitation to treat chondral defects, *Clin. Orthop.* **391** (Suppl.) (2001), 362–369.
[53] J.R. Steadman, W.G. Rodkey and K.K. Briggs, Microfracture to treat full-thickness chondral defects: surgical technique, rehabilitation, and outcomes, *J. Knee Surg.* **15** (2002), 170–176.
[54] J.S. Temenoff and A.G. Mikos, Review: tissue engineering for regeneration of articular cartilage, *Biomaterials* **21** (2000), 431–440.

[55] D. Uebelhart, E.J. Thonar, P.D. Delmas, A. Chantraine and E. Vignon, Effects of oral chondroitin sulphate on the progression of knee osteoarthritis: a pilot study, *Osteoarthritis Cartilage* **6** (Suppl. A) (1998), 39–46.
[56] A.M. Vachon, C.W. McIlwraith, G.W. Trotter, R.W. Norrdin and B.E. Powers, Morphologic study of repair of induced osteochondral defects of the distal portion of the radial carpal bone in horses by use of glued periosteal autografts, *Am. J. Vet. Res.* **52** (1991), 317–327.
[57] G. Vunjak-Novakovik, I. Martin, B. Obradovic, S. Treppo, A.J. Grodzinsky, R. Langer and L.E. Freed, Bioreactor cultivation conditions modulate the composition and mechanical properties of tissue-engineered cartilage, *J. Orthop. Res.* **17** (1999), 130–139.
[58] S. Wakitani, T. Goto, R.G. Young, J.M. Mansour, V.M. Goldberg and A. Caplan, Repair of large full-thickness articular cartilage defects with allograft articular chondrocytes embedded in a collagen gel, *Tissue Eng.* **4** (1998), 429–444.
[59] T. Yamamoto, M. Katoh, R. Fukushima, T. Kurushima and M. Ochi, Effect of glycosaminoglycan production on hardness of cultured cartilage fabricated by the collagen-gel embedding method, *Tissue Eng.* **8** (2002), 119–129.

Biorheology 41 (2004) 445–458
IOS Press

Long-term culture of tissue engineered cartilage in a perfused chamber with mechanical stimulation

J.O. Seidel, M. Pei, M.L. Gray, R. Langer, L.E. Freed and G. Vunjak-Novakovic *
Harvard-MIT Division of Health Sciences and Technology, MIT, E25-330, 45 Carleton Street, Cambridge, MA 02139, USA

Abstract. One approach to functional tissue engineering of cartilage is to utilize bioreactors to provide environmental conditions that stimulate chondrogenesis in cells cultured on biomaterial scaffolds. We report the combined use of a three-dimensional *in vitro* model and a novel bioreactor with perfusion of culture medium and mechanical stimulation in long-term studies of cartilage development and function. To engineer cartilage, scaffolds made of a non-woven mesh of polyglycolic acid (PGA) were seeded with bovine calf articular chondrocytes, cultured for an initial 30-day period under free swelling conditions, and cultured for an additional 37 day period in one of the three groups: (1) free-swelling, (2) static compression (on 24 h/day, strain control, static offset 10%), and (3) dynamic compression (on 1 h/day; off 23 h/day; strain control, static offset 2%, dynamic strain amplitude 5%; frequency 0.3 Hz). Constructs were sampled at timed intervals and assessed with respect to structure, biochemical composition, and mechanical function. Mechanical simulation had little effect on the compositions, morphologies and on mechanical properties of construct interiors discs, but it resulted in distincly different properties of the peripheral rings and face sides. Contructs cultured with mechanical loading maintained their cylindrical shape with flat and parallel top and bottom surfaces, and retained larger amounts of GAG. The modular bioreactor system with medium perfusion and mechanical loading can be utilized to define the conditions of cultivation for functional tissue engineering of cartilage.

Keywords: Chondrogenesis, bioreactor, loading, fluid flow, glycosaminoglycans, collagen

1. Introduction

An evolving discipline called *functional tissue engineering* is focused on the role of biomechanical factors in tissue regeneration, with the overall goal of developing a series of rational design principles to guide tissue development [22]. To reach this goal, the influence of mechanical forces on biological structure and function need to be understood [3,39]. This is particularly important in guiding tissue growth for orthopaedic applications, such as cartilage repair. Cells, biomaterial scaffolds, and biochemical and physical regulatory signals can be utilized in a variety of ways to engineer cartilage, *in vitro* and *in vivo* (e.g. [4,7,8,11,20,22,27,32,34,36,37,45]). In most cases, tissue engineering attempts to recapitulate aspects of the environment present *in vivo* during tissue development.

It has been hypothesized that mechanical forces may provide a critical stimulus in modulating the structure and function of engineered cartilage [8,21,29,34,35]. Numerous physical factors have been

*Address for correspondence: Gordana Vunjak-Novakovic, PhD, Harvard-MIT Division of Health Sciences and Technology, Massachusetts Institute of Technology, MIT E25-330, 45 Carleton Street, Cambridge, Massachusetts 02139, USA. Tel.: +1 617 452 2593; E-mail: gordana@mit.edu.

utilized to modulate tissue engineered cartilage, including fluid flow [36,47], dynamic fluctuations in hydrodynamic shear and pressure [18,45], cyclic hydrostatic pressure [7], cyclic mechanical compression [5,34] and cyclic stretch [48]. *In vivo*, cyclic loading induced chondrogenesis in mesenchymal stem cells [40]. *In vitro*, dynamic but not static compression enhanced synthesis of proteoglycans in cartilage explants [38] and engineered cartilage [5], improved the mechanical function of engineered cartilage [7,28,34] and enhanced chondrogenesis of chick limb bud cells [11,12]. Certain regimes of mechanical stimulation significantly improved the structure and function of engineered cartilage [34], whereas other regimes resulted in denaturation of collagen and degradation of cartilage [42]. The response of cartilage constructs to mechanical stimulation was more pronounced if the constructs were comprised of more mature extracellular matrix (ECM) [5], supporting the notion that cell–matrix interactions and the ECM itself play an important role in the response of chondrocytes to static and dynamic compression. Taken together, these previous studies establish the feasibility of using mechanical stimulation to modulate the biosynthetic responses of chondrocytes in native and engineered cartilage. The current research needs involve long-term cultivation of engineered cartilage with the application of mechanical stimuli, and the assessment of the effects of biophysical factors on tissue structure and mechanical behavior.

We describe a novel bioreactor system with medium perfusion (for enhanced mass transfer of nutrients and metabolites) and mechanical stimulation (unconfined; strain-controlled; static or dynamic; continuous or periodic) of engineered cartilage that enables long term studies under controlled conditions of loading. The strain (frequency and amplitude) and the regime (periods of loading and rest) of mechanical stimulation were selected based on previous studies [5,21,26,34]. We also report the biochemical compositions, histomorphologies and mechanical properties of the resulting cartilaginous constructs cultured for up to 3 months under free-swelling conditions, and with static or dynamic mechanical stimulation.

2. Perfused bioreactor with mechanical stimulation

A perfused bioreactor was designed to allow long-term cultivation of engineered cartilage constructs with mechanical stimulation (Fig. 1) and coupled to a mechanical spectrometer that was previously custom-designed to allow quantitative studies of the biological response of cartilage explants to mechanical deformation [2]. The original device allowed the application of displacement- or load-controlled mechanical loading to cartilage discs (2–4 mm in diameter $\times$ 0.8–2 mm thick) that were unconfined to promote tissue viability.

Two main modifications of the system, developed in the course of the present study were: *perfusion of culture medium* equilibrated with respect to the concentration of dissolved oxygen and carbon dioxide (to enhance exchange of nutrients and metabolites), and the ability to culture *multiple constructs* in a single chamber (five constructs per chamber, five chambers per stimulation regime). These modifications, described below in more detail, enabled long-term studies of the effects of mechanical stimulation on tissue engineered cartilage. In the new design, one set of five chambers could be dynamically stimulated and an additional set of five chambers could be held under static compression. Each chamber held up to five constructs and was independently perfused with culture medium. Free swelling constructs cultured in 6-well tissue culture plates served as controls.

2.1. *Perfused chamber design*

Compression chambers (34 mm in diameter, 25 mm high) were made from polysulfone and equipped with five shallow circular wells (7 mm in diameter $\times$ 0.5 mm deep) to accommodate the cultivation and

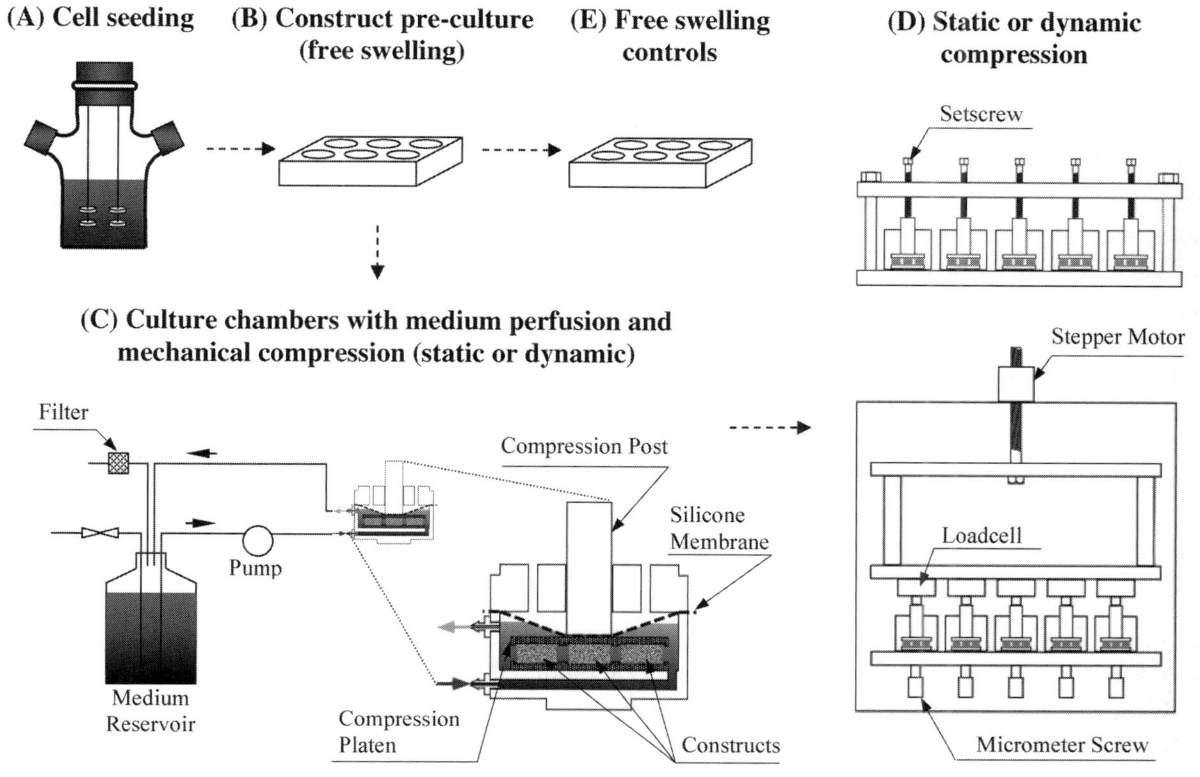

Fig. 1. Experimental system: (A) Cartilagenous constructs were prepared starting form bovine calf articular chondrocytes that were seeded onto fibrous polyglycolic acid scaffolds in well-mixed spinner flasks for 3 days. (B) Constructs were then cultured for 30 days under free-swelling conditions (in 35 mm dishes, at 50 rpm) to allow the formation of cartilaginous tissue matrix. (C, D) Pre-cultured constructs were transferred into the perfused chambers and cultured for up to 37 days with the application of either static or dynamic mechanical compression (groups 1 and 2, respectively). (E) Constructs cultured under free-swelling conditions served as controls (group 3). Constructs were evaluated structurally and functionally after 30, 40, 47 and 67 days of cultivation.

mechanical loading of up to five constructs (Fig. 1C). A compression platen (25 mm in diameter) was placed on top of the constructs and the chamber was covered with a silicone membrane that served as a seal between the chamber and the incubator environment while allowing free motion of the platen when a displacement was applied by the mechanical spectrometer. A compression post placed on the top of the membrane was used to transmit axial compression to the constructs.

Each stimulation chamber had its own recirculation loop that consisted of one channel of a multichanel peristaltic pump and a 60 ml medium reservoir. The chambers had a resident volume of about 4.5 ml and were perfused with tissue culture medium at a rate of 180 ml/h. The medium reservoir was fitted with a gas-permeable filter to allow gas exchange with incubator air (5% CO_2 in air) and the equilibration of medium entering the chambers with respect to concentrations of CO_2 and O_2. Medium could replaced through ports on the medium reservoir without stopping mechanical stimulation of the constructs. The medium in the reservoir was replaced at a rate of 50 ml every 3–4 days.

2.2. Mechanical stimulation

Static axial compression was imposed on constructs through setscrews mounted over the 5 chambers, such that the samples were constrained to the desired compressive strain (Fig. 1D). Dynamic compression was imposed on constructs via a computer-controlled stepper motor (Parker, Compumotor, Rohnert Park, CA). Loads for individual chambers were measured by load cells (50 N, Sensotec, Columbus, OH) mounted above the chambers in contact with the compression posts (Fig. 1D). Displacement was measured with a displacement sensor (Sensotec, Columbus, OH) and used in conjunction with initial sample thickness to determine actual strains. Loads for each load cell and displacement were recorded by a plotter (L2200, Linseis GmbH, Selb, Germany) and stored on disk. The chambers were clamped to a micrometer-adjustable base (L.S. Starrett Co., Athol, MA) that allowed for individual adjustment of the initial displacement, and the mounting and removing of single chambers. The system allowed the application of dynamic compression in the range of 0.003 to 1.0 Hz, either continuously or with periods of loading and rest. Frequency of stimulation was regulated by the computer-controlled stepper motor and verified by analyzing the recorded load and displacement readings.

3. Construct assembly and stimulation

3.1. Cells

Chondrocytes were isolated from the femoropatellar grooves of 2–3 week old bovine calves within 8 h of slaughter. Cartilage was digested using 0.15% (w/v) type II collagenase (300 U/mg, Worthington, Freehold, NJ) in Dubecco's Modified Eagle Medium (DMEM) containing 10% Fetal Bovine Serum (FBS) for 16 h at 37°C [16]. Chondrocytes were resuspended in basal culture medium (DMEM containing 4.5 g/l glucose and supplemented with 10% FBS, 10 mM HEPES, 0.1 mM nonessential amino acids, 0.4 mM proline, 50 mg/ml L-ascorbic acid, 100 U/ml penicillin, 100 μg/ml streptomycin) [4,38].

3.2. Cell–polymer constructs

Polyglycolic acid (PGA) scaffolds (5-mm in diameter × 2-mm thick discs, 97% void volume, Albany International, Mansfield, MA) were seeded with freshly isolated chondrocytes in well-mixed spinner flasks as previously described [44]. Scaffolds were sterilized with ethylene oxide, aerated, wetted in 70% ethanol, rinsed in PBS and transferred to culture medium. Pre-wetted scaffolds were then threaded onto needles suspended from the stopper of a spinner flask (Bellco, Vineland, NJ) (3 scaffolds per needle, 4 needles per flask). Flasks were filled with 120 ml of culture medium, mixed using a nonsuspended magnetic stir bar at 50 rpm and placed in a humidified (37°C, 5% CO_2) incubator for 24 h. Flasks were then inoculated with 60 million freshly isolated chondrocytes (5×10^6 cells per scaffold) and magnetically stirred (50 rpm) for 3 days. After three days constructs were transferred into untreated 6-well tissue culture plates (1 scaffold per well in 7 ml culture medium) and cultured on an orbital shaker (50 rpm) in humidified 5% CO_2 incubator for 30 days to allow for development of a functional extracellular matrix. Culture media were replaced every 3–4 days.

3.3. Mechanical stimulation

After 30 days of free-swelling culture, constructs were divided into three experimental groups. In *group 1*, constructs continued being cultured under free-swelling conditions in orbitally mixed 6-well plates as described above for up to an additional 37 days. In *group 2*, constructs were transferred into tissue compression chambers (5 chambers, with 5 constructs each) and subjected to unconfined static compression at 10% compressive strain for up to 37 days. In *group 3*, constructs were transferred into tissue compression chambers (5 chambers, with 5 constructs each) and subjected to unconfined dynamic compression of 5% dynamic strain amplitude superimposed onto a static offset of 2%, resulting in a maximum strain of 7%; the stimulation frequency was 0.3 Hz; stimulation was applied for 1 h followed by 23 h of rest during which the construct was unloaded but in contact with the compression platen, for up to 37 days.

3.4. Sampling

Constructs for biochemical and histological analyses were sampled after a total culture period of 30, 40, 47 or 67 days, corresponding to stimulation periods of 0, 10, 17 or 37 days. Samples for biomechanical analysis were taken at 0, 17 or 37 days of stimulation, corresponding to 30, 47 or 67 days of total culture period.

4. Structural and functional analyses

4.1. Biochemical analysis

Samples were weighed, frozen, lyophilized, and digested with 0.125 mg/ml papain in 100 mM phosphate buffer (PBS) with 10 mM EDTA and 10 mM cysteine for 18 h at 60°C [16,38], using 1 ml enzyme solution per 5–10 mg dry weight of the sample. The number of chondrocytes was determined from the measured amount of DNA, using Hoechst 33258 dye and a spectrofluorometer (Aminco, Urbana, IL) [25], based on the known DNA content per cell (7.7 pg/cell; [25]) and the mass of a single cell (10 ng/cell; [17]). Glycosaminoglycans (GAG) content was determined spectrophotometrically after reaction with dimethylmethylene blue dye, using bovine chondroitin sulfate in PBS as a standard [13]. Total collagen content was determined from the hydroxyproline content after acid hydrolysis (6 N HCl at 115°C for 18 h) and reaction with p-dimethylaminobenzaldehyde and chloramine-T [46] using a hydroxyproline to collagen ratio of 1 : 10 [23]. Biochemical analyses were done using $n = 3$ samples per time point.

4.2. Mechanical analysis

Compressive equilibrium modulus was determined in confined uniaxial compression as previously described [14,45]. Samples were transferred to 0.15 M PBS supplemented with proteinase inhibitors (1 mM phenylmethanesulfonyl fluoride, 2 mM disodium ethylenediamine tetraacetate, 5 mM benzamidine-HCl, 10 mM N-ethylmaleimide) and equilibrated for 30 min. Disks 3.25 mm in diameter and 1.5–2 mm thick were harvested from middle region of the 5 mm diameter × 3 mm thick engineered constructs. The cored disk was equilibrated for 10–15 min at room temperature in 0.15 M PBS with proteinase inhibitors and transferred into a cylindrical confining chamber [14]. The chamber was mounted in a servo-controlled Dynastat mechanical spectrometer (IMASS, Hingham, MA) and the specimen was compressed using a porous polyethylene platen. Five sequential increments of 2% strain were applied after an initial tare displacement of 2% up to a maximum of 15% strain, i.e. in the range where equilibrium stress varied linearly with applied strain. After stress relaxation, the equilibrium stress was measured and plotted against applied strain; the equilibrium modulus was determined from the slope of the best linear fit. After mechanical testing, the cored center and the peripheral ring were frozen and analyzed with respect to biochemical compositions. Mechanical analyses were done using $n = 5$ samples per time point.

4.3. Histological analysis

Samples for histological evaluation were fixed for 24 h at 4°C in 10% neutral buffered formalin, embedded in paraffin, sectioned to 5 μm (cross sections), and stained with safranin-O/fast green for GAG according to standard protocols [31]. Immunohistochemical staining was performed as previously described [37]. In brief, sections were deparaffinized, hydrated, incubated for 30 min at room temperature with 2 mg/ml testicular hyaluronidase in PBS (pH 5.0) [1], rinsed with PBS, and incubated for 30 min at room temperature with normal goat serum diluted at $1:10$ in PBS and for 1 h at room temperature with the appropriate primary antibody. Primary antibodies were monoclonal antibodies against bovine collagen type I (Sigma Chemical Co., St Louis, MO) and type II (NewMaker, Fremont, CA). Secondary antibody and staining was done using the Vectastain ABC Kit (Vector, Burlingame, CA) and hematoxylin counter-stain. Histological analyses were done using $n = 2$ samples per time point.

4.4. Image analysis

Images for histological evaluation were recorded using a solid-state 3 CCID camera (Hitachi HV-C20) mounted on a Nikon Diaphot inverted microscope. Red, green and blue (RGB) image planes were digitized using a LG-3 frame grabber (Scion Corporation, Frederick, MID). GAG distribution was determined from safranin-O stained sections using the method of Martin et al. [33] with some modifications. Images were processed using IPLab Spectrum (Macintosh version, Sigma Analytics Corporation, Vienna, VA). Histology images were separated into a horizontal and a vertical section, by selecting a horizontal or vertical band running through the center of the construct and covering 50% of the area of the section. Each band then was eroded step-wise from the surface of the construct inwards, in the horizontal or in the vertical direction. The horizontal and vertical distributions of GAG were then determined by calculating the red content R_c in each eroded zone [33]. To compare GAG distributions the plots were fitted by broken-line linear regression with one break point (Eq. (1)):

$$f(x,\overline{p}) = p_1 + p_2 x - p_2(x - p_3)\sigma, \tag{1}$$

where x was the depth into the tissue, p_3 was the distance of the breakpoint from the construct surface, and σ was defined as:

$$\sigma = \begin{cases} 1, & x > p_3, \\ 0, & x \leqslant p_3. \end{cases} \tag{2}$$

4.5. Statistics

Statistical analyses included t-test and ANOVA in conjunction with Tukey's post hoc test using Prism 4 for Macintosh (GraphPad Software, Inc.).

5. Results

5.1. Biochemical and mechanical properties

DNA (% wet weight) was generally higher in bioreactor cultured constructs exposed to either static or dynamic stimulation, than in free swelling controls; with time in culture, the differences between

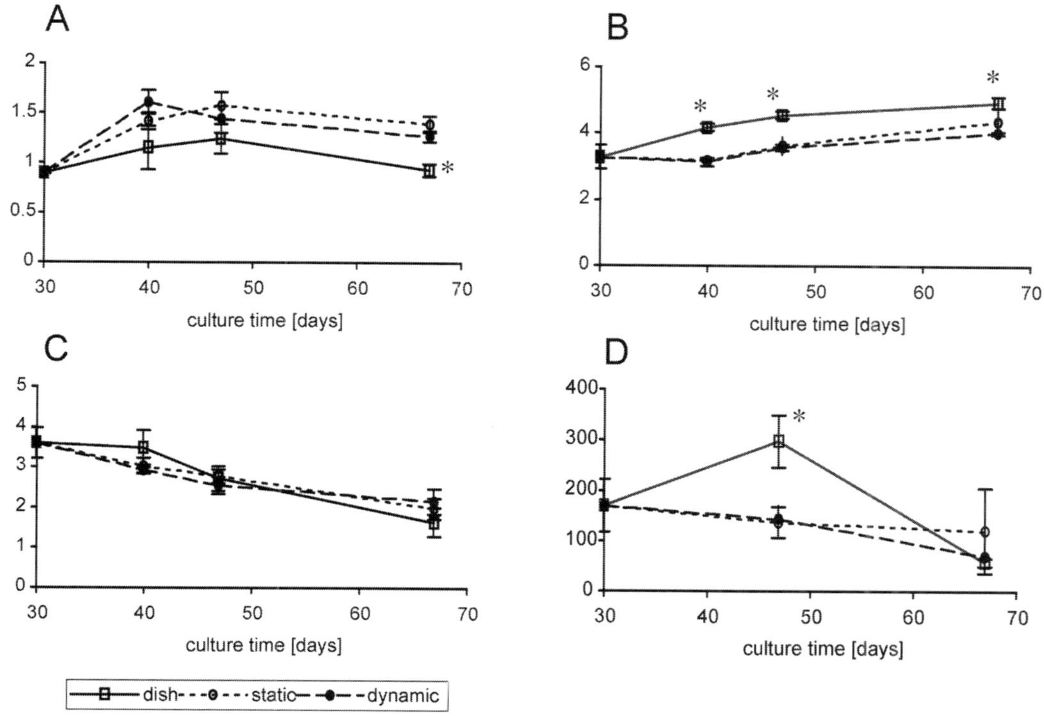

Fig. 2. Biochemical and mechanical properties of engineered constructs as a function of experimental group and time of culture. (A) DNA (% of wet weight, ww), (B) Total collagen (%ww), (C) GAG (%ww), (D) compressive modulus (kPa). Constructs cultured free-swelling, in static compression, and in dynamic compression are shown by solid line, dotted line and dashed lines, respectively. Data represents average ± standard deviation of $n = 5$. (*) indicates significant differences between the free-swelling constructs and constructs in either static or dynamic compression ($p < 0.05$).

groups became statistically significant (Fig. 2A). Collagen (% wet weight) increased and GAG decreased with time of culture (Figs 2B and 2C, respectively). Free swelling constructs contained slightly but significantly more collagen than bioreactor-grown constructs. GAG fractions were comparable for the three groups, at all time points. Aggregate modulus of constructs cultured in bioreactors decreased with time in culture, in parallel with the decrease in GAG fractions. Free swelling constructs showed an increase in modulus at 47 days that did not correlate with GAG concentration (Fig. 2D).

5.2. Histological evaluation

Representative low magnification sections of constructs after 67 days of culture stained for GAG, type I and type II collagen are shown in Fig. 3. GAG and collagen II had a pattern of co-localization previously established for tissue engineered cartilage [19]. Collagen I was present in areas devoid of GAG and collagen II. All three groups showed a decrease in the intensity of safranin-O staining between day 30 and day 67 of culture, indicating a loss in GAG (data not shown). The peripheral regions appeared to be devoid of GAG, whereas interior regions appeared to retain GAG, as indicated by safranin-O staining. GAG distributions in constructs cultured under compression and free swelling conditions were distinctly different. Under free swelling conditions, the region of lower GAG concentration extended over the entire surface of the construct. In constructs cultured in compression (static or dynamic), the region of lower GAG concentration was present only at the lateral sides but not on face surfaces that were in contact with the compression platens. Also, constructs cultured under compression maintained

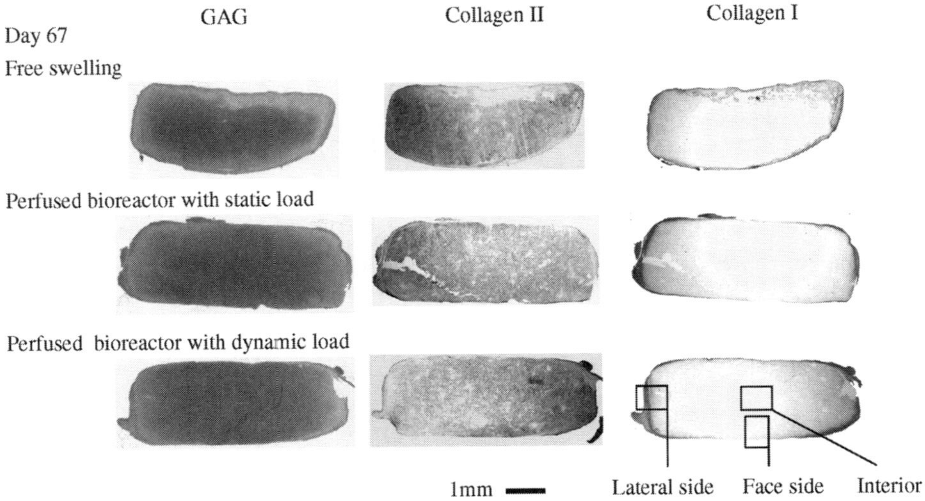

Fig. 3. Construct histomotrphologies. Full cross-sections of constructs cultured for a total of 67 days under free-swelling conditions, with static or dynamic compression, are stained for GAG, type II collagen and type I collagen. Scale bar: 1 mm. Boxes indicate the areas of interest shown at a higher magnification in Fig. 4.

their cylindrical shape with flat, parallel face sides, whereas the free swelling constructs became more irregular.

Representative high magnification sections of constructs after 67 days of culture stained for GAG, type I and type II collagen are shown in Fig. 4. Free swelling constructs showed a pattern of reduced safranin-O staining at the surfaces in contact with culture medium (indicative of GAG leaching that decreased GAG concentration) that was similar on the face and lateral sides (Figs 4A1 and 4A4, respectively). The presence of layers of flattened cells at the surfaces of constructs that stained for collagen type I (Figs 4A3, 4A6) was indicative of dedifferentiation of chondrocytes into a more fibroblastic phenotype. In constructs from perfused bioreactors that were exposed to either static or dynamic stimulation, the lateral sides also contained flattened cells and little GAG, and stained for collagen type I (Figs 4B4–6, 4C4–6). However, the face sides of bioreactor grown constructs contained round cells and stained strongly for GAG and type II collagen and only weakly for type I collagen (Figs 4B1–3, 4C1–3). In all groups, the construct interiors contained round, well differentiated cells embedded in ECM rich in GAG (Figs 4A7, 4B7, 4C7), that was co-localized with type II collagen (data not shown).

5.3. Distributions of ECM components

To quantitate the compositional differences observed histologically after 67 days in culture, two construct regions: the central disc (3.25 mm in diameter $\times$ 1.5–2 mm thick) and the outer ring (7 mm/3.25 mm $\times$ 3 mm) obtained after coring out the central disc were analyzed biochemically for the amounts of DNA, GAG and total collagen (Fig. 5). The outer ring had higher wet weight fractions of DNA (Fig. 5A), and total collagen (Fig. 5C), but lower GAG (Fig. 5B) and GAG/DNA ratio than the central "interior" disc. The compositions of central discs were significantly different from the compositions of the outer rings with respect to all biochemical parameters: DNA, GAG, collagen, and GAG/DNA ratio in all construct groups.

Construct GAG distributions were evaluated using an algorithm for measuring GAG concentrations in tissue sections by image processing (Fig. 6). The gradient in GAG concentration depended on the

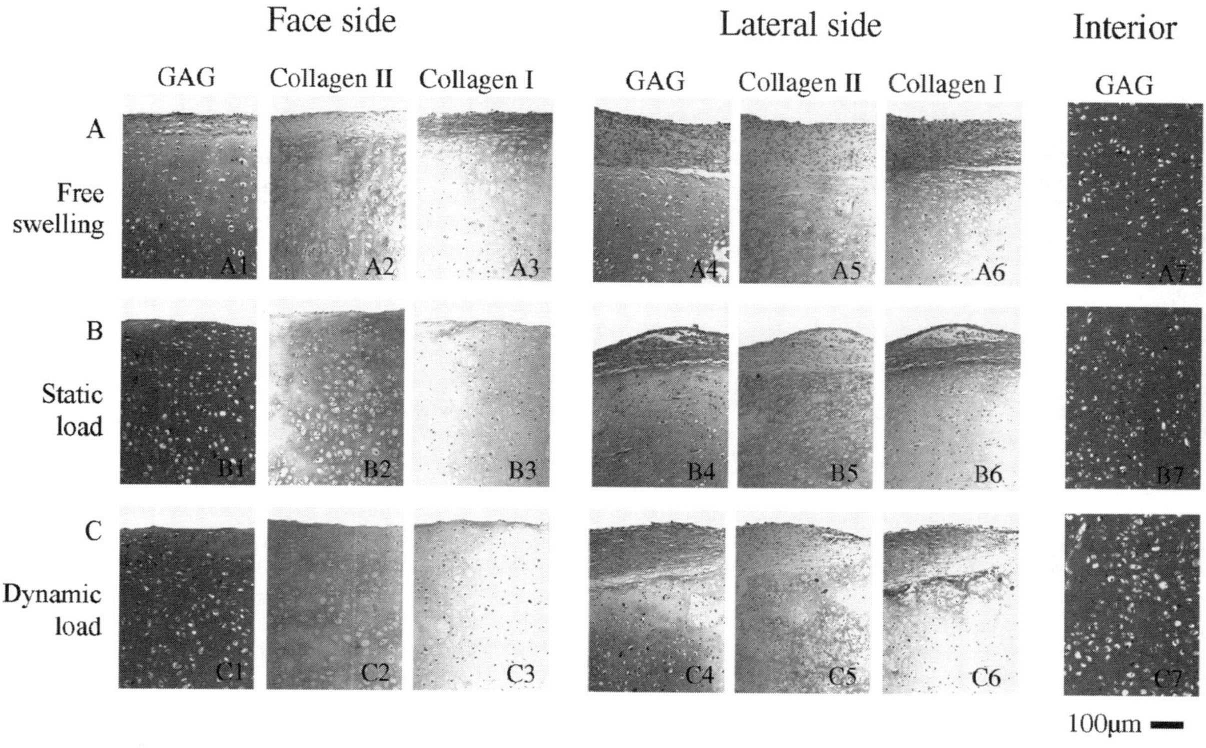

Fig. 4. GAG distributions. High magnification images of the face side region, lateral side region and construct interior after 67 days of culture are shown as indicated in Fig. 3. (A) Free swelling constructs, (B) constructs cultured with static compression, (C) constructs cultured with dynamic compression. Representative histological sections were stained for GAG (A1, B1, C1, A4, B4, C4, A7, B7, C7), type II collagen (A2, B2, C2, A5, B5, C5) and type I collagen (A3, B3, C3, A6, B6, C6). In each group (A, B, or C), images 1–3 are for the face side, images 4–6 are for the lateral side, and image 7 is for the construct interior. Scale bar: 100 μm.

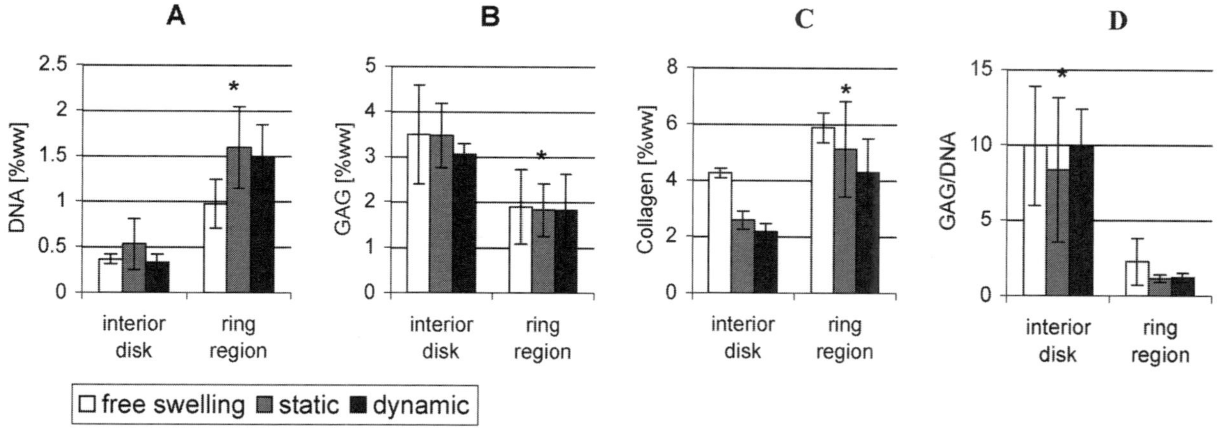

Fig. 5. Biochemical compositions of the construct interior disc and lateral ring regions. (A) DNA (%ww), (B) GAG (%ww), (C) total collagen (%ww), (D) GAG/DNA (μg/μg). Constructs cultured for 67 days under free-swelling conditions, in static and in dynamic compression are shown by white, gray and black bars, respectively. Data represent average $\pm$ SD of $n = 5$. (*) Indicates significant differences between the interior and the ring regions of the construct ($p < 0.05$).

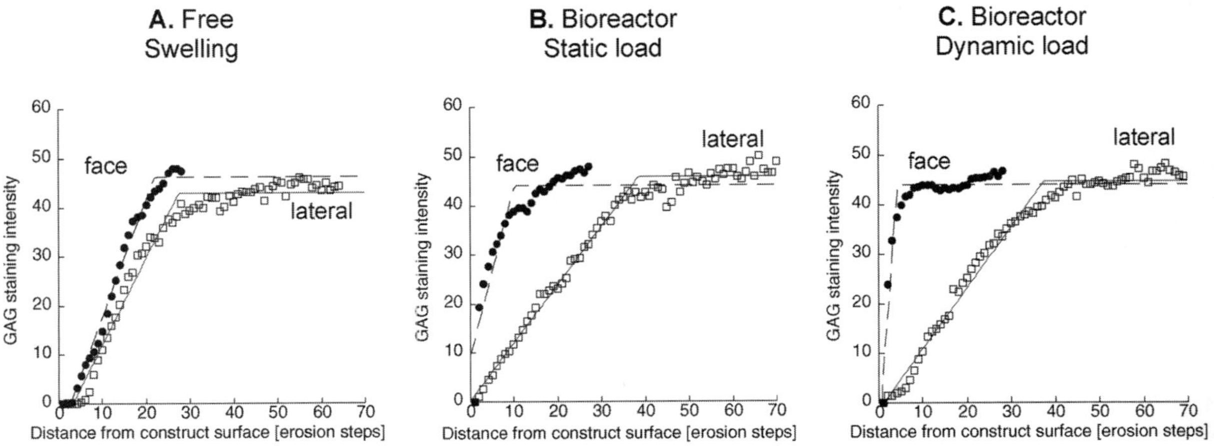

Fig. 6. GAG concentration profiles at construct surfaces. GAG concentration measured by image processing of histological sections is shown as a function of the distance from the construct surface. Representative images were used for 67-day constructs from the three groups: (A) free swelling, (B) static compression in bioreactor, (C) dynamic compression in bioreactor. Concentration profiles at the lateral and in face surfaces are shown by open squares and closed circles, respectively.

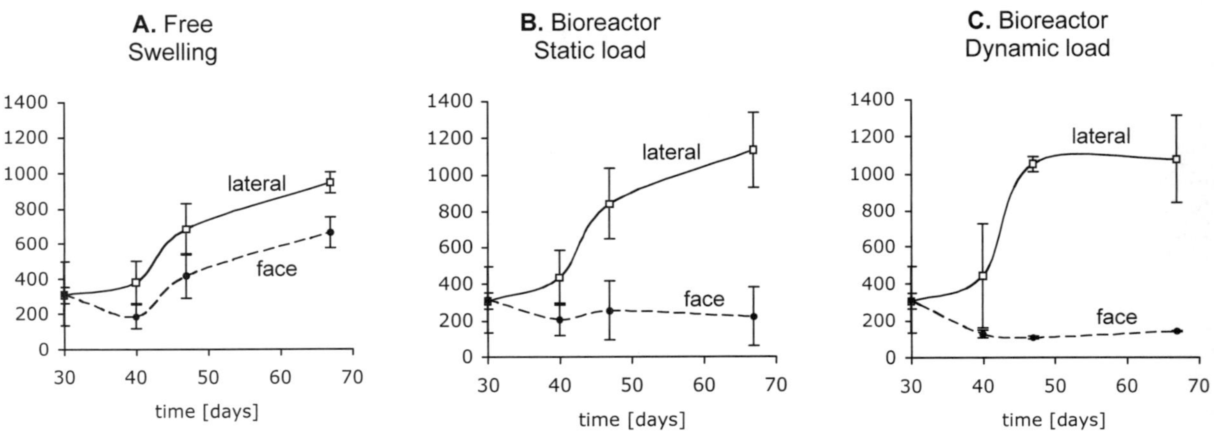

Fig. 7. Time progression of GAG leaching. The depth of GAG leaching zone was determined by a broken line fit, as shown in Fig. 6, as a function of cultivation time, for constructs from the three groups: (A) free swelling, (B) static compression in bioreactor, (C) dynamic compression in bioreactor. Leaching depths at the lateral and face surfaces are shown by open squares and closed circles, respectively. Data represents average $\pm$ SD of $n = 4$.

conditions of cultivation (free swelling, static or dynamic compression) and the position at the construct surface (face or lateral side). The depth of the GAG leaching zone, defined by the broken line fit, was 950 ± 50 μm at lateral sides and 660 ± 90 μm at face sides in free swelling constructs (Fig. 6A), 1070 ± 200 μm for the lateral sides, but only 220 ± 160 μm for the face sides in statically loaded constructs (Fig. 6B), and 1130 ± 230 μm for the lateral sides, but only 140 ± 10 μm for the face sides in dynamically loaded constructs (Fig. 6C). The GAG contents of the construct centers were comparable for the three groups (plateau levels in Figs 6A, 6B, 6C).

Figure 7 shows the progression of the GAG leaching at the face and lateral sides of constructs cultured under free swelling conditions, and with static and dynamic mechanical loading. Free swelling constructs showed a progressive increase in the depth of GAG leaching zone both at the lateral and face side of

Table 1
Overview of culture conditions

	Free swelling controls in 6-well plates	Static compression in bioreactors	Dynamic compression in bioreactors
Construct number	1 per well	5 per chamber	5 per chamber
Medium volume	7 ml/well	60 ml/chmaber	60 ml/chamber
Perfusion rate	none	180 ml/h	180 ml/h
Medium exchange rate	2 ml/construct/day	3 ml/construct/day	3 ml/construct/day
Strain	none	10% static offset	2% static offset 5% dynamic strain
Loading frequency	none	none	0.3 Hz
Loading regime	none	Continuous	1 h on / 23 h off

the construct (Fig. 7A). In constructs cultured with mechanical loading, the depth of the leaching zone increased with time at the lateral sides, but decreased at the face sides of the constructs; the trend was similar but more pronounced for dynamically stimulated constructs (Figs 7B, 7C).

6. Discussion

We report the design and application of a bioreactor system that was custom-built for studies of tissue engineered cartilage in response to mechanical stimulation, by upgrading an existing mechanical spectrometer [2]. The system was capable of supporting controlled long term (5-week) studies of tissue engineered cartilage cultured with the application of various regimes of mechanical stimulation. Important advantages were the capability for environmental control via perfusion and aseptic exchange of medium without interrupting cultivation or mechanical stimulation. The system could potentially be adapted to fit other mechanical spectrometer setups that currently use non-perfused dishes [10,15,34].

Ideally, a bioreactor for cartilage tissue engineering should provide an *in vitro* environment for rapid and orderly chondrogenesis. Hydrodynamic factors present during *in vitro* culture can modulate engineered cartilage in at least two ways: by enhanced *mass transport* between the cells and culture medium (e.g., oxygen, nutrients, growth factors), and by direct *physical stimulation* of the cells (e.g., shear, pressure). *In vivo*, mass transfer within articular cartilage involves a combination of diffusion and convective flow driven by concentration gradients and hydrodynamic pressures associated with dynamic loading. It is possible that the observed effects of dynamic compression *in vitro* (e.g., [5,34]) and *in vivo* [41] were also due in part to fluid flow. Loading can cause additional effects on chondrocytes, either by direct effects on their shape and interfibrillar spacing, or via related changes in hydrostatic pressure, fluid flow, and streaming potentials. These effects can modulate the synthesis, breakdown and structural adaptation of the ECM, and thereby mediate the catabolic and anabolic metabolism of chondrocytes.

Controlled *in vitro* studies are important for improving our understanding of mechanotransduction in chondrocytes and cartilage. However, the results should be interpreted with caution, because the loading regimes utilized *in vitro* can only remotely mimic the complexity of joint loading *in vivo*. In our system perfusion of medium, that was continuously equilibrated to maintain oxygen and pH levels, was utilized to enhance mass transport at construct surfaces, whereas the interstitial fluid flow was generated by mechanical compression (Fig. 1, Table 1).

The decrease of GAG content with prolonged culture time, that was observed in all experimental groups (Fig. 2), was not expected. Many previous studies of engineered cartilage have shown that GAG contents increased with time [19,43,45], in some cases over periods that were as long as 7 months [18]. Other studies reported that GAG contents decreased with time, in a manner similar to that observed in the present study, for bovine chondrocytes cultured on fibrous PGA scaffolds [9] and in agarose gel [35]. In recent studies, Mauck [34,35] provided further evidence that the ECM synthesis rates decrease with the development of mature ECM and the associated increase in fixed charge density. Increased catabolic activity was also thought to be responsible for ECM degradation and GAG loss by diffusion [24]. Our findings that DNA and collagen contents increased with time and that GAG content correlated with the compressive modulus (Fig. 2) were consistent with previous studies [45].

In the present study, mechanical stimulation was applied after 4 weeks of free swelling culture to constructs that already contained substantial amounts of cartilaginous ECM (>3% ww GAG, 3% ww collagen, Figs 2B, 2C), as the presence of ECM was previously linked to a more pronounced response to mechanical stimulation [5]. Interestingly, mechanical stimulation had little effect on the compositions, morphologies and on mechanical properties of construct interiors discs (Figs 2, 4, 5), but it resulted in distinctly different properties of the peripheral rings (Figs 3, 4) and face sides (Figs 6, 7).

Constructs cultured with mechanical loading maintained their cylindrical shape with flat and parallel top and bottom surfaces (Fig. 3). The platens served as a diffusional barrier, retaining GAG inside the construct and preventing its loss into culture medium at the face side (Fig. 4). The platens also prevented the formation of the fibrous capsule at the face side (Fig. 4), presumably due to the lack of hydrodynamic shear forces. Round cell morphology and the presence of ECM rich in GAG and type II collagen indicated the better maintenance of differentiated cell phenotype in the face side, as compared to the lateral side of constructs grown in static or dynamic compression, or the entire surfaces of free-swelling constructs. The regular geometry and cartilaginous nature of the face surfaces may be important for the survival and function of tissue constructs following their implantation into articular joint surfaces.

The distribution of ECM was distinctly different in the centers and the ring sections of the constructs, in a manner that depended on the conditions of cultivation. Fluid shear at the construct surfaces resulted in the loss of GAG (Figs 3, 4, 5B, 6), lack of type II collagen (Figs 3, 4) and accumulation of DNA (Fig. 5A). In particular, the lack of GAG and type II collagen and the presence of type I collagen were also observed at the all surfaces of free swelling constructs, and the lateral surfaces of mechanically stimulated constructs, but not at the face surfaces of mechanically stimulated constructs.

The loss of GAG may be further augmented by loading-induced interstitial fluid flow. In unconfined dynamic compression, loading-induced fluid flow is highest in the lateral region of the constructs and lowest at the construct centers [6,30]. These areas were previously associated with increased rates of GAG synthesis [6], but they also appear to be associated with the GAG loss (Figs 3, 4 and 5). One could postulate that the loss of GAG due to interstitial fluid flow may be a contributing factor for the up-regulation of GAG synthesis. Studies my Mauck [35] have shown increased rates of GAG accumulation in chondrocyte-seeded agarose hydrogels as a result of mechanical stimulation at early times in culture (up to 28 days), but also found that after an initial accumulation of GAG over the first 35 days in free swelling culture, GAG content decrease during the following 28 days. These findings and the progression of GAG loss (Figs 2, 7), clearly indicate the need to further study the effects of duration of construct cultivation and mechanical stimulation on structural and functional changes, both before and after the loading is applied. Factors that may act synergistically with mechanical loading include: supplemental growth factors, use of gel as a cell delivery vehicle to promote mechanotransduction and reduce GAG loss [5,34,35], and the application of loading at an earlier point during cultivation.

In summary, the duration of culture (before and after the application of mechanical stimulation) and the regime of loading can be utilized to modulate the development of engineered cartilage. The modular bioreactor system with medium perfusion and mechanical loading can be utilized for controlled studies of native and engineered cartilage, and to define the conditions of cultivation for functional tissue engineering of cartilage.

Acknowledgements

This work was supported by the National Aeronautics and Space Administration (NASA, Grant NCC8-174). We would like to thank Sue Kangiser for her help with the manuscript preparation.

References

[1] T. Aigner, W. Bertling, H. Stoss, G. Weseloh and K. von der Mark, Independent expression of fibril-forming collagens I, II, and III in chondrocytes of human osteoarthritic cartilage, *J. Clin. Invest.* **91** (1993), 829–837.

[2] R.G. Allen, R. Eisenberg and M.L. Gray, Quantitative measurement of the biological response of cartilage to mechanical deformation, in: *Methods in Molecular Medicine*, J.R. Morgan and M.L. Yarmush, eds, Vol. 18, Humana Press, Totowa, NJ, 1999, pp. 521–542.

[3] G.H. Altman, H.H. Lu, R.L. Horan, T. Calabro, D. Ryder, D.L. Kaplan, P. Stark, I. Martin, J.C. Richmond and G. Vunjak-Novakovic, Advanced bioreactor with controlled application of multi-dimensional strain for tissue engineering, *J. Biomech. Eng.* **124** (2003), 734–741.

[4] M.D. Buschmann, Y.A. Gluzband, A.J. Grodzinsky, J.H. Kimura and E.B. Hunziker, Chondrocytes in agarose culture synthesize a mechanically functional extracellular matrix, *J. Orthop. Res.* **10** (1992), 745–758.

[5] M.D. Buschmann, Y.A. Gluzband, A.J. Grodzinsky and E.B. Hunziker, Mechanical compression modulates matrix biosynthesis in chondrocyte/agarose culture, *J. Cell Sci.* **108** (1995), 1497–1508.

[6] M.D. Buschmann, Y.J. Kim, M. Wong, E. Frank, E.B. Hunziker and A.J. Grodzinsky, Stimulation of aggrecan synthesis in cartilage explants by cyclic loading is localized to regions of high interstitial fluid flow, *Arch. Biochem. Biophys.* **366** (1999), 1–7.

[7] S.E. Carver and C.A. Heath, Semi-continuous perfusion system for delivering intermittent physiological pressure to regenerating cartilage, *Tissue Eng.* **5** (1999), 1–11.

[8] T.H. Davisson, S. Kunig, A.C. Chen, R.L. Sah and A. Ratcliffe, Static and dynamic compression modulate biosynthesis in tissue engineered cartilage, *J. Orthop. Res.* **20** (2002), 842–848.

[9] T.H. Davisson, F. Wu and A. Ratcliffe, The kinetics of growth for tissue engineered cartilage: the effects of donor age, donor stie, and culture method, in: *49th Annual Meeting of the Orthopaedic Research Society*, New Orleans, 2003.

[10] O. Demarteau, M. Jakob, D. Schafer, M. Heberer and I. Martin, Development and validation of a bioreactor for physical stimulation of engineered cartilage, *Biorheology* **40** (2003), 331–336.

[11] S.H. Elder, J.H. Kimura, L.J. Soslowsky, M. Lavagnino and S.A. Goldstein, Effect of compressive loading on chondrocyte differentiation in agarose cultures of chick limb-bud cells, *J. Orthop. Res.* **18** (2000), 78–86.

[12] S.H. Elder, S.A. Goldstein, J.H. Kimura, L.J. Soslowsky and D.M. Spengler, Chondrocyte differentiation is modulated by frequency and duration of cyclic compressive loading, *Ann. Biomed. Eng.* **29** (2001), 476–482.

[13] R.W. Farndale, D.J. Buttler and A.J. Barrett, Improved quantitation and discrimination of sulphated glycosaminoglycans by the use of dimethylmethylene blue, *Biochim. Biophys. Acta* **883** (1986), 173–177.

[14] E.H. Frank and A.J. Grodzinsky, Cartilage electromechanics I. Electrokinetic transduction and the effects of electrolyte pH and ionic strength, *J. Biomech.* **20** (1987), 615–627.

[15] E.H. Frank, M. Jin, A.M. Loening, M.E. Levenston and A.J. Grodzinsky, A versatile shear and compression apparatus for mechanical stimulation of tissue culture explants, *J. Biomech.* **33** (2000), 1523–1527.

[16] L.E. Freed, J.C. Marquis, A. Nohria, J. Emmanual, A.G. Mikos and R. Langer, Neocartilage formation in vitro and in vivo using cells cultured on synthetic biodegradable polymers, *J. Biomed. Mater. Res.* **27** (1993), 11–23.

[17] L.E. Freed, G. Vunjak-Novakovic, R. Biron, D. Eagles, D. Lesnoy, S. Barlow and R. Langer, Biodegradable polymer scaffolds for tissue engineering, *Bio/Technology* **12** (1994), 689–693.

[18] L.E. Freed, R. Langer, I. Martin, N. Pellis and G. Vunjak-Novakovic, Tissue engineering of cartilage in space, *Proc. Natl. Acad. Sci. USA* **94** (1997), 13885–13890.

[19] L.E. Freed, A.P. Hollander, I. Martin, J.R. Barry, R. Langer and G. Vunjak-Novakovic, Chondrogenesis in a cell–polymer-bioreactor system, *Exp. Cell Res.* **240** (1998), 58–65.

[20] L.E. Freed and G. Vunjak-Novakovic, Tissue engineering bioreactors, in: *Principles of Tissue Engineering*, R.P. Lanza, R. Langer and J. Vacanti, eds, Academic Press, San Diego, 2000, pp. 143–156.

[21] A.J. Grodzinsky, M.E. Levenston, M. Jin and E.H. Frank, Cartilage tissue remodeling in response to mechanical forces, *Annu. Rev. Biomed. Eng.* **2** (2000), 691–713.

[22] F. Guilak, D.L. Butler and S.A. Goldstein, Functional tissue engineering: the role of biomechanics in articular cartilage repair, *Clin. Orthop.* **391**(Suppl.) (2001), S295–S305.
[23] A.P. Hollander, T.F. Heathfield, C. Webber, Y. Iwata, R. Bourne, C. Rorabeck and R.A. Poole, Increased damage to type II collagen in osteoarthritic articular cartilage detected by a new immunoassay, *J. Clin. Invest.* **93** (1994), 1722–1732.
[24] G. Jin, R.L. Sah, Y.S. Li, M. Lotz, J.Y. Shyy and S. Chien, Biomechanical regulation of matrix metalloproteinase-9 in cultured chondrocytes, *J. Orthop. Res.* **18** (2000), 899–908.
[25] Y.J. Kim, R.L. Sah, J.Y.H. Doong and A.J. Grodzinsky, Fluorometric assay of DNA in cartilage explants using Hoechst 33258, *Anal. Biochem.* **174** (1988), 168–176.
[26] Y.J. Kim, R.L. Sah, A.J. Grodzinsky, A.H. Plaas and J.D. Sandy, Mechanical regulation of cartilage biosynthetic behavior: physical stimuli, *Arch. Biochem. Biophys.* **311** (1994), 1–12.
[27] J. Kisiday, M. Jin, B. Kurz, H. Hung, C. Semino, S. Zhang and A.J. Grodzinsky, Self-assembling peptide hydrogel fosters chondrocyte extracellular matrix production and cell division: implications for cartilage tissue repair, *Proc. Natl. Acad. Sci. USA* **99** (2002), 9996–10001.
[28] D.A. Lee and D.L. Bader, Compressive strains at physiological frequencies influence the metabolism of chondrocytes seeded in agarose, *J. Orthop. Res.* **15** (1997), 181–188.
[29] J.H. Lee, J. Kisiday and A.J. Grodzinsky, Tissue-engineered versus native cartilage: linkage between cellular mechano-transduction and biomechanical properties, *Novartis Found. Symp.* **249** (2003), 52–64.
[30] M.E. Levenston, E.H. Frank and A.J. Grodzinsky, Electrokinetic and poroelastic coupling during finite deformations of charged porous media, *J. Appl. Mech.* **66** (1999), 1–12.
[31] R.D. Lillie and H.M. Fullmer, *Histopathologic Techniques and Practical Histochemistry,* McGraw Hill, New York, 1976.
[32] H. Madry, R. Padera, J. Seidel, R. Langer, L.E. Freed, S.B. Trippel and G. Vunjak-Novakovic, Gene transfer of a human insulin-like growth factor I cDNA enhances tissue engineering of cartilage, *Hum. Gene Ther.* **13** (2002), 1621–1630.
[33] I. Martin, B. Obradovic, L.E. Freed and G. Vunjak-Novakovic, A method for quantitative analysis of glycosaminoglycan distribution in cultured natural and engineered cartilage, *Ann. Biomed. Eng.* **27** (1999), 656–662.
[34] R.L. Mauck, M.A. Soltz, C.C.B. Wang, D.D. Wong, P.G. Chao, W.B. Valhmu, C.T. Hung and G.A. Ateshian, Functional tissue engineering of articular cartilage through dynamic loading of chondrocyte-seeded agarose gels, *J. Biomech. Eng.* **122** (2000), 252–260.
[35] R.L. Mauck, S.L. Seyhan, G.A. Ateshian and C.T. Hung, Influence of seeding density and dynamic deformational loading on the developing structure/function relationships of chondrocyte-seeded agarose hydrogels, *Ann. Biomed. Eng.* **30** (2002), 1046–1056.
[36] D. Pazzano, K.A. Mercier, J.M. Moran, S.S. Fong, D.D. DiBiasio, J.X. Rulfs, S.S. Kohles and L.J. Bonassar, Comparison of chondrogensis in static and perfused bioreactor culture, *Biotechnol. Prog.* **16** (2000), 893–896.
[37] M. Pei, J. Seidel, G. Vunjak-Novakovic and L.E. Freed, Growth factors for sequential cellular de- and re-differentiation in tissue engineering, *Biochem. Bioph. Res. Co.* **294** (2002), 149–154.
[38] R.L.Y. Sah, Y.J. Kim, J.Y.H. Doong, A.J. Grodzinsky, A.H.K. Plaas and J.D. Sandy, Biosynthetic response of cartilage explants to dynamic compression, *J. Orthop. Res.* **7** (1989), 619–636.
[39] J.F. Stoltz and X. Wang, From biomechanics to mechanobiology, *Biorheology* **39** (2002), 5–10.
[40] M. Tagil and P. Aspenberg, Cartilage induction by controlled mechanical stimulation in vivo, *J. Orthop. Res.* **17** (1999), 200–204.
[41] I. Takahashi, G.H. Nuckolis, K. Takahashi, O. Tanaka, I. Semba, R. Dashner, L. Shum and H.C. Slavkin, Compressive force promotes Sox9, type II collagen and aggrecan and inhibits IL-1b expression resulting in chondrogenesis in mouse embryonic limb bud mesenchymal cells, *J. Cell Sci.* **111** (1998), 2067–2076.
[42] M. Thibault, A.R. Poole and M.D. Buschmann, Cyclic compression of cartilage/bone explants in vitro leads to physical weakening, mechanical breakdown of collagen and release of matrix fragments, *J. Orthop. Res.* **20** (2002), 1265–1273.
[43] G. Vunjak-Novakovic, L.E. Freed, R.J. Biron and R. Langer, Effects of mixing on the composition and morphology of tissue-engineered cartilage, *AIChE J.* **42** (1996), 850–860.
[44] G. Vunjak-Novakovic, B. Obradovic, P. Bursac, I. Martin, R. Langer and L.E. Freed, Dynamic cell seeding of polymer scaffolds for cartilage tissue engineering, *Biotechnol. Prog.* **14** (1998), 193–202.
[45] G. Vunjak-Novakovic, I. Martin, B. Obradovic, S. Treppo, A.J. Grodzinsky, R. Langer and L.E. Freed, Bioreactor cultivation conditions modulate the composition and mechanical properties of tissue engineered cartilage, *J. Orthop. Res.* **17** (1999), 130–138.
[46] J.F. Woessner, The determination of hydroxyproline in tissue and protein samples containing small proportions of this imino acid, *Arch. Biochem. Biophys.* **93** (1961), 440–447.
[47] F. Wu, N. Dunkelman, A. Peterson, T. Davisson, R. De La Torre and D. Jain, Bioreactor development for tissue-engineered cartilage, *Ann. N. Y. Acad. Sci.* **875** (1999), 405–411.
[48] Q.-q. Wu and Q. Chen, Mechanoregulation of chondrocyte proliferation, maturation and hypertrophy: ion-channel dependent transduction of matrix deformation signals, *Exp. Cell Res.* **256** (2000), 383–391.

Biorheology 41 (2004) 459–467
IOS Press

Spectral and lifetime fluorescence imaging microscopies: New modalities of multiphoton microscopy applied to tissue or cell engineering

D. Dumas [a,*], N. Gaborit [b], L. Grossin [b], B. Riquelme [c], C. Gigant-Huselstein [a], N. de Isla [a], P. Gillet [b], P. Netter [b] and J.F. Stoltz [a]

[a] *Laboratoire de Mécanique et Ingénierie Cellulaire et Tissulaire, UMR CNRS 7563 LEMTA et IFR 111 CNRS, UHP-INPL-CHU, 54505 Vandoeuvre-lès-Nancy, France*

[b] *Physiopathologie et Pharmacologie Articulaires, UMR 7561 CNRS-UHP et IFR 111 CNRS, UHP-INPL-CHU Nancy 1, 54505 Vandoeuvre les Nancy, France*

[c] *Grupo de Optica Aplicada, Instituto de Fisica de Rosario (CONICET-UNR), Argentina*

Abstract. Spectral and multiphoton imaging is the preferred approach for non-invasive study allowing deeper penetration to image molecular processes in living cells. But currently available fluorescence microscopic techniques based on fluorescence intensity, such as confocal or multiphoton excitation, cannot provide detailed quantitative information about the dynamic of complex cellular structure (molecular interaction). Due to the variation of the probe concentration, photostability, cross-talking, its effects cannot be distinguished in simple intensity images. Therefore, Time Resolved fluorescence image is required to investigate molecular interactions in biological systems. Fluorescence lifetimes are generally absolute, sensitive to environment, independent of the concentration of the probe and allow the use of probes with overlapping spectra but that not have the same fluorescence lifetime. In this work, we present the possibilities that are opened up by Fluorescence Lifetime Imaging Microscopy, firstly to collect images based on fluorescence lifetime contrast of GFP variants used as a reporter of gene expression in chondrocytes and secondly, to measure molecular proximity in erythrocyte (glycophorin/membrane) by Fluorescence Resonance Energy Transfer (FLIM-FRET).

Keywords: GFP variants, glycophorin, FRET, lifetime imaging, spectral, chondrocytes, erythrocytes

1. Introduction

There is a growing interest for screening programs (drug efficiency) in the potential of spectral imaging methods for discriminating fluorophores with closely related emission spectra. Unlike a fluorescence image, where each point represents the intensity at a different locations (x, y, z) in space, a spectral image is a sequence of image representing the intensity of the 2D plane of the sample at different wavelength (spectral information). Combination of fluorescence spectroscopy with image cytometry is useful

*Address for correspondence: Dumas Dominique, Service d'Imagerie et Biophysique Cellulaire et tissulaire de Lorraine, Laboratoire de Mécanique et Ingénierie Cellulaire et Tissulaire, UMR CNRS 7563 et IFR 111, Faculté de Médecine, B.P. 184, 54505 Vandoeuvre-les-Nancy Cedex, France. Tel.: +33 03 83 68 34 65; Fax: +33 03 83 59 26 43; E-mail: dumas@hemato.u-nancy.fr.

0006-355X/04/$17.00

in biology by using fluorescent probes to identify and map several fluorophores and to improve the signal-to-noise ratio. The ability to identify a fluorophore signal based on spectral characteristics rather than intensity has significant advantages when trying to separate signals from undesirable background fluorescence (cartilage, blood cells, etc.), and when trying to separate signals from closely related fluorophore. Spectral imaging has also been used successfully to detect of micro-environmental changes (intracellular ion concentration), the measurement of auto-fluorescence from natural pigments (skin, etc.). Fluorescence resonant energy transfer (FRET) is a powerful technique for measuring intermolecular distances on a scale of a few nanometers, orientation and dynamic properties. However, due to the spectral overlap of donor and acceptor excitation/emission, the acceptor is often also excited directly by the donor excitation. Most of pair of fluorophore used in FRET experiment are difficult to separate spectrally by currently available methods based on fluorescence intensity.

Multi-photon excitation has evolved as a powerful way of performing three-dimensional fluorescence scanning microscopy in biological science. Compared to single-photon confocal microscopy, the multiphoton microscopy significantly improves performance, as the result of the pulsed excitation in the (infra) red region of the spectrum. This multiphoton process allows two or more photons of long wavelength light to create the same excitation as one photon of shorter wavelength light provided the photons arrive simultaneously. In order for a two-photon event to occur, a significantly density of incident photons is required. This phenomenon is made possible by the use of ultra-fast, mode-locked near-infrared lasers which can deliver a very high intensity combined with time-related concentration of a very brief laser flash (100–200 femtosecond pulse duration and very repetition rate (76 MHz) [1]. When confocal images of optical sections that are deep within a thick sample are obtained, the fluorescence signal is attenuated by light scatter. Since the red or near infrared illuminated light used for 2-P excitation has approximatively twice the wavelength of that employed for 1-P excitation which is less scattered in biological tissue, allowing deeper penetration into highly scattering thick specimen [2]. Under the appropriate power, near-infrared light is far less damaging to many living samples. Because in multiphoton mode spatial resolution is the result of absorption (and excitation) confinement to the focal event (smaller of one femtoliter) [3], the photobleaching patterns and photo-generation of toxic products are restricted to the focal plane [4].

Spectral and multiphoton imaging is the preferred approach to image molecular processes in living cells based on fluorescence intensity. However, the intensity signal is only a good measure of the dye concentration if the quantum efficiency (QE) stays constant, and QE is often affected by the cell micro-environment (pH, viscosity, temperature, probe concentration, oxygen, ...). To overcome the problems associated with intensity-based measurements (photobleaching, concentration and environment dependence, lost of dye, ...) as given by the spectral imaging, the resolution and contrast of image can be enhanced by a time-resolved technique. The fluorescence lifetime is defined as the average time that a molecule remains in an excited state prior to returning to the ground state and offers many opportunities for studies of dynamic event of living cells. Fluorescence lifetime can be measured with different strategies either in the frequency domain or in the temporal domain [5]. In time-domain methods, pulsed light used as the excitation source, and fluorescence lifetimes are measured from the fluorescence signal directly or by photon counting. Lifetime imaging systems have been demonstrated using wide-field [6], confocal [7] and multiphoton [8] imaging modes. Time-correlated single photon counting (TCSPC) has been used in combination with two-photon laser scanning microscope to deliver clear images of deep tissue layers. Importantly, the lifetime value measured for each pixel and dispayed a contrast by FLIM (fluorescence lifetime imaging microscopy) is a property of individual fluorescent molecules and is therefore largely independent of fluorophore concentration and photobleaching. These measurements are generally absolute. Fluorescence lifetime imaging microscopy is a very sensitive detector for any environment changes surrounding the fluorophore. In previous study based on GFP fluorescence intensity in the presence of strong auto-fluorescence background [9], gene expression into chondrocytes *in vivo* after electro-permeabilization of GFP was located in definite areas (superficial layer, middle and deep areas) indicating that the target tissue was transduced (electroporation) in its whole depth. In addition by FLIM, the emission of specific fluorophores or wild-GFP can be contrasted against a strong auto-fluorescence background as encountered in articular cartilage.

In the case of multi-labelling experiments, FLIM provides a discrimination of molecules on their different lifetimes (contrast image) whereas their emission spectra overlap (GFP variants). The emerging of new assay based on FRET measured by Time Resolved fluorescence (TR-FRET) have been developed for screening programs (drug targets, signal transduction, cellular function). Similarly, FLIM can also be used to visualize the lifetime distribution in particular the decrease of the donor fluorescence lifetime in a pair of fluorophore undergoing fluorescence resonance energy transfer (FRET).

In this work, we present a description of the combined spectral and temporal system and present images which demonstrate that the spectra and the lifetime of biological samples can be resolved and discuss a range of possible future applications in screening programs applied to bio-engineering (signalling cascades, cellular functions, protein–protein interactions, ...)

Firstly, we developed a FLIM-FRET method to characterize the hemoagglutination based on the interaction of pairs fluorophores combined respectively to secondary antibody directed against the primary agglutinin mAb–Alexa488™ or located in the RBC membrane (DiI).

Secondly, we used FLIM to deliver contrast lifetime images of different GFP variants used as a reporter of gene expression in chondrocytes because the discrimination of these GFP mutants by spectral imaging is hampered, by a high degree of spectral overlap with chondrocyte autofluorescence, low quantum efficiencies and extinction coefficients, or rapid photobleaching.

2. Material and methods

2.1. Sample preparation

2.1.1. FRET experiment on red blood cells

Fresh blood samples were collected from healthy donors by venepuncture in sterile vials containing EDTA as anticoagulant, stored at 5°C and analyzed within 24 hours. RBC was washed three times with phosphate buffered saline (PBS). A 2% suspension of RBC in PBS has been labeled with 3,3′-dioctadecyloxacarbocyanine perchlorate (DiI, 2.15 mM, Molecular Probes). The DiI-RBC were agglutinated in the presence of mAb and revealed by secondary antibody combined to Alexa488™. This is a donor (Alexa-488™)–acceptor (DiI) couple in a Forster resonance energy transfer process (FRET), whereby the fluorescence of the donor is quenched by the acceptor whose fluorescence is enhanced provided these fluorophores are in close proximity and the fluorescence of the donor overlaps the absorption spectrum of the acceptor.

As primary antibodies we used four anti-GP specific mouse monoclonal antibody (mAb) (Table 1) provided by Dr. Blanchard (EFS Nantes, France). Briefly, 50 μl of 2% RBC suspension were incubated at 37°C with 50 μl of primary antibody. After 30 min of incubation, the samples were washed twice. Then, 50 ml of diluted secondary antibody combined with Alexa-488™ (goat anti-mouse IgG, Molecular Probes A-11001) were added and incubate at 20°C for 30 min. Finally, after washing twice with PBS,

Table 1

Anti-glycophorin specific mouse monoclonal antibody. RBC samples were prepared in similar form to the recommended protocol by 4th International Workshop on Monoclonal Antibodies against Human Red Blood Cells and Related Antigens, Paris, July 2001

Antibody	Code	Concentration μg/ml	Culture supernatant	Clone	Epitope
anti-GPA+B	3-16	55	Mouse IgG1	NaM167-8A2	GPA: 1–9 (weak) GPB: 1–9
anti-GPA	3-17	43	Mouse IgG2b	NaM228-2E2	GPA: 34–38 GPB: (weak)
Anti-GPA+B	3-46	65	Mouse IgG1	NaM26-3F4	GPA: 49–52 (weak) GPB: 34–37
Anti-GPA+B	3-47	5.3	Mouse IgG2a	NaM10-6G4	GPA: 38–43

the RBC was suspended in PBS containing 0.5% bovine serum albumin (BSA) for subsequent FRET measurement (spectral and FLIM).

2.1.2. GFP variants

Isolation and culture of chondrocytes. Normal articular cartilage was obtained from Wistar male rats (180–200 g, Charles River, France) killed under anesthesia (ketamine 50 mg kg^{-1}, (Rhône-Mérieux) and acepromazine (1.25 mg kg^{-1}) (Sanofi Santé Animale). After joint surgery, articular cartilage pieces were aseptically dissected from femoral head caps and chondrocytes were obtained by sequential digestion with pronase (2 mg/ml) and collagenase B (1.5 mg/ml) (Roche Molecular Biochemical). The cells were washed two times in phosphate-buffered saline (PBS) and cultured to confluence in 25-cm^2 flasks at 37°C in a humidified atmosphere containing 5% CO_2. The medium used was DMEM/Ham's F-12 supplemented with L-glutamine (2 mM), gentamicin (50 μg/ml), amphotericin B (0.5 μg/ml), and heat-inactivated fetal calf serum (10%) (Invitrogen). Cells used at passage 2 or 3. In all experiments, chondrocytes were cultured under low FCS conditions (5%, v/v) and used at confluence.

Transfection of rat chondrocytes. Transfection of rat chondrocytes was performed with polyethylenimine (PEI) reagent (Exgen 500, Euromedex). Briefly, chondrocytes were seeded in 24-well plates in DMEM/F12 with 5% FCS at 37°C in 5% CO_2. For each well, at 70% confluence, chondrocytes were transfected with the expression vectors (500 ng of plasmid DNA; respectively, pcDNA3.1/CT-GFP-TOPO, pSecTag/eGFP/FRT/V5-His-TOPO, and pSecTag2A/EYFP) according to manufacturer's recommendations during 3 hours. The mixture was removed and chondrocytes were seeded in DMEM/F12 during 48 hours. After 48 h transfection, chondrocytes were fixed with 4% paraformaldehyde, and placed between two slides in a mounting medium for microscopy analysis.

2.2. Single-photon confocal microscopy

For the image series and spectra measurement, a Leica TCS SP2 equipped with an acousto-optical beamsplitter (AOBS) was used. Excitation was achieved by the 488 nm line from an Ar laser (GFP variant, Alexa 488^{TM}) and the 543 nm line from an HeNe laser (DiI). The bandwidths of the detected fluorescence wavelengths have been optimized for each probe (Alexa488^{TM}/diI) or GFP variants (GFP, eGFP, YFP) to the maximum emission. All photomultipliers used were adjusted to the same sensitivity (gain/offset). We employed a 1.32 numerical aperture oil immersion lens (Leica 63×, Planapo, Wetzlar, Germany). Three-dimensional images of reconstituted tissues were either compiled into a single view projection using LCS3D image processing software (Leica Microsystems, Mannheim, Germany) or compiled into a 3D projection using Imaris reconstruction software (Bitplane).

2.3. Two-photon microscopy

For two-photon excitation, we used a mode-locked Ti:sapphire laser (Coherent, Mira 900F) pumped by an 8 W argon-ion laser (Coherent, Verdi) which delivers a pulse width from 120 fs with a repetition rate of 76 MHz. The power of the excitation light in the focus was measured as about 5 mW. The detection was performed in a non-descanned mode (NDD-PMT).

2.4. FRET

Fluorescence resonance energy transfer, as described by Förster's theory, explains FRET as a dipole–dipole interaction between neighboring molecules and derives the dependence of the energy transfer efficiency E on their actual proximity R [10]. The direct non-linear dependence between E and R is described as: $E = 1/[1 + (R/R_0)^6]$ with R_0 the distance between the donor (Alexa488^{TM}) and acceptor (DiI) for $E = 0.5$. The properties of Alexa488^{TM}/DiI pair including acceptor quantum yield, spectral overlap between DiI-emission and Alexa488^{TM}-excitation and the relative spatial Alexa488^{TM}–DiI orientation are reflected by R_0. E can be obtained directly from the experiment as $E = 1 - I_{\text{Alexa488}^{TM}\text{–DiI}}/I_{\text{Alexa488}^{TM}}$ with $I_{\text{Alexa488}^{TM}\text{–DiI}}$ the fluorescence intensity of the donor in the presence of the acceptor and $II_{\text{Alexa488}^{TM}}$ the fluorescence intensity of the donor alone. In single-photon mode at 488 nm, the donor (Alexa488^{TM}) emission at 520 nm or, in case of FRET, the acceptor (DiI) emission at 570 nm could be separated easily using an acousto-optical beamsplitter (AOBS). First cells were excited at 488 nm and donor (520 nm) and acceptor (570 nm) emission was determined. To determine the amount of DiI present in sample, cells were excited at 543 nm. The spectra analysis cannot solve the problems due to the donor bleaching and FRET was measured by the decrease in donor fluorescence intensity (spectra) but also in lifetime (FLIM) by the presence of acceptor.

2.5. FLIM

For lifetime imaging, SPC-730 TCSPC Imaging module (Becker&Hickl, Berlin) was interfaced (signals, Pixel Clock, Frame Sync) to the scan controller of the Leica TCS-AOBS multiphoton laser scanning microscope. The PMT was typically built-in the microscope attached to the non-descanned port without discrimination for the different emission wavelengths of the different probes. The decay analysis measured by time-correlated single photon counting was performed using the SPCImage software (Becker&Hickl GmbH). The instrument response function was measured on erythrosine (full-width-half-maximum $= 220 \pm 15$ ps) and used for deconvolution.

3. Results and discussion

3.1. FLIM-FRET on red blood cells

For FRET revealed in single-photon confocal microscopy, the excitation wavelength was adjusted to minimize the direct excitation of the acceptor (DiI) by the donor excitation at 488 nm. For 3-17 mAb (Anti-GPA), the proximity requirement for the occurrence of FRET cannot be met upon excitation at 488 nm (energy transfer efficiency, $E = 0\%$) as shown in Fig. 1 which displays the fluorescence spectrum without acceptor emission at 565 nm. The FRET from Alexa-488^{TM} to DiI was found for 3-16 mAb ($E = 67\%$) with detection of fluorescence from acceptor. For 3-46 ($E = 8\%$) and 3-47 mAb

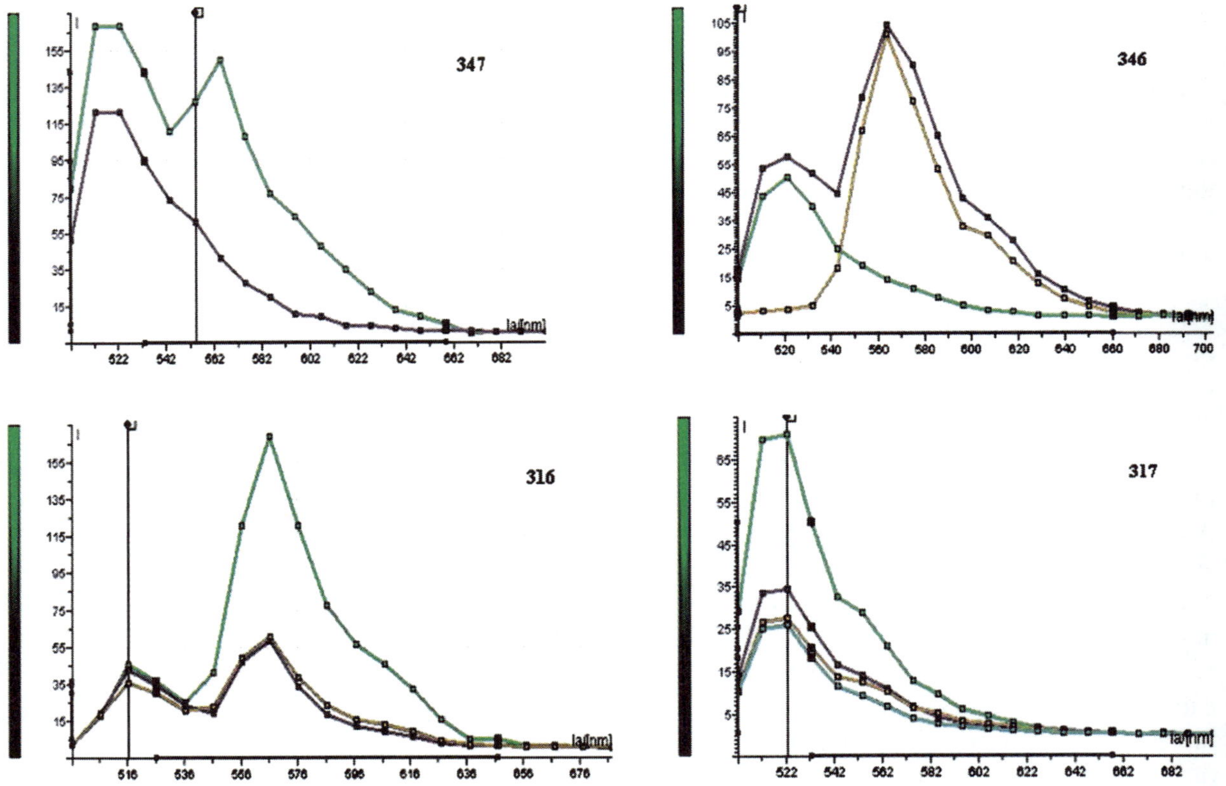

Fig. 1. Fluorescence spectra of RBC stained with DiI, then agglutinated with different anti-glycophorin A (3-17 and 3-47) or anti-glycophorin B (3-16 and 3-46) antibodies and revealed by secondary IgG antibody combined to Alexa-488™.

($E = 5\%$), the spectra analysis cannot solve the problems of spectral contaminations that arise in steady-state (intensity) measurements, as specific detection of fluorescence alone from donor (without acceptor excitation) or acceptor (without detection of fluorescence from unpaired donor).

The reduced lifetime of Alexa-488™ (indirectly combined to 3-16 and 3-46 mAb) indicates a molecular interaction with DiI incorporated in the RBC membrane (Fig. 2). The proximity requirement for the occurrence of FRET was not met for the 3-47 and 3-17 mAb.

3.2. *FLIM of GFP variants*

Under single-photon excitation, the fluorescence spectra of wild-type GFP and variants shows respectively a strong emission band at 509 nm for GFP, at 507 nm for eGFP and at 530 nm for EYFP. These spectra are similar under two-photon excitation at respectively 880 nm for GFP, 900 nm for eGFP and 920 nm for YFP (Fig. 3) and the cell autofluorescence exhibits negligible emission. The 2P-fluorescence specifically due to GFP and its mutants in transfected cells was observed to be slightly weaker than in 1P-fluorescence in which autofluorescence was increased (Fig. 4). By FLIM, the fluorescence decays of GFP, eGFP and YFP were best described as a single-exponential fit with $\tau_{\mathrm{GFP}} = 2.4 \pm 0.1$ ns, $\tau e_{\mathrm{GFP}} = 2.1 \pm 0.1$ ns and $\tau_{\mathrm{YFP}} = 3.6 \pm 0.2$ ns. These results are in agreement with Pepperkok et al. [11]. We used the τ information to create a mask for segmentation of lifetime images, as shown for eGFP (Fig. 5) and the emission of specific eGFP can be contrasted against an auto-fluorescence background.

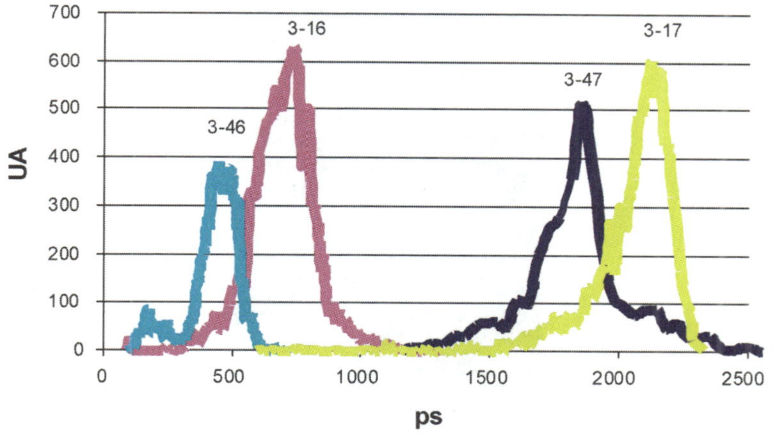

Fig. 2. Fluorescence lifetime microscopy. Temporal histogram depicting the Alexa-488™ fluorescence lifetime distribution. When the Alexa-488™ fluorophore is combined to secondary mAb (IgG), its average fluorescence lifetime is in the nanosecond range (1.7–2.2 ns) but upon transferring energy (FRET), it may drop in picoseconds (350–750 ps).

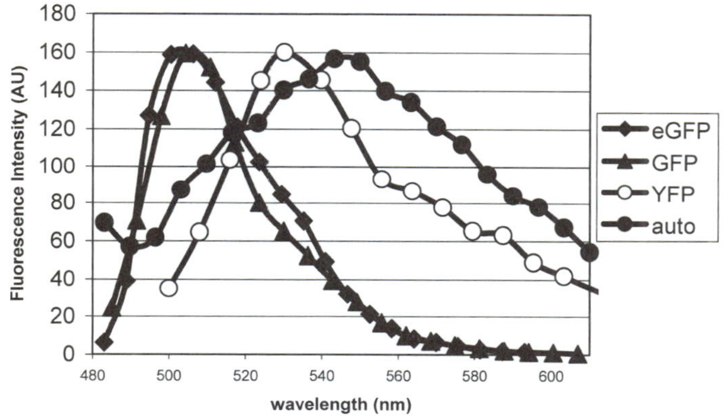

Fig. 3. Normalized fluorescence spectra in two-photon fluorescence of GFP (λ_{ex}: 880 nm), eGFP (λ_{ex}: 900 nm) and YFP (λ_{ex}: 920 nm) used as a reporter of gene expression in chondrocytes and showing the overlap with the undesirable chondrocyte fluorescence (×10).

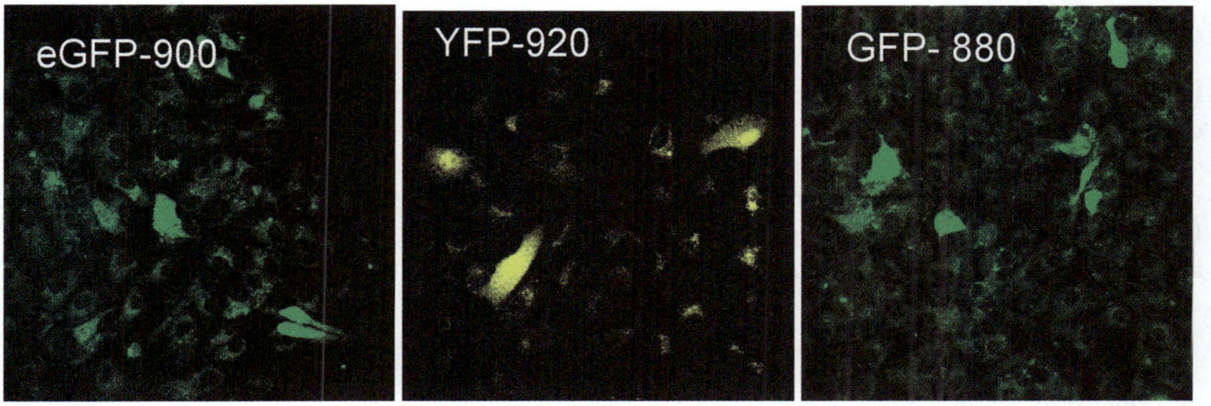

Fig. 4. Multiphoton excitation and expression of GFP (880 nm), EGFP (900 nm) and YFP (920 nm) vector transfected *in vitro* in chondrocytes from articular cartilage. The fluorescence detection channel was set to the following ranges: GFP and eGFP: 490–510 nm, YFP: 520–540 nm.

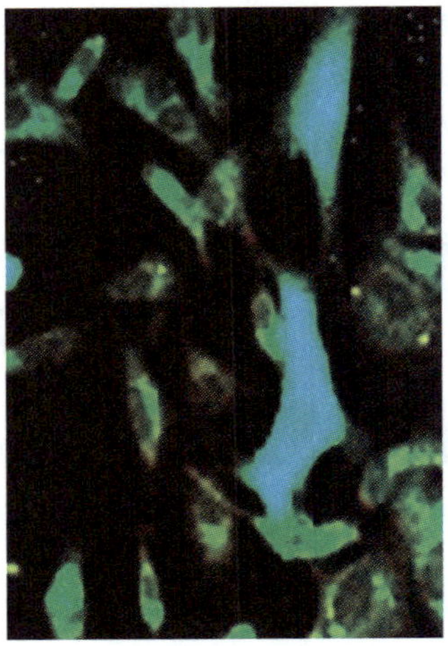

Fig. 5. Fluorescence lifetime imaging of EGFP vector transfected *in vitro* in chondrocytes from articular cartilage. Following Two-photon excitation (900 nm), the fluorescence decay of eGFP consisted in a major component (2.1 ± 0.1 ns) used to generate a mask for segmentation and to contrast the eGFP expression specifically against the cell autofluorescence.

4. Conclusion

4.1. RBC FRET

In view of the fluorescence properties depicted in RBC for the pair of fluorophore Alexa-488^{TM} (donor) and DiI (acceptor), it may be expected that upon agglutination of RBC, effective FRET should be observed, whereby the fluorescence of the donor is quenched (with reduced lifetime) by the acceptor in close proximity whose fluorescence is enhanced (at 565 nm). Figure 2 demonstrates upon excitation at 780 nm the FLIM-FRET from Alexa488^{TM} to DiI for the 3-16 and 3-46 mAbs only with a lifetime distribution in the picosecond range. Similarly, effective FRET was not observed for the 3-17 and 3-47 mAbs. The interaction of RBC surface glycophorin and the receptor clustering and changes in their conformation could be resolved by these methods.

4.2. GFP variants

For transfection (PEI) in chondrocytes, several variants of GFP have been used that have differences in spectral characteristics to make them more suitable for specific purposes, as to improve the sensitivity of GFP detection by tuning the excitation (multiphoton) and emission spectrum, to provide a higher quantum yield (affected by pH, temperature, ions, . . .). The ability to distinguish fusion proteins labelled with different GFP variants opens up the possibility of multi-labelling imaging experiments in live cells. Discrimination of these co-expressed GFP variants by spectral imaging (wavelength) could be hampered by a high degree of spectral overlap with autofluorescence, low quantum efficiencies and extinction coefficients, or rapid bleaching.

5. Conclusion

To conclude, Fluorescence Lifetime Imaging Microscopy (FLIM) is powerful technique as a good indicator of close proximity in protein–protein interactions by FRET. Combined intensity (spectral) and lifetime (FLIM) imaging was used to discriminate the signal of molecules on their different lifetimes whereas their emission spectra overlap as independent phenomena of the fluorophore concentration and photobleaching.

Acknowledgements

The author is grateful to the Association pour la Recherche sur le Cancer (ARC) and "Région de Lorraine" for financial support.

References

[1] W. Denk, J.H. Stricklen W.W. Webb, Two-photon fluorescence scanning microscopy, *Science* **2** (1990), 248–273.
[2] D. Dumas, L. Grossin, G. Cauchois, M. Gentils and J.F. Stoltz, Comparison of wide-field/deconvolution and confocal microscopy in bioengineering. Interest of multi-photon microscopy in the study of articular cartilage, *Biorheology* **40** (2003), 253–259.
[3] A. Diaspro and M. Robello, Two-photon excitation of fluorescence for three-dimensional optical imaging of biological structures, *J. Photochem. Photobiol.* **55** (2000), 1–8.
[4] H.J. Koester, D. Baur, R. Uhl and S.W. Hell, Ca^{2+} fluorescence imaging with pico-anf femtosecond two-photon excitation: signal and photodamage, *Biophys. J.* **77** (1999), 2226–2236.
[5] D. Dumas, S. Muller, J.J. Padilla, V. Latger, S. Woodard, M.C. Carré, W. Blondel, F. Baros, M.L. Viriot and J.F. Stoltz, New trends in opical bioengineering: applications to cell biology, *Recent Res. Devel. Optical Engg.* **2** (1999), 295–315.
[6] B. Pinsky, J. Ladasky, J.R. Lakowicz, K. Berndt and R.A. Hoffman, Phase-resolved fluorescence lifetime measurements for flow cytometry, *Cytometry* **14** (1993), 123–135.
[7] R. Sanders, A. Draaijer, H.C. Gerritsen, P.M. Houpt and Y.K. Levine, Quantitative pH imaging in cells using confocal fluorescence lifetime imaging microscopy, *Analytical Biochemistry* **227** (1995), 302–308.
[8] T. French, P.T. So, D.J. Weaver, T. Coelho-Sampaio, E. Gratton, E.W. Voss and J. Carrero, Two-photon fluorescence lifetime imaging microscopy of macrophage-mediated antigen processing, *Journal of Microscopy* **185** (1997), 339–353.
[9] L. Grossin, C. Cournil-Henrionnet, L.M. Mir, B. Liagre, D. Dumas, S. Etienne, C. Guimgamp, P. Netter and P. Gillet, Direct gene transfer into articular cartilage by in vivo electroporation, *FASEB J.* **17** (2003), 829–835.
[10] I. Majoul, M. Straub, R. Dufen, S.H. Hell and H-D. Söling, Fluorescence resonance energy transfer analysis of protein–protein interactions in single living cells by mulifocal multiphoton microscopy, *Reviews in Molecular Biotechnology* **82** (2002), 267–277.
[11] R. Pepperkok, A. Squire, S. Geley and P.I. Bastiaens, Simultaneous detection of multiple green fluorescent proteins in live cells by fluorescence lifetime imaging microscopy, *Current Biology* **9** (1999), 269–272.

Biorheology 41 (2004) 469–476
IOS Press

Human mesenchymal stem cells suppress induction of cytotoxic response to alloantigens

D. Angoulvant [a], A. Clerc [b], S. Benchalal [b] C. Galambrun [c], A. Farre [b], Y. Bertrand [c] and A. Eljaafari [a,b,*]

[a] *INSERM E 0226, Faculté de Medecine Lyon Nord, Université Claude Bernard, Lyon, France*
[b] *Cell Therapy Department, EFS Rhone-Alpes, Lyon, France*
[c] *Hematology Department, Debrousse Hospital, Lyon, France*

Abstract. Mesenchymal stem cells (MSC) fail to induce allogeneic responses in mixed lymphocyte reaction assays. Because MSC express HLA class I molecules, here we investaged whether they could be recognized as allogeneic targets by cytolytic T lymphocytes (CTL). With this aim, CTL precursor (CTLp) frequencies were measured following stimulation of T cells with either allogeneic mononuclear cells (MNC) or MSC originated from the same human bone marrow donor. Lysis of MSC was measured at day 10 of culture in standard chromium release assays. In addition, allogeneic PHA blast T cells or B-EBV lymphoblastoid cell lines (LCLs) generated from the same donor were used as positive controls of lysis. Our results showed that when allogeneic MNC were used to stimulate T cells, a high CTLp frequency was detected towards MSC targets. However, when MSC were used as stimulators, CTLp frequencies were markedly altered whatever the targets used, i.e.: MSC, PHA blast T cells or EBV-B LCLs. Moreover, when graded concentrations of MSC were added together with MNC upon stimulation of alloreactive T cells, we observed a dose-dependent decrease in CTLp frequencies towards MSC targets. This inhibition of MSC lysis was partially overcomed by adding exogenous rh-IL-2 from the beginning of cultures. In addition, this suppressive effect was totally reproduced when, instead of MSC, supernatant harvested from MSC cultures was added to allogeneic MNC, upon stimulation of alloreactive T cells. In conclusion, our results demonstrate that MSC which can be recognized as targets by pre-activated alloreactive CTLs, may be able to suppress differentiation of CTL precursors into CTL effectors through secretion of suppressive factors.

1. Introduction

The bone marrow microenvironment consists of stromal cells, also termed mesenchymal stem cells (MSC), and of extracellular matrix components which act in concert to regulate growth and differentiation of hematopoietic stem cells (HSC) [6]. MSC can be easily isolated and expanded from bone marrow, due to their adherence to plastic and replicative capacities [17]. MSC have also the ability to differentiate *in vitro and in vivo* into multiple tissue cell types such as osteocytes, chondrocytes, adipocytes, myocytes, cardiomyocytes, neurons, and tenocytes [7,8,14–16,19,20].

Due to these properties, MSC are good candidates for tissue engineering, for correction of acquired or inherited tissue disorders, or for regeneration of the bone marrow microenvironment, following radiation

[*]Corresponding author: Dr. Assia Eljaafari, Unite de Therapie Cellulaire el Tissus, CHU Brabois, Rue Morvan, 54511 Vandœuvre les Nancy, France. E-mail: Assia.Eljaafari@medecine.uhp-nancy.fr.

or intense chemotherapy [5,9,11,12]. Whereas autologous MSC can theoretically be used, allogeneic MSC should be more helpful, with the goal to prepare tissues or cells, in advance. Recent studies have investigated MSC allostimulatory capacities. Interestingly, it has been shown that human MSC do not express HLA class II molecules at their surface membrane [2]. Furthermore, MSC also present a lack of CD40, B7-1 and B7-2 molecules, which are involved in T cell co-activation or costimulation [3]. Thus, as expected from their phenotype, it has been shown that human MSC do not induce proliferative allore-sponses. MSC were also shown to inhibit T cell proliferative responses against alloantigens, induced by third party stimulating cells, and to prolong allogeneic skin graft survival [1,4,13]. Besides the secretion of immunosuppressive cytokines, like TGFβ [4] by MSC, low expression of costimulatory molecules and absence of human leukocyte antigen (HLA) class II expression on human MSC were shown to likely contribute to the immunomodulatory effect of MSC, and to the absence of induction of CD4+ T cell alloresponse, respectively [18]. But, as yet, no study has investigated the possibility that MSC could induce a cytotoxic allo-response, while HLA class I molecules are present on human MSC membrane. This expression could indeed result in the induction of a cytotoxic alloresponse, which would lead to MSC disappearance following their injection *in vivo* or following transplantation of tissues engineered from MSC. We aimed to test this hypothesis through the analysis of the capacity of MSC to induce a cytotoxic T lymphocytes (CTL) alloresponse and their susceptibility to lysis.

2. Materials and methods

2.1. Bone marrow derived allogeneic mesenchymal stem cells (MSC)

Human bone marrow mononuclear cells (1.8–2 $\times$ 10^6/ml), resuspended in alpha-minimal essential medium (GIBCO) supplemented with fetal bovine serum (10%) and penicillin streptomycin (1%), were inoculated at 5 $\times$ 10^4/cm^2 into tissue culture flask and incubated at 37°C in a humidified atmosphere supplemented with 5% CO_2. The cultures were fed by complete replacement of culture medium twice a week. Confluent MSC layers were trypsinised after 21 days before second passage. MSC quality was assessed by flow cytometry analysis as described below. MSC were used as stimulator and target cells.

2.2. MSC surface phenotype

MSC were detached using trypsin/EDTA and resuspended in complete medium. MSC (5 $\times$ 10^6/ml). They were then stained for 20 minutes at 4°C with combinations of saturating amounts of fluorescein isothiocyanate (FITC)-conjugated or phycoerythrin (PE)-conjugated monoclonal antibodies. The FITC- or PE-conjugated antibodies recognizing CD45, CD34, CD90, CD73, CD86, CD80, CD40, HLA DR and HLA ABC were purchased from Beckton Dickinson. Each fluorescence analysis included appropriate FITC- and PE-conjugated mice IgG1, as negative isotype controls. The percentage of positive cells was determined by substracting the percentage of fluorescent cells in the control from the percentage of cells positively stained with the appropriate antibody. Cells were analyzed by using a FACScan laser flow cytometry system (Becton Dickinson) equipped with a Macintosh personal computer and Cell Quest software.

2.3. Allogeneic peripheral blood mononuclear cells (PBMC)

Peripheral blood mononuclear cells were isolated by means of the Ficoll ditrizoate gradient separation and used as responder or stimulator cells.

2.4. *Targets*

Generation of PHA blasts T cells from PBMC was performed by incubating PBMC (1×10^6/ml) with PHA M 1% for 72 hours. Cells were then cultured in the presence of interleukin-2 (IL-2) for 7 additional days. B lymphocytes were isolated from PBMC and infected with Epstein Barr Virus (EBV) to obtain B-EBV lymphoblastoid cell lines used as target cells.

2.5. *CTLp assay*

Allogeneic MNC and MSC, originated from the same donor, used as stimulator cells were irradiated (30 Gy) before being cultured with responder PBMC. Cells were mixed at different stimulator-to-responder cell ratios in V-bottomed 96-well culture plates for 10 days in 0.2 ml of RPMI 1640 medium (Life Technologies) containing 10% heat-inactivated AB serum, 2 mM glutamine (Life Technologies), 1% NaHCO3 (Life Technologies), 100 U/ml penicillin and 100 μg/ml streptomycin (Life Technologies). Briefly, limiting dilution assays were performed by adding a fixed concentration of irradiated stimulating cells and graded concentrations of responder MNC, starting from 20 000 cells/well. When MNC were used as stimulators, the concentration was 20 000/well. When MSC were used to stimulate T cells, the concentration was 10 000/well. Each dilution was repeated 16 fold. After three and six days, 100 μl of medium was discarded and replaced with 100 μl of medium containing 20 U/ml recombinant human (rh) IL-2. The same operation was repeated after 6 days. CTLp cytotoxicity assay was performed after 10 days with ^{51}Cr-labelled target cells, which were either allogeneic ^{51}Cr-labeled MSC, PHA blast T cells or B-EBV lymphoblastoid cell lines (LCLs) originated from the same donor. CTLp frequency was then measured by using a statistical method, based on the numeration of negative wells.

2.6. *CTLp modified assay*

The CTLp assay was modified, by using a fixed concentration of responder cells, and graded concentrations of stimulating cells. 16 wells were performed for each dilution. Cells were cultured for 10 days in the presence of IL-2, and cytolytic activity was measured as described above. Results were expressed as number of positive wells.

3. Results

3.1. *MSC are good allogeneic targets for CTL effectors*

Despite the lack of MHC class II, human MSC express MHC class I molecules at their surface membrane. Thus, their ability, to induce an allogeneic response was tested in CTLp assay. With this aim, MNC were plated at graded concentrations and were stimulated either with allogeneic MNC or allogeneic MSC originated from the same bone marrow donor. Cultures were fed with IL-2 at day 3 and day 6, and CTLp frequencies were measured at day 10, using as targets either PHA blast T cells, MSC, or B-EBV LCLs, which were established from the same bone marrow donor.

Figure 1 shows that whatever the target used, CTLp frequency was high following stimulation of responder cells with allogeneic MNC. However, when allogeneic MSC were used to stimulate responder cells, no lysis was observed, even when B-EBV LCLs were used as targets. Thus, these data showed that, on the one hand MSC can be recognized as allogeneic targets, but that on the other hand, MSC were not able to induce CTLp differentiation into CTL effectors.

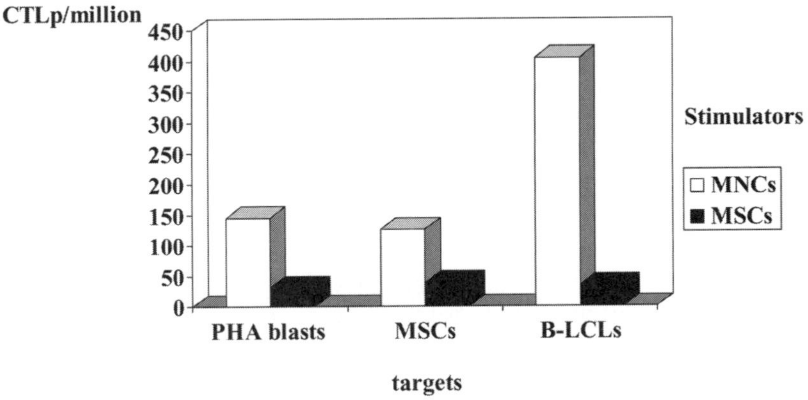

Fig. 1. MSC are good allogeneic targets for CTL effectors. Either allogeneic MSC or MNC were used to stimulate CTL precursors. Their differentiation into CTL effectors was tested by using different targets, issued from the same bone marrow donor.

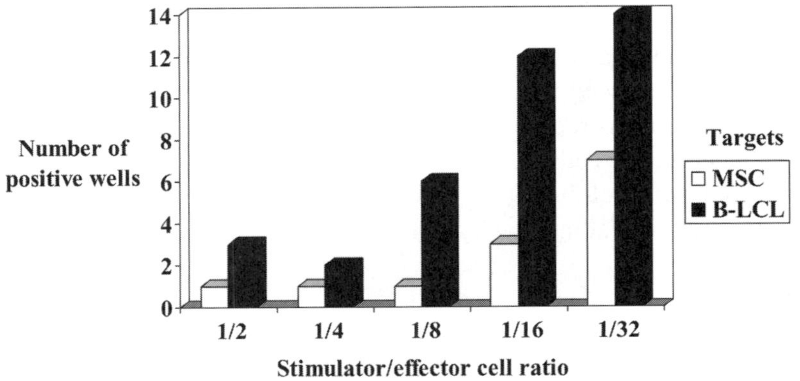

Fig. 2. MSC failure to induce an allogeneic response is not related to a poor stimulatory capacity. Allogeneic MSC were used at different concentrations to stimulate responder CTLp. Either B-EBV LCLs or MSC were used as targets.

3.2. MSC failure to induce an allogeneic response is not related to a poor stimulatory capacity

Since MSC were not able to induce CTLp differentiation into CTL effectors, we addressed whether this could be related to a poor stimulatory capacity. With this aim, we performed a CTLp assay by using a modified method. In this assay, responder MNC were plated at only one concentration (the highest one, 20 000 cells/well) and allogeneic MSC were added at graded concentrations, starting from a concentration of 1 stimulator/2 effector cell ratio. Targets were either MSC or B-EBV cell lines. Results were expressed as number of positive wells. Surprisingly and as shown in Fig. 2, we observed that decreasing the concentration of stimulating MSC resulted in the recovery of target cell lysis. Since low concentrations of MSC resulted in high CTL induction, these results ruled out the hypothesis that MSC could be poor stimulators of CTL allo-response.

3.3. MSC inhibit the induction of a CTL allo-response, by inhibiting CTLp differentiation

Since MSC are likely to be good stimulators at low concentrations, we then hypothesized that high concentrations of MSC might inhibit CTLp stimulation. To address this issue, we performed a CTLp

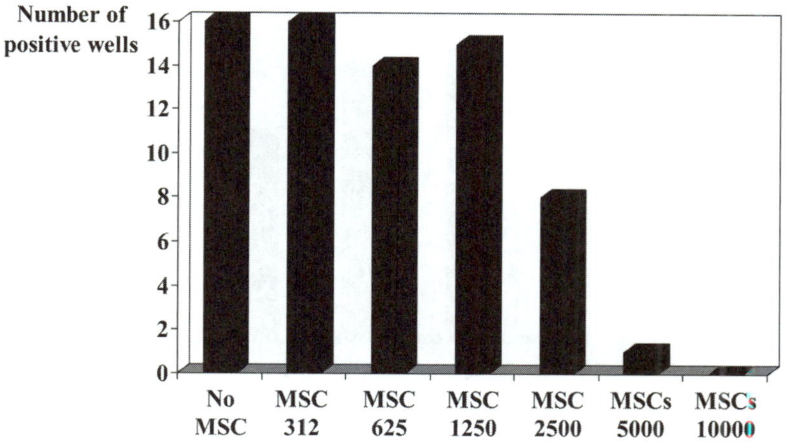

Fig. 3. MSC inhibit MNC-mediated CTL allo-response. MSC were added at various concentrations into wells containing blood CTLp and allogeneic stimulating MNC issued from the same bone marrow donor than MSC donor. Targets used in this assay were MSC.

modified assay in which responder cells were stimulated with a unique concentration of allogeneic MNC, but either in the absence or in the presence of graded concentrations of MSC originated from the same allogeneic bone marrow donor. As expected and as shown in Fig. 3, we observed that increasing the concentrations of MSC resulted in inhibiting the CTL response mediated by allogeneic MNC. Thus, these data demonstrated that high concentrations of MSC prevent MNC-mediated CTLp differentiation into CTL effectors.

3.4. IL-2 restores MNC-mediated CTLp response

Because IL-2 is known to be able to reverse the immunomodulatory effect of some suppressive factors, we then performed the same CTLp modified assay, as described above, but we added rh IL-2 from the beginning of cultures, instead of adding it at day 3, as usual. Figure 4 shows that target cell lysis was partially recovered when rh IL-2 was added together with MSC upon stimulation, at day zero. Thus, these results strongly suggested that inhibition by MSC of MNC allo-stimulatory capacity was likely to be related to an immune suppression mechanism.

3.5. MSC inhibition of CTL allogeneic response is mediated by soluble factors

Because it has been shown that inhibition of allogeneic T cell proliferation by MSC is partly related to their ability to secrete soluble factors, such as TGFb or HGF [4], we then investigated whether MSC supernatant could replace the presence of MSC and inhibit CTLp differentiation. With this aim, CTLp assays were performed using as effectors peripheral blood T cells and HLA-mismatched stimulating MNC. Then, supernatant harvested from 24 hour MSC cultures was added from the beginning of cultures. rh IL-2 was then added at day 3. Results shown in Fig. 5, clearly demonstrated that MSC supernatant was, by itself, able to inhibit MNC-mediated CTL allo-response. Moreover, using antibodies specific to TGFβ and HGF resulted in the partial recovery of target cell lysis, which suggested that these cytokines might be involved, at least partly, in the inhibition of CTLp differentiation.

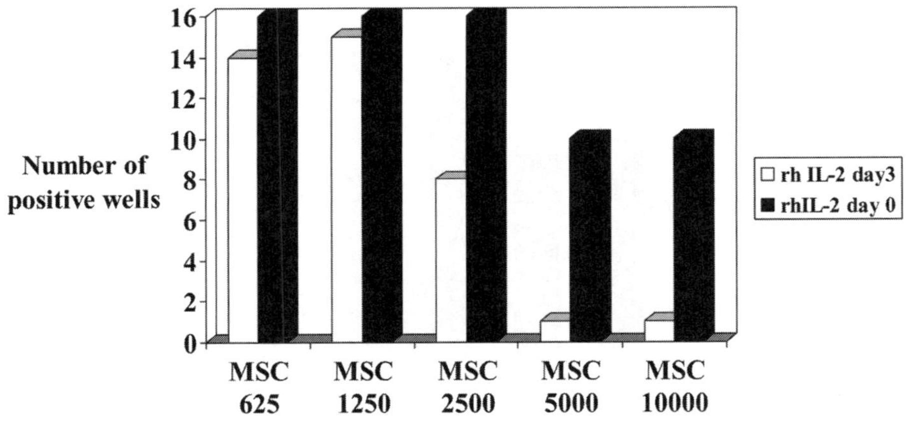

Fig. 4. IL-2 restores the MNC-mediated CTLp response that is inhibited by MSC. CTL precursors were stimulated by allogeneic MNC, at a 1/1 effector/stimulator ratio. Allogeneic MSC issued from the same bone marrow donor were added at various concentrations. IL-2 was added either at day zero or day 3.

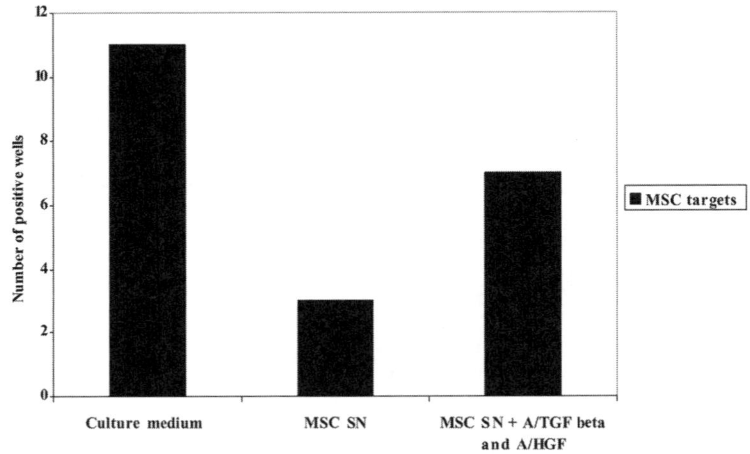

Fig. 5. MSC inhibition of CTL precursor response is mediated by soluble factors. CTLp assays were performed by adding allogeneic stimulating MNC. Supernatant (SN) harvested from MSC 24 hour culture was added at day zero, in the presence or absence of anti-TGF beta and anti-HGF specific mAbs.

4. Discussion

In this study, we investigated the induction of CTL alloresponse by allogeneic MSC. Since human MSC express MHC class I molecules at their cell surface, they should be able to induce a CTL alloresponse, through a direct or indirect mechanism of recognition by CTL.

Whereas, MSC were shown to represent good targets for pre-activated CTLs, we observed that allogeneic MSC were unable to activate CTL precursors in order to induce their differentiation into CTL effectors, as assessed by the absence of lysis of various allogeneic targets, including B-EBV LCL (Fig. 1). However, this unability was dependent on the concentration of MSC used to stimulate CTLp, since as shown in Fig. 2, low doses of MSC were potent CTLp stimulators. That high concentrations of MSC might inhibit CTLp differentiation was then suggested by our next experiment showing that MSC inhibited the stimulation of CTLp that was mediated by allogeneic blood MNC (Fig. 3). Once again, this inhibition was dose-dependent, since occuring when high concentrations of MSC were used, only. These concentrations were nevertheless always inferior to that of stimulating MNC, in order to favour stimulation-of responder T cells by MNC. The possibility that MSC could compete with MNC for binding to CTLp T cell receptor (TcR), was ruled out by adding rh IL-2 to CTLp cultures from the beginning of the assay. Indeed, in the presence of rh IL-2, MNC-mediated CTLp differentiation was observed even in the presence of high MSC concentrations. Thus, these results indicated that TcR can be efficiently stimulated by allogeneic MNC, even in the presence of MSC.

We then showed that the overcoming effect of rh IL-2 was likely to be due to the reversal of the suppressive activity of soluble factors secreted by MSC. Indeed, replacing MSC by supernatant harvested from 24 hour MSC cultures, resulted in inhibition of CTLp differentiation. Moreover, these suppressive factors were likely to include TGFβ and HGF, as judged by the reappearance of target cell lysis in the presence of blocking Abs specific to these cytokines.

In conclusion, we show herein that MSC are good tragets for pre-activated CTLs but, are able to inhibit induction of CTL allo-response, depending on the concentration of MSC used. Moreover, our results suggest that this inhibitory effect may be at least partly related to secretion of inhibitory factors, such as TGFβ and HGF, as reported by Di Nicola et al. [4] in another model. These results extend thus to the CTL function, previous reports demonstrating that MSC inhibit T cell response to allo-Ag. Therefore, our results suggest that it might be possible to graft MSC together with tissues or HSC and avoid an allo-response *in vivo*. However, they point out that the MSC dose, as well as the time of injections are important factors needing to be determined, first.

References

[1] A. Bartholomew, C. Sturgeon, M. Siatskas, K. Ferrer, K. McIntosh, S. Patil, W. Hardy, S. Devine, D. Ucker, R. Deans, A. Moseley and R. Hoffman, Mesenchymal stem cells suppress lymphocyte proliferation in vitro and prolong skin graft survival in vivo, *Exp. Hematol.* **30** (2002), 42–48.

[2] R.J. Deans and A.B. Moseley, Mesenchymal stem cells: Biology and potential clinical uses: review, **28** (2000), 875–884.

[3] S.M. Devine and R. Hoffman, Role of mesenchymal stem cells in hematopoietic stem cell transplantation. Review, *Curr. Opin. Hematol.* **7** (2000), 358–363.

[4] M. Di Nicola, C. Carlo-Stella, M. Magni, M. Milanesi, P.D. Longoni, P. Matteucci, S. Grisanti and A.M. Gianni, Human bone marrow stromal cells suppress T-lymphocyte proliferation induced by cellular or nonspecific mitogenic stimuli, *Blood* **99** (2002), 3838–3843.

[5] G. Ferrari, G. Cusella-De Angelis and M. Colletta et al., Muscle regeneration by bone marrow-derived myogenic progenitors, **279** (1998), 1528–1530.

[6] A. Fridenstein, Stromal bone marrow cells and the hematopietic micŕoenvironment, *Arkh. Pathol.* **44** (1982), 3–11.

[7] K. Fukuda, Development of regenerative cardiomyocytes from mesenchymal stem cells for cardiovascular tissue engineering, *Artif. Organs* **25** (2001), 187–193.

[8] M.C. Galmiche, V.E. Koteliansky, J. Briere, P. Herve and P. Charbord, Stromal cells from human long-term marrow cultures are mesenchymal cells that differentiate following a vascular smooth muscle differentiation pathway, *Blood* **82** (1993), 66–76.

[9] S. Gronthos and P.J. Simmons, The bioloy and application of human bone marrow stromal cell precursors, *J. Hematother.* **5** (1995), 15–23.

[10] H.W. Hirte, D.A. Clark, G. O'Connell, J. Rusthoven and J. Mazurka, Reversal of suppression of lymphokine-activated killer cells by transforming growth factor-beta in ovarian carcinoma ascitic fluid requires interleukin-2 combined with anti-CD3 antibody, *Cell. Immunol.* **142** (1992), 207–216.

[11] E.M. Horwitz, P.L. Gordon, W.K. Koo, J.C. Marx, M.D. Neel, R.Y. McNall, L. Muul and T. Hofmann, Isolated allogeneic bone marrow-derived mesenchymal cells engraft and stimulate growth in children with osteogenesis imperfecta: Implications for cell therapy of bone, *Proc. Natl. Acad. Sci. USA* **99** (2002), 8932–8937.

[12] E.M. Horwitz, D.J. Prockop, L.A. Fitzpatrick, W.W. Koo, P.L. Gordon, M. Neel, M. Sussman, P. Orchard, J.C. Marx, R.E. Pyeritz and M.K. Brenner, *Nat. Med.* **5** (1999), 309–313.

[13] M. Krampera, S. Glennie, J. Dyson, D. Scott, R. Laylor, E. Simpson and F. Dazzi, Bone marrow mesenchymal stem cells inhibit the response of naive and memory antigen-specific T cells to their cognate peptide, *Blood* **101** (2003), 3722–3729.
[14] R.F. Pereira, K.W. Halford, M.D. O'Hara, D.B. Leeper, B.P. Sokolov, M.D. Pollard, O. Bagasra and D.J. Prockop, Cultured adherent cells from marrow can serve as long-lasting precursor cells for bone, cartilage, and lung in irradiated mice, *Proc. Natl. Acad. Sci. USA* **92** (1995), 4857–4861
[15] M.F. Pittenger, A.M. Mackay, S.C. Beck, R.K. Jaiswal, R. Douglas, J.D. Mosca, M.A. Moorman, D.W. Simonetti, S. Craig and D.R. Marshak, Multilineage potential of adult human mesenchymal stem cells, *Science* **284** (1999), 143–147.
[16] D.J. Prockop, Marrow stromal cells as stem cells for nonhematopoietic tissues, *Science* **276** (1997), 71–74.
[17] M. Reyes, T. Lund, T. Lenvik, D. Aguiar, L. Koodie and C.M. Verfaillie, Purification and ex vivo expansion of postnatal human marrow mesodermal progenitor cells, *Blood* **98** (2001), 2615–2625.
[18] W.T. Tse, J.D. Pendleton, W.M. Beyer, M.C. Egalka and E.C. Guinan, Suppression of allogeneic T-cell proliferation by human marrow stromal cells: implications in transplantation, *Transplantation* **75** (2003), 389–397.
[19] S. Watikani, T. Saito and A.I. Caplan, Myogenic cells derived from rat bone marrow mesenchymal stem cells exposed to 5-azacytidine, *Muscle Nerve* **18** (1995), 1417–1426.
[20] D. Woodbury, E.J. Schwartz, D.J. Prockop and I.B. Black, Adult rat and human bone marrow stromal cells differentiate into neurons, *J. Neurosci. Res.* **61** (2000), 364–370.

Part V

Experimental, clinical and pharmacological approaches

Biorheology 41 (2004) 479–491
IOS Press

Post-traumatic osteoarthritis: The role of accelerated chondrocyte senescence

James A. Martin, Thomas Brown, Anneliese Heiner and Joseph A. Buckwalter *
Department of Orthopaedics, University of Iowa, Iowa City, IA 52242, USA

Abstract. Joint injuries frequently lead to progressive joint degeneration that causes the clinical syndrome of post-traumatic osteoarthritis. The pathogenesis of osteoarthritis remains poorly understood, but patient age is a significant risk factor for progressive joint degeneration. We have found that articular cartilage chondrocytes show strong evidence of senescence with increasing age, including synthesis of smaller more irregular aggrecans; increased expression of lysosomal beta-galactosidase and telomere erosion; and decreased proteoglycan synthesis, response to the anabolic cytokine IGF-I, proliferative capacity, and mitochondrial function. These observations help explain the strong association between age and joint degeneration, but they do not explain how joint injury increases the risk of joint degeneration in younger individuals. We hypothesized that excessive loading of articular surfaces due to acute joint trauma or post-traumatic joint instability, incongruity or mal-alignment increases release of reactive oxygen species, and that the increased oxidative stress on chondrocytes accelerates chondrocyte senescence thereby decreasing the ability of the cells to maintain or restore the tissue. To test this hypothesis, we exposed human articular cartilage chondrocytes from young adults to mechanical and oxidative stress. We found that shear stress applied to cartilage explants in a triaxial pressure vessel increased release of reactive oxygen species and oxidative stress induced chondrocyte senescence (as measured by expression of lysosomal beta-galactosidase, nuclear and mitochondrial DNA damage and decreased mitochondrial function). These observations support the hypothesis that joint injury accelerates chondrocyte senescence and that this acceleration plays a role in the joint degeneration responsible for post-traumatic osteoarthritis.

1. Introduction

Clinical experience and epidemiologic studies show that joint injuries including direct and indirect joint impact loading; meniscal, ligament and joint capsule tears; and intra-articular fractures, increase the risk of the progressive joint degeneration that causes post-traumatic osteoarthritis [33,57]. The causes of this increased risk are poorly understood. The relationships between severity of acute articular surface injury and risk of joint degeneration have not been well defined, and the mechanisms responsible for progressive loss of grossly normal articular surfaces following various types of joint injuries have received little attention. Furthermore, the risk of post-traumatic osteoarthritis varies among joints and among individuals [25,50,57], and only recently have investigators started to explore the reasons for the decreased capacity for repair of articular surface injuries with increasing age [60].

Basic scientific investigations have shown how mechanical forces damage articular surfaces, and defined the responses of articular surfaces to injury [12,16,17,23]. Other work has demonstrated the effects of residual articular surface incongruity following intra-articular fractures on articular surface contact stress [57]. Although residual joint instability, incongruity and mal-alignment increase the risk of joint

*Address for correspondence: Joseph A. Buckwalter, 01008 Pappajohn Pavilion, Department of Orthopaedics, University of Iowa College of Medicine, Iowa City, IA 52242, USA. Tel.: +1 319 356 2595; Fax: +1 319 356 8999; E-mail: joseph-buckwalter@uiowa.edu.

degeneration [17,57], there has been little experimental study of how these post-traumatic joint abnormalities cause articular surface degeneration [57], and the relationship between acute high intensity loading of articular surfaces and subsequent joint degeneration has not been explained.

Development of effective methods of preventing or slowing the progression of post-traumatic osteoarthritis depend on advances in understanding of the pathogenesis of this disease. This manuscript first reviews current understanding of how mechanical forces damage articular surfaces and how articular surfaces respond to injury. It then presents the results of experimental studies that show how chondrocyte senescence increases the risk of joint degeneration. The final section presents experimental studies demonstrating that intense mechanical loading causes chondrocytes to release reactive oxygen species and that oxidative stress accelerates chondrocyte senescence.

2. Mechanical injuries of articular surfaces

Disrupting a normal articular surface with a single impact requires substantial force, presumably because of the ability of articular cartilage and subchondral bone to dampen and distribute loads. A transarticular load of 2170 newtons applied to canine patellofemoral joints caused fractures in the zone of calcified cartilage visible by light microscopy, and caused articular cartilage fissures that extended from the articular surface to the transitional or superficial radial zone of the articular cartilage [10,12,75]. A study of the effects of blunt trauma on articular surfaces showed that articular cartilage could withstand impact loads of up to 25 newtons per square millimeter (25 MPa) without apparent damage. Impact loads exceeding this level caused chondrocyte death and cartilage fissures [69]. The authors suggested that reaching a stress level that could cause cartilage damage required a force greater than that necessary to fracture the subchondral bone. Another study [35] measured the pressure on human patellofemoral articular cartilage during impact loading, and found that impact loads less than the level necessary to fracture bone caused stresses greater than 25 MPa in some regions of the articular surface. With the knee flexed 90 degrees, 50% of the load necessary to cause a bone fracture produced joint pressures greater than 25 MPa for nearly 20% of the patellofemoral joint. At 70% of the bone fracture load, nearly 35% of the contact area of the patellofemoral joint pressures exceeded 25 MPa, and at 100% of the bone fracture load, 60% of the patellofemoral joint pressures exceeded 25 MPa.

Although the magnitude of the applied force is important in determining the extent of articular surface damage [36,67], the rate of loading also is important. Slowly applied loads and suddenly applied loads differ considerably in their effects. The articular cartilage extracellular matrix consists of water and a macromolecular framework formed primarily by collagens and large aggregating proteoglycans [15]. The collagens give the tissue its form and tensile strength, and the interaction of aggregating proteoglycans with water give the tissue its stiffness to compression, resilience, and durability. Both the solid component of the matrix, primarily the collagen–proteoglycan macromolecular framework, and the fluid component, have roles in bearing loads applied to articular surfaces. Following all but the most rapidly applied forces, loading of articular surfaces causes movement of fluid and realignment of the matrix macromolecules, which dampen and distribute loads within the cartilage and to the subchondral bone [65,66]. When this occurs slowly, the fluid movement and re-alignment of the macromolecular framework allows the cartilage to deform. These changes decrease the stress applied to the matrix macromolecular framework. When loading occurs too rapidly for fluid movement through the matrix, as with sudden impact or torsional loading of the joint, the matrix macromolecular framework sustains a greater force, and it is likely the chondrocytes are subjected to more damaging levels of mechanical stress. If this force is great enough it ruptures the matrix macromolecular framework [47], damages or kills the cells [49], and exceeds the ability of articular cartilage to prevent subchondral bone damage by dampening and distributing loads. For these reasons, rapidly applied loads may cause more articular surface damage than more slowly applied loads of the same magnitude.

3. Types of articular surface mechanical injury

Based on the type of tissue damage, articular surface injuries caused by mechanical forces can be classified into three types: (1) damage to the cells and matrices of articular cartilage and subchondral bone without visible disruption of the joint surface, (2) visible mechanical disruption of the articular surface limited to articular cartilage that takes the form of chondral fissures, flap tears or chondral defects, and (3) visible mechanical disruption of articular cartilage and bone, that is, intra-articular fractures [11,16, 19,20,23]. Each type of tissue damage stimulates a different repair response.

3.1. Cell and matrix damage

Articular cartilage damage that leaves the articular surface grossly intact occurs with almost every joint injury [10,30,44,56,75,79]. The intensity and type of joint loading that can cause chondral and subchondral damage without visible articular surface disruption has not been well defined. Physiologic levels of joint loading do not appear to cause joint injury, but impact loading above that generated by normal activities like walking or lifting light objects, but less than that necessary to produce cartilage disruption, can cause alterations of the cartilage matrix and damage chondrocytes [10,45,46,56]. Experimental evidence shows that following impact loading, loss of proteoglycans or alteration of their organization occurs before other signs of cartilage injury. The loss of proteoglycans may be due either to increased degradation or decreased synthesis. Significant loss of matrix proteoglycans decreases cartilage stiffness and increases its permeability. These alterations may cause greater loading of the remaining macromolecular framework, including the collagen fibrils, increasing the vulnerability of the tissue to further damage from loading. These injuries may cause other matrix abnormalities besides loss of proteoglycans, such as distortions of the collagen fibril meshwork or disruptions of the collagen fibril proteoglycan relationships, and swelling of the matrix [30], and they may injure or kill chondrocytes [49,56].

The ability of chondrocytes to sense changes in matrix composition and synthesize new molecules makes it possible for them to repair damage to the macromolecular framework [58,61]. It is not clear at what point this type of injury becomes irreversible and leads to progressive loss of articular cartilage. Presumably, the chondrocytes can restore the matrix as long as the loss of matrix proteoglycan does not exceed what the cells can rapidly produce, if the fibrillar collagen meshwork remains intact and if enough chondrocytes remain capable of responding to the matrix damage. When these conditions are not met the cells cannot restore the matrix, the chondrocytes will be exposed to excessive mechanical and metabolic stresses and the tissue will degenerate.

3.2. Cartilage disruption

Chondrocytes respond to injuries that disrupt articular cartilage, but that do not extend into subchondral bone. Following this type of injury they proliferate and increase synthesis of matrix macromolecules near the injury; but, the newly synthesized matrix and proliferating cells do not fill the tissue defect, and soon after injury the increased proliferative and synthetic activity ceases [10,16]. This leaves a permanent articular surface defect that can alter joint mechanical function and increase the risk of joint degeneration.

3.3. Cartilage and subchondral bone disruption (intra-articular fractures)

Unlike injuries limited to cartilage, injuries that extend into subchondral bone cause hemorrhage and fibrin clot formation, and activate the inflammatory response [10,16,18]. Soon after injury, blood escaping from the damaged bone blood vessels forms a hematoma that temporarily fills the injury site. Fibrin forms within the hematoma and platelets bind to fibrillar collagen. A continuous fibrin clot fills the bone defect and extends for a variable distance into the cartilage defect. Platelets within the clot release vasoactive mediators and growth factors or cytokines, including transforming growth factor beta and platelet derived growth factor. Bone matrix also contains growth factors such as transforming growth factor beta, bone morphogenic protein, platelet derived growth factor, insulin like growth factor I, insulin-like growth factor II, and possibly others. Release of these growth factors may have an important role in the repair of osteochondral defects. In particular, they probably stimulate vascular invasion and migration of undifferentiated cells into the clot, and influence the proliferative and synthetic activities of the cells. Shortly after entering the tissue defect, the undifferentiated mesenchymal cells proliferate and synthesize a new matrix. Within two weeks of injury, some mesenchymal cells assume the rounded form of chondrocytes and begin to synthesize a matrix that contains type II collagen and a relatively high concentration of proteoglycans. These cells produce regions of hyaline-like cartilage in the chondral and bone portions of the defect. Six to eight weeks following injury the repair tissue within the chondral region of osteochondral defects contains many chondrocyte-like cells in a matrix consisting of type II collagen, proteoglycans, some type I collagen and noncollagenous proteins. Unlike the cells in the chondral portion of the defect, the cells in the bony portion of the defect produce immature bone, fibrous tissue and hyaline like cartilage. The chondral repair tissue typically has a composition and structure intermediate between hyaline cartilage and fibrocartilage; it rarely, if ever, replicates the elaborate structure of normal articular cartilage [10,16,18].

3.4. Age related loss of chondrocyte function and articular cartilage degeneration

Articular cartilage matrix undergoes significant structural, molecular, and mechanical changes with age, including surface fibrillation, alteration of proteoglycan structure and composition, increased collagen cross linking, and decreased tensile strength and stiffness [15,17,26]. Deterioration of chondrocyte function accompanies these changes in the matrix. The cells synthesize smaller aggrecans and less functional link proteins leading to the formation of smaller more irregular proteoglycan aggregates [13,21,22, 24,60,74]. The mitotic and synthetic activity of human chondrocytes decline with age [9,60,63]. Human chondrocytes become less responsive to anabolic mechanical stimuli (Personal Communication – Alan Grodzinsky, 2002) and rat articular cartilage chondrocytes show an age related decline in response to the anabolic cytokine, insulin-like growth factor I (IGF-I) [61,64]. Given these aging changes in chondrocyte function, it is not surprising that after age 40 the incidence of the joint degeneration responsible for the clinical syndrome of osteoarthritis increases with each passing decade [16,17], and the effectiveness of articular surface repair declines [20]. The risk of developing post-traumatic osteoarthritis following an intra-articular fracture of knee increases three to four fold after 50 years of age [43,72,77,78], and *in vitro* compressive injury causes a more severe decrease in glycosaminoglycan synthesis in chondrocytes from people over 50 years of age than in chondrocytes from people less than 35 years of age [29].

Perichondrial arthroplasties and drilling of articular surfaces are much less successful in patients over age 50 than in patients under age 30 [52,70]. The age related loss of chondrocyte function and resulting increased risk of joint degeneration and decreased effectiveness of articular cartilage repair appear to be caused at least in part by age related loss of chondrocyte function [5,27,37–39,58–60,62,63].

4. Mechanisms of age related loss of chondrocyte function

In 1879, Minot noted that "With each successive generation of cells the power of growth diminishes. This loss of power I term senescence" [71]. In the 1960s, Hayflick's demonstration that cells lose their ability to replicate after a certain number of cells divisions confirmed Minot's observation and defined the phenomenon of replicative cell senescence, the "Hayflick Limit" [37–39]. Although loss of the ability of cells to divide is the recognized measure of senescence, cell function may deteriorate before cells reach cell cycle arrest [27]. Abnormal protein synthesis, altered growth factor responses, and longer population doubling times, appear well in advance of growth arrest in continuously dividing somatic cell populations. This suggests that cell populations drift toward senescence early in their replicative lifespans; that is, they develop progressive cell or phenotypic senescence well in advance of arrest of the cell cycle. Telomere erosion and oxidative damage are two possible explanations of cell senescence.

4.1. Telomere erosion

Recent evidence of an association between shorter telomeres in white blood cells and death from heart disease and infectious diseases suggests that shorter telomere length increases the risk of disease [28]. Our studies of human articular cartilage chondrocytes shows that with increasing age chondrocyte telomere length decreases [63]. Telomeres, DNA sequences at the ends of chromosomes, are necessary for chromosomal replication but they also may help maintain normal chromosome function by preventing enzymatic degradation and clumping of chromosomes. The growth of most somatic cell populations is sharply limited by telomere erosion, which proceeds in increments with each round of cell division [1–3]. The telomere erosion hypothesis suggests that cell aging is regulated by an intrinsic genetic "clock" associated with erosion of telomeres [8,27]. Most authors have assumed that chondrocyte mitotic activity in normal human articular cartilage occurs rarely after skeletal maturity [15–17]; however, even a slow rate of cell division would be sufficient to cause telomere erosion over many decades. Furthermore, chondrocyte mitotic activity increases several-fold following cartilage injury [16,20]; therefore, repetitive intense joint loading or joint trauma could accelerate telomere erosion and help explain the association between repetitive excessive joint loading and the development of joint degeneration [14].

4.2. Oxidative damage

Although the evidence for an association between telomere erosion and chondrocyte senescence is strong, cumulative cell damage from oxidative stress provides an alternative or complimentary explanation for chondrocyte senescence [31,76]. Stress-induced senescence is manifested in age-related degeneration of mitochondria [34]. Mitochondria provide metabolic energy in the form of ATP and NADH via respiration, supporting the anabolic and catabolic activities that cells require for tissue maintenance. Mitochondria also protect cells from the toxic effects of oxygen and its free-radical derivatives. Thus, mitochondrial damage both limits energy production and exposes cells to oxidative damage.

Numerous studies have shown age-related declines in the respiratory activity of tissues such as bone, skeletal muscle, liver, and brain [4]. The decline is due to progressive age-dependent decrease in the number and activity of mitochondria [6,7,51], a phenomenon related to alterations in the mitochondrial genome. Deletions, duplications and point mutations that disrupt expression of electron transport proteins accumulate with age [4,48,53]. These events lead to increasing free radical production as a consequence of faulty electron transport. Oxidative damage to mitochondria can initiate apoptosis through caspase activation but may also lead to irreversible growth arrest similar to replicative senescence [76].

We found that increasing number of human chondrocyte cell divisions *in vitro* causes degradation of the mitochondrial genome and a decline in the function and numerical density of chondrocyte mitochondria [60]. This observation suggests that mitochondrial degeneration and loss is associated with chondrocyte senescence. Although it remains to be seen if mitochondrial degeneration increases with aging *in vivo*, our data suggest that functional lifespan of chondrocytes may be limited by mitochondrial degeneration. These findings also suggest that metabolic stress plays a role in chondrocyte senescence by contributing to free-radical production that could accelerate mitochondrial degeneration.

In addition to their effects on mitochondrial DNA, free radicals also damage chromosomal DNA. Intriguingly, multiple lines of evidence indicate that oxidation plays a role in telomere erosion. Polyguanosine sequences in telomere DNA are hypersensitive to oxidative de-polymerization [55]. In telomerase-deficient cells, oxidative stress promotes telomere attrition and the antioxidant ascorbic acid slows the rate of age-related telomere loss [32,73]. These findings implicate oxidative damage as an accelerant of telomere-dependent aging processes and suggest a role for oxidative stress in the age-related telomere erosion and senescence of articular cartilage chondrocytes.

5. Acceleration of chondrocyte senescence

Although high levels of mechanical force applied to articular surfaces can kill chondrocytes [49,56], many cells survive even the force necessary to cause mechanical disruption of articular cartilage and intra-articular fractures [30,49,75]. In most instances the joint degeneration that causes post-traumatic osteoarthritis develops over several years following joint injury, and in some patients it does not develop for more than a decade after injury. Clinical experience shows that the development and progression of post-traumatic osteoarthritis is not necessarily due to rapid wear of an irregular articular surface. Even when surgeons restore the congruity, alignment and stability of a fractured joint, the joint surface may degenerate [57]. Furthermore, apparently normal articular surfaces often degenerate following injuries that cause residual joint instability [57]. These observations indicate that chondrocytes that survive an acute joint injury may have decreased ability to maintain and repair the tissue and that repetitive abnormally high joint surface loading due to joint instability causes chondrocytes to lose their ability to maintain the tissue. We hypothesized that accelerated chondrocyte senescence due to oxidative damage contributes to the progressive degeneration of anatomically restored fractured articular surfaces and the articular surfaces of joints with post-traumatic instability, mal-alignment or incongruity.

5.1. Mechanical stress

To determine if application of mechanical force to articular cartilage can cause oxidative damage, we tested the effects of high shear stress (2 MPa axial compression at 1 Hz cyclic loading) as compared with compressive stress (simultaneous 2 MPa axial compression and 2 MPa transverse compression at 1 Hz cyclic loading) on cartilage explants. Implants were placed in a triaxial compression vessel that can apply

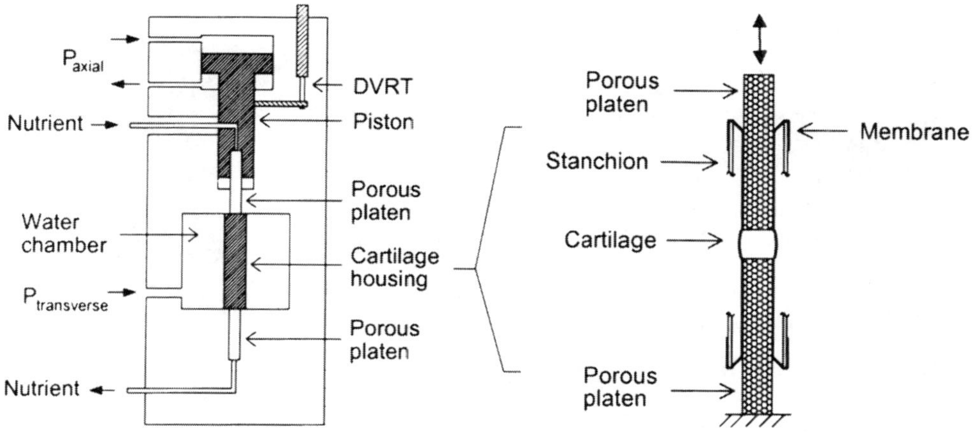

Fig. 1. Schematic drawing of the triaxial pressure vessel. The cartilage housing secures cartilage samples within the pressure vessel.

variable shear stress independently of compressive stress [40–42] (Fig. 1). The triaxial compression vessel subjects cylindrical cartilage explants to two independent compression inputs: (1) axial, from platen contact against the cylinder ends, and (2) radially transverse, from external fluid compression of an impermeable but distensible membrane jacketing the curved cylinder wall. The cartilage explant (4 mm diameter) is mounted in a sub-component denoted as the cartilage housing (Fig. 1). The cartilage housing holds the explant in position within the main housing (Fig. 1), and chemically isolates the explant from the water in the transverse pressurization chamber, by means of a thin (0.8 mm) silicone sheath, stretched inside-out over the ends of a perforated cylindrical stanchion. Pressurization of the water in this chamber applies radial transverse compression to the cartilage explant. Axial loading is applied by a low-friction piston, driven by air pressure.

To maintain free exchange of nutrients and waste products, the top and bottom surfaces of the cartilage explant are constantly bathed with culture medium. This is accomplished by peristaltically pumping medium through the porous platens at a very low flow rate (<0.001 ml/minute). The medium flows around the explant (between the explant and silicon tubing) and fills the space below the explant. Medium flow is interrupted by application of transverse compression (50% of duty cycle) but resumes when the transverse compression is released. Pilot studies using histologic analysis of $^{35}SO_4$ incorporation revealed that these nutrient conditions supported the viability of cells in explants treated with axial + transverse compression for 1 hour (data not shown).

Dihydroethidium was used to detect release of reactive oxygen species [54]. This cell-permanent probe is non-fluorescent in reduced form, but intercalates in DNA and fluoresces when oxidized to ethidium. Tibial and talar articular cartilage explants were harvested from the ankle joint of a 52 year old donor and cultured for seven days in 5% oxygen atmosphere to mimic conditions in ankle joints and to minimize oxidative stress prior to mechanical stress treatment. Dihydroethidium (5 μM) was added to the medium 15 minutes before stress treatment. Random explants were treated with 2 mM n-acetyl cysteine (NAC), a potent anti-oxidant. The explants were then exposed to high and low shear stress conditions. After stress treatment the explants were immediately snap-frozen in cryoembedding medium. Ten μM-thick sections were mounted on slides and examined using epi-fluorescent optics. Figure 2 shows that high shear stress increased release of reactive oxygen species and that n-acetyl cysteine blocked the stress induced release. Other experiments showed that high shear stress suppressed aggrecan synthesis (data not shown).

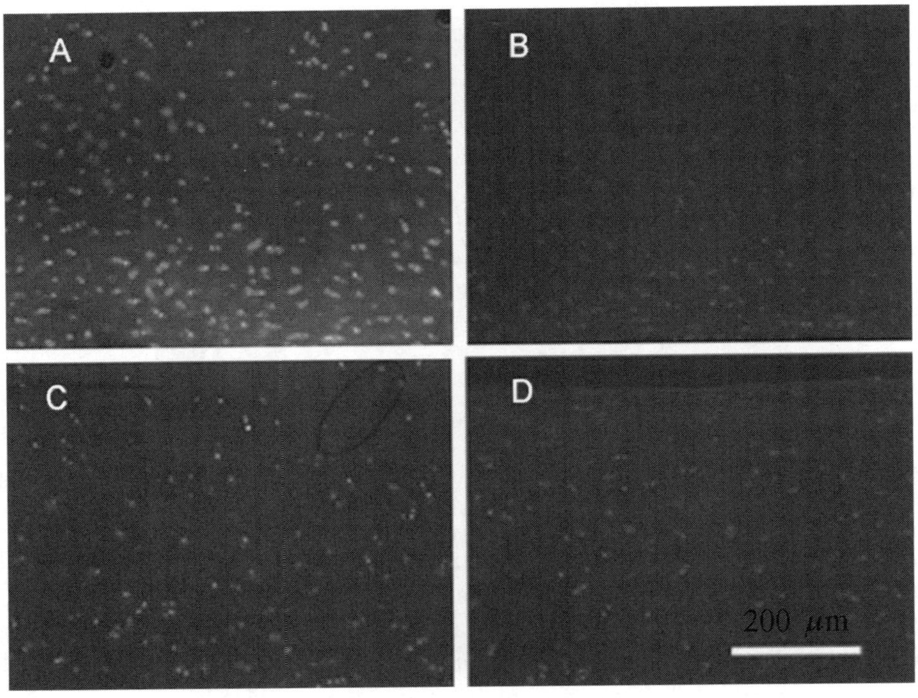

Fig. 2. Shear-stress induced release of reactive oxygen species. Fluorescence micrographs of cryosections of human articular cartilage explants stained with dihydroethidium in the absence of n-acetyl cysteine (A&C) or in the presence of n-acetyl cysteine (B&D). Explants were exposed to high-shear stress (A&B) or low-shear stress (C&D). The articular surface is at the top of each image and the bar shown in D represents 200 μm.

5.2. *Oxidative stress*

To determine if oxidative damage to young chondrocytes can accelerate senescence we exposed tibial articular chondrocytes from a 21 year old donor to sub-lethal levels of tert-butylhydroperoxide, a potent oxidant. Treatment of chondrocytes for 4 days with 20 μM tert-butylhydroperoxide (4 hours per day) caused increased expression of senescence-associated β-galactosidase and DNA oxidation, and decreased mitochondrial function (Fig. 3). A second experiment tested the cumulative effects of tert-butylhydroperoxide treatments applied weekly over a period of several weeks. This study showed that 3 or more treatments significantly increased β-galactosidase expression (Fig. 4) and induced population growth arrest (Fig. 5). These results indicate that the cellular damage caused by sub-lethal peroxide doses is cumulative and accelerates chondrocyte senescence.

6. Conclusions

Joint injuries cause increased mechanical stress on articular cartilage and certain patterns of mechanical stress, in particular high levels of shear stress, increase release of reactive oxygen species and decrease proteoglycan synthesis. In addition to high levels of mechanical stress, joint injuries can cause synovitis and recent work shows that chondrocytes produce potentially damaging levels of oxygen-free radicals in response to inflammatory cytokines [68]. Oxidative stress on chondrocytes accelerated chondrocyte senescence as demonstrated by increased expression of beta-galactosidase, DNA damage, and

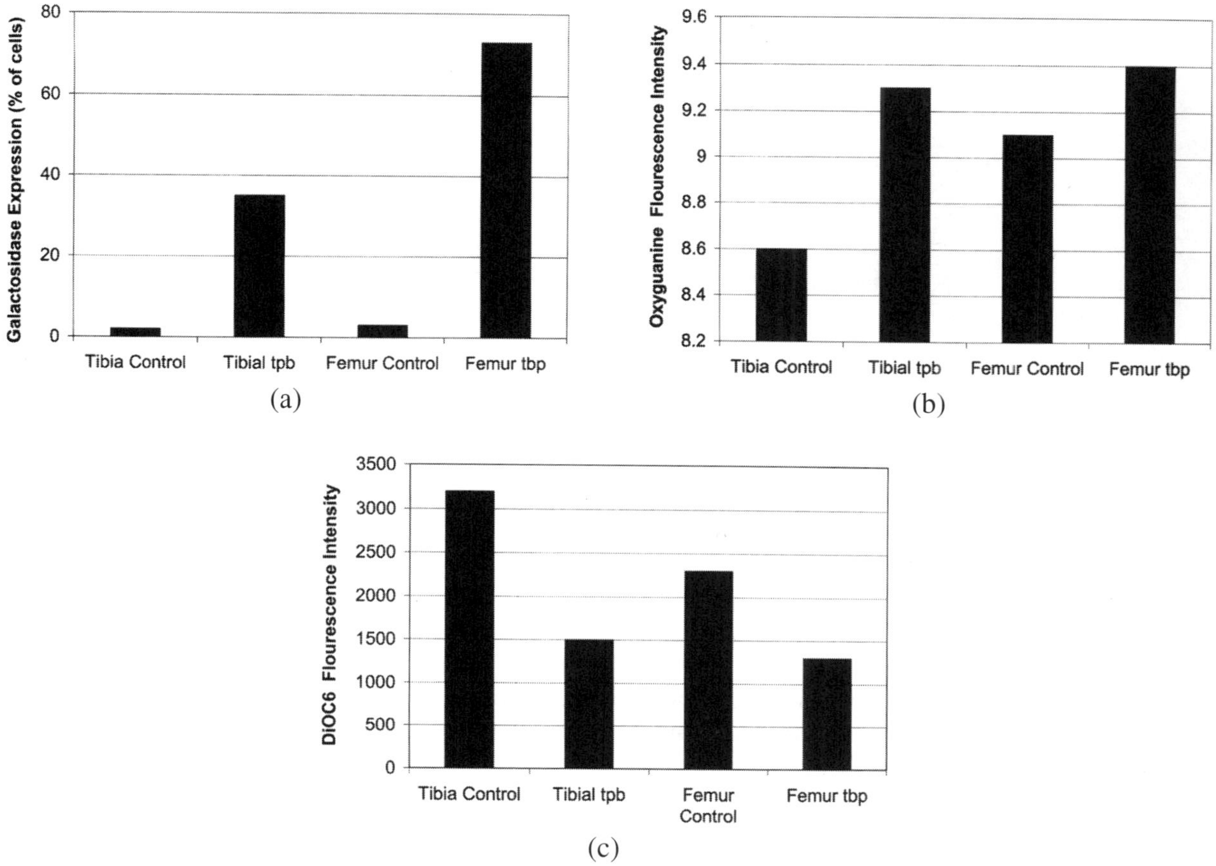

Fig. 3. Histograms showing the effects of oxidative stress from tert-butylhydroperoxide (tbp) on articular cartilage chondrocytes from a 21 year old donor. (a): Expression of senescence associated beta-galactosidase. (b): Oxyguanine fluorescence intensity (a measure of DNA damage). (c): DiOC6 fluorescence intensity (a measure of mitochondrial electron transport proteins).

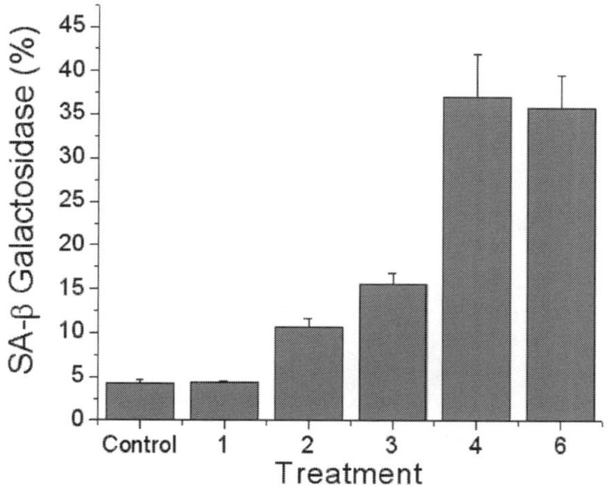

Fig. 4. The effects of repeated tert-butylhydroperoxide exposure on the percent of cells expressing beta-galactosidase. A succession of three exposures increased the percent of cells expressing beta-galactosidase.

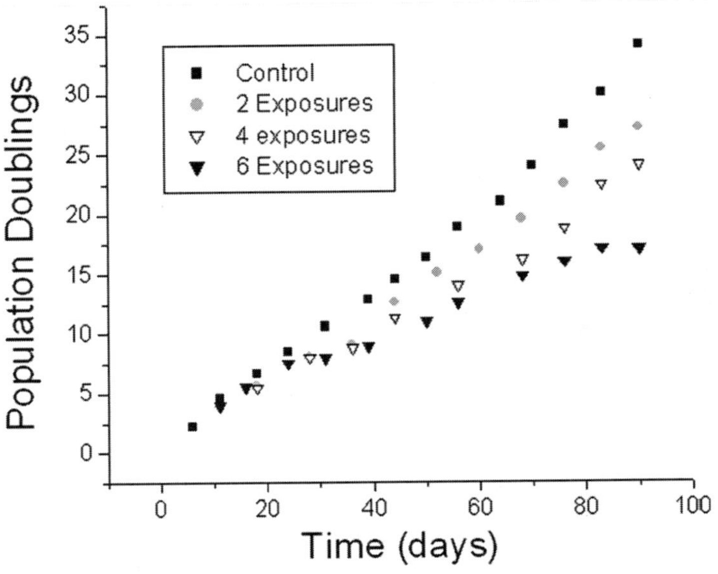

Fig. 5. The effects of repeated tert-butylhydroperoxide exposure on human chondrocyte population doublings. Two exposures slowed the rate of cell proliferation. Six exposures caused early replicative senescence.

a decrease in mitochondrial electron transport proteins. These data strongly suggest that chondrocyte senescence contributes to the risk of post-traumatic articular cartilage degeneration by decreasing the ability of the cells to maintain and repair the tissue. The most likely mechanisms of chondrocyte senescence are telomere erosion and oxidative damage. In patients who develop post-traumatic osteoarthritis these mechanisms may both contribute to accelerated chondrocyte senescence: articular cartilage injury stimulates chondrocyte proliferation that increases metabolic stress on the cells and causes telomere erosion, and oxidative stress due to excessive mechanical loading causes telomere erosion and mitochondrial damage. For these reasons accelerated chondrocyte senescence may be an important component of the pathogenesis of post-traumatic osteoarthritis. Strategies that may decrease the risk of post-traumatic osteoarthritis include minimizing oxidative damage, treatments that provide the optimal mechanical environment for restoration of articular surfaces following injury, and, in patients with severe joint disruption, introducing new cell populations soon after injuries.

Acknowledgement

This work was supported by grant P50 AR48939 from the National Institute of Arthritis and Musculoskeletal and Skin Diseases.

References

[1] R. Allsopp, E. Chang, M. Kashefi-Aazam, E. Rogaec, M. Piatyszek, J. Shay and C. Harley, Telomere shortening is associated with cell division in vitro and in vivo, *Exp. Cell Res.* **220** (1995), 194–200.

[2] R. Allsopp and C. Harley, Evidence for a critical telomere length in senescent human fibroblasts, *Exp. Cell Res.* **219** (1995), 130–136.

[3] R. Allsopp, H. Vaziri, M. Piatyszek, S. Goldstein, E. Younglai, A. Futcher, C. Greider and C. Harley, Telomere length predicts replicative capacity of human fibroblasts, *Proc. Natl. Acad. Sci. USA* **89** (1992), 10114–10118.

[4] N. Arnheim and G. Cortopassi, Deleterious mitochondrial DNA mutations accumulate in aging human tissues, *Mutation Res.* **275** (1992), 157–167.
[5] A. Barbero, S.P. Grogan, S. Ploegert, O. Demarteau, M. Koerner, T. Schaffner, D. Schaffner, P. Mainil-Varlet and I. Martin, Age related changes in chondrocyte biology, in: *4th International Cartilage Repair Symposium*, Toronto, Canada, 2002.
[6] K.B. Beckman and B.N. Ames, Endogenous oxidative damage of mtDNA, *Mutat. Res.* **424** (1999), 51–58.
[7] K.B. Beckman and B.N. Ames, Oxidative decay of DNA, *J. Biol. Chem.* **272**(32) (1997), 19633–19636.
[8] E. Blackburn, Structure and function of telomeres, *Nature* **350** (1991), 569–572.
[9] M.C. Bolton, J. Dudhia and M.T. Bayliss, Age-related changes in the synthesis of link protein and aggrecan in human articular cartilage: implications for aggregate stability, *Biochem. J.* **337** (1999), 77–82.
[10] J.A. Buckwalter, Articular cartilage injuries, *Clin. Ortho. Rel. Res.* **402** (2002), 21–37.
[11] J.A. Buckwalter, Evaluating methods of restoring cartilagenous articular surfaces, *Clin. Ortho. Rel. Res.* **367S** (1999), 224–238.
[12] J.A. Buckwalter, Mechanical injuries of articular cartilage, in: *Biology and Biomechanics of the Traumatized Synovial Joint*, G. Finerman, ed., American Academy of Orthopaedic Surgeons, Park Ridge, IL, 1992, pp. 83–96.
[13] J.A. Buckwalter, H. Choi, L. Tang, L. Rosenberg and R. Ungar, The effect of link protein concentration of proteoglycan aggregation, in: *Trans. 32nd Meeting Ortho. Res. Soc.*, 1986, pp. 11–98.
[14] J.A. Buckwalter and N.E. Lane, Athletics and osteoarthritis, *Am. J. Sports Med.* **25** (1997), 873–881.
[15] J.A. Buckwalter and H.J. Mankin, Articular cartilage I. Tissue design and chondrocyte–matrix interactions, *J. Bone Joint Surg.* **79A**(4) (1997), 600–611.
[16] J.A. Buckwalter and H.J. Mankin, Articular cartilage II. Degeneration and osteoarthrosis, repair, regeneration and transplantation, *J. Bone Joint Surg.* **79A**(4) (1997), 612–632.
[17] J.A. Buckwalter, J.A. Martin and H.J. Mankin, Synovial joint degeneration and the syndrome of osteoarthritis, *Instructional Course Lectures* **49** (2000), 481–489.
[18] J.A. Buckwalter, J.A. Martin, M. Olmstead, K.A. Athanasiou, M.P. Rosenwasser and V.C. Mow, Osteochondral repair of primate knee femoral and patellar articular surfaces: Implications for preventing post-traumatic ostearthritis, *Iowa Ortho. J.* **23** (2003), 66–74.
[19] J.A. Buckwalter and V.C. Mow, Cartilage repair in osteoarthritis, in: *Osteoarthritis: Diagnosis and Management*, 2nd edn, R.W. Moskowitz, D.S. Howell, V.M. Goldberg and H.J. Mankin, eds, Saunders, Philadephia, PA, 1992, pp. 71–107.
[20] J.A. Buckwalter, V.C. Mow and E.B. Hunziker, Concepts of cartilage rpair in osteoarthritis, in: *Osteoarthritis: Diagnosis and Medical/Surgical Management*, 3rd edn, R. Moskowitz, V.M. Goldberg, D. Howell, R. Altman and J.A. Buckwalter, eds, Saunders, Philadelphia, PA, 2001, pp. 101–114.
[21] J.A. Buckwalter and L.C. Rosenberg, Electron microscopic studies of cartilage proteoglycans, *Elec. Microsc. Rev.* **1** (1988), 87–112.
[22] J.A. Buckwalter and L.C. Rosenberg, Electron microscopic studies of cartilage proteoglycans: Direct evidence for the variable length of the chondroitin sulfate rich region of the proteoglycan subunit core protein, *J. Biol. Chem.* **257** (1982), 9830–9839.
[23] J.A. Buckwalter, L.C. Rosenberg, R. Coutts, E. Hunziker, A.H. Reddi and V.C. Mow, Articular cartilage: Injury and repair, in: *Injury and Repair of the Musculoskeletal Soft Tissues*, S.L. Woo and J.A. Buckwalter, eds, American Academy of Orthopaedic Surgeons, Park Ridge, IL, 1988, pp. 465–482.
[24] J.A. Buckwalter, P.J. Roughley and L.C. Rosenberg, Age-related changes in cartilage proteoglycans: quantitative electron microscopic studies, *Micros. Res. Tech.* **28** (1994), 398–408.
[25] J.A. Buckwalter and C.L. Saltzman, Ankle osteoarthritis: distinctive characteristics, *Instr. Course Lect.* **48** (1999), 233–241.
[26] J.A. Buckwalter, S.L.-Y. Woo, V.M. Goldberg, E.C. Hadley, F. Booth, T.R. Oegema and D.R. Eyre, Soft tissue aging and musculoskeletal function, *J. Bone Joint Surg.* **75A** (1993), 1533–1548.
[27] J. Campisi, Replicative senescence and immortalization, in: *The Molecular Basis of Cell Cycle and Growth Control*, G.S. Stein, ed., Wiley/Liss, New York, 1999, pp. 348–373.
[28] R.M. Cawthon, K.R. Smith, E. O'Brien, A. Sivatchenko and R.A. Kerber, Association between telomere length in blood and mortality in people aged 60 years and older, *Lancet* **361** (2003), 393–395.
[29] D.D. D'Lima, A. Bergula, M. Lotz and C.W. Colwell, Effect of age and osteoarthritis on glycosaminoglycan synthesis, in: *4th International Cartilage Repair Symposium*, Toronto, Canada, 2002.
[30] J.M. Donohue, D. Buss, T.R. Oegema and R.C. Thompson, The effects of indirect blunt trauma on adult canine articular cartilage, *J. Bone Joint Surg.* **65-A**(7) (1983), 948–956.
[31] P. Dumont, Q.M. Chen, M. Burton, E.S. Gonos, C. Frippiat, J.B. Mazarati, F. Eliaers, J. Remacle and O. Toussaint, Induction of replicative senescence biomarkers by sublethal oxidative stress in normal human fibroblasts, *Free Rad. Biol. Med.* **28** (2000), 361–373.
[32] K. Furumoto, E. Inoue, N. Nagao, E. Hiyama and N. Miwa, Age-dependent telomere shortening is slowed down by enrichment of intracellular vitamin C via suppression of oxidative stress, *Life Sci.* **63**(11) (1998), 935–948.

[33] A.C. Gelber, M.C. Hockberg, L.A. Mead, N.Y. Wang, F.M. Wigley and M.J. Klag, Joint injury in young adults and the risk for susequent knee and hip osteoarthritis, *Ann. Int. Med.* **133** (2000), 321–328.
[34] T. Golden and S. Melov, Mitochondrial DNA mutations, oxidative stress, and aging, *Mech. Ageing Dev.* **122** (2001), 1577–1589.
[35] R.C. Haut, Contact pressures in the patellofemoral joint during impact loading on the human flexed knee, *J. Ortho. Res.* **7** (1989), 272–280.
[36] R.C. Haut, T.M. Ide and C.E. Camp, Mechanical responses of the rabbit patello-femoral joint to blunt impact, *J. Biomech. Eng.* **117** (1995), 402–408.
[37] L. Hayflick, *How and Why We Age*, Ballantine Books, New York, 1996, p. 377.
[38] L. Hayflick, The limited in vitro lifetime of human diploid cell strains, *Exp. Cell Res.* **37** (1965), 614–636.
[39] L. Hayflick and P.S. Moorhead, The serial cultivation of human diploid cell strains, *Exp. Cell Res.* **25** (1961), 585–621.
[40] A.D. Heiner and J.A. Martin, Cartilage responses to a novel triaxial mechanostimulatory culture system, *Journal of Biomechanics*, In press.
[41] A.D. Heiner and J.A. Martin, The triaxial compression vessel: A novel mechanically active culture system for cartilage explants, in: *Trans. 24th Annual Meeting of the American Society of Biomechanics*, 2000, pp. 15–16.
[42] A.D. Heiner, J.A. Martin and T.D. Brown, Cartilage responses to stress in a novel triaxial mechanical stress culture system, in: *Trans. 4th Combined Orthopaedic Research Societies Meeting*, 2001, pp. 141.
[43] S.E. Honkonen, Degenerative arthritis after tibial plateau fractures, *J. Orthop. Trauma* **9** (1995), 272–277.
[44] G.D. Jay, K.A. Elsaid and C.O. Chichester, Early damage to the articular cartilage matrix occurs following traumatic knee injuries without fracture, *Acad. Emerg. Med.* **10**(5) (2003), 496.
[45] J.E. Jeffrey, D.W. Gregory and R.M. Aspden, Matrix damage and chondrocyte viability following a single impact load on articular cartilage, *Arch. Biochem. Biophys.* **10** (1995), 87–96.
[46] J.E. Jeffrey, L.A. Thomson and R.M. Aspden, Matrix loss and synthesis following a single impact load on articular cartilage in vitro, *Biochem. Biophys. Acta* **15** (1997), 223–232.
[47] V. Kafka, Surface fissures in articular cartilage: new concepts, hypotheses and modeling, *Clin. Biomech.* **17** (2002), 73–80.
[48] C.M. Kang, B.S. Kristal and B.P. Yu, Age-related mitochondrial DNA deletions: effect of dietary restriction, *Free Radical Biology & Medicine* **24**(1) (1998), 148–154.
[49] H. Kim, M. Lo and R. Pillarisetty, Chondrocyte apopotosis following intraarticular fracture in humans, *Osteoarthritis Cartilage* **10** (2002), 747–749.
[50] H. Koepp, W. Eger, C. Muehleman, A. Valdellon, J.A. Buckwalter, K.E. Keuttner and A.A. Cole, Prevalence of articular cartilage degeneration in the ankle and knee joints of human organ donars, *J. Ortho. Science* **4** (1999), 407–412.
[51] G. Kopsidas, S.A. Kovalenko, J.M. Kelso and A.W. Linnane, An age-associated correlation between cellular bioenergy decline and mtDNA rearrangements in human skeletal muscle, *Mutation Res.* **421** (1998), 27–36.
[52] T. Kumai, Y. Takakura, I. Higashiyama and S. Tamai, Arthroscopic drilling for the treatment of osteochondral lesions of the talus, *J. Bone Joint Surg.* **81A** (1999), 1229–1235.
[53] C.M. Lee, R. Weindrich and J.M. Aiken, Age-related alterations of the mitochondrial genome, *Free Radical Biology & Medicine* **22**(7) (1997), 1259–1269.
[54] W. Li, F. Miller, M. Brown, P. Chatterjee, G. Aylsworth, J. Shao, A. Spector, L. Oberley and N. Weintraub, Enhanced H_2O_2-induced cytotoxity in "epitheliod" smooth muscle cells implications for neointimal regression, *Arterio. Thromb. Vasc. Biol.* **20** (2000), 1473–1479.
[55] L. Liu, J.R. Trimarchi, P. Navarro, M.A. Blasco and D.L. Keefe, Oxidative stress contributes to arsenic-induced telomere attrition, chromosome instability and apoptosis, *J. Biol. Chem.* (2003).
[56] A.M. Loening, I.E. James, M.E. Levenston, A.M. Badger, E.H. Frank, B. Kurz, M.E. Nuttall, H.H. Hung, S.M. Blake, A.J. Grodzinsky and M.W. Lark, Injurious mechanical compression of bovine articular cartilage induces chondrocyte apoptosis, *Arch. Biochem. Biophys.* **381** (2000), 205–212.
[57] J.L. Marsh, J.A. Buckwalter, R. Gelberman, D. Dirschl, S. Olson, T.D. Brown and A. Llinias, AOA symposium – articular fractures: Does an anatomic reduction really change the result?, *J. Bone Joint Surg.* **84A** (2002), 1259–1271.
[58] J.A. Martin and J.A. Buckwalter, Human chondrocyte senescence and osteoarthritis, *Biorheology* **39** (2002), 145–152.
[59] J.A. Martin and J.A. Buckwalter, Human chondrocyte senesence, in: *4th International Cartilage Repair Symposium*, Toronto, Canada, 2002.
[60] J.A. Martin and J.A. Buckwalter, The role of chondrocyte senescence in the pathogenesis of osteoarthritis and in limiting cartilage repair, *J. Bone Joint Surg. Am.* **85-A** (Suppl. 2) (2003), 106–110.
[61] J.A. Martin and J.A. Buckwalter, The role of chondrocyte–matrix interactions in maintaining and repairing articular cartilage, *Biorheology* **37** (2000), 129–140.
[62] J.A. Martin and J.A. Buckwalter, Roles of articular cartilage aging and chondrocyte senescence in the pathogenesis of osteoarthritis, *Iowa Ortho. J.* **21** (2001), 1–7.
[63] J.A. Martin and J.A. Buckwalter, Telomere erosion and senescence in human articular cartilage chondrocytes, *J. Gerontology Biol. Sci.* **56A** (2001), B172–B179.

[64] J.A. Martin, S.M. Ellerbroek and J.A. Buckwalter, The age-related decline in chondrocyte response to insulin-like growth factor-I: The role of growth factor binding proteins, *J. Ortho. Res.* **15** (1997), 491–498.

[65] V.C. Mow and M.P. Rosenwasser, Articular cartilage: Biomechanics, in: *Injury and Repair of the Musculoskeletal Soft Tissues*, S.L. Woo and J.A. Buckwalter, eds, American Academy of Orthopaedic Surgeons, Park Ridge, IL, 1988, pp. 427–463.

[66] V.C. Mow, L.A. Setton, F. Guilak and A. Ratcliffe, Mechanical factors in articular cartilage and their role in osteoarthritis, in: *Osteoarthritic Disorders*, K.E. Kuettner and V.M. Goldberg, eds, American Academy of Orthopaedic Surgeons, Rosemont, IL, 1995, pp. 147–171.

[67] W.N. Newberry, J.J. Garcia, C.D. Mackenzie, C.E. Decamp and R.C. Haut, Analysis of acute mechanical insult in an animal model of post-traumatic osteoarthrosis, *J. Biomech. Eng.* **120** (1998), 704–709.

[68] M. Oh, K. Fukuda, S. Asada, Y. Yasuda and S. Tanaka, Concurrent generation of nitric oxide and superoxide inhibits proteoglycan synthesis in bovine articular chondrocytes: involvement of peroxynitrite, *J. Rheumatol.* **25** (1998), 2169–2174.

[69] R.U. Repo and J.B. Finlay, Survival of articular cartilage after controlled impact, *J. Bone Joint Surg.* **59-A**(8) (1977), 1068–1075.

[70] H. Seradge, J.A. Kutz, H.E. Kleinert, G.D. Lister, T.W. Wolff and E. Atasoy, Perichondrial resurfacing arthroplasty in the hand, *J. Hand Surg.* **9A** (1984), 880–886.

[71] J.A. Simpson and E.S.C. Weiner, eds, *The Oxford English Dictionary*, 2nd edn, Clarendon Press, Oxford, 1989.

[72] D.G. Stevens, R. Beharry, M.D. McKee, J.P. Wadell and E.H. Schemitsch, The long-term functional outcome of operatively treated tibial plateau fractures, *J. Orthop. Trauma* **15** (2001), 312–320.

[73] V.A. Szalai, M.J. Singer and H.H. Thorp, Site-specific probing of oxidative reactivity and telomerase function using 7,8-dihydro-8-oxoguanine in telomeric DNA, *J. Am. Chem. Soc.* **124**(8) (2002), 1625–1631.

[74] L.H. Tang, J.A. Buckwalter and L.C. Rosenberg, The effect of link protein concentration on articular cartilage proteoglycan aggregation, *J. Orthop. Res.* **14** (1996), 334–339.

[75] R.C. Thompson, T.R. Oegema, J.L. Lewis and L. Wallace, Osteoarthritic changes after acute transarticular load: an animal model, *J. Bone Joint Surg.* **73A** (1991), 990–1001.

[76] O. Toussaint, E. Medrano and T.V. Zglinicki, Cellular and molecular mechanisms of stress-induced premature senescence (sips) of human diploid fibroblasts and melanocytes, *Experimental Gerontology* **35** (2000), 927–945.

[77] G. Volpin, G.S. Dowd, H. Stein and G. Bentley, Degenerative arthritis after intra-articular fractures of the knee: long-term results, *J. Bone Joint Surg.* **72B** (1990), 634–638.

[78] V. Wright, Post-traumatic osteoarthritis – a medico-legal minefield, *Br. J. Rheumatol.* **29**(6) (1990), 474-478.

[79] H. Zang, M.S. Vrahas, R.V. Baratta and D.M. Rosler, Damage to rabbit femoral articular cartilage following direct impacts of uniform stresses: an in vitro study, *Clin. Biomechanics* **14** (1999), 543–548.

Biorheology 41 (2004) 493–502
IOS Press

Hyaluronan injection affects neither osteoarthritis progression nor loading of the OA knee in dogs

Kenneth D. Brandt *, Gerald N. Smith, Jr. and Stephen L. Myers
Rheumatology Division, Indiana University School of Medicine, 1110 West Michigan Street, Room 545, Indianapolis, IN 46202, USA

Abstract. We previously reported that intraarticular injections of hyaluronan (HA), administered prophylactically to dogs in whom knee osteoarthritis had been induced by transection of the anterior cruicate ligament, did not significantly modify the intraarticular pathology but decreased the proteogylcan concentration of the articular cartilage by as much as 30%. Because the cartilage proteoglycan concentration is directly related to the stiffness of the tissue, these results raised the possibility that intraarticular HA therapy could exacerbate OA. In the present study, using a different HA formulation, with a longer interval between intraarticular HA injection and examination of joint tissues, we found that neither prophylactic nor therapeutic administration of HA had an effect on the severity of OA pathology, the magnitude of vertical ground reaction forces generated by the unstable hind limb (a surrogate for joint pain), or the cartilage proteoglycan concentration. The data suggest that the suppression of proteoglycan synthesis induced by HA is temporary and fully reversible and that HA injections do not result in overloading of the OA extremity. A significant correlation was noted between the severity of chondropathy and the magnitude of the vertical ground reaction forces generated by the unstable limb.

Keywords: Intra-articular injection, gait, limb loading

1. Introduction

Intra-articular (IA) injection of hyaluronan (HA) is currently used as symptomatic treatment for patients with knee OA whose joint pain has not responded adequately to analgesics or nonsteroidal anti-inflammatory drugs (NSAIDs) and nonpharmacologic measures, such as education in principles of joint protection, periarticular muscle strengthening, use of thermal modialities and orthotics. Several studies have described relief of joint pain after such treatment [2,19,22,23] and three preparations of HA have recently been approved for use by the Food and Drug Administration in the United States. The rationale for this therapeutic approach has been claimed to be "viscosupplementation", i.e., augmentation of the lubricating potential of the synovial fluid in the OA joint [30]. However, a number of biologic effects of HA that could be relevant to palliation of joint pain have also been reported. For example, exogenous HA has been shown to stimulate synthesis of HA by synovial cells from OA joints *in vitro* [35]; to inhibit PGE-2 synthesis induced by interleukin-1 [37,39]; to affect leukocyte adherence, proliferation, migration and phagocytosis [15]; and to protect against cytotoxicity caused by reactive oxygen species [29].

*Address for correspondence: Kenneth D. Brandt, MD, Rheumatology Division, Indiana University School of Medicine, 1110 West Michigan Street, Room 545, Indianapolis, IN 46202, USA. Tel.: +1 317 274 4225; Fax: +1 317 274 7792; E-mail: kbrandt@iupui.edu.

Even though HA is cleared rapidly from the joint after IA injection (within 20 hours in animal models) [12], in some patients improvement in joint pain and function persists for months after this treatment [37,39]. The basis for the prolonged beneficial effect on symptoms noted in some patients after IA HA injection is unknown. Some have suggested that it represents only a placebo response, which is known to be large and often prolonged after invasive procedures such as arthrocentesis and IA injection [7]. Nonetheless, given the growing concerns over the adverse effects of NSAIDs, especially in the elderly (the population at greatest risk for OA), an alternative treatment for OA pain that permits a reduction in the dose, or discontinuation, of NSAIDs holds considerable appeal.

Although the objective of current pharmacologic treatment of OA is to reduce joint pain and improve mobility, a variety of drugs exist that have been shown to favorably influence the development or progression of pathologic changes in animal models of OA. Such agents have been classified as DMOADs (Disease-Modifying OA Drugs). In some experimental models of OA, HA has been reported to exert a DMOAD effect [1,3,20,31,40] but in other cases, either no effect [13,14] or deleterious effects have been noted [13,14].

To date, only one study has examined the DMOAD potential of HA in humans: Listrat et al. [21] reported that patients with knee OA who received a series of 3 IA injections of HA at 3-month intervals showed significantly less progression of the disease one year later, as judged by arthroscopy, than controls who received conventional therapy but no IA injections. However, arthroscopy does not permit accurate measurement of the thickness of the articular cartilage or the size of large chondral defects and has not been validated as an outcome measure in studies of OA progression.

We recently examined the effect of IA HA injection on the changes in articular cartilage that occur during the early stages of OA in the canine anterior cruciate ligament transection (ACLT) model [33]. Although treatment had no effect on morphologic changes in the OA joint, 7 weeks after the last injection the cartilage uronic acid concentration in the OA knee was significantly lower than that in cartilage from the OA knee of control dogs which had received IA injections of saline. Although no direct mechanical measurements of the cartilage were made in that study, this finding suggested that IA HA injection led, at least temporarily, to cartilage softening [26].

In clinical practice, HA treatment of OA is not used prophylactically, i.e., prior to the appearance of structural changes in the unstable joint – as it was in our previous canine study – but is used therapeutically, i.e., in patients in whom structural damage in the OA joint is already apparent (e.g., by radiography). We, therefore, performed further studies to examine the effects of HA on structural changes of OA with various treatment regimens. In addition, using gait analysis as a surrogate for pain in the unstable knee, we examined the effects of HA injection on the vertical ground reaction forces generated by the arthritic hind limb.

2. Methods

OA was induced in 30 adult dogs of mixed breed by unilateral ACLT [34]. The dogs were divided into 3 groups: In the "control group" the operated knee was injected weekly with saline during weeks 1–5 and 13–17 after ACLT. The "therapeutic HA group" received weekly IA saline injections during weeks 1–5 and weekly HA injections during weeks 13–17. The "prophylactic HA group" received IA HA injections during weeks 1–5 and IA saline injections during weeks 13–17. Each injection was one ml. The HA preparation contained 10 mg/ml of sodium hyaluronate (molecular weight $= 0.5\text{–}7.3 \times 10^6$ Da; Hyalgan®, Lot #064600, Fidia, Abano Terme, Italy).

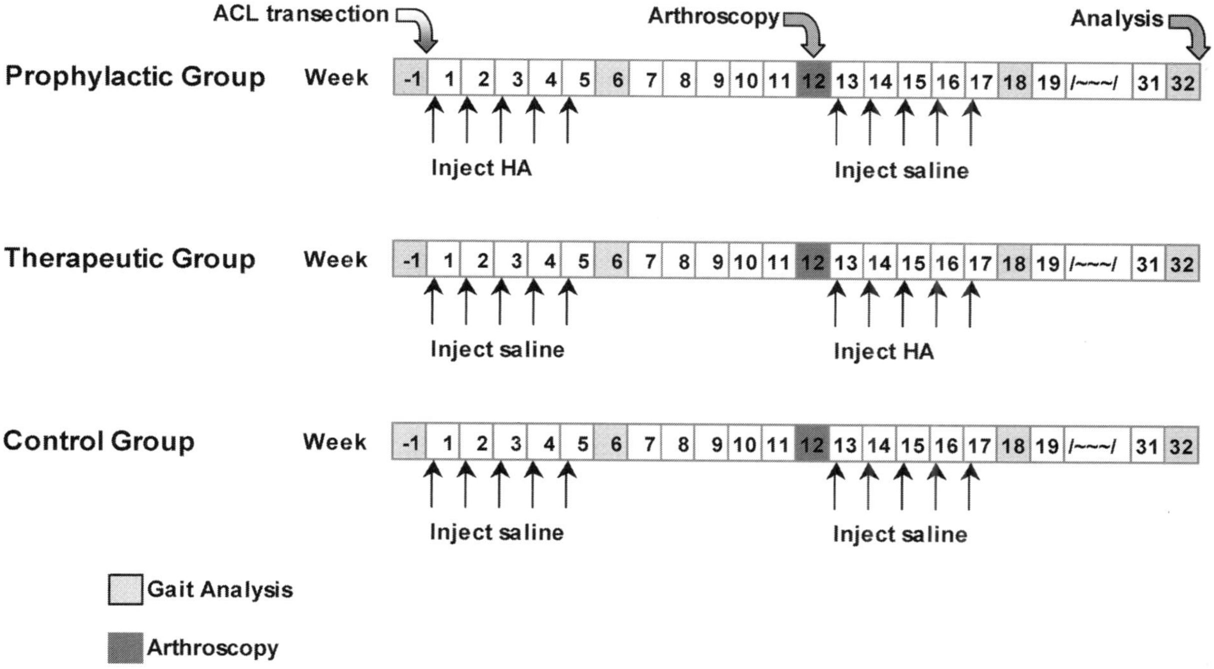

Fig. 1. The experimental protocol.

The progression of joint damage was evaluated by arthroscopy 12 weeks after ACLT and by gross examination 32 weeks after surgery at the time of sacrifice [34]. At that time, samples of cartilage and synovium were taken for examination of histological changes of OA and determination of water content, proteoglycan (PG) concentration and determination of the net rate of cartilage PG synthesis [34].

The extent to which the unstable limb was loaded during gait was determined by forceplate analysis of the peak vertical ground reaction force (VGRF) generated before surgery, after each series of IA injections (weeks 6 and 18), and in the week before euthanasia (week 32) [28]. The experimental protocol is shown in Fig. 1.

3. Results

3.1. Chondropathy

Twelve weeks after ACLT, arthroscopic examination revealed modest changes of OA in the cruciate-deficient knee in most dogs. Chondropathy scores ranged from 0 to 8 (possible range = 0–65), with a mean score ($\pm$ standard deviation) of 2.5 ± 1.3 for the controls, 2.5 ± 2.5 for the therapeutic HA group, and 2.1 ± 1.3 for the prophylactic treatment group.

The severity of cartilage changes was progressive; 32 weeks after ACLT, chondropathy scores were uniformly more severe than those obtained 20 weeks earlier at arthroscopy (Fig. 2). The gross chondropathy scores at the termination of the experiment were 14.0 ± 5.2 for the controls, 17.6 ± 6.8 for the therapeutic HA group, and 13.3 ± 5.0 for the prophylactic HA group. However, the differences between treatment groups were not statistically significant at either time point. The extent of osteophytosis was also unaltered by HA injection.

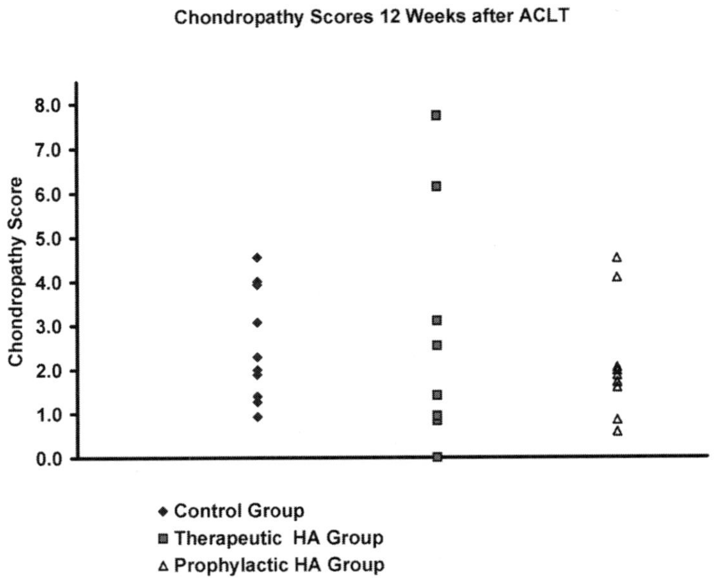

Fig. 2. Chondropathy scores 32 weeks after ACLT.

3.2. Histologic changes

Histologic changes typical of those seen in articular cartilage from the OA knee at sacrifice are depicted in Fig. 3, which shows disruption of surface integrity, loss of staining with Safranin O in the superficial zone, chondrocyte clusters and early fibrillation. However, scores for the histologic severity of cartilage changes, based on the Mankin grading scale [24], did not differ significantly among the 3 treatment groups (Table 1).

Histologic grading of synovitis showed a somewhat greater difference between treatment groups (Table 1) than the grading of cartilage changes. The mean score in the group of animals that was dosed therapeutically was higher than that in the group dosed prophylactically or in the control group (3.2 ± 1.2 vs. 2.5 ± 1.4 and 2.4 ± 1.4, respectively, range $= 0$–4). However, this difference also was not statistically significant.

On the basis of the above results, we concluded that a series of IA HA injections, administered either immediately after transection of the ACL or after structural changes of OA are clearly established, had no effect on the progression of OA pathology in the canine knee.

3.3. Forceplate analyses

Forceplate analyses have shown that a normal dog bears about 30% more of its total body weight on its forelimbs than on its hind limbs. After ACLT, VGRF generated by the unstable hind limb are decreased by an additional 30% and then slowly increase to about 75% of the baseline value by 32 weeks after surgery. Figure 4 depicts changes in VGRF after ACLT in our 2 treatment groups and the unoperated control group. Based on the magnitude of the VGRF generated by the unstable limb, HA injection, administered either prophylactically or therapeutically, did not alter loading of the arthritic knee.

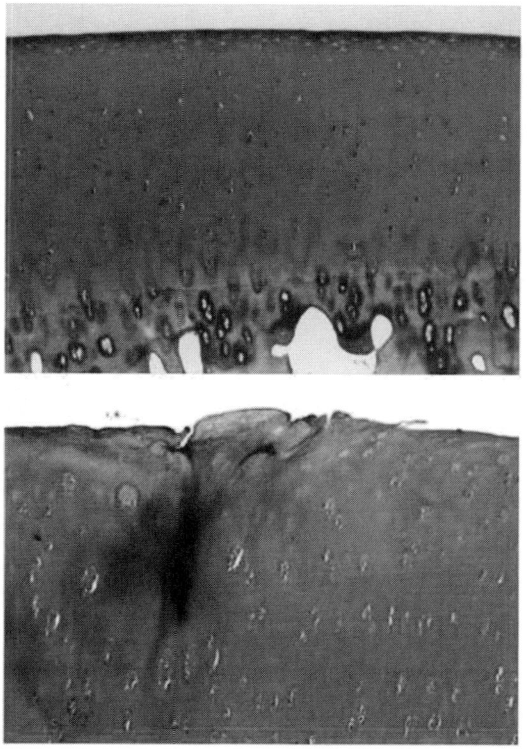

Fig. 3. Upper panel = sample from the contralateral, morphologically normal knee. Lower panel = representative histologic section of cartilage from the medial femoral condyle 32 weeks after ACLT. Safranin O – fast green stain, original magnification 100×. In the section from the cruciate-deficient knee, note the disruption of integrity of the cartilage surface; loss of staining the superficial zone, suggesting depletion of proteoglycans; cloning of chondrocytes and fibrillation.

Table 1
Histopathologic changes of OA

Treatment group	Cartilage histology score mean ± SD		Synovitis score mean ± SD	
	OA knee	CK	OA knee	CK
Control group	4.7 ± 1.6	0.0	2.4 ± 1.4	0.7 ± 0.7
Therapeutic HA group	4.4 ± 1.7	0.1	3.2 ± 1.2	0.3 ± 0.5
Prophylactic HA group	4.7 ± 1.9	0.0	2.5 ± 1.4	0.3 ± 0.7

Articular cartilage histopathology was graded according to the Mankin scale [24] 32 weeks after ACLT.
Synovitis was graded as described in the text, on a scale ranging from 0 to 4.
CK = contralateral knee.

3.4. Cartilage biochemistry

Cartilage water content and the net rate of PG synthesis, as measured by incorporation of $^{35}SO_4$ into proteoglycans [33], were unaffected by either mode of HA therapy.

In our previous study [33], knees that were injected prophylactically with HA exhibited a reduction in the PG (uronic acid) concentration of the articular cartilage 7 weeks after the last injection. In the present study, we likewise measured the PG concentration of the cartilage obtained at the time of sac-

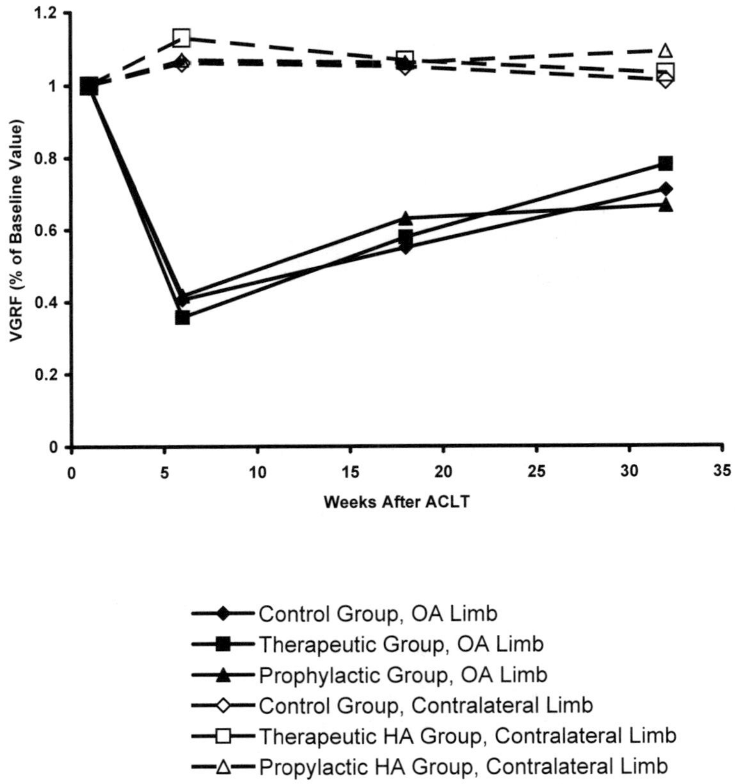

Fig. 4. Vertical ground reaction forces (VGRF) generated by the hind limbs at baseline (i.e., 1 week before ACLT) and 6 weeks, 18 weeks and 32 weeks after ACLT. IA injections of HA, performed either prophylactically or therapeutically, did not alter VGRF, in comparison with the control group, which received IA injections of saline.

rifice. However, in this study the interval between the last HA injection and analysis of the cartilage in the group treated prophylactically was 27 weeks, i.e., 20 weeks longer than in our prior study. In comparison, in the group that received therapeutic HA injections in the present study, the interval between the last injection and procurement of samples of articular cartilage for biochemical analysis was only 15 weeks.

Table 2 shows the results of the cartilage uronic acid analyses in the present study. In the control group, as expected, the uronic acid concentration of cartilage from the OA knee was about 20% greater than that of cartilage from the contralateral knee, reflecting the phenomenon of hypertrophic repair [38], which we have described previously in this model [7,9]. The cartilage uronic acid concentrations in our prophylactic HA treatment group were essentially comparable to those in the OA knee of the control group. However, among the dogs that received HA therapeutically, the mean uronic acid concentration of cartilage from the OA knee was significantly lower ($p < 0.05$) than that in the other two groups, and was no greater than that in the contralateral knee, consistent with the *absence* of hypertrophic repair (Table 2).

Finally, we examined the relationship between the magnitude of VGRF and the severity of chondropathy in cartilage obtained at sacrifice, 32 weeks after ACL transection. As indicated in Fig. 5, a moderately strong relationship ($r = 0.35$) existed, which was statistically significant ($p = 0.04$).

Table 2
Proteoglycan (uronic acid) concentration of articular cartilage 32 weeks after ACLT

Treatment group	Cartilage uronic acid concentration, μg/mg dry weight, mean ± SD		
	OA knee	Contralateral knee	OA knee: contralateral knee*
Control	5.6 ± 0.6	4.7 ± 0.5	1.19
Therapeutic HA	5.3 ± 0.4	5.2 ± 0.3	1.03*
Prophylactic HA	5.8 ± 0.6	4.5 ± 0.6	1.24

* $P < 0.05$.

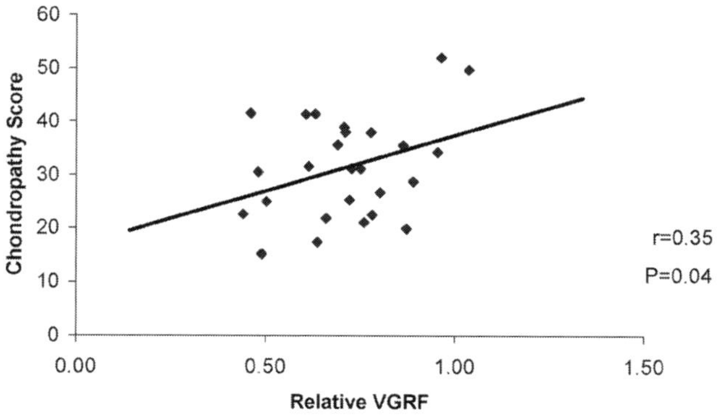

Fig. 5. Relationship between the severity of chondropathy and VGRF generated by the unstable hind limb. The severity of cartilage damage increases with increased loading of the unstable hind limb. Relative VGRF = VGRF measured at the experimental time point (32 weeks after ACLT) divided by VGRF measured at baseline.

4. Discussion

In neither the present study, which was 32 weeks in duration, nor our previous study, which was only 12 weeks in duration [33], did IA HA injection affect the progression of structural changes of OA. However, in the previous study, 5 weekly IA injections of HA, administered immediately after ACLT, resulted in as much as a 30% reduction in the PG concentration of articular cartilage 7 weeks later. Because the compressive stiffness of cartilage is related directly to its PG concentration [26], these data suggest that the articular cartilage might have become softer and, therefore, more susceptible to damage after IA injection of HA, especially in the presence of persistent knee instability resulting from ACLT. This raised the possibility that, if the PG depletion persisted, the lengthier period of joint instability afforded by the experimental design of the present study might have resulted in the acceleration of structural damage in the OA joint.

In the present study, however, no loss of PG was apparent in the OA cartilage 15 weeks after the last HA injection in the therapeutic HA group (Table 2) or 27 weeks after the last HA injection in the prophylactic HA group. Indeed, OA cartilage from the prophylactic treatment group, like that from the control group, showed no evidence of PG depletion; indeed, the PG concentration of cartilage from the OA knee was about 25% greater than that in the contralateral knee.

As noted above, the interval between the last HA injection and analysis of the articular cartilage was some 12 weeks shorter in our therapeutic HA group than in our prophylactic HA group (i.e., 15 weeks, 27 weeks, respectively). Presumably, the 15 week interval was long enough to permit restoration of the cartilage PG concentration to a normal level, but not sufficiently long to permit the chondrocytes to achieve the supranormal PG concentration associated with hypertrophic repair (as was observed in our prophylactic HA group). This explanation would be consistent with the results of our earlier study, which showed that the PG concentration of the cartilage matrix was significantly lower than that in the contralateral knee 7 weeks after HA injection [33].

The possibility must also be considered, however, that the absence of PG depletion after HA injection in the present study was due to the fact that the HA preparation employed here was not identical to that used in our earlier study. Because the latter was no longer available to us for use in the present study, an alternate HA preparation was employed, the molecular weight of which ($0.5–0.73 \times 10^6$ Da) was only about half as great as that of the HA used in the previous study (1.5×10^6 Da). However, in a 12 week pilot study we found that a series of 5 IA injections of the HA preparation used in the present study reduced the PG concentration of the cartilage from the cruciate-deficient knee by $18 \pm 3\%$ [25]. In contrast, we have found repeatedly that cartilage from saline-injected OA knees exhibits a 15–30% *increase* in its uronic acid concentration, reflecting hypertrophic repair. We have also shown that sham ACLT does not result in pathologic changes in the operated knee or affect the PG concentration of the articular cartilage 12 weeks after surgery [25].

We interpret our finding that IA HA injections – regardless of whether they were administered prophylactically or therapeutically – had no effect on VGRF generated by the unstable knee as evidence that the injections had little effect on the dogs' perception of joint pain. To the extent that loading of the arthritic limb is a surrogate for the pain associated with load-bearing on the OA knee, our data suggest that IA HA therapy has little effect on the perception of joint pain in this model. Finally, the results shown in Fig. 5 can be interpreted as indicating that the severity of cartilage damage in this model increases with an increase in loading of the unstable limb during gait.

The concept of analgesic arthropathy, i.e., that relief of pain in patients with OA results in detrimental overloading of the arthritic joint, thereby accelerating disease progression, was first suggested in reports describing rapid progression of hip OA in patients treated with NSAIDs [12,15,17]. Whether NSAIDs accelerate OA progression remains controversial [5,13]. However, in studies of humans with medial compartment knee OA, the decrease in joint pain resulting from treatment with an analgesic/NSAID was accompanied by an increase in the adductor moment and net quadriceps moment, substantially increasing the load on the damaged medial compartment; cessation of analgesic/NSAID therapy was followed by an increase in knee pain and a reduction in the forces acting on the medial compartment [16].

Studies of gait in the ovine meniscectomy model of OA suggested that IA HA injection increased loading of the affected joint, an effect associated with an increase in osteophytosis, and reductions in the PG concentration and net rate of PG synthesis in the OA cartilage [13,14]. These observations, and our finding that the severity of joint pathology is correlated with the magnitude of loading of the OA knee, suggest that relief of joint pain in the absence of concomitant inhibition of the processes driving structural damage in the joint may accelerate disease progression.

Although we have not found evidence that IA HA alters the rate of OA progression in the cruciate-deficient canine knee, others, working with different models and using a variety of outcome measures, have reported both positive and negative effects after IA injection of a variety of HA preparations [9]. These discrepancies may be attributed to fundamental differences in the animal models employed in these studies, such as differences in gait, and/or to differences among the various exogenous HA preparations used, or in the properties of HA in normal joint fluid among various species. In rabbits, IA HA injection after either ACLT or meniscectomy has been reported to have positive effects on the progression of structural damage and the expression of MMP-3 and IL-1β [5,18,27,32]. However, in rabbits with less severe OA, IA HA injection was reported to result in an increase of MMP-3 expression [32].

In the ovine partial medial meniscectomy model of OA, HA injection resulted in an increase in osteophytosis with no effect on the progression of cartilage damage although, as noted above, cartilage PG concentration and the rate of PG synthesis were reduced [36,41].

Clearly, if IA HA injection has an effect on progression of cartilage damage in our canine model of OA, it is not robust, based on the present findings and our previous studies, in which we did not observe either slowing of the progression of structural damage or improvement in the concentration or molecular weight of synovial fluid HA after IA HA therapy [13].

Acknowledgements

We thank Elizabeth Mickler and Marjorie Albrecht for their expert technical assistance in the conduct of this study and Deborah Jenkins and Kathie Lane for their clerical assistance.

References

[1] G. Abatangelo, P. Botti, M. Del Bue, G. Gei, J.C. Samson, R. Cortivo, A. De Galateo and M. Martelli, Intraarticular sodium hyaluronate injections in the Pond-Nuki experimental model of osteoarthritis in dogs. I. Biochemical results, *Clin. Orthop. Rel. Res.* **241** (1989), 278–285.
[2] M.E. Adams, M.H. Atkinson, A.J. Lussier, J.I. Schulz, K.A. Siminovitch, J.P. Wade and M. Zummer, The role of viscosupplementation with hylan G-F 20 (Synvisc) in the treatment of osteoarthritis of the knee: a Canadian multicenter trial comparing hylan G-F 20 alone, hylan G-F 20 with non-steroidal anti-inflammatory drugs (NSAIDs) and NSAIDs alone, *Osteoarthritis Cartilage* **3** (1995), 213–225.
[3] R.D. Altman, R. Moskowitz and the Hyalgan Study Group, Intraarticular sodium hyaluronate (Hyalgan^{R}) in the treatment of osteoarthritis of the knee: a randomized clinical trial, *J. Rheum.* **25** (1998), 2203–2212.
[4] T. Bitter and H.M. Muir, A modified uronic acid carbazole reaction, *Anal. Biochem.* **4** (1962), 330–334.
[5] J.S. Arora and R.H. Maudsley, Indomethacin arthropathy of the hips, *Proc. Roy. Soc. Med.* **61** (1968), 669.
[6] K.D. Brandt, E.M. Braunstein, D.M. Visco, B. O'Connor, D. Heck, B. Katz and M. Albrecht, Anterior (cranial) cruciate ligament transection in the dog: A bona fide model of osteoarthritis, not merely of cartilage injury and repair, *J. Rheum.* **18** (1991), 436–446.
[7] K.D. Brandt, Insights into the natural history of osteoarthritis provided by the cruciate-deficient dog: an animal model of osteoarthritis, *Ann. NY Acad. Sci.* **732** (1994), 199–205.
[8] K.D. Brandt, G.N. Smith Jr. and L.S. Simon, Intraarticular injection of hyaluronic acid as treatment for knee osteoarthritis. What is the evidence?, *Arthritis Rheum.* **43** (2000), 1192–1203.
[9] E.M. Braunstein, K.D. Brandt and M. Albrecht, MRI demonstration of hypertrophic articular cartilage repair in osteoarthritis, *Skeletal Radiology* **19** (1990), 335–339.
[10] H. Coke, Long term indomethacin therapy of coxarthrosis, *Ann. Rheum. Dis.* **26** (1967), 346–347.
[11] M. Doherty and A. Jones, Indomethacin hastens large joint OA in humans—How strong is the evidence?, *J. Rheum.* **22** (1995), 2013–2016.
[12] J.R.E. Fraser, W.G. Kimton, B.K. Pierscioner and R.N.P. Cahill, The kinetics of hyaluronan in normal and acutely inflamed synovial joints: exploratory observations with experimental arthritis in sheep, *Semin. Arthritis Rheum.* **22** (Suppl. 1) (1993), 9–17.
[13] P. Ghosh, R. Read, S. Armstrong, D. Wilson, R. Marshall and P. McNair, The effects of intraarticular administration of hyaluronan in a model of early osteoarthritis in sheep. I. Gait analysis and radiological and morphological studies, *Semin. Arthritis Rheum.* **22**(6) (Suppl. 1) (1993), 18–30.
[14] P. Ghosh, R. Read, Y. Numata, S. Smith, S. Armstrong and D. Wilson, The effects of intraarticular administration of hyaluronan in a model of early osteoarthritis in sheep. II. Cartilage composition and proteoglycan metabolism, *Semin. Arthritis Rheum.* **22**(6) (Suppl. 1) (1993), 31–42.
[15] P. Ghosh, The role of hyaluronic acid (hyaluronan) in health and disease: interactions with cells, cartilage and components of the synovial fluid, *Clin. Exp. Rheumatol.* **12** (1994), 75–82.
[16] D. Hurwitz, L. Sharma and T.P. Andriacchi, Effects of knee pain on joint loading in patients with osteoarthritis, *Curr. Opin. Rheumatol.* **11** (1999), 422–426.
[17] E.C. Huskisson, H. Berry, P. Gishen, R.W. Jubb, J. Whitehead and the LINK Study Group, Effects of antiinflammatory drugs on the progression of OA of the knee, *J. Rheumatol.* **22** (1995), 1941–1946.

[18] H. Iwata, Pharmacologic and clinical aspects of intraarticular injection of hyaluronate, *Clin. Orth. Rel. Res.* **289** (1993), 285–291.
[19] A.C. Jones, M. Pattrick, S. Doherty and M. Doherty, Intra-articular hyaluronic acid compared to intra-articular triamcinolone hexacetonide in inflammatory knee osteoarthritis, *Osteoarthritis Cartilage* **3** (1995), 269–273.
[20] T. Kikuchi, H. Yamada and M. Shimmei, Effect of high molecular weight hyaluronan on cartilage degeneration in a rabbit model of osteoarthritis, *Osteoarthritis Cartilage* **4** (1996), 99–110.
[21] V. Listrat, X. Ayral, F. Patarnello, J.-P. Bonvarlet, J. Simonnet, B. Amor and M. Dougados, Arthroscopic evaluation of potential structure modifying activity of hyaluronan (Hyalgan) in osteoarthritis of the knee, *Osteoarthritis Cartilage* **5** (1997), 153–160.
[22] L.S. Lohmander, N. Dalen, G. Englund, M. Hamalainen, E.M. Jensen, K. Karlsson, M. Odensten, L. Ryd, I. Sernbo, O. Suomalainen and A. Tegnander, Intra-articular hyaluronan injections in the treatment of osteoarthritis of the knee: a randomised, double blind, placebo controlled multicentre trial. Hyaluronan Multicentre Trial Group, *Ann. Rheum. Dis.* **55** (1996), 424–431.
[23] A. Lussier, A.A. Cividino, C.A. McFarlane, W.P. Olszynski, W.J. Potashner and R. De Medicis, Viscosupplementation with hylan for the treatment of osteoarthritis: findings from clinical practice in Canada, *J. Rheumatol.* **23** (1996), 1579–1585.
[24] H.J. Mankin, H. Dorfman, Z. Lipiello and A. Zarins, Biochemical and metabolic abnormalities in articular cartilage from osteoarthritic human hips. II. Correlation of morphology with biochemical and metabolic data, *J. Bone Joint Surg. Am.* **53A** (1997), 523–537.
[25] E.A. Mickler, M. Albrecht, R. Herman and G.N. Smith, Jr., Intra-articular hyaluronan injection does not alter synovial fluid hyaluronan or cartilage proteoglycan concentration in stable knees, *Conn. Tiss. Res.* **41** (2000), 302 [abstract].
[26] V.C. Mow and L.A. Setton, Mechanical properties of normal and osteoarthritic cartilage, in: *Osteoarthritis*, K.D. Brandt, M. Doherty and L.S. Lohmander, eds, Oxford University Press, Oxford, 1998, pp. 108–122.
[27] S.L. Myers, K.D. Brandt, B.L. O'Connor, D.M. Visco and M.E. Albrecht, Synovitis and osteoarthritic changes in canine articular cartilage after anterior cruciate ligament transection, *Arthritis Rheum.* **33** (1990), 1406–1415.
[28] B.L. O'Connor, D.M. Visco, P.I. Rogers, L.A. Mamlin and K.D. Brandt, Serial force plate analyses of dogs with unilateral knee instability, with or without interruption of the sensory input from the ipsilateral limb, *Osteoarthritis Cartilage* **7** (1999), 567–573.
[29] D. Presti and J.E. Scott, Hyaluronan mediated protective effect against cell damage caused by enzymatically generated hydroxyl radicals is dependent on hyaluronan molecular mass, *Cell. Biochem. Function.* **12** (1994), 281–288.
[30] N. Rydell and E.A. Balazs, Effect of intraarticular injection of hyaluronic acid on the clinical symptoms of osteoarthritis and on granulation tissue formation, *Clin. Orthop. Rel. Res.* **80** (1971), 25–32.
[31] A. Schiavinato, E. Lini, D. Guidolin, G. Pezzoli, P. Botti, M. Martelli, R. Cortivo and A. De Galateo and G. Abatangelo, Intraarticular sodium hyaluronate injections in the Pond-Nuki experimental model of osteoarthritis in dogs. II. Morphological findings, *Clin. Orthop. Rel. Res.* **241** (1989), 286–299.
[32] C. Shimizu, M. Yoshioka, R.D. Coutts, F.L. Harwood, T. Kubo, Y. Hirasawa and D. Amiel, The effects of hyaluronan during the development of osteoarthritis, *Osteoarthritis Cartilage* **5** (1997), 251–260.
[33] G. Smith, Jr., S.L. Myers, K.D. Brandt and E.A. Mickler, Effect of intra-articular hyaluronan injection in experimental canine osteoarthritis, *Arthritis Rheum.* **41** (1998), 976–985.
[34] G.N. Smith, Jr., S.L. Myers, K.D. Brandt, E.A. Mickler and M. Albrecht, Diacerhein treatment reduces the severity of osteoarthritis in the canine cruciate-deficiency model of osteoarthritis, *Arthritis Rheum.* **42** (1999), 549–554.
[35] M.M. Smith and P. Ghosh, The synthesis of hyaluronic acid by human synovial fibroblasts is influenced by the nature of the hyaluronate in the extracellular environment, *Rheumatol. Int.* **7** (1987), 113–122.
[36] K. Takahashi, R.S. Goomer, F. Harwood, T. Kubo, Y. Hirasawa and D. Amiel, The effects of hyaluronan on matrix metalloproteinase-3 (MMP-3), interleukin-1β (IL-1β), and tissue inhibitor of metalloproteinase-1 (TIMP-1) gene expression during the development of osteoarthritis, *Osteoarthritis Cartilage* **7** (1999), 182–190.
[37] K. Tobetto, T. Uasui and T. Ando, Inhibitory effects of hyaluronan on (^{14}C) arachidonic acid release from labeled human synovial fibroblasts, *Japan J. Pharmacol.* **60** (1992), 79–84.
[38] E.M. Vignon, D. Arlot, D. Hartman, B. Moyer and G. Willie, Hypertrophic repair of particular cartilage in experimental osteoarthrosis, *J. Rheum.* **142** (1983), 82–88.
[39] T. Yasui, M. Adatsuka, K. Tobetto, M. Hayaishi and T. Anto, The effect of hyaluronan on interleukin-1-a-induced prostaglandin-E_2 production in human osteoarthritis synovial cells, *Agents Actions* **37** (1992), 155–156.
[40] T. Yoshimi, T. Kikuchi, T. Obara, T. Yamaguchi, Y. Sakakibara, H. Itoh, H. Iwata and T. Miura, Effects of high-molecular-weight sodium hyaluronate on experimental osteoarthrosis induced by the resection of rabbit anterior cruciate ligament, *Clin. Orthop. Rel. Res.* **298** (1994), 296–304.
[41] M. Yoshioka, C. Shimizu, F.L. Harwood, R.D. Coutts and D. Amiel, Long-term effects of hyaluronan on experimental osteoarthritis in the rabbit knee, *Osteoarthritis Cartilage* **6** (1998), 1–10.

Biorheology 41 (2004) 503–508
IOS Press

Lubricin reduces cartilage–cartilage integration

Dirk B. Schaefer [a], David Wendt [a], Matteo Moretti [a], Marcel Jakob [a], Gregory D. Jay [b], Michael Heberer [a] and Ivan Martin [a,*]

[a] *Departments of Surgery and of Research, University Hospital Basel, Switzerland*
[b] *Department of Medicine, Section of Emergency Medicine and Division of Engineering, Brown University, Providence, RI*

Abstract. Cartilage integration *in vivo* does not occur, such that even cartilage fissures do not heal. This could be due not only to the limited access of chondrocytes to the wound, but also to exogenous factors. In this paper, we tested the hypothesis that lubricin, a lubricating protein physiologically present in the synovial fluid, reduces the integrative cartilage repair capacity. Disk/ring composites of bovine articular cartilage were prepared using concentric circular blades and cultured for 6 weeks with or without treatment with 250 μg/ml lubricin applied three times per week. Following culture, the percentage of contact area between the disks and the rings, as assessed by light microscopy, were equal in both groups. The adhesive strength of the integration interface, as assessed by push-out mechanical tests, was markedly and significantly lower in lubricin-treated specimens (2.5 kPa) than in the controls (28.7 kPa). Histological observation of Safranin-O stained cross-sections confirmed the reduced integration in the lubricin treated composites. Our findings suggest that the synovial milieu, by providing lubrication of cartilage surfaces, impairs cartilage–cartilage integration.

1. Introduction

It is generally agreed that injured articular cartilage has a limited intrinsic repair capacity. Clinical observations and animal experiments indicate that even thin fissures in articular cartilage can persist for years without healing [1,6]. The poor integrative repair capacity of opposing cartilage surfaces has been related to the limited number of chondrocytes that are capable of migrating or proliferating in the cartilage–cartilage gaps [7]. The insufficient cell recruitment could be due to the fact that articular cartilage is avascular and chondrocytes are entrapped in their own extracellular matrix [2] and/or to the large extent of cell death that occurs following cartilage incisions [8]. The limited cellularity within purely cartilaginous wounds could also be explained by the anti-adhesive properties of the defect surface conferred by dermatan sulfate and other proteoglycans [9]. For example, decorin and biglycan are known to inhibit adhesion of cells to macromolecules, such as fibronectin, in the extracellular matrix [17,18, 21]. In this context, molecules locally present in synovial fluid, which provide lubrication of the articular surface, are also likely to play a role in cartilage–cartilage integration.

Lubricin, also known as PRG4 [10], articular cartilage superficial zone protein (SZP) [22], and megakaryocyte stimulating factor precursor [13] is a mucinous glycoprotein found in the synovial

*Address for correspondence: Ivan Martin, Institute for Surgical Research and Hospital Management, University Hospital Basel, Hebelstrasse 20, ZLF, Room 405, 4031 Basel, Switzerland. Tel.: +41 61 265 2384; Fax: +41 61 265 3990; E-mail: imartin@uhbs.ch.

0006-355X/04/$17.00

fluid [23]. Lubricin provides boundary lubrication of congruent articular surfaces under conditions of high contact pressure and near zero sliding speed [16]. These lubricating properties have also been demonstrated *in vitro* [12]. Cells capable of synthesizing lubricin have been found in synovial tissue and within the superficial zone of articular cartilage within diarthrodial joints [16].

Mechanisms of integration of cartilage–cartilage interfaces have been investigated using different *in vitro* model systems, in order to decouple the biological processes that modulate the repair from the complex loading patterns in synovial joints [20]. In particular, a disk–ring composite model has been described, that well mimics aspects of clinical methods of chondral transplantation [19]. In the present study, we used the previously developed disk–ring model system to test the hypothesis that lubricin reduces cartilage–cartilage integration.

2. Materials and methods

2.1. Generation and culture of disk/ring composites

Articular cartilage was harvested from the femoral–patellar grooves of 3–6 week old calves under aseptic conditions within 6 hours after death and rinsed thoroughly in phosphate buffered saline (PBS) supplemented with 100 U/ml penicillin and 100 μg/ml streptomycin. Top and bottom surfaces were removed from the cartilage explants, and surfaces cut flat at an approximate thickness of 2 mm. Disks (5.0 mm o.d.) and rings (10.0 mm o.d.) were cored out simultaneously from a cartilage explant using two concentric circular blades. Disks were immersed for 30 min in PBS, with or without 250 μg/ml of lubricin extracted and purified from pooled bovine synovial fluid, as previously described [12]. This concentration falls between the threshold and maximal values of 200 μg/ml and 250 μg/ml, respectively, previously determined for lubrication by lubricin in a glass–rubber model [12]. Each disk was then placed back into the corresponding ring from which it was cored and the composites cultured for up to 6 weeks on an orbital shaker at 16 rpm in Dulbecco's Modified Eagle Medium (4.5 g/l glucose with nonessential amino acids) supplemented with 10% fetal bovine serum, 10 mM HEPES, 2 mM glutamine, 1 mM sodium pyruvate, 100 U/ml penicillin, 100 μg/ml streptomycin (all previous from Gibco; Grand Island, NY), 0.1 mM ascorbic acid 2-phosphate (Sigma; Buchs, CH), and 1 U/ml insulin (Novo Nordisk A/S; Bagsvaerd, Denmark). At each change of medium (3 times/week) samples initially treated with lubricin were incubated for 5 min with 20 μl lubricin solution applied on the top surface of each construct. A total of 34 disk–ring composites were generated and cultured.

2.2. Assessment of disk/ring composites

After 6 weeks of culture, the contact area was measured for each composite as the percentage of the disk/ring interface that did not transmit light when analyzed using a light transmission microscope. Mechanical properties of the disk/ring interface were assessed from a push-out test in which a plunger displaced the disk from the cartilage ring using a custom-made mechanical testing device [19]. A computer activated micro-stepper motor controlled the displacement of the plunger at 0.50 mm/min, and a load cell coupled to the plunger measured the push-out force. For each specimen, load measurements were first recorded until the disk was fully displaced from the ring, then load measurements were repeated with the disk removed. The second set of data, which functioned as a baseline, was subtracted from the first data set to remove any contribution to the force measurements due to friction between the plunger and the cartilage ring. The adhesive strength was evaluated as the maximum force to failure per unit of interfacial area. The interfacial area was determined for each composite by measuring with an electronic caliper the thickness of the disk and the ring at four different locations. For each experimental group, 6 to 7 composites were tested mechanically. Values are presented as averages ± the standard error of the mean, and statistical differences among experimental groups were assessed using nonparametric Mann–Whitney tests and considered significant at values of $p < 0.05$. Composites were also assessed histologically by Safranin-O stain of horizontal cross-sections.

3. Results

After 6 weeks of culture, the contact area at the disk/ring interface reached an approximate average of 90% for both experimental groups. For one of the composites, following lubricin treatment, the disk was completely displaced from the surrounding ring, and was therefore not included in the data analysis.

Mechanical testing resulted in characteristic force–displacement profiles, with forces increasing until a complete displacement of the disk occurred (Fig. 1). As compared to lubricin-treated specimens, control composites displayed peaks in the force–displacement profiles which reached higher levels and were better defined. The adhesive strength of control composites (28.7 ± 8.3 kPa) was markedly and significantly higher ($p = 0.004$) than in lubricin treated specimens (2.5 ± 0.9 kPa) (Fig. 2).

Histological analysis demonstrated a uniform and intense staining for Safranin-O of both disks and rings in composites treated and not treated with lubricin. In both groups the disk/ring interface could be identified in all explants. Lubricin-treated composites displayed a pattern of interrupted contact zones, whereas control composites had few interruptions (Fig. 3).

4. Discussion

In this study, we verified the hypothesis that the glycoprotein lubricin, physiologically present in the synovial fluid of diarthrodial joints, reduces the integrative capacity of articular cartilage *in vitro*. In particular, using a previously developed *in vitro* model system (i.e., dislocation of a cartilage disk from a cartilage ring [19] we found that treatment with lubricin reduced the adhesive strength by more than 10-fold.

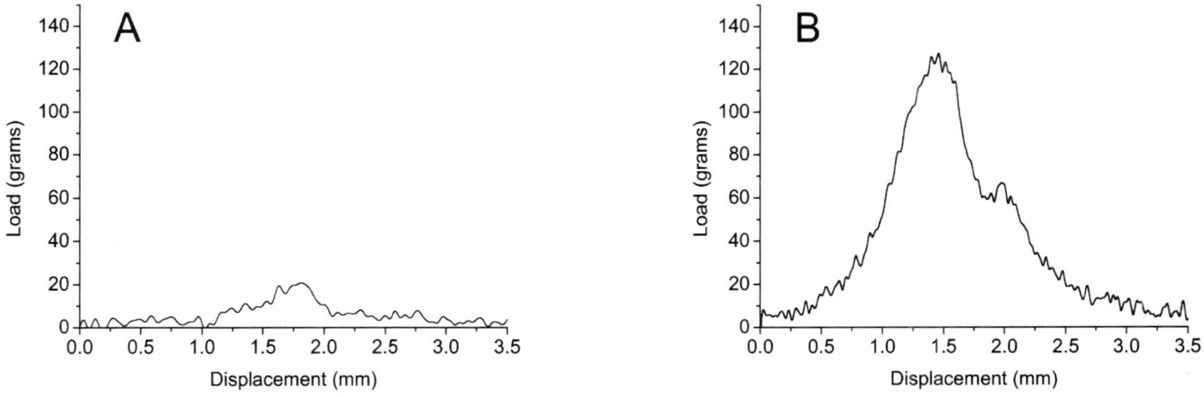

Fig. 1. Force–displacement curves during push-out mechanical tests. (A) Profiles acquired from lubricin treated composites were significantly lower in magnitude than untreated control samples, indicating negligible adhesive strength. (B) Profiles acquired from untreated composites displayed a higher peak, indicating the dislocation of the disk from the ring.

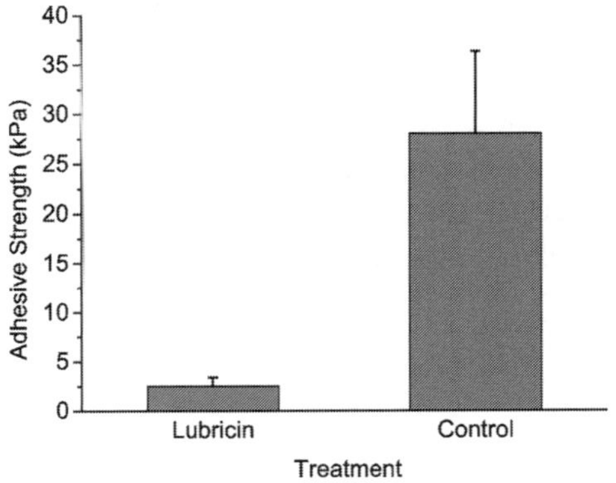

Fig. 2. Adhesive strength of lubricin treated and control composites. As assessed by this parameter, cartilage treatment with lubricin resulted in a 10-fold decrease of integrative repair capacity.

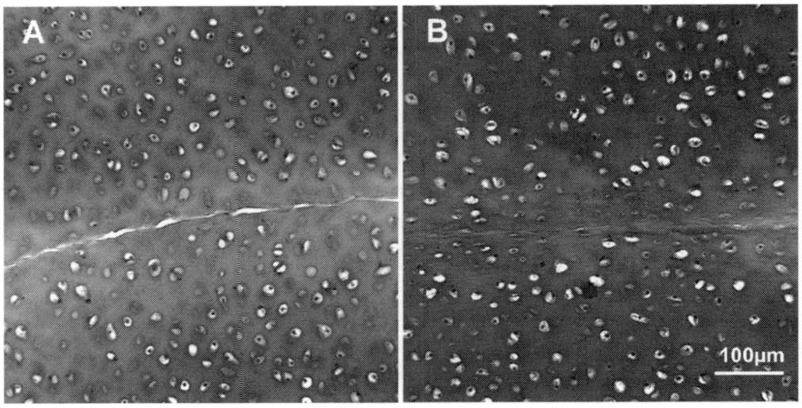

Fig. 3. Representative fields of disk/ring cartilage interfaces. Lubricin treated and control samples show a homogenous distribution of GAG at the disk/ring interface. (A) Lubricin treatment resulted in a pattern of interrupted contact zones, whereas those areas were rarely observed in (B) control samples (Safranin-O staining, 100×).

Lubricin has been proposed to be a key factor for joint lubrication. Homozygous knock-out mice lacking the orthologous gene PRG4, displayed significantly inferior joint lubrication as compared to the wild-type [14]. The boundary lubricating mechanism of lubricin has been related to the parallel orientation of lubricin molecules at the surface of articular cartilage, resulting in a repulsive hydration force if the distance between two opposing surfaces is less than 30 Å [11,12]. In an oscillating glass–rubber model it has been shown that the lubricating properties of purified bovine lubricin are dose-dependent, such that the lubrication ability occurs only at concentrations higher than 200 μg/ml [12]. In our study, a concentration of 250 μg/ml was repeatedly applied at the cartilage–cartilage interface. The use of lower concentrations for more limited times would aid in understanding if the reduction of cartilage–cartilage integration by lubricin is associated with its lubrication properties and what is the duration of its effect.

Our finding that lubricin reduces the integrative capacity of articular cartilage provides one plausible explanation for the limited ability of cartilage defects or even small cartilage fissures to heal, and suggests that factors specific to the intra-articular milieu are not optimal for joint repair. These results are consistent with a recent report using a different *in vitro* model system, that synovial fluid, and in particular SZP, reduced the integration of opposing cartilage surfaces [4]. Furthermore, the comparable histological staining and contact areas of the treated and non-treated constructs in our study are also consistent with the finding that SZP did not appreciably affect chondrocyte biosynthesis [4].

Using the disk/ring *in vitro* model system, it was previously demonstrated that cartilage treatment with trypsin may enhance its integration capacity by digesting cartilage proteoglycans [19]. The finding that trypsin can also eliminate the lubricating activity of molecules of the synovial fluid [15] would provide an additional motivation for the possible introduction of specific proteases in cartilage repair procedures. Indeed, when rabbit patellar cartilage was implanted in a full-thickness articular cartilage defect, integration and remodeling were improved by pre-treatment with trypsin [3].

In conclusion, lubricin at a concentration necessary for boundary lubrication markedly reduced cartilage–cartilage integration *in vitro*. Together with the recent finding that specific mechanical factors regulate the expression of lubricin [5], our results indicate that attempts to improve cartilage integration have to address an orchestrated control over a variety of physicochemical factors.

Acknowledgements

This work was supported in part by a program to promote academic science at the University of Basel to D.B.S., by Fidia Advanced Biopolymers (Abano Terme, IT) to M.M. and by NIA K08AG/AR01008 to G.D.J.

References

[1] J.A. Buckwalter and H.J. Mankin, Articular cartilage: degeneration and osteoarthritis, repair, regeneration, and transplantation, in: *Instructional Course Lectures*, W.D. Cannon, ed., AAOS, Rosemont, 1998, pp. 487–504.

[2] A.I. Caplan, M. Elyaderani, Y. Mochizuki, S. Wakitani and V.M. Goldberg, Principles of cartilage repair and regeneration, *Clin. Orthop.* **342** (1997), 254–269.

[3] J. Chen, S. Maniwa and M. Ochi, Influence of trypsin on the biological bonding of cartilaginous surface to bone in rabbits, *Arch. Orthop. Trauma Surg.* **120** (2000), 587–591.

[4] C. Englert, K.B. McGowan, T.J. Klein, A. Giurea, B.L. Schuhmacher and R.L. Sah, Inhibition of integrative cartilage repair by synovial fluid components, *Trans. Orthop. Res.* **29** (2003), 189–189.

[5] S. Grad, K. Gorna, S. Gogolewski, M. Wimmer and M. Alini, Surface motion but not axial compression induces superficial zone protein expression in chondrocytes seeded onto 3D scaffolds, *Trans. Orthop. Res.* **29** (2003), 594.

[6] E.B. Hunziker, Articular cartilage repair: basic science and clinical progress. A review of the current status and prospects, *Osteoarthritis Cartilage* **10** (2001), 432–463.

[7] E.B. Hunziker, Biologic repair of articular cartilage. Defect models in experimental animals and matrix requirements, *Clin. Orthop.* **367** (1999), S135–S146.

[8] E.B. Hunziker and T.M. Quinn, Surgical removal of articular cartilage leads to loss of chondrocytes from the wound edges, *Trans. Orthop. Res.* **46** (2000), 185.

[9] E.B. Hunziker and L.C. Rosenberg, Repair of partial-thickness defects in articular cartilage: cell recruitment from the synovial membrane, *J. Bone Joint. Surg. Am.* **78** (1996), 721–733.

[10] S. Ikegawa, M. Sano, Y. Koshizuka and Y. Nakamura, Isolation, characterization and mapping of the mouse and human PRG4 (proteoglycan 4) genes, *Cytogenet. Cell. Genet.* **90** (2000), 291–297.

[11] J. Israelachvili, Intermolecular and surface forces with applications to colloidal and biological systems, in: *Intermolecular and Surface Forces with Applications to Colloidal and Biological Systems*, Academic Press, New York, 1965, pp. 201–207.

[12] G.D. Jay, Characterization of a bovine synovial fluid lubricating factor. I. Chemical, surface activity and lubricating properties, *Connect. Tissue Res.* **28** (1992), 71–88.

[13] G.D. Jay, D.E. Britt and C.J. Cha, Lubricin is a product of megakaryocyte stimulating factor gene expression by human synovial fibroblasts, *J. Rheumatol.* **27** (2000), 594–600.

[14] G.D. Jay, J.D. Carp, K. Rhee, J.R. Torr and M.L. Warm, Analysis of the frictional characteristics of CACP knockout mice joints with the modified Stanton pendulum technique, *Trans. Orthop. Res.* **28** (2003), 136.

[15] G.D. Jay and C.J. Cha, The effect of phospholipase digestion upon the boundary lubricating ability of synovial fluid, *J. Rheumatol.* **26** (1999), 2454–2457.
[16] G.D. Jay, U. Tantravahi, D.E. Britt, H.J. Barrach and C.J. Cha, Homology of lubricin and superficial zone protein (SZP): products of megakaryocyte stimulating factor (MSF) gene expression by human synovial fibroblasts and articular chondrocytes localized to chromosome 1q25, *J. Orthop. Res.* **19** (2001), 677–687.
[17] K. Lewandowska, H.U. Choi, L.C. Rosenberg, L. Zardi and L.A. Culp, Fibronectin-mediated adhesion of fibroblasts: inhibition by dermatan sulfate proteoglycan and evidence for a cryptic glycosaminoglycan-binding domain, *J. Cell. Biol.* **105** (1987), 1443–1454.
[18] Y. Mitani, A. Honda and H.E. Jasin, Polymorphonuclear leukocyte adhesion to articular cartilage is inhibited by cartilage surface macromolecules, *Rheumatol. Int.* **20** (2001), 180–185.
[19] B. Obradovic, I. Martin, R.F. Padera, S. Treppo, L.E. Freed and G. Vunjak-Novakovic, Integration of engineered cartilage, *J. Orthop. Res.* **19** (2001), 1089–1097.
[20] E.S. Reindel, A.M. Ayroso, A.C. Chen, D.M. Chun, R.M. Schinagl and R.L. Sah, Integrative repair of articular cartilage in vitro: adhesive strength of the interface region, *J. Orthop. Res.* **13** (1995), 751–760.
[21] G. Schmidt, H. Robenek, B. Harrach, J. Glossl, V. Nolte, H. Hormann, H. Richter and H. Kresse, Interaction of small dermatan sulfate proteoglycan from fibroblasts with fibronectin, *J. Cell. Biol.* **104** (1987), 1683–1691.
[22] B.L. Schumacher, J.A. Block, T.M. Schmid, M.B. Aydelotte and K.E. Kuettner, A novel proteoglycan synthesized and secreted by chondrocytes of the superficial zone of articular cartilage, *Arch. Biochem. Biophys.* **311** (1994), 144–152.
[23] D.A. Swann, H.S. Slayter and F.H. Silver, The molecular structure of lubricating glycoprotein-I, the boundary lubricant for articular cartilage, *J. Biol.Chem.* **256** (1981), 5921–5925.

Biorheology 41 (2004) 509–519
IOS Press

Short-term changes in cell and matrix damage following mechanical injury of articular cartilage explants and modelling of microphysical mediators

V. Morel and T.M. Quinn *
Center for Orthopaedic Research, Swiss Federal Institue of Technology, 1015 Lausanne, Switzerland

Abstract. The short-term responses of articular cartilage to mechanical injury have important implications for prevention and treatment of degenerative disease. Cell and matrix responses were monitored for 11 days following injurious compression of cartilage in osteochondral explants. Injury was applied as a single ramp compression to 14 MPa peak stress at one of three strain rates: 7×10^{-1}, 7×10^{-3} or 7×10^{-5} s^{-1}. Responses were quantified in terms of the appearance of macroscopic matrix cracks, changes in cell viability, and changes in cartilage wet weights. Loading at the highest strain rate resulted in acute cell death near the superficial zone in association with cracks, followed over the 11 days after compression by a gradual increase in cell death and loss of demarcation between matrix zones containing viable versus nonviable cells. In contrast, loading at the lowest strain rate resulted in more severe, nearly full-depth cell death acutely, but with no apparent worsening over the 11 days following compression. Between days 4 and 11, all mechanically injured explants significantly increased in wet weight, suggesting loss of matrix mechanical integrity independent of compression strain rate. Results demonstrate that short-term responses of cartilage depend upon the biomechanical characteristics of injurious loading, and suggest multiple independent pathways of mechanically-induced cell death and matrix degradation. Modifications to an existing fiber-reinforced poroelastic finite element model were introduced and the model was used for data interpretation and identification of microphysical events involved in cell and matrix injury. The model performed reasonably well at the slower strain rates and exhibited some capacity for anticipating the formation of superficial cracks during injurious loading. However, several improvements appear to be necessary before such a model could reliably be used to draw upon *in vitro* experimental results for prediction of injurious loading situations *in vivo*.

1. Introduction

Nonphysiological mechanical loading contributes to the initiation and progression of degenerative disease of articular cartilage. The processes by which abnormal loading conditions can evolve into long-term degradative disease may depend upon a number of interacting mechanical, biological and biochemical factors. From a biomechanical perspective alone, multiple pathways may be involved due to the wide range of nonphysiological loads which may be applied to a joint, including acute trauma and loads associated with obesity [30] or joint misalignment [3]. Previous *in vitro* studies have shown that relatively high strain rate loading can give rise to acute cell death and matrix disruption [11,29], which over the days following injury can lead to cell death in initially uninjured tissue regions [16] perhaps related to

*Address for correpondence: Dr. Thomas Quinn, Center for Orthopaedic Research, AA.B0.19, 1015 Lausanne, Switzerland. Tel.: +41 21 693 83 50; Fax: +41 21 693 86 60; E-mail: Thomas.Quinn@epfl.ch.

the release of biochemical mediators such as toxic oxygen species after cell disruption [14] and induced apoptosis [8]. It has also been shown that acute effects of cartilage injurious loading depend upon the rate at which compression is applied [13,26], with the gel diffusion rate (the inherent mechanical relaxation rate of the tissue) representing a threshold separating different cell and matrix responses to injury [22]. It is possible that the short-term sequelae of mechanical injuries may also depend upon the biomechanical nature of the injury itself, which may have important implications for disease progression and treatment options.

In vitro experimental models of cartilage injury have the advantage of geometrical simplicity, making them useful for establishing relationships between specific micromechanical events within the poroelastic tissue [1,2,12,17] and localized biological changes such as cell death or alterations in cell activities [28]. *In vivo* loading of articular cartilage is in general much more complex; nevertheless the ability to anticipate and avoid cartilage injury arising due to hemiarthroplasty, joint misalignment, or mechanical inhomogeneities associated with tissue repair and tissue engineering treatments would be clinically useful. Accurate modelling of articular cartilage mechanics is therefore of significant interest for bringing research results to bear on clinical problems [2]. A model for cartilage mechanics based upon the underlying molecular physics of the tissue, including the separate contributions of the collagen network and the proteoglycan gel, would seem to be of greatest versatility for these purposes since it could be based simply upon tissue composition and structure, and thereby independent of the loading environment.

The aims of this study were therefore (1) to investigate the effects of compression strain rate on the temporal evolution of spatial patterns of cell and matrix injury between 1 hour and 11 days following mechanical injury of cartilage explants, and (2) to evaluate the performance of an existing model for cartilage mechanics as a tool for interpretation of experimental data. A fibril-reinforced poroelastic finite element model [17] was chosen because it was consistent with our desire to separately account for collagen network and proteoglycan gel contributions to tissue mechanics, and could be modified to include existing models of molecular physics.

2. Material and methods

2.1. Experimental

Osteochondral cores of 4 mm diameter were drilled from adult bovine humeral heads. The full-thickness cartilage layer was trimmed to 2.7 mm diameter and the bone to 1 mm thickness. The explants were incubated in DMEM supplemented with fetal bovine serum, L-proline, non-essential amino acids and an antibiotic/antimycotic cocktail for up to 17 days with daily media changes as previously described [22]. After 6 days of culture, a single ramp compression to a maximal stress of 14 MPa was applied at a strain rate of one of 7×10^{-1}, 7×10^{-3} or 7×10^{-5} s^{-1}. Stress was defined relative to the explant articular surface area at dissection, and strain was defined relative to the cartilage thickness measured under a dissecting microscope just before compression. Explants were weighed on an analytical balance immediately before and after compression (after removal of surface fluid with a tissue) to determine water loss. One hour, 4 days and 11 days after compression, explants were removed from culture, weighed again, examined with the naked eye for the presence of macroscopic cracks, and assessed histologically for chondrocyte viability. As previously described [22], $\sim$1 mm thick vertical slices of cartilage from the middle of explants were obtained manually with a razor blade, incubated for 5 minutes in a solution of calcein AM and ethidium homodimer-1 and examined under a confocal fluorescence microscope to visualize spatial distributions of live and dead cells. To estimate the proportion of the tissue containing viable cells, a human user identified areas of images containing cartilage (regardless of cell viability) and containing viable cells with a normal appearance. Before compression, each explant was randomly assigned to an experimental group characterized by compression strain rate and time of removal from culture. 6 dissections contributed explants to this study. Data are reported as mean ± SEM (n). Significant differences ($p < 0.05$) between groups were identified by ANOVA with post hoc Tukey tests.

2.2. Numerical

Cartilage mechanics was modelled using a fibril reinforced poroelastic finite element (FE) model [17] with modifications introduced in an effort to describe individual matrix constituents in a manner consistent with other existing models of molecular physics. Finite element calculation was carried out using the commercial FE package Abaqus 5.8. Cartilage tissue was modelled as a combination of a poroelastic component corresponding to the water-rich proteoglycan (PG) gel and a network of nonlinear springs corresponding to the collagen network.

The PG gel was modelled by 4-node poroelastic elements (CAX4P) requiring specification of elastic moduli, hydraulic permeability, and Poisson's ratio. These material properties were all defined as functions of solid volume fraction σ which varied with depth inside cartilage (due to tissue inhomogeneity) and also with mechanical deformation. For the initial condition in numerical simulations, the uncompressed solid volume fraction σ_0 versus depth was chosen to approximate previous results of serial sectioning experiments [18]; specifically, σ_0 for the PG gel varied linearly from 0.16 at the articular surface to 0.25 at the attachment to bone. The PG gel was assumed to contain all tissue water and the solid part was considered to be effectively incompressible. Its equilibrium bulk modulus and hydraulic permeability were estimated using existing "unit cell" models based on molecular physics. Within this description, individual glycosaminoglycan (GAG) molecules were modelled as solid cylinders of radius $a = 0.55$ nm [7,23] associated with a sheath of surrounding fluid extending to a radius b (Fig. 1) such that $\sigma = a^2/b^2$. Therefore, the evolution of PG gel material properties during matrix deformation simulations was governed by changes in σ at the tissue scale and, equivalently, b at the molecular scale. Solutions of a Poisson–Boltzmann equation [5] for the electrostatic potential within the unit cell fluid

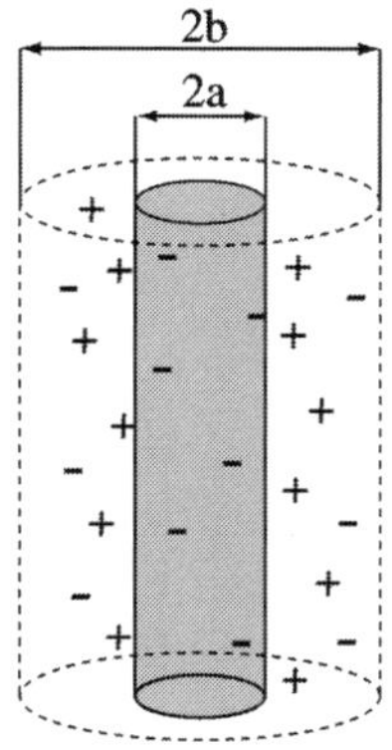

Fig. 1. A "unit cell" representing a single GAG molecule within the water-rich proteoglycan gel. The inner solid cylinder represents the negatively charged GAG molecule itself and has constant radius a. The outer radius b of the unit cell is determined by the solid volume fraction σ of the PG gel and can change with mechanical deformation.

were obtained using a 20-node finite difference model implemented in a spreadsheet. Boundary conditions were imposed by the negative electrical charges fixed to the GAG rod and the unit cell symmetry. Results were indistinguishable from those previously obtained [5]. Also as suggested by this previous work, the total bulk modulus was taken to be the sum of this electrostatic component and a nonelectrostatic part accounting for other contributions (e.g., osmotic pressure of the GAG rods themselves [9]) estimated at 0.25 MPa [5]. Hydraulic permeability (κ) of the PG gel was estimated using Kuwabara's analytical unit cell model for fluid flow through an array of cylinders [15], which has been shown to provide a very good approximation to more precise finite element calculations [27]. Kuwabara's analytical result may be written

$$\kappa = \frac{b^2}{3\mu}\left(-\frac{1}{2}\ln\sigma - \frac{3}{4} + \sigma - \frac{1}{4}\sigma^2\right), \tag{1}$$

where $\mu = 7 \times 10^{-4}$ Pa·s is the viscosity of water at 37°C. The equilibrium Poisson's ratio ν was chosen to vary linearly with fluid content: $\nu = 0.4\sigma + 0.08$. This relation corresponds to near-incompressibility for $\sigma = 1$ and a material that can be compressed axially without radial expansion for $\sigma = 0$; in simulations this resulted in values of ν which were similar to those used previously [17].

The collagen network was modelled by springs that resist tension only (SPRINGA) with elastic constants depending upon depth and strain as determined previously from unconfined compression experiments [17]. This previous work modelled the collagen network using radially-directed springs only and we extended this to include axial springs since for our osteochondral explants large gradients in radial displacement could occur near the site of attachment to bone. For axial fibers, we assumed that the elastic constant varied with distance from subchondral bone in the same way as the elastic constant for radial fibers varied with distance from the articular surface in that previous work [17].

The FE model was axisymmetric with dimensions typical of those used in experiments: a cartilage disk of 1.35 mm radius and 1 mm thickness. The FE mesh was constructed to include 20 elements between the explant center and radial edge, and 10 elements in the axial direction. Therefore, 20×10 poroelastic elements contributed to the model, together with 20×11 radially oriented springs and 10×21 axially oriented springs. Boundary conditions consisted of no radial displacement at the axial center, no displacement at the point of attachment to subchondral bone, zero pore pressure at the explant disk radial edge and no axial fluid flow through axial surfaces. Nodes at the articular surface were subjected to an axial displacement corresponding to the ramp compressive strain rates and maximal strains applied during experiments (Fig. 2); radial movement of articular surface nodes was unrestricted.

3. Results

3.1. Experimental

Consistent with previous findings [22], the occurrence of superficial matrix cracks decreased with decreasing strain rate, while mechanically-induced water loss and maximum compressive strains increased. 83% of explants were fissured at 7×10^{-1} s^{-1} strain rate, 26% at 7×10^{-3} s^{-1} and none at 7×10^{-5} s^{-1}. For these same strain rates, normalized water losses (volumes of water lost divided by initial cartilage volumes) during compression were 0.12 ± 0.01 ($n = 30$), 0.35 ± 0.02 ($n = 27$), and 0.61 ± 0.02 ($n = 22$), respectively. Similarly, maximum axial strains were 0.57 ± 0.02, 0.71 ± 0.02, and 0.87 ± 0.01, respectively.

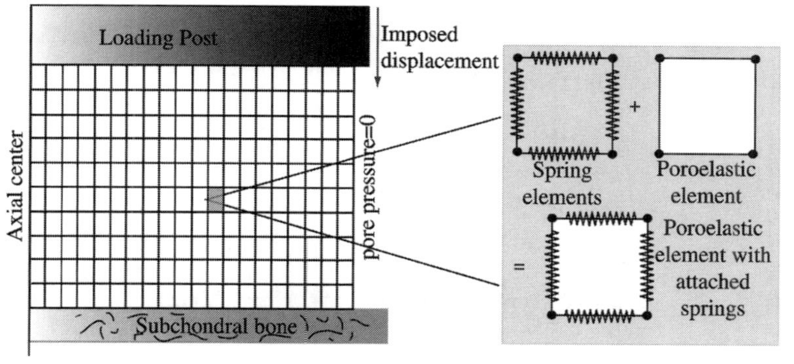

Fig. 2. A finite element model for cartilage mechanics during injurious compression included the superposition of poroelastic elements representing the PG gel and spring elements (oriented along the explant axial and radial directions) representing the collagen network. Boundary conditions were imposed corresponding to cartilage attachment to bone and unconfined compression by a rigid, nonporous surface.

One hour after compression, explants exhibited the same wet weight as before compression regardless of strain rate (Fig. 3a). The area fraction of cartilage containing viable chondrocytes was ~0.95 for uncompressed controls and the intermediate strain rate (7×10^{-3} s^{-1}). For the most rapid strain rate (7×10^{-1} s^{-1}), this fraction was slightly (but not significantly) lower at 0.87 ± 0.02 ($n = 8$). The lowest strain rate (7×10^{-5} s^{-1}) was most injurious in terms of cell viability 1 hour after compression with an area fraction of viable cells of 0.54 ± 0.14 ($n = 7$; Fig. 3b).

4 days after compression, explant wet weights remained unchanged compared to pre-compression values (Fig. 3a). Similarly, area fractions of histological images containing viable cells did not change significantly compared to their values 1 hour after compression (Fig. 3b).

11 days after compression and for every applied strain rate, explant wet weights were significantly greater than pre-compression values, and also with respect to uncompressed controls (which did not significantly change in wet weight over the course of the experiment) (Fig. 3a). No significant changes in the area fraction containing viable cells were found for uncompressed controls, nor for the slowest and intermediate strain rates. However, for the most rapid strain rate (7×10^{-1} s^{-1}), the area fraction of histological images containing viable cells was significantly decreased compared to its values 1 hour and 4 days after compression (Fig. 3b).

3.2. Numerical

Tissue-scale results from the finite element model were compared to experimental results to provide a preliminary indication of model performance. At the highest strain rate (7×10^{-1} s^{-1}), the model gravely overestimated stresses during tissue compression (not shown), and subsequent results for this strain rate were considered to be too inaccurate to warrant further attention. For strain rates $\leqslant 7 \times 10^{-3}$ s^{-1}, the model followed the experimental stress–strain relationship reasonably closely (Figs 5a,b). However, for the slowest strain rate (7×10^{-5} s^{-1}), the FE calculation could not proceed beyond an axial strain of 65% due to numerical convergence problems and therefore could not simulate the full extent of the displacements imposed experimentally. Normalized water loss during compression at 7×10^{-3} s^{-1} strain rate was estimated to be 0.59, which exceeded the value of 0.35 ± 0.02 ($n = 27$) measured in experiments. Similarly, for 7×10^{-5} s^{-1} strain rate, normalized water loss predicted by the model at 65% compression was 0.64, which exceeded the experimental value of 0.61 ± 0.02 ($n = 22$) attained at ~87% axial strain.

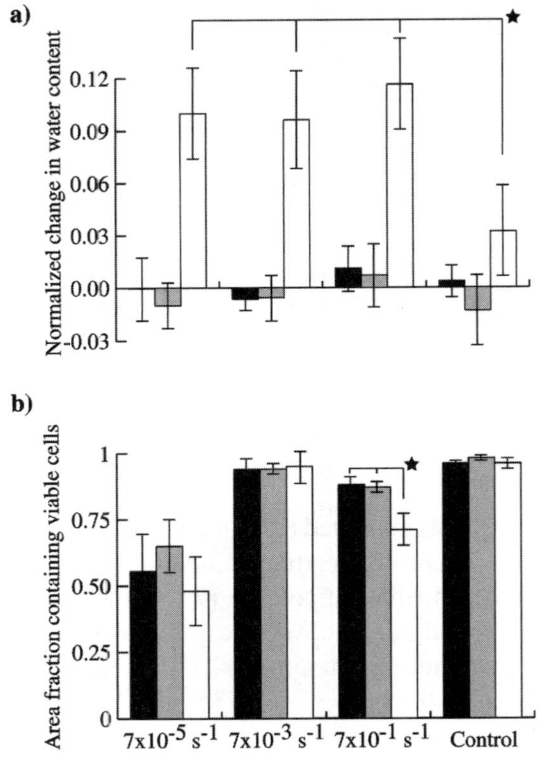

Fig. 3. Effects of strain rate on short-term changes in cartilage matrix water content and chondrocyte viability. Data were acquired 1 hour (black bars), 4 days (gray bars) and 11 days (white bars) after compression at strain rates of 7×10^{-5}, 7×10^{-3}, and $7 \times 10^{-1}\ s^{-1}$, and compared to uncompressed controls. (a) Normalized changes in water content as determined from osteochondral explant wet weights. (b) Area fraction of cartilage explants containing viable cells as measured from confocal images. Stars are placed vertically above data which are significantly different ($p < 0.05$) from data marked by the associated lines.

Modelling results of microphysical phenomena within the tissue during compression indicated that pore pressure contributed increasingly to the total stress as strain rate increased (Fig. 5c,d). For the lowest strain rate, maximal pore pressure did not exceed 3 MPa but rose to many times this value for the higher strain rates. The location within explants at which maximal pore pressure occurred was always the axial center, with relatively little dependence upon axial position (Fig. 6a,b). In contrast, matrix radial displacements and strains depended strongly upon axial position due to attachment to underlying bone. In the superficial zone, where fissures were observed in experiments, radial strain increased dramatically with strain rate. Furthermore, radial strain varied more strongly with radial position and exhibited a clear maximum at the explant axial center during ramp compression at $7 \times 10^{-3}\ s^{-1}$ strain rate (Fig. 6c) as compared to $7 \times 10^{-5}\ s^{-1}$ strain rate (Fig. 6d).

4. Discussion

Consistent with previous findings [22,26], low strain rate ($7 \times 10^{-5}\ s^{-1}$) injurious ramp compression was associated with chondrocyte death throughout the full cartilage depth in the absence of macroscopic cracks, while high strain rate compression ($7 \times 10^{-1}\ s^{-1}$) induced cell death primarily in the superficial zone and in close association with cracks. In the present study, these differences in the acute response

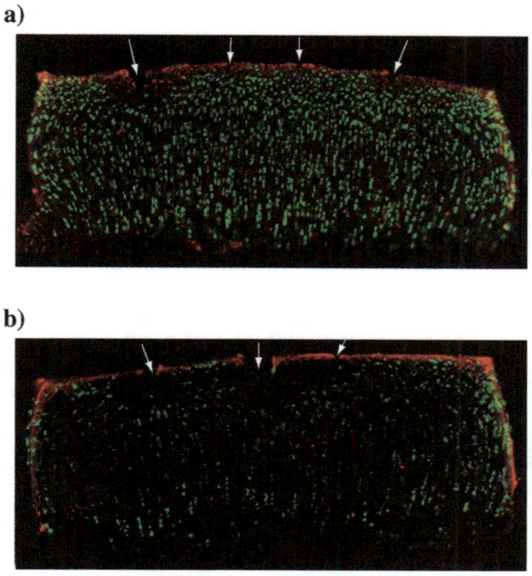

Fig. 4. Confocal images of full-thickness cartilage from ostechondral explants compressed at a strain rate of 7×10^{-1} s^{-1}. (a) One hour after compression, dead cells were concentrated near macroscopically visible cracks in the articular surface (arrows). (b) 11 days after compression the area fraction containing viable cells had significantly decreased. In addition, regions containing viable cells also appeared to contain large numbers of nonviable cells which had not been evident for up to 4 days after compression.

to injury were found to have distinctly different sequelae over the 11 days following injury. While more injurious in terms of immediate decreases in cell viability (Fig. 3), loading at the lowest strain rate appeared to induce distributions of cell death which were more stable over the subsequent days: tissue zones containing nonviable cells did not increase in size over the 11 days following compression and the demarcation between zones containing viable and nonviable cells remained clearly defined. In contrast, loading at the highest strain rate induced less immediate cell death but was associated with significant increases of cell death over the 11 days following compression, and loss of a clear demarcation between matrix zones containing viable versus nonviable cells (Fig. 4). It is possible that increases in cell death with time following high strain rate injury were due to apoptosis, which has been associated with cartilage mechanical injury [8,19] and can become more severe with time following injury [10]. This short-term induction of cell death may be mediated by intercellular signalling as previously observed for cyclically impacted explants [16], including release of biochemical factors such as toxic oxygen species [14] after physical disruption of cells. Alternatively, apoptosis could be induced by disruption of the collagen network at sites of matrix cracks and modified interactions between chondrocytes and the extracellular matrix [6]. Interestingly, violent mechanical disruption of cells and matrix may not occur at all during slow strain rate injury, so that such signalling cascades leading to induced cell death over the days following injury are not initiated. Therefore, different mechanical regimes of injurious loading can induce distinct patterns of acute cell and matrix injury, as well as differences in the short-term evolution of these microscopic changes.

For all strain rates applied, significant increases in explant wet weights occurred between days 4 and 11 following injury, while wet weights of uncompressed controls did not change significantly throughout the experiment (Fig. 3a). These findings suggest short-term changes in matrix mechanical integrity resulting from injury, which were independent of the applied strain rate. This was in contrast with the short-term evolution of cell viability, which appeared to depend strongly upon strain rate. Taken to-

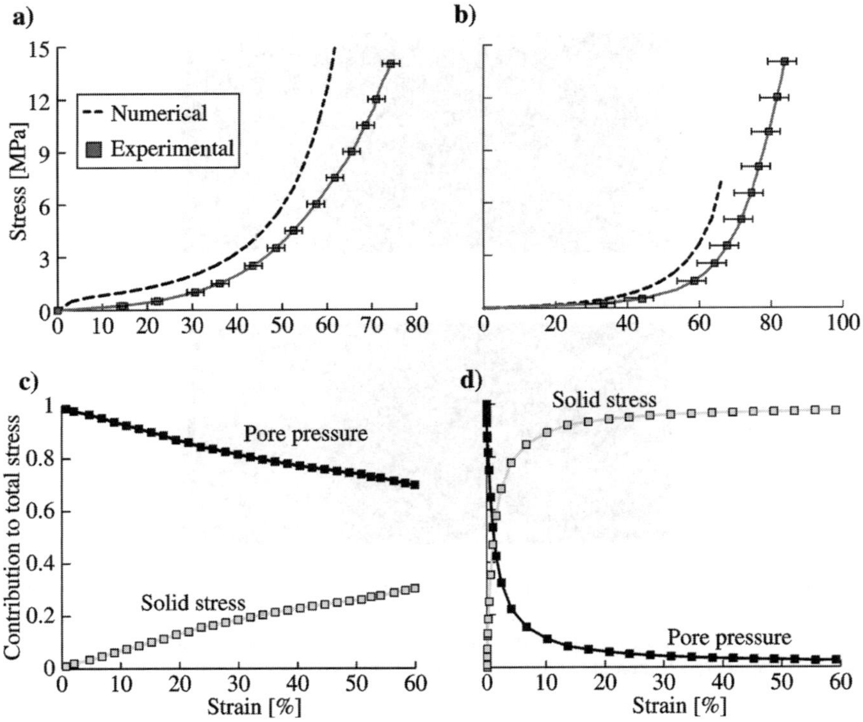

Fig. 5. Finite element modelling results compared to experimental stress versus strain data, and relative contributions of solid and fluid stresses as predicted by the model. (a) Stress versus strain for injurious ramp compression of cartilage explants at 7×10^{-3} s^{-1} strain rate as measured in experiments (black squares) and as predicted by the finite element model (dotted line). (b) Stress versus strain at 7×10^{-5} s^{-1} strain rate. (c) Model predictions for the relative contributions of the mean axial solid stress (grey squares) and pore pressure (black squares) to the total measured stress at the articular surface during compression at 7×10^{-3} s^{-1} strain rate. (d) Model predictions for the relative contributions of solid stress and pore pressure at 7×10^{-5} s^{-1} strain rate.

gether, these findings suggest that biochemical and biomechanical pathways of cartilage matrix degradation induced by mechanical injury [24,25,31] may act independently of pathways leading to induced cell death.

In an effort to better interpret experimental data, a fiber-reinforced poroelastic finite element model was used for comparison with experimental data. This modelling approach attempts to retain a separation of the individual contributions of the collagen network and the PG gel to cartilage mechanics [17]. We introduced the modification that the elastic modulus and hydraulic permeability of the PG gel were defined using existing models of molecular physics; however, PG gel Poisson's ratio and collagen network mechanics were still based largely on semi-empirical previous work [17]. At the intermediate and slow strain rates applied in experiments, the model provided reasonable results for explant stress versus strain during ramp compression. Despite consistently overpredicting the rate and extent of water loss from our radially unconfined axially compressed explants, model results provided some useful insights into the individual roles of fluid pressure and solid stress in the mechanics of injurious compression (Figs. 5c,d). Furthermore, predictions for radial strain in the superficial zone indicated increasingly large strains as explant compressive strain rate increased, with an increasing concentration of superficial zone radial strain near the explant axial center; these trends corresponded closely to the mechanical conditions and locations within explants where macroscopic cracks were observed experimentally. Indeed, the relatively

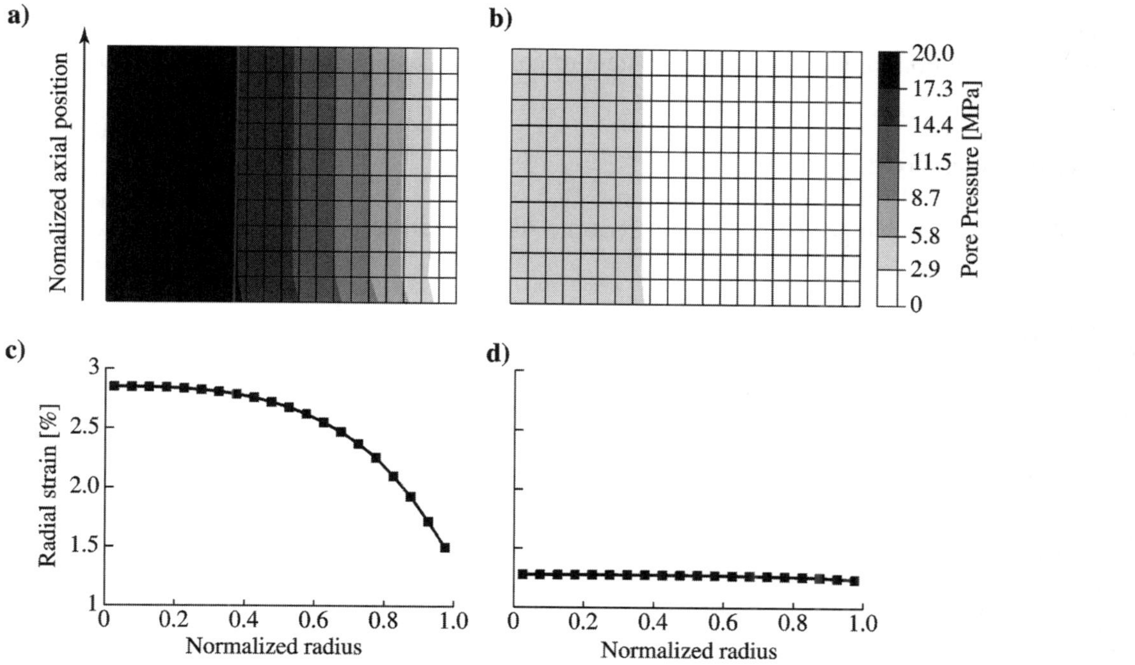

Fig. 6. Finite element modelling results for intratissue stress distributions. In all cases, calculations were performed at 65% axial strain; beyond this limit the FE calculation did not converge for the lower strain rate. (a) Distribution of pore pressure within explants for ramp compression at 7×10^{-3} s^{-1} strain rate. (b) Pore pressure distribution at 7×10^{-5} s^{-1} strain rate. (c) Radial strain at the articular surface for ramp compression at 7×10^{-3} s^{-1} strain rate. (d) Radial strain at the articular surface for 7×10^{-5} s^{-1} strain rate.

poor performance of the model at the highest strain rate studied may in part be due to the very common formation of tissue cracks experimentally, the mechanics of which are beyond the model's intended scope.

Insofar as the fiber-reinforced poroelastic approach [17] represented an important step forward in the sophistication of cartilage mechanics modelling, it is useful to identify some of the model's shortcomings in order to highlight possible directions for future work. Under no external load, the collagen network of cartilage is in tension while it counteracts the swelling pressure of the PG gel [20]. However, the model includes no prestress equilibrium between its components, and cannot readily be interpreted as a superposition of models for the collagen network and the PG gel. Furthermore, treatment of the PG gel by itself as a poroelastic material possessing a Poisson's ratio is questionable. This description gives to the PG gel the properties of a solid while a solution of PG behaves more like a gas. The collagen network is described by a discrete set of springs rather than a true network of fibers; this approach therefore ignores changes in network mechanical properties due to geometric reorientation of fibers during tissue deformation, which plays an important role in supporting shear stresses in cartilage [4]. Also, lack of attention to the distribution of water between PG and collagen components during matrix compression [21] may have contributed further sources of error. Some of the model's limitations were likely also due to our choice of a lateral free-slip boundary condition between the loading post and the articular surface, a choice which is difficult to validate [1,12]. Despite these limitations, progressive improvement of modelling approaches which draw connections between tissue mechanics and molecular structure have great promise as tools for in cartilage biomechanics. Identification of microphysical events leading to cell and matrix injury within specific injurious loading experimental protocols could be used for the prediction, prevention, and treatment of cartilage degenerative disease in a much larger context.

Acknowledgement

This work was supported by a grant from the AO Foundation, Switzerland.

References

[1] C.G. Armstrong, W.M. Lai and V.C. Mow, An analysis of the unconfined compression of articular cartilage, *J. Biomech. Eng.* **106**(2) (1984), 165–73.
[2] T.S. Atkinson, R.C. Haut and N.J. Altiero, Impact-induced fissuring of articular cartilage: an investigation of failure criteria, *J. Biomech. Eng.* **120**(2) (1998), 181–7.
[3] J.A. Buckwalter, Osteoarthritis and articular cartilage use, disuse, and abuse: experimental studies, *J. Rheumatol. Suppl.* **43** (1995), 13–5.
[4] P. Bursac, C.V. McGrath, S.R. Eisenberg and D. Stamenovic, A microstructural model of elastostatic properties of articular cartilage in confined compression, *J. Biomech. Eng.* **122**(4) (2000), 347–53.
[5] M.D. Buschmann and A.J. Grodzinsky, A molecular model of proteoglycan-associated electrostatic forces in cartilage mechanics, *J. Biomech. Eng.* **117**(2) (1995), 179–92.
[6] L. Cao, V. Lee, M.E. Adams, C. Kiani, Y. Zhang, W. Hu and B.B. Yang, Beta-integrin–collagen interaction reduces chondrocyte apoptosis, *Matrix Biol.* **18**(4) (1999), 343–55.
[7] W.D. Comper and R.P. Williams, Osmotic flow caused by chondroitin sulfate proteoglycan across well-defined nuclepore membranes, *Biophys. Chem.* **36**(3) (1990), 215–22.
[8] D.D. D'Lima, S. Hashimoto, P.C. Chen, C.W., Colwell, Jr. and M.K. Lotz, Human chondrocyte apoptosis in response to mechanical injury, *Osteoarthritis Cartilage* **9**(8) (2001), 712–9.
[9] S. Ehrlich, N. Wolff, R. Schneiderman, A. Maroudas, K.H. Parker and C.P. Winlove, The osmotic pressure of chondroitin sulphate solutions: experimental measurements and theoretical analysis, *Biorheology* **35**(6) (1998), 383–97.
[10] S.P. Grogan, B. Aklin, M. Frenz, T. Brunner, T. Schaffner and P. Mainil-Varlet, *In vitro* model for the study of necrosis and apoptosis in native cartilage, *J. Pathol.* **198**(1) (2002), 5–13.
[11] J.E. Jeffrey, D.W. Gregory and R.M. Aspden, Matrix damage and chondrocyte viability following a single impact load on articular cartilage, *Arch. Biochem. Biophys.* **322**(1) (1995), 87–96.
[12] Y.J. Kim, R.L. Sah, A.J. Grodzinsky, A.H. Plaas and J.D. Sandy, Mechanical regulation of cartilage biosynthetic behavior: physical stimuli, *Arch. Biochem. Biophys.* **311**(1) (1994), 1–12.
[13] B. Kurz, M. Jin, P. Patwari, D.M. Cheng, M.W. Lark and A.J. Grodzinsky, Biosynthetic response and mechanical properties of articular cartilage after injurious compression, *J. Orthop. Res.* **19**(6) (2001), 1140–6. 21639380 0736-0266 Journal Article.
[14] B. Kurz, M. Kehn, C. Domm, E.H. Frank, A.J. Grodzinsky and M. Shuenke, Mechanical induction of apoptosis in articular cartilage is inhibited by superoxid dismutase mimetic, in: *48th Meeting of the Orthopaedic Research Society*, Dallas, 2002, p. 409.
[15] S. Kuwabara, The forces experienced by randomly distributed parallel circular cylinders or spheres in a viscous flow at small reynolds numbers, *J. Phys. Soc. Japan* **14** (1959), 527–32.
[16] A. Levin, N. Burton-Wurster, C.T. Chen and G. Lust, Intercellular signaling as a cause of cell death in cyclically impacted cartilage explants, *Osteoarthritis Cartilage* **9**(8) (2001), 702–11.
[17] L.P. Li, M.D. Buschmann and A. Shirazi-Adl, A fibril reinforced nonhomogeneous poroelastic model for articular cartilage: inhomogeneous response in unconfined compression, *J. Biomech.* **33**(12) (2000), 1533–41.
[18] H. Lipshitz, 3rd, R. Etheredge and M.J. Glimcher, Changes in the hexosamine content and swelling ratio of articular cartilage as functions of depth from the surface, *J. Bone Joint Surg. Am.* **58**(8) (1976), 1149–53.
[19] A.M. Loening, I.E. James, M.E. Levenston, A.M. Badger, E.H. Frank, B. Kurz, M.E. Nuttall, H.H. Hung, S.M. Blake, A.J. Grodzinsky and M.W. Lark, Injurious mechanical compression of bovine articular cartilage induces chondrocyte apoptosis, *Arch. Biochem. Biophys.* **381**(2) (2000), 205–12.
[20] A. Maroudas, Balance between swelling pressure and collagen tension in normal and degenerate cartilage, *Nature* **260** (1976), 808–809.
[21] A. Maroudas, E. Wachtel, G. Grushko, E.P. Katz and P. Weinberg, The effect of osmotic and mechanical pressures on water partitioning in articular cartilage, *Biochim. Biophys. Acta* **1073** (1991), 285–294.

[22] V. Morel and T.M. Quinn, Cartilage injury by ramp compression near the gel diffusion rate, *J. Orthop. Res.*, in press.
[23] A.G. Ogston, B.N. Preston and J.D. Wells, On the transport of compact particles through solutions of chain-polymers, *Proc. R. Soc. Lond. A* **333** (1973), 297–316.
[24] P. Patwari, M.N. Cook, M.A. DiMicco, S.M. Blake, I.E. James, S. Kumar, A.A. Cole, M.W. Lark and A.J. Grodzinsky, Proteoglycan degradation after injurious compression of bovine and human articular cartilage in vitro: interaction with exogenous cytokines, *Arthritis Rheum.* **48**(5) (2003), 1292–301.
[25] E.A. Pickvance, T.R. Oegema, Jr. and R.C. Thompson, Jr., Immunolocalization of selected cytokines and proteases in canine articular cartilage after transarticular loading, *J. Orthop. Res.* **11**(3) (1993), 313–23.
[26] T.M. Quinn, R.G. Allen, B.J. Schalet, P. Perumbuli and E.B. Hunziker, Matrix and cell injury due to sub-impact loading of adult bovine articular cartilage explants: effects of strain rate and peak stress, *J. Orthop. Res.* **19**(2) (2001), 242–9.
[27] T.M. Quinn, P. Dierickx and A.J. Grodzinsky, Glycosaminoglycan network geometry may contribute to anisotropic hydraulic permeability in cartilage under compression, *J. Biomech.* **34**(11) (2001), 1483–90.
[28] T.M. Quinn, A.J. Grodzinsky, M.D. Buschmann, Y.J. Kim and E.B. Hunziker, Mechanical compression alters proteoglycan deposition and matrix deformation around individual cells in cartilage explants, *J. Cell. Sci.* **111**(Pt 5) (1998), 573–83.
[29] R.U. Repo and J.B. Finlay, Survival of articular cartilage after controlled impact, *J. Bone Joint Surg. [Am]* **59**(8) (1977), 1068–76.
[30] I.Y. Syed and B.L. Davis, Obesity and osteoarthritis of the knee: hypotheses concerning the relationship between ground reaction forces and quadriceps fatigue in long-duration walking, *Med. Hypotheses* **54**(2) (2000), 182–5.
[31] M. Thibault, A.R. Poole and M.D. Buschmann, Cyclic compression of cartilage/bone explants in vitro leads to physical weakening, mechanical breakdown of collagen and release of matrix fragments, *J. Orthop. Res.* **20**(6) (2002), 1265–73.

Biorheology 41 (2004) 521–534
IOS Press

Induction of heat shock protein 70 (Hsp70) by proteasome inhibitor MG 132 protects articular chondrocytes from cellular death *in vitro* and *in vivo* [1]

Laurent Grossin, Stéphanie Etienne, Nadège Gaborit, Astrid Pinzano, Christel Cournil-Henrionnet, Catherine Gerard, Elisabeth Payan, Patrick Netter, Bernard Terlain and Pierre Gillet *

UMR 7561 CNRS – Université Nancy I, Laboratoire de Pharmacologie, Faculté de Médecine de Nancy, BP 184, F54505 Vandœuvre, France

Abstract. The aim of this work was to determine whether Hsp70 overexpression via proteasome inhibitor MG132 was able to protect chondrocytes towards mono-iodoacetate (MIA) cytotoxicity both *in vitro* and *in vivo*. *In vitro*, overexpression of Hsp70 via MG132 was significantly able to protect chondrocytes from MIA toxicity (MTT/LDH analyses). Hsp70 essentially mediated this chondroprotective effect as demonstrated by antisense strategy. *In vivo*, chondrocytic overexpression of Hsp70, after a preventive intra-articular injection of MG132 in rat knee, was sufficient to decrease the severity of OA-induced MIA lesions, as demonstrated histologically and biochemically. In conclusion, intracellular overexpression of Hsp70, through proteasome inhibition, could be an interesting tool in protecting chondrocytes from cellular injuries, either necrotic or apoptotic in nature, and thus might be a novel chondroprotective modality in rat experimental OA.

1. Introduction

Osteoarthritis (OA) is a degenerative joint disease, characterized by hyaline cartilage fibrillations and erosions, subchondral bone sclerosis and osteophyte formation at joint margins. These changes result from poorly understood events occurring over a long period, thus leading to both cartilage extracellular (ECM) matrix degradation and ECM synthesis inhibition. Furthermore, cartilage hypocellularity due to cell death also contributes to the development of OA, either apoptotic in nature, due to programmed cell death [1], or necrotic as an end result, ensuing from cellular injury. In fact, chondrocyte apoptosis is observed during OA, in both human OA [2] and OA experimental models [15].

An indicator of cellular suffering is the induction of heat shocks proteins (Hsps). Hsps are a set of conserved proteins in Prokaryotes and Eukaryotes, in which Hsps expression is induced by a lot of cellular stresses. Hsps also play a protective role as molecular chaperones in cells, by facilitating folding,

[1]This work received support from the GIP HMR AVENTIS (grant FR99RHU037), Region Lorraine and CPRC of the CHU of Nancy.

*Address for correspondence: UMR 7561 CNRS – Université Nancy I, Laboratoire de Pharmacologie, Faculté de Médecine de Nancy, BP 184, F54505 Vandœuvre, France. E-mail: pierre.gillet@medecine.uhp-nancy.fr.

intracellular transport, assembly, and disassembly of proteins. Hsps as well protect cells from oxidative damage both *in vivo* and *in vitro* [22] and additionally protect cell from either necrosis [31] or apoptosis [24]. An increase in the expression of Hsp70 in chondrocytes has been related to the severity of the joint lesions in humans. Therefore, Hsp70 could be an indicator of early stages of OA [29] and cellular suffering, and Hsp70 was thus suggested to have an important influence on OA progression. The same authors [28] found that the enhancement of Hsp70 expression protects cells from cell death induced by stress [18], mechanical, and biological factors [23].

In several cell types, proteasome inhibitors, like MG132, induce heat shock proteins. However, data in chondrocytes are currently scanty. The 26S proteasome is a multisubunit complex, which recognizes and degrades proteins, including poly-ubiquitinated, misfolded and short-lived regulatory proteins. Inhibition of the degradation of intracellular proteins by an inhibitor results in the accumulation of altered proteins, which, as an end result, induces the overexpression of Hsps [17].

The aim of the present work was to determine the chondroprotective properties of Hsp70 overexpression towards mono-iodoacetate (MIA) induced cytotoxicity *in vitro* and *in vivo*. To this end, we have used the peptide aldehyde inhibitor of proteasome MG132 as a potent inducer of Hsp70 expression in chondrocytes. This study assessing the chondroprotective properties of the overexpression of Hsp70 *via* preventive MG132 exposure was realized both *in vitro* in rat chondrocytes exposed to MIA and *in vivo* during the MIA-induced OA model in rat knee. It is noteworthy that MIA, when injected intra-articularly in rat knee, reproduces OA-like lesions, functional joint impairment and chondrocytic apoptosis and necrosis whose severity are clearly related to the dose of MIA used [13].

2. Material and methods

2.1. Cell culture

2.1.1. Reagents

Iodoacetic acid (MIA) was purchased from Sigma (Saint Quentin Fallavier, France) and proteasome inhibitor (Z-Leu-Leu-leucinal, MG132) was purchased from Calbiochem (VWR, Fontenay-sous-Bois, France).

2.1.2. Isolation and culture of chondrocytes

Articular chondrocytes were obtained from femoral head caps of male Wistar rats (175–200 g) (Charles River Laboratories, Saint-Aubin-les-Elbeuf, France, France). After sacrifice, hips were quickly dissected under sterile conditions and chondrocytes isolated by sequential digestion with pronase (2 mg/ml) and collagenase B (1.5 mg/ml) (Roche, Meylan, France). Cells were then plated in DMEM/Ham's F12 medium supplemented with 10% (v/v) heat-inactivated fetal calf serum (FCS), 2 mM L-glutamine, gentamicin, amphotericin B (Invitrogen, Cergy-Pontoise, France). Cells were cultured at confluence at 37°C in a humidified atmosphere containing 5% CO_2, and were used at passage 2. In all experiments, chondrocytes were cultured under low FCS conditions (5%, v/v) and were used in monolayers at confluence.

2.1.3. Mono-idoacetate (MIA) cytotoxicity

Cell stress was induced using different concentrations of MIA (2; 5 and 10 μM). For Hsps induction, cells were treated with DMSO (0.1% (v/v) final concentration) as vehicle control or with Z-Leu-Leu-leucinal (MG132) [31] during 4 h followed by an incubation for 14 h in DMEM/F12. MG132 was used at 0.15, 0.5 and 1.5 μM diluted in DMEM/F12 solution (0.1% (v/v) DMSO).

2.2. *Assessment of cytotoxicity*

2.2.1. *Assay for lactate dehydrogenase activity*

Cells were seeded in 24-well plates at a density of 2×10^5 cells per well. Cytotoxicity Detection Kit (Roche) measures the lactate dehydrogenase (LDH) activity released from the cytosol of damaged cells into the supernatant and evaluates the percentage of cell death. One hundred microliters of supernatant were incubated with 100 μl of the reaction mixture. After 10 min of incubation in the dark, the reaction was stopped by 50 μl of an HCl 1 N solution. Absorbance was measured using plate reader at 490 nm (MR5000, Dynatech). Cytotoxicity was determined in comparison using a positive test with a Triton X-100 2% (v/v) diluted in DMEM/F12 solution.

2.2.2. *MTT assay*

Cells were seeded in 96-well plates at a density of 5×10^4 cells per well. The assay is based on the cleavage of the yellow tetrazolium salt MTT (3-(4,5-dimethylthiazole-2-yl)-2,5-diphenyltetrazolium bromure) to purple formazan crystals by dehydrogenase activity in intact mitochondria. Then, the formazan crystals are solubilized and absorbance of the resulting colored solution is measured at 550 nm (Dynatech). Cells were incubated with 100 μl of DMEM/F12 and 25 μl of MTT solution (5 mg/ml in PBS). After 4 h, the supernatant was removed and 100 μl of a solubilization solution (sodium dodecyl sulfate + di-methyl formamide) were added into each well. The plate was incubated overnight before measurement.

2.3. *Reverse transcription and polymerisation chain reaction*

Cells were seeded in 6-well plates at a density of 5×10^5 cells per well. Hsp70 expression was increased by stimulation with MG132 in the range of 0.15 to 1.5 μM for 4 h. Chondrocytes were washed twice with PBS, and total RNAs were extracted by the guanidium thiocyanate-phenol-chloroform method [7] using Trizol™ (Invitrogen). Total RNA amount was quantified by spectrophotometry at 260 nm. cDNA was obtained by reverse transcription (Omniscript® Reverse Transcriptase, Qiagen, Courtaboeuf, France) using 1 μg of total RNA and oligodT primers (polydT$_{12–18}$, Invitrogen). Then, PCR amplification was performed using specific primers for Hsp70i (1946 bp product), Hsp27 (636 bp product) and GAPDH (816 bp product) (Table 1). Samples were first heated at 94°C for 15 min to activate the HotStar*Taq* polymerase (Qiagen). Amplification was performed for 30 cycles at 94°C for 45 s, 62°C for 45 s and 72°C for 1 min using a thermocycler (Eppendorf, VWR). The final extension step lasted 10 min (72°C). Amplification products were resolved by electrophoresis in a 1% (w/v) agarose gel, stained by ethidium bromide (0.5 μg/ml), and photographed under UV light by a computer-assisted camera (GelDoc, Biorad, Marnes-la-Coquette, France).

2.4. *Western blot analysis*

Cells were seeded in 6-well plates at a density of 5×10^5 cells per well. After MG132 exposure, cells were washed with PBS, scrapped and lysed by sonication in 200 μl of PBS. Proteins concentrations were measured by the method of Bradford [3]. Chondrocyte lysates were diluted with the gel loading buffer in a ratio of 1 : 1 (152 mM Tris-HCl, pH 6.8, 4% (w/v) SDS, 20% (v/v) glycerol, 10% (v/v) 2-mercapto-ethanol and 0.004% (v/v) bromophenol blue), and heated at 95°C for 5 min. Electrophoresis was performed using 10% (w/v) SDS-polyacrylamide gel (20 μg protein per lane) according to Laemmli [19].

Table 1

Primers used to evaluate expression of the genes GAPDH and Hsp70i by RT-PCR

	Forward	Reverse
GAPDH	5′-ACGGATTTGGCCGTATTGGCCGC-3′	5′-TGGTCCTCAGTGTAGCCCAGGAT-3′
Hsp70i	5′-AGCGACATGGCCAAGAAAACAGCG-3′	5′-CCAGAAAAGCCTCCAATCCACCTCC-3′
Hsp27	5′-GCCAAGACCATGACCGAG-3′	5′-CTTCCACTTGGCTCCAGAC-3′

Separated proteins were transferred onto PVDF membranes (Immobilon-P, Biorad) and non-specific binding sites were blocked overnight at 4°C with 3% (w/v) BSA in TBST solution (9 mM Tris-HCl, pH 9.0, 154 mM NaCl, 0.05% v/v Tween 20). Membranes were incubated with primary antibodies, rabbit polyclonal anti-Hsp70 (1/20000) (Stressgen, Tebu, Le Perray en Yvelines, France) or rabbit polyclonal anti-Hsp27 (1/2000) (Stressgen) for 1 h. After 6 washes with TBST solution, membranes were further incubated for 1 h with the secondary antibody (goat anti-rabbit IgG) conjugated to alkaline phosphatase (1/40000). Membranes were washed again with TBST, incubated for 5 min with 5-bromo-4-chloro-3-indolyl phosphate/nitroblue tetrazolium (BCIP/NBT, Roche) and reaction was stopped with methanol. Protein expression was evaluated by densitometry analysis of each band (NIH Image) in arbitrary units.

2.5. *Antisense strategy*

The antisense oligonucleotide against Hsp70i was performed in order to determine if this oligonucleotide was able to inhibit the protective effect of MG132 towards the toxicity of MIA. Optimized antisense oligomers, 5′-CACCTTGCCGTGCTGGGAA-3′ (10 μM), located in the starting-coding region of the rat Hsp70, were mixed with polyethylenimine reagent (Exgen 500, Euromedex, Mundolsheim, France) (3%, v/v) and then incubated 15 min at room temperature. An equal volume of complete medium was mixed to a final volume of 200 μl for the 6-well plates and 30 μl for 96-well plates and added to the cells. After incubation for 4 h at 37°C, medium was changed and cells were incubated for an additional 20 h sequence. A second transfection was done, in the same conditions (PEI 3% v/v, 10 μM antisense oligomers). After 90 min, the supernatant was removed and experiments were started. Control using nonsense oligonucleotide (5′-TGGATCCGACATGTCAGA-3′) were done concurrently.

2.6. *Experimental in vivo procedures in the rat*

2.6.1. *Induction of MIA induced OA and treatment procedure*

Male Wistar rats weighing 150 g (Charles River Laboratories) were housed in solid-bottomed plastic cages designed to allow easy access to standard laboratory food and water *ad libitum*. Rats were kept in a 12 : 12 light-dark cycle (light-on period, 6:00 am–6:00 pm) in a controlled temperature chamber. Several doses of MIA (0.03 or 0.3 mg, Sigma), dissolved in 50 μl of sterile physiologic saline, were injected through the intra-patellar ligament [18]. MG132 was dissolved in 50 μl of a sterile solution of PEG (polyethylene glycol, Sigma) and this solution was injected 3 h before MIA. For each study, two control groups of rats received concomitantly sterile saline solution and sterile solution of PEG, respectively.

2.6.2. *Histological assessment*

Fifteen days after injection with either MIA, saline, or MG132, rats were killed and the whole knee joints were fixed immediately with 10% (v/v) formalin, decalcified, and embedded in paraffin. Sagittal sections were prepared and stained with hematoxylin–eosine–safran, toluidine blue or Sirius red. The severity of OA lesions was graded on a scale of 0–21, by 2 independent observers, using a modified form of histological/histochemical scale of Mankin [5,11]. This scale evaluates the severity of OA lesions based on proteoglycan integrity (scale 0–4), cellular changes (scale 0–3), collagen fibers integrity (scale 0–3), invasion of the tidemark by blood vessels (scale 0–1), bone modifications (scale 0–2) and structural changes (scale 0–5, were 0 = normal cartilage structure and 5 = erosion of the cartilage down to the subchondral bone).

2.6.3. Gross evaluation

Rat knee joints were isolated on day 15 and carefully dissected to assess the severity of macroscopic chondral alterations after injections of either MIA alone, MG132 (15 μM) and MIA (0.3 mg) or saline. All the compartments of the knee (medial and lateral tibial plateaus, medial and lateral condyles, patellae, and femoral groove) were observed. Macroscopic lesions were graded as follow: 0 = normal appearance; 1 = slight yellowish discoloration of the chondral surface; 2 = little cartilage erosions in load-bearing areas; 3 = large erosion extending down to the subchondral bone; 4 = large erosions with large areas of subchondral bone exposure. The 6 compartment scores were combined (for a maximum possible score of 24) and the mean was determined for each group.

2.6.4. Measurement of proteoglycan biosynthesis

Proteoglycan synthesis was assessed as described by van der Kraan et al. [30], with some modifications. Fifteen days after injections, rats were killed as described earlier and the patellae from the injected right knees were carefully dissected out. The patellae from the injected left knees were used to determine sulphated GAG concentrations. They were placed in 1 ml of medium consisting of RPMI 1640 with HEPES and sodium bicarbonate, supplemented with L-glutamine (2 mM), penicillin (100 IU/ml), streptomycin (100 μg/ml) (Invitrogen), and $Na_2{}^{35}SO_4$ (0.7 μCi/ml; Amersham, Les Ulis, France). After incubation for 3 h (37°C, 5% CO_2), patellae were washed with 1 ml of cold physiologic saline and fixed overnight in 0.5% cetylpyridinium chloride (Sigma) in phosphate buffered formalin. After decalcification in 5% formic acid for 6 h, the central area of the patella was sampled with a 2 mm-diameter biopsy punch (Stiefel, Nanterre, France). The punched-out part of the patellae and the remaining peripheral part of the patellae were separately dissolved overnight in 0.5 ml of Soluene-350 (Packard, Rungis, France). The amount of ^{35}S-sulfate incorporated in each compartment was counted using scintillation fluid (Hionic Fluor; Packard). Results were expressed as the differences between the MIA-injected, the MG132-injected and the MIA + MG132-injected patellae and control patellae in the mean percentage of ^{35}S incorporated.

2.6.5. Immunostaining

Bilateral injections of MG132 (15 μM) were realized through infra-patellar ligament and rats were sacrificed at 24 h or 48 h. Patellae and knees were isolated and fixed in 4% paraformaldehyde, decalcified in 10% EDTA and embedded in paraffin. Sections (5 μm) of paraffin-embedded samples were deparaffinized in toluene, rehydrated in a graded series of ethanol, and incubated with trypsin 0.1% in $CaCl_2$ 0.1% for 30 min at 37°C. Slices were then washed in PBS, and endogenous peroxydase was inhibited in 3% hydrogen peroxide/TNT buffer (Tris-HCl 0.1 M, pH7.5; NaCl 0.15 M; 0.05% Tween 20) for 10 min. They were further incubated with TNB buffer (Tris-HCl 0.1 M, pH 7.5; NaCl 0.15 M; 0.5% blocking reagent) and overlaid with antibodies (rabbit polyclonal anti-Hsp70; 1 : 300) (Stressgen) for 1 h at room temperature in a humidified chamber. Each slide was washed 3 times in TNT buffer, pH 7.4, then incubated in DNP amplification reagent (Kit Nen amplification, Novocastra, Tebu) for 7 min. After 3 washing in TNT buffer in a shaking water bath, slides were incubated with anti-DNP-HRP antibodies from the kit for 30 min in a humidified chamber and then washed again in TNT buffer. Color was developed with DAB containing hydroxide peroxide and slides were counterstained for 3 min in 1% methyl green. The counterstaining gives green-stained nuclei and specific staining appears as brown-stained cytoplasm. The percentage of positive chondrocytes was measured by counting the positive cells and the total number of cells in the different cartilage layers. For each specimen, 3 microscopic fields were examined (magnification, ×40), in each layer of cartilage (the superficial, intermediate and deep layers). Each slide was subjected to a double-blind evaluation with a maximal variation of 5% observed.

3. Results

3.1. Overexpression of Hsps and cytoprotection in MG132-treated cells

3.1.1. Hsp70i and Hsp27 mRNA expression in chondrocytes in response to MG132 treatment

RT-PCR led to amplicons of the expected size: 1946, 636 and 816 bp for Hsp70i, Hsp27 and GAPDH respectively (Fig. 1). In each sample, expression of the gene encoding for Hsp70i was normalized according to that GAPDH mRNA (Fig. 1). Hsp70i and Hsp27 mRNA were not detected in control chondrocytes. A MG132 exposure stimulated the mRNA expression of Hsp70i in a dose-dependent manner, the lowest dose of MG132 also being able to induce mRNA overexpression. The same pattern of expression was observed, in a lesser extent, for Hsp27.

3.1.2. Hsp70i and Hsp27 proteins expression in chondrocytes in response to MG132 treatment

As expected, MG132 also induced Hsp70i (70 kDa) protein expression in chondrocytes (Fig. 2). The protein was detected 6 h after MG132 exposure, the signal intensity increasing until 24 h. At 48 h, Hsp70 level decreased, probably related to the short half-life of the protein. In fact, at 48 h, Hsp70i protein level was around 3.3-fold less expressed, when compared to 24 h with 1.5 μM MG132 treatment.

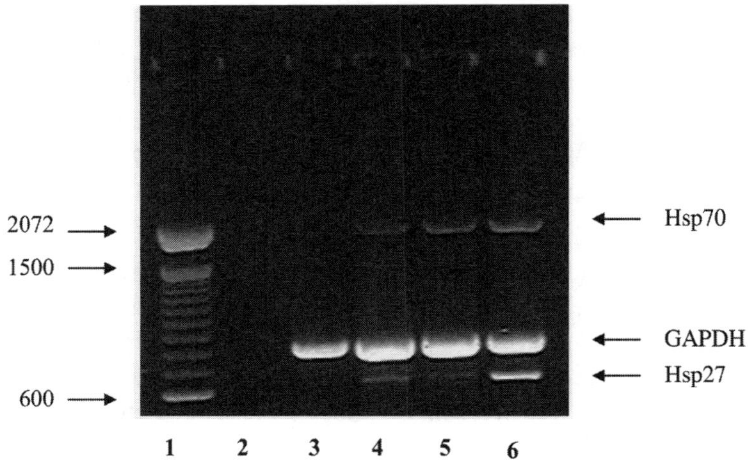

Fig. 1. Electrophoretic profile of RT-PCR products of Hsp70i, Hsp27 and GAPDH using specific primers. Chondrocytes are submitted to MG132 exposure at several concentrations (0.15 μM, 0.5 μM, 1.5 μM) for 6 h. Then total RNA are extracted and submitted to RT-PCR reaction. PCR products generated by amplification are visualized on 1% agarose gel, stained with ethidium bromide (0.5 μg/ml). Lane 1, molecular weight markers; Lane 2, negative PCR control; Lane 3, control cells without MG312; Lane 4, cells treated with 0.15 μM MG132; Lane 5, cells treated with 0.5 μM MG132 (lane 5); Lane 6, cells treated with 1.5 μM MG132. Induction of Hsps proteins were observed with each concentration of proteasome inhibitor (MG132), as soon as 6 h after exposure and the maximum of induction was depicted with 1.5 μM of MG132, without any signs of cytotoxicity.

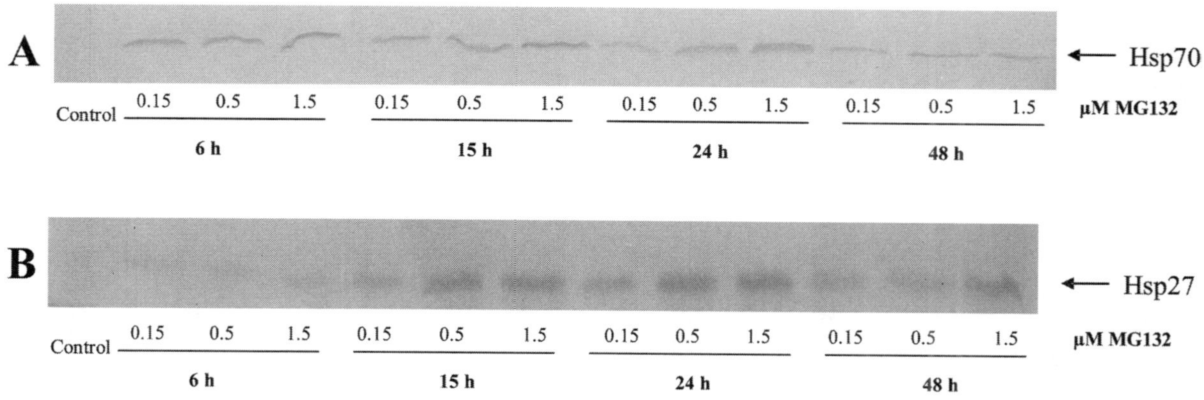

Fig. 2. Expression levels of the proteins Hsp70i in chondrocytes treated with MG132. Cells were treated either with the vehicle (0.1% DMSO in medium) or 0.15; 0.5 or 1.5 μM MG132 for 4 h, and then incubated in fresh medium for 2, 11, 20 or 44 h. Cells were lysed and the lysates were used to perform SDS/PAGE. For each sample, 20 μg of total proteins were loaded on the gel. Specific primary antibody against the inducible form of Hsp70i was used. Level of expression of the protein were estimated by densitometry. For Hsp70i, induction was already detected at 6 h after exposure and the maximum amount of proteins was observed at 24 h with the highest concentration of MG312 (1.5 μM). For Hsp27, a weak induction was observed at 6 h and as for Hsp70, the maximum induction was reached at 24 h after proteasome inhibitor exposure.

3.1.3. MG132 protection against MIA-induced cell death

Cells were pre-treated or not with different concentrations of MG132 (0.15, 0.5 or 1.5 μM) and then were incubated with different concentrations of MIA (2, 5 and 10 μM) for 6 h, 24 h and 48 h, respectively. Cell survival was quantified by measuring lactate dehydrogenase release. Figure 3 shows the results obtained after MIA treatment (A: 6 h, B: 24 h, C: 48 h). A pretreatment with MG132 has a significantly advantage against cell injury during MIA treatment. Moreover, cytoprotection was MG132 dose-dependent, and, in an other hand, clearly related to MIA dose and time exposure. These results were confirmed by MTT test and they showed the same effect of MG132 on MIA toxicity (data not shown). Moreover, data showed that MG132 was not chondrotoxic and not a pro-apoptotic in our experimental conditions as assessed by TUNEL method (data not shown).

3.1.4. Effects of antisense oligonucleotides

To determine whether the proteasome inhibitor was related to the level of Hsp70i expression and affected protection, rat chondrocytes were pre-treated with 10 μM antisense oligonucleotides specific against the heat-inducible Hsp70 mRNA prior to 1.5 μM MG132 stimulation. The resulting increases in Hsp70i protein in response to MG132 stimulation were blocked by Hsp70 oligonucleotides, shown by western-blot. Pre-treatment of cells with antisense oligonucleotides decreased the protection of MG132 towards MIA (10 μM, 6 h), as observed by the MTT test (Fig. 4). In fact, addition of 10 μM Hsp70 antisense oligonucleotides reduced significantly ($p < 0.005$) MG132 protection towards MIA to 18% protection in MG132-treated cells when compared with a 60% to 70% protection in cells pretreated with MG132 alone or with 10 μM nonsense oligonucleotides.

4. *In vivo* experimental studies in the rat

4.1. Increased Hsps expression after MG132 stimulation

Immuno-histochemical analysis of cartilage with Hsp70i antibody after injection of MG132 or saline is shown in Fig. 5. The results showed a higher level of Hsp70 protein in the lower hypertrophic chon-

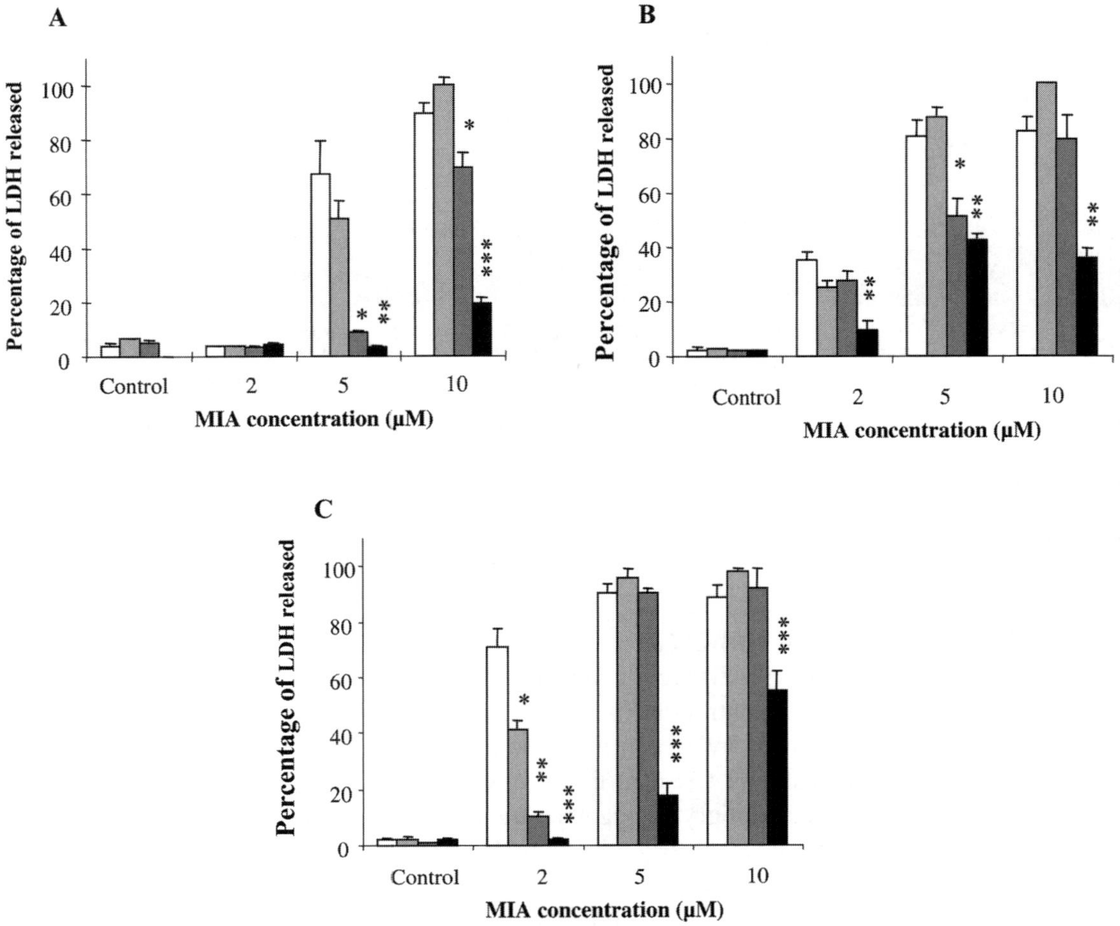

Fig. 3. Protective properties of MG132 towards MIA cytotoxicity. The release of LDH from cells pretreated with various doses of MG132 (□, chondrocytes without preventive exposure to MG132; ▒, chondrocytes preventively exposed to proteasome inhibitor MG132 (0.15 μM); ▓, chondrocytes preventively exposed to proteasome inhibitor MG132 (0. 5 μM); ■, chondrocytes preventively exposed to proteasome inhibitor MG132 (1.5 μM)) or treated with different concentrations of MIA (2 μM; 5 μM and 10 μM) for 6 h (panel A), 24 h (panel B) or 48 h (panel C) was measured. Results are expressed as percentage of LDH released from the cells over total LDH for each sample. There were normalized to the amount of LDH released by control cells. Errors bars show the mean ± SEM ($n = 3$ per group in 3 repeated experiments).$^{*}p < 0.05$; $^{**}p < 0.005$; $^{***}p < 0.0005$ (Student's t-test).

drocytes either in control than treated rats (100% of positive cells). In control rats, there was a weak expression of Hsp70 in the superficial layer (around 20% of positive cells). In contrast MG132 *ia* injection into rat knee joint induced the expression of Hsp70 in chondrocytes of the superficial and middle layer ($n = 8$ patellae). Moreover, after one injection of MG132, Hsp70 proteins in the superficial layer were expressed 24 h after injection and expression was maximal at 48 h. In fact, the percentage of positive cells for Hsp70i in the superficial zone was significantly higher in response to MG132 injection (44 ± 12% for 24 h and 63 ± 18% at 48 h) than in normal cartilage (27 ± 4%) (Fig. 5).

4.1.1. Macroscopic study after MIA ia injection and effect of MG132 pre-treatment

Fifteen days after the *ia* injection of 0.3 mg of MIA, preceded or not by a preventive *ia* injection of MG132 (0.35 μg), articular cartilage of the different compartments was observed macroscopically

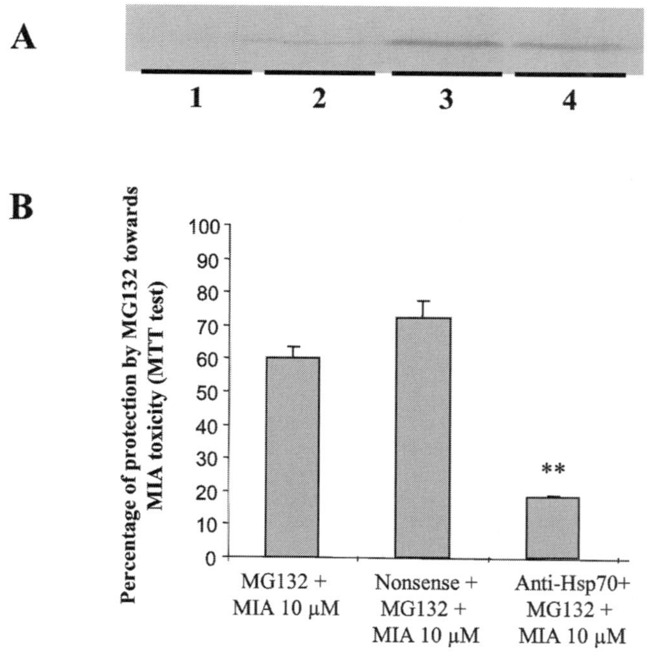

Fig. 4. Antisense experiments. Panel A: Western blot analysis of Hsp70i after treatment of chondrocytes with Hsp70i antisense oligonucleotides. Cells were treated with 10 μM nonsense or 10 μM antisense oligonucleotides in the presence or absence of 1.5 μM MG132. Lane 1, no exposure to neither MG132 or oligonucleotides; Lane 2, exposure to antisens oligonucleotide prior to MG132 exposure; Lane 3, exposure to non-sens oligonucleotide prior to MG132 exposure; Lane 4, exposure to MG132 exposure; Panel B: Toxicity of MIA in cells treated with Hsp70i antisense oligonucleotides. Cells were first treated with either 10 μM nonsense or 10 μM antisense oligonucleotides and then stimulated with 1.5 μM MG132 for 4 h. After an incubation for 14 h at 37°C–5% CO_2, cells were treated by 10 μM MIA for 6 h before MTT test. Errors bars show the mean ± SEM ($n = 3$ per group). $^{**}p < 0.005$ (Student's t-test).

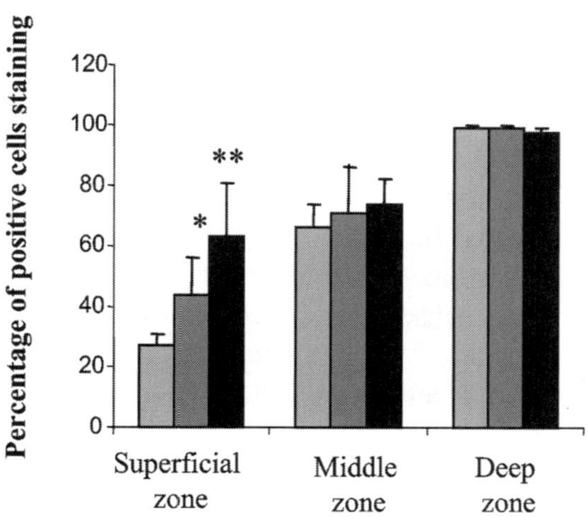

Fig. 5. Immunohistochemistry of Hsp70 in articular cartilage. Rats received an intra-articular injection of MG132 0.35 μg (15 μM in 50 μl) into the knee joints and were sacrificed at 24 and 48 h. Percentage of positive cell staining for Hsp70i in cartilage from the different layers in the presence or absence of MG132. ▫, control patellae of rats, receiving an intra-articular injection of saline solution; ▪, patellae of rats, submitted to an intra-articular injection of MG132 (15 μM) for 24 hours; ■, Patellae of rats, submitted to an intra-articular injection of MG132 (15 μM) for 48 hours. Bars show the mean ± SEM ($n = 8$ patellae). $^{*}p < 0.05$; $^{**}p < 0.005$ (Student's t-test).

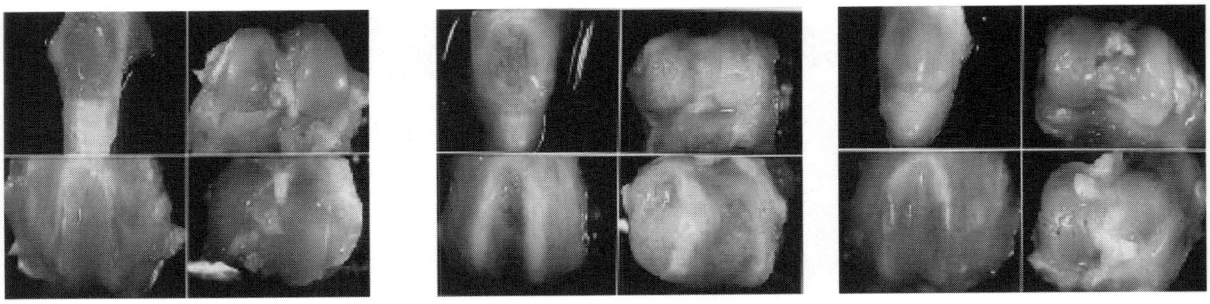

A B C

Fig. 6. Macroscopic changes in the knee joint compartments after intra-articular injection of MIA 0.3 mg with or not a previous injection of MG132 0.35 μg. The knees of 3 rats per group were carefully dissected on day 15, under a magnifying glass. The different articular surfaces shown in each part of the figure are the patella (upper left), the femoral groove (lower left), the femoral condyles (upper right) and the tibial plateaus (lower right). Panel A: No macroscopic lesion can be seen in the normal rat knee. Panel B: After intra-articular injection of 0.3 mg MIA, large subchondral bone exposures are present in the different compartments and osteophytes are prominent in the peripatella and femoral groove. Panel C: A previous intra-articular injection of MG132 0.35 μg, allowed to reduce the lesions.

(Fig. 6). Injection of MIA alone led to lesions affecting all articular cartilage of the knee joint, especially in the femoro-tibial joint, with large chondral erosions and subchondral bone exposure in both tibial plateaus and femoral condyles (mean score: 18.5 ± 0.96). Large osteophytes were present in the nonweight-bearing areas of femoral condyles and tibial plateaus. Patella and femoral groove were also affected, but in a lesser extend. In rat knees, previously injected with MG132, lesions were less severe on day 15. Hyaline cartilage of central weight-bearing areas of each articular surface was thin and yellowish, with focal erosions in femoral condyles and tibial plateaus. Patella and femoral groove seemed to be not affected. We have not assessed macroscopically the protective effect of MG132 on the lowest dose of MIA (0.03 mg) because macroscopic lesions observed at this dose-regimen were not important enough to observe a protective effect of MG132 [18] (mean score: 6.33 ± 2.67, $p < 0.005$).

4.1.2. Histological study after MIA ia injection and effect of MG132 pre-treatment

On day 15 after *ia* injection of 0.03 or 0.3 mg of MIA preceded or not by a preventive *ia* injection of MG132 (0.35 μg), the whole knee joints that had been injected with either MIA, saline or MG132 were fixed, decalcified and embedded in paraffin. Sagittal sections were prepared and stained with hematoxylin–eosine–safran, toluidine blue and sirius red. Histologic scores for the lesions on cartilage patellae, condyles and plateaus were realised (Fig. 7). Histological scoring for of the lesions was higher in knees injected with MIA (0.3 mg and 0.03 mg) than those injected with saline or MG132. Morphologic changes characteristic of experimental MIA-induced OA [13] were observed: subchondral bone exposure, subchondral bone sclerosis, synovial membrane hyperplasia and fibrosis, slight infiltration of mononuclear leukocytes, hypocellularity and clusters of chondrocytes, and decrease of proteoglycan integrity. Interestingly, a minimal toxicity of MG132 (lesions graded 4.5) was present, when compared with saline alone: synovial membrane was hyperplasic and fibrotic, but, in contrast, cartilage cellularity and matrix integrity were not affected. In contrast, in the MG132-pretreated group, the lesions were significantly less severe (lesions graded 9.62 ± 1.99 for MIA 0.03 mg and 12.00 ± 3.17 for MIA 0.3 mg) compared with those in the MIA alone group (0.03 mg, lesions graded 18.96 ± 1.37; 0.3 mg, lesions graded 20.50 ± 0.29).

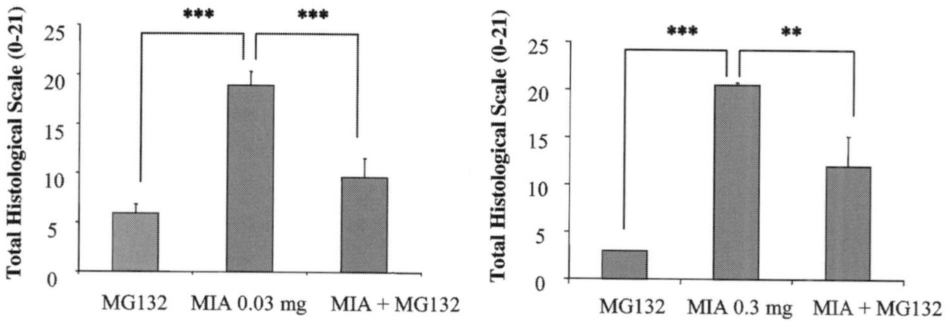

Fig. 7. Histologic changes in the knee joints after intra-articular injection of MIA 0.3 mg and 0.03 mg with or not a previous injection of MG132. First, rats received or not an intra-articular injection of MG132 0.35 μg (15 μM in 50 μl) into the knee joints followed by an injection of MIA (0.3 mg and 0.03 mg) and were sacrificed at 15 days after MIA injection. Histologic grading of cartilage from the patellae, femoral condyles and tibial plateaus. Bars show the mean and SEM ($n = 12$ patellae per group). $^{**}p < 0.005$; $^{***}p < 0.0005$ (Student's t test).

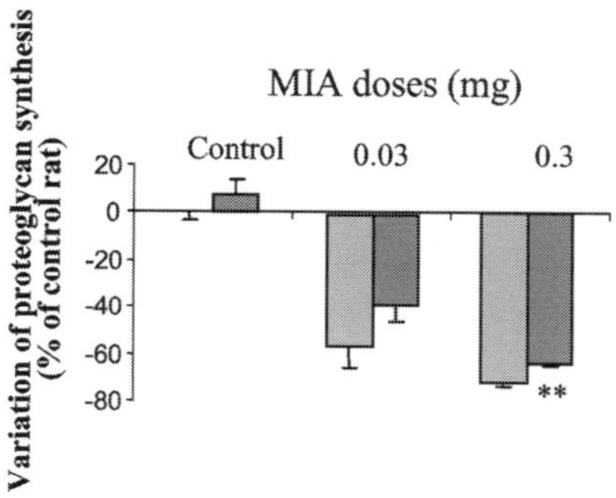

Fig. 8. Chondrocytes metabolism. Rats received or not an intra-articular injection of MG132 0.35 μg (15 μM in 50 μl) into the knee joints followed by an injection of MIA (0.3 mg and 0.03 mg) and were sacrificed at 15 days after MIA injection. Proteoglycan biosynthesis in patellar cartilage, measured by $Na_2{}^{35}SO_4$ incorporation after intra-articular injection of MG132 and/or MIA in rat knee joint. Results are expressed as the percentage of change in ^{35}S incorporation in the central part of the patella *versus* the control rats ($n = 6$ patellae per group). □, Rats only treated with a single injection of MIA (0.3 mg; 0.03 mg); ■, Rats were submitted to an intra-articular injection of MG132 (15 μM). Bars show the mean ± SEM. $^{*}p < 0.05$; $^{**}p < 0.005$, *versus* rats injected with MIA only (no injection of MG132), Student's t test.

4.1.3. Effects of MIA on proteoglycan biosynthesis and chondroprotective effect of MG132

Fifteen days after injection of either 0.03 mg or 0.3 mg of MIA into rat knee joint, ^{35}S-sulphate incorporation in the central patella was strongly inhibited by $56.32 \pm 8.47\%$ and $72.41 \pm 1.29\%$ respectively (Fig. 8). A previous injection of MG132 0.35 μg reduced the strong inhibition of anabolism (-62%, $p < 0.005$, 0.3 mg). Moreover, the MG132 alone did not affect significantly proteoglycan anabolism.

5. Discussion

The aim of this study was to evaluate the protective effect of MG132, a transient inhibitor of the proteasome pathway, in chondrocytes submitted to cytotoxic potencies of mono-iodoacetate (MIA). In a first step, these experiments were performed *in vitro* to determine that this cytoprotection was mediated *via* Hsp induction. Secondarily, according to these data, *in vivo* studies were then conducted in the MIA-induced OA model in the rat to confirm the chondroprotective effect of Hsp overexpression ancillary to the temporary inhibition of the proteasomal degradation pathway.

In fact, numerous reports have described both pro and antiapoptotic properties of a variety of proteasome inhibitors. In general, inhibition of the 26S proteasome elicits either (1) an anti-apoptotic effect in terminally differentiated cells [20,25,27] and non-rapidly dividing cells, or (2) a pro-apoptotic process in fast-dividing cells [10]. Besides, accumulation of abnormal or short-lived protein inside cell cytoplasm, due to the inhibition of the proteasome pathway, induces *de facto* the overexpression of stress proteins, such as Hsp70 [6].

The results presented herein demonstrate that chondrocyte exposure to a transient proteasome inhibitor led to such an overexpression of Heat shock proteins, mainly Hsp70 and Hsp27, without any signs of chondrotoxicity *in vitro* and *in vivo*. Additionally, we have reported in this study that a preventive treatment of chondrocytes with several doses of proteasome inhibitor confers a relative cytoprotection versus the cytotoxic effect generated by MIA exposure. This resistance against cellular stress was mainly mediated *via* the induction of Hsp70, as demonstrated by oligonucleotide antisense strategy, despite the induction, in a lesser extent, of all the members of Heat Shock Proteins family and relative cofactors (personal data). Thus, the persistence of a weak protection, even when Hsp70 expression was knock-out, was probably related to others Hsps (e.g., Hsp27) and also to other factors, such as HSF1 [17].

To the best of our knowledge, this is the first report in which a transient inhibition of the proteasome pathway was able to protect chondrocytes cells from lethal stresses, even if the precise molecular cytoprotective mechanism was not clearly understood. This effect is believed to be similar to other heat shock proteins in behaving as a molecular chaperone and thus preventing denaturation and misolding of cytoplasmic proteins [21]. This protective effect might also be due to the ability of proteasome inhibition to assist to the transfer of newly synthesized proteins into the mitochondria, thus helping to maintain overall mitochondria integrity [26], or, in an other way, by preventing apoptosis by inhibiting caspases activation [23]. *In vitro*, the chondroprotective effect of MG132 against MIA toxicity seems to be essentially mediated by Hsp70 whereas Hsp27 does not take an important part in this effect. In contrast, the induction of Hsp27 *in vivo* should be of importance due to its ability to inhibit cell death process induced by other pathways [4,8,12].

We have previously reported (internal data) that a single intra-articular injection of MG132 in rat knee was able to induce Hsps overexpression (Hsp70 and Hsp27) in articular chondrocytes: this induction was detectable up to 48–72 hours after a single injection, and then decreasing to its basic level. Additionally, cartilage biosynthetic activity seems to be not affected by this single injection thus suggesting that an intra-articular (*ia*) injection of MG132 was not deleterious for chondrocytes viability and metabolism. With this in mind, *in vivo* experiments were then performed in order to determine whether the induction of Hsps was sufficient to protect chondrocytes from cellular stresses generated by MIA exposure which reproduces OA-like lesions in rat knee 10 to 15 days after *ia* injection [14]. The results presented here clearly indicate that a preventive injection of transient proteasome inhibitor is able to decrease the severity of macrosopical and histological lesions secondary to MIA exposure. This beneficial influence was also confirmed by measuring the biosynthetic activity of chondrocytes (proteoglycan synthesis), which is partially restored after MG132 exposure, when compared with controls only submitted to MIA injection.

The interest of a reversible proteasome inhibitor in altering the course of chondral lesions in experimental OA has not been previously assessed. This original concept appears as an interesting way to improve or reinforce chondrocytes to withstand against cellular stresses leading to cell death, and therefore, to cartilage degradation. This chondroprotective potency during MIA-induced OA needs therefore to be confirmed during in others OA models, e.g., anterior cruciate ligament transection which also closely mimics OA progression. Interestingly, in this model, apoptotic events also play a pivotal role in the establishment and progression of the disease. In this case, the ambivalent effect of proteasome inhibitor on proliferating and nondividing cells could be of a particular interest since a proliferation of synovial cells (synovitis) is associated to chondral erosion, thus mimicking OA flare.

With this in mind, *ia* injection of MG132 should both improve chondrocytes resistance to apoptotic events and simultaneously sensitize highly proliferating synovial cells to apoptotic process [16]. Moreover, an increase of Hsp70 in synoviocytes and chondrocytes could present another interest, due to its ability to inhibit the NFκB related transduction signal, pathway leading to the overexpression of proinflammatory cytokines (IL-1β, TNFα, ...) [9]. Finally, single or repeated *ia* injections of a transient proteasome inhibitor could increase chondrocytes viability in rendering articular cells less sensitive to inflammatory mediators and ancillary apoptotic signals. The combination of these effects could delay the degradation of the extracellular matrix occurring during osteoarthritis.

Acknowledgment

We gratefully thank Michel Thiery for his good care to the animals.

References

[1] T. Aigner and H.A. Kim, Apoptosis and cellular vitality: issues in osteoarthritic cartilage degeneration, *Arthritis Rheum.* **46** (2002), 1986–1996.

[2] F.J. Blanco, R. Guitian, E. Vazquez-Martul, F.J. de Toro and F. Galdo, Osteoarthritis chondrocytes die by apoptosis. A possible pathway for osteoarthritis pathology, *Arthritis Rheum.* **41** (1998), 284–289.

[3] M.M. Bradford, A rapid and sensitive method for the quantitation of microgram quantities of protein utilizing the principle of protein-dye binding, *Anal. Biochem.* **72** (1976), 248–254.

[4] J.M. Bruey, C. Ducasse, P. Bonniaud, L. Ravagnan, S.A. Susin, C. Diaz-Latoud, S. Gurbuxani, A.P. Arrigo, G. Kroemer, E. Solary and C. Garrido, Hsp27 negatively regulates cell death by interacting with cytochrome c, *Nat. Cell Biol.* **2** (2000), 645–652.

[5] J.A. Buckwalter and H.J. Mankin, Articular cartilage repair and transplantation, *Arthritis Rheum.* **41** (1998), 1331–1342.

[6] K.T. Bush, A.L. Goldberg and S.K. Nigam, Proteasome inhibition leads to a heat-shock response, induction of endoplasmic reticulum chaperones, and thermotolerance, *J. Biol. Chem.* **272** (1997), 9086–9092.

[7] P. Chomczynski and N. Sacchi, Single-step method of RNA isolation by acid guanidinium thiocyanate–phenol–chloroform extraction, *Anal. Biochem.* **162** (1987), 156–159.

[8] C.G. Concannon, A.M. Gorman and A. Samali, On the role of Hsp27 in regulating apoptosis, *Apoptosis* **8** (2003), 61–70.

[9] X.Z. Ding, C.M. Fernandez-Prada, A.K. Bhattacharjee and D.L. Hoover, Over-expression of hsp-70 inhibits bacterial lipopolysaccharide-induced production of cytokines in human monocyte-derived macrophages, *Cytokine* **16** (2001), 210–219.

[10] H.C. Drexler, Activation of the cell death program by inhibition of proteasome function, *Proc. Natl. Acad. Sci. USA* **94** (1997), 855–860.

[11] M.G. Ehrlich, H.J. Mankin, H. Jones, A. Grossman, C. Crispen and D. Ancona, Biochemical confirmation of an experimental osteoarthritis model, *J. Bone Joint Surg. Am.* **57** (1975), 392–396.

[12] C. Garrido, J.M. Bruey, A. Fromentin, A. Hammann, A.P. Arrigo and E. Solary, HSP27 inhibits cytochrome c-dependent activation of procaspase-9, *FASEB J.* **14** (1999), 2061–2070.

[13] C. Guingamp, P. Gegout-Pottie, L. Philippe, B. Terlain, P. Netter and P. Gillet, Mono-iodoacetate-induced experimental osteoarthritis: a dose-response study of loss of mobility, morphology, and biochemistry, *Arthritis Rheum.* **40** (1997), 1670–1679.

[14] R.E. Guzman, M.G. Evans, S. Bove, B. Morenko and K. Kilgore, Mono-iodoacetate-induced histologic changes in subchondral bone and articular cartilage of rat femorotibial joints: an animal model of osteoarthritis, *Toxicol. Pathol.* **31** (2003), 619–624.

[15] S. Hashimoto, L. Creighton-Achermann, K. Takahashi, D. Amiel, R.D. Coutts and M. Lotz, Development and regulation of osteophyte formation during experimental osteoarthritis. *Osteoarthritis Cartilage*, **10** (2002), 180–187.

[16] A. Kawakami, T. Nakashima, H. Sakai, A. Hida, S. Urayama, S. Yamasaki, H. Nakamura, H. Ida, Y. Ichinose, T. Aoyagi, I. Furuichi, M. Nakashima, K. Migita, Y. Kawabe and K. Eguchi, Regulation of synovial cell apoptosis by proteasome inhibitor, *Arthritis Rheum.* **42** (1999), 2440–2448.

[17] D. Kim, S.H. Kim and G.C. Li, Proteasome inhibitors MG132 and lactacystin hyperphosphorylate HSF1 and induce hsp70 and hsp27 expression, *Biochem. Biophys. Res. Commun.* **254** (1999), 264–268.

[18] T. Kubo, Y. Arai, K. Takahashi, T. Ikeda, S. Ohashi, I. Kitajima, O. Mazda, M. Takigawa, J. Imanishi and Y. Hirasawa, Expression of transduced HSP70 gene protects chondrocytes from stress, *J. Rheumatol.* **28** (2001), 330–335.

[19] U.K. Laemmli, F. Beguin and G. Gujer-Kellenberger, A factor preventing the major head protein of bacteriophage T4 from random aggregation, *J. Mol. Biol.* **47** (1970), 69–85.

[20] A.B. Meriin, V.L. Gabai, J. Yaglom, V.I. Shifrin and M.Y. Sherman, Proteasome inhibitors activate stress kinases and induce Hsp72. Diverse effects on apoptosis, *J. Biol. Chem.* **273** (1998), 6373–6379.

[21] D.D. Mosser, A.W. Caron, L. Bourget, A.B. Meriin, M.Y. Sherman, R.I. Morimoto and B. Massie, The chaperone function of hsp70 is required for protection against stress-induced apoptosis, *Mol. Cell. Biol.* **20** (2000), 7146–7159.

[22] B.S. Polla, S. Kantengwa, D. Francois, S. Salvioli, C. Franceschi, C. Marsac and A. Cossarizza, Mitochondria are selective targets for the protective effects of heat shock against oxidative injury, *Proc. Natl. Acad. Sci. USA* **93** (1996), 6458–6463.

[23] L. Ravagnan, S. Gurbuxani, S.A. Susin, C. Maisse, E. Daugas, N. Zamzami, T. Mak, M. Jaattela, J.M. Penninger, C. Garrido and G. Kroemer, Heat-shock protein 70 antagonizes apoptosis-inducing factor, *Nat. Cell Biol.* **3** (2001), 839–843.

[24] J.D. Robertson, K. Datta, S.S. Biswal and J.P. Kehrer, Heat-shock protein 70 antisense oligomers enhance proteasome inhibitor-induced apoptosis, *Biochem. J.* **344** (1999), 477–485.

[25] R. Sadoul, P.A. Fernandez, A.L. Quiquerez, I. Martinou, M. Maki, M. Schroter, J.D. Becherer, M. Irmler, J. Tschopp and J.C. Martinou, Involvement of the proteasome in the programmed cell death of NGF-deprived sympathetic neurons, *EMBO J.* **15** (1996), 3845–3852.

[26] A. Samali, J.D. Robertson, E. Peterson, F. Manero, L. van Zeijl, C. Paul, I.A. Cotgreave, A.P. Arrigo and S. Orrenius, Hsp27 protects mitochondria of thermotolerant cells against apoptotic stimuli, *Cell Stress Chaperones* **6** (2001), 49–58.

[27] C. Stefanelli, F. Bonavita, I. Stanic, C. Pignatti, G. Farruggia, L. Masotti, C. Guarnieri and C.M. Caldarera, Inhibition of etoposide-induced apoptosis with peptide aldehyde inhibitors of proteasome, *Biochem. J.* **332** (1998), 661–665.

[28] K. Takahashi, T. Kubo, Y. Arai, J. Imanishi, M. Kawata and Y. Hirasawa, Localization of heat shock protein in osteoarthritic cartilage, *Scand. J. Rheumatol.* **26** (1997), 368–375.

[29] K. Takahashi, T. Kubo, R.S. Goomer, D. Amiel, K. Kobayashi, J. Imanishi, R. Teshima and Y. Hirasawa, Analysis of heat shock proteins and cytokines expressed during early stages of osteoarthritis in a mouse model, *Osteoarthritis Cartilage* **5** (1997), 321–329.

[30] P.M. van der Kraan, E.L. Vitters, H.M. van Beuningen, W.B. van den Berg, Proteoglycan synthesis and osteophyte formation in 'metabolically' and 'mechanically' induced murine degenerative joint disease: an in-vivo autoradiographic study, *Int. J. Exp. Pathol.* **73** (1992), 335–350.

[31] M. Vayssier, N. Banzet, D. Francois, K. Bellmann and B.S. Polla, Tobacco smoke induces both apoptosis and necrosis in mammalian cells: differential effects of HSP70, *Am. J. Physiol.* **275** (1998), L771–L779.

Biorheology 41 (2004) 535–542
IOS Press

Modulation of collagen synthesis in normal and osteoarthritic cartilage

Jérôme Gouttenoire, Ulrich Valcourt, Marie-Claire Ronzière, Elisabeth Aubert-Foucher, Frédéric Mallein-Gerin and Daniel Herbage *

Institut de Biologie et Chimie des Protéines, UMR 5086 CNRS-UCB Lyon I, IFR128 Biosciences Lyon-Gerland, 7 passage du Vercors, 69367 Lyon Cedex 07, France

Abstract. In osteoarthritic cartilage, chondrocytes are able to present heterogeneous cellular reactions with expression and synthesis of the (pro)collagen types characteristic of prechondrocytes (type IIA), hypertrophic chondrocytes (type X), as well as differentiated (types IIB, IX, XI, VI) and dedifferentiated (types I, III) chondrocytes. The expression of type IIA procollagen in human osteoarthritic cartilage support the assumption that OA chondrocytes reverse their phenotype towards a chondroprogenitor phenotype. Recently, we have shown that dedifferentiation of mouse chondrocytes induced by subculture was associated with the alternative splicing of type II procollagen pre-mRNA with a switch from the IIB to the IIA form. In this context, we demonstrated that BMP-2 favours expression of type IIB whereas TGF-β1 potentiates expression of type IIA induced by subculture. These data reveal the specific capability of BMP-2 to reverse the program of chondrocyte dedifferentiation. This interesting feature needs to be tested with human chondrocytes since cell amplification is required for the currently used autologous chondrocyte transplantation.

Keywords: Chondrocyte phenotype, type IIA and IIB procollagen, BMP-2, TGF-β1, cartilage repair

1. Introduction

Articular cartilage is composed of a fibrous collagen network embedded in a proteoglycan rich ground substance. Collagen fibers contain at least three different types of collagen (II, IX, and XI) interacting with small proteoglycans (decorin, biglycan) and matricellular proteins [9,11,13]. In osteoarthritis, mechanical insult, joint instability and presence of catabolic or anabolic cytokines trigger different cellular responses [15,16,33]: chondrocytes of articular cartilage proliferate in clusters and increase their synthesis of matrix molecules and degradative enzymes such as matrix metalloproteinases (MMP-2, -3, -8, -13), aggrecanases 1 and 2 (ADAMTS 4 and 5) and cathepsins, and decrease their synthesis of inhibitors such as the tissue inhibitors of metalloproteinases (TIMPs). As the balance between synthesis and degradation is disrupted in favour of catabolism, cartilage degeneration is accelerated. Furthermore, several authors have described increased apoptosis in osteoarthritic cartilage but the extent of cell death and its impact on the pathology is still controversial [4]. Another characteristic result of chondrocyte activation in osteoarthritis is the expression and synthesis of diverse molecular markers classically described in dedifferentiated chondrocytes or in chondrocytes at different stages of differentiation (Fig. 1), but not in normal articular chondrocytes [2,3,15,33]. In this paper, we limit our interest to the modulation

*Address for correspondence: Dr. Daniel Herbage, Laboratory of Cartilage Biology and Engineering, IBCP, 7 Passage du Vercors, 69367 Lyon Cedex 07, France. Tel.: +33 (0)4 72 72 26 15; Fax: +33 (0)4 72 72 26 02; E-mail: d.herbage@ibcp.fr.

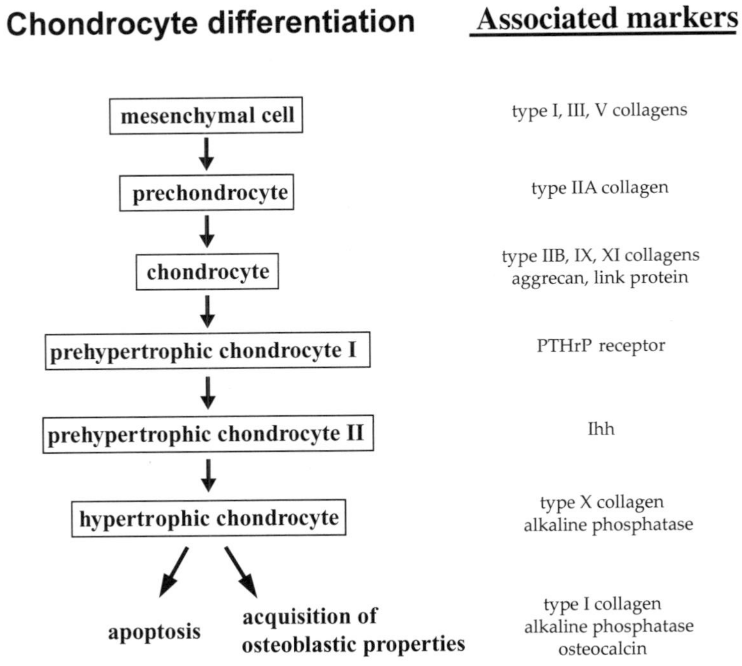

Fig. 1. Schematic representation of the chondrocyte differentiation pathway with associated markers.

of synthesis of the different types of collagen with specific attention given to the alternative splicing of type II procollagen pre-mRNA.

2. Cartilage synthesis and organization in the extracellular matrix of articular cartilage

Type II collagen, a homotrimer composed of three $\alpha 1$(II) chains and the most abundant collagen in cartilage, forms the backbone of heteropolymeric fibrils with types IX and XI collagens. Whereas type XI collagen is located within the fibrils, type IX collagen is located at the surface of the fibrils in an antiparallel orientation. Type IX collagen contains a large non-collagenous (NC) globular domain with a glycosaminoglycan chain (chondroitin sulfate), likely playing a role in inter-fibril and fibril-non-collagenous components interactions [9,11,13]. The presence of type III collagen was detected in normal articular cartilage [43], probably associated with type II collagen in single fibrils [45]. Very recently, expression analysis of the last fibrillar collagen gene described (COL27A1) indicates that it is primarily expressed in cartilage, suggesting that the corresponding protein is a new component of the extracellular matrix of cartilage [28].

Several other non-fibrillar collagens have been found in cartilage. Type VI collagen, an ubiquitous matrix component of most tissues, forms a filamentous network with a typical 100 nm cross-striation [17,29]. Type XII, XIV and XVI collagens, members of the FACIT collagen subfamilly are also present in cartilage [20,42]. In territorial cartilage, type XVI collagen was recently shown [20] to reside in a discrete population of thin, weakly banded collagen fibrils containing types II and XI collagens. Type IX collagen is also present in this population of fibrils, but at different locations from those of collagen XVI.

Types I and X collagens are only detected in young articular cartilage in the superficial and deep zones, respectively. The detection of type X collagen at the surface of normal adult human pig and rat articular cartilage by immunohistochemistry and biochemistry [30] was not confirmed [8,40,41]. The distribution and organization of these collagens differ between the different zones of the articular cartilage (superficial, transitional, upper and lower radial and calcified) and also within the three matrix compartments (pericellular, territorial and interterritorial) [18,19].

3. Modification of the collagen network composition and properties in osteoarthritic cartilage

An interesting hypothesis to explain the pathogenesis of osteoarthritis is a fatigue failure in the collagen network [14]. The concept that a collagen-network failure precedes the development of fibrillation in osteoarthritis derives from studies of hydration and mechanical properties of articular cartilage. Indeed, in the early stages of the disease, an increased hydration and swelling of the cartilage and a reduction of its tensile strength precede the biochemical changes (decrease in glycosaminoglycan content). Recently, several interesting data have been obtained by the TeKoppele's group [7,10,39] concerning the relationship between the age-related changes in collagen network, particularly the accumulation of advanced glycation end products (AGEs), and the alterations of the tensile properties of articular cartilage. They suggest that increased stiffness by AGE crosslinking may contribute to the age-related failure of the collagen network [39] and predispose human articular cartilage to increased stress concentration, fracture and aging-associated biomechanical dysfunction [10]. As expected, several studies have shown that the increased swelling and instantaneous deformation of osteoarthritic cartilage is correlated with collagen degradation [7] and there is a close correspondence between the cleavage of type II collagen and the sites where MMP-1 and -13 are detected [44]. Telopeptide cleavages of collagen II by stromelysin-1 (MMP-3) [44] could explain fibril depolymerization and the increased collagen solubilization in human osteoarthritic cartilage that we observed in a previous study [29]. Indeed, if only 22% of total collagen was soluble in normal cartilage after extraction with 4 M guanidinium chloride followed by two pepsin treatments, more than 38% was extracted in osteoarthritic samples. The different types of collagen present in the extracts were determined by electrophoresis and immunobloting. Type II, IX, XI and VI collagens were detected in both tissues whereas type I collagen was present in osteoarthritic samples, demonstrating the phenotypic modulation of chondrocytes. We confirmed these results by immunohistochemistry (Fig. 2). Whereas type II collagen remains the main collagen present throughout the cartilage, type I collagen was present in the clusters of the fibrillated areas in late stage of osteoarthritis. It is worth noting that a strong labeling with antibodies against type IX collagen was localized in the same area. Fragments of type IX collagen are probably released in this region where high proteolytic activities have been detected. As we observed previously *in vitro* that these chains were potent activators of monocytes/macrophages and fibroblasts, with increased synthesis of collagenase and interleukin-1 [12], a possible activation of the chondrocytes present in the clusters could amplify the catabolic pathway. In the same study, we observed by electron microscopy that type VI collagen was deposited in the pericellular region in osteoarthritic tissue, as abundant broad-banded aggregates (with the typical 100 nm periodicity) rarely observed in normal cartilage. Similar severe disturbance of the distribution and expression of this collagen (increase in the lower middle and upper deep zones and significant loss in the upper zones) was described by Hambach et al. [17]. Several authors have shown the presence of "abnormal" collagens in osteoarthritic cartilage. Immunohistochemical [26] and *in situ* hybridization [1] analyses indicate low and independent expression and synthesis of type I (low level and absence in early phase of the disease) and III (high level) collagens with focal pericellular distribution in osteoarthritic lesions. More consistently, type X collagen usually present in hypertrophic cartilage of

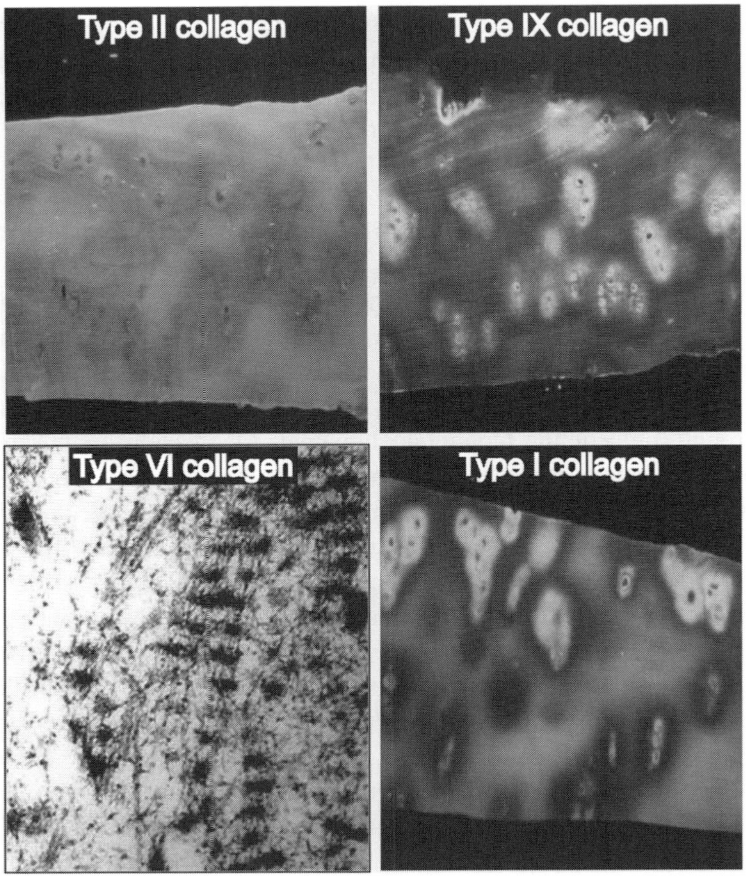

Fig. 2. Immunohistochemical localization of types II, IX and I collagens and electron microscopic visualization of the typical fibrous aggregates of type VI collagen in the upper layer of osteoarthritic cartilage (for more details see [29]).

the growth plate, was expressed by chondrocytes in the upper zone of moderately and severely affected human osteoarthritic cartilage as shown by immunolocalization [8,24,40,41]. Of possible importance in the pathogenesis of osteoarthritis is the report by Bailey et al. [6] that the quality of the collagen of subchondral bone in osteoarthritic femurs is modified by the formation of type I homotrimer. The presence of the latter could account for narrower disorganised collagen fibres, decreased mineralization and a reduction in mechanical stability of the osteoarthritic femoral heads.

4. Biological significance/consequence of the reexpression of type IIA procollagen in osteoarthritic cartilage

The description by Ryan and Sandell [31] of the differential expression of a cystein-rich domain in the amino-terminal propeptide of type II procollagen by alternative splicing of pre-mRNA, was the first example of alternative splicing in a fibrillar collagen gene, thus providing a second mechanism for generating diversity in these collagen genes. Type II procollagen is expressed as two spliced forms, type IIA and IIB (Fig. 3a). Type IIA procollagen is a mRNA spliced form that contains an additional 207 bases exon (exon 2) encoding the 69 amino acid cysteine-rich domain of the NH_2-propeptide. Type IIA procollagen

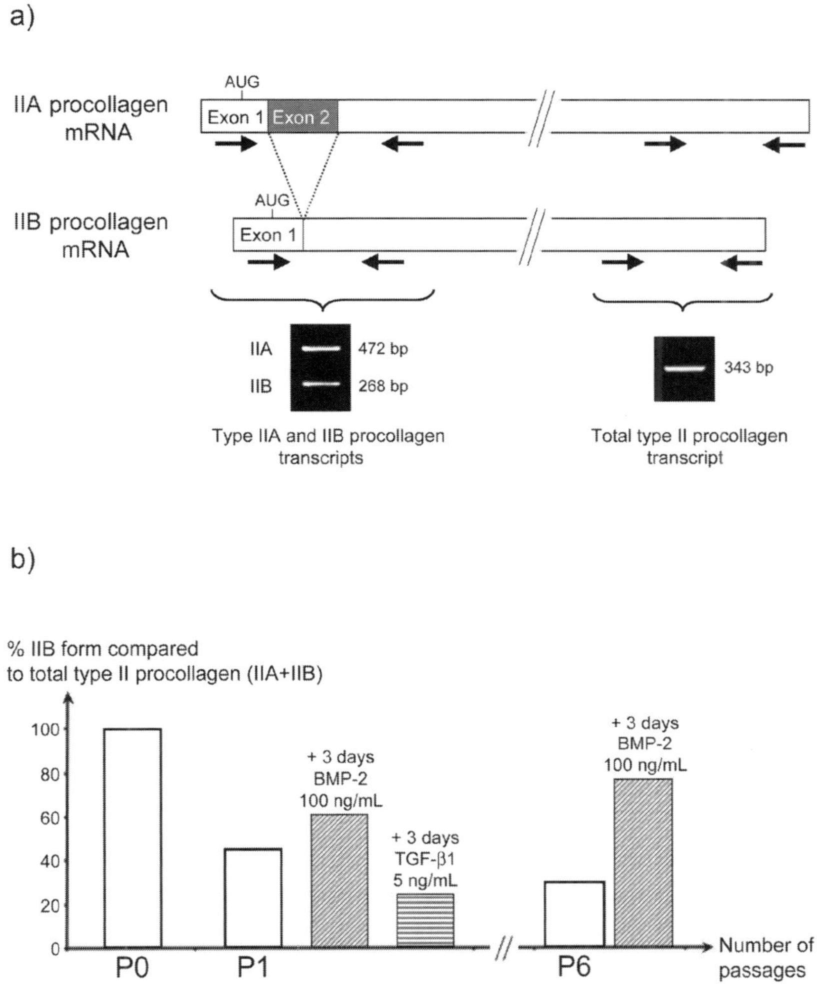

Fig. 3. Alternative splicing of type II procollagen pre-mRNA. (a) To examine the expression of spliced forms of type II procollagen pre-mRNA by RT-PCR, we used a sense oligonucleotide located in exon 1 and an antisense oligonucleotide located in exon 5 of the mouse *Col2a1* gene, to amplify 2 PCR products: a long form corresponding to type IIA mRNA (containing exon 2) and a shorter form corresponding to type IIB mRNA (lacking exon 2). The total amount of type II procollagen (IIA + IIB) transcripts was determined by designing a sense oligonucleotide in exon 50 and an antisense primer in exon 52 of the mouse *Col2a1* gene. (b) Relative amounts of type IIA and IIB procollagen mRNAs expressed in mouse chondrocytes after several passages in monolayer cultures (P0 to P6), and after treatment with BMP-2 or TGF-β1 at P1. These experiments show a decrease in the percentage of type IIB form when the number of passages in culture increases. Treatment of P1 chondrocytes with 100 ng/ml BMP-2 stimulates type IIB form reexpression. However, a 5 ng/ml TGF-β1 treatment favours the decrease of the type IIB form expression. After 6 passages, the redifferentiation effect of BMP-2 is still clearly observed (for more details see [38]).

mRNA precedes type IIB mRNA expression during formation of the endochondral skeleton [23,27,33]. This form is present in the somites, neuroepithelia and prechondrogenic mesenchyme of mouse and human embryos [23]. In tissues that undergo chondrogenesis, the mRNA spliced form switches from type IIA to IIB upon differentiation into chondroblasts. An interesting finding by Zhu et al. [46] is that type IIA procollagen could potentially play a role in induction and differentiation of the skeleton. Indeed, immunoelectron microscopy revealed type IIA procollagen fibrils labeled with antibodies against the NH_2-propeptide, suggesting that the NH_2-propeptide remains attached to the collagen molecule in the extracellular matrix whereas the COOH-propeptide is removed. The cleaved form of the type IIA NH_2-propeptide was also detected in extracts of 7 days chick embryonic limb [24]. Furthermore, using recombinant NH_2-propeptide and a solid phase binding assay, Zhu et al. [46] showed no reaction with bFGF and IGF-1 but binding with TGF-β1 and BMP-2. These results suggest that type IIA-propeptide could regulate the concentration and biodisponibility of these chondrogenic/osteogenic factors at different stages of cartilage development or repair under normal and pathological conditions.

Absent in healthy human adult articular cartilage, several authors have described the expression of type IIA procollagen in human osteoarthritic cartilage [5,24] and in rabbit [22] and mouse [32] models of osteoarthritis. It was thus postulated that a significant proportion of adult articular chondrocytes starts to re-express a chondroprogenitor phenotype during osteoarthritic cartilage degeneration.

A different interpretation is possible considering the recent results obtained in our laboratory concerning the modulation of synthesis of type IIA and IIB procollagen [38]. Recently, using rib cage embryonic mouse chondrocytes [36,37], we have shown (Fig. 3b) that dedifferentiation of these cells by serial subcultures (up to passage 6) was associated with a decrease in the gene expression of total type II procollagen. More precisely, expression of the type IIB form, characteristic of P0 chondrocytes, was decreased in P1 chondrocytes whereas the type IIA form was expressed only after subculture. Thus, alternative splicing of type II procollagen pre-mRNA occurs during dedifferentiation of mouse and rat [21] chondrocytes in monolayer. We further investigated the responsiveness of dedifferentiated P1 chondrocytes to BMP-2 and TGF-β1, during a 3 and 6 day-culture period. BMP-2 favours expression of type IIB form whereas TGF-β1 potentiates expression of type IIA form induced by subculture. These observations reveal the specific capability of BMP-2 to reverse the dedifferentiation state of chondrocytes even after 6 passages. This interesting feature could explain the positive effect of BMP-2 on the healing of full-thickness defects of articular cartilage [34]. According to these results, the presence of type IIA form of procollagen in osteoarthritic cartilage could represent a sign of chondrocyte dedifferentiation. Furthermore, the recent localization of BMP-2 [25] in human osteoarthritic cartilage and osteophyte (BMP-2 bound to the NH_2-propeptide of type IIA procollagen ?) suggests its possible involvement in the attempt of repair that occurs in the early stages of osteoarthritis. In contrast, TGF-β1, also present in osteoarthritic cartilage, may rather function as an inducer of osteophyte formation [35] than a chondro-inducer (Fig. 4).

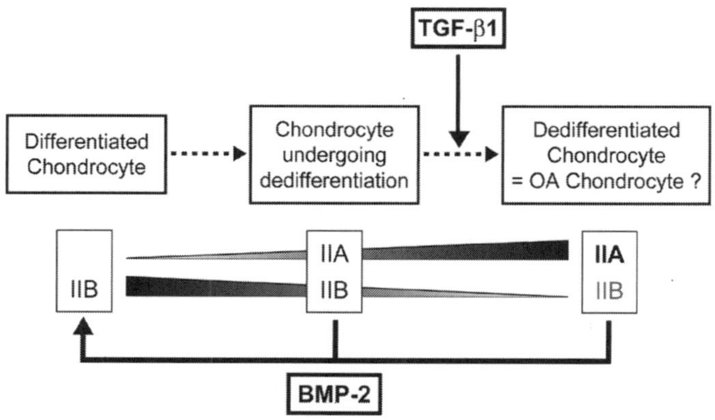

Fig. 4. Schematic representation of the effects of BMP-2 or TGF-β1 observed on alternative splicing of type II procollagen pre-mRNA in mouse chondrocytes grown on plastic. In regard of the dedifferentiation induced by subculture, BMP-2 acts as a redifferentiation factor whereas TGF-β1 acts as a stimulator of dedifferentiation. Interestingly, the reexpression of type IIA procollagen by dedifferentiated chondrocytes mimics the phenotype expressed by human osteoarthritic chondrocytes.

Acknowledgements

J.G. was supported by a doctoral fellowship from the "Association pour la Recherche sur le Cancer". This work was funded by grants from the "Rhône-Alpes Région (085113) Programme Thématique Prioritaire", from "CNRS-INSERM, Ingénierie Tissulaire, 2000–2003" and from the "Fondation de l'Avenir pour la Recherche Médicale Appliquée, Paris (ET2-320)".

References

[1] T. Aigner, W. Bertlingt, H. Stöss, G. Weseloh and K. von der Mark, Independent expression of fibril-forming collagens I, II and III in chondrocytes of human osteoarthritic cartilage, *J. Clin. Invest.* **91** (1993), 829–837.

[2] T. Aigner and J. Dudhia, Phenotypic modulation of chondrocytes as a potential therapeutic target in osteoarthritis: a hypothesis, *Ann. Rheum. Dis.* **56** (1997), 287–291.

[3] T. Aigner, K. Glückert and K. von der Mark, Activation of fibrillar collagen synthesis and phenotypic modulation of chondrocytes in early human osteoarthritic cartilage lesions, *Osteoarthritis Cart.* **5** (1997), 183–189.

[4] T. Aigner, M. Hemmel, D. Neureiter, P.M. Gebhard, G. Zeiler, T. Kirchner and L. McKenna, Apoptotic cell death is not a widespread phenomenon in normal aging and osteoarthritic human articular cartilage, *Arthritis Rheum.* **44** (2001), 1304–1312.

[5] T. Aigner, Y. Zhu, H.H. Chansky, F.A. Matsen III, W.J. Maloney and L.J. Sandell, Reexpression ot type IIA procollagen by adult articular chondrocytes in osteoarthritic cartilage, *Arthritis Rheum.* **42** (1999), 1443–1450.

[6] A.J. Bailey, T.J. Sims and L Knott, Phenotypic expression of osteoblast collagen in osteoarthritic bone: production ot type I homotrimer, *Int. J. Biochem. Cell Biol.* **343** (2002), 176–182.

[7] R.A. Bank, M. Soudry, A. Maroudas, J. Mizrahi and J.M. TeKoppele, The increased swelling and instantaneous deformation of osteoarthritic cartilage is highly correlated with collagen degradation, *Arthritis Rheum.* **43** (2000), 2202–2210.

[8] N. Boos, A.G. Nerlich, I. Wiest, K. von der Mark, R. Ganz and M. Aebi, Immuno-histochemical analysis of type X collagen expression in osteoarthritis of the hip joint, *J. Orthop. Res.* **17** (1999), 495–502.

[9] J.A. Buckwalter and H.J. Mankin, Articular Cartilage. Part I: tissue design and chondrocyte-matrix interactions, *J. Bone Joint Surgery* **79-A** (1997), 600–612.

[10] A.C. Chen, M.M. Temple, D.M. Ng, N. Verzijl, J. DeGroot, J.M. TeKoppele and R. L. Sah, Induction of advanced glycation end products and alterations of the tensile properties of articular cartilage, *Arthritis Rheum.* **46** (2002), 3212–3217.

[11] M.A. Cremer, E.F. Rosloniec and A.H. Kang, The cartilage collagens: a review of their structure, organization and role in the pathogenesis of experimental arthritis in animals and in human rheumatic disease, *J. Mol. Med.* **76** (1998), 275–288.

[12] J.M. Dayer, S. Ricard-Blum, M.T. Kaufmann and D. Herbage, Type IX collagen a potent inducer of PGE2 and IL1 production by human monocyte macrophage, *FEBS Lett.* **198** (1986), 208–212.

[13] D. Eyre, Collagen of articular cartilage, *Arthritis Res.* **4** (2002), 30–35.

[14] M.A.R. Freeman, Is collagen fatigue failure a cause of osteoarthrosis and prosthetic component migration? A hypothesis. *J. Orthop. Res.* **17** (1999), 3–8.

[15] N. Fukui, C.R. Purple and L.J. Sandell, Cell biology of osteoarthritis: the chondrocyte's response to injury, *Curr. Rheum. Rep.* **3** (2001), 496–505.

[16] M.B. Goldring, Osteoarthritis and cartilage: the role of cytokines, *Curr. Rheum. Rep.* **2** (2000), 459–465.

[17] L. Hambach, D. Neureiter, G. Zeiler, T. Kirchner and T. Aigner, Severe disturbance of the distribution and expression of type VI collagen chains in osteoarthritic articular cartilage, *Arthritis Rheum.* **41** (1998), 986–996.

[18] E.B. Hunziker, M. Michel and D. Studer, Ultrastructure of adult human articular cartilage matrix after cryotechnical processing, *Microscop. Res. Tech.* **37** (1997), 271–284.

[19] E.B. Hunziker, T.M. Quinn and H.-J. Häuselmann, Quantitative structural organization of normal adult human articular cartilage, *Osteoarthritis Cart.* **10** (2002), 564–572.

[20] A. Kassner, U. Hansen, N. Miosge, D.P. Reinhardt, T. Aigner, L. Bruckner-Tuderman, P. Bruckner and S. Grässel, Discrete integration of collagen XVI into tissue-specific collagen fibrils or beaded microfibrils, *Matrix Biol.* **22** (2003), 131–143.

[21] T.K. Kim, J.S. Park, M.C. Lee, S.C. Seong and H. J. Kim, Alternative splicing of type II procollagen gene in the dedifferentiation of rat epiphyseal chondrocytes serially cultured in monolayer, *Connect. Tiss. Res.* **43** (2002), 56–62.

[22] T.P. Lefkoe, A.M. Nalin, J.M. Clark, R.A. Reife, J. Sugai and L.J. Sandell, Gene expression of types IIA and IX correlates with ultrastructural events in early osteoarthrosis: new application of the rabbit meniscectomy model, *J. Rheum.* **24** (1997), 1155–1163.

[23] V.C.H. Lui, L.J. Ng, J. Nicholls, P.P.L. Tam and K.S.E. Cheah, Tissue-specific and differential expression of alternatively spliced a1(II) collagen mRNAs in early human embryos, *Dev. Dyn.* **203** (1995), 198–211.

[24] H.-D. Nah, B. Swoboda, D.E. Birk and T. Kirsch, Type IIA procollagen: expression in developing chicken limb cartilage and human osteoarthritic articular cartilage, *Dev. Dyn.* **220** (2001), 307–322.
[25] T. Nakase, T. Miyaji, T. Tomita, M. Kaéneko, K. Kuriyama, A. Myoui, K. Sugamoto, T. Ochi and H. Yoshikawa, Localization of bone morphogenetic protein-2 in human osteoarthritic cartilage and osteophyte, *Osteoarthritis and Cart.* **11** (2003), 278–284.
[26] A.G. Nerlich, I. Wiest and K. von der Mark, Immunohistochemical analysis of interstitial collagens in cartiloage of different stages of osteoarthrosis, *Virchows Archiv. B Cell Pathol.* **63** (1993), 249–255.
[27] L.J. Ng, P.P.L. Tam and K.S.E. Cheah, Preferential expression of alternatively spliced mRNAs encoding type II procollagen with a cysteine-rich amino-propeptide in differentiating cartilage and nonchondrogenic tissues during early mouse development, *Dev. Biol.* **159** (1993), 403–417.
[28] J.M. Pace, M. Corrado, C. Missero and P.H. Byers, Identification, characterization and expression analysis of a new fibrillar collagen gene, COL27A1, *Matrix Biol.* **22** (2003), 3–14.
[29] M.-C. Ronzière, S. Ricard-Blum, J. Tiollier, D.J. Hartmann, R. Garrone and D. Herbage, Comparative analysis of collagens solubilized from human foetal, and normal and osteoarthritic adult articular cartilage, with emphasis on type VI collagen, *Biochim. Biophys. Acta* **1038** (1990), 222–230.
[30] G.J. Rucklidge, G. Milne and S.P. Robins, Collagen type X: a component of the surface of normal human, pig and rat articular cartilage, *Biochem. Biophys. Res. Commun.* **224** (1996), 297–302.
[31] M.C. Ryan and L.J. Sandell, Differential expression of a cysteine-rich domain in the amino-terminal propeptide of type II (cartilage) procollagen by alternative splicing of mRNA, *J. Biol. Chem.* **265** (1990), 10334–10339.
[32] H. Salminen, E. Vuorio and A.-M. Säämänen, Expression of Sox9 and type IIA procollagen during attempted repair of articular cartilage damage in a transgenic mouse model of osteoarthritis, *Arthritis and Rheum.* **44** (2001), 947–955.
[33] L.J. Sandell and T. Aigner, Articular cartilage and changes in arthritis. An introduction: cell biology of osteoarthritis, *Arthritis Res.* **3** (2001), 107–113.
[34] A. Scharstuhl, H.L. Glansbeek, H.M. van Beuningen, E.L. Vitters, P.M. van der Kraan and W.B. van den Berg, Inhibition of endogenous TGF-beta during experimental osteoarthritis prevents osteophyte formation and impairs cartilage repair, *J. Immunol.* **169** (2002), 507–514.
[35] R.S. Sellers, D. Peluso and E.A. Morris, The effect of recombinant human BMP-2 on the healing of full-thickness defects of articular cartilage, *J. Bone Joint Surg.* **79-A** (1997), 1452–1463.
[36] U. Valcourt, M.C. Ronzière, P. Winkler, V. Rosen, D. Herbage and F. Mallein-Gerin, Different effects of bone morphogenetic proteins 2, 4, 12, 13 on the expression of cartilage and bone markers in the MC615 chondrocyte cell line, *Exp. Cell Res.* **251** (1999), 264–274.
[37] U. Valcourt, J. Gouttenoire, A. Moustakas, D. Herbage and F. Mallein-Gerin, Functions of TGF-b family type I receptors and smad proteins in the hypertrophic maturation and osteoblastic differentiation of chondrocytes, *J. Biol. Chem.* **277** (2002), 33545–33558.
[38] U. Valcourt, J. Gouttenoire, E. Aubert-Foucher, D. Herbage and F. Mallein-Gerin, Alternative splicing of type II procollagen pre-mRNA in chondrocytes is oppositely regulated by BMP-2 and TGF-β1, *FEBS Lett.* **545** (2003), 115–119.
[39] N. Verzijl, J.DeGroot, C. Ben Zaken, O. Braun-Benjamin, A. Maroudas, R.A. Bank, J. Mizrahi, C.G. Schalkwijk, S.R. Thorpe, J.W. Baynes, J.W.J. Bijlsma, F.P.J.G. Lafeber and J.M.TeKoppele, Crosslinking by advanced glycation end products increases the stiffness of the collagen network in human articular cartilage, *Arthritis Rheum.* **46** (2002), 114–123.
[40] K. von der Mark, T. Kirsch, A. Nehrlich, A. Kuss, G. Weseloh, K. Glückert and H. Stöss, Type X collagen synthesis in human osteoarthritic cartilage. Indication of chondrocyte hypertrophy, *Arthritis Rheum.* **35** (1992), 806–811.
[41] G.D. Walker, M. Fisher, J. Gannon, R.C. Thompson and T.R. Oegema, Expression of type X collagen in osteoarthritis, *J. Orthop. Res.* **13** (1995), 4–12.
[42] S.L. Watt, G.P. Lundstrum, A.M. McDonough, D.R. Keene, R.E. Burgeson and N.P. Morris, Characterization of collagen types XII and XIV from fetal bovine cartilage, *J. Biol. Chem.* **267** (1992), 20093–20099.
[43] S.F. Wotton and V.C. Duance, Type III collagen in normal articular cartilage, *Histochem. J.* **26** (1994), 412–416.
[44] W. Wu, R.C. Billinghurst, I. Pidoux, J. Antoniou, D. Zukor, M.Tanzer and A.R. Poole, Sites of collagenase cleavage and denaturation of type II collagen in aging and osteoarthritic articular cartilage and their relationship to the distribution of matrix metalloproteinase 1 and matrix metalloproteinase 13, *Arthritis Rheum.* **46** (2002), 2087–2094.
[45] R.D. Young, P.A. Lawrence, V.C. Duance, T. Aigner and P. Monaghan, Immunolocalization of collagen types II and III in single fibrils of human articular cartilage, *J. Histochem. Cytochem.* **48** (2002), 423–432.
[46] Y. Zhu, A. Oganesian, D.R. Keene and L.J. Sandell, Type IIA procollagen containing the cystein-rich amino propeptide is deposited in the extracellular matrix of prechondrogenic tissue and binds to TGF-β1 and BMP-2, *J. Cell Biol.* **144** (1999), 1069–1080.

Biorheology 41 (2004) 543–547
IOS Press

Type II collagen peptides for measuring cartilage degradation

Y. Henrotin [a,b,*], M. Deberg [a], J.-E. Dubuc [b], E. Quettier [b], S. Christgau [c] and J.-Y. Reginster [a]
[a] *Bone and Cartilage Research Unit, Institute of Pathology, CHU Sart-Tilman, 4000 Liège, Belgium*
[b] *Princess Paola Hospital, 6900 Marche-en-Famenne, Belgium*
[c] *Nordic Bioscience, Herlev Hovedgade 207, Herlev, Denmark*

Abstract. This paper describes two new immunoassays for a peptide of the triple helix of type II collagen (Coll 2-1) and its nitrated form (Coll 2-1 NO_2). In healthy subjects aged between 20 and 65 years old, Coll 2-1 and Coll 2-1 NO_2 levels in serum were in means 125.13 ± 3.71 and 0.16 ± 0.08 nmol/l, respectively. These levels did not significantly vary with age. However, up to 45 years of age, Coll 2-1 NO_2 levels in women were significantly higher than in men. In patients with knee osteoarthritis (OA), Coll 2-1 in serum was found to be elevated compared to healthy controls (267.45 ± 26.42 nmol/l vs 126.78 ± 6.61 nmol/l). Further, we have demonstrated that an increase of the urinary levels of Coll 2-1 or Coll 2-1 NO_2 over 1 year was predictive of joint space narrowing progression in OA patients. In conclusion, these preliminary results indicate that Coll 2-1 could be a predictive marker of knee OA progression.

Keywords: Ageing, osteoarthritis, biological markers, nitric oxide, oxygen

1. Introduction

Type II collagen is one of the most important component in cartilage, representing 15–25% of the wet weight, 50% of the dry weight and 90–95% of the total collagen. Type II collagen is a fibrillar collagen formed by the association of three α chains in triple helix with short linear sequences at the N- and C-terminal ends, termed telopeptides [13]. It is synthesized as procollagen α chains which are secreted from the cell into the extracellular matrix where it forms fibrils. This is accompanied by the removal of C- and N-propeptides (CPII and NPII) by specific C- and N-proteases, as collagen fibrils are formed. CPII and NPII are considered as markers of type II collagen synthesis. These are maximal in foetal cartilage, being reduced almost threefold at birth and reaching very low levels (mean 3.9% of newborn) in normal adult cartilage [10]. In osteoarthritis (OA), CPII is significantly elevated in cartilage when compared to normal adult tissue. Nevertheless, the increase in CPII in OA cartilage is not reflected in serum. In rheumatoid arthritis (RA), CPII concentration in serum exhibits a marked increase, whereas it has been reported to be reduced in OA [10].

Type II collagen degradation can be appreciated by measuring a neoepitope generated by collagenases in the triple helix domain [1,9,11] or fragments of the telopeptides [3–5]. C-telopeptide fragment (CTX-II) is found to be elevated in urine of RA and OA patients compared to healthy controls [3,5]. Furthermore, high baseline level of CTX-II in urine is associated with increased risk of progression of joint

*Address for correspondence: Prof. Y. Henrotin, Bone and Cartilage Research Unit, Institute of pathology, CHU Sart-Tilman, 4000 Liège, Belgium. Tel.: +32 4 366 24 67; Fax: +32 4 366 47 34; E-mail: yhenrotin@ulg.ac.be.

destruction over 1 year in early RA [5]. In OA, urinary CTX-II concentration is positively correlated with the radiographic parameters of joint destruction (i.e. joint space narrowing) [3]. Recently, Garnero et al. [4] reported that a decrease of serum N-propeptide (PIIANP) coupled to an increase of CTX-II, was highly predictive of progression of joint damage in knee OA.

Herein, we describe a new strategy for quantifying type II collagen network degradation. This approach is based on the detection in biological fluids (serum and urine) of a nitrated peptide released from the triple helix of type II collagen during its processing. Peptide nitration results from the reaction of aromatic amino acids with peroxynitrite ($ONOO^-$), a strong oxidant formed by the reaction of nitric oxide ($^{\bullet}NO$) and superoxide anion ($O_2^{\bullet -}$). In OA joint, $^{\bullet}NO$ and $O_2^{\bullet -}$ are produced by a large number of cells including macrophages and chondrocytes [7,8]. Tyrosine, tryptophan and phenylalanine are particularly sensitive to nitration [14]. Type II collagen contains two tyrosines, one located in the triple helix and the other in the telopeptide at the C-terminal end. We have developed two new specific immunoassays, one for the peptide ^{108}HRGYPGLDG^{116} (Coll 2-1) and the other for the nitrated form of this peptide ^{108}HRGY(NO_2)PGLDG^{116} (Coll 2-1 NO_2) [2]. These immunoassays have been validated in sera or urines of healthy subjects and osteoarthritic patients [6].

2. Materials and methods

Coll 2-1 and Coll 2-1 NO_2 were injected into rabbits. By this way, we have got two antisera, named D3 and D37, with a high specificity and affinity for Coll 2-1 and Coll 2-1 NO_2, respectively. These antisera were used to design two immunoassays, one specific for Coll 2-1 and the other for Coll 2-1 NO_2. The limit of detection of Coll 2-1 and Coll 2-1 NO_2 was 17 nmol/l and 25 pmol/l, respectively. For both immunoassays, the inter- and intra-assay CV were below 10%. For validation, we used 242 sera from healthy subjects, 13 sera from patients with knee OA and urines from 75 patients with primary knee OA corresponding to the placebo arm of a study evaluating the effects of glucosamine sulphate [12]. All OA subjects had a low level of C-reactive protein (<2 mg/l) indicating that they were not affected by systemic inflammation.

3. Results and discussion

In serum from healthy subjects ($n = 242$), the mean concentrations of Coll 2-1 and Coll 2-1 NO_2 were 125.13 ± 3.71 nmol/l and 0.16 ± 0.08 nmol/l, respectively. Further, these values were not significantly affected by ageing (Table 1). Nevertheless, in donors younger than 45 years old, Coll 2-1 NO_2 concentration in serum was significantly higher in women than in men (0.22 ± 0.02 nmol/l vs 0.16 ± 0.01 nmol/l, $p < 0.01$), suggesting that oestrogen could influence the level of Coll 2-1 nitration. Furthermore, Coll 2-1 NO_2 levels were higher in pre-menopausal (0.22 ± 0.02 nmol/l) than in post-menopausal women (0.12 ± 0.01 nmol/l). The Coll 2-1 level (Fig. 1A) was elevated in serum of knee OA patients ($n = 13$) compared to normal age-matched subjects ($n = 80$) (267.45 ± 26.42 nmol/l vs 126.78 ± 6.61 nmol/l), confirming that type II collagen processing is accelerated in OA. In contrast, Coll 2-1 NO_2 level (Fig. 1B) was not elevated in our population compared to normal age-matched subjects (0.12 ± 0.01 nmol/l vs 0.13 ± 0.01 nmol/l), probably because joint inflammation was not flared up at the sample time, as suggested by the very low level of C-reactive protein.

These biochemical markers were also measured in urines of a cohort of 75 patients with knee OA corresponding to the placebo arm of a large 3-year, randomized, double-blind study comparing the efficacy

Table 1

Coll 2-1 and Coll 2-1 NO_2 concentrations in serum of healthy donors

Age (years)	Coll 2-1 (nmol/l)		Coll 2-1 NO_2 (nmol/l)	
	Men ($n = 170$)	Women ($n = 72$)	Men ($n = 170$)	Women ($n = 72$)
20–25	142.15 ± 26.14	148.23 ± 18.88	0.13 ± 0.01	0.19 ± 0.02*
26–30	93.35 ± 9.10	94.67 ± 10.48	0.21 ± 0.02	0.27 ± 0.02*
31–35	123.87 ± 17.66	103.91 ± 13.20	0.16 ± 0.01	0.16 ± 0.01
36–40	114.57 ± 10.59	125.31 ± 16.59	0.11 ± 0.01	0.14 ± 0.02
41–45	140.46 ± 9.96	127.69 ± 15.41	0.17 ± 0.02	0.18 ± 0.04
46–50	122.01 ± 10.40	148.18 ± 24.33	0.10 ± 0.01	0.11 ± 0.02
51–55	128.76 ± 10.96	152.74 ± 17.66	0.14 ± 0.04	0.13 ± 0.02
56–60	122.26 ± 15.97	188.99 ± 37.33	0.11 ± 0.01	0.12 ± 0.01
61–65	117.00 ± 8.65	143.32 ± 9.33	0.13 ± 0.01	0.12 ± 0.01

Results are expressed as means ± SEM. The difference between men and women is significant for the subgroups 20–25 and 26–30: $^*p < 0.05$.

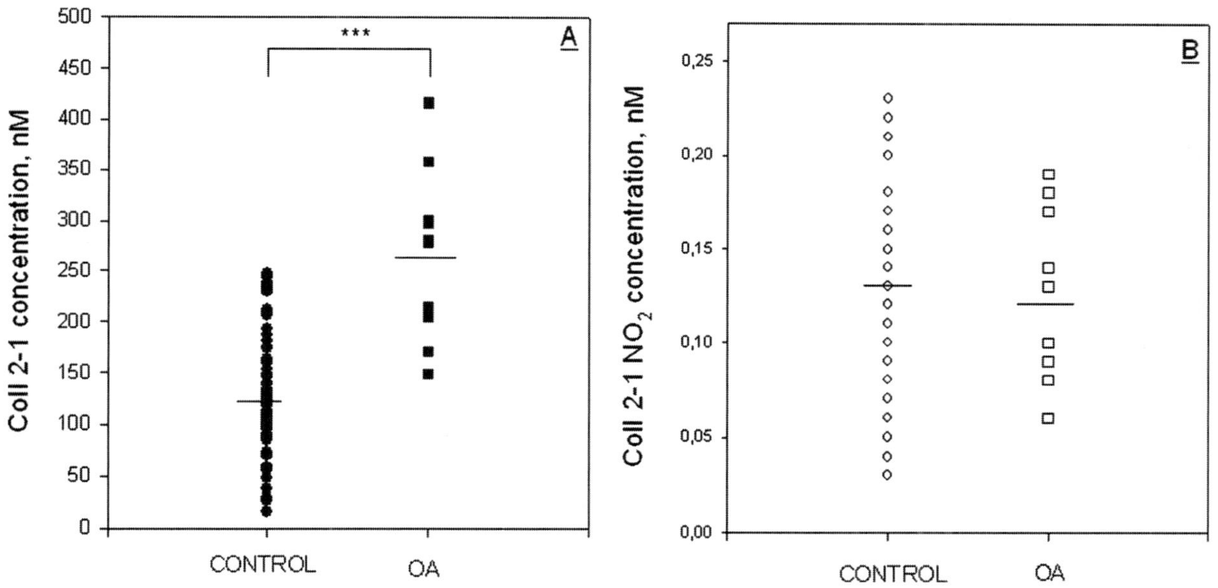

Fig. 1. Coll 2-1 and Coll 2-1 NO_2 levels in sera of age-matched healthy donors and of patients with knee osteoarthritis. The Coll 2-1 level is significantly different between healthy and OA subjects with $^{***}p < 0.001$.

of glucosamine sulphate with placebo [12]. At baseline, significant correlations were found between the urinary levels of Coll 2-1 and Coll 2-1 NO_2 and the global Western Ontario and Mc Master Universities score (WOMAC osteoarthritis index which is a validated, disease-specific questionnaire addressing the severity of joint pain, stiffness and limitation) (Coll 2-1: $r = 0.28$, $p = 0.01$; Coll 2-1 NO_2: $r = 0.27$, $p = 0.02$) and its pain (Coll 2-1: $r = 0.27$, $p = 0.01$; Coll 2-1 NO_2: $r = 0.30$, $p = 0.01$) and function (Coll 2-1: $r = 0.29$, $p = 0.01$; Coll 2-1 NO_2: $r = 0.27$, $p = 0.02$) subscales. Furthermore, there was a significant negative correlation between the first year change of these markers in urine, and the 3-year change of medial joint space width (Coll 2-1: $r = -0.31$, $p = 0.035$; Coll 2-1 NO_2: $r = -0.31$, $p = 0.035$). It was also observed that the subjects of the highest quartile, corresponding to subject with the highest 1-year increase of these markers, were also those with the highest joint space narrowing over 3 years [6].

4. Conclusion

This paper describes two new immunoassays for quantifying in serum and urine a peptide of the triple helix of type II collagen in its native and nitrated form. The preliminary data suggest that Coll 2-1 and Coll 2-1 NO_2 are potential markers of OA clinical activity and that the increase of these peptides over one year seems to be predictive of the radiological progression of knee OA.

Acknowledgements

The author thank the "Service de Transfusion Sanguine" for providing healthy subjects and the Department of Clinical Biology (Prof. Gielen) for the measurements of C-reactive protein.

References

[1] R.C. Billinghurst, L. Dahlberg, M. Ionescu, A. Reiner, R. Bourne, C. Rorabeck, P. Mitchell, J. Hambor, O. Diekmann, H. Tschesche, J. Chen, H. Van Wart and R. Poole, Enhanced cleavage of type II collagen by collagenases in osteoarthritic articular cartilage, *J. Clin. Invest.* **99** (1997), 1534–1545.

[2] M. Deberg, S. Christgau, P. Cloos, D.B. Henriksen, C. Sanchez, A. Labasse, J.Y. Reginster and Y. Henrotin, Development of new immunoassays for the quantification of inflammatory related cartilage degradation, *Arthritis Rheum.* **46** (2002), S496 (P1319).

[3] P. Garnero, M. Piperno, E. Gineyts, S. Christgau, P.D. Delmas and E. Vignon, Cross sectional evaluation of biochemical markers of bone, cartilage and synovial tissue metabolism in patients with knee osteoarthritis: relations with disease activity and joint damage, *Ann. Rheum. Dis.* **60** (2001), 619–626.

[4] P. Garnero, X. Ayral, J.C. Rousseau, S. Christgau, L. Sandell, M. Dougados and P.D. Delmas, Uncoupling of type II collagen synthesis and degradation predicts progression of joint damage in patients with knee osteoarthritis, *Arthritis Rheum.* **46** (2002), 2613–2624.

[5] P. Garnero, E. Gineyts, S. Christgau, B. Finck and P. Delmas, Association of baseline levels of urinary glucosyl-galactosyl-pyridinoline and type II collagen C-telopeptide with progression of joint destruction in patients with early rheumatoid arthritis, *Arthritis Rheum.* **46** (2002), 21–30.

[6] Y. Henrotin, M. Deberg, S. Christgau, D.B. Henriksen, L. Seidel and J.Y. Reginster, Type II collagen derived fragment (Coll 2-1) is a new marker predictive of osteoarthritis progression, *Osteoporosis Int.* **13** (2002), S17 (021).

[7] Y. Henrotin, G. Deby-Dupont, C. Deby, M. De Bruyn, M. Lamy and P. Franchimont, Production of active oxygen species by isolated human chondrocytes, *Br. J. Rheumatol.* **32** (1993), 562–567.

[8] Y.E. Henrotin, S.X. Zheng, G.P. Deby, A.H. Labasse, J.M.R. Crielaard and J.Y. Reginster, Nitric oxide downregulates interleukin 1β (IL-1β) stimulated IL-6, IL-8, and prostaglandin E2 production by human chondrocytes, *J. Rheumatol.* **25** (1998), 1595–1601.

[9] A.P. Hollander, T.F. Heathfield, C. Webber, Y. Iwata, R. Bourne, C. Rorabeck and A.R. Poole, Increased damage to type II collagen in osteoarthritic articular cartilage detected by a new immunoassay, *J. Clin. Invest.* **93** (1994), 1722–1732.

[10] F. Nelson, L. Dahlberg, S. Laverty, A. Reiner, I. Pidoux, M. Ionescu, G.L. Fraser, E. Brooks, M. Tanzer, L.C. Rosenberg, P. Dieppe and R. Poole, Evidence for altered synthesis of type II collagen in patients with osteoarthritis, *J. Clin. Invest.* **102** (1998), 2115–2125.

[11] I.G. Otterness, J.T. Downs, C. Lane, M.L. Bliven, H. Stukenbrok, D.N. Scampoli, A.J. Milici and P.S. Mèzes, Detection of collagenase-induced damage of collagen by 9A4, a monoclonal C-terminal neoepitope antibody, *Matrix Biology* **18** (1999), 331–341.

[12] J.Y. Reginster, R. Deroisy, L.C. Rovati, R.L. Lee, E. Lejeune, O. Bruyere, G. Giacovelli, Y. Henrotin, J.E. Dacre and C. Gossett, Long-term effects of glucosamine sulphate on osteoarthritis progression: a randomised, placebo-controlled trial, *The Lancet* **357** (2001), 251–256.

[13] E.J.M.A. Thonar, K. Masuda, D.H. Manicourt and K.E. Kuettner, Structure and function of normal adult articular cartilage, in: *Osteoarthritis: Clinical and Experimental Aspects*, J.Y. Reginster, J.P. Pelletier, J. Martel-Pelletier and Y. Henrotin, eds, Springer, Berlin, 1999, pp. 1–19.

[14] A. Van der Vliet, D. Smith, C.A. O'Neill, H. Kaur, V. Darley-Usmar, C.E. Cross and B. Halliwell, Interactions of peroxynitrite with human plasma and its constituents: oxidative damage and antioxidant depletion, *Biochem. J.* **303** (1994), 295–301.

Biorheology 41 (2004) 549–561
IOS Press

Articular chondrocytes cultured in hypoxia: Their response to interleukin-1β and rhein, the active metabolite of diacerhein

G. Martin [a], P. Bogdanowicz [a], F. Domagala [b], H. Ficheux [b] and J.-P. Pujol [a,*]

[a] *Laboratory of Connective Tissue Biochemistry, Faculty of Medicine, 14032 Caen Cedex, France*
[b] *Laboratoires Negma, Toussus-le-Noble, 78771 Magny-Les-Hameaux Cedex, France*

Abstract. In the present report, we show that bovine articular chondrocytes cultured in low oxygen tension, i.e. in conditions mimicking their hypoxic *in vivo* environment, respond to IL-1β (10 ng/ml) by an increased DNA binding activity of NF-κB and AP-1 transcription factors. Incubation of the cells with 10^{-5} M rhein for 24 h was found to reduce this activity, particularly in the case of AP-1. Mitogen activated kinases (ERK-1 and ERK-2) were activated by exposure of the chondrocytes to 1-h treatment with IL-1β. This effect was greater in hypoxia (3% O_2) than in normoxia (21% O_2). Rhein was capable of reducing the IL-1β-stimulated ERK1/ERK2 pathway whatever the tension of oxygen present in the environment. The mRNA steady-state levels of collagen type II (COL2A1) and aggrecan core protein were found to be significantly increased by a 24-h treatment with 10^{-5} M rhein. This stimulating effect was also observed in the presence of IL-1β, suggesting that the drug could prevent or reduce the IL-1β-induced inhibition of extracellular matrix synthesis. IL-1-induced collagenase (MMP1) expression was significantly decreased by rhein in the same conditions. In conclusion, rhein can effectively inhibit the IL-1-activated MAPK pathway and the binding of NF-κB and AP-1 transcription factors, two key factors involved in the expression of several pro-inflammatory genes by chondrocytes. In addition, the drug can reduce the procatabolic effect of the cytokine, by reducing the MMP1 synthesis, and enhance the synthesis of matrix components, such as type II collagen and aggrecan. These results may explain the anti-osteoarthritic properties of rhein and its disease-modifying effects on OA cartilage, in spite of absence of inhibition at prostaglandin level.

Keywords: Cartilage, chondrocytes, interleukin-1, osteoarthritis, rhein

Abbreviations

AP-1, activating protein-1; ATP, adenosine triphosphate; BAC, bovine articular chondrocytes; BSA, bovine serum albumin; poly dI-dC, poly(deoxy inositinic-deoxy cytidilic); COX, cyclooxygenase; DMEM, Dulbecco's modified Eagle's medium; DTT, dithiothreitol; EDTA, ethylene diamine tetracetic acid; Egr-1, early growth response factor; EGTA, ethylene glycol-bis (2 aminoethyl ether)-N,N,N′,N′-tetracetic acid; ERKs, extracellular signal-regulated kinases; ESE-1, epithelium specific Ets factor; FCS, fetal calf serum; GAPDH, glyceraldehyde phosphate dehydrogenase; Hepes, N-2-hydroxyethylpiperazine-N-ethanesulfonic acid; IL-1β: interleukin-1β; iNOS, inducible nitric oxide synthase; MMPs, matrix metalloproteases; MOPS, 3-N-morpholino-propanesulfonic acid; NF-κB, nuclear factor kappa B; NO, nitric oxide; OA, osteoarthritis; PGE_2, prostaglandin E_2; PMSF, phenylmethanesulfonyl fluoride; SDS-PAGE, sodium dodecyl sulfate-polyacrylamide gel electrophoresis; TGF-β, transforming growth factor-β; TNF-α: tumor necrosis factor α.

*Address for correspondence: Tel.: +33 2 31 06 82 18; Fax: +33 2 31 06 82 24; E-mail: pujol@ibba.unicaen.fr.

1. Introduction

Mechanical stress on articular cartilage is considered as a fundamental stimulus for chondrocyte metabolic activities. Cell and tissue deformation, as well as cyclic fluctuations in hydrostatic pressure and fluid movement, are associated with cartilage loading and involved in regulating chondrocyte metabolic activity [76]. Several studies have emphasized the importance of joint motion and weight-bearing in the homeostasis of normal healthy cartilage. Many investigators have used *in vitro* systems that allow the analysis of cellular response to loading more directly. Generally, forces have been applied to cartilage explants in different ways: either statistically [29,40], as an impact load [15,32,36,39] or as a repetitive cycle of forces, either applied continuously [40,69,70,74] or intermittently [12,35,69, 77]. These protocols tended to simulate more closely the *in vivo* conditions cartilage might experience during normal or pathological joint function. However, there is another factor from the cartilage tissue specific environment which has been rather neglected so far by most of the investigators who used either explant or isolated cell culture protocols, that is the low oxygen tension present within cartilage. Indeed, chondrocytes are living in hypoxic conditions, the level of oxygen in articular cartilage ranging from 2 to 7%, related to 21% in normal atmosphere. Several studies have shown that signaling pathways and gene expression are highly dependent on the environmental oxygen level of chondrocytes [17, 30,49,62,80]. Furthermore, hypoxic conditions may potentially control the intensity of chondrocyte response to cytokines such as interleukin-1, which is known to play a major role in the physiopathology of osteoarthritis [52,60].

Osteoarthritis (OA) is a multifactorial degenerative joint disease in which the cartilaginous matrix is destroyed. Disruption of the balance between cytokine-induced catabolic and anabolic processes leads to cartilage erosion that can reach the level of subchondral bone [53]. We, and others, have largely documented that Interleukin-1β (IL-1β) is a key mediator of joint breakdown, through its ability to induce expression of several metalloproteases (MMPs) [52,60]. Il-1β can also down-regulate the synthesis of matrix components, including type II collagen and aggrecans, limiting the repair potential of articular cartilage [6,27,59]. IL-1β induces the expression of the nitric oxide synthase (iNOS) [42,51,68] and cyclooxygenase (COX) [3,33] enzymes, resulting in increased production of NO and prostaglandin E2 (PGE_2).

IL-1β triggers several signaling cascades in chondrocytes, including sequential activation of various cytoplasmic protein kinases such as mitogen activated protein kinases (MAPKs). The MAPK family, composed of extracellular signal-regulated kinases (ERK 1 and ERK 2), c-Jun N-terminal kinases (JNKs) and p38 kinase [18,66], controls cellular growth and differentiation via the regulation of several gene expressions. For example, MAPKs may activate the DNA binding activity of NF-κB and AP-1, two transcription factors implicated in the transcription of many pro-inflammatory genes, including COX [41], NOS [24,41,47,78] and MMPs [4,19,20]. Indeed, high NF-κB and AP-1 binding activity has been found in the synovium of patients with rheumatoid arthritis and osteoarthritis, suggesting that activation of NF-κB and AP-1 may be involved in the pathogenesis of the joint diseases.

Conventional NSAIDs inhibit cyclooxygenase (COX), suppressing the formation of prostaglandins from arachidonic acid. As COX-2 expression in response to various stimuli is known to be regulated by MAPK cascades in several cell types [1,44,71], it has been suggested that one or more of the MAPK may mediate the effects of NSAIDs on OA joint cells. In this regard, it is worth noting that it has been recently found that *in vivo* inhibition of MAPK 1/2 in rabbit experimental OA was associated with a reduction in the development of structural changes [56].

In contrast to conventional NSAIDs, diacerein and rhein, its active metabolite, cannot inhibit the production of prostaglandin E_2 by several cell types, including articular chondrocytes [25,58]. Nevertheless, this anthraquinone that has been shown to reduce the severity of OA, both in man [22,38,43,50, 55,72] and in animal models [11,31,67], exerting structure-modifying effects on cartilage. The molecule is currently used in the treatment of OA [22] and has also been found to suppress the evolution of adjuvant-induced arthritis in rats [73]. The cellular mechanisms underlying these beneficial effects are not completely understood. *In vitro*, diacerein and rhein are capable of inhibiting the IL-1β-induced stimulation of collagenase expression by cultured rabbit chondrocytes whereas they can increase the synthesis of collagen and [8,61,79]. These compounds also stimulate the expression of TGF-β1 [26], an effect that may be responsible for the action of rhein, since TGF-β counteracts the IL-1-induced modulation of MMP expression and production of matrix molecules [63,64]. In human normal and osteoarthritic chondrocytes, rhein has been shown to reduce IL-1β induction of iNOS synthesis and to enhance COX2 production and activity [54].

In this study, we investigated the effects of rhein on the major signaling pathways of chondrocytes that could be responsible for the drug beneficial action in OA, with special attention to the activation of MAPKs, NF-kB and AP-1 induced by IL-1 in bovine articular chondrocytes cultured in hypoxic environment. Moreover, we also studied the effects of rhein on the expression of the major cartilage matrix genes and collagenase (MMP-1), by evaluating the steady-state levels of corresponding mRNAs.

2. Materials and methods

2.1. Reagents

Mouse anti-phospho MAP kinase (specific for threonine and tyrosine phosphorylated residues of ERK1/ERK2) and rabbit anti-MAP kinase 1/2 monoclonal antibodies (ERK1/2-CT) were purchased from Upstate Biotechnology, Inc (Lake Placid, NY, USA). Rabbit polyclonal anti-IκBα and rabbit polyclonal anti-c-jun was obtained from Santa Cruz Biotechnology Inc. (Santa Cruz, CA, USA). Primary antibodies were revealed with anti-rabbit or anti-mouse horseradish peroxidase-labeled secondary antibodies (Santa Cruz), using an ECL + Plus Western Blot detection kit (Amersham Pharmacia Biotech Europe, Orsay, France). The consensus oligonucleotide probes: AP1 (5′-5′CGCTTGATGAGTCAGCCCGGAA-3′) and NF-κB (5′-AGTTGAGGGGACTTTCCCAGGC-3′) were supplied by Life Technologies, Inc. (Cergy Pontoise, France). All other chemicals were of the highest purity available and ordered from Sigma-Aldrich Co. (St Quentin-Fallavier, France). Human recombinant IL-1β was a generous gift from Dr Soichiro Sato, Shizuoka, Japan.

2.2. Culture and treatment of articular chondrocytes

Chondrocytes were isolated from the knee joints of freshly slaughtered calves by enzymatic treatment [7]. Briefly, slices of cartilage were dissected out and kept in Earle's balanced salt solution. Chondrocytes were released by digestion with type XIV protease (4 mg/ml) for 1.5 h and type I collagenase (1 mg/ml) overnight in DMEM at 37°C. The cells were centrifuged and seeded in DMEM (Life Technologies, Inc.) supplemented with 10% heat-inactivated fetal calf serum (FCS) and antibiotics: penicillin (100 IU/ml), streptomycin (100 μg/ml), and fungizone (0.25 μg/ml). The cells were allowed to recover for 48 h at 37°C in a humidified atmosphere supplemented with 5% CO_2.

For experiments in hypoxic environment, cartilage samples were dissected in a special plastic chamber previously equilibrated at 3% oxygen and the subsequent procedure was performed in the same conditions. The O_2 level of the culture medium was reduced from 21% to 3% O_2, by bubbling a gas mixture without oxygen for 1.5 hour. Because of its antioxidant properties, phenol red, generally used as a pH indicator, was not added to the medium.

2.3. *Preparation of cytoplasmic and nuclear extracts*

At the end of the experiments, the chondrocytes were rinsed twice with ice-cold phosphate-buffered saline (PBS) and scraped in a hypotonic buffer [5,65]. The cell suspension was centrifuged at 10,000 $\times g$ for 10 min and the resulting supernatants used as cytoplasmic extracts. The pellet was incubated in hypertonic buffer for 4 h at 4°C and centrifuged at 10,000 $\times$ g for 30 min to provide nuclear extracts. Hypotonic buffer contained: 10 mM Hepes (pH 7.9), 0.1% NP-40, 1.5 mM $MgCl_2$, 10 mM KCl, 1 mM EGTA, 1 mM EDTA, 5 mM NaF, 0.5 mM DTT, 0.5 mM PMSF, Leupeptin, Pepsatin A and Aprotinin at 10 μg/ml, 1 mM Na_3VO_4. Hypertonic buffer was composed of: 20 mM Hepes (pH 7.6), 25% glycerol, 1 mM $MgCl_2$, 420 mM NaCl, 0.2 mM EDTA, 0.25 mM DTT, 0.5 mM PMSF, Leupeptin, Pepstatin A and Aprotinin at 10 μg/ml, 1 mM Na_3VO_4. The protein amount was determined by the Bradford's colorimetric procedure (Bio-Rad S.A., Ivry sur Seine, France).

2.4. *Western blot analysis*

Cellular extracts (15 μg protein) were subjected to sodium dodecylsulfate-polyacrylamide gel electrophoresis (SDS-PAGE) under reducing conditions [34] and electrophoretically transferred to polyvinylidene difluoride transfer membranes (PVDF, NEN Life Sciences Products, Zawentem, Belgium). The membranes were blocked for 1 h at room temperature in Tris-buffered saline (pH 7.6) with 0.1% Tween-20 (TBS-T) and 10% nonfat dry milk. Then, they were rinsed twice in TBS-T and incubated overnight at 4°C with the primary antibody. After washing with TBS-T, they were incubated for 1 h with the appropriate secondary antibody. The membranes were developed with the ECL+Plus chemoluminescence detection kit. To check for the presence of equal amounts of protein or to analyse the expression of several proteins, blots were stripped (100 mM 2-mercaptoethanol, 2% SDS, 62.5 mM Tris-HCl (pH 6.7) for 30 min at 50°C with stirring, and used again.

2.5. *Electrophoretic mobility shift assays (EMSA)*

Nuclear extracts (10 μg protein) were incubated in binding buffer for 30 min at 25°C with the cDNA probes, previously labeled with ^{32}P-γ-ATP (25 fmoles) using T4 polynucleotide kinase (Life Technologies, Inc.). Final binding reaction for NF-κB was performed, in: 20 mM Hepes (pH 7.5), 50 mM KCl, 4 mM $MgCl_2$, 0.2 mM EDTA, 0.5 mM DTT, 0.05% NP40, 20% glycerol, 1 mg/ml BSA, 0.025 mM poly dI-dC. Binding reaction for AP-1 was composed of Hepes 20 mM, Tris-HCl pH 8.2 mM, DTT 1 mM, NaCl 80 mM, glycerol 10% (v/v), PMSF 0.2 mM, EDTA 0.4 mM, EGTA 0.2 mM. The samples were then submitted to 8% PAGE in 0.5 $\times$ TBE (45 mM Tris (pH 7.8), 45 mM boric acid and 1 mM EDTA) and visualized by autoradiography [65].

2.6. *RNA extraction and Northern blot*

Total RNA from the chondrocyte cultures were extracted by the guanidium isothiocyanate–phenol–chloroform procedure [16]. RNA was fractionated by electrophoresis on 1% agarose/MOPS/formadelyde gel and transferred to a nylon membrane (Biodyne B, NEN, Boston, USA). RNA samples were fixed on the membrane by UV exposure. Type II collagen, aggrecan core and GAPDH probes were generated by reverse transcription-polymerase chain reaction, using the following primers: type II collagen, forward primer: GACCCCATGCAGTACATG, reverse primer: GACCGTCTTGCCCCACTT; aggrecan core protein: f.p.: CCCTGGACTTTGACAGGGC, r.p.: AGGAAACTCGTCCTTGTCTCC; GAPDH: f.p.: TGGTATCGTGGAAGGACTCATGAC, r.p.: ATGCCAGTGAGCTTCCCGTTCAGC. The probes were ^{32}P-labeled, using a random priming kit (Invitrogen. Inc). Prehybridization (1 h at 42°C) and hybridization (18 h at 48° to 60°C) were performed in 2/3 Na_2HPO_4 pH 7.2–7.4, 1/3 SDS 20% (v/v). Blots were washed several times in 2 × SSC (standard saline citrate) and 0.1% SDS at 42°C. Final washes were in 0.1 × SSC plus 0.1% SDS at 42°C. The ^{32}P-labeled cDNA-mRNA hybrids were visualized by autoradiography. The relative density of detected signals was measured by densitometric scans of X-ray film (Kodak, X-Omat, NY, USA) and quantified by the Image Quant program (Molecular Dynamics, Sunnyvale, USA).

3. Results

3.1. *Effect of normoxia and hypoxia on interleukin 1β-induced NF-κB and AP-1 DNA binding*

Treatment of chondrocytes with IL-1β for 1 h and 24 h, either in normal atmosphere (21% O_2) or in hypoxic environment (3% O_2), induced the binding of NF-κB complexes to the corresponding consensus DNA sequence (Fig. 1). Two protein complexes were observed, which are dimers composed of the subunits p65 and p50, as we and others have already demonstrated, using specific antibodies (not shown).

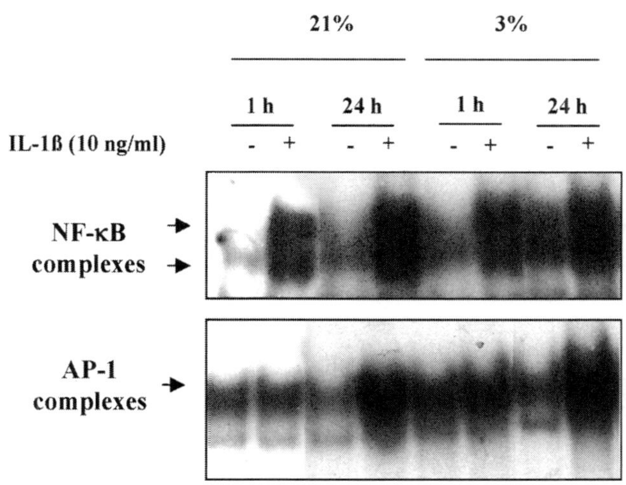

Fig. 1. Effect of IL-1β treatment on NF-κB and AP-1 DNA binding activity in normoxia and hypoxia. Chondrocyte cultures in either normoxia (21% O_2) or hypoxia (3% O_2) were incubated for 1 h or 24 h. EMSA were performed as described under "Material and methods". Arrows indicate the position of the DNA binding complexes. The migration of the free probe is not shown. The autoradiography shown is representative of three independent experiments.

The binding was found to be greater for 24 h of IL-1β treatment than after 1 h. Furthermore, the effect of IL-1 was slightly superior in low-oxygen environment, compared to normoxia, specially after 1 h incubation of chondrocytes with the cytokine.

IL-1 was also capable of strongly enhancing the binding of proteins to the consensus AP-1 sequence (Fig. 1). The effect was particularly visible after a 24-h treatment of the chondrocyte cultures with the cytokine, suggesting that the activation of AP-1 in these conditions requires more time than that of NF-κB. The inducing effect exerted by IL-1β on AP-1 activation was always greater in hypoxic than in normoxic cultures.

3.2. *Effect of rhein on IL-1β-induced NF-κB and AP-1 activation*

Taking that the activating effect exerted by IL-1β on NF-κB and AP-1 transcription factors was more significant in hypoxic cultures of chondrocytes, a situation close to the environment of these cells within the cartilage, we chose those conditions to determine the effect of rhein. Pretreatment of chondrocyte cultures for 18 h with 10^{-5} M rhein slightly decreased interleukin-1β-induced NF-κB binding to the corresponding consensus probe, whereas it significantly reduced that of AP-1 (Fig. 2). The drug alone had no effect on the bands corresponding to both NF-κB- and AP-1-oligonucleotide probe complexes revealed by EMSA. Degradation of IκB-α protein from the cytoplasm and translocation of NF-κB subunits to the nucleus have been described as the sequential mechanism which is responsible for activation of the transcription factor and its binding to specific sequences of target genes. As expected, IL-1β induced a decrease of the amount of IκB-α protein detected by Western blotting in cytoplasmic extracts of hypoxic cultures. Rhein alone was capable of increasing the amount of IκB-α and to reduce the IL-1 effect when added in combination with the cytokine (Fig. 3). This effect suggests that the drug could prevent, or rather delay, the degradation of IkB-α in hypoxic chondrocyte cultures. These data are to be correlated with the slight reduction observed in the formation of the NF-κB-oligonucleotide complexes, as detected by EMSA (Fig. 2).

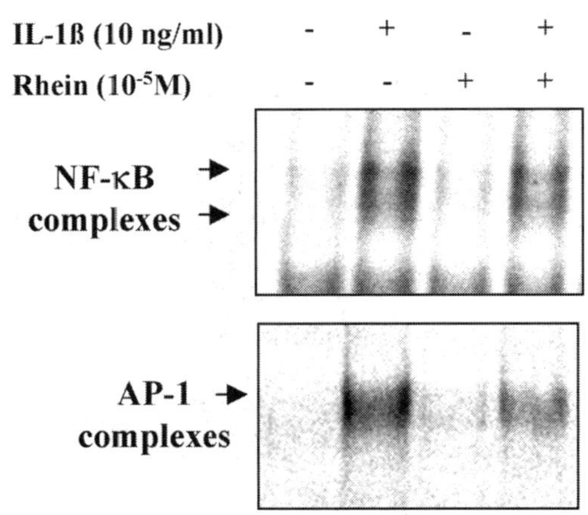

Fig. 2. Effect of 10^{-5} M rhein on IL-1-β-induced NF-κB and AP-1 DNA binding activity and IκBα degradation (C). Chondrocytes were cultured in hypoxia (3% O_2) and incubated with rhein (10^{-5} M) for 18 h in 1% FCS-containing medium. Then, the cells were further incubated for 24 h in the presence of IL-1β (10 ng/ml) and the drug. EMSA were performed as described under "Material and methods". The autoradiography shown is representative of three independent experiments.

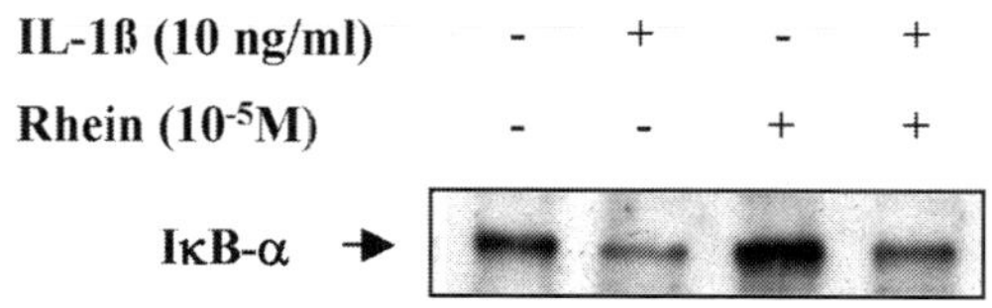

Fig. 3. Effect of 10^{-5} M rhein on IL-1-β-induced IκBα degradation. Chondrocytes were cultured in hypoxia (3% O_2) and incubated with rhein (10^{-5} M) for 18 h in 1% FCS-containing medium. Then, the cells were further incubated for 24 h in the presence of IL-1β (10 ng/ml) and the drug. Western-blot were performed as described under "Material and methods". The autoradiography shown is representative of three independent experiments.

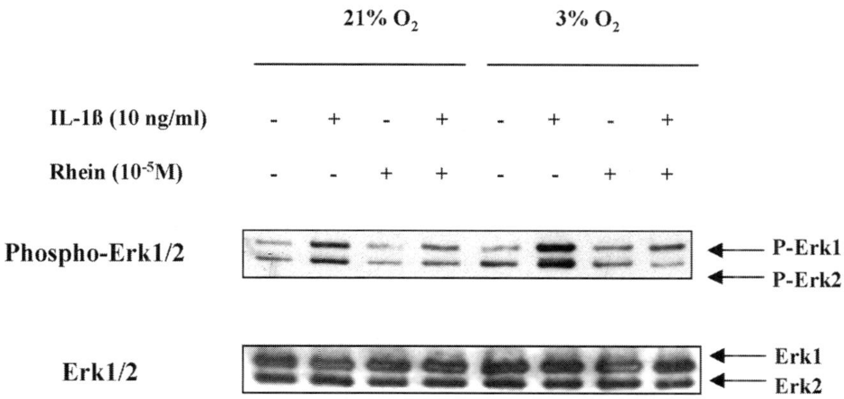

Fig. 4. Effect of rhein on IL-1β-induced activity of ERK 1/2. Chondrocytes were cultured either in normoxia (21% O_2) or hypoxia (3% O_2) and incubated for 18 h with 10^{-5} M rhein. Then, the cells were further incubated for 1 h in the presence of IL-1β (10 ng/ml). The cytoplasmic extracts were prepared and subjected to Western blot analysis, using anti-phospho-Erk 1/2 and anti-Erk 1/2, as described under "Material and methods". The protein-antibody complexes were visualized by chemiluminescence, using HRP-conjugated secondary antibody. The results shown are representative of three independent experiments.

3.3. Effect of interleukin-1β and rhein on ERK 1/2 activation

To further investigate the signaling pathways, upstream NF-κB and AP-1 DNA binding, that could potentially be affected by rhein, we first determined the effect of IL-1β treatment on the activity of MAP kinases. Using Western blotting, we found that treatment of cells with IL-1β for 1 h caused an increased phosphorylation of ERK1 and ERK2, reflecting the activation of these kinases (Fig. 4). The corresponding bands were always of greater intensity when the chondrocytes were cultured in a 3% O_2 environment, indicating that IL-1β stimulation is more effective in hypoxia than in normoxia. Incubation of the cultures with 10^{-5} M rhein for 18 h reduced the IL-1β-stimulated MAP kinase activity in both experimental conditions, a greater effect being observed however in low oxygen tension.

3.4. Effect of rhein on the type II collagen and aggrecan core protein mRNA steady-state levels

The expression of matrix specific genes by chondrocytes cultured in hypoxia was found to be modulated by the presence of rhein in the culture medium. Indeed, Northern blots showed that the type II collagen mRNA levels were strongly increased in cells pre-treated with 10^{-5} M rhein for 18 h (Fig. 5). Furthermore, this stimulating effect of the drug was still visible in chondrocytes further exposed to IL-1β for 24 h, indicating that rhein is capable of preventing the inhibitory effect of IL-1β on type II collagen expression. Similar effects were found for the aggrecan core protein mRNA levels (Fig. 5).

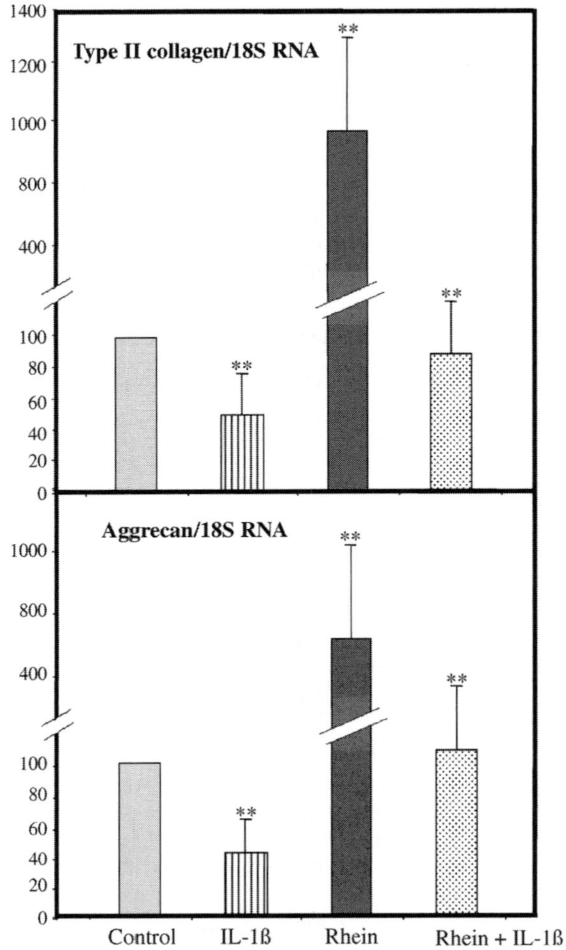

Fig. 5. Effect of 10^{-5} M rhein on IL-1β decrease of type II collagen and aggrecan core protein mRNA levels. Chondrocyte cultures were treated for 18 h with 10^{-5} M rhein, in hypoxia. Then, they were further incubated for 24 h with IL-1β (10 ng/ml). Total RNA was extracted and used for Northern blotting with COL2A1 and core protein cDNA probes. Hybridization with a probe corresponding to 18 sRNA is shown as an internal control of RNA loading. Histograms show the means ± S.E.M. values obtained by densitometric scanning of the three experimental filters. Significance of the differences between treated and control samples was estimated by the Student's t test ($^{*}p < 0.05$; $^{**}p < 0.01$; $^{***}p < 0.001$).

3.5. *Effect of rhein on the collagenase (MMP 1) mRNA levels*

Since IL-1-induced activation of MAP kinases and AP-1 may up-regulate the expression of metalloproteases, including MMP 1, it was of interest to determine if the mRNA steady-state levels of collagenase 1 were altered by pre-treatment of chondrocytes with rhein. Northern blots performed on RNA extracted from chondrocytes incubated in the same experimental conditions clearly show that rhein significantly reduced the strong IL-1β-induced collagenase expression (Fig. 6).

4. Discussion

The aim of this study was to determine the signaling pathways of chondrocytes that be modulated by rhein and may explain the anti-inflammatory and anti-osteoarthritic effects of the drug. Since it has

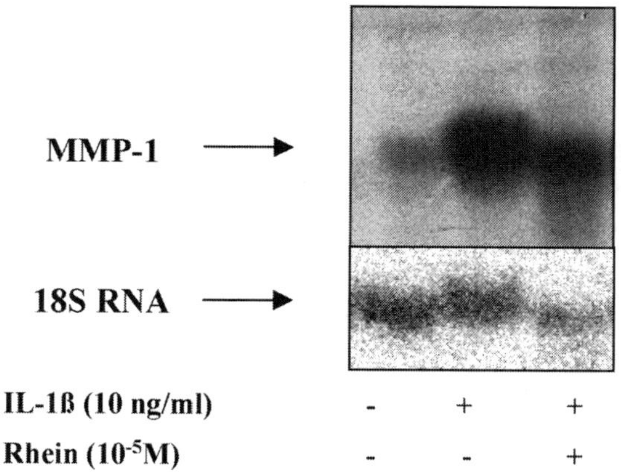

Fig. 6. Effect of rhein on IL-1β-induced collagenase I gene transcription. Chondrocytes were treated in the same conditions as previously described, in the presence or absence of 10^{-5} M rhein. Northern-blots were performed, using a cDNA probe specific for MMP1. Hybridization with a 18s RNA probe was also performed as an internal control of RNA loading.

been demonstrated earlier that transduction mechanisms and gene expression were significantly affected by the oxygen level present in the chondrocyte environment, we performed the experiments in hypoxia to mimick the situation of these cells within cartilage [17,30,49,62,80]. Rhein was found to reduce interleukin-1-stimulated ERK kinase signal transduction, as well as the activation of AP-1 and NF-κB transcription factors in bovine articular chondrocytes. This effect was associated with down-regulation of the IL-1-induced collagenase (MMP1) gene expression and partially reduced the inhibition exerted by the cytokine on the synthesis of type II collagen and aggrecan core protein. In this regard, our data are in agreement with a recent study showing that inhibition of interleukin-1-stimulated MAP kinases, AP-1 and NF-κB transcription factors, down-regulates matrix metalloproteinase gene expression, including MMP-3 and MMP-13, in bovine articular chondrocytes [37]. That report clearly demonstrates that activation of MAP kinases is responsible for subsequent stimulation of AP-1 and NF-κB DNA binding, leading to MMP induction. The promoters of MMPs are known to contain AP-1-responsive elements and the role of AP-1 in the regulation of several MMP genes has been largely documented [13,57]. Implication of NF-κB, whose sites are not systematically found in all MMP promoters, has also been supported by recent evidence. For example, NF-κB activation is involved in the regulation of MMP-1 and MMP-13 genes in chondrosarcoma cells [48]. Moreover, blocking NF-κB activity by overexpression of its inhibitor, IκBα, was also found to reduce inflammation and tissue destruction by MMP-1 and MMP-3 in rheumatoid synovial tissue and dermal fibroblasts [9,10]. One possibility whereby NF-κB could play a role in MMP regulation may be a crosstalk with AP-1 in the nucleus, as suggested for TNF-α regulation of the vascular cell adhesion molecule-1 gene expression [2].

We also found that the IL-1β-induced ERK 1/2 response was exacerbated in chondrocyte cultures maintained in hypoxia, related to what is observed in normoxia. This observation is in agreement with a previous report demonstrating that human cartilage explants cultured in 5% pO_2 responded to IL-1α by a MAP kinase activation which was enhanced compared to 21% pO_2 [23]. Furthermore, using MAPK inhibitors, the authors revealed that the IL-1α-stimulated inhibition of ^{35}S-sulfate incorporation utilises the ERK 1/2 pathway and that transcription of aggrecan, link protein and decorin is dependent upon this mechanism. These data explain the inhibitory effect exerted by rhein on the IL-1β-induced down-regulation of aggrecan core protein expression observed in our study, and more globally the synthesis of proteoglycan [8,61].

In our study, the inhibitory effect of rhein on IL-1β-induced NF-κB DNA binding was generally less significant that its action on AP-1 binding. During this work, a study has also reported a preventing effect of diacerhein and rhein on IL-1β-induced NF-κB activation in bovine chondrocytes [46]. Our results are in agreement with most of those data, excepted that we cannot conclude that rhein was capable of completely suppress IKB-α degradation. Anyway, the inhibitory effect exerted by rhein on IL-1-induced NF-κB activation has probably great impact on gene expression since the implication of NF-κB in the transcription of several gene promoters, including those of pro-inflammatory cytokines or mediators, has been largely documented. For example, it has been shown that diacerhein and rhein inhibited IL-1β-induced iNOs expression by preventing NF-κB activation [54]. In turn, studies in different cells, including chondrocytes, have shown that NO can activate NF-κB by stimulating its activity induced by cytokines and other stimuli [47].

Rhein inhibition of IL-1β-induced NF-κB activation could also provide an explanation for its effect on the down-regulation of type II collagen synthesis caused by the cytokine. Indeed, it has been recently demonstrated that inhibition of COL2A1 transcription by IL-1β was mediated by the $-577/+125$ bp promoter region and was associated with induction of Egr-1, NF-κB and ESE-1 binding activities in nuclear extracts [28]. However, the authors found that NF-κB acted indirectly by inducing ESE-1 promoter activity and subsequent ESE-1 mRNA expression. Another responsible mechanism contributing to the effect of rhein on collagen synthesis would be its inhibition of NO production, since it has been reported that NO could down-regulate type II collagen synthesis at a post-transcriptional level [14].

In our study, the IL-1β-induced AP-1 DNA binding was clearly shown to be reduced by pre-treatment of cells with rhein. This effect could also be secondary to decreased NO production by rhein since NO was reported to induce AP-1-DNA binding [47]. Nevertheless, the mechanism whereby AP-1 activation is inhibited by rhein requires further investigation. We have shown earlier that c-jun expression does not seem to be modulated by the drug [21,61]. Most probably, the binding activity of the AP-1 complex itself, formed by dimers of c-Jun and c-Fos proteins, is modified following rhein treatment of the chondrocytes. Work is in progress to determine if the Jun kinase phosphorylating activity could be one of the targets used by rhein to enhance AP-1 DNA binding.

To conclude, we have seen here that rhein acts at an early step of the IL-1 signaling cascade, leading from MAP kinase activation to modulation of the transcription factor NF-κB and AP-1 activities and subsequent regulation of transcription of genes, including iNOS, MMPs, type II collagen and aggrecan. By its ability to inhibit the IL-1β-induced NF-κB and AP-1 activation, rhein could contribute to reduce the deleterious effects exerted by the cytokine on the cartilage matrix. Taken that NF-kB- and AP-1-dependent gene expression is also playing a great role in inflammatory diseases, such as rheumatoid arthritis, rhein could also be of some interest in the treatment of chronic inflammatory diseases.

Acknowledgement

This study was supported by a grant from Negma Laboratories, France.

References

[1] S.R. Adderley and D.J. Fitzgerald, Oxidative damage of cardiomyocytes is limited by extracellular regulated kinases 1/2-mediated induction of cyclooxygenase-2, *J. Biol. Chem.* **274** (1999), 5038–5046.

[2] M. Ahmad, P. Theofanidis and R.M. Medford, Role of activating protein-1 in the regulation of the vascular cell adhesion molecule-1 gene expression by tumor necrosis factor-alpha, *J. Biol. Chem.* **273** (1998), 4616–4621.

[3] M.G. Attur, I.R. Patel, R.N. Patel, S.B. Abramson and A.R. Amin, Autocrine production of IL-1 beta by human osteoarthritis-affected cartilage and differential regulation of endogenous nitric oxide, IL-6, prostaglandin E_2, and IL-8, *Proc. Assoc. Am. Physicians* **110** (1998), 65–72.

[4] D.T. Auble and C.E. Brinckerhoff, The AP-1 sequence is necessary but not sufficient for phorbol ester induction of collagenase in fibroblasts, *Biochemistry* **30** (1991), 4629–4635.

[5] T. Bellido, V.J.C. Borda, P. Roberson and S.C. Manolagas, Activation of the Janus kinase/STAT (signal transducers and activator of transcription), *Endocrinology* **138** (1997), 3666–3676.

[6] H.P. Benton and J.A. Tyler, Inhibition of cartilage proteoglycan synthesis by interleukin-1, *Biochem. Biophys. Res. Commun.* **154** (1988), 421–428.

[7] P.D. Benya, S.A. Padilla and M.E. Nimni, The progeny of rabbit articular chondrocytes synthesize collagen types I and III and type I trimer, but not type II. Verification by cyanogen bromide peptide analysis, *Biochemistry* **16** (1977), 865–872.

[8] M. Boittin, F. Redini, G. Loyau and J.P. Pujol, Effet de la diacerhéine (Art 50) sur la synthèse de la matrice et la sécrétion de la collagénase par les chondrocytes articulaires de lapin en culture, *Rev. Rhum. (Ed. Fr.)* **60** (1993), 68S–76S.

[9] M. Bond, A.H. Baker and A.C. Newby, Nuclear factor kappaB activity is essential for matrix metalloproteinase-1 and -3 upregulation in rabbit dermal fibroblasts, *Biochem. Biophys. Res. Commun.* **264** (1999), 561–567.

[10] J. Bondeson, B. Foxwell, F. Brennan and M. Feldmann, Defining therapeutic targets by using adenovirus: blocking NF-kappaB inhibits both inflammatory mediators, *Proc. Natl. Acad. Sci. USA* **96** (1999), 5668–5673.

[11] K.D. Brandt, G. Smith, S.Y. Kang, S. Myers, B. O'Connor and M. Albrecht, Effects of diacerhein in an accelerated canine model of osteoarthritis, *Osteoarthritis Cartilage* **5** (1997), 438–449.

[12] N. Burton-Wurster, M. Vernier-Singer, T. Farquhar and G. Lust, Effect of compressive loading and unloading on the synthesis of total protein, proteoglycan, and fibronectin by canine cartilage explants, *J. Orthop. Res.* **11** (1993), 717–729.

[13] G. Buttice, S. Quinones and M. Kurkinen, The AP-1 site is required for basal expression but is not necessary for TPA-response of the human stromelysin gene, *Nucleic Acids Res.* **19** (1991), 3723–3731.

[14] M. Cao, A. Westerhausen-Larson, C. Niyibizi et al., Nitric oxide inhibits the synthesis of type-II collagen without altering Col2A1 mRNA abundance: Prolyl hydroxylase as a possible target, *Biochem. J.* **324** (1997), 305–310.

[15] C.T. Chen, N. Burton-Wurster, C. Borden, K. Hueffer, S. Bloom and G. Lust, Chondrocyte necrosis and apoptosis in impact damaged articular cartilage, *J. Orthop. Res.* **19** (2001), 703–711.

[16] P. Chomczynski and N. Sacchi, Single-step method of RNA isolation by acid guanidium thiocyanate–phenol–chloroform extraction, *Anal. Biochem.* **162** (1987), 156–159.

[17] C.C. Clark, B.S. Tolin and C.T. Brighton, The effect of oxygen tension on proteoglycan synthesis and aggregation in mammalian growth plate chondrocytes, *J. Orthoped. Res.* **9** (1991), 477–484.

[18] M.H. Cobb, MAP kinase pathways, *Prog. Biophys. Mol. Biol.* **71** (1999), 479–500.

[19] W. Conca, P.B. Kaplan and S.M. Krane, Increases in the levels of procollagenase mRNA in cultured fibroblasts induced by human recombinant interleukin-1 or serum follow c-jun expression and are dependent on new protein synthesis, *J. Clin. Invest.* **83** (1989), 1753–1758.

[20] W. Conca, P.E. Auron, M. Aonun-Wathne, N. Bennett, P. Seckinger, Welgus et al., An interleukin 1-beta point mutant demonstrates that jun/fos expression is not sufficient for fibroblast metalloproteinase expression, *J. Biol. Chem.* **266** (1991), 16265–16268.

[21] T.F. Cruz, J. Tang, S. Pronost and J.P. Pujol, Mécanismes moléculaires impliqués dans l'inhibition de l'expression de la collagénase par la diacérhéine, *Rev. Prat.* **4** (1996), 515–519.

[22] M. Dougados, M. Nguyen, L. Berdah, B. Mazières, E. Vignon and M. Lequesne, for the ECHODIAH investigators study group, Evaluation of the structure-modifying effects of diacerein in hip osteoarthritis, *Arthritis Rheum.* **44** (2001), 2539–2547.

[23] J. Dudhia, M.T. Bayliss, C.P.D. and Wheeler-Jones, The IL-1α induced p42/p44 mapk response in cartilage is exacerbated in 5% pO_2, in: *XVIIth Meeting of the Federation of European Connective Tissue Societies* (FECTS), Patras, July 2000, p. F32.

[24] W.C. Eberhardt, C. Pliiss, R. Hummel and J. Pfeilschifter, Molecular mechanisms of inducible nitric oxide synthase gene expression by interleukin-1β and cAMP in rat mesangial cells, *J. Immunol.* **160** (1998), 4961–4969.

[25] S. Franchi-Micheli, L. Lavacchi, C.A. Friedmann and L. Ziletti, The influence of rhein on the biosynthesis of prostaglandin-like substances *in vitro*, *J. Pharm. Pharmacol.* **35** (1983), 262–264.

[26] N. Félisaz, K. Boumediene, C. Ghayor, J.F. Herrouin, P. Bogdanowicz, P. Galéra and J.P. Pujol, Stimulating effect of diacerein on TGF-β 1 and β 2 expression in articular chondrocytes cultured with and without interleukin-1, *Osteoarthritis Cartilage* **7** (1999), 255–267.

[27] M.B. Goldring, L. Birkead, T. Kimura and S.M. Krane, Interleukin-1 suppresses expression of cartilage specific types II and IX collagens and increases types I and III collagens in human chondrocytes, *J. Clin. Invest.* **82** (1988), 2026–2037.

[28] M.B. Goldring, L. Tan, B.K. Choy and M. Osaki, Multiple interacting IL-1β-induced transcription factors converge to suppress constitutive *COL2A1* promoter activity, *Osteoarthritis Cartilage* **10** (Suppl. A) (2002), S12.

[29] A.J. Grodzinsky, Y.J. Kim, M.D. Buschmann, A.M. Garcia, T.M. Quinn and E.B. Hunziker, Response of the chondrocyte

to mechanical stimuli, in: *Osteoarthritis*, K.D. Brandt, M. Koherty and L.S. Lohmander, eds, Oxford University Press, Oxford, 1998, pp. 123–136.

[30] G.E. Henderson and R.M. Mason, Effect of oxygen tension on ^{35}S glycosaminoglycan synthesis and UDP-sugar pool size in articular cartilage, *Biochem. Soc. Trans.* **19** (1991), 364S.

[31] S.Y. Hwa, D. Burkhardt, C. Little and P. Ghosh, The effect of orally administered diacerein on cartilage and subchondral bone in an ovine model of osteoarthritis, *J. Rheumatol.* **28** (2001), 825–834.

[32] J.E. Jeffrey, D.W. Gregory and R.M. Aspden, Matrix damage and chondrocyte viability following a single impact load on articular cartilage, *Arch. Biochem. Biophys.* **322** (1995), 87–96.

[33] I. Knott, M. Dieu, M. Burton, A. Houbion, J. Remacle and M. Raes, Induction of cycloxygenase by interleukin-1: comparative study between synovial cells and chondrocytes, *J. Rheumatol.* **21** (1994), 462–466.

[34] U.K. Laemmli, Cleavage of structural proteins during the assembly of the head of bacteriophage T4, *Nature* **227** (1970), 680–685.

[35] T. Larson, R.M. Aspden and D. Heinegard, Effects of mechanical load on cartilage matrix biosynthesis *in vitro*, *Matrix* **11** (1991), 388–394.

[36] A. Levin, N. Burton-Wurster, C.T. Chen and G. Lust, Intercellular signaling as a cause of cell death in cyclically impacted cartilage explants, *Osteoarthritis Cartilage* **9** (2001), 702–711.

[37] A. Liacini, J. Sylvester, W.Q. Li and M. Zafarullah, Inhibition of interleukin-1-stimulated MAP kinases, activating protein-1 (AP-1) and nuclear factor kappa B (NF-κB) transcription factors down-regulates matrix metalloproteinase gene expression in articular chondrocytes, *Matrix Biol.* **21** (2002), 251–262.

[38] M. Lingetti, P.L. D'ambrosio, F. Grezia, P. Sorrentino and E.A. Lingetti, Controlled study in the treatment of osteoarthritis with diacerhein (Artrodar), *Curr. Ther. Res.* **31** (1982), 408–412.

[39] A.M. Loening, I.E. James, M.E. Levenston, A.M. Badger, E.H. Franck, B. Kurz et al., Injurious mechanical compression of bovine articular cartilage induces chondrocyte apoptosis, *Arch. Biochem. Biophys.* **381** (2000), 205–212.

[40] E. Lucchinetti, C.S. Adams, W.E. Horton, Jr and P.A. Torzilli, Cartilage viability after repetitive loading: a preliminary report, *Osteoarthritis Cartilage* **10** (2002), 71–81.

[41] W.J. Lukiw and N.G. Bazan, Strong nuclear factor-κB-DNA binding parallels cyclooxygenase-2 gene transcription in aging and in sporadic Alzheimer's disease superior temporal lobe neocortex, *J. Neurosci. Res.* **53** (1998), 583–592.

[42] R. Maier, G. Bilbe, J. Rediske and M. Lotz, Inducible nitric oxide synthase from human articular chondrocytes: cDNA cloning and analysis of mRNA expression, *Biochim. Biophys. Acta* **1208** (1994), 145–150.

[43] R. Marcolongo, A. Fioravanti, S. Adami, E. Tozzi, M. Mian and A. Zampieri, Efficacy and tolerability of diacerhein in the treatment of osteoarthrosis, *Curr. Ther. Res.* **43** (1988), 878–887.

[44] H. Matsuhura, M. Sakaue, K. Subbaramaiah, H. Kamitani, E.E. Eling, A.J. Dannenberg et al., Regulation of cyclooxygenase-2 by interferon γ and transforming growth factor β in normal human epidermal keratinocytes and squamous carcinoma cells, *J. Biol. Chem.* **274** (1999), 29138–29148.

[45] A.F. Mendes, M.M. Caramona, A.P. Carvalho and M.C. Lopes, Role of mitogen-activated protein kinases and tyrosine kinases on interleukin-1β-induced NF-κB activation and iNOS expression in bovine articular chondrocytes, *Nitric Oxide* **6** (2002), 35–44.

[46] A.F. Mendes, M.M. Caramona, A.P. Carvalho and M.C. Lopes, Diacerhein and rhein prevent interleukin-1β-induced nuclear factor-κB activation by inhibiting the degradation of inhibitor κB-α, *Pharmacology Toxicology* **91** (2002), 22–28.

[47] A.F. Mendes, A.P. Carvalho, M.M. Caramona and M.C. Lopes, Role of nitric oxide in the activation of NF-kB, AP-1 and NO II expression in articular chondrocytes, *Inflamm. Res.* **51** (2002), 369–375.

[48] J.A. Mengshol, M.P. Vincenti, C.L. Coon, A. Barchowsky, C.E. Brinckerhoff, Interleukin-1 induction of collagenase 3 (matrix metalloproteinase 13) gene expression in chondrocytes requires p38, c-Jun N-terminal kinase, and nuclear factor kappaB: differential regulation of collagenase I and collagenase 3, *Arthritis Rheum.* **43** (2000), 801–811.

[49] C.L. Murphy and A. Sambanis, Effect of oxygen tension on chondrocyte extracellular matrix accumulation, *Connect. Tiss. Res.* **42** (2001), 87–96.

[50] M. Nguyen, M. Dougados, L. Berdah and B. Amor, Diacerhein in the treatment of osteoarthritis of the hip, *Arthritis Rheum.* **37** (1994), 529–536.

[51] R.M.J. Palmer, M.S. Hickery, I.G. Charles, S. Moncada and M.T. Bayliss, Induction of nitric oxide synthase in human chondrocytes, *Biochem. Biophys. Res. Commun.* **193** (1993), 398–405.

[52] J.P. Pelletier and J. Martel-Pelletier, Evidence for the involvement of interleukin-1 in human, osteoarthritic cartilage degradation: protective effect of NSAID, *J. Rheumatol.* **16** (1989), 19–27.

[53] J.P. Pelletier, J. Martel-Pelletier and D.S. Howell, Etiopathogenesis of osteoarthritis, in: *Arthritis and Allied Conditions: A Textbook of Rheumatology*, D.J. Mc Carty and W.J. Koopman, eds, Lea and Febiger, Philadelphia, 1996.

[54] J.P. Pelletier, F. Mineau, J.C. Fernandes, N. Duval and J. Martel-Pelletier, Diacerhein and rhein reduce the interleukin 1β stimulated inducible nitric oxide synthesis level and activity while stimulating cyclooxygenase-2 synthesis in human osteoarthritic chondrocytes, *J. Rheumatol.* **25** (1998), 2417–2424.

[55] J.P. Pelletier, M. Yaron, B. Haraoui, P. Cohen, M. Nahir, D. Choquette et al., Efficacy and safety of diacerein in osteoarthritis of the knee: a double-blind, placebo-controlled trial, *Arthritis Rheum.* **43** (2000), 2339–2348.
[56] J.P. Pelletier, J.C. Fernandes, J. Brunet, F. Moldovan, D. Schrier, C. Flory and J. Martel-Pelletier, *In vivo* selective inhibition of mitogen-activated kinase 1/2 in rabbit experimental osteoarthritis is associated with a reduction in the development of structural changes, *Arthritis Rheum.* **48** (2003), 1582–1593.
[57] A.M. Pendas, M. Balbin, E. Llano, M.G. Jimenez and C. Lopez-Otin, Structural analysis and promoter characterization of the human collagenase-3-gene (MMP13), *Genomics* **40** (1997), 222–233.
[58] P. Pomarelli, M. Berti, M.T. Gatti and P. Mosconi, A non-steroidal anti-inflammatory drug that stimulates prostaglandin release, *Farmaco* **35** (1980), 834–842.
[59] J.P. Pujol, M. Brisset, C. Jourdan, J. Bocquet, V. Jouis, R. Béliard et al., Effect of a monocyte cell factor (MCF) on collagen production in articular chondrocytes: Role of prostaglandin E_2, *Biochem. Biophys. Res. Commun.* **119** (1984), 499–508.
[60] J.P. Pujol and G. Loyau, Interleukin-1 and osteoarthritis, *Life Sci.* **41** (1987), 1187–1198.
[61] J.P. Pujol, N. Félisaz, K. Boumediene, C. Ghayor, J.F. Herrouin, P. Bogdanowicz and P. Galéra, Effects of diacerhein on biosynthesis activities of chondrocytes in culture, *Biorheology* **37** (2000), 177–184.
[62] R. Rajpurohit, C.J. Koch, Z. Tao, C.M. Teixeira and I.M. Shapiro, Adaptation of chondrocytes to low oxygen tension: relationships between hypoxia and cellular metabolism, *J. Cell. Physiol.* **168** (1996), 424–432.
[63] F. Rédini, P. Galéra, A. Mauviel, G. Loyau and J.P. Pujol, Transforming growth factor β stimulated collagen and glycosaminoglycan biosynthesis in cultured rabbit chondrocytes, *FEBS Lett.* **234** (1988), 172–176.
[64] F. Rédini, M. Daireaux, A. Mauviel, P. Galéra, G. Loyau and J.P. Pujol, Characterization of proteoglycans synthesis by rabbit articular chondrocytes in response to transforming growth factor-β (TGF-β), *Biochem. Biophys. Acta* **1093** (1991), 196–206.
[65] H.B. Sadowski and M.Z. Gilman, Cell-free activation of a DNA binding protein by epidermal growth factor, *Nature* **362** (1993), 79–83.
[66] H.J. Schaeffer and M. Weber, Mitogen-activated protein kinases: Specific messages from ubiquitous messengers, *Mol. Cell. Biol.* **19** (1999), 2435–2444.
[67] G.N. Smith, S.L. Myers, K.D. Brandt, E.A. Mickler and M.E. Albrecht, Diacerhein treatment reduces the severity of osteoarthritis in the canine cruciate-deficiency model of osteoarthritis, *Arthritis Rheum.* **42** (1999), 545–554.
[68] J. Stadler, M. Stefanovic-Racic, T.R. Billiar, R.D. Curran, L.A. McIntyre, H.I. Georgescu et al., Articular chondrocytes synthesize nitric oxide in response to cytokines and lipopolysaccharide, *J. Immunol.* **147** (1991), 3915–3920.
[69] J. Steinmeyer, B. Ackermann and R.X. Raiss, Intermittent cyclic loading of cartilage explants modulates fibronectin metabolism, *Osteoarthritis Cartilage* **5** (1997), 331–341.
[70] J. Steimeyer, S. Knue, R.X. Raiss and I. Pelzer, Effects of intermittently applied cyclic loading on proteoglycan metabolism and swelling behaviour of articular cartilage explants, *Osteoarthritis Cartilage* **7** (1999), 155–164.
[71] K. Subbaramaiah, W.J. Chung and A.J. Dannenberg, Ceramide regulates the transcription of cyclooxygenase-2, *J. Biol. Chem.* **273** (1998), 32943–32949.
[72] A.L. Taccoen and L. Berdah, Diacetylrhein, a new therapeutic approach of osteoarthritis, *Rev. Rhum. (Ed. Fr.)* **60** (1993), 83S–86S.
[73] T. Tamura and K. Ohmori, Diacerein suppresses the increase in plasma nitric oxide in rat adjuvant-induced arthritis, *Eur. J. Pharmacol.* **419** (2001), 269–274.
[74] P.A. Torzilli and R. Grigiene, Continuous cyclic load reduces proteoglycan release from articular cartilage, *Osteoarthritis Cartilage* **6** (1998), 260–266.
[75] P.A. Torzilli, R. Grigiene, J. Borelli Jr and D.L. Hellet, Effect of impact load on articular cartilage: cell metabolism and viability, and matrix water content, *J. Biomech. Eng.* **121** (1999), 433–441.
[76] J.P. Urban, Present perspectives on cartilage and chondrocyte mechanobiology, *Biorheology* **37** (2000), 185–190.
[77] G.P. Van Kampen, G.H. Korver and R.J. van de Stadt, Modulation of proteoglycan composition in cultured anatomically intact join cartilage by cyclic loads of various magnitudes, *Int. J. Tissue. React.* **16** (1994), 171–179.
[78] Q. Xie, Y. Kashiwabara and C. Nathan, Role of transcription factor NF-kB/Rel in induction of nitric oxide synthase, *J. Bio. Chem.* **269** (1994), 4705–4708.
[79] M. Yaron, I. Shirazi and I. Yaron, Anti-interleukin-1 effects of diacerhein and rhein in human osteoarthritic synovial tissue and cartilage cultures, *Osteoarthritis Cartilage* **7** (1999), 272–280.
[80] G. Ysart and R. Mason, Responses of articular cartilage explant cultures to different oxygen tensions, *Biochem. Biophys. Acta* **1221** (1994), 15–20.

Part VI

Lectures of NEGMA-LERADS prize

Biorheology 41 (2004) 565
IOS Press

Introduction to Negma Lerads Prize 2003 on mechanobiology of cartilage and chondrocyte

I am glad to open this ceremony concerning the Negma Lerads Prize winners. This prize was created last year to stimulate researches in articular cartilage and chondrocyte. The choice of the winners was entirely in the hands of an international scientific panel who juged the scientific merit, the scope and objectives, the methods used, the medical potential of research and the track record in the field. Among the numerous proposal received, two were unanimously selected by the board.

I am glad to introduce the first winner Jane Millward-Sadler. Dr. Millward-Sadler is a young postdoctoral research fellow in the department of Pathology of the University of Edinburgh. She defended her PhD thesis in 1996 at the University of Newcastle upon Tyne. The selected is entitled concern "*human articular chondrocyte signalling pathways activated following mechanical stimulation*". During the last 4 years, Dr. Millward-Sadler published 9 full papers in refereed journals and has given more than 10 presentations at meetings on cartilage and chondrocyte. She also presented a communication during the last symposium in Paris.

The second winner is Clark T. Hung. Clark is Associate Professor of Biomedical Engineering at Columbia University in New York since 2002. He defended his PhD thesis at the University of Philadelphia in 1995. His main interest is in cartilage bioengineering. He is principal investigator in many grants on cartilage physiomechanics and engineering. As did Dr. Millward-Sadler, he had presented a paper during the second symposium in Paris.

I am glad to congratule the two winners. The prize will be hand over to the two winners by Dr. Burger a representative of Negma Lerads, a French pharmaceutical Company.

Congratulations to the two winners and I hope we will receive more proposals for a second prize in two years.

J.F. Stoltz

Biorheology 41 (2004) 567–575
IOS Press

ATP in the mechanotransduction pathway of normal human chondrocytes

S.J. Millward-Sadler [a,*], M.O. Wright [a], P.W. Flatman [b] and D.M. Salter [a]

[a] *Osteoarticular Research Group, Division of Pathology, School of Molecular and Clinical Medicine, College of Medicine and Veterinary Medicine, Edinburgh University Medical School, Teviot Place, Edinburgh, UK*

[b] *Membrane Biology Group, School of Biomedical Sciences, College of Medicine and Veterinary Medicine, Edinburgh University Medical School, Teviot Place, Edinburgh, UK*

Abstract. Extracellular nucleotides have been shown to have diverse effects on chondrocyte function, generally acting via P2 purinoceptors. We have previously shown that mechanical stimulation at 0.33 Hz of normal human chondrocyte cultures causes cellular hyperpolarisation, while chondrocytes derived from osteoarthritic (OA) cartilage depolarise. Experiments have been undertaken to establish whether ATP is involved in the response of the chondrocyte to mechanical stimulation.

Chondrocytes, isolated from normal and OA cartilage obtained, with consent, from human knee joints following surgery, were cultured in non-confluent monolayer. Cells were mechanically stimulated at 0.33 Hz for 20 minutes at 37°C in the presence or absence of inhibitors of ATP signalling, or were stimulated by the addition of exogenous ATP or derivatives, and electrophysiological measurements recorded. Samples of medium bathing the cells were collected before and after mechanical stimulation, and the concentration of ATP in the cell medium was measured. Total RNA was extracted from cultured chondrocytes, reverse-transcribed and used for RT-PCR with primers specific for P2Y2 purinoceptors.

ATP, UTP 2-methylthioadenosine and $\alpha\beta$-methylene adenosine $5'$-triphosphate all induced a hyperpolarisation response in normal human articular chondrocytes. No significant change was observed in the membrane potentials of chondrocytes isolated from OA cartilage following the addition of these nucleotides to the medium. In normal chondrocytes, the hyperpolarisation induced by ATP was blocked by the presence of apamin, indicating the involvement of small-conductance calcium-activated potassium channels. Following mechanical stimulation of normal chondrocytes, an increase was observed in ATP concentration in the cell culture medium bathing the cells. The presence within the culture medium of suramin or pyridoxal-phosphate-6-azophenyl-$2'$,$4'$-disulphonic acid (PPADS) prior to and during the period of mechanical stimulation abolished the hyperpolarisation response in normal chondrocytes. The presence of mRNA for P2Y2 purinoceptors was demonstrated in both normal and OA chondrocytes by RT-PCR.

These results suggest that ATP has a role in the response of normal chondrocytes to mechanical stimulation, via P2Y2 purinoceptors. This response appears to be different in chondrocytes derived from OA cartilage, and may be important in the progression of this disease.

Keywords: Osteoarthritis, mechanical stimulation, signal transduction, purinoceptor

1. Introduction

Articular cartilage covers the ends of long bones in synovial joints, and protects the underlying bone from shearing and compressive forces. Physiological loading results in the maintenance of normal cartilage structure and matrix production, while abnormal loading, whether under- or overloading, leads to proteoglycan depletion and the onset of osteoarthritis (OA) [1,2,29].

*Address for correspondence: Tel.: +44 131 650 2954; Fax: +44 131 650 6528; E-mail: Jane.Sadler@ed.ac.uk.

The mechanisms by which chondrocytes recognise and respond to mechanical stimulation have been the focus of much research, and these mechanisms are beginning to be understood. Mechanical forces *in vivo* lead to the compression of cartilage. This results in a number of physicochemical changes including production of hydrostatic pressure gradients, fluid flow, streaming potentials, alterations in matrix water content, fixed charge density, mobile ion concentrations and changes in pH and osmotic pressure [26, 30]. Recognition of the mechanical signal by the chondrocyte results in the activation of intracellular signalling pathways, leading to the regulation of gene expression and chondrocyte metabolism.

In normal cartilage, chondrocytes exist within individual lacunae, separated from other cells by the pericellular matrix surrounding them, with no evidence of cell processes connecting neighbouring cells. Intercellular communication, therefore, is limited to the diffusion of small molecules through the matrix that can act in an autocrine and paracrine manner within the local environment. Studies within our group have shown that normal chondrocytes, when mechanically stimulated *in vitro*, release substance P and interleukin-4, both of which act in an autocrine/paracrine manner to activate signalling pathways that lead to the upregulation of aggrecan mRNA and concomitant downregulation of matrix metalloproteinase 3 (MMP3) mRNA [18–20]. The response of chondrocytes derived from OA cartilage to the same stimulus involves substance P and an IL1β autocrine/paracrine loop, and does not result in alteration of aggrecan and MMP3 mRNA levels [18,19,21].

Another candidate for paracrine signalling within chondrocytes is ATP. ATP, besides being involved in energy metabolism and synaptic transmission, has also been shown to be involved in autocrine/paracrine signalling in a variety of cell types. It is released from cells not only following cell damage, but also by a variety of non-lytic pathways such as transmembrane transport or exocytosis [3,11,22]. Following release, ATP binds to specific plasma membrane receptors, and is quickly degraded, thereby ending its actions. The receptors for the nucleotide can be classified according to whether they are ligand-gated (P2X) or G-protein-coupled (P2Y), and can be functionally identified using analogues of ATP and also pharmacological agents [9]. Compressive loading of bovine chondrocytes in chondrons or in agarose pellets leads to the release of ATP [10,13], and human chondrocytes have been shown to express both P1 (adenosine) and P2 (purine) receptors by RT-PCR [16]. Exogenous ATP has been shown to increase proteoglycan synthesis from bovine chondrocyte pellets, and to increase intracellular calcium levels in a dose-dependent manner [8,14,15].

The aim of this study was to identify whether ATP-purinoceptor signalling was involved in the mechanotransduction pathways activated in response to mechanical stimulation at 0.33 Hz in human articular chondrocytes isolated from both normal and OA cartilage.

2. Materials and methods

2.1. Isolation and culture of chondrocytes

Human articular cartilage was obtained, with patients consent, at operation from knee joint arthroplasty and limb resection specimens. Joints were opened and graded macroscopically for the presence or absence of OA using the Collins/McElligott system [7]. Articular cartilage was removed from either macroscopically normal or abnormal parts of the joint for isolation of chondrocytes by enzymatic release [31]. Isolated cells were seeded in Iscoves modified Dulbeccos medium supplemented with 10% FCS, 2 mM L-glutamine, 100 IU/ml penicillin and 100 μg/ml streptomycin, at a concentration of 5×10^4 cells/ml into 58 mm diameter tissue culture dishes (NUNC). Primary, non-confluent cultures of chondrocytes were used in all experiments.

Morphologically, the cells from both normal and OA cartilage were typically flattened with a polygonal cell shape and did not show the fibroblastic appearance of dedifferentiated chondrocytes. In addition, RT-PCR and immunohistochemistry showed expression of cartilage specific molecules, including type II collagen and aggrecan [31].

2.2. *RNA extraction of chondrocytes*

Total RNA was extracted from cultured chondrocytes as described in the Micro RNA Isolation Kit (Stratagene, Amsterdam, Netherlands), using a denaturing buffer of 4 M guanidine thiocyanate, 0.75 M sodium citrate, 10% (w/v) lauroyl sarcosine and 7.2 μl/ml β-mercaptoethanol. The quantity of RNA isolated was determined by the absorbance reading at 260 nm on a spectrophotometer (Genequant; Amersham Biosciences, Chalfont St Giles, UK).

2.3. *RT-PCR and gel analysis for expression of P2Y2 mRNA*

Template cDNA was synthesised using 0.5 μg RNA, Superscript II (Invitrogen Ltd, Paisley, UK) and oligo dT [12–18] (Amersham Biosciences) according to the manufacturers instructions. The primers used for the PCR reactions for P2Y2 purinoceptor [4] were:

sense: 5′-CGTCATCCTTGTCTGTTACGTGCT-3′,
antisense: 5′-CTACAGCCGAATGTCCTTAGTG-3′.

A typical 20 μl PCR reaction contained 20 mM ammonium sulphate, 75 mM Tris-HCl, pH 8.8, 0.01% (v/v) Tween-20, 1.5 mM $MgCl_2$, 1 μM of each primer, 2 μl cDNA, 100 μM dNTPs, 0.1% (w/v) bovine serum albumin, 0.25 units Taq polymerase (Biogene, Kimbolton, UK). The PCR cycle programme was: 94°C for 3 minutes; 40 cycles of 94°C for 1 minute, 60°C for 1 minute, 72°C for 1 minute 30 seconds. PCR products were analysed by electrophoresis using a 1% (w/v) agarose gel stained with ethidium bromide.

2.4. *Mechanical stimulation and electrophysiological recording*

The technique used has previously been described in detail [31]. In brief, flexible plastic tissue culture dishes were placed in a sealed pressure chamber with inlet and outlet ports. The chamber was pressurised with nitrogen gas from a cylinder, the frequency of pressurisation being dictated by an electronic timer controlling the inlet and outlet valves. The standard regime of cyclical pressurisation used in this study consisted of pressure pulses of 16 kPa above atmospheric pressure, at a frequency of 0.33 Hz (2 seconds on, 1 second off) for 20 minutes. Mechanical stimulation in this system has been shown to induce deformation and strain on the base of the plastic tissue culture dish and its adherent cells [31], a pressure of 1 atmosphere resulting in 3700 microstrain on the base of the dish.

Membrane potentials of cells were recorded using a single electrode bridge circuit and calibrator as previously described [27]. Microelectrodes with tip resistances of 40 to 60 MΩ and tip potentials of approximately 3 mV were used to impale the cells. Membrane potentials of isolated cells were measured, and results were accepted if, on cell impalement, there was a rapid change in voltage to the membrane potential level that remained constant for at least 60 s. Experiments were performed at 37°C. The membrane potentials of 5–10 cells were measured prior to and following the addition of nucleotides or inhibitors, and also following the period of mechanical stimulation.

2.5. Experimental protocol

The medium covering the chondrocytes was replaced with serum-free medium for a minimum of 30 minutes prior to the control readings of membrane potentials being recorded. In separate experiments, 100 μM ATP, UTP (Roche Diagnostics Ltd, Lewes, UK), 2-methylthioATP or $\alpha\beta$-methylene ATP was added to the medium for 5 minutes prior to the membrane potential of a further 5–10 cells being recorded. The reagent was still in contact with the cells when the post-stimulation membrane potentials were being recorded. In a separate experiment, normal chondrocytes were incubated with apamin (4.9 μM), a blocker of SK channels, for 10 minutes prior to the addition of ATP. For mechanical stimulation experiments, the medium was replaced with serum-free medium as before, and the membrane potentials of 5–10 cells were recorded. The chondrocytes were then mechanically stimulated for 20 minutes at 0.33 Hz, and the membrane potentials of a further 5–10 cells were recorded post-stimulation. In separate experiments, normal chondrocytes were incubated with suramin (100 μM) or pyridoxal-phosphate-6-azophenyl-2′,4′-disulphonic acid (PPADS; 30 μM), known antagonists of P2 receptors, for 10 minutes prior to mechanical stimulation or addition of 10 pg/ml recombinant interleukin-4 (R&D Systems Ltd, Abingdon, UK).

All nucleotides and inhibitors were obtained from RBI, Sigma, Poole, UK unless otherwise stated.

2.6. ATP measurements

Cell culture medium (1 ml) was collected from dishes of normal chondrocytes prior to and immediately following the period of mechanical stimulation, and 5 mM EDTA was immediately added to slow the hydrolysis of ATP. The ATP concentration was then determined in a 10 μl sample according to the method of Brown [5] using the luciferin/luciferase assay (LKB, Finland) with and without deproteination of the sample. The sample was deproteinised with perchloric acid followed by neutralisation with potassium hydroxide. This latter refinement was found to be unnecessary. ATP standards were obtained from Roche Biochemicals.

2.7. Statistics

The mean, standard deviation, and standard error of the mean were calculated for each set of data. In the statistical comparison of means, when the F ratio of the variances of the two means reached significance, the non-parametric Mann–Whitney U test was performed to determine statistical significance. If the F ratio did not reach significance, the Student t-test was used.

3. Results

3.1. Human articular chondrocytes express P2Y2 receptors

Reverse-transcriptase-PCR (RT-PCR) on total RNA isolated from both normal and OA primary cultured chondrocytes, using primers specific for the purinoceptor P2Y2 [4], resulted in amplification of a 483 bp region of DNA (Fig. 1), corresponding to 881–1379 bp of the P2Y2 receptor (Genbank Accession number U07225).

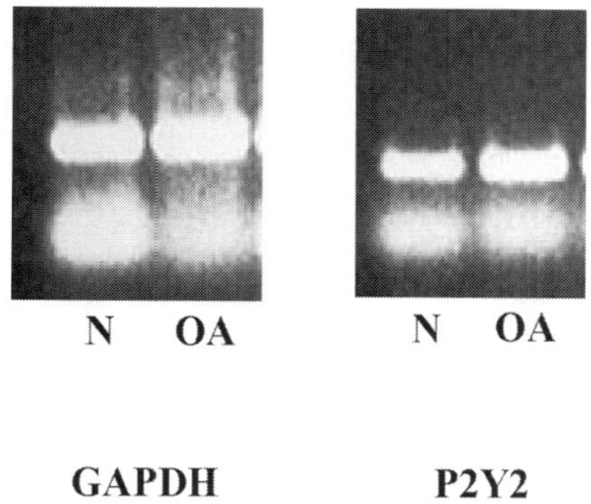

Fig. 1. Expression of the ATP receptor P2Y2 in human articular chondrocytes. RT-PCR was carried out on RNA extracted from primary cultures of normal (lanes 1 and 3) and osteoarthritic (lanes 2 and 4) human articular chondrocytes using primers specific for GAPDH (lanes 1 and 2) or the P2Y2 purine receptor (lanes 3 and 4).

Table 1

Effect of the addition of ATP or its analogues on the membrane potentials of normal and OA chondrocytes. Membrane potentials of chondrocytes were measured prior to and following addition of the agent (100 μM ATP, ADP, UTP or $\alpha\beta$-methylene ATP). Membrane potentials of normal chondrocytes were measured prior to and following the addition of 100 μM ATP in the presence or absence of 4.9 μM apamin. Values are mean $\pm$ SEM, $n = 3$

Peptide	Chondrocyte	Inhibitor	Membrane potential reading ($\pm$ SEM)			% change	P value
(100 μM)	type		resting	with inhibitor	post-stimulation		
ATP	Normal		26.8 $\pm$ 1.1		38.2 $\pm$ 1.2	43	<0.0001
	Normal	Apamin (4.9 μM)	28.0 $\pm$ 0.9	27.6 $\pm$ 1.7	27.8 $\pm$ 0.6		ns
	Osteoarthritic		36.4 $\pm$ 0.62		36.1 $\pm$ 0.45		ns
UTP	Normal		25.6 $\pm$ 0.6		31.8 $\pm$ 1.16	24	0.0014
	Osteoarthritic		35.9 $\pm$ 0.50		37.3 $\pm$ 0.52		ns
2meSATP	Normal		23.0 $\pm$ 1.58		30.6 $\pm$ 1.6	33	0.0097
	Osteoarthritic		nd		nd		
$\alpha\beta$meATP	Normal		23.6 $\pm$ 0.93		27.4 $\pm$ 1.17	16	0.03
	Osteoarthritic		nd		nd		

ns: not significant; nd: not determined.

3.2. Effect of ATP and analogues on chondrocyte membrane potential

Following addition of 100 μM ATP, normal chondrocytes hyperpolarised by 29%, and this hyperpolarisation response was blocked by the presence of apamin (4.9 μM), an inhibitor of SK channels (Table 1). In separate experiments, the addition of 100 μM 2-methylthioATP, UTP or $\alpha\beta$-methylene ATP caused normal human articular chondrocytes to hyperpolarise by 33%, 25% and 16%, respectively. In the case of chondrocytes isolated from OA cartilage, the addition of 100 μM ATP or UTP did not cause a significant change in the membrane potential of the cells.

3.3. ATP is involved in the response of normal chondrocytes to mechanical stimulation at 0.33 Hz

When normal chondrocytes were mechanically stimulated for 20 minutes at 0.33 Hz, the cells hyperpolarised by 61%. Addition of recombinant IL4 to the culture medium caused normal chondrocytes to hyperpolarise by 46%. Incubation of chondrocytes derived from normal cartilage with either 30 μM PPADS or 100 μM suramin (both P2 purinoceptor antagonists) prevented the hyperpolarisation response

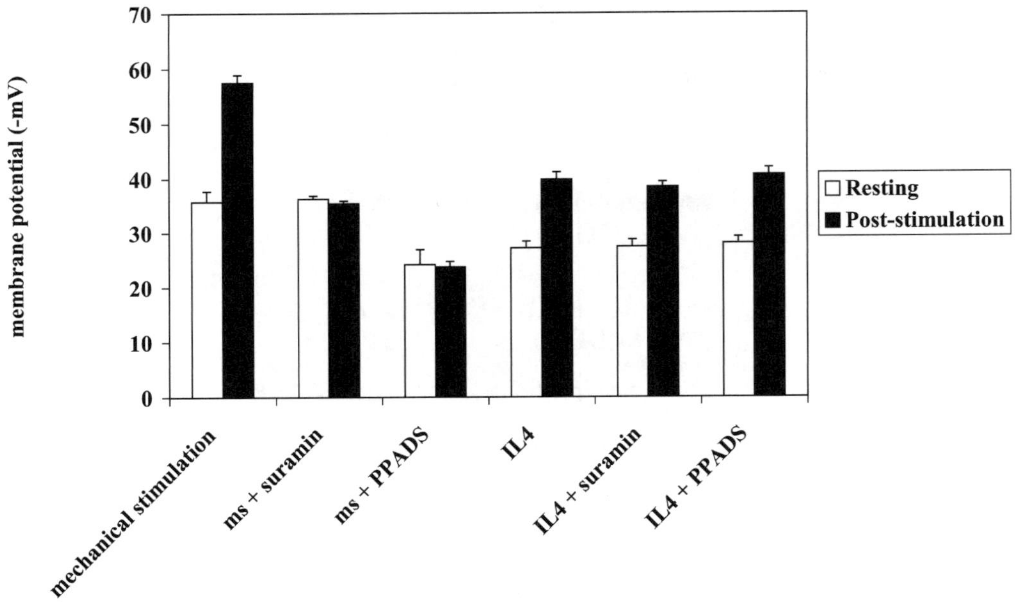

Fig. 2. Effect of inhibitors of ATP signalling on the response of normal chondrocytes to mechanical or IL4 stimulation. Chondrocytes were mechanically stimulated or stimulated by the addition of 10 pg/ml recombinant IL4 in the presence or absence of suramin (100 μM) or pyridoxal-phosphate-6-azophenyl-2′,4′-disulphonic acid (PPADS; 30 μM) to determine the effect of these inhibitors on the cell membrane hyperpolarisation response. Error bars represent $\pm$ SEM, $n = 3$.

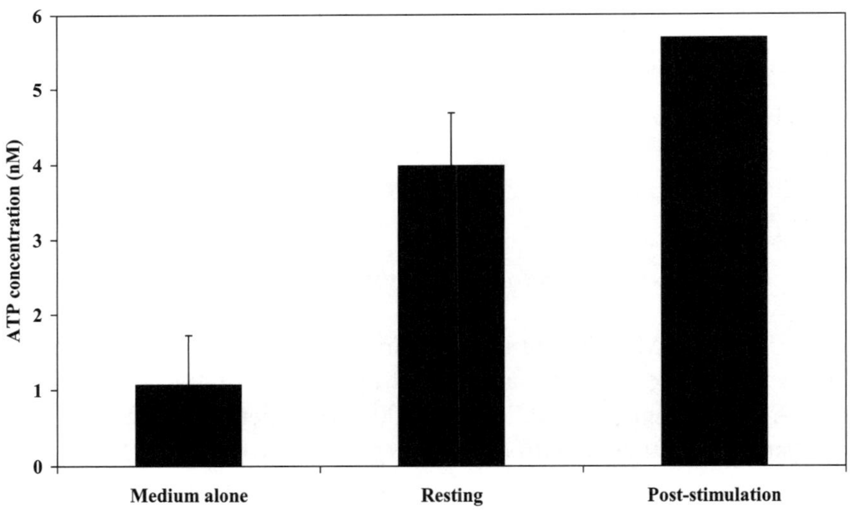

Fig. 3. Effect of mechanical stimulation on the concentration of ATP in the cell culture medium. ATP concentration in the cell culture medium was measured prior to and following a period of 20 minutes mechanical stimulation at 0.33 Hz. Error bars represent $\pm$ SEM, $n = 3$.

to mechanical stimulation (Fig. 2). However, the presence of PPADS or suramin had no effect on the hyperpolarisation caused by the addition of IL4 (Fig. 2). Following mechanical stimulation, the ATP concentration in the cell culture medium bathing the chondrocytes rose by 43% (Fig. 3).

4. Discussion

Mammalian chondrocytes in culture have been shown to express purinoceptors, and release ATP following mechanical stimulation [10,13,16]. In this study we show that chondrocytes derived from both normal and OA human cartilage express the mRNA for P2Y2 receptors, and that normal chondrocytes release ATP into the culture medium following mechanical stimulation. Further, the presence of suramin and PPADS in the culture medium (both antagonists of P2 receptor) inhibit the electrophysiological response of normal chondrocytes to mechanical stimulation but not to the addition of interleukin-4 (IL4). Normal chondrocytes respond to ATP, UTP, α, β-methylene ATP and methylthio ATP, and the response to ATP can be inhibited by apamin, an SK channel blocker. The addition of ATP or UTP to chondrocytes derived from OA cartilage had no effect on the cell membrane potential, despite the cells expressing mRNA for the P2Y2 receptor.

ATP acts through purinoceptors to activate downstream signalling pathways. These receptors are of two main types: ligand-gated, or P2X receptors, and G-protein coupled or P2Y receptors. The type of receptor involved in ATP signalling can be classified according to pharmacological data. The greater effect of ATP and UTP over $\alpha\beta$-methylene ATP on the membrane potential response of normal chondrocytes suggests that the nucleotides are acting on P2Y receptors [23,32], and the mRNA for P2Y2 purinoceptors has been shown to be expressed by chondrocytes isolated from both normal and OA cartilage. The observation that suramin blocks the electrophysiological response to mechanical stimulation also supports P2Y2 activation. However, although suramin is widely used as a P2Y antagonist, it does not completely discriminate between P2X and P2Y receptors [24]. The inhibition of the membrane hyperpolarisation response by PPADS following mechanical stimulation of normal chondrocytes is also not consistent with P2Y2 receptor activation, as PPADS is a non-competitive inhibitor of P2X receptors [17]. If signalling was solely through P2Y2 receptors, an equipotent rather than differential membrane response to ATP and UTP would be expected [23,32]. The problem with clearly identifying which type of purinoceptor is activated following stimulation is that there are currently no antagonists that are specific to a single subtype of P2 receptor. This means that it is necessary to use a range of different pharmacological agonists and antagonists in an attempt to elucidate which receptor type is being stimulated, due to the gradient of responses found against the different chemical inhibitors utilised. The results presented here would suggest that ATP may signal through more than one receptor type in chondrocytes, and it may not be solely a P2Y2 receptor that is activated following mechanical stimulation. This could explain why the response to ATP is lost in chondrocytes isolated from OA cartilage despite the expression of P2Y2 mRNA by these cells. Alternatively, the chondrocytes from OA cartilage may be desensitised to ATP signalling as ATP is found in the synovial fluid of both rheumatoid and OA patients, probably as a result of cell death [25]. Despite the presence of mRNA in the chondrocyte there may be little expression of the P2Y2 receptor at the cell surface in OA chondrocytes due to receptor down-regulation. ATP or UTP binding to the P2Y2 receptor results in the internalisation of the receptor–ligand complex. Under normal conditions the ligand is degraded and the receptor is rapidly recycled to the cell surface. However, epitope-tagging studies have shown that with long-term agonist stimulation, cell-surface receptor expression is dependent on *de novo* protein synthesis [28]. Alternatively, P2Y2 receptors may be desensitised by phosphorylation of the intracellular portion of the receptor. Increasing concentrations of UTP have been shown to induce P2Y2 receptor desensitisation, as does kinase activation by phorbol esters [12], suggesting that phosphorylation of the intracellular portion of the receptor is important in reversible desensitisation. Further studies would need to be performed to determine the expression and phosphorylation state of receptor protein in both normal and OA chondrocytes, and to investigate the possibility of expression of other purinoceptor subtypes in human articular chondrocytes.

We have previously studied the effect of mechanical stimulation at 0.33 Hz on normal and OA chondrocytes and have demonstrated that there is an upregulation of aggrecan and a decrease in MMP3 mRNA following this stimulus in normal chondrocytes, which is IL4-mediated. This response to mechanical stimulation is lost in cells isolated from OA cartilage [19]. This study suggests that ATP is involved in the mechanotransduction pathways that lead to this anabolic response in normal chondrocytes, and acts upstream of the release of IL4. It seems likely that on release of ATP following mechanical stimulation, the ATP binds to the cell surface purinoceptors, acting in an autocrine/paracrine manner, to activate further intracellular signalling pathways, leading to the upregulation of matrix production, and a decrease in metalloproteinase activity. This would be consistent with the observation that ATP is chondroprotective in normal cartilage, decreasing IL1-stimulated proteoglycan loss [6]. The finding that OA chondrocytes are no longer responsive to ATP-mediated signalling may be significant in the onset and progression of the disease. We have previously shown that the transduction pathway in OA chondrocytes differs from that of normal cells [21]. It may be that the signalling pathway in OA chondrocytes has already deviated from that activated in normal chondrocytes prior to the release of ATP following mechanical stimulation. Further work is required to elucidate ATP signalling and mechanotransduction mechanisms in articular cartilage, to answer such questions.

References

[1] M.E. Adams and K. Brandt, Hypertrophic repair of canine articular cartilage in osteoarthritis after anterior cruciate ligament transection, *J. Rheumatol.* **18** (1991), 428–435.

[2] F. Behrens, E.L. Kraft and T.R. Oegema Jr, Biochemical changes in articular cartilage after joint immobilization by casting or external fixation, *J. Orthop. Res.* **7** (1989), 335–343.

[3] M.R. Boarder and S.M.O. Hourani, The regulation of vascular function by P2 receptors: multiple sites and multiple receptors, *TIPS* **19** (1998), 99–107.

[4] W.B. Bowler, M.A. Birch, J.A. Gallagher and G. Bilbe, Identification and cloning of human P2U purinoceptor present in osteoclastoma, bone and osteoblasts, *J. Bone Min. Res.* **10** (1995), 1137–1145.

[5] Brown, ATP and ATPase determinants in red blood cells, in: *Red Cell Membranes – A Methodological Approach*, J.C. Ellory and J.D. Young, eds, Academic Press, London, 1982, pp. 223–238.

[6] C.J. Brown, A.M. Caswell, S. Rahman, R.G. Russell and D.J. Buttle, Proteoglycan breakdown from bovine nasal cartilage is increased, and from articular cartilage is decreased, by extracellular ATP, *Biochim. Biophys. Acta* **1362** (1997), 208–220.

[7] D.H. Collins and T.F. McElligott, Sulphate uptake by chondrocytes in relation to histological changes in osteoarthritic human articular cartilage, *Ann. Rheum. Dis.* **19** (1960), 318–330.

[8] L.J. Croucher, A. Crawford, P.V. Hatton, R.G.G. Russell and D.J. Buttle, Extracellular ATP and UTP stimulate cartilage proteoglycan and collagen accumulation in bovine articular chondrocyte pellet cultures, *Biochim. Biophys. Acta* **1502** (2000), 297–306.

[9] G.R. Dubyak and C. el-Moatassim, Signal transduction via P2-purinergic receptors for extracellular ATP and other nucleotides, *Am. J. Physiol.* **265** (1993), C577–606.

[10] M.K. Elfervig, R.D. Graff, G.M. Lee, S.S. Kelley, A. Sood and A.J. Banes, ATP induces Ca^{2+} signalling in human chondrons cultured in three-dimensional agarose films, *OA Cart.* **9** (2001), 518–526.

[11] D. Ferrari, P. Chiozzi, S. Falzoni, S. Hanau and F. Di Virgilio, Purinergic modulation of interleukin-1β release from microglial cells stimulated with bacterial endotoxin, *J. Exp. Med.* **185** (1997), 579–582.

[12] R.C. Garrad, M.A. Otero, L. Erb, P.M. Theiss, L.L. Clarke, F.A. Gonzalez, J.T. Turner and G.A. Weisman, Structural basis of agonist-induced desensitization and sequestration of the P2Y2 nucleotide receptor, *J. Biol. Chem.* **278** (1998), 29437–29444.

[13] R.D. Graff, E.R. Lazarowski, A.J. Banes and G.M. Lee, ATP release by mechanically loaded porcine chondrons in pellet culture, *Arthritis Rheum.* **43** (2000), 1571–1579.

[14] A.D. Kaplan, D.M. Kilkenny, D.J. Hill and S.J. Dixon, Extracellular nucleotides act through P2U purinoceptors to elevate $[Ca^{2+}]_i$ and enhance basic fibroblast growth factor-induced proliferation in sheep chondrocytes, *Endocrinology* **137** (1996), 4757–4766.

[15] M. Koolpe and H.P. Benton, Calcium-mobilising purine receptors on the surface of mammalian articular chondrocytes, *J. Orthop. Res.* **15** (1997), 204–212.
[16] M. Koolpe, D. Pearson and H.P. Benton, Expression of both P1 and P2 purine receptor genes by human articular chondrocytes and profile of ligand-mediated prostaglandin E2 release, *Arthritis Rheum.* **42** (1999), 258–267.
[17] G. Lambrecht, T. Friebe, U. Grimm, U. Windscheif, E. Bungardt, C. Hildebrandt, H.G. Baumert, G. Spatz-Kumbel and E. Mutschler, PPADS, a novel functionally selective antagonist of P2 purinoceptor-mediated responses, *Eur. J. Pharmacol.* **217** (1992), 217–219.
[18] S.J. Millward-Sadler, A. Mackenzie, M.O Wright, H.-S. Lee, K. Elliot, L. Gerrard, C.E. Fiskerstrand, D.M. Salter and J.P. Quinn, Tachykinin expression in cartilage and function in human articular chondrocyte mechanotransduction, *Arthritis Rheum.* **48** (2003), 146–156.
[19] S.J. Millward-Sadler, M.O. Wright, L.W. Davies, G. Nuki and D.M. Salter, Mechanotransduction via integrins and interleukin-4 results in altered aggrecan and matrix metalloproteinase 3 gene expression in normal but not osteoarthritic human articular chondrocytes, *Arthritis Rheum.* **43** (2000), 2091–2099.
[20] S.J. Millward-Sadler, M.O. Wright, K. Nishida, H. Caldwell, G. Nuki and D.M. Salter, Integrin-regulated secretion of interleukin-4: a novel pathway of mechanotransduction in human articular chondrocytes, *J. Cell Biol.* **145** (1999), 183–189.
[21] S.J. Millward-Sadler, M.O. Wright, H.-S. Lee, H. Caldwell, G. Nuki and D.M. Salter, Altered electrophysiological responses to mechanical stimulation and abnormal signalling through alpha5beta1 integrin in chondrocytes from osteoarthritic cartilage, *OA Cart.* **8** (2000), 272–78.
[22] C.H. Mitchell, D.A. Carre, A.M. McGlinn, R.A. Stone and M.M. Civan, A release mechanism for stored ATP in ocular ciliary epithelial cells, *Proc. Natl. Acad. Sci. USA* **95** (1998), 7174–7178.
[23] D.M. Morse, J.L. Smullen and C.W. Davies, Differential effects of UTP, ATP and adenosine on ciliary activity of human nasal epithelial cells, *Am. J. Physiol.* **280** (2001), C1485–1497.
[24] V. Ravelic and G. Burnstock, Receptors for purines and pyrimidines, *Pharmacol. Rev.* **50** (1998), 413–492.
[25] L.M. Ryan, J.W. Rachow and D.J. McCarty, Synovial fluid ATP: a potential substarte for the production of inorganic pyrophosphate, *J. Rheumatol.* **18** (1991), 716–720.
[26] R.L. Sah, Y.J. Kim, J.Y. Doong, A.J. Grodzinsky, A.H. Plaas and J.D. Sandy, Biosynthetic response of cartilage explants to dynamic compression, *J. Orthop. Res.* **7** (1989), 619–636.
[27] D.M. Salter, J.E. Robb and M.O. Wright, Electrophysiological responses of human bone cells to mechanical stimulation: Evidence for specific integrin function in mechanotransduction, *J. Bone Miner. Res.* **12** (1997), 1133–1141.
[28] S.M. Sromek and T.K. Harden, Agonist-induced internalization of the P2Y2 receptor, *Mol. Pharmacol.* **54** (1998), 485–94.
[29] M. Tammi, A.M. Saamanen, A. Jauhiainen, O. Malminen, I. Kiviranta and H. Helminen, Proteoglycan alterations in rabbit knee articular cartilage following physical exercise and immobilization, *Connect. Tiss. Res.* **11** (1983), 45–55.
[30] J.P.G. Urban, The chondrocyte: A cell under pressure, *Br. J. Rheumatol.* **33** (1994), 901–908.
[31] M.O. Wright, P. Jobanputra, C. Bavington, D.M. Salter and G. Nuki, The effects of intermittent pressurisation on the electrophysiology of cultured human articular chondrocytes: evidence for the presence of stretch-activated membrane ion channels, *Clin. Sci.* **90** (1996), 61–71.
[32] M. Yamaya, K. Sekizawa, Y. Kakuta, T. Ohrui, T. Sawai and H. Sasaki, P2U-purinoceptor regulation of chloride secretion in cultured human tracheal submucosal glands, *Am. J. Physiol.* **270** (1996), L979–984.

Biorheology 41 (2004) 577–590
IOS Press

Functional tissue engineering of chondral and osteochondral constructs

Eric G. Lima [a], Robert L. Mauck [a,*], Shelley H. Han [a], Seonghun Park [b], Kenneth W. Ng [a], Gerard A. Ateshian [a,b] and Clark T. Hung [a,**]

[a] *Department of Biomedical Engineering, Columbia University, New York, NY, USA*
[b] *Department of Mechanical Engineering, Columbia University, New York, NY, USA*

Abstract. Due to the prevalence of osteoarthritis (OA) and damage to articular cartilage, coupled with the poor intrinsic healing capacity of this avascular connective tissue, there is a great demand for an articular cartilage substitute. As the bearing material of diarthrodial joints, articular cartilage has remarkable functional properties that have been difficult to reproduce in tissue-engineered constructs. We have previously demonstrated that by using a functional tissue engineering approach that incorporates mechanical loading into the long-term culture environment, one can enhance the development of mechanical properties in chondrocyte-seeded agarose constructs. As these gel constructs begin to achieve material properties similar to that of the native tissue, however, new challenges arise, including integration of the construct with the underlying native bone. To address this issue, we have developed a technique for producing gel constructs integrated into an underlying bony substrate. These osteochondral constructs develop cartilage-like extracellular matrix and material properties over time in free swelling culture. In this study, as a preliminary to loading such osteochondral constructs, finite element modeling (FEM) was used to predict the spatial and temporal stress, strain, and fluid flow fields within constructs subjected to dynamic deformational loading. The results of these models suggest that while chondral ("gel alone") constructs see a largely homogenous field of mechanical signals, osteochondral ("gel bone") constructs see a largely inhomogeneous distribution of mechanical signals. Such inhomogeneity in the mechanical environment may aid in the development of inhomogeneity in the engineered osteochondral constructs. Together with experimental observations, we anticipate that such modeling efforts will provide direction for our efforts aimed at the optimization of applied physical forces for the functional tissue engineering of an osteochondral articular cartilage substitute.

Keywords: Functional tissue engineering, deformational loading, articular cartilage, osteochondral constructs, finite element models

1. Introduction

Damage to articular cartilage is a common condition affecting the joints of millions of people in the U.S. alone. This damage is complicated by the poor regenerative capacity of adult articular cartilage and the disability and pain that accompanies these injuries [30]. Nearly 10% of the U.S. population aged 30 and older has clinical osteoarthritis (OA) of the hip or knee, with total direct costs estimated at $28.6 billion dollars per year [15]. There exists a range of clinical options (short of total joint replacement) with variable degrees of success, for the repair of focal lesions and damage to the articular surface. These

[*] Present address: Cartilage Biology and Orthopaedics Branch, National Institute of Arthritis, Musculoskeletal & Skin Diseases, NIH, Bethesda, MD, USA.

[**] Address for correspondence: Clark T. Hung, PhD, Columbia University, Department of Biomedical Engineering, 351 Engineering Terrace, MC8904, 1210 Amsterdam Avenue, New York, NY 10027, USA. Tel.: +1 212 854 6542; Fax: +1 212 854 8725; E-mail: cth6@columbia.edu.

approaches include tissue adhesives (e.g., [3,21]), enzymatic treatments [13], laser solder welding [61], autograft cell/tissue transfer via periosteal grafts [41], autologous osteochondral grafting such as mosaicplasty [20] and the Carticel method [8]. While promising in many respects, these approaches are limited by the amount of tissue available for such procedures, as well as donor site morbidity associated with its harvest. The poor intrinsic healing capacity along with the lack of effective clinical repair strategies has generated great interest in the engineering of replacement cartilage tissues.

The most prevalent approach to cartilage tissue engineering involves the combination of chondrocytes (or mesenchymal progenitor cells) with scaffolds (fibrous meshes or hydrogels) for long-term culture. Such constructs have been grown both statically (i.e., in Petri dishes or in the backs of nude mice) and in sophisticated bioreactors designed to provide a nutrient rich environment for construct development [2,14,16,17,43,45,51,52]. In our studies, the underlying premise has been that the application of physiologic deformational loading, as part of the culture environment, will promote development of cartilage tissue engineered constructs having functional load-bearing tissue properties. The rationale for such an approach is based on the well-characterized response of chondrocytes to a variety of physical stimuli that arise with loading of diarthrodial joints (see review in [18]). More specifically, we have used the application of uni-axial unconfined dynamic deformational loading (1 Hz, 10% peak-to-peak deformation) of chondrocyte-seeded agarose constructs to encourage tissue growth *in vitro*. This mode of deformational loading was chosen as it engenders mechanical signals in gels similar to that seen in the native tissue with joint contact (Fig. 1, [44]). To carry out such studies, we employed agarose hydrogels, which not only provide a 3D environment that maintains the chondrocyte phenotype and allows for the development of a functional extracellular matrix [7,9,10], but also can withstand deformational loading from the outset of encapsulation, without any significant pre-incubation period. With time in culture, chondrocyte-seeded agarose hydrogels increase in tissue properties and respond to short-term applied deformational loading in a manner similar to native tissue [11,26,27]. Furthermore, long-term deformational loading of these constructs increases the rate and extent of development of tissue properties compared to free swelling controls [31–35]. As these engineered constructs approach the mechanical and biochemical properties of the native tissue, significant new challenges become apparent [4,42]. In particular, integration of these constructs into the damaged articular surface remains a difficult problem [4,42,60].

One possible solution to the problem of integration may be the development of engineered osteochondral constructs [5,24,46–48] similar to those used for autologous transplantation [1,20]. A number of different approaches have explored the generation of such constructs, employing a diverse spectrum of scaffolds and cell populations [5,24,46–48]. In a previous study, we suggested the production of osteochondral constructs composed of a cell-seeded agarose layer integrated with a trabecular subchondral bony layer, with the hypothesis that the bone layer would provide a suitable interface for *in vivo* bony ingrowth [24]. Osteochondral constructs of this composition (seeded with primary bovine chondrocytes) were shown to increase in biochemical and mechanical properties with time in free swelling culture [24], indicating that the culture method is conducive to cartilage-specific tissue growth.

Dynamic deformational loading has been shown to improve the mechanical properties of chondrocyte-seeded agarose constructs. This approach may similarly prove useful for the culture of osteochondral constructs. A number of interesting issues arise, however, from the intrinsic differences in the stiffness of each layer in such a bilayered construct. First, the trabecular bone region of such constructs is expected to be much stiffer than the region comprised totally of agarose. Consequently one would expect all deformation to occur solely in the agarose gel region. Secondly, the gel is at least partially

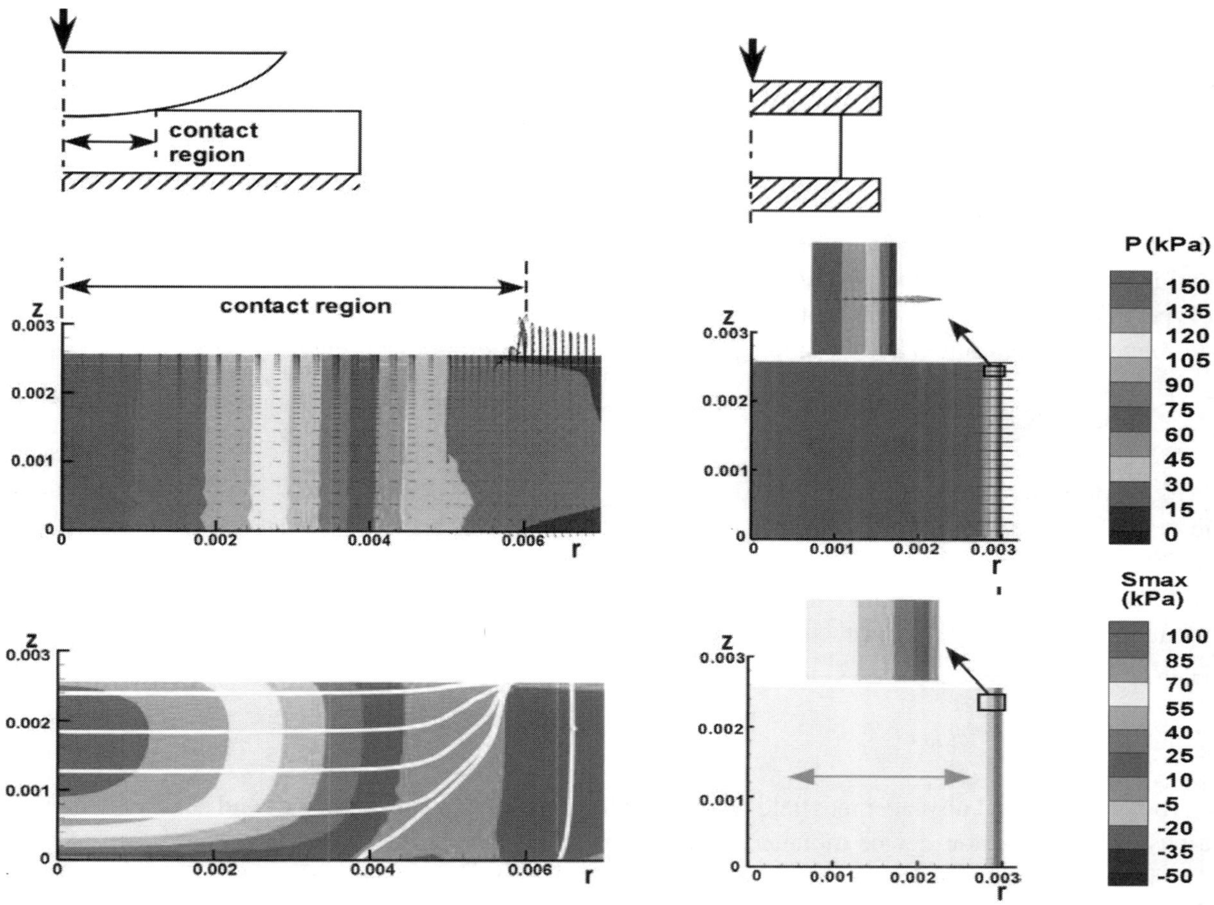

Fig. 1. Finite element prediction of fluid pressure and fluid flow (top) and maximum stress (bottom) occurring within an isotropic articular cartilage layer loaded in contact (left) and a cartilage disk loaded in unconfined compression. Note the similarity between the two loading conditions in mechanical signals generated near the point of contact. Adapted from [44].

infused into the bone along the gel bone interface. Such fixation would be expected to restrict the lateral expansion of the gel layer with axial deformation, and control the movement of fluid through these regions of differing permeability. Given these differences in boundary conditions brought about by the inclusion of a bony layer, the goals of the current study were three-fold. First, we were interested in examining the bulk mechanical properties of acellular agarose constructs with and without a bony layer to determine if inclusion of a bony layer would alter the overall mechanical behavior. Second, we tested the hypothesis that deformation would occur solely within the gel layer of osteochondral constructs. Third, we used biphasic finite element models to compare the differing mechanical stimuli that are predicted to arise with dynamic deformational loading of osteochondral constructs compared to those that occur in gel alone constructs in response to the same mechanical loading protocol. To gain an appreciation of the implication of these spatially-ranging mechanical stimuli on cell activities, gene expression was performed on dynamically-loaded chondral (gel-alone) and osteochondral (gel-bone) constructs.

2. Materials and methods

2.1. Osteochondral construct preparation

Cylindrical trabecular bone cores, ∅5 mm × 4 mm thick, were harvested from the epiphysis of metatarsal bones of 3 month-old calves (Fig. 2). Cores were cut to size with a diamond saw, cleaned of marrow with a water pick, and mass and volume measurements were taken to determine the apparent density. Cores were grouped by density, and infused with a molten 3% (in PBS) agarose hydrogel (Type VII, Sigma) in a custom mold to produce constructs with an ~2 mm gel-only region, an ~2 mm gel-bone interface region, and a ~2 mm bone-only region, as described previously [24].

2.2. Bulk mechanical testing

Five constructs per group were tested in unconfined compression as in [35], with stress relaxation tests to 10% strain (of upper gel thickness) followed by dynamic testing (with an applied displacement of 20 μm) at frequencies ranging from 0.005–1 Hz. The Young's modulus was determined from the equilibrium reaction force (taken at >1200 seconds) and the specimen geometry, while the dynamic modulus was calculated from the maximum peak-to-peak stress normalized to the maximum peak-to-peak strain.

2.3. Displacement fields

To obtain an axial displacement field, osteochondral constructs were cut in half and compressed using a custom compression device mounted on the stage of an inverted microscope, as described in [53]. The initial thickness of the specimen was measured optically (1.66 μm/pixel) and a 5% tare strain (of the gel thickness) was applied. After equilibrium, an image was acquired, and the specimen was further compressed another 5% strain increment with a second image acquired after equilibrium. Image analysis was performed using an automated digital image correlation technique to produce an axial displacement field [53,54,57].

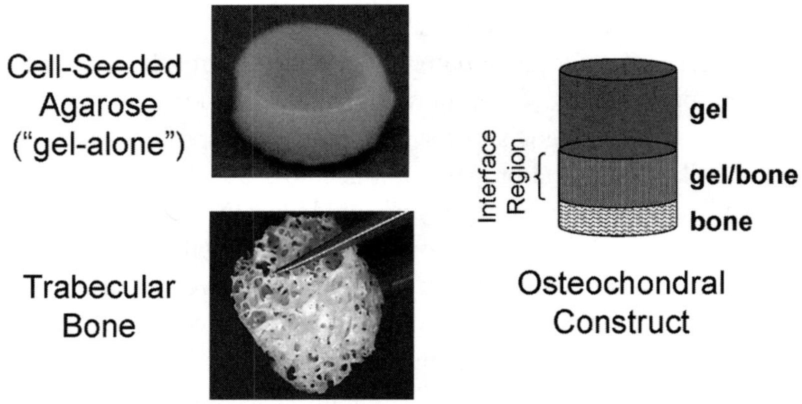

Fig. 2. Schematic for making bilayered osteochondral constructs composed of a gel layer (agarose) interpenetrated into a subchondral bony substrate. Constructs can be formed with a defined geometry and gel penetration depth.

2.4. Biphasic finite element models

Finite element meshes were defined with a commercial software package (I-DEAS, SDRC, Plano, TX) to model gel-alone agarose constructs (2 mm thickness × 5 mm diameter) and osteochondral composite constructs (4 mm thickness × 5 mm diameter), with a variable gel bone interface region (0–100% of bone layer thickness), Fig. 5. Meshes were defined as axisymmetric, and contained 600 elements per mesh with eight nodes per element, with the distribution of elements biased towards the free edge of the construct. Each region was assumed homogenous, with a linear isotropic elastic solid matrix, with the agarose gel region having $E_Y = 10$ kPa, $\nu = 0.3$, $k = 1 \times 10^{-12}$ m^4/N s, the gel-bone region having $E_Y = 1000$ MPa, $\nu = 0.3$, $k = 0.25 \times 10^{-12}$ m^4/N s, and the bone region having $E_Y = 1000$ MPa, $\nu = 0.3$, $k = 1 \times 10^{-8}$ m^4/N s. The degree gel penetration was set at 0, 25, 50, 75, and 100% of the underlying bony thickness. A custom FEM program (as described in [44]) incorporating biphasic theory [38] was used to solve the problem of unconfined deformational loading of such constructs in both stress relaxation (to 10% gel thickness deformation) and with an applied sinusoidal deformation of 10% of the gel thickness at a frequency of 1 Hz. The resulting fluctuations in mechanical signals generated within each construct were output and analyzed at equilibrium for stress relaxation testing ($t = 1000$ s) or at the point of maximal compressive axial deformation ($t = 0.5$ s) for the case of dynamic deformational loading.

2.5. Gene expression

Gene expression was analyzed for previously frozen chondrocyte-seeded (60×10^6 cells/ml) chondral and osteochondral constructs that had been dynamically loaded for 6 hours to 10% of the gel thickness as in [35]. The gel and bone regions of the osteochondral constructs were carefully separated under RNase free conditions using a scalpel. Two samples were pooled and homogenized using a mortar and pestle for the disc and gel region and a mechanical homogenizer for the bone region. RNA was isolated from the samples by two successive guanidine hydrochloride extractions [22] followed by a TRIzol® (Invitrogen, Life Technologies) extraction. Total RNA concentration was measured using a spectrophotometer at a 260 nm wavelength (μQuant, Bio-Tek Instruments). A reverse transcription polymerase chain reaction (RT-PCR) was carried out using the SuperScript™ First-Strand Synthesis System for RT-PCR (Invitrogen, Life Technologies) to determine relative mRNA expression levels for aggrecan and Type II Collagen using glyceraldehyde phosphate dehydrogenase (GAPDH) as an internal control. 500 ng of total RNA in a 20 μl volume was used for the reverse transcription reaction. One microliter of cDNA was used for each PCR reaction using primer sets for Type II Collagen, aggrecan, and GAPDH. Primer sequences were aggrecan (sense): CACTGTTACCGCCACTTCCC, (antisense): GACATCGTTCCACTCGCCCT; Type II Collagen (sense): ATGACAATCTGGCTCCCAAC, (antisense): GCCCTATGTCCACACCGATT; GAPDH (sense): GGTGATGCTGGTGCTGAGTA, (antisense) ATCCACAGTCTTCTGGGTGG. Thirty cycles of PCR were carried out using annealing temperatures of 51°C, 61°C, and 58°C respectively. PCR products were separated on 2% agarose gels containing ethidium bromide, and the bands were digitized and quantified (ImageJ, NIH). Aggrecan and Type II Collagen mRNA levels were normalized by GAPDH for each sample ($n = 6$–8 per group). Data was analyzed by first normalizing the gene expression of dynamically-loaded constructs by the average expression of the corresponding free-swelling control. Three groups were studied: gel-alone (chondral), bone region (lower portion of osteochondral constructs) and gel region (upper portion of osteochondral constructs). One-way analysis of variance with Fisher's LSD test was used to detect significant difference ($\alpha = 0.05$) among the groups using Statistica software (SYSTAT, IL).

3. Results

Bone cores were found to vary in apparent density, and were grouped into high (0.46 ± 0.05), medium (0.36 ± 0.03), and low (0.24 ± 0.03) density groups. Mechanical testing of acellular agarose osteochondral constructs revealed no statistical difference in the Young's modulus with decreasing density of the underlying trabecular bone compared to gel-alone constructs (Fig. 3). However, the dynamic modulus decreased significantly ($p < 0.05$) in the case of the lowest density subchondral bone osteochondral constructs compared to gel-alone constructs (Fig. 3).

Microscopic evaluation of the axial displacement field of osteochondral constructs in unconfined compression (Fig. 4A) revealed that deformation occurred solely in the gel region (Fig. 4B). A FEM simulation of stress relaxation generated a similar deformation profile at equilibrium (Fig. 4C).

Finite element meshes were constructed to model dynamic deformation of both uniform gel alone constructs and osteochondral constructs with mixed material properties (Fig. 5). The application of dynamic compression to the axial boundaries of these constructs showed quantitative and qualitative differences in the resulting magnitude and distribution of mechanical signals. In particular, radial strain in gel-alone constructs was largely homogeneous, while radial strains were larger in magnitude and heterogeneously distributed in osteochondral constructs (Fig. 6), with the highest strain magnitude observed at the surface of the constructs. Fluid pressurization was observed to be largely uniform through the thickness and radial direction in gel-alone constructs (Fig. 7, top left), while there was a higher, and much less uniform distribution of pressure in osteochondral constructs (Fig. 7, top right, bottom left and right). Fluid flux was directed radially and was greatest at the edges of gel-alone constructs (Fig. 7, top left). In osteochondral constructs, high fluid flows were observed at the radial edge (highest at the interface of the gel layer with the subchondral bone), with some flow through the bottom of the gel into the bone-gel region (Fig. 7, top right, bottom left and right). Significant variation of fluid pressurization occurred in

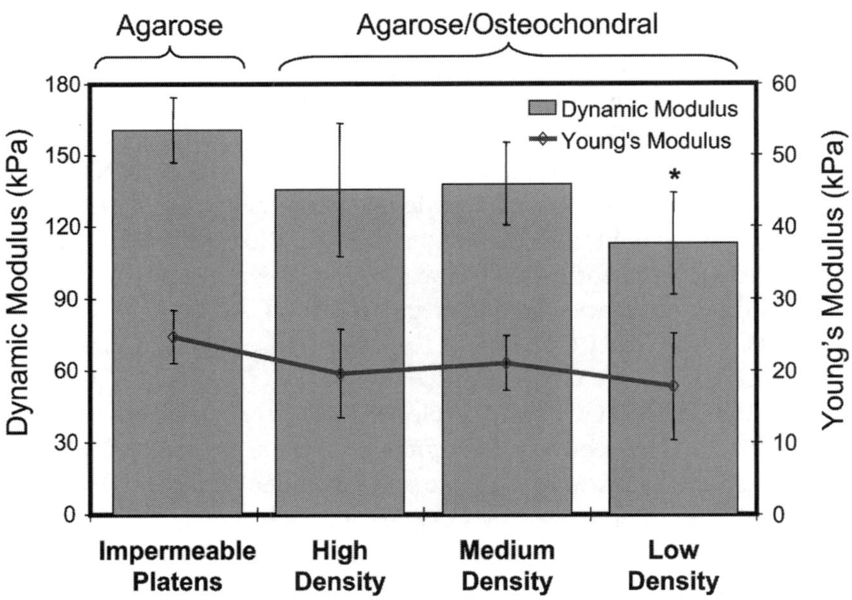

Fig. 3. Dynamic modulus (at 0.5 Hz) and Young's modulus of acellular osteochondral constructs with varying density trabecular bone substrates. * Indicates significant decrease in dynamic modulus compared to gel-alone construct impermeable platen ($p < 0.05$, $n = 5$).

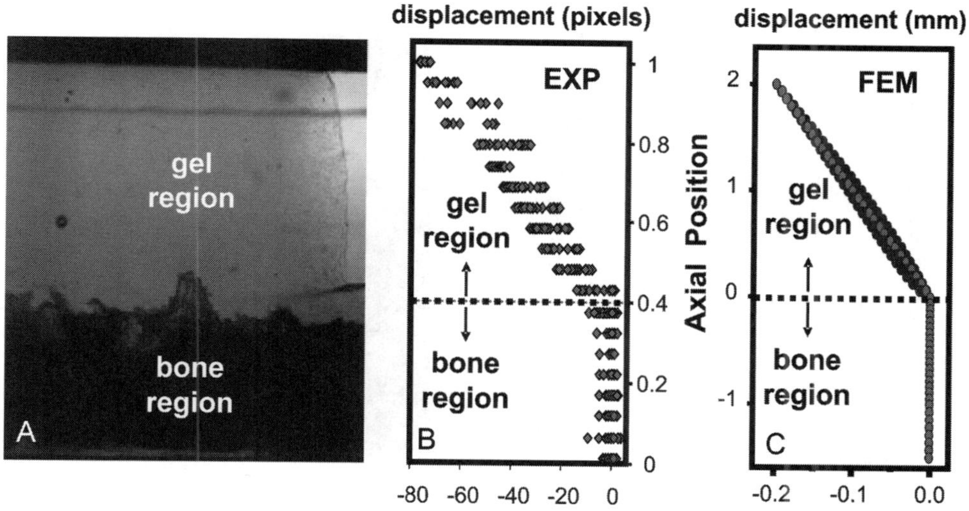

Fig. 4. (A) Cross sectional image of cell-free agarose mounted in a custom microscopy device for measuring local deformation in cartilage and tissue-engineered constructs [53,57]. (B) Measured axial deformation (in pixels) in cell-free osteochondral construct in response to a static axial deformation of 5% of the gel layer thickness. (C) Predicted axial deformation under the same loading conditions (at equilibrium) using the FEM model formulation. Due to the much stiffer bone region, axial deformation occurs solely in the upper gel region.

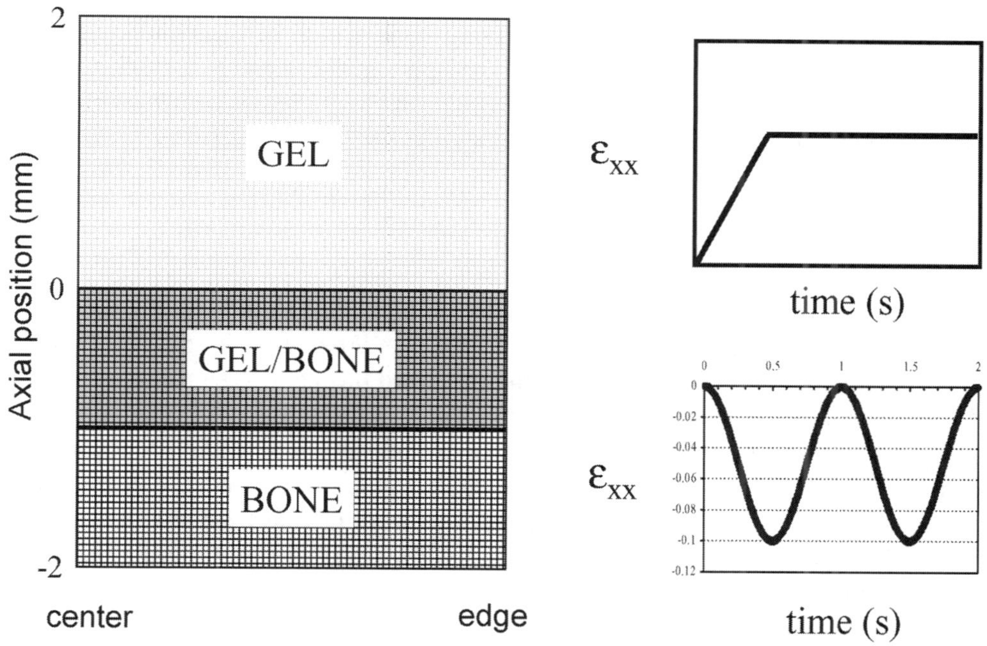

Fig. 5. Finite Element Model (FEM) of osteochondral construct and dynamic loading regime used for analysis.

the bone–gel region as well, and was highest at the gel/bone-gel interface, and decreased to ambient levels at the bone-gel/bone interface. As the penetration of gel into the bony substrate increased form 0% (top right) to 50% (bottom left) to 100% (bottom right) a progressively smaller proportion of fluid flowed through the gel/gel-bone interface, though the distribution of pressure and flow in the superior (chondral) gel region was largely unaffected.

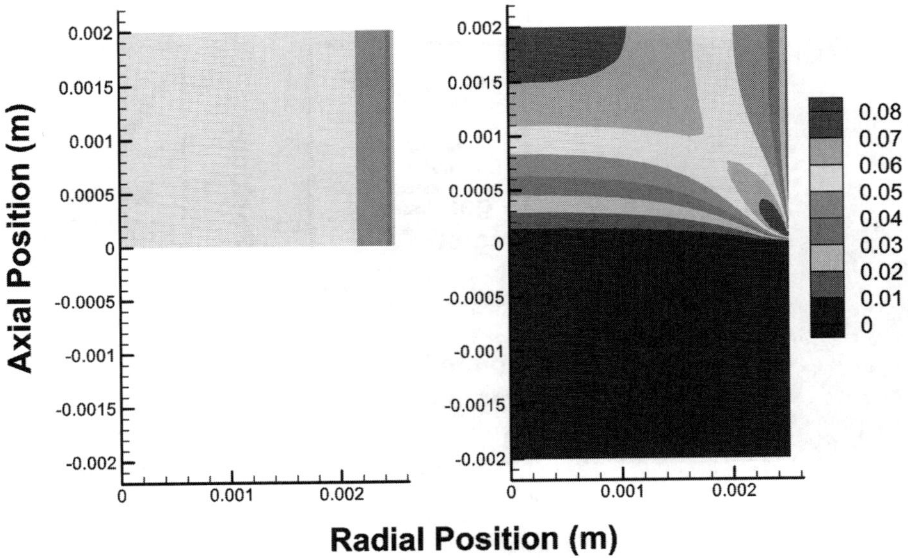

Fig. 6. Radial strain at the peak axial compression of a dynamically loaded (1 Hz, 10% deformation of gel layer thickness) gel alone (left) or gel/bone osteochondral construct (right) as predicted by the biphasic FEM model.

In studies monitoring gene expression of extracellular matrix proteins, the aggrecan and Type II Collagen mRNA levels of chondrocytes in the underlying bone region of osteochondral constructs was significantly elevated compared to that of levels in the upper gel regions ($p < 0.05$, $n = 6$–8), Fig. 8. Additionally, the relative gene expression of chondrocytes in the gel region of these gel-bone constructs was similar to those of chondrocytes residing in gel only constructs.

4. Discussion

Constitutive models coupled with finite element techniques are a useful tool for analyzing complex events occurring in structures that have non-uniform shapes and/or compositions. This preliminary study used biphasic finite element models to compare the predicted mechanical signals that arise in chondral and osteochondral constructs subjected to the same dynamic deformational loading regime. To construct these models, gel constructs were assumed homogenous and isotropic, while osteochondral constructs were given three distinct material regions. The top region of the osteochondral construct, the gel layer, was given the material properties of 2% agarose. Previous work has shown that biphasic theory can adequately predict the mechanical response of agarose hydrogels, particularly during the transient phase of loading [33], and to an even greater extent when one uses a strain-dependent permeability (unpublished observations). The middle region of the osteochondral construct was assigned the permeability of agarose (normalized by the decrease in area brought about by the presence of impermeable bone) and the stiffness of trabecular bone. Finally, the bottom bony region was given the material properties and permeability (very high) of free-draining, cleaned trabecular bone. Such osteochondral constructs, composed of three distinct, but adhered layers, would be expected to alter the mechanical signals generated with deformational loading due to the differing boundary conditions, especially at the gel-gel/bone interface. Similar analyses can be performed for alternate underlying substrate materials (e.g., natural coralline [19], bioactive scaffolds [29], hydrogels [40] and polymers [47]).

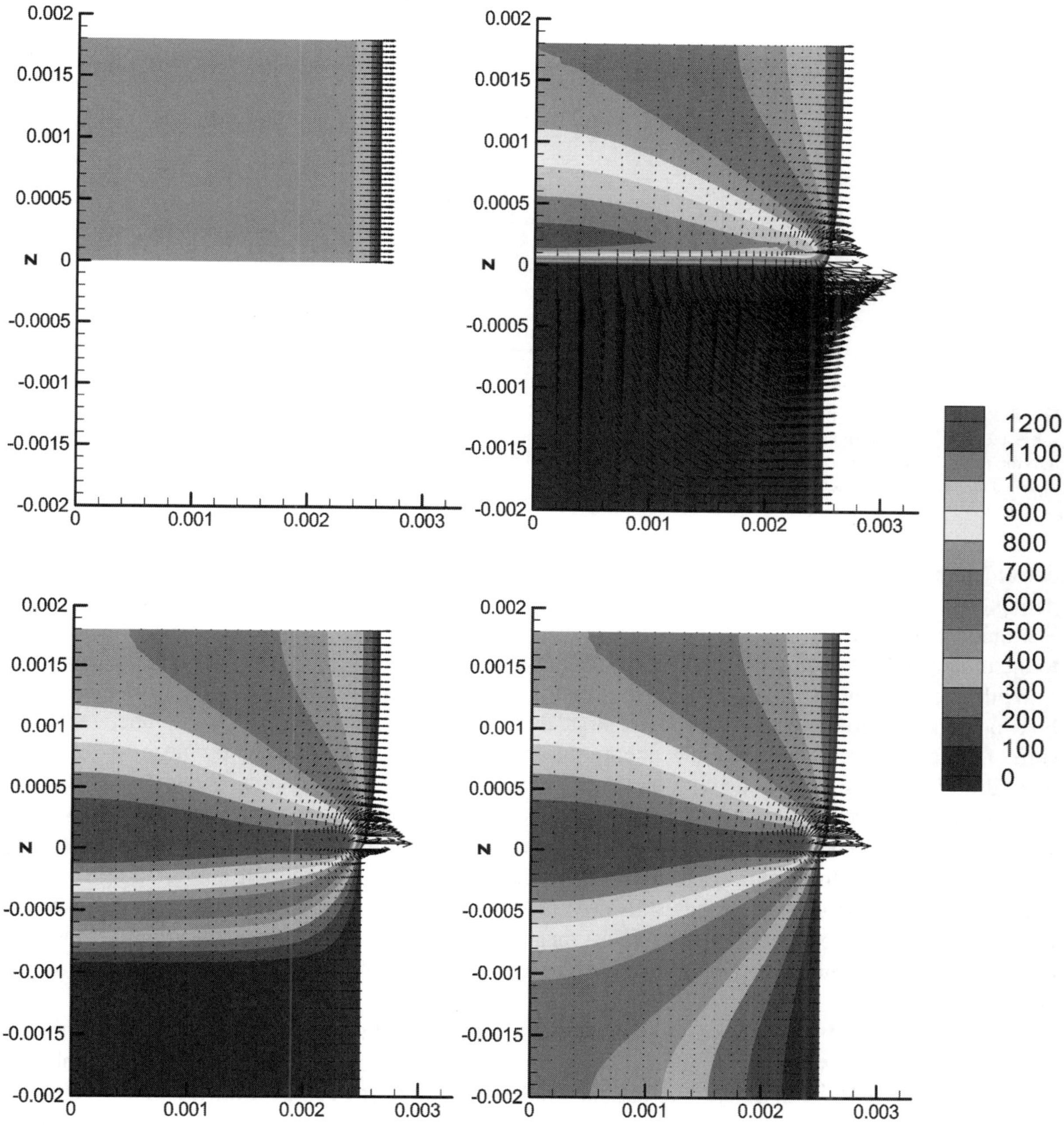

Fig. 7. Hydrostatic pressure (color mapping) and fluid flow (arrows) at the peak axial compression of a dynamically loaded (1 Hz, 10% deformation of gel layer thickness) gel alone (top left) or gel/bone osteochondral construct (top right, bottom left and right) as predicted by the biphasic FEM model.

The results of this study do indeed suggest that the two differing constructs will see disparate mechanical signals through their axial depth and radial position. Experimentally, we first observed that static deformation of an osteochondral construct resulted in deformation occurring solely in the gel region, as one would expect of a construct composed of two layers with such dissimilar material properties. This

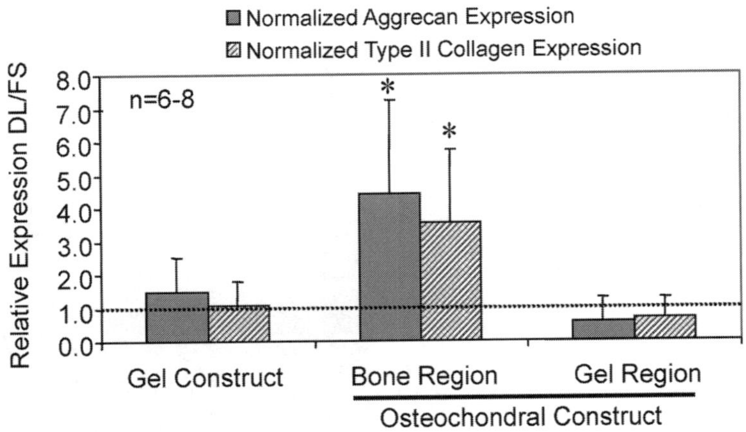

Fig. 8. Relative aggrecan and Type II Collagen gene expression of chondrocyte-seeded gel alone and osteochondral constructs subjected to 6 hours of dynamic loading (1 Hz, 10% deformation). For analysis, osteochondral constructs have been divided into the lower bone region and upper gel regions. Data is normalized to respective free-swelling control constructs and regions. ($^*p < 0.05$ versus gel construct and gel region).

deformation profile was linear, as has previously been seen for freshly cast hydrogel constructs, validating our assumption of homogenous material properties in the gel region [56,57]. Using the Biphasic FEM model we were able to predict the form of this deformation by applying the same static deformation to the model and examining the equilibrium response.

While the construct stiffness measured at equilibrium exhibits no relationship with the underlying bone porosity, experimental findings demonstrate that the construct dynamic modulus (a functional measure of the tissue properties that reflects both the matrix stiffness and hydraulic permeability) exhibits a trend of decreasing magnitude with increasing porosity. The latter likely reflects the degree of interstitial fluid pressurization that accompanies dynamic material testing of constructs, which is maximum when the gel is loaded between two impermeable surfaces (permitting only radial fluid exudation during loading) [44]. Since the material properties at equilibrium reflect only the stiffness of the solid construct (when fluid pressurization has subsided) [33], the similar Young's modulus observed over the bone porosity range seems reasonable. These measured bulk (or average) material properties indicate that the presence of an underlying bony substrate can alter the physical environment within the constructs.

FEM analyses can be performed to appreciate the spatial distribution of physical stimuli within the constructs. Using the same FEM model described above, but altering the loading pattern to apply dynamic loading at 1 Hz with a deformation magnitude 10% that of the gel layer thickness of osteochondral (or chondral) constructs reveals more interesting results. In chondral constructs, fluid pressurization and radial strain are similar through the depth and radial position, with only the expected edge effects (both signals rapidly go to zero) [25,36,39]. Fluid flow is also as expected highest at the radial periphery [12]. Dynamic deformational loading of osteochondral constructs, on the other hand, creates hydrostatic pressures, fluid flows, and radial strains that are quantitatively and qualitatively different from those occurring in gel-alone constructs. In particular, at the peak of compression, the gel region of the osteochondral construct experiences significant inhomogeneity in its hydrostatic pressure, with the highest values occurring at the middle of the construct just above the bone interface. Furthermore, radial normal strain is highest at the surface and minimal at the gel/bone interface (where by adhesion of the two layers the bone keeps the gel from expanding laterally). Finally, as mentioned in the results, osteochondral constructs see the highest levels of fluid flow directed out of the radial edge of the gel region, with a decreasing amount of fluid driven into the bone gel region as the penetration thickness increases.

These findings highlight the differences in mechanical environment that would be seen by cells seeded in the various regions of a dynamically loaded osteochondral construct compared to a gel-alone control. In support of this notion, chondrocyte aggrecan and Type II Collagen gene expression were found to be significantly elevated in the lower bone regions compared to upper gel regions. This finding may suggest that differing regional load-induced signals will likely alter the developing mechanical properties and matrix distribution in osteochondral constructs compared to gel-alone controls [9], and may be harnessed to modulate the development of tissue inhomogeneity in engineered constructs [55,58]. In articular cartilage, tensile stiffness is highest in the most superficial articular cartilage layers [58], and the tension compression non-linearity that derives from this fact is believed to play a role in the frictional response and load bearing capacities of the tissue [6,23,37,49]. In this study, the fact that the highest radial strain with dynamic loading is predicted to occur at the articular surface of the osteochondral construct suggests that such constructs may develop enhanced tensile stiffness in this region (to counteract the strain), and therefore may more closely match important features of the native tissue. Experimental work using cartilage and agarose gels have also shown that cells have maximum biosynthetic activity in regions of high fluid flows [9,12,18,25]. In this study, the prediction of fluid flows through the gel/bone interface may be taken advantage of to encourage subchondral plate formation at the gel–bone interface, and to possibly enhance gel layer-to-bone layer integration strength. The incorporation of pre-seeded bone cells on the underlying substrate may also facilitate development of a subchondral plate [50,59]. Current studies are exploring these possibilities by examining the development of both bulk and local mechanical properties of chondrocyte-seeded agarose bone osteochondral constructs subjected to long-term daily dynamic deformational loading [28].

Acknowledgements

This study was supported by the NIH [R01 AR46568 (CTH), R01 AR46532 (GAA)] and a predoctoral fellowship from the Whitaker Foundation (RLM).

References

[1] C.S. Ahmad, Z.A. Cohen, W.N. Levine, G.A. Ateshian and V.C. Mow, Biomechanical and topographic considerations for autologous osteochondral grafting in the knee, *Am. J. Sports Med.* **29**(2) (2001), 201–6.

[2] T. Ahsan, L. Chin, V.W. Wong, A.C. Chen, R.L. Sah, R.A. Bank, N. Verzijl and A. Ratcliffe, Effects of long term growth on tissue engineered cartilage, *Trans. Orthop. Res. Soc.* **28** (2003), 309.

[3] T. Ahsan, L.M. Lottman, F. Harwood, D. Amiel and R.L. Sah, Integrative cartilage repair: inhibition by beta-aminopropionitrile, *J. Orthop. Res.* **17**(6) (1999), 850–7.

[4] T. Ahsan and R.L. Sah, Biomechanics of integrative cartilage repair, *Osteoarthritis Cartilage* **7**(1) (1999), 29–40.

[5] P. Angele, R. Kujat, M. Nerlich, J. Yoo, V. Goldberg and B. Johnstone, Engineering of osteochondral tissue with bone marrow mesenchymal progenitor cells in a derivatized hyaluronan-gelatin composite sponge, *Tissue Eng.* **5**(6) (1999), 545–54.

[6] G.A. Ateshian, M.A. Soltz, R.L. Mauck, I.M. Basalo, C.T. Hung and W.M. Lai, The role of osmotic pressure and tension–compression nonlinearity in the frictional response of articular cartilage, *Transport in Porous Media* **50** (2003), 5–33.

[7] P.D. Benya and J.D. Shaffer, Dedifferentiated chondrocytes reexpress the differentiated collagen phenotype when cultured in agarose gels, *Cell* **30**(1) (1982), 215–24.

[8] M. Brittberg, A. Nilsson, A. Lindahl, C. Ohlsson and L. Peterson, Rabbit articular cartilage defects treated with autologous cultured chondrocytes, *Clin. Orthop.* (326) (1996), 270–83.
[9] M.D. Buschmann, Y.A. Gluzband, A.J. Grodzinsky and E.B. Hunziker, Mechanical compression modulates matrix biosynthesis in chondrocyte/agarose culture, *J. Cell Sci.* **108** (Pt 4) (1995), 1497–508.
[10] M.D. Buschmann, Y.A. Gluzband, A.J. Grodzinsky, J.H. Kimura and E.B. Hunziker, Chondrocytes in agarose culture synthesize a mechanically functional extracellular matrix, *J. Orthop. Res.* **10**(6) (1992), 745–58.
[11] M.D. Buschmann and A.J. Grodzinsky, A molecular model of proteoglycan-associated electrostatic forces in cartilage mechanics, *J. Biomech. Eng.* **117**(2) (1995), 179–92.
[12] M.D. Buschmann, Y.J. Kim, M. Wong, E. Frank, E.B. Hunziker and A.J. Grodzinsky, Stimulation of aggrecan synthesis in cartilage explants by cyclic loading is localized to regions of high interstitial fluid flow, *Arch. Biochem. Biophys.* **366**(1) (1999), 1–7.
[13] A.I. Caplan, M. Elyaderani, Y. Mochizuki, S. Wakitani and V.M. Goldberg, Principles of cartilage repair and regeneration, *Clin. Orthop.* (342) (1997), 254–69.
[14] S.C. Chang, J.A. Rowley, G. Tobias, N.G. Genes, A.K. Roy, D.J. Mooney, C.A. Vacanti and L.J. Bonassar, Injection molding of chondrocyte/alginate constructs in the shape of facial implants, *J. Biomed. Mater. Res.* **55**(4) (2001), 503–11.
[15] D.T. Felson and Y. Zhang, An update on the epidemiology of knee and hip osteoarthritis with a view to prevention, *Arthritis Rheum.* **41**(8) (1998), 1343–55.
[16] L.E. Freed, R. Langer, I. Martin, N.R. Pellis and G. Vunjak-Novakovic, Tissue engineering of cartilage in space, *Proc. Natl. Acad. Sci. USA* **94**(25) (1997), 13885–90.
[17] L.E. Freed, N. Pellis, N. Searby, J. de Luis, C. Preda, J. Bordonaro and G. Vunjak-Novakovic, Microgravity cultivation of cells and tissues, *Gravit. Space Biol. Bull.* **12**(2) (1999), 57–66.
[18] F. Guilak, R.L. Sah and L.A. Setton, Physical regulation of cartilage metabolism, in: *Basic Orthopaedic Biomechanics*, W.C. Hayes, ed., Lippincott-Raven, Philadelphia, 1997, pp. 179–207.
[19] S.M. Haddock, J.C. Debes, E.A. Nauman, K.E. Fong, Y.P. Arramon and T.M. Keaveny, Structure–function relationships for coralline hydroxyapatite bone substitute, *J. Biomed. Mater. Res.* **47**(1) (1999), 71–8.
[20] L. Hangody, G. Kish, Z. Karpati, I. Szerb and I. Udvarhelyi, Arthroscopic autogenous osteochondral mosaicplasty for the treatment of femoral condylar articular defects. A preliminary report, *Knee Surg. Sports Traumatol. Arthrosc.* **5**(4) (1997), 262–7.
[21] M.C. Harper, Viscous isoamyl 2-cyanoacrylate as an osseous adhesive in the repair of osteochondral osteotomies in rabbits, *J. Orthop. Res.* **6**(2) (1988), 287–92.
[22] C.D. Hoemann, J. Sun, V. Chrzanowski and M.D. Buschmann, A multivalent assay to detect glycosaminoglycan, protein, collagen, RNA, and DNA content in milligram samples of cartilage of hydrogel-based repair cartilage, *Anal. Biochem.* **300**(1) (2002), 1–10.
[23] C.Y. Huang, V.C. Mow and G.A. Ateshian, The role of flow-independent viscoelasticity in the biphasic tensile and compressive responses of articular cartilage, *J. Biomech. Eng.* **123**(5) (2001), 410–7.
[24] C.T. Hung, E.G. Lima, R.L. Mauck, M.A. LeRoux, E. Takai, H.H. Lu, R.G. Stark, X.E. Guo and G.A. Ateshian, Anatomically shaped osteochondral constructs for articular cartilage repair, *J. Biomechanics* **36** (2003), 1853–64.
[25] Y.J. Kim, R.L. Sah, A.J. Grodzinsky, A.H. Plaas and J.D. Sandy, Mechanical regulation of cartilage biosynthetic behavior: physical stimuli, *Arch. Biochem. Biophys.* **311**(1) (1994), 1–12.
[26] D.A. Lee and D.L. Bader, Compressive strains at physiological frequencies influence the metabolism of chondrocytes seeded in agarose, *J. Orthop. Res.* **15**(2) (1997), 181–8.
[27] D.A. Lee, T. Noguchi, S.P. Frean, P. Lees and D.L. Bader, The influence of mechanical loading on isolated chondrocytes seeded in agarose constructs, *Biorheology* **37**(1–2) (2000), 149–61.
[28] E.G. Lima, R.L. Mauck, S. Gasinu, G.A. Ateshian and C.T. Hung, Functional tissue engineering of free swelling and dynamically loaded osteochondral constructs, *Trans. Orthop. Res. Soc.* **29** (2004), 13.
[29] T. Livingston, P. Ducheyne and J. Garino, In vivo evaluation of a bioactive scaffold for bone tissue engineering, *J. Biomed. Mater. Res.* **62**(1) (2002), 1–13.
[30] H.J. Mankin, The response of articular cartilage to mechanical injury, *J. Bone Joint Surg.* **64-A**(460-66) (1982).
[31] R.L. Mauck, S.B. Nicoll, S.L. Seyhan, G.A. Ateshian and C.T. Hung, Synergistic action of growth factors and dynamic loading for articular cartilage tissue engineering, *Tiss. Eng.* **9**(4) (2003), 597–611.
[32] R.L. Mauck, S.L. Seyhan, G.A. Ateshian and C.T. Hung, Influence of seeding density and dynamic deformational loading on the developing structure/function relationships of chondrocyte-seeded agarose hydrogels, *Ann. Biomed. Eng.* **30**(8) (2002), 1046–56.
[33] R.L. Mauck, M.A. Soltz, C.C. Wang, D.D. Wong, P.H. Chao, W.B. Valhmu, C.T. Hung and G.A. Ateshian, Functional tissue engineering of articular cartilage through dynamic loading of chondrocyte-seeded agarose gels, *J. Biomech. Eng.* **122**(3) (2000), 252–60.
[34] R.L. Mauck, C.C.-B. Wang, Q. Cheng, N. Gabriel, H.H. Lu, F.H. Chen, G.A. Ateshian and C.T. Hung, Optimization of parameters of for articular cartilage tissue engineering with deformational loading, *Trans. Orthop. Res. Soc.* **28** (2003).

[35] R.L. Mauck, C.C.-B. Wang, E.S. Oswald, F.H. Chen, G.A. Ateshian and C.T. Hung, The role of cell seeding density and nutrient supply for articular cartilage tissue engineering with deformational loading, *Osteoarthritis and Cartilage* **11** (2003), 879–890.
[36] V.C. Mow, N.M. Bachrach, L.A. Setton and F. Guilak, Stress, strain, pressure, and flow fields in articular cartilage and chondrocytes, in: *Cell Mechanics and Cellular Engineering*, R.M. Hochmuth, ed., Springer-Verlag, New York, 1994, p. 345–79.
[37] V.C. Mow and X.E. Guo, Mechano-electrochemical properties of articular cartilage: their inhomogeneities and anisotropies, *Annu. Rev. Biomed. Eng.* **4** (2002), 175–209.
[38] V.C. Mow, S.C. Kuei, W.M. Lai and C.G. Armstrong, Biphasic creep and stress relaxation of articular cartilage in compression? Theory and experiments, *J. Biomech. Eng.* **102**(1) (1980), 73–84.
[39] V.C. Mow and C.C. Wang, Some bioengineering considerations for tissue engineering of articular cartilage, *Clin. Orthop.* (367 Suppl.) (1999), S204–23.
[40] K.W. Ng, C.C.-B. Wang, X.E. Guo, G.A. Ateshian and C.T. Hung, Characterization of inhomogeneous bi-layered chondrocyte-seeded agarose constructs of differing agarose concentrations, *Trans. Orthop. Res. Soc.* **28** (2003), 960.
[41] S.W. O'Driscoll, F.W. Keeley and R.B. Salter, The chondrogenic potential of free autogenous periosteal grafts for biological resurfacing of major full-thickness defects in joint surfaces under the influence of continuous passive motion. An experimental investigation in the rabbit, *J. Bone Joint Surg. Am.* **68**(7) (1986), 1017–35.
[42] B. Obradovic, I. Martin, R.F. Padera, S. Treppo, L.E. Freed and G. Vunjak-Novakovic, Integration of engineered cartilage, *J. Orthop. Res.* **19**(6) (2001), 1089–97.
[43] K.T. Paige, L.G. Cima, M.J. Yaremchuk, J.P. Vacanti and C.A. Vacanti, Injectable cartilage, *Plast. Reconstr. Surg.* **96**(6) (1995), 1390–8; discussion 1399–400.
[44] S. Park, C.T. Hung and G.A. Ateshian, Mechanical response of bovine articular cartilage under dynamic unconfined compression loading at physiological stress levels, *Osteoarthritis Cartilage* **12**(1) (2003), 65–73.
[45] D. Pazzano, K.A. Mercier, J.M. Moran, S.S. Fong, D.D. DiBiasio, J.X. Rulfs, S.S. Kohles and L.J. Bonassar, Comparison of chondrogensis in static and perfused bioreactor culture, *Biotechnol. Prog.* **16**(5) (2000), 893–6.
[46] D. Schaefer, I. Martin, G. Jundt, J. Seidel, M. Heberer, A. Grodzinsky, I. Bergin, G. Vunjak-Novakovic and L.E. Freed, Tissue-engineered composites for the repair of large osteochondral defects, *Arthritis Rheum.* **46**(9) (2002), 2524–34.
[47] D. Schaefer, I. Martin, P. Shastri, R.F. Padera, R. Langer, L.E. Freed and G. Vunjak-Novakovic, In vitro generation of osteochondral composites, *Biomaterials* **21**(24) (2000), 2599–606.
[48] J.K. Sherwood, S.K. Riley, R. Palazzolo, S.C. Brown, D.C. Monkhouse, M. Coates, L.G. Griffith, L.K. Landeen and A. Ratcliffe, A three-dimensional osteochondral composite scaffold for articular cartilage repair, *Biomaterials* **23**(24) (2002), 4739–51.
[49] M.A. Soltz and G.A. Ateshian, A Conewise Linear Elasticity mixture model for the analysis of tension–compression nonlinearity in articular cartilage, *J. Biomech. Eng.* **122**(6) (2000), 576–86.
[50] E. Takai, E.G. Lima, H.H. Lu, G.A. Ateshian, X.E. Guo and C.T. Hung, Devitalized bone as a mineralized substrate for osteochondral tissue engineered hydrogel constructs, *Trans. Orthop. Res. Soc.* **28** (2003), 307.
[51] G. Vunjak-Novakovic, I. Martin, B. Obradovic, S. Treppo, A.J. Grodzinsky, R. Langer and L.E. Freed, Bioreactor cultivation conditions modulate the composition and mechanical properties of tissue-engineered cartilage, *J. Orthop. Res.* **17**(1) (1999), 130–8.
[52] G. Vunjak-Novakovic, B. Obradovic, I. Martin and L.E. Freed, Bioreactor studies of native and tissue engineered cartilage, *Biorheology* **39**(1–2) (2002), 259–68.
[53] C.C. Wang, J.M. Deng, G.A. Ateshian and C.T. Hung, An automated approach for direct measurement of two-dimensional strain distributions within articular cartilage under unconfined compression, *J. Biomech. Eng.* **124**(5) (2002), 557–67.
[54] C.C. Wang, X.E. Guo, D. Sun, V.C. Mow, G.A. Ateshian and C.T. Hung, The functional environment of chondrocytes within cartilage subjected to compressive loading: a theoretical and experimental approach, *Biorheology* **39**(1–2) (2002), 11–25.
[55] C.C. Wang, C.T. Hung and V.C. Mow, An analysis of the effects of depth-dependent aggregate modulus on articular cartilage stress-relaxation behavior in compression, *J. Biomech.* **34**(1) (2001), 75–84.
[56] C.C.-B. Wang, T.N. Kelly, R.L. Mauck, G.A. Ateshian and C.T. Hung, Temporal and spatial development of construct stiffness in chondrocyte-seeded agarose disks cultured in free-swelling and dynamical-loading configurations, *ASME-BED* (Summer Meeting) (2003), 279–280.
[57] C.C.-B. Wang, M.A. Soltz, R.L. Mauck, W.B. Valhmu, G.A. Ateshian and C.T. Hung, Comparison of the equilibrium axial strain distribution in articular cartilage explants and cell-seeded alginate disks under unconfined compression, *Trans. Orthop. Res. Soc.* **25** (2000), 131.
[58] C.C.B. Wang, N.O. Chahine, C.T. Hung and G.A. Ateshian, Optical determination of anisotropic properties of bovine articular cartilage in compression, *J. Biomech.* **36**(3) (2003), 339–53.

[59] Y. Weng, Y. Cao, C.A. Silva, M.P. Vacanti and C.A. Vacanti, Tissue-engineered composites of bone and cartilage for mandible condylar reconstruction, *J. Oral Maxillofac. Surg.* **59**(2) (2001), 185–190.

[60] J.Z. Wu, W. Herzog and E.M. Hasler, Inadequate placement of osteochondral plugs may induce abnormal stress–strain distributions in articular cartilage – finite element simulations, *Med. Eng. Phys.* **24**(2) (2002), 85–97.

[61] B.J. Zuger, B. Ott, P. Mainil-Varlet, T. Schaffner, J.F. Clemence, H.P. Weber and M. Frenz, Laser solder welding of articular cartilage: tensile strength and chondrocyte viability, *Lasers Surg. Med.* **28**(5) (2001), 427–34.

Part VII

Perspectives and concluding remarks

Biorheology 41 (2004) 593–596
IOS Press

Current perspectives on cartilage and chondrocyte mechanobiology

Mikko J. Lammi
Department of Anatomy, University of Kuopio, P.O. Box 1627, 70211 Kuopio, Finland
Tel.: +358 17 163 027; Telefax: +358 17 163 032; E-mail: mikko.lammi@uku.fi

Abstract. It is well known that physiological forces are essential for the maintenance of normal composition and structure of articular cartilage. Although some of the mechanisms of mechanotransduction are known today, there are certainly many others left unrevealed. In order to understand the complicated systems present in articular cartilage, we have to bring together the data from all fields of cartilage mechanobiology. The 3rd Symposium on Mechanobiology of Cartilage and Chondrocyte was a good effort towards that goal.

1. Introduction

Regular joint loading is known to be needed for the maintenance of normal healthy articular cartilage [9], while joint immobilisation rapidly causes cartilage atrophy [7,8]. Further, *in vitro* metabolic studies have indicated that mechanical forces are able to increase the production of extracellular matrix molecules [18]. In general, the net responses of various loading conditions on extracellular matrix production in cartilage explants or cell cultures have been investigated quite a lot. Recently, the focus of cartilage mechanobiology has more and more shifted on attempts to understand the cellular mechanisms of chondrocytes that respond to mechanical forces, although theoretical and clinical aspects also remain important topics of cartilage mechanobiology. Here I will present an overview of the topics presented at the 3rd International Symposium on Mechanobiology of Cartilage and Chondrocyte held in Brussels in May 2003.

2. Mechanobiology and chondrocytes

Although the effects of loading on extracellular matrix synthesis in cartilage tissue or cell cultures have been investigated a lot, there has been a lack of systematic effort to find the optimal ratio between loading and rest for the most efficient matrix synthesis. The work presented by Dr. Bader is an important addition to our knowledge of the temporal regulation of chondrocyte metabolism in agarose constructs subjected to mechanical loading. A growing number of possible mechanotransduction pathways activated in chondrocytes by the load has been recently presented, of which integrin-interleukin-4 route is perhaps the best characterized so far [15]. In the symposium we heard Dr. Salter's and Dr. Millward-Sadler's talks that included NMDA receptors and P2Y2 purinoceptors in the group of potential key players of chondrocyte mechanotransduction. Thymosin β_1 has previously been shown to be induced by mechanical loading [2], and new data show its involvement, e.g., in the assembly of filamentous actin.

0006-355X/04/$17.00

In chondrocytes, a number of transport proteins are involved in the maintenance of constant intracellular milieu, and many pathways by which load regulates intracellular composition and the macromolecule turnover have been proposed [21]. Dr. Urban presented how changes in cartilage pH, lactate concentration, osmolarity and oxygen consumption are related with chondrocyte's metabolism. Previously, intracellular calcium transients have been shown to occur during swelling of articular chondrocytes [22]. Dr. Edllich showed that shear stress created by fluid flow also stimulates calcium transients in bovine articular chondrocytes [5]. Steady fluid flow was more stimulatory than oscillatory, and saw-tooth waveform more efficient than regular sinusoidal oscillatory fluid flow. Previously, loading of cartilage explants larger than the loading head resulted in increased ^{35}S-sulphate incorporation in the superficial zone, the area obviously subjected to some level of tensile forces during the compression of the central part of the tissue [14]. The finding presented by Dr. Levenston that oscillatory tension decreased both proline and sulphate incorporation of chondrocytes seeded in fibrin constructs and increased their proliferation rates indicates the complexity of the cellular responses to mechanical forces.

3. Tissue engineering

Investments in cartilage tissue engineering have increased remarkably during the last years. Different approaches have been tested to find mixtures of supplements that would preserve the differentiated state of chondrocytes even after efficient multiplication of the cells. TGF-βs, IGF, FGFs, and BMPs are among the most investigated differentiation modifiers tested, and as presented by Dr. Mandl, FGF-2 in the absence of serum appears to decrease dedifferentiation of human chondrocytes during monolayer expansion stage [11]. Dr. Herbage showed that BMP-2 can potentially reverse dedifferentiation that occurs in passaged chondrocyte cultures [19]. An interesting new approach utilising a technique of organ printing was presented by Dr. Forgacs. Although a considerable progress in cartilage tissue engineering has been achieved, it appears that mechanical loading of the tissue constructs may play an important, if not essential, role in our efforts to manufacture constructs that not only contain an adequate amount of extracellular matrix constituents [13], but also have a correct three-dimensional structure that is most likely critical for the durability of the transplanted construct.

The scaffolds that are used for cartilage tissue engineering include, for instance, hyaluronan derivatives, agarose, alginate, collagen constructs of either collagen type I or II, and biodegradable materials. These different materials are widely tested at the moment, and only the future will show which scaffold types and 3D matrices will have the best properties for functional tissue engineering of cartilage. Hyaluronan and collagen type II matrices certainly have one clear advantage, as they are the natural constituents of the cartilage. A lack of sufficient source for articular chondrocytes to be used in transplants has led to an intense reseach on multipotential adult stem cells. Bone marrow mesenchymal stem cells have been considered a very potential source with certain success [1]. However, certain other sources appear also promising. As Dr. Guilak showed in his talk, adipose tissue-derived stromal cells is a potential source of progenitor cells for musculoskeletal tissue engineering [20].

Attempts of cartilage repair have shown that integration of transplanted construct is still a difficult problem. Dr. Van de Breevaart showed how hyaluronidase and collagenase enzymatic treatments improve the integration of cartilage transplants. The use of heat prestress or proteasome inhibitor MG13, reported to induce heat shock protein 70 and increase cell vitality by Dr. Gillet, might improve the integration by reducing the cellular death of the transplant.

4. Theoretical aspects and imaging

Articular cartilage is subjected to compressive and tensile loads during locomotion. In osteoarthrosis, cartilage is lost due to mechanical wear too extreme for the tissue. Theoretical approaches to determine the cartilage properties are important to understand the mechanisms responsible for the failure of the tissue in osteoarthrosis. Dr. Verteramo presented data on zonal and directional tensile properties of bovine articular cartilage showing that at high tensile strain rates cartilage stiffness appeared to be greater than at lower rates. In his talk, Dr. Li concluded that collagen viscoelasticity does not have a significant effect on the response of cartilage in compression, however, its viscoelastic behaviour can be observed in tensile testing. Dr. Poschetti used poroelastic model to define permeability dependence on deformation. Dr. Huyghe proposed how to model changes in collagen fibre content and orientation in connective tissues due to fibre stretch.

A great progress in imaging of cartilage and its constituents has been achieved during the last years. The talk by Dr. Grodzinsky illustrated how molecular force probe and atomic force microscopy techniques can be utilized to obtain high-resolution images of the cartilage matrix molecules such as proteoglycan aggregates, and these methods can also provide us with the data of the electromechanical behavior of polyelectrolyte brushes such as glycosaminoglycans [4,16]. Drs. Luo and An presented data how single molecule mechanical properties of collagen types I and II and hyaluronan can be measured with an optical tweezer technique [6,17]. Magnetic resonance imaging of articular cartilage is used frequently today, and Dr. Jurvelin showed that acoustic parameters obtained by ultrasound measurements can used for clinical diagnostic imaging [10], as well as to analyse material properties of articular cartilage. Quantitative electrical impedance device to map focal cartilage lesions was also introduced by Dr. Frank. FTIR imaging also provides a new effective tool for cartilage research.

5. Approaches of osteoarthrosis research

Osteoarthrosis is a disease affecting both the articular cartilage and the bone. Dr. Bailey described the changes that can be observed in the osteoarthrotic subchondral and trabecular bone, and Dr. Weinans demonstrated with micro-CT images how the overall architecture of subchondral bone changes in human osteoarthrotic joints [3]. These results clearly show osteoarthrosis must be examined at level of the whole joint. With ageing, many indicators show phenotypic senescence of chondrocytes [12], and Dr. Buckwalter demonstrated that events associated with joint trauma, such as increased oxidative and mechanical stress, can accelerate chondrocyte phenotypic senescence. A number of drugs claimed to be disease-modifying are available to treat osteoarthrosis, although for some medications the mechanism of actions are not well understood. Therefore, the profound analysis of the mechanism how rhein affects osteoarthrotic cartilage, presented by Dr. Pujol, is valuable addition to our knowledge of the disease. Dr. Gillet reported that proteasome-inhibitor MG132, an effective inducer of heat shock protein 70, could protect articular cartilage chondrocytes from mono-iodoacetate induced cytotoxicity. Intra-articular hyaluronan injections are used to alleviate joint pain in patients with osteoarthrosis, and it has been suggested that it may have disease-modifying properties. However, Dr. Brandt showed that hyaluronan injections did not affect the progression of osteoarthrosis in cruciate ligament-deficient dogs.

6. Conclusions

In summary, a great number of researchers working in cartilage mechanobiology were gathered together in this symposium presenting interesting new findings on a wide spectrum of cartilage mechanobiology. I express my warmest thanks to Professor J.F. Stoltz for organising such an up-to-date symposium, and look forward to the possible fourth symposium. I also thank Negma-Lerad Laboratories and especially Dr. Burger for their outstanding support to organise the symposium.

References

[1] F.B. Barry, Mesenchymal stem cell therapy in joint disease, *Novartis Found. Symp.* **249** (2003), 86–96.

[2] E.J. Blain, D.J. Mason and V.C. Duance, The effect of cyclical compression loading on gene expression in articular cartilage, *Biorheology* **40** (2003), 111–117.

[3] J.S. Day, M. Ding, J.C. van der Linden, I. Hvid, D.R. Sumner and H. Weinans, A decreased subchondral trabecular bone tissue elastic modulus is associated with pre-arthritic cartilage damage, *J. Orthop. Res.* **19** (2001), 914–918.

[4] D. Dean, J. Seog, C. Ortiz and A.J. Grodzinsky, Molecular-level theoretical model for electrostatic interactions within polyelectrolyte brushes: applications to charged glycosaminoglycans, *Langmuir* **19** (2003), 5526–5539.

[5] M. Edlich, C.E. Yellowley, C.R. Jacobs and H.J. Donahue, Oscillating fluid flow regulates cytosolic calcium concentration in bovine articular cartilage, *J. Biomech.* **34** (2001), 59–65.

[6] T. Fujii, Y.L. Sun, K.N. An and Z.P. Luo, Mechanical properties of single hyaluronan molecules, *J. Biomech.* **35** (2002), 527–531.

[7] J. Haapala, J.P. Arokoski, M.M. Hyttinen, M. Lammi, M. Tammi, V. Kovanen, H.J. Helminen and I. Kiviranta, Remobilization does not fully restore immobilization induced articular cartilage atrophy, *Clin. Orthop.* (1999), 218–229.

[8] M.O. Jortikka, R.I. Inkinen, M.I. Tammi, J.J. Parkkinen, J. Haapala, I. Kiviranta, H.J. Helminen and M.J. Lammi, Immobilisation causes longlasting matrix changes both in the immobilised and contralateral joint cartilage, *Ann. Rheum. Dis.* **56** (1997), 255–261.

[9] I. Kiviranta, M. Tammi, J. Jurvelin, A.-M. Säämänen and H.J. Helminen, Moderate running exercise augments glycosaminoglycans and thickness of articular cartilage in the knee joint of young beagle dogs, *J. Orthop. Res.* **6** (1988), 188–195.

[10] M.S. Laasanen, J. TšyrŠs, A.I. Vasara, M.M. Hyttinen, S. Saarakkala, J. Hirvonen, J.S. Jurvelin and I. Kiviranta, Mechano-acoustic diagnosis of cartilage degeneration and repair, *J. Bone Joint Surg. Am.* **85**(Suppl. 2) (2003), 78–84.

[11] E.W. Mandl, S.W. van der Veen, J.A. Verhaar and G.J. van Osch, Serum-free medium supplemented with high-concentration FGF2 for cell expansion culture of human ear chondrocytes promotes redifferentiation capacity, *Tissue Eng.* **8** (2002), 573–580.

[12] J.A. Martin and J.A. Buckwalter, The role of chondrocyte senescence in the pathogenesis of osteoarthritis and in limiting cartilage repair, *J. Bone Joint Surg. Am.* **85A**(Suppl. 2) (2003), 106–110.

[13] R.L. Mauck, S.L. Seyhan, G.A. Ateshian and C.T. Hung, Influence of seeding density and dynamic deformational loading on the developing structure/function relationships of chondrocyte-seeded agarose hydrogels, *Ann. Biomed. Eng.* **30** (2002), 1046–1056.

[14] J.J. Parkkinen, M.J. Lammi, H.J. Helminen and M. Tammi, Local stimulation of proteoglycan synthesis in articular cartilage explants by dynamic compression in vitro, *J. Orthop. Res.* **10** (1992), 610–620.

[15] D.M. Salter, S.J. Millward-Sadler, G. Nuki and M.O. Wright, Integrin-interleukin-4 mechanotransduction pathways in human chondrocytes, *Clin. Orthop.* (2001), S49–S60.

[16] J. Seog, D. Dean, A.H.K. Plaas, S. Wong-Palms, A.J. Grodzinsky and C. Ortiz, Direct measurement of glycosaminoglycan intermolecular interactions via high-resolution force spectroscopy, *Macromolecules* **35** (2002), 5601–5615.

[17] Y.L. Sun, Z.P. Luo, A. Fertala and K.N. An, Direct quantification of the flexibility of type I collagen monomer, *Biochem. Biophys. Res. Commun.* **295** (2002), 382–386.

[18] J.P. Urban, The chondrocyte: a cell under pressure, *Br. J. Rheumatol.* **33** (1994), 901–908.

[19] U. Valcourt, J. Gouttenoire, E. Aubert-Foucher, D. Herbage and F. Mallein-Gerin, Alternative splicing of type II procollagen pre-mRNA in chondrocytes is oppositely regulated by BMP-2 and TGF-β1, *FEBS Lett.* **545** (2003), 115–119.

[20] M.Q. Wickham, G.R. Erickson, J.M. Gimble, T.P. Vail and F. Guilak, Multipotent stromal cells derived from the infrapatellar fat pad of the knee, *Clin. Orthop.* (2003), 196–212.

[21] R.J. Wilkins, J.A. Browning and J.P. Urban, Chondrocyte regulation by mechanical load, *Biorheology* **37** (2000), 67–74.

[22] C.E. Yellowley, J.C. Hancox and H.J. Donahue, Effects of cell swelling on intracellular calcium and membrane currents in bovine articular chondrocytes, *J. Cell. Biochem.* **86** (2002), 290–301.

Biorheology 41 (2004) 597
IOS Press

Concluding remarks

During this third symposium, the communications represented a broad spectrum of the bioengineering and medical research in mechanobiology of cartilage and chondrocyte. The communications presented are in part overviews and in part taken from the cutting edges of this new approach. Data drawn from diverse fields, when brought together, reinforced potentially important phenomena. A number of exciting new finds were presented or reviewed. Of the new engineering developments, the one that seems destined for earliest exploitation is the construct of biocartilages and the researches on stem cells.

I am grateful to all the participants for their outstanding contributions. In order to draw up a new balance of the research in mechanobiology of cartilage and chondrocyte, I am looking forward to meet you again for a fourth symposium in two or three years.

Before closing, I would like to thank Negma Lerads for its partnership and support. This is an example of how scientific societies, like the European society on Cell and Tissue Engineering and Therapy and pharmaceutical companies should work together in a constructive way. I also want to thank Martine Burger who gave so generously of her time in the preparation of this symposium.

Brussels, 17 May 2003

J.F. Stoltz
President of the symposium

Author Index